Analytical Separation Science

Analytical Separation Science

By

Bob W. J. Pirok

University of Amsterdam, The Netherlands
Email: B.W.J.Pirok@uva.nl

and

Peter J. Schoenmakers

University of Amsterdam, The Netherlands
Email: P.J.Schoenmakers@uva.nl

ROYAL SOCIETY
OF **CHEMISTRY**

Print ISBN: 978-1-83767-103-8
PDF ISBN: 978-1-83767-482-4
EPUB ISBN: 978-1-83767-483-1

A catalogue record for this book is available from the British Library

The Royal Society of Chemistry is a charity, registered in England and Wales, Number 207890, and a company incorporated in England by Royal Charter (Registered No. RC000524), registered office: Burlington House, Piccadilly, London W1J 0BA, UK, Telephone: +44 (0) 20 7437 8656.

For further information see our website at www.rsc.org

For general enquiries, please contact books@rsc.org

For EU product safety enquiries, please email books@rsc.org or contact Royal Society of Chemistry Worldwide (Germany) GmbH, Römischer Hof, Unter den Linden 10, 10117 Berlin.

To Lotte

To Bonnie

Preface

We have been teaching analytical separation science for 5 years (Bob) to 25 years (Peter). Students have always appreciated our classes (or at least, that is what they have told us). However, they were not abundantly pleased with the series of research papers we asked them to read. Over the years, the demand for a comprehensive text to supplement our slides has grown louder, and we finally decided to respond. We wrote this book for our own students – and for students worldwide who are interested in analytical separation science.

We aim to cover the field of analytical separation science from how it was established to where it is still developing. Each chapter is divided into dedicated modules, intended for Bachelor students (B modules), Master students (M), and Advanced (PhD) students and experienced Analysts (A). Readers can progress through the levels.

The book reflects how we teach. We aim to help our students master the foundations of the science. For example, we want them to understand why gas chromatographers use open-tubular capillary columns, while liquid chromatographers use columns packed with very small particles, resulting in a book that admittedly contains more (usually simple) equations than chromatograms. This is how we teach *Essentials of Analytical Chemistry* in the joint BSc Chemistry programme of the Amsterdam Universities (the University of Amsterdam and Vrije Universiteit). It is also how we teach *Separation Science*, *Advanced Separation Science*, and *Chemometrics and Statistics* in our joint MSc programme in Analytical Chemistry.

Colleagues in the team also teach applied courses, such as Bioanalytical Science, Environmental Science, Forensics, and Protein Analysis, where students learn how Analytical Separation Science is applied in practice. We encourage teachers using this book to incorporate their own experiences and the context of their course into their teaching. For example, teachers with experience in Food Science should use examples from food analysis, while those in a Pharmaceutical Science programme may discuss drug analysis,

Analytical Separation Science
By Bob W. J. Pirok and Peter J. Schoenmakers
© Bob W. J. Pirok and Peter J. Schoenmakers 2025
Published by the Royal Society of Chemistry, www.rsc.org

and so on. This underlines the importance of our field. Separation science is needed everywhere. A sound foundation saves time and money and makes the world a bit "greener" by saving solvents and energy.

We are committed to maintaining a website (ass-ets.org) to support readers of the book (students and analysts), as well as teachers. We will continue adding material to the website and examples that may be used in class. If you have an example that clearly supports the text of this book and can be added to the website, please contact us.

Several "Heroes of Analytical Separation Science" (and some "Heroes of Chemometrics") are honoured in this book. We want our students to know that all the science in this book has been developed by people with a face and a soul. Initially, we only considered scientists who were deceased and whom we found most exceptional. This resulted in an amazing set of Heroes, but unfortunately, almost all of them were white males. Therefore, we decided to also include amazing female scientists from our field who are formally retired. We are deeply indebted to all these Heroes and many others. If you believe other scientists deserve to be inducted into the Hall of Fame on our website, then you are invited to make your case.

The heroes mentioned in this book are not the only role models we want to present to our students. Anna Baglai is one of our personal heroes. She did not have the chance to complete a brilliant career, as she passed away much, much too soon. However, through her perseverance and tireless optimism, she had a massive impact on us and her broader community. We trust that she will inspire future generations and convince them that science is open to everybody.

Box 1 Hero of analytical separation science: Anna Baglai.

Anna Baglai (1988–2022, The Netherlands) Anna Baglai was originally from Kharkiv and studied in Kyiv, Ukraine. She then joined our group in Amsterdam, where she learned about analytical separation science. In 2018, she obtained her PhD in comprehensive two-dimensional liquid chromatography, with a special focus on stationary-phase-assisted modulation (SPAM). One of her eye-catching results can be found in this book, in Figure 7.26. With her vibrant personality, she won the hearts of all her colleagues. Scores of separation scientists who interacted with Anna at a scientific conference or elsewhere are bound to remember her.

We have made many friends in the analytical separation science community that we cherish. We hope that they will appreciate this book. Many top scientists and graduate students proofread chapters from this book. Their critical reviews provided us with invaluable advice and, in many cases, saved us from embarrassment. Any mistakes left are entirely our fault. We are deeply grateful to Tijmen Bos, Ab Buijtenhuijs, Ken Broeckhoven, Jan Christensen, Gert Desmet, Sebastiaan Eeltink, Attila Felinger, Paul Ferguson, James Grinias, Fabrice Gritti, Gerben van Henten, Rick van den Hurk, Hans-Gerd Janssen, Lourdes Ramos, Govert Somsen, Dwight Stoll, André Striegel, Peter Tranchida, Gabriel Vivó-Truyols, Caroline West, Johan Westerhuis, and Annika van der Zon. We are also grateful to the students in our MSc programme, particularly Janne Bolwerk, Boaz Geurtsen, Rebecca Gibkes, and Ebru Kara Chasan Memet, who reviewed earlier versions of some of our chapters.

See you at ass-ets.org

Bob Pirok
Peter Schoenmakers
Amsterdam

Foreword

What is better than a book? Three books. With *Analytical Separation Science*, the authors have written a book that, by virtue of a clever B-, M- and A-level labelling system, can be read at three degrees of depth and, hence, will be your first guide and reference point throughout your entire career in the analytical sciences: from your very first steps at university or college to the time you reach the senior expert level in your professional life.

The book arrives at precisely the right moment. The demand for skilled chemists in industries such as pharmaceuticals, environmental analysis, food safety, and petrochemicals peaks as never before. However, while analytical separation techniques, such as chromatography and electrophoresis, are used as never before, and analytical problems in industry or life-science research have become ever more complex, academic research on the fundamentals of these techniques is gradually receding. This inevitably reflects in the chemistry curricula of colleges and universities, leaving fresh graduates with both a lack of hands-on experience and understanding, leading to inefficiencies, errors and even regulatory non-compliance.

Analytical Separation Science is a monumental book written by two passionate thought leaders in the field who live and breathe separation science. The book is very detailed and comprehensive in nature, covering all fundamental and practical aspects ranging from the sample preparation stage to the statistical data interpretation. It is full of details, beautifully and richly illustrated and filled to the brim with the latest insights. However, this content is not conveyed to you in an encyclopaedic way but in a carefully designed didactic way.

Maximizing a deep understanding of the subject matter, the book helps you choose the right separation method through sound reasoning and in the most informed way. It will help you find your way through the panoply of options and the vast amount of information. Ultimately, the book will bring

Analytical Separation Science
By Bob W. J. Pirok and Peter J. Schoenmakers
© Bob W. J. Pirok and Peter J. Schoenmakers 2025
Published by the Royal Society of Chemistry, www.rsc.org

you to a level of understanding that will allow you to follow future developments and even generate numerous ideas for new research, missing knowledge, and gaps in our understanding.

Gert Desmet
Brussels

Author Biographies

Bob Pirok

Bob Pirok obtained his PhD in 2019 in Amsterdam, with the distinction *cum laude*. Earlier, he worked for several years in the industry. He is currently an associate professor at the University of Amsterdam, where he focuses on the application of chemometrics in analytical chemistry, with a special interest in method development and data analysis for (multi-dimensional) chromatography. He currently teaches separation science, industrial analysis, chemometrics and statistics. He is also a visiting research professor at Gustavus Adolphus College (St. Peter, MN, USA) in the group of Prof. Dwight Stoll.

He has received six international awards, including the Csaba Horváth Young Scientist Award in 2017, the *Journal of Chromatography* Award in 2018, and the HTC Innovation Award in 2024 for his work on the automation of method development in LC. He has successfully established a growing research group after being awarded several grants from the Dutch Research Council (NWO).

He was selected as an Early Career Board member for *Analytical Chemistry* in 2021 and is an Editorial Advisory Board member for the *Journal of Separation Science* and *LC-GC International*. *The Analytical Scientist* featured him in the 2021 and 2022 editions of the Power List. The Power List recognizes the 100 most influential analytical scientists worldwide as an inspiration to their fellows.

Analytical Separation Science
By Bob W. J. Pirok and Peter J. Schoenmakers
© Bob W. J. Pirok and Peter J. Schoenmakers 2025
Published by the Royal Society of Chemistry, www.rsc.org

Peter Schoenmakers

Peter Schoenmakers has been a full-time professor of Analytical Chemistry at the University of Amsterdam (UvA) since 2002. He has been teaching various subjects, but mostly analytical chemistry, and separation science in particular. He developed and directed national analytical-science honours programs for BSc and MSc students.

Earlier, he obtained a Master's Degree in chemical engineering from the Technical University of Delft, The Netherlands, and conducted his PhD research with Professor Leo de Galan in Delft and with Professor Barry Karger in Boston, MA, USA. Thereafter, he worked for Philips in Eindhoven (The Netherlands) and for Shell in Amsterdam and in Houston, TX, USA.

He has conducted research on many aspects of chromatography, with a recent emphasis on multi-dimensional liquid chromatography and a focus on applications involving macromolecules. Many of his projects have involved collaborations with various industries and research laboratories.

Peter Schoenmakers has published nearly 350 papers on chromatography and other aspects of analytical chemistry. He has received numerous professional awards and was awarded an ERC Advanced grant for the project STAMP (Separation Technology for A Million Peaks) in 2016. From 2005 to 2020, he served as an editor of the *Journal of Chromatography A*. Through royal recognition, he became a Knight in the Order of the Dutch Lion.

Contents

Analytical Separation Science
By Bob W. J. Pirok and Peter J. Schoenmakers
© Bob W. J. Pirok and Peter J. Schoenmakers 2025
Published by the Royal Society of Chemistry, www.rsc.org

2   Gas Chromatography   87

4 Separation of Large Molecules   346

5 Capillary Electrophoresis   410

8 Sample Preparation   560

9 Data Analysis: Chemometrics and Statistics 611

10 Method Development and Optimization 757

1 Fundamentals of Chromatography

The basic principles of analytical separation science are described in the first five modules of this chapter, which are devoted to chromatography, retention, selectivity, efficiency, and resolution and peak capacity, respectively. This knowledge is needed for a basic understanding of analytical separation science and for studying the remainder of this book. Thermodynamic equilibria, phase ratios, flow profiles, and column permeability are then discussed in somewhat greater depth (M). A large module is devoted to a closer look at chromatographic band broadening in packed and open-tubular columns at the M (and sometimes A) level. Non-linear adsorption isotherms also affect observed peak shapes and are highly relevant for preparative separations. The final module of this chapter describes three generic categories of analytical separations: target, group-type, and non-target or fingerprinting analyses (Figure 1.1).

1.1 Chromatography (B)

1.1.1 Chromatographic Separation

Chromatography is a family of separation techniques in which separation is based on the differential distribution of molecules between two (or more) phases that move at different velocities. Most forms of chromatography involve a stagnant stationary phase and a moving mobile phase. Separation is established due to different sample components, also referred to as **solutes** or **analytes**, being distributed differently between the two phases. This process is

Analytical Separation Science
By Bob W. J. Pirok and Peter J. Schoenmakers
© Bob W. J. Pirok and Peter J. Schoenmakers 2025
Published by the Royal Society of Chemistry, www.rsc.org

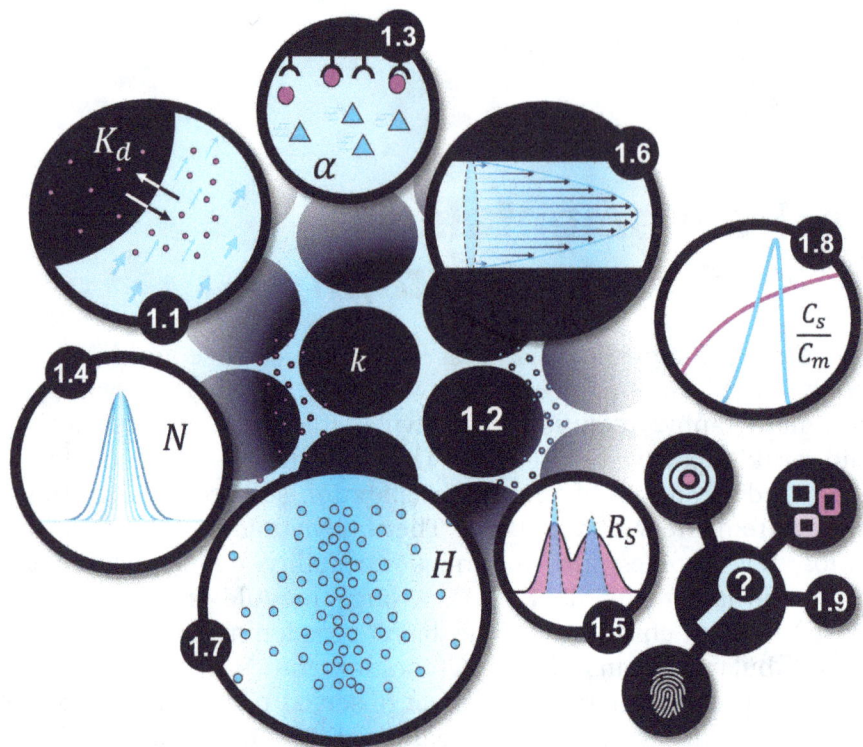

Figure 1.1 Graphical overview of the modules in Chapter 1.

depicted on the left side of Figure 1.2A. Here, analyte i is distributed between the two phases, with a quantity (weight) of the molecules in the mobile phase ($q_{i,m}$) and a quantity in the stationary phase ($q_{i,s}$). The distribution of analyte molecules is determined by their physicochemical interactions (or lack thereof) with the mobile and stationary phases. Analytes that spend time in the stationary phase are considered **retained**. An analyte with such a high affinity for the stationary phase that it exclusively resides there will be retained indefinitely. Conversely, analytes that do not enter the stationary phase will elute at the same velocity (u_0) as the mobile phase and are referred to as **unretained** analytes. The degree of retention is quantified as the **retention factor** and is given by

$$k_i = \frac{q_{i,s}}{q_{i,m}} \tag{1.1}$$

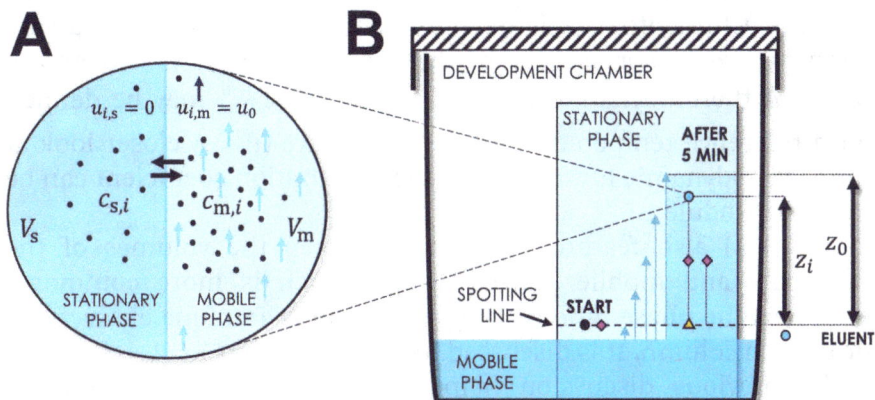

Figure 1.2 (A) The concept of a dynamic equilibrium in a partition process that governs the chromatographic separation. (B) Schematic representation of thin-layer chromatography.

The weights ($q_{i,\mathrm{m}}$ and $q_{i,\mathrm{s}}$) given in eqn (1.1) are the product of the concentration (c_i) of analyte i in each phase and the volume of these phases (V):

$$k_i = \frac{q_{i,\mathrm{s}}}{q_{i,\mathrm{m}}} = \frac{c_{i,\mathrm{s}}}{c_{i,\mathrm{m}}} \cdot \frac{V_\mathrm{s}}{V_\mathrm{m}} = K_{\mathrm{d},i} \cdot \phi \tag{1.2}$$

where ϕ is the phase ratio. Eqn (1.2) is an important extension, because it demonstrates, as shown in Figure 1.2A, a **thermodynamic equilibrium** with a distribution coefficient of $K_{\mathrm{d},i}$ that equals $c_{i,\mathrm{s}}/c_{i,\mathrm{m}}$. The phase ratio may be difficult to measure accurately, especially in liquid chromatography (LC; see Chapter 3).

At high concentrations of the analyte one or both phases may become saturated, and the relation between $c_{i,\mathrm{s}}$ and $c_{i,\mathrm{m}}$ may no longer be linear. In this case, K_d also depends on the analyte concentration. However, analytical separations are usually conducted at very low concentrations of analytes, so that K_d can be considered independent of the concentration. The relation between $c_{i,\mathrm{s}}$ and $c_{i,\mathrm{m}}$ is known as the **adsorption isotherm**. It is more extensively discussed in Module 1.8. For now, it suffices to understand that chromatographic separations are typically carried out in the region where the adsorption isotherm is linear. It is important to note that $K_{\mathrm{d},i}$ (eqn 1.2) is deliberately referred to as a **distribution coefficient** and not as a distribution constant or equilibrium constant. Indeed, even when independent of the analyte concentration, $K_{\mathrm{d},i}$ still depends on temperature and pressure. Chromatographic separations may be conducted at varying temperatures and pressures, and the word constant is thus not appropriate. A thermodynamic equilibrium constant $K_{d,i}^0$ may be defined for a reference temperature (T^0) and pressure (P^0). A closer look at the thermodynamic foundation of the distribution coefficient can be found in Module 1.6.

Eqn (1.2) also features the ratio between the volumes of the stationary and mobile phases, V_s/V_m, which is more commonly known as the **phase ratio** ϕ. The phase ratio is a unique characteristic of each column. It is discussed further in Section 1.6.2.

The previous discussion helps to quantify how the different molecules are distributed across the two phases. However, it does not explain how this distribution leads to separation, which requires considering that the two phases move at different velocities.

Referring back to Figure 1.2A, at any given time, a fraction of the analyte molecules is in the mobile phase $(f_{i,m})$, while the remaining fraction $(f_{i,s} = 1 - f_{i,m})$ is in the stationary phase. This is quantified by eqn (1.3a) and (1.3b).

$$f_{i,m} = \frac{q_{i,m}}{q_{i,m} + q_{i,s}} = \frac{1}{1 + q_{i,s}/q_{i,m}} = \frac{1}{1 + k_i} \tag{1.3a}$$

$$f_{i,s} = \frac{q_{i,s}}{q_{i,m} + q_{i,s}} = \frac{q_{i,s}/q_{i,m}}{1 + q_{i,s}/q_{i,m}} = \frac{k_i}{1 + k_i} \tag{1.3b}$$

The average linear velocity of the analyte can then be calculated from

$$u_i = f_{i,m} \cdot u_{i,m} + f_{i,s} \cdot u_{i,s} = f_{m,i} \cdot u_0 \tag{1.4}$$

where the final equality arises from the realization that $u_{s,i} = 0$ and from the introduction of the common symbol u_0 for the average mobile-phase velocity.

Separation can be established for compounds that distribute differently (*i.e.* show different values of $K_{d,i}$ and hence k_i and $f_{i,m}$) because they will move at different linear velocities. Eqn (1.4) can be rewritten by substituting $f_{i,m}$ from eqn (1.3a).

$$u_i = \frac{1}{1 + k_i} \cdot u_0 \tag{1.5}$$

So far, we have assumed that u_0 is a constant, which is often a good approximation. However, depending on the chromatographic system and conditions, u_0 may actually vary with position (along the length of the column) and time. Such situations will be considered when appropriate in later modules.

1.1.2 Thin-layer Chromatography

Thin-layer chromatography (TLC) is a simple and low-cost chromatographic method. It is used for tentative confirmation of the presence of specific analytes in relatively simple mixtures, assessment of compound purity, and monitoring the progress of chemical reactions. Paper chromatography is a classical implementation of TLC, but in most contemporary implementations, a thin layer of a **stationary-phase sorbent**, typically a silica gel, is formed with an inert binder on an inert plate made of glass, plastic, or sometimes aluminium foil. The stationary phase is then dried and activated, usually by heating. TLC plates are relatively cheap and widely available commercially.

The separation principle for TLC is depicted in Figure 1.2B. First, an environment ("development chamber") is formed by filling a transparent container with a small amount of mobile phase (typically with a depth of less than 10 mm), which is then closed and allowed to saturate with mobile-phase vapour. Optionally, this saturation is accelerated by partially immersing a piece of filter paper in the layer of the mobile phase.

Next, the TLC plate is prepared. Small aliquots of one or more samples are deposited next to each other using a syringe or a mechanical pipette along a so-called spotting line close to the bottom of the plate. The spotting line should be just above the mobile phase level when the plate is placed in the container, allowing sufficient migration length for the analytes. The lid is again placed on top of the chamber to minimize solvent evaporation.

After the plate has been positioned in the development chamber, the mobile phase will start migrating towards the top of the plate due to capillary action. When the mobile-phase front reaches the sample spots, the chromatographic process begins. Depending on their $K_{d,i}$ values, different analytes in the sample will migrate at different average velocities. After a sufficient amount of time – usually when the solvent front reaches a level close to the top of the plate – the plate is removed, and the separation on the plate is assessed. Visualization can be aided by spraying a reagent and/or using UV light (see Section 8.3.3).

In TLC, different analytes have the same migration time but different migration distances (z) due to different $K_{d,i}$ values. Eqn (1.5) can be written for TLC as

$$z_i = \frac{1}{1 + k_i} \cdot z_0 \tag{1.6}$$

where z_i is the migration distance of analyte i, and z_0 is the migration distance of the mobile-phase solvent. Traditionally, retention is expressed by the **retardation factor** (R_f), which is the ratio between z_i and z_0:

$$R_f = \frac{z_i}{z_0} \tag{1.7}$$

From eqn (1.6) and (1.7), the retention factor that corresponds to the TLC experiment can be calculated as

$$k_{i,\text{TLC}} = \frac{z_0 - z_i}{z_i} = \frac{1 - R_f}{R_f} \tag{1.8}$$

While TLC is useful for assessment and comparison of relatively simple mixtures, quantification is difficult, as it relies on measuring spot intensities, which cannot be easily performed with high precision. An advantage of TLC is that several parallel separations can be visually observed. Such results may be convincing, for example, as evidence in court. Another advantage is that (the combination of all) analytes that do not migrate can be visualized for each sample, which cannot be achieved in column high-pressure liquid chromatography (HPLC) using stainless-steel columns.

1.1.3 Column Chromatography

Column chromatography is a streamlined form of chromatography that is more effective for quantification and more amenable to automation. This book almost exclusively deals with **column chromatography**, although the fundamentals are similar for other techniques, such as TLC. Column chromatography uses separation

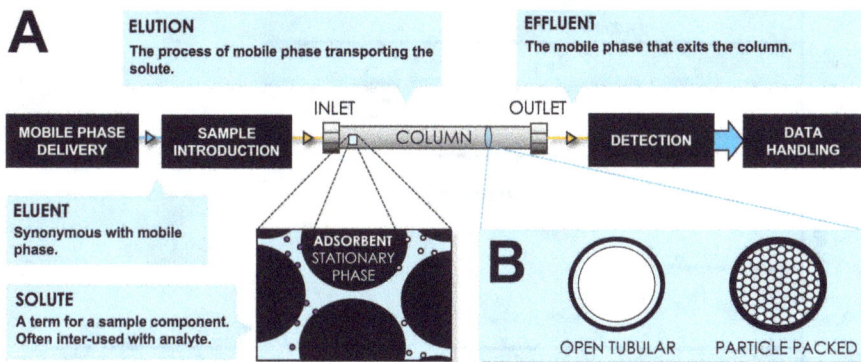

Figure 1.3 (A) Schematic representation of a chromatographic system. The yellow connections that connect the sample introduction system with the inlet of the column and the outlet of the column with the detection system contribute to the extra-column volume (V_{ec}), together with the volume of effluent in the detector (V_{det}). (B) Schematic representation of open-tubular and particle-packed columns.

channels that are typically cylindrically shaped and contain the stationary phase. A schematic representation of a chromatographic system using a column is depicted in Figure 1.3A. The scheme applies to all forms of column chromatography, although the techniques used for each of the functions in the boxes (mobile phase delivery, sample introduction, *etc.*) will differ. The remainder of this chapter will focus on the processes occurring in the column.

Chromatography is mostly conducted using either columns with a layer of stationary phase coated on the inner wall, so-called **wall-coated open-tubular columns** (WCOTs, Figure 1.3B, left), or **packed columns** filled with stationary-phase particles (Figure 1.3B, right). The stationary phase remains in the column, and the mobile phase is transported through it. Analyte bands that reach the detector can be measured to obtain a **chromatogram**, which presents the signal intensity of each analyte band ("peak") as a function of time (Figure 1.4).

For columns packed with porous particles, unretained analytes are considered to sample the entire volume of the column occupied by the mobile phase (V_m), including both the mobile-phase volume inside the pores of the porous particles (V_{pores}) and the (interstitial) volume outside the particles (V_{int}). When analyte molecules do not enter any of the pores (for example, as they are too large), they are considered **excluded** from the pores. This is exploited in size-exclusion chromatography, which will be discussed in detail in

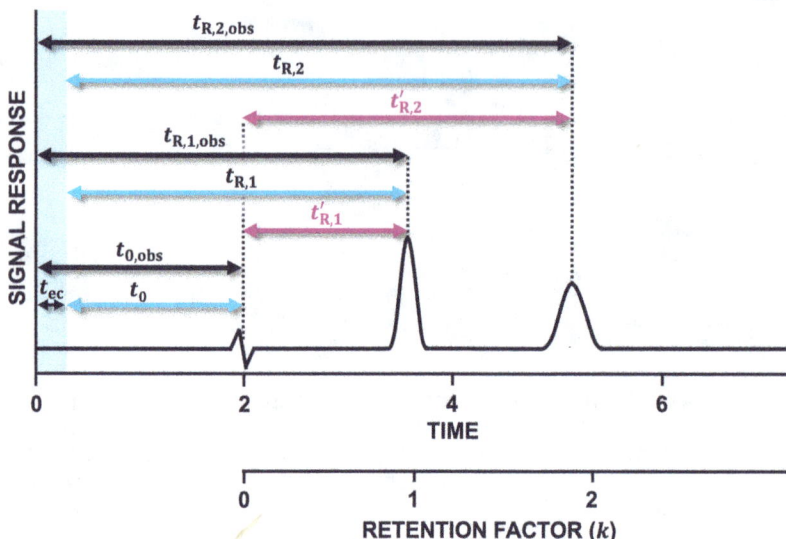

Figure 1.4 Schematic example of a chromatogram that presents different parameters related to the concept of retention.

Module 4.2. In the remainder of this chapter, it is assumed that analytes are not excluded.

1.2   Retention (B)

In contrast to TLC, where migration distances (z_i) are considered, all analytes that elute from a chromatographic column have spent an equal amount of time in the mobile phase, provided that the mobile-phase velocity is constant. Differences arise from the time analyte molecules spend in the stationary phase. Analytes are transported through a column of length L at an average migration velocity u_i, which is defined as

$$u_i = \frac{L}{t_{R,i}} \tag{1.9}$$

Here, $t_{R,i}$ is the time at which the analyte exits the column, which is referred to as the **retention time** of the analyte. It is the sum of the time that the analyte has spent in the mobile phase (t_m) and in the stationary phase $(t_{i,s})$. The latter is equal to the more commonly used

net retention time $(t'_{R,i})$. The process by which an analyte exits the column is also referred to as its **elution**, *i.e.* the analyte **elutes** from the column.

$$t_{R,i} = t_{i,s} + t_m = t'_{R,i} + t_m \tag{1.10}$$

Unretained analytes that never spend time in the stationary phase $(t'_{R,i} = t_{i,s} = 0)$ will migrate through the column at the same velocity as the mobile-phase molecules. The average linear velocity of the mobile phase molecules is u_0

$$u_0 = \frac{L}{t_m} = \frac{L}{t_0} \tag{1.11}$$

The time that mobile-phase molecules reside in the mobile phase (t_m) is often also denoted as t_0 and is interchangeably referred to as the dead time, hold-up time or unretained time. It relates to the distance travelled by the solvent front in TLC.

Eqn (1.9) and (1.11) can now be substituted into eqn (1.5) to relate the retention time to the retention factor k_i of analyte i in column chromatography:

$$t_{R,i} = (1 + k_i) \cdot t_0 \tag{1.12}$$

Thus, the **retention factor** k_i in column chromatography can be defined as

$$k_i = \frac{t_{R,i} - t_0}{t_0} = \frac{t'_{R,i}}{t_0} = \frac{t_{R,i}}{t_0} - 1 \tag{1.13}$$

This treatment and the term "retention time" are generally valid for analytes that are retained in the column. However, analytes that are fully or partially excluded from the pores of the packing material may elute before the mobile phase molecules. A more general term, **elution time,** is appropriate for situations that may occur for ions (such as electrostatic exclusion, see Section 3.7.4) or for macromolecules (such as size exclusion, Module 4.2). In the case of exclusion, retention factors would take on negative values, as shown in eqn (1.13); however, they lose their fundamental meaning (eqn (1.1)) and are not commonly used in such cases.

Eqn (1.12) and (1.13) are fundamental equations for quantifying retention in chromatography and will be used extensively throughout this book. The major advantage of the retention factor is that it is a dimensionless quantity, which can be used to compare different chromatographic systems. It should be noted that eqn (1.12) and (1.13) assume a constant retention factor during the chromatographic experiment. If conditions vary during the analysis, the retention factor, as defined in these equations, is not meaningful.

At this point, it is useful to reconsider the scheme of our chromatographic system shown in Figure 1.3A. The retention time is equal to the time the analyte spends in the mobile and stationary phases (eqn (1.10)). It is, therefore, not exactly equal to the time at which the analyte reaches the detector. Indeed, the schematic shown in Figure 1.3A depicts connections between the sample introduction system and the inlet of the column and between the outlet of the column and the detector. The time the analyte spends in these connections is referred to as the extra-column time (t_{ec}), during which no separation takes place. Since t_{ec} may be very small relative to t_0 and $t_{R,i}$, t_{ec} is often ignored in practice. Nevertheless, especially when retention times and factors of different systems are compared, an accurate assessment of the retention factor requires extending eqn (1.13) to

$$k_i = \frac{(t_{R,i,\text{obs}} - t_{ec}) - (t_{0,\text{obs}} - t_{ec})}{(t_{0,\text{obs}} - t_{ec})} = \frac{(t_{R,i,\text{obs}} - t_{ec})}{(t_{0,\text{obs}} - t_{ec})} - 1 = \frac{t_{R,i,\text{obs}} - t_{0,\text{obs}}}{t_{0,\text{obs}} - t_{ec}} \qquad (1.14)$$

where the subscript **obs** denotes the observed time. In the remainder of this book, when we use t_0, $t_{R,i}$ or $t'_{R,i}$, we will assume that t_{ec} is either negligible or properly accounted for.

Eqn (1.12) and (1.14) also demonstrate that the retention factor can be determined from a chromatogram. This is shown in Figure 1.4, where two different analytes are shown to elute at two different retention times, $t_{R,1}$ and $t_{R,2}$. If a signal corresponding to an unretained analyte is observed, then the dead time, t_0 (or t_m), may be determined from the chromatogram and the retention factor can be calculated. Note that this is not always the case and that a perturbance signal, such as that displayed in Figure 1.4, does not always mark the exact t_0. The two different x-axes shown in Figure 1.4 depict the relation between time and retention factor.

Throughout this section, we related the linear velocity of the analyte to the migration time. Neither property, however, is uniquely defined. As shown in Figure 1.3, a mobile-phase delivery system pushes the mobile phase through the column. Almost all chromato-graphic systems rely on pressure to transport the mobile phase. Electro-driven chromatographic separations (capillary electro-chromatography; see Section 5.1.4.5) are rare. The volumetric **flow rate** (F) is the amount of mobile phase that passes through the column in chromatography, and it is typically expressed in mL min^{-1}, although miniaturized ("micro") chromatographic systems operate at flow rates in the μL min^{-1} range, and even "nano" separation systems operating in the nL min^{-1} exist today. The mobile-phase flow rate may increase along the length of the column if the mobile phase is compressible. In a stationary state (with a constant volumetric flow, pressure, and temperature), the mass flow (mass time^{-1}, *e.g.* g min^{-1}) is constant.

The higher the flow rate on a given column, the higher the average linear velocity and, thus, the shorter the retention time. This relation between the linear velocity and the volumetric flow rate in **open-tubular chromatography** is given by

$$u_{0,ot} = \frac{F}{A_{cross}} = \frac{4F}{\pi(d_c - 2d_f)^2} \tag{1.15}$$

where A_{cross} is the cross-sectional area of the column, d_c is its internal diameter and d_f is the film thickness of the stationary phase. For **packed-column chromatography**, the relation is

$$u_{0,pc} = \frac{F}{A_{cross}} = \frac{4F}{\varepsilon_{tot} \cdot \pi \cdot d_c^2} \tag{1.16}$$

where ε_{tot} depicts the total column **porosity**, which is the total fraction of the column volume that is occupied by the mobile phase. In the case of porous particles, this is the mobile-phase volume inside the pores of the particles (V_{pores}), plus the interstitial volume between the particles (V_{int}), divided by the total volume of the empty column (V_c)

$$\varepsilon_{tot} = \varepsilon_{pores} + \varepsilon_{int} = \frac{V_{pores}}{V_c} + \frac{V_{int}}{V_c} = \frac{V_0}{V_c} \tag{1.17}$$

The complement of the porosity, $1 - \varepsilon_{tot}$, represents the fraction of the column volume occupied by the (solid) packing material and

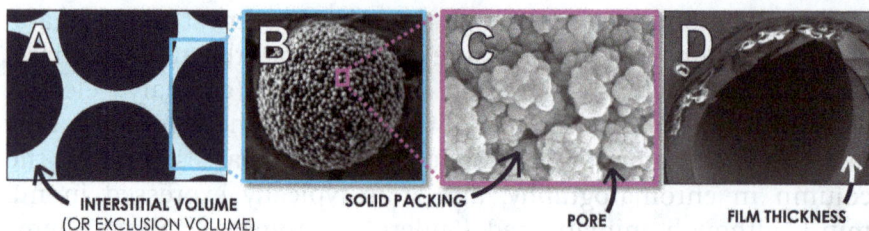

Figure 1.5 (A) Schematic depiction of a packed column. (B) Scanning electron microscopy (SEM) photograph of a stationary-phase particle used for particle-packed columns.[1] (C) SEM photo of the porous outer surface of a stationary-phase particle.[2] (D) SEM photo of an open-tubular capillary with a stationary-phase film.[3] Panel B reproduced from ref. 1 with permission from Elsevier, Copyright 2017. Panel C reproduced from ref. 2 with permission from Elsevier, Copyright 2016. Panel D reproduced from ref. 3 with permission from the Royal Society of Chemistry.

the stationary phase. Note that we are discussing the porosity of the columns. Sometimes, the porosity of the particles is also used ($\varepsilon_{particle} = V_{pores}/(V_c - V_{int})$) to characterize a packing material.

Estimating the dead time, t_0, and, consequently, the porosity is important for measuring retention factors. In Module 3.11, we will see that measuring t_0 is also often useful for troubleshooting in LC (*e.g.* to diagnose pump malfunctioning). The different volumes contributing to the porosity are illustrated in Figure 1.5. In case of size-exclusion chromatography (Module 4.2), the interstitial volume is commonly referred to as the total exclusion volume ($V_{int} = V_{excl}$ and $\varepsilon_{int} = \varepsilon_{excl}$). For uniform, perfectly packed spherical particles, ε_{int} is roughly 0.4, whereas ε_{pores} ($=V_{pores}/V_c$) is roughly 0.2 to 0.3 for typical LC particles but may be as high as 0.4 (see Module 4.2). The total porosity, thus, typically is 60% to 70% of the total column volume ($0.6 < \varepsilon_{tot} < 0.7$). For size-exclusion chromatography columns, ε_{pores} can be up to 0.4, so that $\varepsilon_{tot} \approx 0.8$ and $V_{excl} = V_{int} \approx 0.5\, V_0 \approx 0.4\, V_c$. The total porosity can be estimated by measuring the elution time of a t_0 marker and solving eqn (1.16) for ε_{tot}.

1.3 Selectivity (B)

To assess the capability of a chromatographic method to distinguish between two analytes, it can be useful to consider the differences in their retention times.

$$\Delta t_{R,j,i} = t_{R,j} - t_{R,i} = t'_{R,j} - t'_{R,i} \qquad (1.18)$$

Here, $t_{R,j}$ and $t'_{R,j}$ depict the retention time and the net retention time of the last eluting peak of a pair, respectively, and $t_{R,i}$ and $t'_{R,i}$ denote the corresponding properties of the first eluting peak, respectively. Consequently, $\Delta t_{R,j,i}$ is always positive. It is indicative of our ability to separate the two compounds, but the value is highly dependent on the specific column and conditions used, and depending on these conditions, smaller or larger values of $\Delta t_{R,j,i}$ may be required to achieve good separation (see Module 1.5).

To compare different combinations of stationary and mobile phases, it is more useful in practice to use the ratio $\alpha_{j,i}$ of net retention times or (dimensionless) retention factors. Eqn (1.19) can be used to define the **selectivity** ($\alpha_{j,i}$) solely based on retention factors.

$$\alpha_{j,i} = \frac{t'_{R,j}}{t'_{R,i}} = \frac{k_j}{k_i} \qquad (1.19)$$

In chromatography, α is the parameter that pertains mostly to the selectivity of the phase system and its ability to chemically separate two different analytes. Its value is largely determined by the chemistry of the mobile and stationary phases in relation to that of the analytes. A value of $\alpha_{j,i} = 1$ implies exactly co-eluting analytes (no separation), whereas a high value of $\alpha_{j,i}$ means that the two analytes can easily be separated. Consequently, eqn (1.19) is another fundamental equation for chromatographic separations that will be frequently used throughout this book.

It is also of interest to express the selectivity in terms of distribution coefficients by substituting eqn (1.2) into eqn (1.19) to obtain eqn (1.20).

$$\alpha_{j,i} = \frac{K_{d,j}}{K_{d,i}} \qquad (1.20)$$

While less useful in practice than eqn (1.19), eqn (1.20) clearly shows that the selectivity is exclusively affected by thermodynamic factors. Indeed, distribution coefficients are only influenced by the analyte, the phase system (*i.e.* mobile-phase and stationary-phase chemistries), temperature and, to a lesser extent, pressure. They are not influenced by "physical parameters", such as the flow rate or the column length (except indirectly through changes in pressure). Parameters that only affect the phase ratio affect the retention factors but do not affect α. These include the inner diameter of open-tubular columns and the surface area of packing materials. However, especially in the latter case, the chemistry of different phases may be somewhat different, resulting in changes in α.

1.4   Efficiency (B)

Thus far, each analyte has been regarded as an assembly of molecules that migrates through the chromatographic column at a given velocity (u_i). However, in practice, each individual molecule of analyte i travels at a slightly different speed through the column. These variations are caused by several different processes that ultimately lead to the broadening of all chromatographic peaks. **Band broadening** is an important factor to consider, as it may cause two theoretically separated peaks to still overlap. Band broadening is quantitatively described in terms of column **efficiency**, which is the topic of this module.

1.4.1   Diffusion

Chromatography, like any other separation principle, is a process of transporting mass from one location to another. Transport under the influence of an external gradient (*e.g.* pressure or voltage) is referred to as **convection. Diffusion** refers to the movement of individual molecules while the fluid is at rest. In chromatography, the mobile phase and the analytes contained in it are transported through the column by convective movement, while diffusion is needed to move analyte molecules between moving and non-moving zones in the column. Separation is established when molecules from one analyte are transported more rapidly (or in a given time, as in TLC, across a greater distance) than those pertaining to another analyte because they spend relatively more time in the mobile phase. In earlier sections, we addressed how thermodynamic equilibria play a central role in achieving separation, but the rates at which such equilibria are achieved have not yet been considered. To understand why such rates are essential to achieving separation, diffusion must be addressed in more detail.

Diffusion is an irreversible process, also known as Brownian motion. It follows from the second law of thermodynamics, as it leads to an increase in entropy. The process is schematically depicted in Figure 1.6, where we assume that a droplet of a concentrated solution of analytes (*i.e.* soluble molecules in the same or a fully miscible solvent) is immersed in a liquid. At this point (Figure 1.6A), there exists a high concentration of the analyte in one fraction of the liquid and a zero or extremely low concentration in the remaining bulk of the liquid. The analyte molecules will consequently diffuse into the remaining liquid (Figure 1.6B) until all analytes are equally or "homogenously" distributed throughout the entire liquid. At that point, the system is in equilibrium (Figure 1.6C). Individual molecules will still be moving, but the average distances they travel become zero in all directions.

The driving force for molecular diffusion is a chemical potential gradient, *i.e.* the difference in chemical potential between liquids that contain different concentrations of the analyte. Diffusional transport in the direction of the gradient is mathematically expressed by **Fick's first law**

$$J_{i,x} = -D_i\frac{\mathrm{d}c_i}{\mathrm{d}x} \tag{1.21}$$

where $J_{i,x}$ is the molar flow of the analyte per unit area, also known as the flux of analyte i, in the x-direction, D_i is, by definition, the diffusion coefficient of the analyte, and c_i is its concentration. The

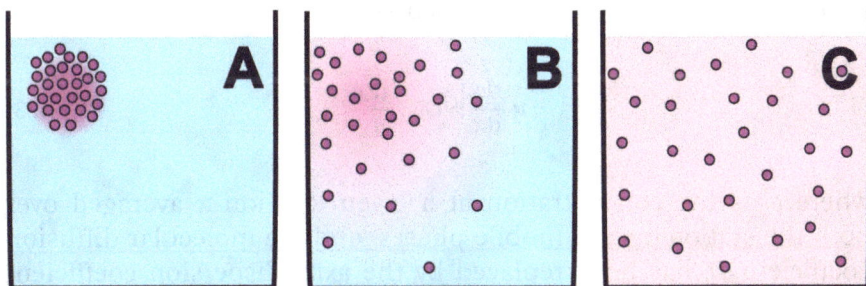

Figure 1.6 Schematic representation of molecular diffusion. (A) Upon injection of an analyte-containing droplet in a larger volume of solvent (blue), there exists a sharp concentration gradient between the two fractions. (B) The analytes (purple) will diffuse into the surrounding bulk liquid until (C) they are homogenously dispersed throughout the bulk liquid. The concentration gradient in each panel is illustrated by a blue-to-pink transition.

negative sign in eqn (1.21) is of importance, as it indicates that the flux is always in the direction from a high to a low concentration (opposite direction from the concentration gradient dc_i/dx). Fick's law helps us understand the process shown in Figure 1.6.

Molecular diffusion is both a blessing and a curse for chromatographers and chromatographic separations. When the mass transfer between mobile and stationary phases is considered, a high rate of molecular diffusion allows the separation process to take place by allowing different distributions between the mobile phase and stationary phase for different analytes to be established rapidly. However, inside a chromatographic column, each analyte will also diffuse within the mobile phase (and to a lesser extent in the stationary phase) in the axial direction (*i.e.* the direction of the convective flow), since every analyte band is formed by concentration gradients. Diffusion from the centre of the band leads to mixing and zone broadening instead of separation.

Consequently, it is paramount to consider the time component, or the rate, at which a diffusion process takes place. To quantify this, **Fick's second law** can be used, with t depicting the time

$$\frac{dc_i}{dt} = D_i \frac{d^2 c_i}{dx^2} \tag{1.22}$$

Eqn (1.22) expresses the rate of concentration change in relation to the spatial rate of change in the direction of the concentration gradient. For chromatographic separations, where there is also a (convective) flow component with a velocity u, the rate of concentration change in time can be expressed as

$$\frac{dc_i}{dt} = -u \frac{dc_i}{dx} + D_{ax,i} \frac{d^2 c_i}{dx^2} \tag{1.23}$$

where c_i is the concentration at a given location x averaged over both the stationary and mobile phases, and the molecular diffusion coefficient D_i has been replaced by the axial dispersion coefficient $D_{ax,i}$ because several processes other than molecular diffusion also contribute to axial dispersion, as will be explained in the following sections.

1.4.2 Band Broadening

Eqn (1.21)–(1.23) help to understand molecular diffusion and the processes that occur in a chromatographic column. When a volume of a homogeneous sample mixture is introduced into the system, the analyte molecules are equally dispersed in the sample solvent ("sample plug"), and their concentration is much higher than that in the surrounding mobile phase (Figure 1.7). Diffusion will immediately take place in the direction of flow (as well as in the opposite direction), which is referred to as **axial dispersion.**

There are several processes that contribute to the broadening of peaks or – more correctly – analyte bands in the chromatographic system. Figure 1.7 illustrates band broadening with the difference in the relative distance of two zones depicted as Δz and the width of a peak as w. As the two analyte zones travel through the chromatographic system, their relative distance Δz increases with travelled distance z when the linear velocity and other conditions, such as temperature or mobile-phase composition, are constant. Moreover, the width of each zone w will increase with $\sqrt{z}$ (see Section 1.7.4).

Note that the width of each analyte band is roughly equal at a given distance z. In Figure 1.7, this translates to $w_i \approx w_j$, assuming that the two peaks are very close to each other.

Of course, diffusion can also take place in the perpendicular direction, referred to as **radial diffusion,** but this is less relevant, given the fact that the analyte concentration gradient is much less

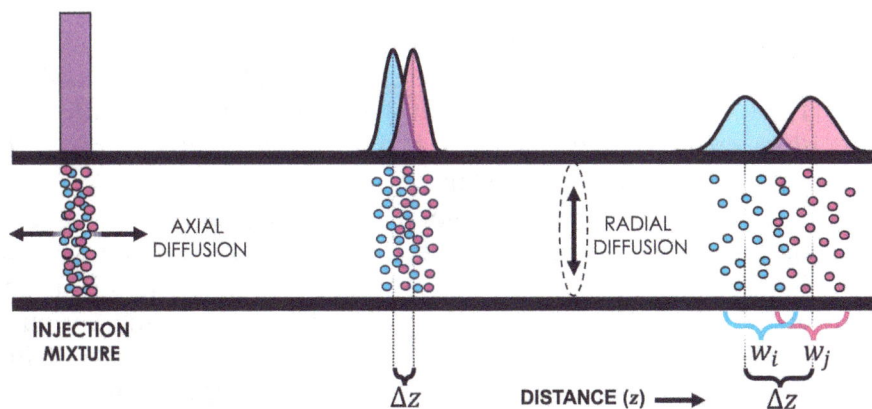

Figure 1.7 Schematic depiction of axial diffusion of two analyte bands within a cylindrical chromatographic column with a flow from left to right.

steep because, ideally, the sample plug is spread evenly across the entire cross section of the column. This is one of the functions of the porous filter ("frit") at the inlet of packed columns. In addition, the column walls limit the room for dispersion. Relevant exceptions to this apply and are discussed in Module 1.7, but the present description suffices for the purpose of explaining column efficiency. As the plug migrates through the column, the band broadening continues.

Molecular diffusion processes are statistical by nature. Indeed, molecules diffuse in a random manner continuously as they migrate through the column. Consequently, the solute molecules arrive at the detector with different average linear velocities, exhibiting a certain spread. The spreading of the analyte molecules in the mobile phase, and, consequently, the peak in the chromatogram, typically takes the form of a normal or **Gaussian distribution.**

If $f(t)$ is the signal (detector response), as a function of time, and t_R is the retention time of the peak, then a Gaussian peak can be described by

$$f(t) = h_{peak}\, e^{-\frac{1}{2}\left(\frac{t-t_R}{\sigma}\right)^2} \tag{1.24}$$

Here, σ is the standard deviation of the peak (in time units), a measure of its width, and h_{peak} is the height at the peak maximum, which can be related to the area of the peak (A_{peak}) by

$$h_{peak} = \frac{A_{peak}}{\sigma\sqrt{2\pi}} \tag{1.25}$$

To quantify band broadening, it is relevant that eqn (1.25) allows σ to be computed from the peak area and height. When applying eqn (1.25), attention must be paid to the units in which the different quantities are expressed. For example, if h_{peak} is expressed in mV and σ in s, A_{peak} should be expressed in mV s.

At this point, it is important to note that while, ideally, chromatographic peaks are of Gaussian shape, they usually are not in practice. This is due to finite sample concentrations, inhomogeneities in or on the stationary phase, dead volumes in the system, and various other factors. In LC, peaks tend to be less symmetrical than in gas chromatography (GC). Nevertheless, to a first approximation, a Gaussian peak shape can be assumed for a chromatographic peak.

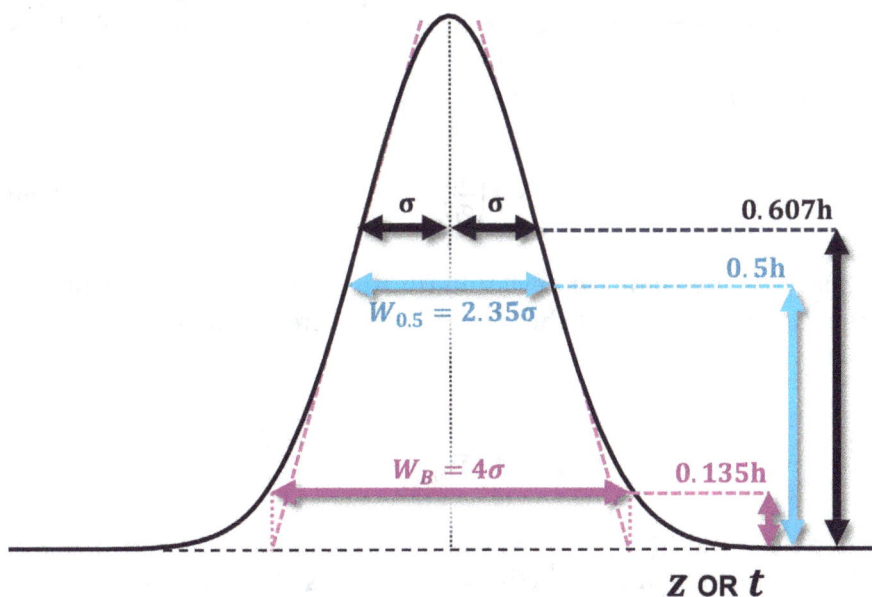

Figure 1.8 Some characteristics of a Gaussian peak. The peak width can conveniently be measured from various heights of the peak.

One particularly attractive characteristic of the Gaussian distribution is that σ can be computed from the peak width at various heights of the peak. This is graphically illustrated in Figure 1.8. It can be seen from this figure that the variance of a genuinely Gaussian peak can be determined by measuring the peak width at some fixed fraction of the peak height. The peak width at the base is usually very hard to measure accurately unless the peak is sufficiently resolved to draw tangent lines from the slopes to the baseline (dotted lines in Figure 1.8). Measurement of the peak width at half-height ($W_{0.5} = 2.35\sigma$) is often considered the most practical approach. However, most measurements of peak width lack precision and accuracy. Several alternative methods to characterize the peak shape, peak width and peak symmetry in chromatography are discussed in Chapter 9.

1.4.3 The Plate Number

The ability to measure the peak width to estimate σ is indispensable to studying the extent of band broadening or, conversely, the efficiency of a chromatographic system. This latter property is commonly expressed in terms of the number of theoretical plates (N), which can be derived from a chromatogram recorded under constant conditions as follows

$$N = \left(\frac{t_R}{\sigma_t}\right)^2$$
(1.26)

where σ_t is the standard deviation (in time units) of the distribution underlying the peak. If the peak is considered to be Gaussian (see Section 1.4.2), eqn (1.26) can also be written as

$$N = 5.54\left(\frac{t_R}{W_{0.5}}\right)^2 = 16\left(\frac{t_R}{W_{0.135}}\right)^2 \approx 16\left(\frac{t_R}{W_B}\right)^2$$
(1.27)

where $W_{0.5}$ and $W_{0.135}$ are the peak widths at half-height and at 13.5% of the peak height, respectively, and W_B is the approximate peak width at the baseline, found from tangent lines drawn from the peak slopes (see Figure 1.8). The peak height (h_{peak}) and the peak area of a well-resolved peak may also be used to obtain

$$N = 2\pi\left(\frac{t_R \cdot h_{peak}}{A_{peak}}\right)^2$$
(1.28)

As with eqn (1.25), units are important when applying eqn (1.28). N is dimensionless, so if t_R is expressed in seconds and h in mV, the area should be expressed in mV s. Integrators or data stations may use different units. Sometimes these are obscure ("counts"), and a conversion or calibration may be needed.

The word **plate** is derived from distillation technology used in industry and shares a number of analogies (Figure 1.9).

The plates in a chromatographic column are theoretical (or virtual), in contrast to the physical plates (equilibrium stages, referred to as "plates") present in industrial distillation columns (see inset on the left of Figure 1.9). The "height equivalent of a theoretical plate" (HETP), usually abbreviated as "plate height" (H), can be calculated from the number of plates and the column length (L)

$$H = \frac{L}{N}$$
(1.29)

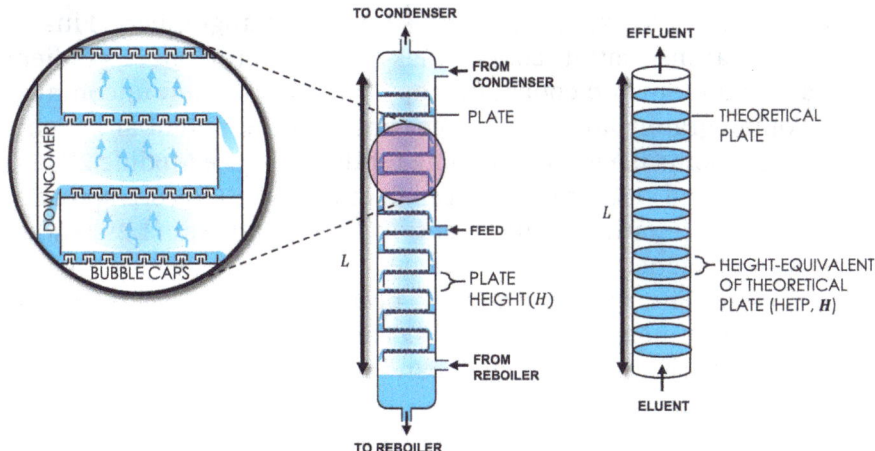

Figure 1.9 Illustration of an industrial distillation column (left) with actual plates (bubble trays) and a chromatographic column (right) with theoretical (virtual) plates. Adapted from ref. 4 with permission from the Author.

A low plate height is of great relevance because distillation "towers", which are often the highest columns in chemical plants or oil refineries, can be lower, while yielding the same performance (N) if the plate height (H) is lower. The same is true for chromatographic columns. Columns and retention times can be shorter if H can be reduced.

Ironically, "distillers" (chemical engineers) also speak about "theoretical plates", but these are defined as trays in full equilibrium, which is not achieved in practice, despite constructions such as the bubble caps in Figure 1.9 (left). The bubble caps are intended to maximize the contact between the vapour rising through the trays and the liquid streaming down in a zigzag manner. An oil distillation column may contain, say, 40 plates and be 25 m tall, with a plate height of say 0.5 m. The chromatography column on the right in Figure 1.9 could be a 0.05 m-long LC column with a plate height of 0.000005 m (5 μm), yielding 10 000 theoretical plates. An open-tubular GC column could be 25 m long (such as an industrial distillation column), but it would be coiled (if only to fit it in an oven), and it may easily produce 100 000 plates. There are more differences between industrial distillation and column chromatography. In the former, there is a continuous feed somewhere in the middle of the column, whereas the latter is a batch process with injections from one end. A distillation column has void areas at the top and bottom that would cause nightmares for chromatographers. Finally, an industrial distillation column has a static temperature gradient (hot at the bottom and cool at the top). This is extremely uncommon in chromatography, although discussions on the potential benefits of such gradients have been ongoing for decades (see Module 2.7).

Eqn (1.26) and (1.27) reveal that high efficiencies (N) correspond to narrow peaks (relative to the retention time), and eqn (1.29) shows that high efficiencies may be obtained if low plate heights (H) can be achieved. The narrower a peak becomes, the easier it can be separated from neighbouring peaks. These concepts can be used to understand the factors that affect band broadening.

1.4.4 The *van Deemter* Equation

In chromatography, the plate height varies with the linear velocity (u_0) of the mobile phase as it moves through the column. This variation can be characterized by an *H vs. u* curve and described by a plate-height equation. The first such equation was described in a seminal paper by van Deemter *et al.*, as far back as 1956.[5] The general, simplified form of the **van Deemter equation**, as it is still referred to today, is

$$H = A + \frac{B}{u} + C \cdot u \tag{1.30}$$

The equation shows a number of contributions to the plate height, depicted as the A, B and C terms. A common mental picture for the A-term or **eddy diffusion** term in the *van Deemter* equation is that it arises from the path-length differences as the different molecules move through the column. This is especially relevant in the case of particle-packed columns, where short paths between particles and longer paths around particles can be imagined (Figure 1.10A, upper right). Each solute molecule can take a different path, with a slightly different length, through the column, resulting in spreading of the solute band. An alternative explanation is that each molecule has a slightly different linear velocity at any one time (known as the Gidding–Eyring stochastic theory).[6,7] van Deemter (Box 1.1) assumed A to be independent of u, but this is a bit simplistic (see Module 1.7).

The B-term is due to the molecular diffusion, as discussed in Section 1.4.1. Analyte molecules unavoidably diffuse away from a region of high concentration (the top of the peak) towards regions of low concentration.

Jan Jozef van Deemter (1918–2004, The Netherlands)

Jan Jozef van Deemter obtained his PhD from the University of Amsterdam (The Netherlands) in 1950 while he was already employed (since 1947) at the Royal Dutch Shell Laboratory in Amsterdam (KSLA). Together with colleagues F.J. Zuiderweg and A. Klinkenberg, he developed a detailed model for dispersion in chromatographic columns, which was published back in 1956. We still use the concepts from his work and – most importantly – his name lives on in the numerous times chromatographers talk about the *van Deemter* equation.

Several different processes underlie the *C*-term, with mass transfer as the common denominator. One important process arises from incomplete equilibration and is depicted in Figure 1.10 (C_s). In the leading edge of the peak, the eluent progressing through the column contains a slightly above-equilibrium concentration of the analyte.

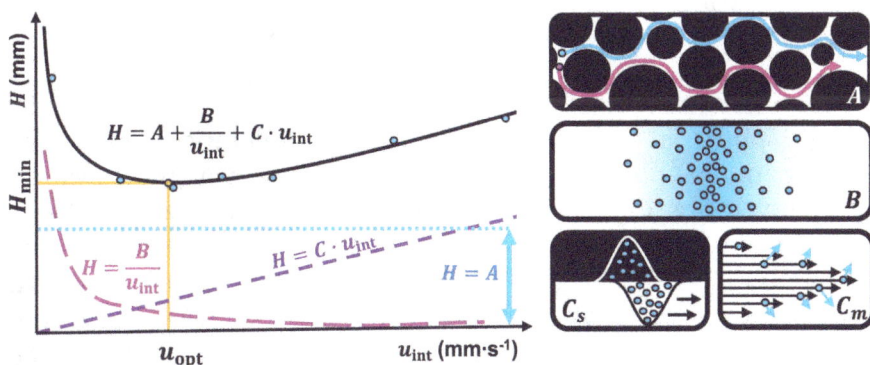

Figure 1.10 Graphical illustration of the van Deemter equation describing the plate height as a function of the mobile-phase linear velocity and schematic illustrations of the three different contributions. Adapted from ref. 4 with permission from the Author.

At the tailing edge of the peak, the concentration of the analyte is slightly above the equilibrium value in the stationary phase.

The *van Deemter* equation has several fundamental applications that enable the study of column technology and band broadening. The physicochemical processes that are described by each of the terms are much more complex in reality. Examples include the flow profile (C_m; see Figure 1.10), which was not considered in the original work by *van Deemter et al.* Another example is that *van Deemter et al.* and most of their followers used the interstitial linear velocity (u_{int}) as their measure of the linear velocity. These and other aspects are discussed in more detail in Module 1.7. Here, we will discuss just the most practical application of the *van Deemter* equation, which allows a convenient determination of the optimal flow rate for a chromatographic system.

In the basic treatment shown below, the (average) interstitial velocity (u_{int}) is used. In a packed-column LC system, u_{int} can be estimated in various experimental ways, for example, by measuring the elution time of a small amount of a UV-active salt, such as KBr, using a low-ionic-strength mobile phase (with the charge on the stationary-phase surface causing it to be excluded from the pores),[8] or the elution time of a high-MW polymer that does not interact with the stationary phase and is too large to enter any of the pores (see Module 4.2, size-exclusion chromatography). Peaks for such analytes are expected to elute around the exclusion time, t_{excl}. Alternatively, an interstitial porosity of $\varepsilon_{int} = 0.4$ may be assumed to find u_{int} from

$$u_{int} = \frac{L}{t_{excl}} = \frac{L \cdot F}{\varepsilon_{int} \cdot V_c} = \frac{4F}{\varepsilon_{int} \cdot \pi \cdot d_c^2} \approx \frac{10F}{\pi \cdot d_c^2} \tag{1.31}$$

The interstitial velocity (u_{int}) is the average velocity of the mobile phase as it moves through the interstitial space. The average mobile-phase velocity ($u_0 = L/t_0$) is a weighted average of the velocity in the interstitial space and the zero velocity of stagnant mobile phase in the pores ($u_{pores} = 0$), resulting in $u_0 = \left(\varepsilon_{int} \cdot u_{int} + \varepsilon_{pores} \cdot u_{pores}\right)/\left(\varepsilon_{int} + \varepsilon_{pores}\right) = \varepsilon_{int} \cdot u_{int}/\varepsilon_{tot}$. If u_0 is used, different values for the B and C coefficients will be found, but the resulting optimal flow rate and minimum plate height will be identical. In open-tubular columns, which are commonly used in GC, there are no particles, so there is no issue, and u_0 can be conveniently used.

If we measure the peak width (Section 1.4.2) of a selected compound at various flow rates (*i.e.* linear velocities), we can fit the *van Deemter* equation (eqn (1.30)) to the obtained data using a regression method. This is illustrated in Figure 1.10, where the resulting H vs. u curve shows the typical shape of the *van Deemter* curve, featuring a minimum plate height (H_{min}) at a specific velocity ($u_{int,opt}$). Setting the derivative dH/du equal to zero yields

$$u_{int,opt} = \sqrt{\frac{B}{C}} \qquad (1.32)$$

and

$$H_{min} = A + 2\sqrt{B \cdot C} \qquad (1.33)$$

In the *van Deemter* curve (Figure 1.10), three regions can be recognized:

- The B-branch, where $u_{init} < u_{int,opt}$. Here, the system performs below its potential ($N < N_{max}$), while the retention times are long. Less performance in more time is a lose–lose situation. Thus, the B-branch is the place to avoid.
- The optimum, where $u_{int} \approx u_{int,opt}$. Here, the system performs best ($N \approx N_{max}$), but the retention times are relatively long. If we need the performance (difficult separations), this is the place to be.
- The C-branch, where $u_{int} > u_{int,opt}$. Here, we sacrifice some performance ($N < N_{max}$), but we gain time. This is the region where compromises can be struck.

The latter kind of compromise is illustrated in Figure 1.11. Working at two or three times the optimum velocity may be attractive because this leads to a minor loss in performance (about 9% or 21%, respectively, in the example shown in Figure 1.11), while the speed (u/H) increases by about 81% (a drop in the time per plate, H/u, by 45%) and 136% (H/u decrease by 58%), respectively.

At higher speeds, the return on investment drops. In the present example, when $u = 19 \cdot u_{opt}$, the performance has dropped as much as the time per plate (see inset). After this point, increasing the velocity is no longer effective. Working at velocities higher than a few times the optimum value may be impractical in any case because the

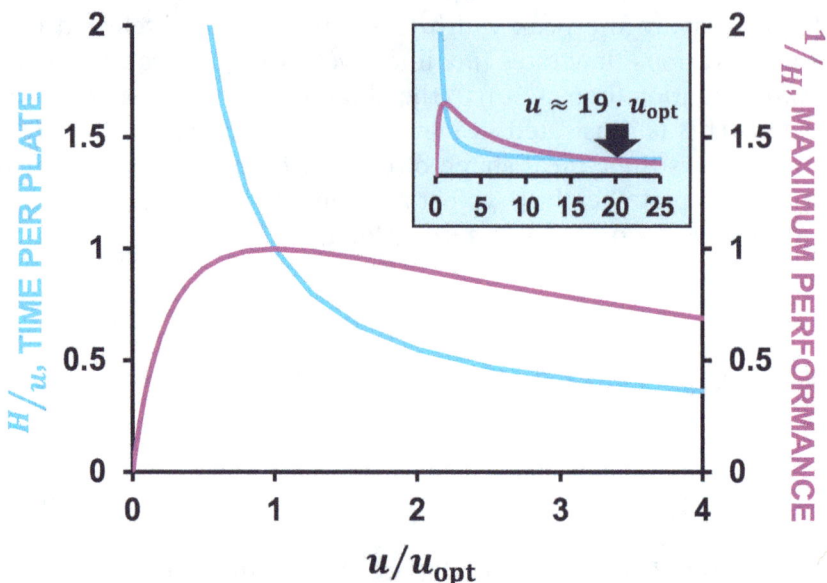

Figure 1.11 Illustrating the performance of a chromatographic system in the B-branch and the C-branch of the van Deemter equation. Both the time per plate (blue) and the performance (pink) are normalized to unity at $u = u_{opt}$. Figure constructed with $A = 1.2$, $B = 2$, $C = 0.08$.

price to pay for the increase in performance is an increase in the pressure drop across the column (see Module 1.6).

The *van Deemter* equation is not the only plate-height equation relevant for chromatographic separations, and many other useful models exist. One especially relevant example is the Golay equation, which applies to open-tubular chromatography and is discussed in Section 1.7.7.

1.4.5   Peak Dilution

In all forms of chromatography, peaks are diluted. Chromatography is not a method to separate a mixture into its pure constituents, unlike distillation. Instead, chromatography yields a series of analyte bands that are **diluted** in the mobile phase. The concentration of analyte i at the top of its peak $(c_{i,\max})$ can be determined using the conservation of mass between the injected and detected peaks:

$$c_{i,\max} = \frac{c_{i,\text{inj}} \cdot V_{\text{inj}}\sqrt{N}}{V_0\sqrt{2\pi}(1 + k_i)} \tag{1.34}$$

where $c_{i,\text{inj}}$ is the injection concentration and V_{inj} is the injection volume. A column with a high plate count and a small volume is desirable for minimizing the dilution of the analyte. High injection concentrations and large injection volumes are favourable, as long as these do not jeopardize the chromatographic separation. A dilution factor can be defined as DF = $c_{i,\text{inj}}/c_{i,\max}$. Extra-column effects may add to the dilution of analytes (see Section 3.3.3, Extra-column Band Broadening).

1.5 Resolution and Peak Capacity (B)

1.5.1 Resolution

Ultimately, the goal of all separations is to isolate different solutes in a mixture. These may represent a targeted set of solutes or the general characterization of the entire mixture. As mixtures grow in complexity, a review of the chromatogram itself will, in practice, often provide insights into the qualitative performance of the method. However, peak overlap can be a major problem in characterizing and quantifying different solutes after separation. In such cases, where (partial) overlap of peaks is a concern, the **resolution** between two successive peaks is a quantitative measure of the degree of separation.

The fundamental equation for the resolution, R_S, between two neighbouring peaks is

$$R_{S,j,i} = \frac{t_{R,j} - t_{R,i}}{2(\sigma_{t,i} + \sigma_{t,j})} \tag{1.35}$$

The elution bands of two adjacent peaks are, in practice, often similar in width (*i.e.* the standard deviations in time units are approximately equal, $\sigma_{t,i} \approx \sigma_{t,j}$). Eqn (1.35) is thus often written as

$$R_{S,j,i} \approx \frac{\Delta t_{R,j,i}}{4\sigma_t} \tag{1.36}$$

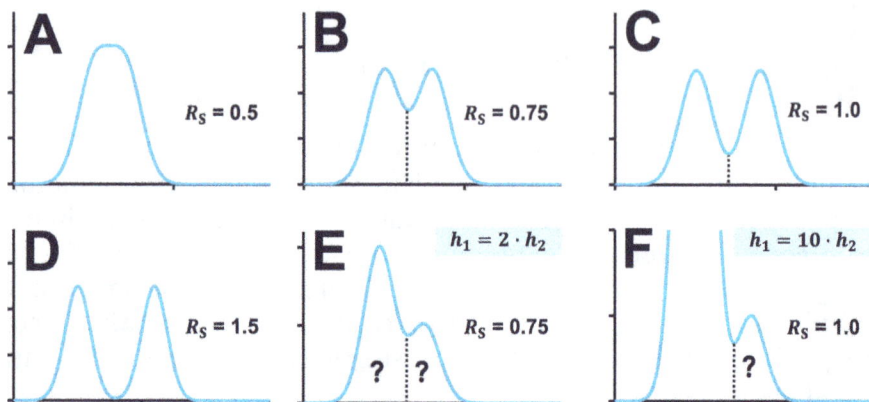

Figure 1.12 Examples of resolution values and the corresponding separation of two Gaussian peaks, with a resolution of 0.5 (A), 0.75
(B), 1.0 (C), and 1.5 (D). A value of 1.5 indicates two baseline-
separated peaks. Lower values indicate partial overlap, but the
significance of co-elution depends highly on the relative
heights of the two peaks (E and F). Peak integration becomes
more challenging as the relative height varies. In practice, this
problem is more serious because – unlike in the present figure –
peaks are rarely observed to be symmetrical.

The resolution of two peaks is depicted in Figure 1.12. By its
definition, a resolution of 1.5 represents two baseline-separated
peaks, as the valley between them is about $3\sigma_t$ away from either top
(Figure 1.12D). Lower values indicate some overlap (Figure 1.12A–
C, E, and F). This may reduce quantitative precision and accuracy,
unless computational methods can be used to accurately deconvolute the two solute bands (see Chapter 9). Note that the significance
of co-elution depends strongly on the relative height of the two
peaks, as illustrated in Figure 1.12E and F, for two partially co-eluting peaks with a resolution of 0.75 or 1.

Eqn (1.35) and (1.36) are helpful in assessing the resolution of two
adjacent peaks, but they do not capture what the chromatographer
can do to improve the resolution. We will thus derive a more useful
form of the resolution equation that is expressed in terms of selectivity, efficiency, and retention. This treatment is valid for separations
under constant (non-programmed) conditions, such as isothermal
GC (constant temperature) or isocratic LC (constant mobile-phase
composition).

We need a few manipulations of the equation for resolution. First,
we substitute eqn (1.26) into eqn (1.35) to obtain

$$R_{S,j,i} = \frac{\sqrt{N}}{2} \cdot \left(\frac{t_{R,j} - t_{R,i}}{t_{R,i} + t_{R,j}} \right) \tag{1.37}$$

where N is the plate count, which is assumed equal for the two adjacent peaks. Next, we can introduce the retention factors using eqn (1.13).

$$R_{S,j,i} = \frac{\sqrt{N}}{2} \left(\frac{k_j - k_i}{2 + k_i + k_j} \right) \tag{1.38}$$

Eqn (1.37) and (1.38) are useful in themselves, as they clearly separate the contributions of the chromatographic efficiency (N) and the (differences in) retention to the resolution. Eqn (1.37) allows the easiest calculations, as it does not require knowledge of t_0.

We can write the bracketed factor in eqn (1.38) as $\dfrac{k_j - k_i}{k_i + k_j} \cdot \dfrac{k_i + k_j}{2 + k_i + k_j}$ and then use eqn (1.19) and the average of the two retention factors $\bar{k} = (k_i + k_j)/2$ to obtain

$$R_{S,j,i} = \frac{\sqrt{N}}{2} \cdot \frac{\alpha_{j,i} - 1}{\alpha_{j,i} + 1} \cdot \frac{\bar{k}}{1 + \bar{k}} \tag{1.39}$$

Finally, for difficult separations (*i.e.* for two peaks that elute closely together), $\alpha_{j,i}$ is close to 1 so that approximately

$$R_{S,j,i} \approx \frac{\sqrt{N}}{4} \cdot \frac{\alpha_{j,i} - 1}{\alpha_{j,i}} \cdot \frac{\bar{k}}{1 + \bar{k}} \tag{1.40}$$

Eqn (1.40) is often used, but it is approximate, unlike the exact eqn (1.39), and is not a simplification. Therefore, we prefer eqn (1.39) and will use this equation when appropriate in the remainder of this book.

Eqn (1.39) and (1.40) are highly useful to the chromatographer, as they clearly demonstrate that co-elution issues can be resolved by adjusting (i) the column efficiency, N, (ii) the selectivity, $\alpha_{j,i}$, and/or (iii) the (average) retention of the solutes, $\bar{k}$. An example is depicted in Figure 1.13, where two peaks have the same retention factors in two different situations, and, thus, the selectivity is identical. Yet in one case (pink, solid lines), the peaks are partially overlapping ($R_{s,j,i} < 1$), while in the other case (blue, dashed lines), they are fully

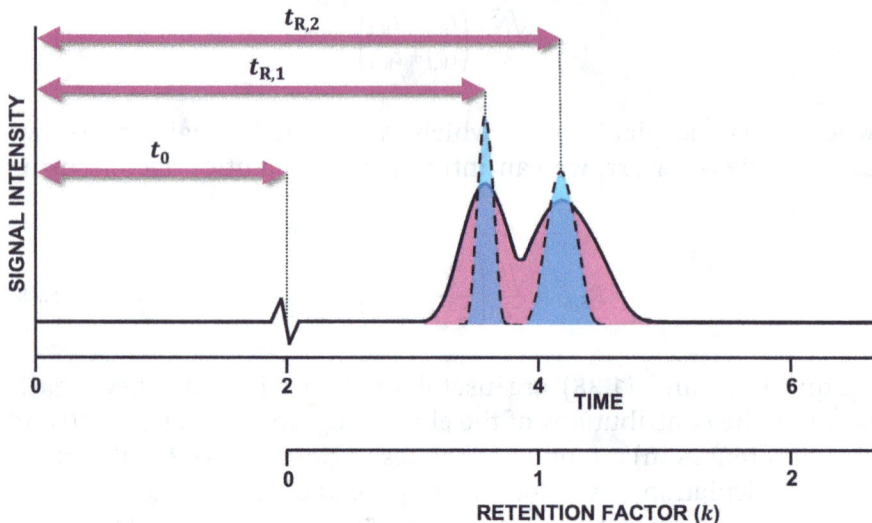

Figure 1.13 Schematic representation of two pairs of neighbouring peaks that, despite exhibiting the same retention factors and, thus, selectivity, exhibit a different resolution (R_s) due to different peak widths. Adapted from ref. 4 with permission from the Author.

separated ($R_{s,j,i} > 1.5$). In this example, the difference in resolution between the two peaks can be explained by a large difference in column efficiency ($N_{blue} \gg N_{pink}$).

We can study eqn (1.39) and (1.40) more closely by examining the three contributing factors. Figure 1.14 illustrates the effect of $\bar{k}$ on the overall resolution. The dramatic negative effect of very low $\bar{k}$ values is evident. Below $\bar{k} = 0.5$, a maximum of one-third of the attainable resolution can be realized. Thus, achieving sufficient retention should be the first goal of the chromatographer. Figure 1.14 shows that the effect of increasing $\bar{k}$ diminishes at higher retention. At $\bar{k} = 4$, 80% of the attainable resolution has been realized, and more retention takes more time and costs sensitivity because peaks get broader and lower with increasing retention time (see eqn (1.26)). Achieving sufficient retention is a vital first step towards obtaining sufficient resolution. In most cases, increasing $\bar{k}$ is not difficult. For example, in LC, $\bar{k}$ can usually be easily altered by adjusting the mobile-phase composition, while in GC, changing the temperature is the most common way to adjust $\bar{k}$.

To explore the effect of selectivity on resolution, we need to inspect eqn (1.39) because, in eqn (1.40), it is explicitly assumed that $\alpha_{j,i} \approx 1$.

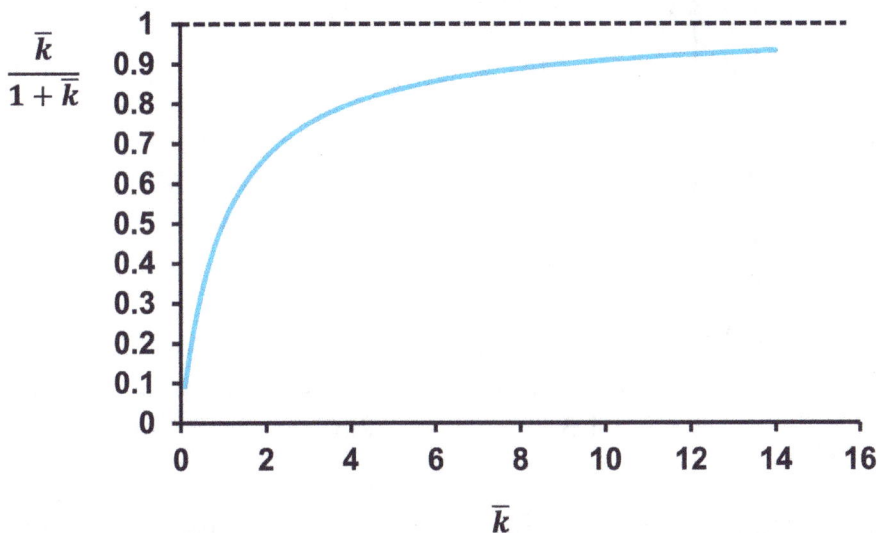

Figure 1.14 Effect of the average retention factor $\bar{k}$ on the retention-related factor in eqn (1.39) and (1.40). In many cases, the retention factor is easy to adjust, for example, by changing the temperature in GC or modifying the composition of the mobile phase in LC. However, retention can only be used to improve the resolution to a limited extent.

The effect of $\alpha_{j,i}$ on the resolution is determined by the factor $(\alpha_{j,i} - 1)/(\alpha_{j,i} + 1)$ and is illustrated in Figure 1.15. Obviously, there is no resolution if there is no selectivity (*i.e.* when $\alpha_{j,i} = 1$). Like the effect of $\bar{k}$, the effect of $\alpha_{j,i}$ diminishes when higher values are achieved. However, unlike the effect of $\bar{k}$, the effect of increasing $\alpha_{j,i}$ is beneficial even when the selectivity is already high. A 10% difference in the retention factor of the two compounds ($\alpha_{j,i} = 1.1$) already entails a significant selectivity; however, it only brings about 5% of the maximum attainable resolution. A difference of a factor of 5 in retention factors ($\alpha_{j,i} = 5$) implies enormous selectivity, but even then, only 67% of the maximum contribution of the selectivity to the resolution is achieved. In LC, modification of the selectivity by changing the nature or composition of the mobile phase is relatively easy, but in practice, the effects of any change are hard to predict. Changing the stationary phase (*i.e.* the column) is less convenient and potentially more costly, but in GC, this is one of the very few options to change the chromatographic selectivity. In chromatograms obtained for complex mixtures (*i.e.* mixtures with

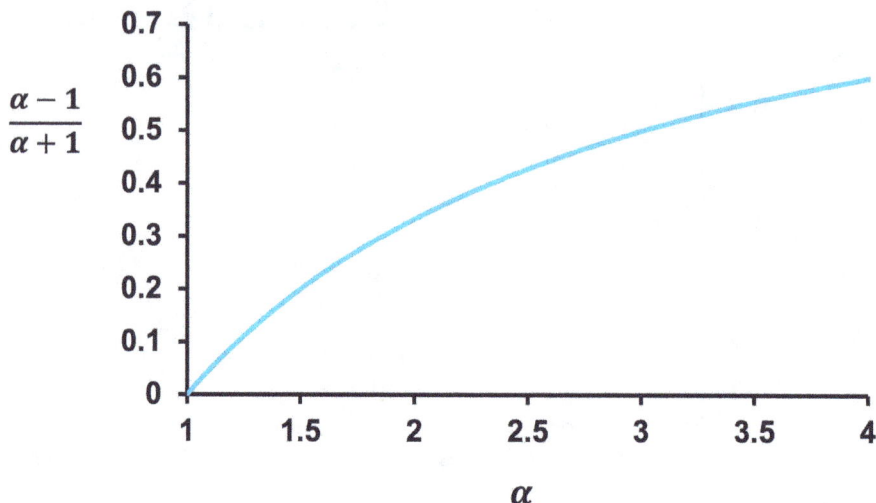

Figure 1.15 Effect of the selectivity factor on the resolution in eqn (1.39). The selectivity factor has a large influence, but it is difficult to predict the separation *a priori.*

more than a few analytes), changing the selectivity may very well improve the separation of one pair of peaks but jeopardize the separation of another peak pair. Because it is hard to predict the selectivity in many situations, multiple columns may need to be scanned experimentally in efforts to increase $\alpha_{j,i}$. Such a process is used, for example, to separate new compounds during drug development in the pharmaceutical industry.

Finally, after achieving sufficient retention and realizing whatever selectivity can be achieved, the chromatographer has to resort to the efficiency (N) to achieve sufficient resolution. As shown in Figure 1.16, the square-root contribution rises monotonously. However, the price of plates may be high, if only because doubling the resolution requires a four-fold increase in N. Increasing N can be achieved most easily by increasing the column length (at the expense of longer analysis times and higher pressures).

In LC, plate counts up to 10 000 can be achieved with relative ease, but higher plate counts may require longer columns, longer analysis times, and higher pressures (see Chapter 3). In GC, plate counts of up to 100 000 are commonly available, but if lower plate counts suffice, much shorter columns may be used, allowing much shorter analysis times.

The best basic strategy to achieve sufficient resolution can be summarized as follows:

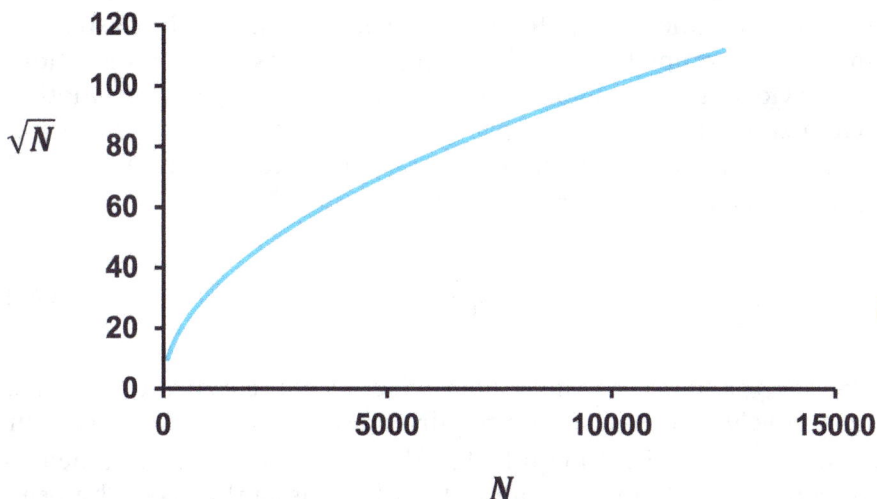

Figure 1.16 Contribution of the column efficiency to the resolution in eqn (1.39) and (1.40).

1. ensure sufficient retention,
2. generate as much selectivity as possible, and
3. choose a column (and a flow rate) that provides the required number of plates.

The latter (N_{req}) can be calculated by rearranging eqn (1.39) to obtain

$$N_{\text{req}} = \left\{ 2\left(R_{S,j,i}\right)_{\text{req}} \cdot \frac{\alpha_{j,i}+1}{\alpha_{j,i}-1} \cdot \frac{1+\bar{k}}{\bar{k}} \right\}^2 \tag{1.41}$$

where $\left(R_{S,j,i}\right)_{\text{req}}$ is the desired resolution. Eqn (1.41) shows that if $\bar{k}$ is very small or if $\alpha_{j,i}$ is close to unity, the required number of plates is exceedingly high. This is why it is imperative to follow the three steps in the indicated order.

1.5.2 Peak Capacity

In our discussions of resolution, we have thus far exclusively focused on the separation of two neighbouring analyte peaks. However, typical samples often contain (many) more analytes. Chromatographic methods are usually assessed on their effectiveness at separating all analytes of interest in a sample. Method development and optimization are the subject of Chapter 10, but it is already of interest to consider two relevant parameters from a fundamental perspective. One parameter is – of course – time; the faster the method is completed, the faster can a response (*i.e.* information) be provided and the sooner a new sample can be analysed. Another parameter is the number of peaks that are present in the chromatogram. The theoretical **peak capacity** (n_p) of a chromatogram obtained under constant conditions can be found from

$$n_p = \frac{\sqrt{N}}{4\,R_{S,req}} \ln\!\left(\frac{1 + k_{last}}{1 + k_{first}}\right) + 1 \qquad (1.42)$$

where k_{first} and k_{last} are the retention factors of the first and last peak, respectively, and $R_{S,req}$ is the required resolution between each pair of successive peaks. In eqn (1.42), the plate count N is assumed to be constant and equal for all (allegedly Gaussian) peaks. The peak capacity can be thought of as the number of peaks that can be fitted in a chromatogram with the same resolution between each pair ($R_{S,req}$). Such a chromatogram is illustrated in Figure 1.17A under non-programmed conditions. About 15 peaks are obtained across the range $0 < k < 15$ with $N = 1000$ according to eqn (1.42). Obviously, a longer analysis time (higher k_{last}) will allow for more peaks in a chromatogram, but – as indicated by eqn (1.39) – greater chromatographic efficiency (*i.e.* a higher number of plates) will cause all peaks to be narrower and thus also create more room for peaks. Later, eluting peaks in non-programmed chromatography are broader and lower, in line with eqn (1.34) (see Section 1.4.5). Under ideal programmed conditions (see, *e.g.* Module 3.4), all peaks are equally broad, and maximum peak capacity corresponds to a chromatogram such as Figure 1.17B. Dilution will also be equal for all peaks, resulting in equal peak heights if the concentrations and detector sensitivities are equal for all analytes.

Statistically, it is of course very unlikely for all peaks to – by pure coincidence – elute right after one another with a resolution of precisely 1.5. Using **statistical overlap theory**, Davis and Giddings[9] pointed out that the theoretical peak capacity is unattainable for a sample consisting of randomly selected analytes. They provided the approximate eqn (1.43), in which p is the number of singular (pure component) peaks and m is the number of analytes.

$$\ln p \approx \ln m - \frac{m}{n_c} \qquad (1.43)$$

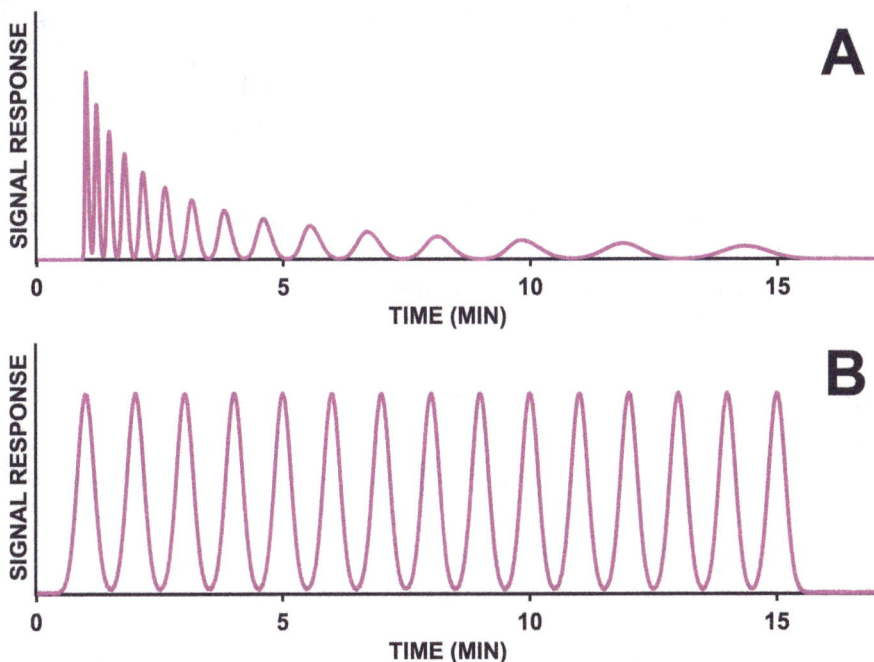

Figure 1.17 Idealized chromatograms in which the available peak capacity is used to the full under (A) non-programmed and (B) ideal programmed conditions.

Their treatment yielded as a rule of thumb that the peak capacity should exceed the theoretically minimum required value (*i.e.* $n_c = m$) by a factor of 20 to provide a greater than 90% probability of an analyte of interest to be baseline separated.

1.6 Equilibria and Migration (M)

The basics of retention have been covered in earlier modules of this chapter. However, several aspects warrant a closer look. This will be provided in the following sections.

1.6.1 Thermodynamics of Retention

We have seen in Module 1.1 that the distribution coefficient describes the retention of an analyte in chromatography. It was mentioned that the distribution coefficient depends on pressure and temperature. This becomes clear when we realize that the distribution coefficient is a thermodynamic property. Fundamentally, it is best expressed in terms of activities, which can be related to the **chemical potential** μ_i of the analyte in solution through

$$\mu_{i,f} = \mu_{i,f}^0 + RT \ln a_{i,f} \tag{1.44}$$

where the subscript f denotes the phase (in our case mobile, m, or stationary, s), the superscript 0 denotes the standard state, R is the gas constant, T is the absolute temperature, and a_i is the activity of analyte i. We can choose the standard state. A common choice is, for example, to choose a pure component i at a given temperature and pressure (*e.g.* 298.15 K and 10^5 Pa). In that case, $\mu_{i,m}^0 = \mu_{i,s}^0 = \mu_i^0$. However, it is also possible to define a different standard state, such as the limit of $\mu_{i,f}$ at infinite dilution (at a given temperature and pressure), in which case $\mu_{i,m}^0 \neq \mu_{i,s}^0$.

In case of equilibrium, we have

$$\mu_{i,m} = \mu_{i,s} \tag{1.45}$$

and from eqn (1.44), it follows that

$$RT \ln K_{d,i}^a = RT \ln \frac{a_{i,s}}{a_{i,m}} = \mu_{i,m}^0 - \mu_{i,s}^0 = \Delta \mu_i^0 \tag{1.46}$$

where $K_{d,i}^a$ is the distribution coefficient in terms of activities. This is not a very practical measure, and it is more convenient to express the distribution coefficient in terms of mole fractions ($K_{d,i}^x$) by introducing an **activity coefficient** γ as follows

$$\mu_{i,f} = \mu_{i,f}^0 + RT \ln(\gamma_{i,f} \cdot x_{i,f}) \tag{1.47}$$

from which we find

$$RT \ln K_{d,i}^x = RT \ln \frac{x_{i,s}}{x_{i,m}} = \Delta \mu_i^0 + RT \ln \frac{\gamma_{i,m}}{\gamma_{i,s}} \tag{1.48}$$

or for very ("infinitely") dilute solutions

$$RT \ln K_{d,i}^x = \Delta\mu_i^0 + RT \ln \frac{\gamma_{i,m}^\infty}{\gamma_{i,s}^\infty} \qquad (1.49)$$

If we conveniently choose the pure component i as the standard state, we have $\Delta\mu_i^0 = 0$ and

$$K_{d,i}^x = \frac{\gamma_{i,m}^\infty}{\gamma_{i,s}^\infty} \qquad (1.50)$$

From eqn (1.50), we see that the distribution coefficient is determined by the activity coefficients, which are measures of the interactions of analyte i with the mobile and stationary phases. The activity coefficient is 1 in the case of pure i. In case i is immersed in a solvent, γ_i is usually greater than 1, especially if the differences in nature between analyte and solvent increase. For example, if we consider an apolar analyte, it will have a high activity coefficient in a polar mobile phase and an activity coefficient closer to 1 in an apolar stationary phase, a familiar situation in what is called reversed-phase liquid chromatography (RPLC, see Module 3.5).

1.6.1.1 Effect of Temperature

Like activities, activity coefficients are thermodynamic properties, and we can write

$$\ln K_{d,i}^x = \ln \frac{\gamma_{i,m}^\infty}{\gamma_{i,s}^\infty} = \frac{-\Delta\mu_i}{RT} = \frac{-\Delta H_i}{RT} + \frac{\Delta S_i}{R} \qquad (1.51)$$

where ΔH_i and ΔS_i are the changes in partial molar enthalpy and entropy, respectively, of transferring analyte i from the mobile to the stationary phase. To move from mole fractions to molar concentrations (c), we need a conversion. If we realize that for phase f (which is either m or s), $c_{i,f} = n_{i,f}/V_f$, at high dilution $x_{i,f}^\infty = n_{i,f}/(n_{i,f} + n_f) \approx n_{i,f}/n_f$ and the number of moles ($n_{i,f}$) relates to the density (ρ_f) and the molecular weight (M_f) of phase f as $n_f = \rho_f V_f/M_f$, we find

$$K_{d,i}^c = \frac{c_{i,s}}{c_{i,m}} = \frac{n_{i,s}}{n_{i,m}} \cdot \frac{V_m}{V_s} \approx \frac{x_{i,s}^\infty}{x_{i,m}^\infty} \cdot \frac{n_s}{n_m} \cdot \frac{V_m}{V_s} = K_{d,i}^{x,\infty} \frac{\rho_s}{\rho_m} \frac{M_m}{M_s} \qquad (1.52)$$

The approximation is valid for highly dilute solutions, as indicated by the subscript ∞ (infinite dilution). Moving on to retention factors requires another conversion

$$k_i = k_{d,i}^c \cdot \phi \approx K_{d,i}^{x,\infty} \frac{\rho_s}{\rho_m} \cdot \frac{M_m}{M_s} \cdot \frac{V_s}{V_m} = K_{d,i}^{x,\infty} \frac{n_s}{n_m} \qquad (1.53)$$

which with eqn (1.51) yields

$$\ln k_i = \ln K_{d,i}^{x,\infty} + \ln \frac{n_s}{n_m} = \frac{-\Delta H}{RT} + \frac{\Delta S_i}{R} + \ln \frac{n_s}{n_m} \qquad (1.54)$$

Eqn (1.54) is an important equation in chromatography. It is known as the **van 't Hoff equation**, named after Jacobus Henricus van 't Hoff, the first Nobel laureate in chemistry. It is typically assumed that ΔH_i and ΔS_i are constants and that the phases are homogeneous. Plots of $\ln k_i$ vs. $1/T$ are often used in retention mechanism studies, but one should be aware of the assumptions made and their limited validity. Also, we have been assuming a closed system in which n_s and n_m are constants. This is not always the case. For example, in GC, the mobile phase expands considerably with increasing temperature, causing the mobile-phase density (ρ_m) and the number of moles of carrier gas in a segment of the column to decrease along its length. For an ideal gas, we may write

$$\ln k_i = \ln K_{d,i}^{x,\infty} + \ln \frac{n_s}{n_m} = \frac{-\Delta H_i}{RT} + \frac{\Delta S_i}{R} + \ln \frac{RT \cdot n_s}{PV} \qquad (1.55)$$

where the pressure P varies along the length of the column (see Module 2.2). The presence of the last term in eqn (1.55) implies that in GC, instead of standard van 't Hoff plots (of $\ln k_i$ vs. $1/T$), plots of $\ln(k_i/T)$ vs. $1/T$ should be used to explore enthalpy and entropy effects.

We can rewrite eqn (1.54) to

$$\ln k_i = \frac{-\Delta G_i}{RT} + \ln \frac{n_s}{n_m} \qquad (1.56)$$

where $\Delta G_i = \Delta H_i - T\Delta S_i$ is the change in partial molar free energy of transferring analyte from the mobile phase to the stationary phase. For the selectivity between the two compounds, this yields

$$\ln \alpha_{j,i} = \ln k_j - \ln k_i = \frac{\Delta G_i}{RT} - \frac{\Delta G_j}{RT} = \frac{\Delta\Delta G_{i,j}}{RT} \qquad (1.57)$$

where $\Delta\Delta G_{i,j}$ is the difference in partial molar free energies of transferring analytes i and j. Clearly, selectivity is a purely thermodynamic property, affected by temperature and – to a lesser extent – by pressure, as well as by the properties of the mobile phase, stationary phase and analytes. It is not affected by the phase ratio or physical parameters, such as flow rate and column dimensions (except indirectly, through changes in pressure).

Eqn (1.57) underlines the great separation power of chromatography. For example, in contemporary (high-performance) LC, we can readily achieve plate counts of $N = 20\,000$. Eqn (1.39) reveals that two compounds with $\alpha_{j,i} = 1.02$ (and $\overline{k} = 3$) can be resolved with better than unit resolution ($R_{S,j,i} > 1$). According to eqn (1.57), the free energy difference for such compounds ($\Delta\Delta G_{i,j}$) is a mere 47 J mol^{-1}. The 100 000 plates that are readily available in contemporary (high-resolution) GC correspond to $\alpha_{j,i} \approx 1.008$ and $\Delta\Delta G_{i,j} \approx$ 20 J mol^{-1}. These are very subtle differences between two analytes and their interactions in chromatographic systems, yet they can be fully separated. Conversely, this great separation power of chromatography makes it extremely challenging to predict selectivity (and resolution) from first principles or computational models. In such processes, errors of 20 J mol^{-1} can make a difference between fully separated and completely coinciding peaks.

1.6.1.2 Effect of Pressure

To understand the effect of pressure on retention, we have to go back to the definition of enthalpy, which is the sum of the internal energy of a system and the product of its volume and the pressure exerted on it. For the partial molar enthalpy, we have

$$H_i = U_i + P \cdot V_i \qquad (1.58)$$

where U_i is the internal energy and V_i is the partial molar volume. Eqn (1.54) now becomes

$$\ln k_i = \frac{-\Delta U_i}{RT} - \frac{P\Delta V_i}{RT} + \frac{\Delta S_i}{R} + \ln \frac{n_s}{n_m} \tag{1.59}$$

where ΔU_i is the change in partial molar energy when component i moves from the mobile phase to the stationary phase and ΔV_i is the change in its partial molar volume. In GC, the mobile phase is compressible, and the phase ratio varies with the pressure (eqn (1.55)). In addition, the energy (ΔU_i) and entropy (ΔS_i) terms may be affected by the pressure. In LC, we may assume that the effect of pressure on all terms on the right-hand side of eqn (1.59) is negligible, except for the second term that explicitly contains P. The effect of pressure on retention in LC can then be obtained by taking the derivative

$$\frac{\partial \ln k_i}{\partial P} = \frac{-\Delta V_i}{RT} \tag{1.60}$$

The change in the partial molar volume of the analyte between the stationary phase and the mobile phase impacts the change in retention with pressure. As the molar volume of the analyte is typically smaller in the adsorbed or absorbed state than it is in the liquid mobile phase, $V_{i,s}$ is smaller than $V_{i,m}$ and $\Delta V_i = V_{i,s} - V_{i,m}$ is negative. Thus, we may expect that the retention factor k_i will increase when the pressure increases. While the effect of pressure on retention was long thought negligible, it has received much more attention in recent years because (i) the pressures used in LC have been increasing with the progression from HPLC to UHPLC and (ii) high-molecular-weight analytes have become targeted more often. Eqn (1.60) shows that the effect of pressure depends on the absolute change in the molar volume, which is likely larger if V_i itself is larger. Changes in the order of 10% in k_i have been reported for small molecules in the HPLC range (*e.g.* a pressure increase from 5 to 25 MPa), while for proteins, retention factors have been found to double or triple with increasing pressure.[10]

It is important to realize that a significant effect of pressure on the retention factor in LC implies that k_i varies with the position in the column. Fekete *et al.* derived the following equation to account for the effect of the pressure drop on the retention time.[11]

$$t_R = t_0 \left[1 + \frac{k_{p=0}RT}{\Delta P \Delta V_i} \cdot e^{\frac{-P_{in}\Delta V_i}{RT}} \cdot \left(e^{\frac{\Delta P \Delta V_i}{RT}} - 1 \right) \right] \tag{1.61}$$

where $k_{P=0}$ is the retention factor extrapolated to zero pressure, P_{in} is the inlet pressure of the column and ΔP is the pressure drop across it. The apparent retention factors (defined as $k_{app} = (t_R - t_0)/t_0$) that follow from eqn (1.61), normalized to the value at $P = 0$ (*i.e.* $k_{app}/k_{P=0}$), are shown in Figure 1.18. It is seen that the effects are relatively minor if the change in molar volume (ΔV_i) is below 0 and −5 mL mol^{-1} (bottom two curves) but become quite substantial if ΔV_i is in the range of −10 to −20 mL mol^{-1} and outright large if ΔV_i becomes as large as −50 or −100 mL mol^{-1} (top two curves). It is not easy to predict values of ΔV_i. Fekete *et al.* experimentally found values at 30 °C of $\Delta V_i \approx$ −15 mL mol^{-1} for a peptide with molecular weights of around 1300 and a value of $\Delta V_i \approx$ −30 mL mol^{-1} for glucagon,[11] a peptide with a molecular weight of about 3000. This suggests that a very rough guess for ΔV_i may be 1% of the molar volume of the pure analyte. Slightly smaller values were reported by McCalley.[12] However, such guesstimates should be treated with extreme caution because the authors used a common mobile phase (25% ACN in water, with 0.1% trifluoroacetic acid

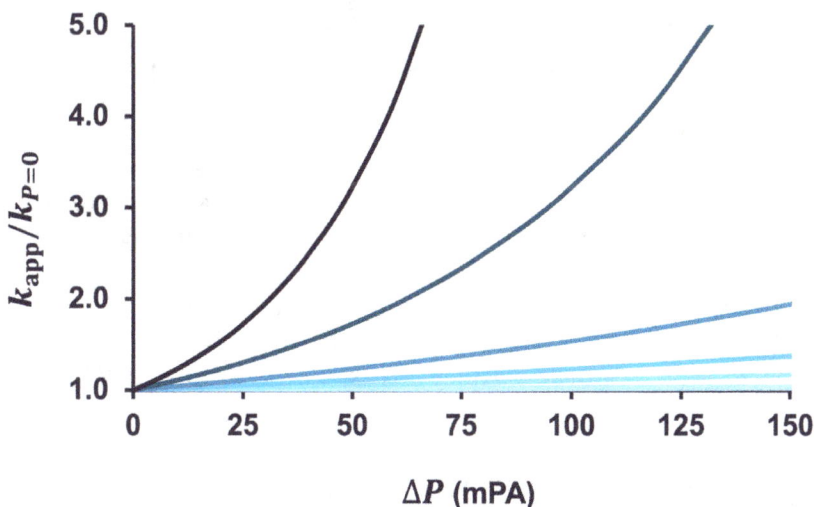

Figure 1.18 Variation in the apparent (isocratic) retention factor with the column pressure drop according to eqn (1.61), assuming a column-outlet pressure of 0.1 MPa. Lines from bottom to top (light to dark) ΔV_i = −1,-5, −10,-20, −50, and −100 mL mol^{-1}.

added) but a rather uncommon stationary phase (featuring butyl chains rather than octadecyl chains). The values found for ΔV will likely be affected by the nature of the stationary phase. The absolute values of ΔV_i measured by Fekete *et al.* decreased significantly when increasing the temperature from 30 to 80 °C (by about half for the largest peptide and about one-third for the smaller ones).[11]

Because the local k value is lower than the average value at the column exit, the analyte band will elute faster than expected based on its overall retention time (and apparent k value) and, therefore, be narrower in time units. This will inflate a plate count determined from eqn (1.27).

In the case of gradient elution, the situation becomes complex because the mobile-phase composition, viscosity and pressure all vary with the position in the column. The changing viscosity implies that there is no simple linear decrease in the pressure along the column length. Numerical approaches can be used to account for all these effects.

1.6.2 Phase Ratio

The **phase ratio** ϕ, defined as V_s/V_m, was briefly mentioned in Section 1.1.1, but thus far, it has not been addressed in detail. However, eqn (1.2) and (1.53) demonstrate that the phase ratio affects the retention factor. Therefore, it is useful to discuss the phase ratio and the parameters that affect it.

The first such parameter is the type of column. Wall-coated open-tubular columns that contain a thin film of a (usually polymeric) stationary phase are predominantly used in GC. Columns packed with particles that are a few micrometres in diameter are typically used in LC. The second parameter is the **surface area** of the stationary phase. For packed columns, the surface area concerns the surface of the stationary-phase particles (Figure 1.5B and C), and for open-tubular columns, it is equal to the surface of the stationary phase coating on the wall. Packed columns tend to offer much higher stationary-phase surface areas. For example, a 25 m-long open-tubular GC column with a 250 μm internal diameter (i.d.) has an internal surface area of about 0.02 m², while a packed LC column may easily feature a stationary-phase surface area of 100 m². The surface area is directly proportional to the phase ratio for (i) solid adsorbents, (ii) chemically modified solid surfaces (*e.g.* chemically bonded stationary phases; see Module 3.2.2.2), and (iii) liquid layers that are deposited on a solid adsorbent with a constant film thickness. The surface area of packing materials is usually reported as a specific surface area per unit weight (*e.g.* m^2 g^{-1}), although it is arguably more useful to consider the surface area per unit volume of packed columns (*e.g.* m^2 mL^{-1}).

The third parameter is the **column porosity,** which is only relevant for packed columns and was discussed in Module 1.2. The fourth parameter is the **film thickness,** which is only well defined for open-tubular columns (Figure 1.5D). Assuming that other parameters are constant, a larger film thickness will increase V_s and (slightly) decrease V_m. Hence, the phase ratio (defined as V_s/V_m) increases with an increasing film thickness. For packed columns, immobilized liquid layers may be formed on the stationary-phase surface (mainly) inside the pores of the particles, in which case the stationary-phase volume increases with the surface area. However, the film thickness of such liquid layers is likely not constant.

The final parameter is the **column inner diameter** (d_c). In principle, this does not affect the phase ratio of packed columns because both V_m and V_s increase quadratically with d_c. In contrast, the column diameter is an important parameter for open-tubular columns. With increasing column diameter, the surface area of the wall increases linearly, yet the column volume increases quadratically. Thus, the phase ratio (V_s/V_m) is inversely proportional to the column inner diameter for such columns. We will see in Section 1.7.7 that the film thickness in open-tubular columns (d_f) typically increases in proportion to the column diameter. If this is the case, $\phi = V_s/V_m$ will be about equal for the different columns (*e.g.* $d_c = 250$ μm with $d_f = 0.25$ μm *vs.* $d_c = 320$ μm with $d_f = 0.32$ μm).

1.6.3 Flow Profiles

We already noted in Module 1.2 that chromatographic separations require a bulk displacement of mobile-phase molecules to transport the solute molecules through the column and from one stationary-phase location to another. Given that flow is of crucial importance to chromatography, it is useful to examine the flow patterns inside the different components of the system and their effects on the mass transfer between the mobile and stationary phases.

For cylindrical channels (*e.g.* connection tubing and open-tubular columns), the flow profile can have a large effect on the efficiency of the chromatographic system. To guide this discussion, we need to first understand the two forces at play in fluid dynamics. When the fluid is pushed by a pump and flows through a channel, it will do so in different velocity layers ("streamlines"). This is depicted in Figure 1.19A. Near the wall, there will be relatively more friction, whereas in the centre the fluid can move relatively unhindered. There are two relevant forces at play here. **Inertial force** represents the inertia of the fluid and its momentum and can be defined as

$$F_{\text{inertia}} = \rho \cdot u^2 \tag{1.62}$$

where ρ is the density of the solution (*i.e.* the mobile phase with solutes) and u is the linear velocity. **Viscous force,** on the other hand, is the friction between fluids on different streamlines. This force increases with the dynamic viscosity of the fluid (η) and with the linear velocity, whereas it decreases with the increasing inner diameter of the channel (d_c), *i.e.*

$$F_{\text{viscous}} = \frac{\eta \cdot u}{d_c} \tag{1.63}$$

When viscous forces are dominant, they constrain the fluid in its defined streamlines. The result is what is known as a **laminar flow**, which yields the parabolic flow profile that is also referred to as Poiseuille flow and is depicted in Figure 1.19A. The relation between

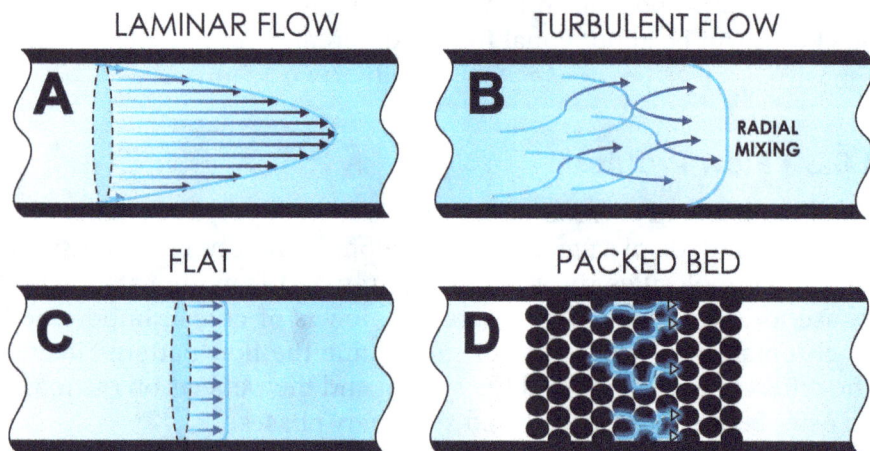

Figure 1.19 Schematic overview of different flow profiles. (A) Laminar flow, (B) turbulent flow, (C) flat flow profile, (D) the flow profile is dampened radially by the presence of stationary-phase particles.

flow and pressure for laminar flow of a (non-compressible) fluid in a cylindrical open tube can be described by the **Hagen–Poiseuille equation**

$$F = \frac{\Delta P \cdot \pi \cdot r_c^4}{8\eta \cdot L} = \frac{\Delta P \cdot \pi \cdot d_c^4}{128\eta \cdot L} \tag{1.64}$$

where ΔP is the pressure drop across the tube, r_c is the radius, d_c is the diameter, and L is the length.

However, as the fluid velocity increases, the inertial forces rise more rapidly than the viscous forces. This can be seen by comparing eqn (1.62) and (1.63), which reveals that the inertial forces increase quadratically with the linear velocity, unlike the viscous forces, which increase linearly. When the inertial forces become too strong (*i.e.* the flow rate becomes too high), the radial pressure differences increase to the point that streamlines start to mix by random fluctuations in the form of eddies, giving rise to a mixing phenomenon known as **eddy diffusion**. This results in **turbulent flow**, in which the radial mixing causes the flow profile to be much flatter than that of laminar flow (Figure 1.19B).

The dimensionless **Reynolds number** can be used to assess whether a laminar or turbulent flow profile can be expected. It is the ratio between the inertial and viscous forces and is given by

$$\mathrm{Re} = \frac{\rho \cdot u \cdot d_c}{\eta} \tag{1.65}$$

At Reynolds numbers below roughly 2300 in open tubes, the viscous forces are dominant, yielding a laminar flow. For packed columns, the Reynolds numbers are calculated based on the particle diameter (d_p) and the interstitial velocity (u_{int})

$$\mathrm{Re} = \frac{\rho \cdot u_{int} \cdot d_p}{\eta} \tag{1.66}$$

and the critical Reynolds number is lower than in open tubes (1000 or lower). In chromatography, Re numbers are usually well below the critical values. We will see later that particle-induced eddy diffusion plays a significant role in the efficiency of packed columns.

In separation sciences, mechanisms other than pressure are also used to facilitate the transport of mobile phases and solute molecules through a separation channel. For example, a high voltage generates an electro-osmotic flow (with a very flat profile, Figure 1.19C) through the column in capillary electrophoresis (see Chapter 5). Finally, it is worth to note that the channels between particles in packed columns can also be considered as channels through which laminar or turbulent flow profiles can develop (Figure 1.19D). However, the packed bed disturbs the laminar flow profile on the scale of the entire column. As a result, packed columns show mostly a flat flow profile except in the vicinity of the column wall.

1.6.4 Column Permeability

In pressure-driven separations, the transport of the mobile phase through a column requires a certain pressure drop across the column (*cf.* eqn (1.64)). This is known as the column pressure or **backpressure**. Column permeability quantifies the pressure required to achieve a given volumetric flow rate. This is of paramount importance to separations, especially when using packed columns. The backpressure is one of the most critical limiting parameters for LC separations. Although HPLC is now often interpreted as high-performance liquid chromatography, it was the availability of precise high-pressure pumps (typically up to 40 MPa or 400 bar) that sparked the dramatic advances in high-pressure liquid chromatography from the 1970s onward. In this section, the parameters affecting the pressure drop across the column will be discussed.

The **Ergun equation** for a packed column (with particle diameter d_p) illustrates the combination of viscous forces (reflected in the first term on the right-hand side of eqn (1.67)) and inertial forces (the second term) discussed in the previous section. It reads

$$\Delta P = \frac{150\,\eta \cdot L}{d_p^2} \frac{(1-\varepsilon_{int})^2}{\varepsilon_{int}^3}\,\varepsilon_{tot} \cdot u_0 + \frac{175\,\rho \cdot L}{d_p} \frac{(1-\varepsilon_{int})}{\varepsilon_{int}^3}\,\varepsilon_{tot}^2 \cdot u_0^2 \qquad (1.67)$$

where, as before, η is the mobile-phase viscosity, ρ is the density, ε_{int} is the interstitial porosity and ε_{tot} is the total porosity. Figure 1.20 shows an example of the results obtained for a typical (150 mm length and 5 μm particles) LC column. At a typical LC flow rate of 1 mm s^{-1}, the viscous forces are seen to strongly dominate (Figure 1.20A).

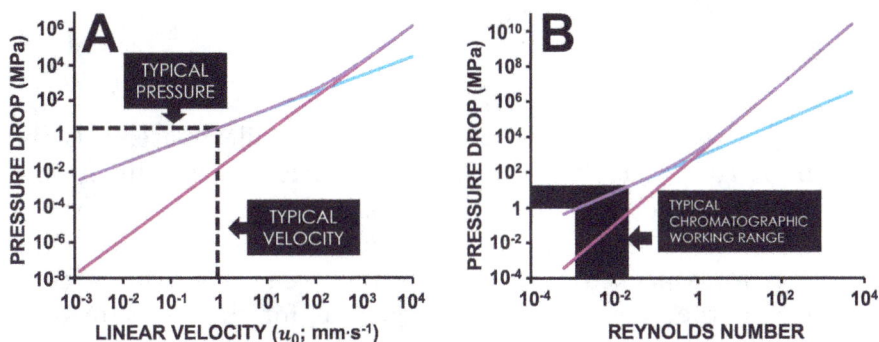

Figure 1.20 Pressure drop calculated from the Ergun equation (eqn (1.67)) for a 0.15 m-long LC column packed with 5 μm particles ($\eta = 10^{-3}$ Pa s; $\rho = 1000$ kg m^{-3}; $\varepsilon_{int} = 0.4$; $\varepsilon_{tot} = 0.7$), plotted against (A) the linear velocity (u_0) in mm s^{-1} and (B) the Reynolds number. Blue lines: pressure drop due to viscous forces; dark-pink lines: pressure drop due to inertial forces; and purple lines: total pressure drop.

The conclusion for this one column can be generalized using the Reynolds number (Re; eqn (1.66)). Viscous forces are seen to dominate in the regime Re <1 (Figure 1.20B). Typical Re values for packed-column chromatography are in the range of 0.001 < Re < 0.1, and values above 0.1 are rarely encountered.[13] In this range, the first term dominates, which can be seen from the ratio between the two terms on the right-hand side of eqn (1.67)

$$\frac{\Delta P_{\text{viscous}}}{\Delta P_{\text{inertial}}} = \frac{150\eta}{175 d_{\text{p}} \cdot \rho \cdot u_0 \cdot \varepsilon_{\text{tot}}} = \frac{6}{7\varepsilon_{\text{tot}}} \cdot \frac{1}{\text{Re}} \tag{1.68}$$

For Re ≤ 0.1, the viscous forces are at least an order of magnitude larger than the inertial forces. Higher Re numbers may be encountered in very fast GC separations and, especially, in supercritical-fluid chromatography (SFC, see Chapter 6).

The first term on the right-hand side of the Ergun equation corresponds to the Darcy equation (sometimes called the Darcy–Weisbach equation), which is more commonly used in chromatography and has been proven valid in the low-Reynolds regime that is appropriate for almost all chromatographic separations.[13] The **Darcy equation** can be written as[14]

$$\Delta P = \frac{\eta \cdot L \cdot u_0}{K_p} = \frac{\psi \cdot \eta \cdot L \cdot u_0}{d^2} \tag{1.69}$$

where η is the (dynamic) viscosity of the mobile phase, L is the column length, K_p is the permeability coefficient, ψ is a dimensionless flow-resistance parameter, and d is the characteristic diameter (the column diameter for an open-tubular column or the particle diameter for a packed column). The Darcy equation can be compared with the Hagen–Poiseuille equation for the pressure drop across a cylindrical tube (with diameter d_c) in the laminar-flow regime if we rewrite eqn (1.64) in terms of the linear velocity (u_0), realizing that $F = \frac{\pi}{4} d_c^2 \cdot u_0$.

$$\Delta P = \frac{128 \cdot \eta \cdot L \cdot F}{\pi \cdot d_c^4} = \frac{32 \cdot \eta \cdot L \cdot u_0}{d_c^2} \tag{1.70}$$

Eqn (1.70) is clearly equivalent to eqn (1.69), and it shows that we may use the Darcy equation with $\psi = 32$ for a cylindrical, open-tubular column. The Darcy equation can be generally applied, but different values for the flow-resistance parameter (ψ) apply in different situations. In chromatography, ψ often takes the place of the more conventional permeability coefficient $K_p = d^2/\psi$ because it makes the effect of the characteristic diameter on the pressure drop explicit. A typical, often quoted value of ψ for a packed column (using $d = d_p$) is 1000, although, for well-packed columns with uniform spherical particles, the value may be as low as 500.[14] This is in line with the **Kozeny–Carman equation**, which in terms of ψ reads

$$\psi = \frac{180(1 - \varepsilon_{int})^2 \varepsilon_{tot}}{S_f \varepsilon_{int}^3} \tag{1.71}$$

where S_f is a sphericity factor.[15] With an interstitial porosity of 0.4 and $S_f = 1$ (the value for spherical particles), this yields a ψ value close to 1000. The Kozeny–Carman equation is also valid in the low-Reynolds regime typically encountered in chromatography (Re ≤ 0.1).

The value of ψ is highly affected by the column geometry (Figure 1.21), especially the (external) porosity, and value ranges for different geometries are provided in Table 1.1. The exact value of ψ for an

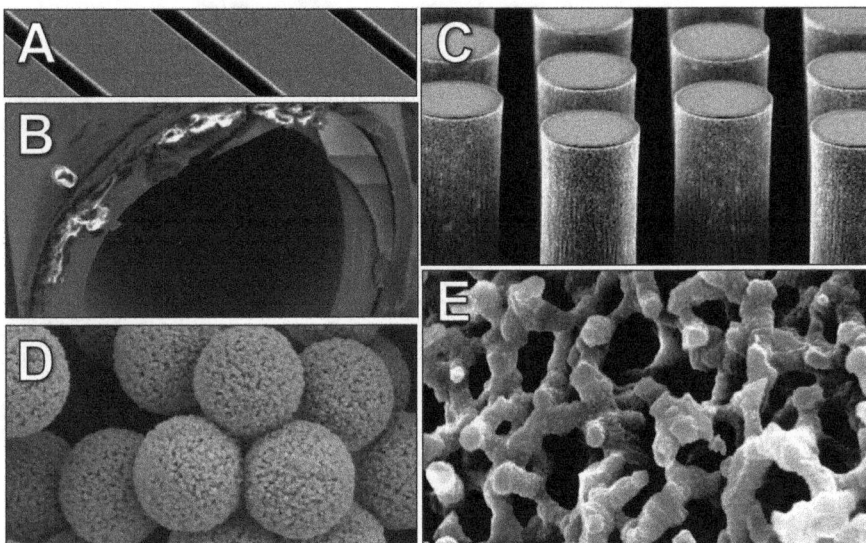

Figure 1.21 SEM photographs of (A) a flat channel, (B) a capillary open-tubular column with stationary-phase film,[3] (C) pillar arrays,[16] (D) solid particles,[2] and (E) monolithic stationary phase.[17] Panel B adapted from ref. 3 with permission from the Royal Society of Chemistry. Panel C adapted from ref. 16 with permission from the Royal Society of Chemistry. Panel D adapted from ref. 2 with permission from Elsevier, Copyright 2016. Panel E reproduced from ref. 17 with permission from Springer Nature, Copyright 2020.

open-tubular column (using $d = d_c$) is 32. Obviously, open-tubular columns have much lower flow-resistance parameters than packed columns, with all other factors equal. For the same number of plates and $d = d_c = d_p$, a good, packed column may need to be about three times longer than an open-tubular column (see Module 1.7). This implies that using the same mobile phase in a packed column requires a backpressure that is 50–100 times higher than needed to operate an open-tubular column.

1.7 Band Broadening (M)

1.7.1 Revisiting the van Deemter Equation

We have learned in Module 1.4 that the *van Deemter* plate-height equation gives rise to an optimal flow rate or linear velocity, together with a maximal column efficiency (minimal plate height). From a

Table 1.1 Overview of the relevant size parameters and typical values for the column resistance factors for different column geometries.

Type	Size parameter	ψ
Cylindrical tube	Column diameter (d_c)	32
Flat channel	Channel height	12
Packed bed of solid spheres	Particle diameter (d_p)	500
Porous silica column	Particle diameter (d_p)	1000
Pillar array	Domain size[a]	50–200
Monolith	Domain size[a]	20–100

[a]The domain size is defined as the diameter of a (average) flow channel plus that of a structural element (pillar or "skeleton" of a monolith). For monolithic columns, it is easier to specify the permeability coefficient $(K_p$; see eqn (1.69)).

fundamental perspective, however, plate-height equations, such as the *van Deemter* equation, are also eminently useful to understand the physicochemical processes that contribute to band broadening. This is the topic of this section.

We must first revisit the *van Deemter* equation,[5] using the interstitial average linear velocity (u_{int}), as done by van Deemter *et al.*

$$H = A + \frac{B}{u_{int}} + C \cdot u_{int} \qquad (1.72)$$

The *van Deemter* equation was derived for packed columns, and the characteristic dimension is the particle size d_p. We will consider open-tubular columns later (Section 1.7.7). In the treatment of *van Deemter* – and subsequent plate-height studies – u_{int} is generally used. It differs from the average mobile-phase velocity u_0 (eqn (1.16)). Unretained molecules spend time in the interstitial volume (*i.e.* the mobile phase outside the packing particles), where they move with average velocity u_{int}, and in the pores of the packing material, where their average lateral movement is zero. The relation between the average interstitial velocity and the average mobile-phase velocity is

$$u_{int} = \frac{F\,L}{\varepsilon_{int}\,V_{col}} = \frac{\varepsilon_{tot}}{\varepsilon_{int}} \cdot \frac{F\,L}{\varepsilon_{tot}\,V_{col}} = \frac{\varepsilon_{tot}}{\varepsilon_{int}} u_0 \qquad (1.73)$$

where ε_{int} is the interstitial porosity, *i.e.* the fraction of the total column volume that is between the particles, and ε_{tot} is the total porosity, which is the total fraction of the column volume that is occupied by the mobile phase. The latter is the sum of the fraction of the total volume between the particles (ε_{int}) and the fraction that is inside the pores (ε_{pores})

$$\varepsilon_{tot} = \varepsilon_{int} + \varepsilon_{pores} \tag{1.74}$$

In practice, eqn (1.72) is used with u_o instead of u_{int} to determine the optimal flow rate for a separation method (in that case $u_{0,opt}$). This is attractive for practical reasons because it is easier to determine the mobile-phase volume ($V_0 = F \cdot t_0$) than the excluded volume (V_{excl}). However, for fundamental plate-height studies, the obtained A, B and C terms will need to be corrected if u_0 is used instead of u_{int}. This can be done by combining eqn (1.72) and (1.73) to find

$$H = A + \frac{B}{u_{int}} + C \cdot u_{int} = A + \frac{B \cdot \varepsilon_{int}}{\varepsilon_{tot} \cdot u_0} + C \cdot \frac{\varepsilon_{tot} \cdot u_0}{\varepsilon_{int}} = A + \frac{B'}{u_0} + C'u_0 \tag{1.75}$$

This shows that the pragmatic values B' and C' obtained using u_0 can be converted into the proper values for the *van Deemter* equation through $B = B'(\varepsilon_{tot}/\varepsilon_{int})$ and $C = C'(\varepsilon_{int}/\varepsilon_{tot})$. The A coefficient, as defined by van Deemter, remains unaltered. However, for packed columns, where the eddy diffusion term is most relevant, it is no longer assumed to be independent of the linear velocity (see Section 1.7.4.3).

1.7.2   Effect of Diffusion Coefficient and Particle Size

A slightly extended form of the *van Deemter* equation is

$$H = a \cdot d_p + \frac{b \cdot D_m}{u_{int}} + c\frac{d_p^2}{D_m} \cdot u_{int} \tag{1.76}$$

where D_m is the diffusion coefficient of the solute in the mobile phase and a, b, and c are proportionality constants that do not vary with either D_m or d_p. Eqn (1.76) shows that two parameters explicitly affect the plate height. In the first place, this is the diffusion coefficient of the analyte in the mobile phase. A higher value of D_m increases the B-term but reduces the C-term. The net effect is that the optimum shifts to the right in the H vs. u_{int} plot. If we combine eqn (1.32) and (1.33) with eqn (1.76), we find that $u_{int,opt}$ increases in proportion with D_m, while H_{min} is not affected by the value of the diffusion coefficient.

$$u_{int,opt} = \sqrt{\frac{B}{C}} = \frac{D_m}{d_p}\sqrt{\frac{b}{c}} \tag{1.77}$$

and

$$H_{min} = A + 2\sqrt{B \cdot C} = a \cdot d_p + 2 \cdot d_p \sqrt{b \cdot c} \tag{1.78}$$

As a result, higher values of D_m imply that separations can be performed faster without a loss in efficiency. Mobile phases based on pressurized carbon dioxide are used in SFC. They exhibit considerably higher diffusion coefficients than mobile phases based on liquid solvents. Therefore, SFC promises to be faster than LC (see Chapter 6).

Another key parameter in eqn (1.76) is the particle size. Eqn (1.77) and (1.78) show that $u_{int,opt}$ is inversely proportional to d_p, while H_{min} increases in proportion to d_p. In addition, the C-branch of the *van Deemter* curves increases sharply with increasing d_p, as illustrated in Figure 1.22A. It is clear that smaller particles imply lower plate heights and smaller penalties for operating at flow rates above the optimum. The main limitation is the increased pressure drop. The overall effects of halving the particle diameter are the following:

- To obtain the same performance (N), the column length may be halved because H_{min} is halved.
- The velocity can be doubled because $u_{int,opt}$ is doubled.
- The analysis time is cut into four, thanks to the combined effects of L and $u_{int,opt}$.
- The pressure drop is increased four-fold, based on the combined effects of L, d_p and $u_{int,opt}$ in Darcy's law (eqn (1.69)).

1.7.3 Reduced Plate-height Equations

The van Deemter equation is useful to fundamentally understand and study the different phenomena that contribute to band broadening. However, one disadvantage of eqn (1.76) is the presence of d_p and D_m. While this clearly demonstrates that both factors impact the plate height, it also complicates the comparison of different chromatographic columns with different particle sizes or when different mobile phases are used. From Figure 1.22A, it is clear that small-particle (*e.g.* 3 µm) columns are advantageous in comparison with larger-particle (*e.g.* 10 µm) columns, provided that the required backpressure does not become a limiting factor. It is impossible to judge from such a figure which column is packed best or which packing material is most efficient.

For a fair comparison of columns packed with particles of different sizes or characterized with different mobile phases, it is useful to introduce **reduced parameters** in the plate-height equation. Reduced parameters are dimensionless and can be seen as a normalization of the variables in the plate-height equation. The **reduced plate height** (*h*) is defined as

$$h = \frac{H}{d_\text{p}} \tag{1.79}$$

Similarly, the linear flow velocity can be normalized for the diffusion coefficient of the solute in the mobile phase by defining the **reduced linear velocity** (ν_int) as

$$\nu_\text{int} = \frac{u_\text{int} \cdot d_\text{p}}{D_\text{m}} \tag{1.80}$$

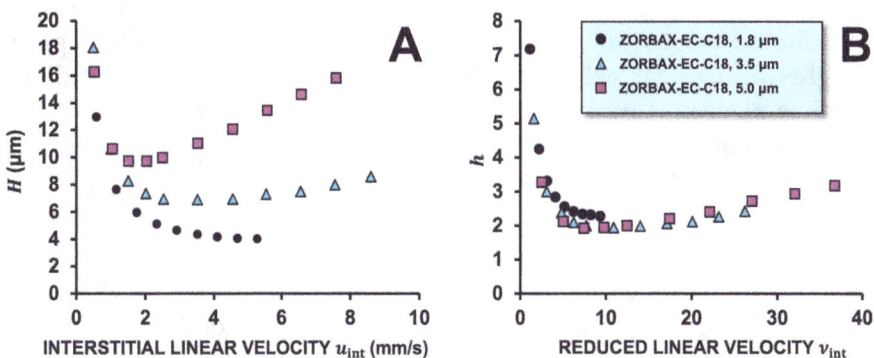

Figure 1.22 (A) Effect of particle size on the plate height as a function of flow rate. For smaller particles, a low plate height can be maintained as u_int increases. (B) The reduced plate heights ($h = H/d_\text{p}$) for the curves shown in (A). Adapted from ref. 18 with permission from Elsevier, Copyright 2017.

Substitution of eqn (1.79) and (1.80) into the *van Deemter* equation (eqn (1.72)) yields the **reduced van Deemter equation**

$$h = a + \frac{b}{v_{\text{int}}} + c \cdot v_{\text{int}} \tag{1.81}$$

Like the variables h and v_{int}, the coefficients a, b, and c in eqn (1.81) are dimensionless, so we no longer need to carefully consider the units in which they are expressed. Moreover, eqn (1.77) and (1.78) are transformed into

$$v_{\text{int,opt}} = \sqrt{\frac{b}{c}} \tag{1.82}$$

and

$$h_{\text{min}} = a + 2\sqrt{b \cdot c} \tag{1.83}$$

These two equations are equivalent to eqn (1.32) and (1.33), except that the coefficients A, B and C in the latter equations are different for each different H vs. u curve (*i.e.* they depend on D_m and d_p), whereas a, b and c in eqn (1.81) are independent of D_m and d_p. This is strikingly obvious from Figure 1.22B. While H vs. u plots following eqn (1.72) are quite different for columns packed with particles of different sizes, the h vs. v plots following eqn (1.81) are seen to overlap completely.

Eqn (1.81) has great practical consequences. If we know a set of reasonable values for a, b, and c, we can estimate the actual plate heights and linear velocities that can be achieved for all kinds of particle sizes and mobile phases, either at the optimum in the *van Deemter* plot (eqn (1.82) and (1.83)) or at any other position on the curve (*e.g.* $v = 3v_{\text{opt}}$). For example, if we assume for a well-packed column $a = 1.2$, $b = 2$, and $c = 0.08$, we obtain $v_{\text{opt}} = 5$ and $h_{\text{min}} = 2$. This instantly reveals that from a well-packed column packed with 10 μm particles, we may expect a minimum actual plate height (H_{min}) of 20 μm. For 5 μm particles, we may expect $H_{\text{min}} = 10$ μm, for 3 μm particles, $H_{\text{min}} = 6$ μm, *etc.* Conversely, the experimentally observed minimum reduced plate height (h_{min}) provides an instant measure of the quality of a packed column. For example, if a column packed with 5 μm particles shows $H_{\text{min}} = 15$ μm (*i.e.* $h_{\text{min}} = 3$), while you have ensured that your LC system and your connections contribute little to the band broadening (see Section 3.3.3), you may be dealing with a mediocre column. If $H_{min} = 25$ μm (*i.e.* $h_{min} = 5$), you either have a genuinely bad column or you are working with a retention mechanism, such as ion-exchange chromatography (see Module 3.7), that is thwarting your efficiency. Note that you do not need to convert to a dimensionless linear velocity and, thus, you do not need to know or estimate the diffusion coefficient to measure h_{min}.

1.7.4 Plate-height Theory – A Closer Look

The plate height H relates to the standard deviation of a chromatographic peak in time units (σ_t) and in length units (σ_z) through

$$H = \frac{L}{N} = \frac{L}{t_R^2}\sigma_t^2 = \frac{1}{L}\sigma_z^2 \tag{1.84}$$

and for independent processes, we may add variances. If we assume three processes A, B and C contributing independently, we may write

$$\sigma_t^2 = \sigma_{t,A}^2 + \sigma_{t,B}^2 + \sigma_{t,C}^2 \tag{1.85}$$

$$H = \frac{L}{t_R^2}\left(\sigma_{t,A}^2 + \sigma_{t,B}^2 + \sigma_{t,C}^2\right) = \frac{1}{L}\left(\sigma_{z,A}^2 + \sigma_{z,B}^2 + \sigma_{z,C}^2\right) = H_A + H_B + H_C \tag{1.86}$$

or

$$H = H_{\text{eddy diffusion}} + H_{\text{longitudinal diffusion}} + H_{\text{mass transfer}} \tag{1.87}$$

Eqn (1.86) represents a summation of contributions of independent processes to the plate height. This forms a justification of the *van Deemter* equation. Below we will take a closer look at the processes that contribute to the plate height.

1.7.4.1 Longitudinal Diffusion

We will start with the *B*-term, which is easiest to grasp conceptually. This contribution is due to molecular diffusion or Brownian motion, as described in Section 1.4.1. Longitudinal diffusion concerns the spreading of molecules in the longitudinal direction (*i.e.* the direction of flow), a process that contributes directly to the band broadening. If we start with an infinitely narrow pulse of analyte molecules and consider some point z (z being a variable in length units that indicates the position along the length of the column), then, after a time t, its concentration profile will be a Gaussian function, the variance of which is given by the **Einstein equation**

$$\sigma_z^2 = 2D_m \cdot t \tag{1.88}$$

where D_m is the diffusion coefficient of the analyte in the mobile phase. One way to describe the diffusion coefficient of the solute is the **Stokes–Einstein equation**

$$D_{m,i} = \frac{RT}{6\pi \cdot r_i \cdot \eta_m \cdot N_A} = \frac{k_B \cdot T}{6\pi \cdot r_i \cdot \eta_m} \tag{1.89}$$

Here, k_B is the Boltzmann constant, N_A is Avogadro's number, T is the temperature, r_i is the radius of the solute molecule (assumed to be a hard sphere), and η_m is the viscosity of the mobile phase. Table 1.2 lists a number of diffusion coefficients for different molecules of various molecular weights in water. As can be seen, D_m decreases as the molecular weight of the solute increases. This is in line with eqn (1.89), which predicts that an increasing solute radius (of a larger and thus heavier molecule) results in a lower diffusion coefficient. It also concurs with the very idea of Brownian motion. For a larger molecule to move in a solvent consisting of smaller molecules, a greater excess of collisions in the same direction is needed and more solvent molecules need to make way.

The most appropriate way to measure molecular diffusion coefficients of analytes in LC mobile phases is the Taylor–Aris method. This is based on measuring the time-based standard deviation (σ_t) of a peak eluting from an uncoated capillary tube. When (i) sufficiently high velocities are used, (ii) capillaries are sufficiently long, not too narrow, and not (tightly) coiled, and (iii) external (injection and detection) band broadening is eliminated or corrected for, the following simple equation applies[19]

$$D_m = \frac{L d_c^2}{96 u \sigma_t^2} \tag{1.90}$$

Table 1.2 Diffusion coefficients in water at 25 °C for different compounds of varying molecular weights.

Solute	Molecular weight	D_{m} $(10^{-9}\,\mathrm{m^2\,s^{-1}})$
Oxygen	32	2.0
Urea	60	1.4
Glucose	180	0.7
Lysozyme	14 000	0.12
Albumin	66 000	0.06
Myoglobin	820 000	0.02
Tobacco mosaic virus	39 000 000	0.004

Substituting eqn (1.88) into eqn (1.84) and realizing that $L = t_{\mathrm{excl}} \cdot u_{\mathrm{int}}$, we obtain an unretained compound

$$H_{\mathrm{B}} = \frac{\sigma_z^2}{L} = \frac{2\gamma D_{\mathrm{m}} t_{\mathrm{excl}}}{t_{\mathrm{excl}}\, u_{\mathrm{int}}} = \frac{B}{u_{\mathrm{int}}} \tag{1.91}$$

so that $B = 2\gamma D_{\mathrm{m}}$. An additional factor γ emerges in eqn (1.91). This is an obstruction factor or, in the words of *van Deemter et al.*, a "labyrinth factor".[5] The plot of H_{B} in Figure 1.10 shows that the *B*-term decreases with increasing linear velocity. This is confirmed by eqn (1.91), and it makes sense because with increasing u, the retention times decrease (eqn (1.9)), so that analyte molecules have less time to diffuse away from the centre of the zone.

Longitudinal diffusion is the only band-broadening process that continues if the flow is stopped. This means that a **peak-parking method** can be used to study the *B*-term. By measuring peak widths with the flow stopped for various durations, the effects of the *B*-term can be isolated. The width of the peak emerging from the column in time units (σ_t) is determined by its width in length units (σ_z) at the end of the column and the speed (u_i) at which the peak of analyte i emerges, *i.e.*

$$u_i = \frac{u_0}{1 + k_i} \tag{1.92}$$

so that with eqn (1.91)

$$\sigma_t^2 = \frac{\sigma_z^2 (1 + k)^2}{u_0^2} = \frac{2D_{\mathrm{eff}} (1 + k)^2}{u_0^2} \left(t_{\mathrm{R}} + t_{\mathrm{park}} \right) \tag{1.93}$$

Therefore, a plot of σ_t^2 vs. the total time spent in the column $(t_R + t_{park})$ will yield a straight line, the slope of which (S_{pp}) reveals the effective diffusion coefficient

$$D_{eff} = \frac{S_{pp} \cdot u_0^2}{2(1+k)^2} \qquad (1.94)$$

This effective diffusion may differ from the molecular diffusion (D_m) because analytes may diffuse differently if they are in the interstitial space, in a confined space in the pores of the packing material, or in (or on) the stationary phase. D_{eff} can be seen as a convoluted average of all these processes.[19]

Of course, we are interested in retained analytes rather than excluded probes. For retained molecules, we need to use $t_R = (L/u_0)(1+k)$ in eqn (1.91), and we need to use the more realistic D_{eff} instead of D_m. This yields

$$H_B = \frac{\sigma_z^2}{L} = \frac{2D_{eff}}{u_0}(1+k) \qquad (1.95)$$

from which it is instantly apparent that the contribution of the longitudinal diffusion is not independent of the retention factor, as initially perceived by van Deemter *et al.*[5] In terms of the reduced parameters $h = H/d_p$ and $v_0 = u_0 \cdot d_p/D_m$

$$h = \frac{2}{v_0}\frac{D_{eff}}{D_m}(1+k) = \frac{2\gamma_{eff}}{v_0}(1+k) \qquad (1.96)$$

where $\gamma_{eff} = D_{eff}/D_m$ is an effective (lack of) obstruction factor. γ_{eff} indicates to what extent diffusion is possible inside the particles (for unhindered diffusion $\gamma_{eff} = 1$ and for hindered diffusion $\gamma_{eff} < 1$). Peak-parking experiments will yield a value for D_{eff}, making the B-term experimentally accessible. Obtaining a value for D_m from, for example, a Taylor–Aris experiment yields insight into γ_{eff}.

1.7.4.2 Resistance to Mass Transfer

Several different processes underlie the C-term, with mass transfer as the common denominator. One process arises from incomplete equilibration and is depicted in Figure 1.10. In the leading edge of the peak, the eluent progressing through the column contains a slightly too high concentration of analytes. At the tailing edge of the peak, the concentration of analytes is slightly too high in the stationary phase. The other processes reflected in the C-term involve variations in the mobile-phase velocity. The parabolic flow profile (see Section 1.6.3) implies that some molecules (near the centre of a channel) move much faster than other molecules (positioned close to the wall of the flow channel). The two processes can be distinguished as C_s (non-equilibrium or stationary-phase contribution) and C_m (flow-profile or mobile-phase contribution) terms, and as a first approximation, the C-term can be described by[19]

$$H_C = C \cdot u_{int} = C_s \cdot u_{int} + C_m \cdot u_{int} = f_s(k)\frac{u_{int} \cdot d_f^2}{D_s} + f_m(k)\frac{u_{int} \cdot d_p^2}{D_m} \qquad (1.97)$$

where d_f is the thickness of a liquid stationary-phase film or (in the case of a stationary phase other than a liquid) a characteristic thickness parameter, D_s is the effective diffusion coefficient of the analyte in the stationary phase, and $f_s(k)$ and $f_m(k)$ are proportionality factors that are functions of the analyte retention factor k (and of the porosity and the Sherwood number, see below). For a packed column, d_p can be substituted for d_f and a particle diffusion coefficient (D_{part}) takes the place of D_s.

During the last fifty years, the theory of band broadening has continued to evolve and more complex equations are now in fashion. We describe a state-of-the-art approach in this section. The contemporary general-plate-height equation yields the following equation for h_{C_m} for fully porous particles in terms of the average reduced interstitial mobile-phase velocity (v_{int})[19]

$$h_{C_m} = \frac{1}{3}\left(\frac{k''}{1+k''}\right)^2\left(\frac{\varepsilon_{int}}{1-\varepsilon_{int}}\right)\frac{v_{int}}{Sh} \qquad (1.98)$$

where Sh is the Sherwood number ($Sh = \kappa_m \cdot d_p/D_m$), which can be seen as the dimensionless form of the mobile zone mass transfer coefficient κ_m. The latter has the dimensions of a velocity (m s^{-1}) and describes the speed at which concentration differences between two phases are diminished.

In plate-height studies, the zone retention factor (k'') is often used instead of k. It is defined as

$$k'' = \frac{t_R - t_{excl}}{t_{excl}} \tag{1.99}$$

where t_{excl} is the elution time of a fully excluded compound. The zone retention factor forms a logical combination with the interstitial velocity u_{int}. The double prime results from an old (and now incorrect) notation of k' for the retention factor (now k). It follows that

$$\frac{1 + k''}{1 + k} = \frac{t_0}{t_{excl}} = \frac{u_{int}}{u_0} = \frac{\varepsilon_{tot}}{\varepsilon_{int}} \tag{1.100}$$

Eqn (1.98) is indeed less convenient in terms of k

$$\begin{aligned}
C_m &= \frac{1}{3} \left\{ \frac{\varepsilon_{tot}(1 + k) - \varepsilon_{int}}{\varepsilon_{tot}(1 + k)} \right\}^2 \left(\frac{\varepsilon_{int}}{1 - \varepsilon_{int}} \right) \frac{\nu_{int}}{Sh} \\
&= \frac{1}{3} \left\{ \frac{\varepsilon_{tot}(1 + k) - \varepsilon_{int}}{\varepsilon_{tot}(1 + k)} \right\}^2 \left(\frac{\varepsilon_{tot}}{1 - \varepsilon_{int}} \right) \frac{\nu_0}{Sh}
\end{aligned} \tag{1.101}$$

but if it is easier to determine k than k'', the two approaches can be used equivalently.

The occurrence of the Sherwood number in eqn (1.98) and (1.101) indicates that the mass-transfer coefficient plays a significant role. If it is high (*i.e.* if the mass transfer is fast), the contribution of h_{C_m} to the overall plate height is small and *vice versa*.

It is difficult to measure κ_m directly in a situation mimicking an LC column, but Desmet *et al.* provide a good indication of the value of Sh in a practical LC system.[20,21]

$$Sh = \frac{13}{1 + 2.1 \cdot \nu_{int}} + 8.6 \cdot \nu_{int}^{0.21} \tag{1.102}$$

For h_{C_s}, we have[19]

$$h_{C_s} = \frac{1}{30} \frac{k''}{(1 + k'')^2} \frac{D_m}{D_{part}} \nu_{int} \tag{1.103}$$

or in terms of k and ν_0,

$$h_{C_s} = \frac{1}{30} \frac{\varepsilon_{tot}(1+k) - \varepsilon_{int}}{\varepsilon_{tot}(1+k)^2} \frac{D_m}{D_{part}} v_0 \tag{1.104}$$

The diffusion coefficient in the particle (D_{part}) can be calculated from D_{eff} measured using the peak-parking method using the effective medium theory.[19]

1.7.4.3 Eddy Diffusion

Although the A-term alphabetically came first in the original van Deemter treatment, it is appropriate to treat it last. It is the most difficult term to describe and determine experimentally. In practice, it is often determined as the remainder of the experimentally observed plate height H after accounting for the other terms, *i.e.* by rewriting eqn (1.87) to

$$H_{eddy\ diffusion} = H - H_{longitudinal\ diffusion} - H_{mass\ transfer} \tag{1.105}$$

In doing so, any uncertainties in the B- and C-terms are reflected in the A-term.

We previously specified that the A-term or **eddy diffusion** term in the *van Deemter* equation is due to path-length differences among different molecules moving through the column. This can also be mathematically approximated with the "random walk" model for eddy diffusion.

We can theoretically approximate a group of solute molecules travelling through the column by imagining that they must travel an n number of steps, each with an average length l, to reach the other end. In other words, if the column has length L, then the total number of steps is

$$n = \frac{L}{l} \tag{1.106}$$

In a column packed with stationary-phase particles of diameter d_p, each analyte molecule can take another route around the particles (as is depicted in Figure 1.10A, upper right), causing each analyte molecule to travel a slightly different distance q. There is thus a variation σ_l in the step length for the population of solute molecules that, if we assume l to be proportional to d_p, is given by

$$\sigma_l = q \cdot l = q \cdot d_p \tag{1.107}$$

And if we square both sides of this equation

$$\sigma_l^2 = q^2 \cdot d_p^2 \tag{1.108}$$

We can use eqn (1.108) because it allows us to sum this variance σ_z^2 for all n steps, yielding

$$\sigma_z^2 = n \cdot \sigma_l^2 = n \cdot q^2 \cdot d_p^2 \tag{1.109}$$

If we now assume the number of steps to be related to the number of particle diameters that would make up the length of the column, we find

$$\sigma_z^2 = \frac{L}{d_p} \cdot q^2 \cdot d_p^2 = L \cdot q^2 \cdot d_p \tag{1.110}$$

We then arrive at the notation of van Deemter to describe the A-term contribution to the plate height.

$$H_A = A = \frac{\sigma_{z,n}^2}{L} = q^2 \cdot d_p = \lambda \cdot d_p \tag{1.111}$$

where λ ($= q^2$) is a "packing-characterization" factor[5] or "tortuosity factor". It is plausible that q^2, which is proportional to the variance in the step size, increases if the variance in the particle size increases. Thus, eqn (1.111) suggests that the A-term contribution to the plate height decreases if the packing particles are more uniform.

Eqn (1.107)–(1.111), as well as the mathematical formulations in the study by van Deemter, suggest that the A-term is independent of the linear velocity (or the flow rate). This is why the plot of H_A shown in Figure 1.10 suggests that the eddy diffusion does not depend on the flow rate.

Giddings and Knox soon realized that eddy diffusion could not be independent of the linear velocity. In other words, they concluded that the A- and C-terms are "coupled". The description of the **coupled** AC-term by Giddings reads

$$H_{AC,Gid} = \left(\frac{1}{A} + \frac{1}{C_m \cdot u_{int}} \right)^{-1} \tag{1.112}$$

This equation implies that at very small values of u_{int}, H_{AC} emerges from the origin ($H = 0$, $u = 0$) as a straight line with slope C_m and bends at higher values of u_{int} to a limiting value of A.

Desmet *et al.* more recently derived a somewhat more complex equation, which reads[20]

$$H_{AC,Des} = C_{AC,Des} \cdot u_{int} \left\{ 1 - \frac{C_{AC,Des} \cdot u_{int}}{2 \cdot A_{AC,Des}} \left(1 - e^{-\frac{2 \cdot A_{AC,Des}}{C_{AC,Des} \cdot u_{int}}} \right) \right\} \tag{1.113}$$

where $A_{AC,Des}$ and $C_{AC,Des}$ are the A and C coefficients, respectively, or, in dimensionless terms,

$$h_{AC,Des} = c_{AC,Des} \cdot v_{int} \left\{ 1 - \frac{c_{AC,Des} \cdot v_{int}}{2 \cdot a_{AC,Des}} \left(1 - e^{-\frac{2 \cdot a_{AC,Des}}{c_{AC,Des} \cdot v_{int}}} \right) \right\} \tag{1.114}$$

Thus far, we have assumed that differences in path length occur at the scale of a particle. Short flow paths along particles and longer paths around these may be imagined. However, Giddings (ref. 22, p. 42) recognized very soon that path-length differences were not restricted to the scale of a single particle. Path-length differences occur (i) across the length of a single flow channel between particles (the *trans-channel* range), (ii) across the scale of a single particle (the *trans-particle* range), (iii) across the range of several particles that may be more tightly or more loosely packed (the *short-range interchannel* range), (iv) in large domains across multiple particles (the *long-range interchannel* range), and (v) across the diameter of the column, for example between the centre regions and regions near the wall (the *trans-column* range). Therefore, the Giddings equation for the A-term must be expanded to a summation across the different length scales

$$H_{AC,Gid} = \sum_{i=1}^{m} \left(\frac{1}{A_i} + \frac{1}{C_{m,i} \cdot u_{int}} \right)^{-1} \tag{1.115}$$

where the index i indicates the m different length scales on which path-length differences or packing inhomogeneities are considered or the Desmet version

$$H_{AC,Des} = \sum_{i=1}^{m} \left[c_{AC,Des} \cdot v_{int} \left\{ 1 - \frac{c_{AC,Des} \cdot v_{int}}{2 \cdot a_{AC,Des}} \left(1 - e^{-\frac{2 \cdot A_{AC,Des,i}}{c_{AC,Des,i} \cdot u_{int}}} \right) \right\} \right] \qquad (1.116)$$

At this point, the eddy diffusion term may have become better understood but impossible to use in practice. Many authors have ignored the summations of eqn (1.115) and (1.116) and resorted to the single-term eqn (1.112) and (1.113).

In a more pragmatic approach, Knox (Box 1.2) suggested the following empirical relationship in terms of the reduced velocity v_{int} with an exponent n_{Knox}.

$$h_{AC,Knox} = A_{Knox} \cdot v_{int}^{n_{Knox}} \qquad (1.117)$$

Knox suggested that 1/3 would be a suitable value for the exponent n_{Knox}, but Desmet *et al.* have refined this to include the effect of the retention factor[20,24]

$$n_{Knox} = 0.55 - \frac{0.11}{(k'')^{0.93}} \qquad (1.118)$$

and

Box 1.2 Hero of analytical separation science: J. H. Knox. Image reproduced from ref. 23 with permission from the Royal Society of Chemistry.

John H. Knox (1927–2018, Scotland)
John Knox was born in Edinburgh and spent most of his professional life as a professor of physical chemistry at that city's university. He is the namesake of a famous 16th-century Scotsman. John H. Knox was one of the brilliant minds of chromatography, and he contributed massively to the theory of dispersion (band broadening) in chromatography. His practical contributions included the development of LC packing materials based on silica and pyrolytic carbon. His student Mary Gilbert and an LC instrument decorated the Scottish 20-pound note for many years.

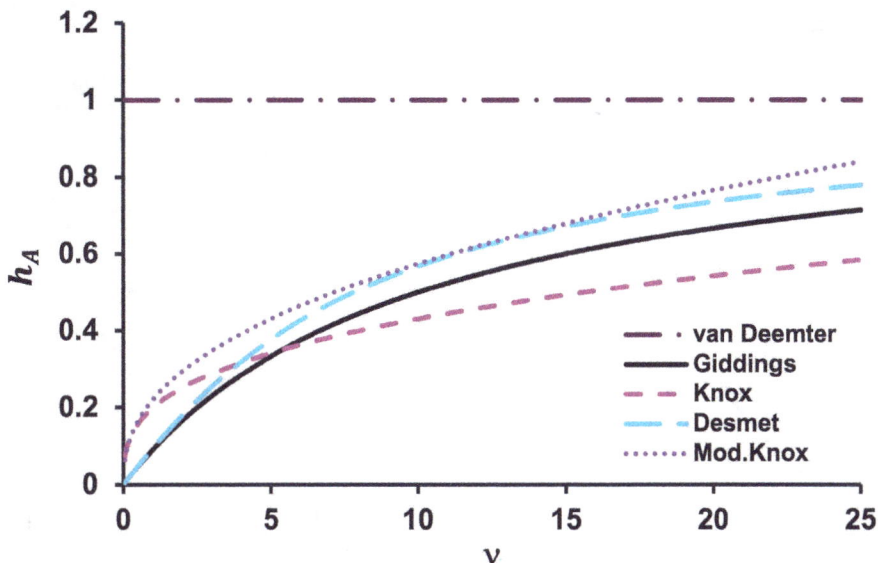

Figure 1.23 Qualitative comparison of the van Deemter *A*-term (with $A = d_p$) and the coupling terms proposed by Giddings (eqn (1.15), one term only, $i = 1$ with $A_1 = 1$ and $C_{m,1} = 0.1$), Desmet (eqn (1.14) with $a_{AC,Des} = 1$ and $c_{AC,Des} = 0.1$), Knox (eqn (1.17), with $A_{Knox} = 0.2$ and $n_{Knox} = 1/3$), and the modified Knox (eqn (1.18) and (1.19) with $k'' = 4.25$).

$$A_{Knox} = 0.19 + \frac{0.25}{(k'')^{0.46}} \qquad (1.119)$$

In Figure 1.23, the different proposals for the coupling term are qualitatively compared. Clearly, the idea of a constant *A*-term as proposed by *van Deemter* has been superseded by contributions that increase with increasing linear velocity. The equations proposed by Giddings and Desmet appear very different, but the behaviour of the two functions is nearly identical. The Desmet model has the best theoretical foundation, and for that reason, it may be preferred.

1.7.4.4 Adsorption/Desorption Kinetics

When the kinetics of adsorption and desorption on and from the stationary phase are slow compared to other kinetic chromatographic processes, an additional *C*-term (C_{ads}) may need to be added to the plate-height equation. Such a C_{ads} term has, for example, been used to describe dispersion in the case of chiral (*i.e.* stereoselective,

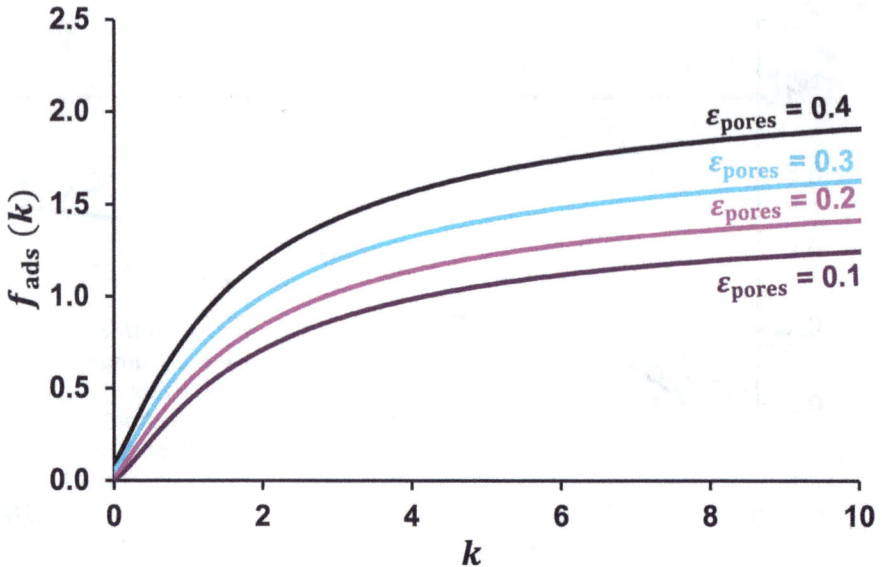

Figure 1.24 Dependence of the slow adsorption/desorption kinetics f_{ads}-factor according to Gritti and Guiochon,[25] calculated for an external porosity of 0.4 and pore porosities as indicated in the figure. For non-porous particles (ε_{pores} = 0), $f_{ads}(k)$ = 0.

see Module 3.13) separations.[25,26] It may also be appropriate for ion-exchange separations, which are also known to be less efficient than typical LC separations of uncharged molecules. Understandably, such a dispersion term is inversely proportional to the adsorption rate constant (r_{ads}, assuming first-order kinetics and, thus, units of s^{-1}), and it is a complex function of the retention factor and the porosity, as illustrated in Figure 1.24.

The contribution of slow adsorption/desorption kinetics to the absolute plate height is

$$H_{ads} = H_{ads}u = \frac{f_{ads}(k)}{r_{ads}}u \qquad (1.120)$$

which shows that the contribution is low as long as the kinetics are fast (*i.e.* r_{ads} is large). If kinetics are slow, this contribution may become dominant. In terms of reduced plate heights, the contribution is

$$h_{ads} = \frac{f_{ads}(k)}{r_{ads}} \cdot \frac{D_m}{d_p^2}v \qquad (1.121)$$

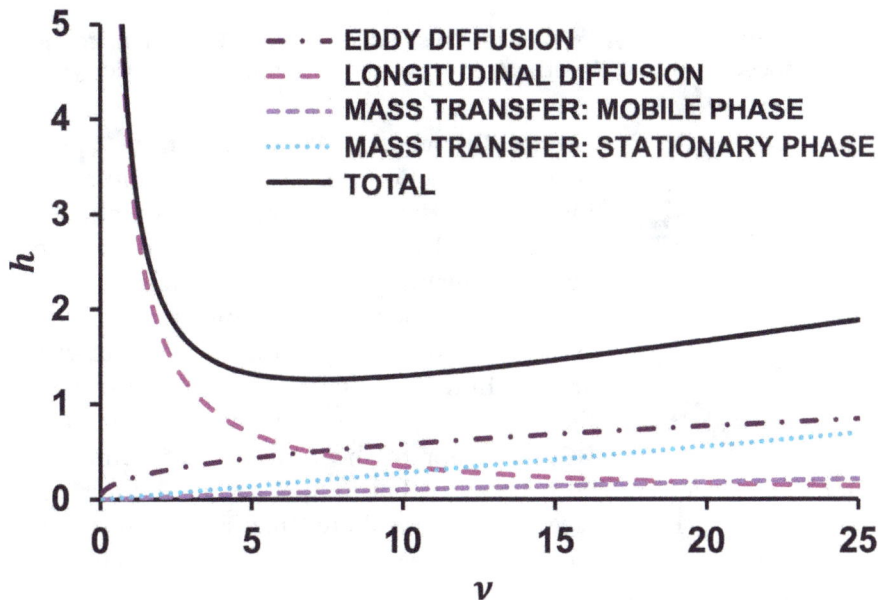

Figure 1.25 Theoretical contributions to the plate height in packed-column LC, assuming a total porosity (ε_{tot}) of 0.7, an interstitial porosity (ε_{int}) of 0.4, $k = 2$ (and, therefore, $k'' = 4.25$), n_{Knox} = 0.414 (eqn (1.118)), A_{Knox} = 0.221 (eqn (1.119)), γ_{eff} = 0.33 (ref. 19 Figure 1.4) and D_m = 5 x 10^{-10} m^2 s^{-1}, value for glucose from Table 1.2, d_p = 1.3 x 10^{-10} m^2 s^{-1} (ref. 19 eqn (1.31) and (1.32)).

This implies that slow kinetics have a relatively greater impact when the diffusion coefficient is high (*i.e.* for low-molecular-weight analytes) and when using small particles. This can be interpreted such that when the kinetics of the chromatographic processes are minimized, the adsorption–desorption kinetics at some point become the limiting factor.

1.7.4.5 Summary

Figure 1.25 shows a summary of all effects described in this section, using eqn (1.96) to describe the longitudinal diffusion term (with γ_{eff} = 0.5), eqn (1.98) for the mobile-phase contribution to the mass-transfer term with the Sherwood number calculated from eqn (1.102), eqn (1.104) for the stationary-phase contribution, and eqn (1.117) to describe the eddy diffusion term.

> **Box 1.3** Hero of analytical separation science: Georges Guiochon. Image reproduced from ref. 28 with permission from Springer Nature, Copyright 2014.
>
>
>
> **Georges Guiochon (1931–2014, France)**
> Georges Guiochon was an amazing and amazingly productive separation scientist, physical chemist, and much more. He graduated from École Polytechnique, near Paris, France, where he also spent the first part of his career, gathering a very strong group of then-young scientists around him. In 1984, he moved to the USA, first to Georgetown University in Washington, DC, and later to the University of Tennessee and Oak Ridge National Laboratory, but in 30 years, he did not lose his charming French accent. He contributed massively with (more than 1000!) often theoretical publications on nearly all aspects of chromatography.

The theoretical models seem to be optimistic in comparison with LC practice, predicting a minimal reduced plate height well below 2 ($h_{min} \approx 1.26$ and $\nu_{opt} \approx 8$).

Much more can be said about band broadening, and the above treatment is by no means complete. The references in the text may help the interested reader to study the subject in more depth. Another good place to start a deeper study is a review by Fabrice Gritti and Georges Guiochon[27] (see also Box 1.3).

1.7.5   Extra-column Band Broadening

The peak width observed in a chromatogram is not solely due to dispersion processes taking place in the column. Generally, if dispersion processes are independent, we may add the variances, so that

$$\sigma_{obs}^2 = \sigma_{col}^2 + \sigma_{ec}^2 \tag{1.122}$$

where σ_{obs} refers to the observed bandwidth, σ_{col} refers to the dispersion due to the column and σ_{ec} refers to extra-column band broadening. The latter may be further specified as

$$\sigma_{ec}^2 = \sigma_{inj}^2 + \sigma_{det}^2 + \sigma_{con}^2 \qquad (1.123)$$

where the subscripts inj, det and con refer to injection, detection and connections, respectively. The variance due to connections (σ_{con}^2) may be further specified, adding the contributions from various capillaries and connectors. Frits in packed columns may also contribute. Extra-column band broadening is not a fundamental but an experimental effect. Ideally, it should be minimized to the extent that it is negligible. Nevertheless, in practice, extra-column band broadening can be very significant, especially in liquid chromatography. This will be discussed in Module 3.3.

1.7.6 Heat Dissipation

Adequate temperature control is an important issue in chromatography. Because of the fabulous discrimination power of chromatography, even very small changes in the partial molar free energy of transfer of analyte i ($\Delta G_i = \Delta H_i - T\Delta S_i$) will affect the retention factor (k_i), and small differences between the ΔG values for two analytes j and i ($\Delta\Delta G_{j,i}$) will affect the selectivity ($\alpha_{j,i}$). We will be discussing temperature control in sections devoted to instrumentation and troubleshooting later in this book, but there is one fundamental aspect that needs mentioning in this section. Almost all the separation systems that we will be discussing in this book rely on pressure to transport a mobile phase through a column. This implies that we put work (W) in the system to the extent

$$W = \pi \cdot t = \Delta P \cdot F \cdot t \qquad (1.124)$$

where π denotes the power. If we express the pressure drop ΔP in Pa $(N\ m^{-2})$, the flow rate F in $m^3\ s^{-1}$, and the time t in s, the work is expressed in N m, *i.e.* J (or W s). For example, if we operate an LC system at 100 MPa and a flow rate of 0.6 mL min^{-1} or 10^{-8} $m^3\ s^{-1}$, we apply a power of 1 W to the system. Inside the column, nearly all the work is converted into frictional heat. What happens to this heat lies somewhere between two extremes. At one extreme, we may aim to keep the temperature in the column constant by removing the heat through the column wall by applying a form of active cooling on the outside. At the other extreme, we may aim to keep all generated heat in the column by insulating. These extremes are the standard thermodynamic isothermal and adiabatic processes, respectively.

The hardware of almost all LC columns that are operated at high pressures consists of stainless steel. If we succeed in keeping the temperature of the outside of the column wall constant, for example, by immersing it in a water bath, we may assume that the inner wall of the steel column also is at the same temperature because metals are good thermal conductors. This is different for the material inside the column (packing particles, stationary phase and mobile phase). To transport heat from the centre of the column to the column wall, we face challenges due to poor thermal conduction and potentially many heat-transfer steps from the mobile-phase to stationary-phase particles and *vice versa*, as well as a final heat transfer to the column wall. As a result, we will not succeed in keeping the entire column at a constant temperature. It will become hotter in the centre than at the column walls. This will cause the viscosity to be higher in the centre than close to the wall, which in turn will result in a higher flow rate in the centre of the column. Moreover, higher temperatures result in lower retention factors, increasing the differences in migration velocities of the analyte between the centre of the column and the wall region. Potentially, the radial temperature gradient has disastrous consequences for the column band broadening, but the damage can be limited by reducing the column diameter. This reduces the distance between the centre and the wall and also the amount of heat generated if the flow rate is adapted, such as to keep the linear velocity constant. Thus, narrow columns are preferred for LC at very high pressures. Capillary LC columns have been used to study operations at extremely high pressures (>150 MPa).

In the adiabatic case, there is no radial heat transport from a relatively warm centre of the column to the relatively cold walls; however, as the frictional heat has no place to go, the mobile phase heats up as it passes through the column. In an eventual steady state, an axial temperature gradient develops and the mobile phase heats up from the inlet temperature to a constant temperature at the column outlet, but the situation is complicated by the thermal conductivity of the (stainless-steel) column, causing a heat flux from the end of the column to the inlet and, again, a temperature difference between the column wall and the mobile phase inside the column. An increase in eluent temperature by more than 10 °C is realistic under practical experimental conditions.[12,29] This implies that the local retention factor decreases along the length of the column. The retention factor of analytes when they exit the column is lower than the apparent retention factor ($k_{app} = t_R/t_0 - 1$) because of the axial temperature gradient and the effect of pressure on retention (see Section 1.6.1.2). Both effects contribute to narrower chromatographic peaks. However, the selectivity may be positively or negatively affected by the combined effects of temperature and pressure.

In practice, it is ill-advised to attempt to keep the temperature of LC columns constant, especially when they are operated at very high pressure drops, which are typically encountered when using very small (sub-2 μm) particles. The standard columns for such situations have an internal diameter of 2.1 mm, which makes them significantly narrower than the conventional 4.6 mm i.d. columns. (Quasi-)adiabatic columns, which are insulated with a vacuum hull, have been explored by Gritti *et al.*[29] With more conventional hardware, it is recommended to keep the column in a still-air oven, avoid any active cooling, and dissipate the produced heat through the column outlet.

1.7.7 Open-tubular Columns – The Golay Equation

Open-tubular columns have a simple and well-defined configuration, which allowed Marcel Golay (Box 1.4) to solve the mass balance and

Box 1.4 Hero of analytical separation science: M. Golay. Image reproduced from ref. 30 with permission from American Chemical Society, Copyright 1989.

Marcel Golay (1902–1989, Switzerland)
Marcel Golay was born in Switzerland in 1902. He was known as a mathematician, physicist and information scientist (the Savitzky–Golay smoothing algorithm carries his name), but he also made great contributions to separation science. Most importantly, he derived an exact equation that described band broadening in open-tubular columns, which we still use today. He played a major role in the transition from packed to open-tubular columns for GC. During a large part of his life (from 1955 onwards), he was associated with Perkin Elmer (Norwalk, CT, USA).

derive an exact form of the van Deemter equation.[31] In this section, we will ignore the compressibility of gases for simplicity, as the effects are usually minor. In that case, the **Golay equation** reads

$$H = \frac{2D_\mathrm{m}}{u} + f(k)\frac{d_\mathrm{c}^2 u}{D_\mathrm{m}} + g(k)\frac{d_\mathrm{f}^2 u}{D_\mathrm{f}} \tag{1.125}$$

where H is the plate height, u is the average linear velocity, d_c is the internal diameter of the column, d_f is the thickness of the (stationary-phase) film, D_m is the diffusion coefficient of the analyte in the mobile phase, and D_f is the diffusion coefficient in the stationary-phase film. $f(k)$ and $g(k)$ are functions of the retention factor, as follows

$$f(k) = \frac{1 + 6k + 11k^2}{96(1 + k)^2} \tag{1.126}$$

$$g(k) = \frac{2k}{3(1 + k)^2} \tag{1.127}$$

Eqn (1.126) and (1.127) are valid for $d_\mathrm{f} \leq 0.1 d_\mathrm{c}$ and in the laminar flow regime. Both conditions are usually met in chromatography (for an extension to turbulent-flow conditions, see ref. 32).

The three terms in eqn (1.125) represent molecular diffusion, mass transfer in the mobile phase and mass transfer in the station-ary phase, respectively. Thus, the Golay equation reflects similar dispersion processes to the *van Deemter* equation, but it specifically applies to open-tubular columns, in which packing particles are absent. We have seen in Section 1.7.3 that it is attractive to introduce dimensionless parameters, *i.e.* the reduced plate height (h), reduced (average) linear velocity (v), and reduced film thickness (δ_f) as follows

$$h = \frac{H}{d_\mathrm{c}} \tag{1.128}$$

$$v = \frac{u d_\mathrm{c}}{D_\mathrm{m}} \tag{1.129}$$

$$\delta_f = \frac{d_f}{d_c}\sqrt{\frac{D_m}{D_f}} \qquad (1.130)$$

If we substitute the above dimensionless parameters into the Golay equation (eqn (1.125)), we obtain

$$h = \frac{2}{\nu} + f(k)\nu + g(k)\delta_f^2\nu \qquad (1.131)$$

This is a very simple plate-height equation, which should be valid (approximately, as we ignored mobile-phase compressibility) for all forms of open-tubular chromatography, whether the mobile phase is a gas, a liquid or a supercritical fluid.

The reduced film thickness should be kept low so that $g(k) \cdot \delta_f^2$ remains significantly smaller than $f(k)$ and the mass-transfer contribution to the plate height is dominated by the mobile-phase term, as shown in Figure 1.26A. It is seen that $\delta_f = 0.3$ yields a much lower contribution to h than $\delta_f = 1.0$, with the latter exhibiting sizeable losses in efficiency. This is in agreement with Figure 1.26B, which shows a plot of the sum of the $f(k)$ and $g(k) \cdot \delta_f^2$ contributions to the Golay equation. Furthermore, Figure 1.26B also suggests that a low retention factor is favourable. On the other hand, δ_f should not be too low if sufficient retention and sample loadability are to be obtained.

The conclusions drawn from Figure 1.26 are confirmed by plotting the Golay equation for different retention factors and film thicknesses, as illustrated in Figure 1.27. In the top two frames, it can be seen that increasing the reduced film thickness from $\delta_f = 0.1$ (top left) to $\delta_f = 0.3$ (top right) causes a minor increase in reduced plate height and, hence, minor losses in efficiency. However, increasing the reduced film thickness to $\delta_f = 0.6$ (bottom left) or $\delta_f = 1$ (bottom right) results in unacceptable losses, even in the region of low retention factors. It is thus not surprising that a value of $\delta_f = 0.3$ has been found optimal.[33]

Figure 1.27 affirms that high efficiencies (low plate heights) can only be maintained in open-tubular chromatography if k values are kept very low. In GC practice, this is achieved by temperature programming. The analytes remain immobilized at the top of the column until the temperature reaches a point where they move through the column quickly (*i.e.* with low effective retention factors).

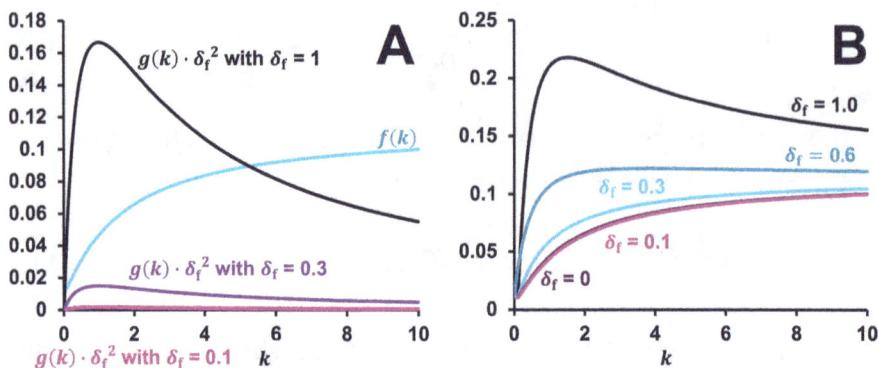

Figure 1.26 (A) $f(k)$ and $g(k) \cdot \delta_f^2$ factors plotted as a function of the retention factor. (B) The sum of the $f(k)$ and $g(k) \cdot \delta_f^2$ contributions to the Golay equation is plotted as a function of the retention factor for different values of the reduced film thickness δ_f. Both plots underline that a dimensionless stationary-phase film thickness of 0.3 is a good compromise between chromatographic efficiency (favoured by low values of δ_f) and sample loadability (favoured by high values of δ_f).

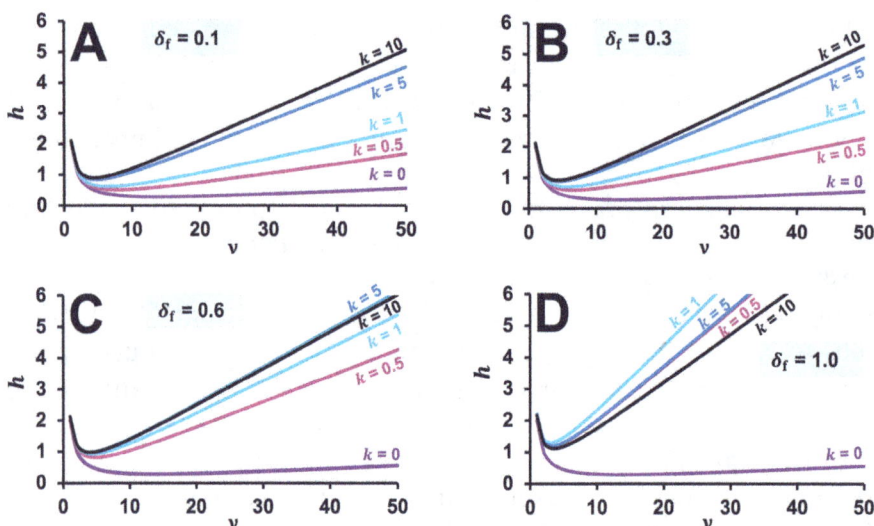

Figure 1.27 Reduced plate height curves based on the dimensionless form of the Golay equation. Plotted for various values of k with a δ_f of 0.1 (A), 0.3 (B), 0.6 (C) and 1.0 (D). It can be seen that a reduced film thickness of 0.3 is acceptable (top right) but that higher values (bottom frames) are not. In addition, it is clear that low retention factors are needed to maintain high efficiencies in open-tubular chromatography.

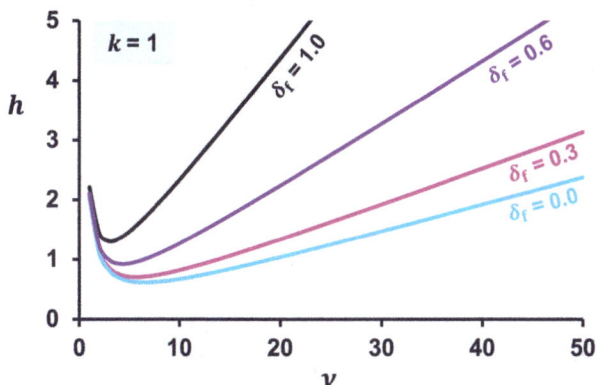

Figure 1.28 Theoretical (Golay) reduced plate-height plots for different dimensionless stationary-phase film thicknesses when the retention factor is 1.

Figure 1.28 shows a different perspective (all lines referring to $k = 1$) that emphasizes the points made above.

The Golay equation yields for the optimum velocity

$$v_{\text{opt}} = \sqrt{\frac{2}{f(k) + g(k)\delta_f^2}} \qquad (1.132)$$

And for the minimum plate height

$$h_{\text{min}} = \sqrt{8\{f(k) + g(k)\delta_f^2\}} \qquad (1.133)$$

It is apparent from Figure 1.27B that the optimum does not shift much at higher k values. This is confirmed in Figure 1.29. The figure shows that high k values can be used in open-tubular chromatography, as long as the column is operated at the optimum linear velocity. At velocities above v_{opt}, k values have a strong negative effect on the efficiency (see Figure 1.27B).

1.7.8 Applicability of Open-tubular Columns in Liquid Chromatography

The Golay equation can be used to calculate the actual values (H, u, D_m, etc.) in quite diverse situations. Using typical values for GC, i.e. $D_m = 5 \times 10^{-6}$ m^2 s^{-1} and, for a silicone-type stationary phase, $D_f = 5 \times 10^{-11}$ m^2 s^{-1}, we find that $\delta_f = 0.3$ corresponds to $d_f =$

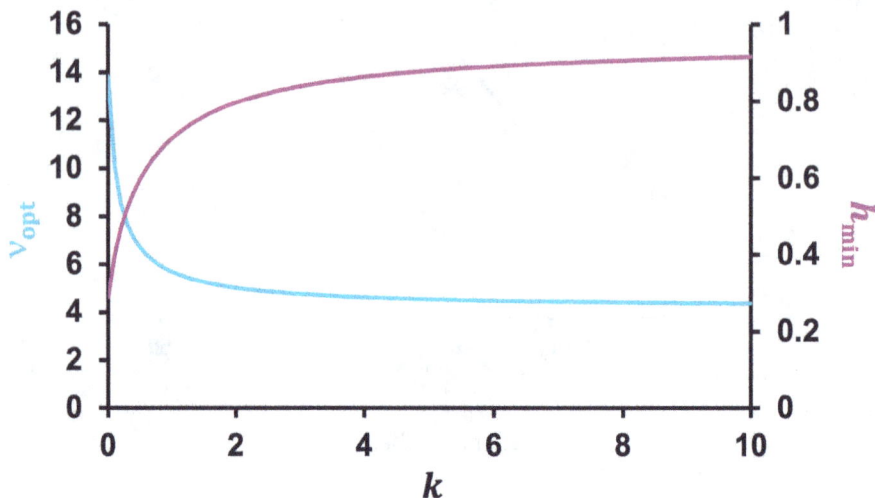

Figure 1.29 Optimum for the reduced linear velocity (eqn (1.132)) and reduced plate height (eqn (1.133)) according to the Golay equation as a function of the retention factor k for a reduced film thickness δ_f of 0.3.

$0.001d_c$. This is an excellent rule of thumb for general-purpose GC columns. The most common GC columns follow this guideline (*e.g.* d_f = 250 μm and D_m = 0.25 μm). For open-tubular LC, D_m = 10^{-9} m² s⁻¹ is a more reasonable value, which implies that a similar silicone film can be about 70 times thicker than in a comparable GC column (d_f = $0.07d_c$). Stationary films that allow faster diffusion (*e.g.* mesoporous coating) can be even thicker. The conditions corresponding to a typical capillary (open-tubular) GC column and a possible open-tubular LC column are summarized in Table 1.3.

The conditions for GC given in Table 1.3 concur with routine practice. For LC, the data in the table confirm that it is practically impossible to perform open-tubular liquid chromatography (OTLC) in a correct manner. While it may be possible – in principle – to deal with the very small volumetric volumes and injection volumes, for example, by installing a liquid-phase splitter before the column, dealing with detection volumes of the order of 40 pL is extremely challenging. A make-up flow would increase the volume available for detection but would dilute the analytes and greatly aggravate the challenge of detecting very low concentrations. Despite a series of valiant attempts, practically useful OTLC systems have not yet been described.

Table 1.3 Typical conditions for open-tubular GC and open-tubular LC, assuming $k = 1$ and $\delta_f = 0.3$ and choosing a column diameter of 5 μm for OTLC. Columns operated at the optimum conditions $v_{opt} = 5.7$ and $h_{min} = 0.7$.

Property	Symbol	Units	GC	LC
Mobile-phase diffusion coefficient	D_m	m² s⁻¹	5×10^{-6}	10^{-9}
Stationary-phase diffusion coefficient	D_f	m² s⁻¹	5×10^{-11}	5×10^{-11}
Column diameter	d_c	μm	250	5
Stationary-phase film thickness ($\delta_f = 0.3$)	d_f	μm	0.24	0.34
Phase ratio ($\beta = V_s/V_m$)	β	—	0.002	0.15
Plate height	H	μm	176	3.5
Average linear velocity	u	mm s⁻¹	110	1.1
Volumetric flow rate	F	μL min⁻¹	335	0.0013
		nL min⁻¹		1.3
Plates per metre	N/L	m⁻¹	5700	285 000
Column length for 100 000 plates	L	m	17.6	0.35
Column hold-up time	t_0	s	155	309
Peak standard deviation (for $k = 1$)	σ_v	μL	5.5	0.00004
		pL		40

1.8 Adsorption Isotherms (M)

In the very first section of this book (Section 1.1.1), we emphasized that the thermodynamic distribution coefficient ($K_{d,i} = c_{i,s}/c_{i,m}$) is not a constant. However, we have so far assumed that by working at very low concentrations ("infinite dilution"), $K_{d,i}$ was independent of the concentration. This corresponds to a straight line in a plot of $c_{i,s}$ vs. $c_{i,m}$, as shown in Figure 1.30B1. Such a plot is called an **adsorption isotherm** or a distribution isotherm, and a constant value of $K_{d,i}$ corresponds to a linear isotherm. The word isotherm refers to a constant temperature. $K_{d,i}$ may be independent of the analyte concentration, but it remains a function of temperature. If we use a simple plate model, in which the contents of each "plate" in the column are transported to the next plate at each step, after which a new equilibrium is established that corresponds to the adsorption isotherm, we can calculate how a peak develops as a function of the distance travelled in the column (z, in number of plate heights). In such a simple approach, it is assumed that the effects of a non-linear isotherm overshadow those of chromatographic band broadening described in Modules 1.4 and 1.7. In Figure 1.30B2, it is seen that

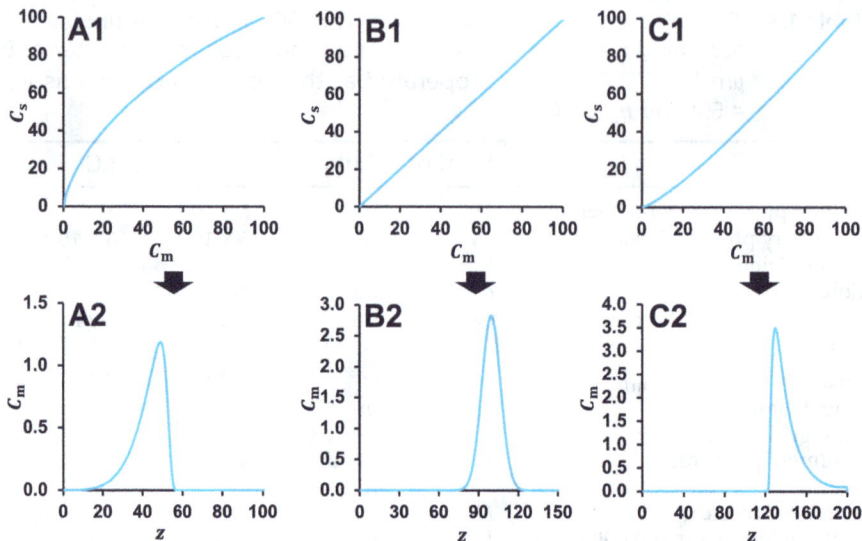

Figure 1.30 Different adsorption isotherms (A1: concave; B1: linear; and C1: convex) and the resulting peak profiles (A2, B2, and C2, respectively) as a function of the location in the column (z).

a linear isotherm results in a symmetrical peak. This is because the rate of migration of a compound can be described using eqn (1.2) as

$$u_i = \frac{u_0}{1 + k_i} = \frac{u_0}{1 + K_{d,i} \cdot \phi} \tag{1.134}$$

where ϕ represents the phase ratio. If $K_{d,i}$ is independent of the concentration, then u_i is also independent, and all parts of the analyte zone, including the high concentration region at the top and the low concentration regions at the edges, migrate at the same average velocity. This is not the case if the adsorption isotherm is not linear. Figure 1.30A1 shows a **concave isotherm** and Figure 1.30A2 shows that this results in a tailing peak (note that the horizontal axis denotes the position in the column in Figure 1.30A2, B2 and C2). This is because the ratio $c_{i,s}/c_{i,m}$, which is the distribution coefficient, is lower at higher concentrations. According to eqn (1.134), this implies that the top of the peak moves faster than the edges. The more rapidly moving centre of the peak catches up with the slow-moving front, while the slow-moving tail falls more and more behind. The opposite effect is observed in the case of a **convex isotherm**, as shown in Figure 1.30C1 and C2. In such a case, fronting peaks are observed. Clearly, chromatographers will wish for linear isotherms. However, these cannot always be ensured.

Non-linear adsorption isotherms are, for example, encountered when the stationary phase is inhomogeneous, featuring strong and weak adsorption sites. A small number of strong adsorption sites results in an isotherm that is steeper at low concentrations (*i.e.* concave) and thus in tailing peaks. If the capacity of the stationary phase is limited, as is, for example, the case in ion-exchange chromatography (Module 3.7) or in normal-phase LC with a solid adsorbent (Module 3.6), concave isotherms and tailing peaks are often encountered. A commonly observed adsorption isotherm on the latter kind of sorbent is the **Langmuir isotherm**, which has the form

$$c_{i,s} = \frac{ac_{i,m}}{1 + bc_{i,m}} \tag{1.135}$$

where $c_{i,s}$ is the surface concentration of the analyte on the adsorbent. The isotherm is illustrated in Figure 1.31. The Langmuir isotherm assumes an ideal surface, with equal and equally spaced adsorption sites, and no interaction between adsorbed molecules. At low concentrations (where $1 \gg bc_{i,m}$), the adsorption isotherm is linear. At higher concentrations, it levels off, ultimately approaching a constant value of $c_{i,s} = a/b$ at very high concentrations, when the stationary-phase surface is covered with a monolayer of analyte. We can rewrite eqn (1.135) in terms of the relative surface coverage

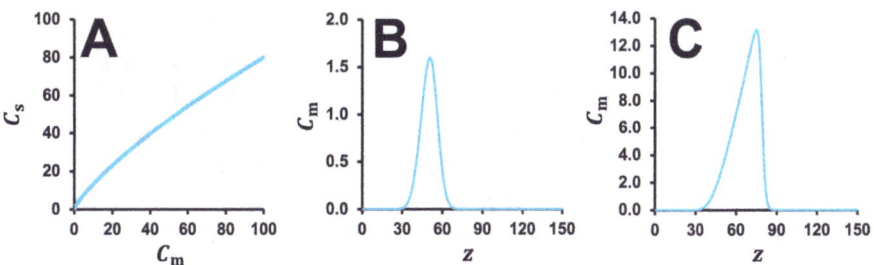

Figure 1.31 Calculated effect of the injected amount on the resulting peak shape in the case of a Langmuir isotherm eqn (1.135) with $a = 3.2$ and $b = 0.03$ (A) One plate volume injected with a concentration of (B) 100 and (C) 1000 units.

$$\frac{c_{i,s}}{c_{i,s}^{\max}} = \frac{K_{d,i}c_{i,m}}{c_L + K_{d,i}c_{i,m}} = \frac{c_{i,s}}{c_L + c_{i,s}} \tag{1.136}$$

where $c_{i,s}^{\max}$ is the surface concentration at saturation, and c_L is a constant with a dimension of surface concentration.

A Langmuir-type isotherm implies that symmetrical peaks can be obtained when injecting small amounts of analytes (Figure 1.31B), whereas strongly asymmetrical peaks are obtained when large amounts of analytes are injected (Figure 1.31C). This is generally referred to as **overloading**. Overloading the stationary phase results in tailing peaks, or triangular peaks, as seen in Figure 1.31C. Overloading the mobile phase, which is less common but may, for example, occur in solubility-limited situations, results in fronting peaks or triangles facing the other way. Overloading is often purposefully accepted in preparative separations, where the goal is not to achieve the best possible separation but rather to obtain the largest possible amount of analytes at a sufficient purity. Understanding the isotherms allows for the calculation of the best possible conditions for preparative separations. Such calculations are of great economic importance when purifying high-value analytes, such as contemporary biopharmaceuticals. Apart from the pure-analyte isotherms, it is also important to know how different compounds affect each other. This can be expressed in a competitive Langmuir isotherm as follows

$$\frac{q_{i,s}}{q_{i,s}^{\max}} = \frac{K_{d,i} \cdot c_{i,m}}{c_L + \Sigma_{j=1}^{n} K_{d,j} \cdot c_{j,m}} \tag{1.137}$$

where the summation involves all analytes present in the sample. Strongly adsorbing analytes with high $K_{d,i}$ values or competing analytes present in high concentrations are seen to reduce the fraction of the surface that is covered with analyte i. In the case of preparative separations based on competitive adsorption (for normal-phase liquid chromatography, see Section 3.6.1 and for ion-exchange chromatography, see Section 3.7.1), the tail of a band may be displaced by the front of the next-eluting band, leading to higher purities than the overlay of the individual peaks would suggest.

The process of predicting peak shapes from adsorption isotherms illustrated in Figures 1.30 and 1.31 can also be reversed.

Computational methods can be used to obtain distribution iso-therms from the peak profile.[34]

Heterogeneous surfaces are often characterized in terms of an empirical bi-Langmuir model with strong adsorption sites (prefix 1) with a high distribution coefficient $^1K_{d,i}$ and a typically low satura-tion capacity $^1q_{i,s}^{max}$ and a larger number (relatively high $^2q_{i,s}^{max}$) of weaker adsorption sites (prefix 2) with a lower distribution coeffi-cient $^2K_{d,i}$ as follows

$$q_{i,s} = \frac{^1q_{i,s}^{max} \cdot {}^1K_{d,i} \cdot c_{i,m}}{c_L + {}^1K_{d,i} \cdot c_{i,m}} + \frac{^2q_{i,s}^{max} \cdot {}^2K_{d,i} \cdot c_{i,m}}{c_L + {}^2K_{d,i} \cdot c_{i,m}} \quad (1.138)$$

In Chapter 3, we will encounter systems in which several reten-tion mechanisms act simultaneously. For example, in reversed-phase LC (see Module 3.5), the strong adsorption sites may be residual silanols. In chiral LC (Module 3.13), one retention mechanism may be enantio-selective, while the other is the general non-chiral retention process.

We have restricted this discussion to Langmuir isotherms. Differ-ent types of isotherms are encountered in chromatography, and the relation between the shape of chromatographic peaks and that of the isotherms can be established along similar lines as described earlier.[35]

1.9 Generic Application Categories (M)

In this book, we focus primarily on (analytical) separation science, including methodology and instrumentation, and less on the numerous applications that are made possible. We mention applications where appropriate, but this textbook features more diagrams than chromatograms. This module is an exception in that it focuses on applications. However, we will describe three application categories, rather than specific applications. van Mispelaar *et al.* first described the classification of chromato-graphic separations into three categories,[36] which is summarized in Table 1.4. In this distinction, three types of applications can be identified, using somewhat more modern words than used by van Mispelaar *et al.*: (1) target(ed) analysis, (2) group-type analysis, and (3) non-target(ed) analysis.

Most of the conventional analytical methods fall under the **target-analysis** category. This includes all methods aimed at determining one or a few specified analytes in a sample. It also includes situations where there is a whole list of targets, such as a list of priority pollutants, illicit drugs, or presumed performance-enhancing drugs proposed by the World Anti-Doping Agency (WADA). The chromatographic behaviour of target compounds may be compared to that of standards, but highly specific detectors, especially mass spectrometry (MS), are the most important tools for these types of analyses (see Table 1.4 for references to other parts of this book).

A less common category are the **group-type separations**, which are not aimed at specific compounds but rather at groups or classes that determine certain properties or together constitute a source of concern. Many such methods have been developed in the oil industry, since the properties of fuels are affected by the relative concentrations of paraffins (or alkanes, P), iso-paraffins (or branched alkanes, I), olefins (or alkenes, O), naphthenes (or cyclo-alkanes, N), and aromatics (A). All these classes are quantified in (paraffins, iso-parraffins, olefins, naphtenes and aromatics) PIONA analysis, but simpler (*e.g.* paraffins, naphthenes, aromatics; PNA) methods are more abundant.[37]

Group-type separations are also important in material characterization. An example is the constituents of a two-component adhesive. Ideally, all molecules possess functional groups that undergo reactions to form a polymeric network. If some molecules have numerous functional groups and a sizeable fraction have none, the formation of a strong network may be thwarted. Thus, it is important to separate the polymeric compounds based on functionality (see Module 4.3). Food labels reflect a number of group-type analyses (such as those for saturated/unsaturated fats and sugars) and some target analyses (for nutrients). However, not all analyses behind the label involve analytical separations. When separations are involved, the key aspect is to find selectivities that allow the separation of different classes. For example, silver ions interact specifically with double bonds. If immobilized on a GC or LC stationary phase, they can be used to distinguish between classes of saturated and unsaturated molecules.

Table 1.4 Classification of applications of analytical separations according to van Mispelaar *et al.*[36]

Class	1. Target analysis, target-compound analysis and lists of compounds	2. Group-type analysis and target-class analysis	3. Non-target analysis, comprehensive analysis and fingerprinting
Examples	• Drug development (*e.g.* pharmacokinetic studies) • Medical diagnosis ("biomarkers") • Healthy (*e.g.* vitamins) or forbidden (*e.g.* toxins) compounds in food • Detection of illicit drugs • Doping analysis in sports	• Mineral-oil fractions (*e.g.* paraffins, naphthenes, and aromatics) • Food (*e.g.* total content of fat; saturated *vs.* unsaturated) • Environmental contaminants (*e.g.* per- and polyfluoroalkyl substances; PFAS) • Polymer characterization (*e.g.* end groups)	• "-omics" fields (metabolomics, proteomics, *etc.*) • Fingerprinting of natural products, such as traditional or herbal medicines, natural food ingredients, and mineral-oil products • Authentication (*e.g.* wine or olive oil)
Methodology	• Hyphenated techniques, especially GC-MS (Module 2.6) and LC-MS (Module 3.10) • Statistics (Chapter 9)	• Separation selectivity aimed at specific sample dimensions • Specific detection (Modules 2.5 and 3.9)	• Comprehensive two-dimensional separations (Modules 7.2, 7.3, and 7.5) • Multivariate statistics (Module 9.9)
Requirements	• High information content • High certainty of identification • Standards for confirmation and quantitation	• Targeted selectivity • Quantitative detection	• High-resolution separations • High peak capacities • Broad dynamic range • High precision (repeatability)

The concept of **sample dimensionality** was introduced by Calvin Giddings in 1995.[38] The dimensionality of a sample is defined as the number of parameters that need to be known to uniquely describe each molecule in the sample. For example, if the sample is a homologous series (such as *n*-alkanes) or a homopolymer (with invariable end-groups), the dimensionality is 1. Only the chain length or molecular weight needs to be specified to pinpoint an individual molecule. In the case of fatty acids, the dimensionality is already large, influenced by the number of carbon atoms, number and position of double bonds, and number, length, and position of branches. In group-type separations, selectivity is aimed at making the relevant sample dimension dominant, such as in the above example, in which silver ions may cause the effect of the presence of double bonds in the molecule to overshadow the effects of chain length, or in critical chromatography, where polymer molecules can be separated based on the types and numbers of functional groups, irrespective of their molecular weight (see Module 4.3).

In many fields, a trend away from target analysis towards **non-target analysis** can be observed. For example, environmental scientists are not just worried about specific pollutants but shifting their focus to the complete **exposome**, which includes all chemicals to which humans are exposed. Likewise, attention has shifted from individual metabolites to metabolomics, from individual proteins to proteomics, *etc.* Genomics is somewhat of an exception in the nomenclature, in that it does not focus on all genes but rather on all base pairs that constitute a specific DNA molecule. Comprehensive non-target analyses pose the greatest challenges for separation scientists. High separation power, including high efficiencies and high peak capacities, is required. In addition, the dynamic range from highly abundant analytes to trace compounds is thought to be extremely important. When comparing different samples, high precision is required to allow good comparisons. This may involve small numbers of samples (*e.g.* flammable liquids in arson investigations) or thousands of samples for biomarker discovery studies.

Acknowledgements

Prof. Gert Desmet, Prof. Ken Broeckhoven and Dr. Fabrice Gritti are kindly acknowledged for their valuable comments and suggestions to improve the quality of this chapter. Gerben van Henten, Janne

Bolwerk and Rebecca Gibkes are kindly acknowledged for suggestions to improve readability.

References

1. M. R. Schure, R. S. Maier, T. J. Shields, C. M. Wunder and B. M. Wagner, *Chem. Eng. Sci.*, 2017, **174**, 445–458.
2. H. Xia, G. Wan, J. Zhao, J. Liu and Q. Bai, *J. Chromatogr. A*, 2016, **1471**, 138–144.
3. T. Sun, X. Jiang, Q. Song, X. Shuai, Y. Chen, X. Zhao, Z. Cai, K. Li, X. Qiao and S. Hu, *RSC Adv.*, 2019, **9**, 28783–28792.
4. B. W. J. Pirok, in *Making Analytical Incompatible Approaches Compatible*, University of Amsterdam, Amsterdam, 2019.
5. J. J. van Deemter, F. J. Zuiderweg and A. Klinkenberg, *Chem. Eng. Sci.*, 1956, **5**, 271–289.
6. F. Dondi, A. Cavazzini and M. Remelli, *Adv. Chromatogr.*, 1998, **38**, 51–74.
7. F. Gritti, *J. Chromatogr. A.*, 2018, **1540**, 55–67.
8. G. E. Berendsen, P. J. Schoenmakers, L. de Galan, G. Vigh, Z. Varga-puchony and J. Inczédy, *J. Liq. Chromatogr.*, 1980, **3**, 1669–1686.
9. J. M. Davis and J. C. Giddings, *Anal. Chem.*, 1983, **55**, 418–424.
10. M. Martin and G. Guiochon, *J. Chromatogr. A*, 2005, **1090**, 16–38.
11. S. Fekete, K. Horváth and D. Guillarme, *J. Chromatogr. A*, 2013, **1311**, 65–71.
12. D. V. McCalley, *TrAC, Trends Anal. Chem.*, 2014, 63.
13. T. Farkas, G. Zhong and G. Guiochon, *J. Chromatogr. A*, 1999, **849**, 35–43.
14. P. A. Bristow and J. H. Knox, *Chromatographia*, 1977, **10**, 279–289.
15. A. Andrés, K. Broeckhoven and G. Desmet, *Anal. Chim. Acta*, 2015, **894**, 20–34.
16. M. Callewaert, J. O. De Beeck, K. Maeno, S. Sukas, H. Thienpont, H. Ottevaere, H. Gardeniers, G. Desmet and W. De Malsche, *Analyst*, 2014, **139**, 618–625.
17. X. Feng, J. Cai, H. Zhao and X. Chen, *Chromatographia*, 2020, **83**, 749–755.
18. M. M. Dittmann and X. Wang, in *Handbook of Advanced Chromatography/mass Spectrometry Techniques*, ed. M. Holçapek W. C. Byrdwell, Elsevier, 2017, pp. 179–225.
19. G. Desmet, K. Broeckhoven, S. Deridder and D. Cabooter, *Anal. Chim. Acta*, 2022, **1214**, 339955.
20. G. Desmet, H. Song, D. Makey, D. R. Stoll and D. Cabooter, *J. Chromatogr. A*, 2020, 461339.
21. S. Deridder and G. Desmet, *J. Chromatogr. A*, 2012, **1227**, 194–202.
22. J. C. Giddings, in *Dynamics of Chromatography; Part I: Principles and theory*, Marcel Dekker, New York, First edit, 1965.
23. John Henderson Knox, FRS1927-2018.
24. K. Broeckhoven and G. Desmet, *Anal. Chim. Acta*, 2022, **1218**, 339962.
25. F. Gritti and G. Guiochon, *J. Chromatogr. A*, 2014, **1332**, 35–45.
26. S. Felletti, M. Catani, G. Mazzoccanti, G. Lievore, A. Buratti, L. Pasti, F. Gasparrini and A. Cavazzini, *J. Chromatogr. A*, 2021, **1637**, 461854.
27. F. Gritti and G. Guiochon, *J. Chromatogr. A*, 2012, **1221**, 2–40.
28. E. R. Adlard, *Chromatographia*, 2015, **78**, 1–1.
29. F. Gritti, M. Gilar and J. A. Jarrell, *J. Chromatogr. A*, 2016, **1444**, 86–98.
30. G. A. Guiochon and L. S. Ettre, *Anal. Chem.*, 1989, **61**, 922A.
31. M. J. E. Golay, in *Gas Chromatography*, ed. D. H. Desty, Academic Press, New York, 1958, pp. 36–55.
32. F. Gritti, *J. Chromatogr. A*, 2017, **1492**, 129–135.
33. P. J. Schoenmakers, *J. High Resolut. Chromatogr.*, 1988, **11**, 278–282.

34. T. Fornstedt, *J. Chromatogr. A*, 2010, **1217**, 792–812.
35. S. Horváth, D. Lukács, E. Farsang and K. Horváth, *Molecules*, 2023, **28**, 1031.
36. V. G. van Mispelaar, H.-G. Janssen, A. C. Tas and P. J. Schoenmakers, *J. Chromatogr. A.*, 2005, **1071**, 229–237.
37. J. Blomberg, P. Schoenmakers and U. A. T. Brinkman, *J. Chromatogr. A*, 2002, **972**, 137–173.
38. J. C. Giddings, *J. Chromatogr. A*, 1995, **703**, 3–15.

2   Gas Chromatography

Gas chromatography (GC) is the preferred separation method for volatile samples, thanks to its inherently high efficiency and the fantastic detectors available. The principles of GC, aspects of mobile and stationary phases, injection and detection are covered in the first half of this chapter (Figure 2.1), followed by a discussion on the near-perfect combination of GC with mass spectrometry. At a higher (A) level, the control of flow and temperature, and compression of the mobile phase under the influence of pressure, deserve significant attention, as do methods for the characterization of the stationary phase in terms of quantitative parameters. Finally, some aspects of troubleshooting in GC are discussed.

2.1   Introduction to Gas Chromatography (B)

As discussed in Chapter 1, faster diffusion leads to faster separations and lower viscosities lead to lower pressure drops. This makes the use of a gas as the mobile phase fundamentally attractive. This chapter will also show that excellent (high-resolution, highly stable and robust) columns and excellent (sensitive; universal, selective or specific) detectors are available for gas chromatography (GC). Moreover, very pure, non-toxic mobile phases can be used. All of this leads to the conclusion that GC should be used *if possible*.

This "*if possible*" illustrates the main bottleneck of GC: target analytes should be sufficiently **volatile** and should not degrade at temperatures below which they reach sufficient volatility. Matrix components that are not sufficiently volatile and not easy to remove may render GC less attractive. Non-volatile analytes may be modified (*e.g.* derivatized through esterification or silylation), but such a step

Analytical Separation Science
By Bob W. J. Pirok and Peter J. Schoenmakers
© Bob W. J. Pirok and Peter J. Schoenmakers 2025
Published by the Royal Society of Chemistry, www.rsc.org

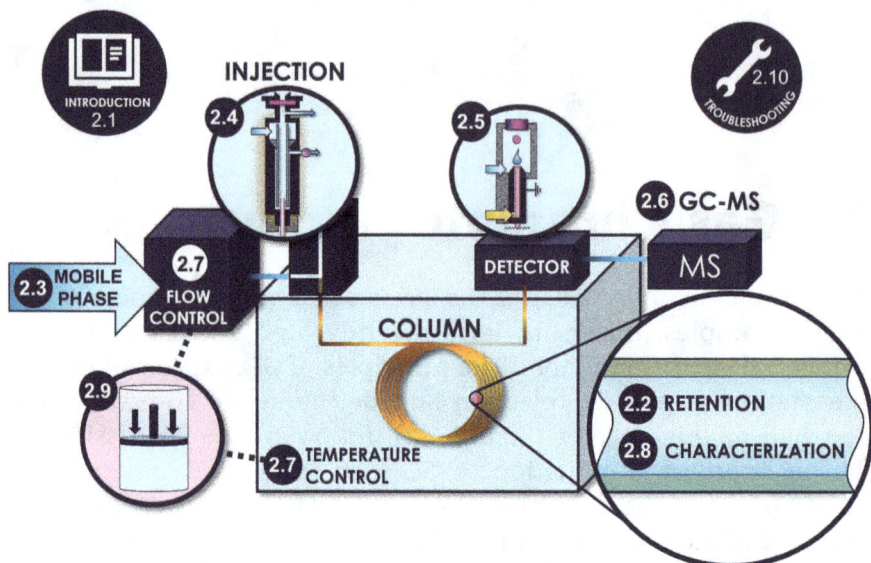

Figure 2.1   Graphical overview of the modules of Chapter 2.

puts GC at a disadvantage in comparison with techniques that allow direct analysis.

2.1.1   Basic Instrumentation and Operation

Like all other separation systems, a **GC instrument** is equipped with an injector to introduce a sample mixture and one (or more) detector(s) to monitor the effluent containing the separated analytes. To limit injection band broadening (see Section 1.7.5), the injection profile needs to be as narrow as possible, with minimal tailing. The ideal injection profile is a "block function" that is as narrow as possible. In practice, this is usually not achievable. A realistic profile is shown in Figure 2.2 (top left). Note that the horizontal (time) axis of the **injection profile** on the left of Figure 2.2 has been expanded in comparison with that of the chromatogram on the right. Achieving a narrow injection profile and injecting a representative sample without bias are far from trivial in GC, which is why injectors are discussed at length in Module 2.4. Detection, on the other hand, is relatively simple in GC. This is due to the inert nature of the mobile phase, which can be exploited to give minimal background. The most common detector in GC is the flame-ionization detector (FID), which measures the number of ions

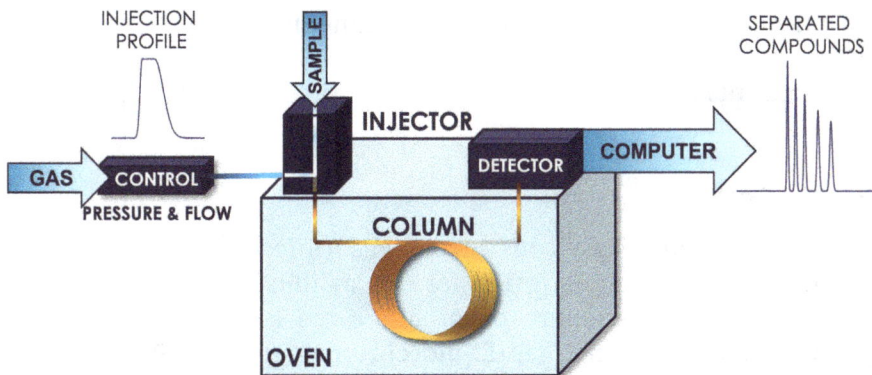

Figure 2.2 Schematic illustration of a basic GC setup. The time axis of the injection profile is expanded in comparison to that of the chromatogram.

formed when the effluent from the GC column is burnt in a small hydrogen flame. Both air and hydrogen must be supplied to form the flame. An FID detector is extremely sensitive (high signals) and shows very little noise, resulting in high signal-to-noise ratios, and thus low detection limits. An FID is also amazingly linear across at least five orders of magnitude. Finally, an FID is a near-universal detector, because it provides a high response for almost all organic compounds, and this response is quite similar for a given class of compounds (*e.g.* hydrocarbons). If more information is sought to identify analytes, GC can readily and very successfully be coupled with a mass spectrometry (MS) instrument. Highly reproducible spectra can be obtained, which allows the creation and application of spectral libraries to quickly suggest candidate structures for each peak in the chromatogram.

The column in GC is typically a highly robust **fused-silica** capillary, with a diameter of a few tenths of a millimetre. A thin layer (few tenths of a micrometre) of stationary phase is coated on the wall of the **open-tubular** column. Because such columns have high permeability (see Section 1.6.4) and the viscosity of gases is very low, long columns can be used and very high plate counts (say, $N = 100\,000$) can routinely be achieved.

In the vast majority of GC analyses, the temperature of the oven that houses the column is increased during the run. Such **temperature programming** complicates the flow control, because the gas expands and the viscosity increases with increasing temperature. Contemporary GC instruments have adequate flow control (see Module 2.7).

Data acquisition is not highly demanding in conventional GC, but more advanced methods, such as the GC-MS combination, produce vast amounts of data and, potentially, information (see Chapter 9).

2.1.2 Selectivity

In GC, the mobile phase consists of a gas. Different stationary-phase formats have been investigated for GC, as illustrated in Figure 2.3. Historically, the **stationary phase** was often a non-volatile liquid, giving rise to the name **gas–liquid chromatography** (GLC). A wide selection of liquids were used, ranging from non-polar (such as squalane, a branched C_{30} alkane) to highly polar ones. Such liquid stationary phases are now largely obsolete, as better polymeric stationary phases have emerged. Occasionally, solid surfaces are used as stationary phases (**gas–solid chromatography**, GSC), especially for the separation of (permanent) gases or highly volatile analytes, which may benefit from a high surface area to provide sufficient retention at attractive temperatures (room temperature or above). Particles can be introduced in a GC column in two formats: as a fully (dense-bed) **packed column** or as a layer along the wall of an open column. The great advantage of open-tubular columns is their low-pressure resistance, which allows using long columns and obtaining high plate counts (as seen in Module 1.4). Attempts were made to create a **"porous-layer open-tubular"** (PLOT) column with an immobilized layer of particles on the inner wall. However, some

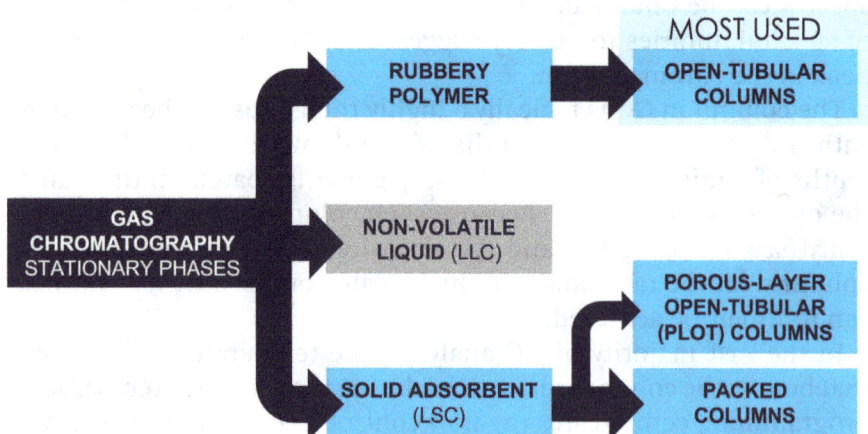

Figure 2.3 Overview of types of stationary phases used in gas chromatography.

kind of bonding between the wall and the particles, and between the particles themselves, is needed and has proven to be challenging.[1]

In the vast majority of cases, a **"wall-coated open-tubular"** column is now used in GC. The stationary phase is a "rubbery" (soft) polymer, *i.e.* a **cross-linked polymer** above its glass-transition temperature. Below this temperature, the polymer is "glassy" (hard), and it is not possible for the analyte to interact effectively with the stationary phase. Intuitively, the migration of analyte molecules in and out of a rubbery polymer will be easier and faster. This was discussed in more quantitative terms in Section 1.7.4.

An ideal stationary phase exhibits the following:

- A low glass-transition temperature (implying a low minimum operating temperature).
- High diffusion coefficients (*i.e.* fast movement of analyte molecules). Higher stationary-phase diffusion coefficients allow for thicker stationary-phase films without jeopardizing the column efficiency (see Section 1.7.4). Thicker films allow larger injections (greater sample capacities), which is favourable for detector sensitivity, and may allow starting elution programs at more convenient (above-ambient) temperatures.
- Low bleed (low volatility and high thermal stability are required because GC detectors are so sensitive that even minute amounts of material evaporating from the stationary phase cause baseline drift and decreased signal-to-background ratios).
- High inertness (showing no chemical or catalytic reactions with analytes or matrix components).
- Good film-forming properties to create a smooth film of constant thickness on the inner wall of the column (and thus maintain a high column efficiency).

Polysiloxanes, *i.e.* polymers with an $[-Si-O-]_n$ backbone, are exceptionally good stationary phases. There are two substituents on every silicon atom in the chain, and these determine the stationary-phase polarity, which is thought to contribute to their low viscosity (making it easier to manufacture good columns) and high diffusivity (contributing to a high efficiency and loadability; see Section 1.7.4). Polysiloxanes are available in high purity, without measurable levels of catalyst residues, and they can be "endcapped" to eliminate reactive end-groups. Silicone films can be cross-linked using radical formers (*e.g.* peroxides), resulting in very stable columns.

"**Polarity**" is a mystical word in GC and, indeed, in (almost) every form of chromatography. There is considerable confusion surrounding the word. Although it can be argued that describing polarity in a single number is senseless,[2] some understanding is necessary for a word that is pervasively used. Conventional wisdom in GC dictates that a stationary phase is polar if it provides increased retention for polar analytes. This only shifts the question to "what is a polar analyte"? One way to define polarity is the **solubility parameter** (δ), which is defined as the square root of the cohesive energy density,

$$\Delta = \sqrt{\frac{-E}{V}} \approx \sqrt{\frac{\Delta H^v}{v}} \tag{2.1}$$

where E is the total (negative) cohesive energy that holds matter together, V is the volume, v is the molar volume, and ΔH^v is the molar heat of evaporation. Another measure for the polarity of compounds is the logarithm of the **octanol–water partition coefficient** of solute i ($\log P_i$ or $\log K_{i,\text{ow}}$),

$$\log P_i = \log K_{i,\text{ow}} = \log \frac{c_{i,\text{octanol}}}{c_{i,\text{water}}} \tag{2.2}$$

where $c_{i,\text{octanol}}$ and $c_{i,\text{water}}$ are the concentrations of the analyte in the respective phases at equilibrium (and 25 °C). When an analyte is distributed across the two liquid phases, octanol and water – an experiment that can feasibly be performed using a classical separation funnel – it will predominantly end up in the more favourable environment. Apolar compounds will mainly occur in the octanol phase and show positive $\log P$ values, whereas polar compounds will end up predominantly in the water phase and show negative $\log P$ values.

It is clear from Table 2.1 that liquids identified as polar show high solubility parameters and low (negative) values of $\log P$. However, the correlation between these two parameters is far from perfect. Retention in reversed-phase LC (RPLC) is thought to correlate well with $\log P$. This correlation has often been explored as the basis for a fast and convenient method for determining $\log P$ values.[3,4] Again, the resulting correlation is imperfect, probably because most

Table 2.1 Some common compounds frequently used as solvents in chromatography, together with two indicators of their polarity, *i.e.* the solubility parameter δ and the octanol–water partition coefficient (log P).

Solvent	δ (cal cm^{-3})$^{\frac{1}{2}}$	log P
Pefluorinated alkanes	~5	6.5[a]
Alkanes	~7	3.9[b]
Ethyl acetate	9.53	0.71[c]
Toluene	9.57	2.73[c]
Tetrahydrofuran	9.88	0.53[d]
Dioxane	10.65	−0.27[e]
Dichloromethane	10.68	1.19[c]
Acetonitrile	13.14	−0.334[c]
Methanol	15.85	−0.69[c]
Water	25.52	*ca.* −1.5[f]

[a]Estimated value for perfluorooctane using the PubChem database (source: https://pubchem.ncbi.nlm.nih.gov/).
[b]Value for *n*-octane (source: https://en.wikipedia.org/wiki/).
[c]Source: https://en.wikipedia.org/wiki/
[d]Source: https://foodb.ca/compounds/FDB021917
[e]Source: https://www.atsdr.cdc.gov/ToxProfiles/tp187-c4.pdf
[f]In liquid–liquid extraction, the octanol layer contains about 20 mol% (or about 3 weight%) of water.

analytes require the presence of significant volume fractions of organic modifiers (*e.g.* methanol and acetonitrile) in the mobile phase to measure retention factors. Obtaining solubility parameters or log P values for typical GC stationary phases is far from straightforward. A suitable quantitative definition of a polarity scale for GC stationary phases is described in Module 2.8.

Retention in GC is determined by two factors, *i.e.* the interactions between the analyte molecules and the stationary phase (related to the polarity of both compounds) and the volatility (pure-component vapour pressure) of the analyte.

2.1.3 Retention Indices

Retention times in gas chromatography are quite variable, as they are not only affected by the analytes and the stationary phase but also by the column length, diameter, and phase ratio (*i.e.* the amount of stationary phase present), the flow rate of the carrier gas, connecting tubing (*e.g.* an eventual retention gap or the transfer line to a mass spectrometer), and the temperature. This makes it virtually impossible to reproduce and standardize retention times, and it renders retention-time tables essentially useless. Retention

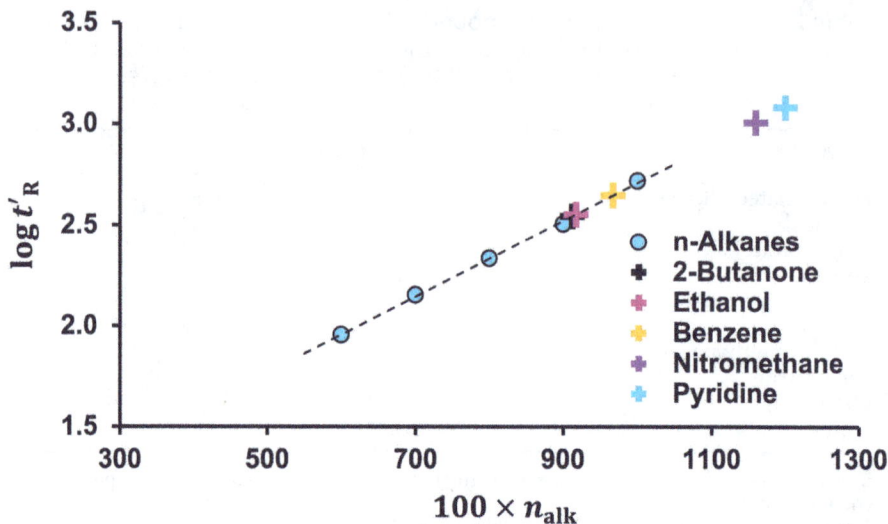

Figure 2.4   Linear relationship between the logarithm of the net retention time and the number of carbon atoms in n-alkanes during isothermal gas chromatography. This straight-line behaviour is often referred to as the "Martin rule" (after Nobel laureate Archer Martin). The horizontal axis depicts one hundred times the number of carbon atoms, yielding a graph that forms the basis for isothermal retention indices. The retention indices of analytes can be found from their position on the horizontal axis, which corresponds to their net retention times.

indices are a way to calibrate retention times, leaving only the effects of the analyte, the stationary phase and the temperature.

Retention indices, as defined by Kováts, are based on the observed linear relationship between the logarithm of the net retention time (t'_R) or the retention factor (k) and the number of carbon atoms (n_{alk}) in a homologous series of n-alkanes in an isothermal GC experiment (Figure 2.4). The retention index of n-alkanes is then set equal to $100 \times n_{alk}$, and the retention index of other analytes is found from their net retention times by interpolation, either graphically or using the following equation:

$$I_x = 100\left(n_{alk,before} + \frac{\log t'_{R,x} - \log t'_{R,before}}{\log t'_{R,after} - \log t'_{R,before}} \right) \qquad (2.3)$$

where x denotes the analyte and "before" and "after" refer to the last n-alkane eluting before the analyte and the first n-alkane eluting after it, respectively.

Retention indices can be tabulated and may aid in the identification of unknown analytes, possibly in combination with information obtained from mass spectra (or ultraviolet or infrared absorption spectra).

The main weakness of the retention-index scheme is that it is designed for isothermal operation, which is increasingly uncommon in GC. In the case of a linear temperature program (*i.e.* temperature increases linearly with time during the run), a plot of either the retention time (t_R) or the net retention time (t'_R) *vs.* n_{alk} becomes approximately linear once the homologues are sufficiently retained (low alkanes may be bunched at the front of the chromatogram) and until the end of the temperature program is approached (after which there is a transition to isothermal analysis). In the range where linearity holds, the following equation may be used.

$$I_{x,linear} \approx 100\left(n_{alk,before} + \frac{t'_{R,x} - t'_{R,before}}{t'_{R,after} - t'_{R,before}}\right)$$
$$= 100\left(n_{alk,before} + \frac{t_{R,x} - t_{R,before}}{t_{R,after} - t_{R,before}}\right)$$

(2.4)

Because of the prevalence of temperature-programmed GC, these types of "linear" retention indices are commonly tabulated (Box 2.1).

Box 2.1 Hero of analytical separation science: Archer J. P. Martin. Photo courtesy of the Nobel Foundation Archive.

A. (Archer) J. P. Martin (1920–2002, United Kingdom)

Archer Martin was awarded the 1952 Nobel Prize in Chemistry (together with Richard Synge) for developing partition chromatography. He was instrumental in developing GC, a technique that took gigantic leaps forward following Martin's seminal papers on the subject. Students should remember him for the Martin rule, which states that the logarithm of the net retention time (or retention factor) in isothermal GC (or isocratic LC) *vs.* the number of carbon atoms forms a straight line for a homologous series. This is the basis for the wellknown retention indices.

2.1.4 Stationary Phases

The best GC stationary phases (high efficiency, high-temperature stability, and low bleed) are non-polar or slightly polar.[5] However, polar analytes and polar sample matrices (solvents) may not be compatible with such phases, giving rise to broad, tailing, or even irregularly shaped peaks. Polar samples typically require polar stationary phases. There are still a significant number of different stationary phases available, but a typical laboratory has only a small selection to cover a range of polarities (if needed) and to meet any special needs. Some important stationary phases are listed in Table 2.2.

Examples of specialty columns include highly polar columns that allow the separation of highly polar analytes without derivatization. Fatty acids are traditionally analyzed as methyl esters. "Free-fatty-acid" (FFA) columns are intended to avoid an esterification step. As mentioned previously, highly volatile analytes may benefit from the large stationary-phase surface area offered by plot columns or packed columns. The latter are still in use for separating (permanent) gases and their impurities.

Chiral separation is of extreme importance for science and society. The function of many molecules in nature is dependent on their stereometric structure, and this is equally true for the action of bioactive compounds, such as drugs, on biological systems. Diastereomeric compounds may be separated with GC, but the direct separation of enantiomers can only be achieved by GC if stereoselectivity is offered by the stationary phase.

If enantiomers are amenable to GC, this is likely the preferred method of separation and analysis, thanks to the high separation efficiency of GC and the availability of excellent detectors, yielding qualitative and quantitative information. However, for preparative separation, LC techniques are preferred. Several types of chiral stationary phases (CSPs) for GC are available, including polysiloxane phases with chiral groups attached to the backbone, chiral metal complexes, and cyclodextrin-containing stationary phases.[7] The latter have proven most versatile for a variety of applications.

Columns that offer other types of shape selectivity, such as those for substitutional isomers or "congeners", also exist. Liquid-crystalline stationary phases have been extensively studied for this purpose, but they have not gained widespread application.[8] In a liquid crystal, significant order exists between molecules or parts thereof. For application in GC, where temperature programming is

Table 2.2 Selection of GC stationary phases and some properties.

| Name Commercial name(s) | Monomeric units (%) | | | | | | | Polarity index[a] | Tempera-ture range (°C)[b] |
	Dime-thylsi-loxane	Diphenylsi-loxane	Trifluoro-methyl methyl siloxane	Cyano-ethyl siloxane	Cyano-propyl siloxane	Ethylene oxide		
Poly(dimethylsiloxane) OV-101, HP-1, DB-1, etc.	100	—	—	—	—	—	46	−60 to 330/350
Poly(dimethylsiloxane-co-diphenyl siloxane) HP-5, DB-5, SE-52	95	5	—	—	—	—	67	−60 to 330/350
Poly(dimethylsiloxane-co-diphenyl siloxane) OV-3	90	10	—	—	—	—	85	−60 to 330/350
Poly(dimethylsiloxane-co-diphenylsiloxane) OV-7, SPB-20, RT_x-20	80	20	—	—	—	—	118	−60 to 330/350
Poly(dimethylsiloxane-co-diphenyl siloxane) OV-17, HP-50, DB-17	50	50	—	—	—	—	177	−60 to 330/350
Poly(trifluoropropyl-methyl siloxane) OV-210	—	—	100	—	—	—	304	0–225/250
Poly(cyanoethyl-phenyl siloxane) OV-225	50	—	—	50	—	—	363	25–220/240
Poly(ethylene glycol) PEG-4000[c]	—	—	—	—	—	100	471	20–230/250

(continued)

Table 2.2 (*continued*)

| Name Commercial name(s) | Monomeric units (%) | | | | | | Polarity index[a] | Tempera-ture range (°C)[b] |
	Dime-thylsi-loxane	Diphenylsi-loxane	Trifluoro-methyl methyl siloxane	Cyano-ethyl siloxane	Cyano-propyl siloxane	Ethylene oxide		
Poly(cyanopropyl-silox-ane) OV-275	—	—	—	—	100	—	844	25–230/250

[a]Average of the first five McReynolds constants[6] (see Module 2.8).
[b]Upper temperature is indicated for continuous (isothermal) operation (before the slash) and for temperature-programmed operation (after the slash).
[c]Average molecular weight: 4000 Da.

the standard mode of operation, an ordered mesophase should exist across a broad range of temperatures.

Liquid-crystalline polymers combine shape selectivity with other required stationary-phase properties, such as high stability and low bleed. One way to synthesize such phases is to connect liquid-crystal forming groups to polysiloxane backbones. Typical analytes separated on liquid-crystalline stationary phases include substituted aromatic or polycyclic aromatic compounds, polychlorinated biphenyls (PCBs) and polychlorinated dibenzo-*p*-dioxins.[8]

A popular area of research in the 21st century has been the development and application of **ionic-liquid columns** for GC.[9] These may offer stable, polar stationary phases for use at high temperatures. The low volatility of ionic liquids results in low bleed in comparison with liquid stationary phases, but their superiority over polar polymeric stationary phases is yet to be proven in practice.

Overall, GC is a high-resolution separation technique. High efficiency (say, 100 000 plates) comes relatively easily in a reasonable timeframe, given (i) the low cost of consumables, (ii) the high degree of automation, and (iii) the excellent retention-time precision obtained through modern instrumentation.[10] In addition, gas chromatographs are increasingly equipped with mass spectrometry (MS) detectors, allowing quantitation of overlapping peaks based on specific ions. The quest for GC selectivity tends to be overshadowed by the available separation and detection power.

> See the website for references to lists of stationary-phase parameters: ass-ets.org

2.1.5 Open-tubular Columns

In most cases, it is not rewarding to make GC columns instead of buying them from a range of companies that produce high-quality columns. By staying within the prescribed temperature ranges, using pure carrier gas and leak-free connections, and ensuring that non-volatile or aggressive components from the sample do not enter the column, good GC columns can be made to last – and remain unaltered – for long periods of time and large numbers of analyses. Non-polar columns tend to be more durable than polar columns, because the latter are not only more interactive, but also more reactive (*e.g.* prone to oxidation).

There are two fundamentally different ways to introduce a polymeric film inside a capillary column.[11] In the dynamic process, a (viscous) plug of polymer is pushed through the column, leaving a thin film on the wall in its wake. In the static process, the column is filled with a solution of the polymer in a volatile solvent and closed at one end. The solvent is then slowly evaporated, leaving a film of polymer. The static process allows for better control of the eventual thickness of the stationary-phase film. The last step typically involves immobilizing the polymeric film by chemical cross-linking, for example, with dicumyl peroxide or ozone[12] or through reactive groups incorporated in the polymer for this purpose.

A gigantic leap forward was taken after 1979 when Dandeneau and Zerenner introduced **fused-silica columns**.[13] The inner surfaces of these columns were more inert than the glass columns used earlier, to reduce peak tailing for polar analytes. The surfaces of fused-silica columns tend to be very smooth and can easily be deactivated with non-polar or polar reagents, to achieve "wettability" for stationary phases or samples.[5] At least as important was the robustness of fused-silica capillaries. Glass columns were extremely breakable, while the new fused-silica columns with an external polymeric coating were not. These days, the common external coating is a poly(imide), which offers excellent thermal stability (up to about 350 °C). Such coatings are not UV transparent, but this is hardly ever an issue in GC.

Temperatures exceeding 350 °C may be of interest in some cases, for example, for "simulated distillation" experiments on (crude) oil samples.[14] Lipsky and Duffy[15,16] proposed aluminium-clad fused-silica columns for very high-temperature GC, but these have been surpassed by more robust all-metal columns.[17]

2.1.6 Basic Gas Chromatography-Mass Spectrometry

A mass spectrometry (MS) instrument is a highly informative GC detector, and gas chromatography–mass spectrometry (GC-MS) can be portrayed as a very successful hyphenated technique, *i.e.* a synergistic coupling between two different instruments. There absolutely is synergy in coupling GC with MS, which means that the combination is much more powerful than a gas chromatograph and a mass spectrometer used independently. GC is a great separation tool and an excellent inlet for the MS. The resulting peaks contain one or a few analytes in addition to an inert carrier gas. MS produces spectra of the introduced analytes. After GC separation, information-rich spectra are obtained, from which many of the analytes can be (tentatively) identified. The GC retention time (or, much more reliably, the retention index) may increase the confidence of the identification.

The process of MS involves various stages. In its simplest form, these can be summarized as (i) ionization of the introduced molecules, (ii) separation of the produced ions, (iii) detection of the separated ions, producing a mass spectrum, and (iv) interpretation of the data. The most popular (small and affordable) type of MS used in conjunction with GC is the (single) quadrupole MS. These are scanning instruments that move through a spectral range of typically 40–400 atomic mass units (amu) within 200 to 500 ms. This means that not all ions are recorded simultaneously, and the intensities may be somewhat distorted when a spectrum is recorded on the slope of a GC peak.

Electron ionization (formerly known as electron-impact ionization, with both names abbreviated to EI) is the most common method to convert the molecules eluting from the GC into ions that can be measured by MS. Electrons are produced by a heated wire (filament). These electrons do not react with the inert carrier gas molecules, but they do react with almost all analytes. EI is a so-called "hard" (high-energy) ionization method. It causes analyte molecules to break apart into a number of fragment ions. Detecting these results in a characteristic mass spectrum that can be conveniently compared with large collections of MS spectra, called libraries. The best-known of these is the NIST library,[18] maintained by the U.S. National Institute of Standards and Technology. MS detection is very sensitive, and GC-MS can be used for reliable quantitative analysis of target compounds. However, frequent (at least daily) calibration or the use of suitable internal standards (see Chapter 9) is required to compensate for variations in MS sensitivity. A more detailed description of GC-MS is provided in Module 2.6.

2.2 Retention (M)

We have seen in Chapter 1 that chromatographic separation arises from differences in the partitioning of analytes between the mobile and stationary phases. We also learned how these processes are expressed in relative concentrations (*e.g.* eqn (1.3)).

In gases, we usually relate such concentrations to partial pressures. In ideal solutions, the vapour pressure of which was quantitatively described by the French chemist François-Marie Raoult (1830–1901), the partial pressure of analyte i is simply proportional to its pure-component vapour pressure (P_i^0) and its mole fraction in the liquid (x_i), *i.e.*

$$P_i = x_i \cdot P_i^0 \tag{2.5}$$

Raoult's law is depicted in Figure 2.5. Ideal solutions are barely ever encountered. A mixture of two very similar compounds (*e.g.* *n*-hexane in *n*-heptane) may come close to ideal behaviour; however, Raoult's law is an oversimplification. A more realistic and more useful approximation was postulated by the English chemist William Henry (1774–1836), *i.e.*

$$P_i = K_H \cdot P_i^0 = \gamma_{i,s}^{\infty} \cdot x_i \cdot P_i^0 \tag{2.6}$$

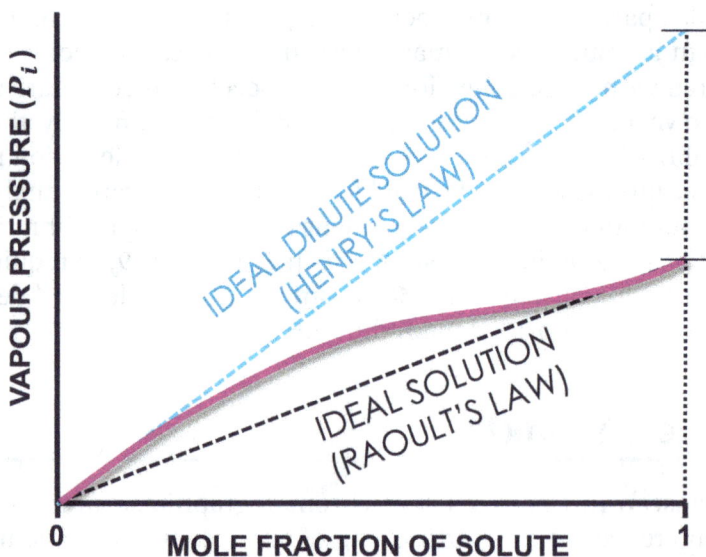

Figure 2.5 Partial vapour pressures of analytes. In the range relevant for GC, at low mole fractions of the analytes, the vapour pressure follows Henry's law, *i.e.* $P_i = K_H \cdot P_i^0 = \gamma_{i,s}^{\infty} \cdot x_i \cdot P_i^0$.

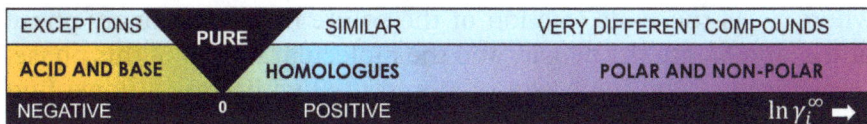

Figure 2.6 Schematic depiction of the values of activity coefficients. Relative to a pure compound (which may be seen as mixing identical compounds), mixing two different components is almost always energetically unfavourable. This is expressed in activity coefficients greater than 1 for similar (but different) compounds and even higher for very different compounds. There are some exceptions to this rule, such as acids and bases, for which energy is released upon mixing, translating to activity coefficients that are less than unity.

where K_H is known as Henry's constant and $\gamma_{i,s}^\infty$ is the **activity coefficient** of analyte i in the stationary phase s at infinite dilution. **Henry's law** is also illustrated in Figure 2.5. The activity coefficient generally reflects the interactions between the stationary phase and the analyte. Chemical compounds tend to be rather xenophobic, as shown in Figure 2.6. They dissolve best in themselves ($\gamma_{i,i}^\infty = 1$) or in solvents with a similar structure ("*like dissolves like*"). The greater the difference between the analyte and the solvent (*e.g.* a polar analyte in a non-polar stationary phase), the larger the value of $\gamma_{i,s}^\infty$. The activity coefficient is always greater than unity, except in exceptional cases (*e.g.* a basic analyte in an acidic solvent or *vice versa*). This rough classification of activity coefficients is illustrated in Figure 2.6.

If we assume ideal-gas behaviour in the mobile phase, then the molar concentration of the analyte is

$$c_{i,m} = \frac{1}{v_{i,m}} = \frac{P_i}{R \cdot T} = \frac{\gamma_{i,s}^\infty \cdot x_{i,s} \cdot P_i^0}{R \cdot T} \tag{2.7}$$

where $v_{i,m}$ is the partial molar volume of the solute in the gaseous phase, R is the gas constant and T is the absolute temperature. In analytical gas chromatography, we work at very low analyte concentrations, justifying the use of eqn (2.7). Moreover, at such low concentrations, the molar concentration of the analyte in the stationary phase is

$$c_{i,s} = x_{i,s} \cdot \frac{\rho_s}{M_s} \tag{2.8}$$

where $x_{i,s}$ is the mole fraction of the solute in the stationary phase and ρ_s and M_s are the density and the molecular weight of this phase, respectively. Combining eqn (2.7) and (2.8) with eqn (1.2), we find the retention factor of the analyte

$$k_i = \frac{c_{i,s}}{c_{i,m}} \frac{V_s}{V_m} = \frac{\rho_s \cdot R \cdot T \cdot V_s}{\gamma_{i,s}^{\infty} \cdot P_i^0 \cdot M_s \cdot V_m} = \frac{R \cdot T \cdot n_s}{\gamma_{i,s}^{\infty} \cdot P_i^0 \cdot V_m} \tag{2.9}$$

where n_s is the number of moles of the stationary phase in the column. Eqn (2.9) is deceptive in that it suggests a linear increase of the retention factor with absolute temperature. However, the pure-component vapour pressure (P_i^0) increases exponentially with T, resulting in a rapid decrease of k_i with increasing temperature.

It follows from eqn (2.9) that the relationship between the retention factor and temperature takes the form

$$\ln\left(\frac{k}{T}\right) = \frac{A_H}{T} + B_S \tag{2.10}$$

where A_H is an enthalpy-related coefficient and the coefficient B_S is related to entropy and the phase ratio. Only two factors are seen to influence the **selectivity** ($\alpha_{j,i}$) in GC (eqn (2.11)),

$$\alpha_{j,i} = \frac{k_j}{k_i} = \frac{\gamma_{i,s}^{\infty} \cdot P_i^0}{\gamma_{j,s}^{\infty} \cdot P_j^0} \tag{2.11}$$

Only the activity coefficient, determined by the interactions of the analyte with the stationary phase, and the vapour pressure of the analyte play a role. This is reflected in the common jargon of GC, where retention is said to depend on **polarity** and **volatility**.

2.3 The Mobile Phase (M)

The Golay equation for band broadening in open-tubular chromatography has been discussed in Section 1.7.7. In this module, some other aspects of open-tubular GC are considered.

The definition of the reduced linear velocity for open-tubular columns (eqn (1.129)) indicates which carrier gas is most favourable in open-tubular GC. Conditions are equivalent at identical values of v, δ_f and k. The latter is kept low by temperature programming. To maintain the same value for δ_f (typically 0.3), a carrier gas that gives rise to higher analyte diffusion coefficients requires thinner stationary-phase films. However, the effect is modest. A twofold increase in D_m requires a 40% decrease in d_f. More importantly, a higher diffusion coefficient allows for faster analysis because the linear velocity u can increase proportionally with D_m if v is to be kept constant. The diffusion coefficient generally decreases with increasing molecular weight of the carrier gas.[19]

Figure 2.7 illustrates the optimal linear velocity for different carrier gases as a function of temperature. The highest diffusion coefficients are encountered with hydrogen, followed by helium. Nitrogen has a much higher molecular weight and the diffusion coefficients are much lower, making it less attractive as a carrier gas for GC. Hydrogen yields the fastest analysis without the increased speed resulting in a higher pressure drop across the column, as it also has the lowest viscosity (see Darcy's law, Section 1.6.4). The obvious disadvantage of hydrogen concerns safety. However,

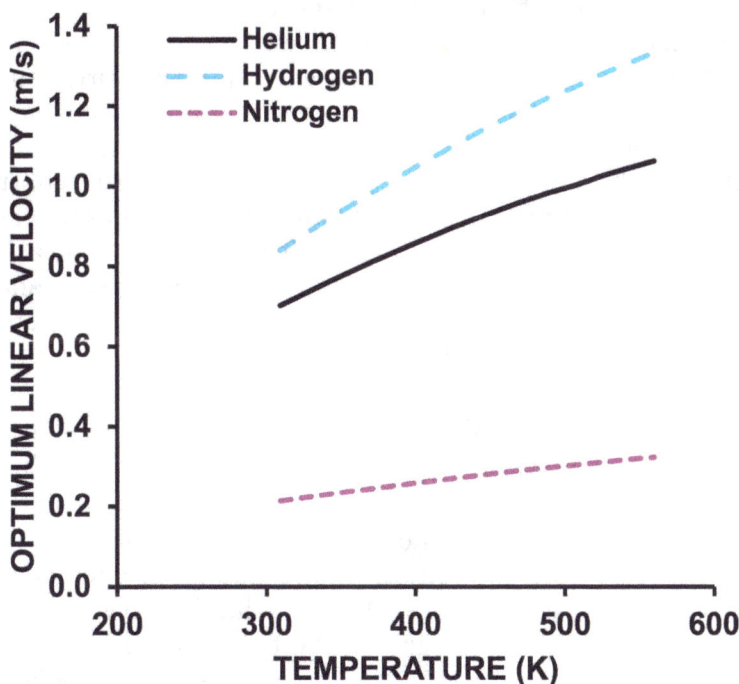

Figure 2.7 Calculated optimal linear velocities on a 250 μm i.d. open-tubular column (with δ_f = 0.3) for *n*-hexane as the analyte.

hydrogen is often available in laboratories for use in flame-based detectors, such as the FID. Hydrogen generators provide a convenient way to produce sufficient quantities of pure hydrogen on the spot. Hydrogen detectors may be installed in the GC oven to enhance laboratory safety. Although hydrogen is less inert than helium, it is not very reactive at the low pressures typically encountered in GC. Nevertheless, care should be taken regarding the inertness of the entire system, avoiding any materials that may exert catalytic effects.

2.4   Injectors for Gas Chromatography (M)

Injectors are a critical aspect of GC. It is difficult to get a representative aliquot of the sample for introduction into the column and subject it to chromatographic separation. A major distinction can be made between (i) gaseous samples, (ii) liquid samples and (iii) multiphase samples. Gaseous samples are easiest to deal with, because no phase transition is required between sampling and analysis. Typically, sampling valves (see Module 3.3) are used to introduce gaseous samples. Reducing valves may be needed to lower the pressure of the gas, but the composition of a mixture of permanent gases should not be affected in the process. This is an issue when injecting pressurized gas–liquid mixtures, such as liquified petroleum gas (LPG),[20] but this will not be discussed here. The fundamental problem with liquid samples introduced in a gas chromatograph is the required phase change and the enormous expansion that accompanies the transformation of a gas to a liquid. The gas volume resulting from the injection of 1 μL of a liquid sample depends on the molecular weight of the solvent and the temperature of the injector. The expansion can be estimated by assuming ideal-gas behaviour,

$$V_{gas} = \frac{V_{inj} \cdot \rho_{solvent}}{M_{solvent}} \cdot \frac{R \cdot T}{P} \tag{2.12}$$

where V_{inj} is the injected volume, which is assumed to consist largely of solvent, V_{gas} is the gas volume after evaporation and expansion of the solvent, $\rho_{solvent}$ is the density of the liquid solvent, $M_{solvent}$ is its molecular weight, R is the gas constant, T is the absolute temperature and P is the pressure. Table 2.3 contains some indicative data. It can be seen that 1 μL of organic liquid typically turns into a few

Table 2.3 Anticipated gas volumes (in microlitres) arising from the injection of 1 μL of the indicated liquid solvents at the indicated injector temperatures [pressure inside the injector: 120 kPa (1.2 bar)].

Temperature (°C)	Water	*n*-Hexane	Dichloromethane
150	1 623	223	459
200	1 814	249	513
250	2 006	276	568
300	2 198	302	622

hundred microlitre volumes of gas. Water has an exceptionally low molecular weight for a liquid, causing 1 μL injections to rapidly turn into about 2 mL of gas. Given the large expansion of the sample upon evaporation, it is understandable that any volume above 2 μL is considered a 'large volume' in GC and is referred to as **large-volume injection (LVI)** Even without entering this large-volume domain, GC injectors should be able to deal with gaseous volumes of at least 1 mL, ensuring that either the entire sample or a representative fraction thereof enters the column and undergoes separation.

Evaporation takes place in an insert ("**liner**") that is typically made of glass or quartz and is easily replaceable. An example is shown in Figure 2.8 on the left side. The volume of the liner should suffice to accommodate the entire volume of the gaseous sample to avoid contact of the analytes with other (metal) surfaces. The surface of the liner must be inert to avoid adsorption or degradation of the analytes. Simple, open liners exhibit the lowest surface area. As a result, deactivated tapered liners minimize the risk of adsorption or degradation. Fritted liners (*i.e.* liners that contain a porous filter made by sintering glass particles with each other and the wall of the liner) are less suitable for highly polar and unstable analytes, but they may give rise to more repeatable injections. Adsorption is usually manifested as a decrease in sensitivity. The injection profile is tailing, but this may not be reflected in tailing peaks in the case of temperature-programmed analysis. The analytes are refocussed at the top of the column at the low initial temperature. Sample degradation results in a decrease in sensitivity, poor quantitation and unwanted artefacts.

In the following sections, the most important devices for introducing liquid samples into a GC are described. The strengths and

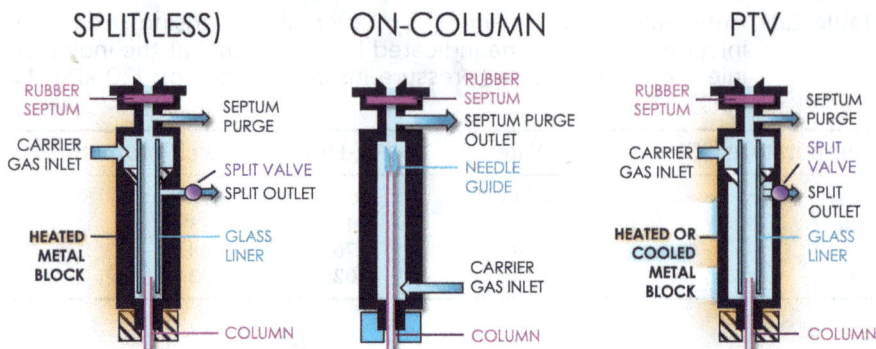

Figure 2.8 Essential aspects of a split/splitless injector (left), an on-column injector (centre) and a PTV injector (right). A split/splitless injector is hot during operation (orange), and the needle is inserted through a septum (pink). The split flow can be switched on or off at programmable times. The on-column injector is cold during operation (blue). Instead of through a septum (as indicated), the (thin) injection needle often enters through an (automatic) valve (not shown) directly into the column or (more commonly) a "retention gap". The temperature of the PTV injector can be rapidly programmed (orange-blue pattern). The split flow and the column flow are also programmable.

weaknesses of the various injectors are summarized at the end of this module in Table 2.4.

2.4.1 Split/Splitless Injector

The most common GC injector is the split/splitless injector (Figure 2.8, left). This injector is typically operated at a high temperature (indicated by the orange colour in the figure) to instantly evaporate the liquid sample. In the **split mode**, most of this sample (often 98% or 99%) is sent to waste through the split outlet (or "split line"), which is possible because (i) gas-phase splitting is an accurate and precise process and (ii) GC detectors are incredibly sensitive (see Module 2.5).

In the **splitless mode**, the split line is closed for a limited time during the injection, allowing (almost) all of the gaseous sample cloud to enter the column. If there is a septum-purge flow, this may also be switched off during this period. However, emptying the entire evaporation space (the liner) follows a typical exponential profile and, thus, takes a long time. In addition, such

Table 2.4 Strengths and weaknesses of different types of injectors for introducing liquid samples into a gas chromatograph.

Injector type	Strengths	Weaknesses
Split/splitless (split mode)	• Simplest and cheapest device for injecting liquid samples	• Discrimination
	• Narrow initial bandwidths	• Poor repeatability
	• Compatible with every column inner diameter	• Risk of thermal degradation
	• Tolerant for high-concentration samples	• Not suitable for trace analysis
	• Easy to use	
	• Robust	
Split/splitless (splitless mode)	• Suitable for trace analysis	• Discrimination
	• (Most of) the solvent does not enter the column	• Only suitable for (relatively) non-volatile analytes
	• Relatively tolerant for "dirty" samples	• Requires optimization (of split time)
	• Robust	• Highest chance of analyte degradation, due to long residence time in the hot injector
On-column (cold)	• No discrimination	• Not compatible with dirty samples
	• No thermal/catalytic degradation	• Not very robust (fragile)
	• High precision in quantitative analysis	• LVI requires optimization
	• Suitable for large-volume injection	

(continued)

Table 2.4 *(continued)*

Injector type	Strengths	Weaknesses
Programmed-temperature vaporizer (programmed)	• Minimal discrimination	• Requires optimization of multiple parameters
	• No thermal/catalytic degradation	• Complex and (relatively) expensive
	• Suitable for complex (e.g. multi-phase) samples	
	• (Most of) the solvent and (all) non-volatile residues do not enter the column	
	• No column/detector/solvent compatibility issues	
	• Suitable for rapid large-volume injection	
	• Robust and precise	
	• May be used for (quantitative) pyrolysis experiments	

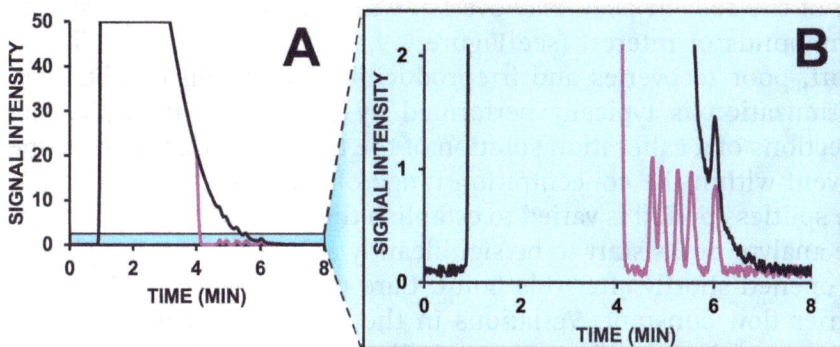

Figure 2.9 Simulated chromatograms illustrating a splitless injection. The dark blue line represents a fully splitless injection. The pink line illustrates an injection in which the split valve is opened after 3.5 min to blow off the last bit of the solvent-peak tail (illustrated in the insert, B, which has a fifty times expanded vertical axis in comparison with the chromatograms in A). Opening the split valve allows for the detection and quantitation of analyte peaks that are much closer to the solvent peak. There is significant noise in the chromatograms because splitless injection is typically used to determine trace-level analytes.

an exponential decay will result in a disturbingly long and high tail of the solvent peak, obstructing the detection of early eluting analytes (especially at low concentrations). Opening the **split valve** to remove the last small fraction (typically much less than 1%) of the solvent cloud greatly reduces this background signal, without affecting the quantitative analysis of the analytes. This is illustrated in Figure 2.8. Following the injection of, say, 1 μL of a sample solution, the split valve is opened after 3.5 min, the consequences of which are visible at the detector a time t_0 later. Figure 2.9A illustrates that the last part of the solvent-peak tail is removed. We can now substantially expand the scale (in this case, by a factor of 50, Figure 2.9B), allowing for the detection of trace-analyte peaks between 4.5 and 5.5 min that would otherwise be obscured by the solvent peak. The peak at 6 min is visible without opening the split valve, but quantitation is significantly improved if the split valve is opened at an appropriate time.

The main advantage of the splitless mode is that lower detection limits can be obtained, facilitating the analysis of trace compounds. A disadvantage is that the time at which the split valve is opened must be optimized and carefully controlled, in conjunction with the flow rate of the carrier gas. If the splitless time is too long, the tail of the solvent peak may overlap with (or fully obstruct) volatile compounds of interest (see Figure 2.9, dark blue line). If it is too short, poor recoveries and irreproducible results will be obtained. Optimization is typically performed by conducting multiple trial injections of a calibration solution of the target analytes in the same solvent within the concentration range of interest. The duration of the splitless period is varied to establish the point where the areas of the analyte peaks start to be significantly affected. The valve should be opened shortly after this point. Care must be taken to keep the carrier flow constant. Variations in the carrier-gas flow, for example, caused by a leaking septum, will compromise the quantitative analyses.

Splitless injection inevitably leads to much broader and less symmetrical injection bands than split injection. Therefore, it becomes important to somehow focus the analyte bands between the injection and their actual migration through the column. Various mechanisms can contribute to analyte focussing. The simplest of these is thermal focussing at the top of the column, prior to the start of a temperature program. Cold trapping is most effective for high-boiling analytes. As a rule of thumb, the initial oven temperature should be 150 °C below the (atmospheric) boiling points of the analytes for cold trapping to be fully effective. It has been argued that a hot injector may affect the temperature at the very top of the column, which is one of the reasons why a short **retention gap** (*i.e.* a short length of deactivated fused-silica capillary without a stationary-phase film) is sometimes used or integrated into the column. However, retention gaps are not commonly used in combination with splitless injection unless incompatibility issues arise between the sample solvent and the stationary phase (see Section 2.4.2 for more on retention gaps).

In **pulsed-splitless** injection, the pressure is briefly increased to push the analyte-containing cloud into the column more quickly. It is claimed that this leads to narrower peaks, but the effect is small because the peak width in volume units is only slightly reduced and bands are eluted at the same speed as in regular splitless injection. Pulsed-splitless injection is considerably more complicated because, in addition to the split-valve-open time, the pulse pressure and the pulse duration need to be established. The pertaining advantage of pulsed-splitless injection is that analytes spend a shorter period of time in the hot injector. This may increase the recovery of thermolabile analytes. However, a better solution for such analytes is to apply cold-on-column or PTV injection – if at all possible. These injectors are described in Sections 2.4.2 and 2.4.3.

2.4.1.1 Discrimination

The split/splitless injector is still popular in GC. It is simple in design and operation. Except for the split time in the splitless mode, little optimization is needed. The injector is robust and reliable. However, there is one major drawback called **discrimination**. This is illustrated in Figure 2.10.

When injecting a mixture of *n*-alkanes with equal concentrations (in mass per volume units), we do not obtain peaks of equal area for all analytes. Instead, we observe a large decrease in peak area for the larger homologues, *i.e.* for the *n*-alkanes with higher boiling points. The response factor (peak area divided by the injected mass of the analyte) is seen to decrease substantially with increasing carbon number (Figure 2.10A). This is a general problem of hot injection in GC, and the process is illustrated in Figure 2.10B. A (relatively) cold syringe is inserted into the injector, typically for a short time. While the (metal) needle may have time to heat up, the (glass) barrel of the syringe will not. Neither will the fingers of the analyst (or the autosampler) holding the plunger. Thus, some kind of temperature gradient will be formed along the syringe. In the

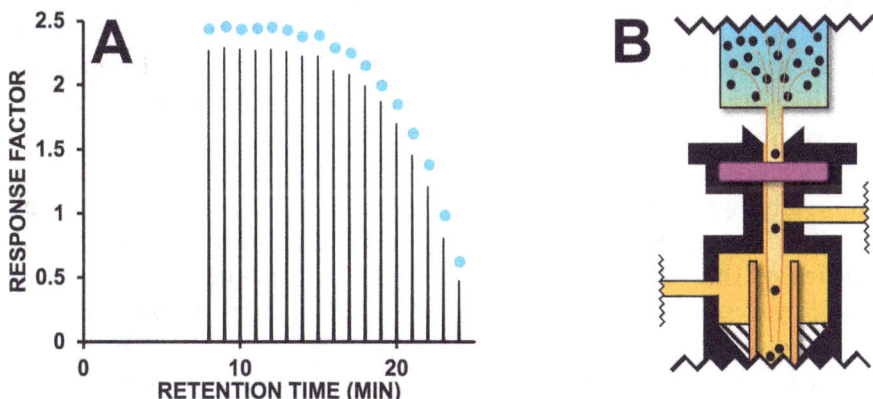

Figure 2.10 Illustration of the problem of discrimination. (A) A typical (temperature-programmed) chromatogram and response factors (blue dots) were obtained with hot split injection for a mixture of *n*-alkanes with similar concentrations. (B) Discrimination due to the needle volume heating up too fast relative to the bulk injection solvent in the (glass) barrel of the syringe (top), which is placed in the injector (bottom).

hottest part (the needle in the hot part of the injector), all analytes may rapidly evaporate. In colder parts of the syringe, solvents and low-boiling analytes may also evaporate, but high-boiling analytes will be left in the syringe. Some form of distillation is taking place along the length of the syringe, resulting in the discrimination phenomenon illustrated in Figure 2.10. Common terms for this phenomenon are "linear discrimination" and "needle discrimination". Response factors vary with the analyte boiling point and are less repeatable due to variations in the temperature profile. The use of several internal standards with different boiling points may help correct the effect in quantitative analysis, but this is neither perfect nor attractive. A more fundamental solution involves cold-injection procedures.

2.4.2 On-column Injector

The best way to ensure that all analyte molecules reach the column is to introduce an aliquot of the liquid sample into the column prior to evaporation. This is achieved in the **on-column injector** (Figure 2.8, middle). Initially, the adjective "cold" was often added, as in cold-on-column injector, to emphasize the fundamental nature of the approach. The sample is introduced through a very thin needle that enters a significant length (typically >30 mm) into the column or, more commonly, into a **retention gap** (Figure 2.11), which is an empty deactivated length of fused-silica tubing attached to the inlet of the column. The internal diameter of this retention gap must be at least 320 μm, and the outside diameter of the needle must be significantly smaller. Pushing such an extremely thin needle through a septum is a risky operation. Therefore, on-column injectors are now almost always equipped with a valve that opens (ideally automatically) when the needle approaches. The needle must then be guided into a closely fitting opening in the injection block. The nature of the (deactivation layer within the) retention gap should match the nature of the solvent. For example, both polar and nonpolar retention gaps exist. It is essential that the temperature at the point of injection is below the boiling point of the solvent (at the inlet pressure). If this is not the case, backflush of the solvent and sample vapours into the cold carrier gas inlets may occur.

Because the injection takes place at a low temperature, thermal focussing does not apply. Therefore, other focussing mechanisms become vitally important. One such mechanism arises from a jump in phase ratio at the point where the analytes move from the

RETENTION GAP | STATIONARY PHASE

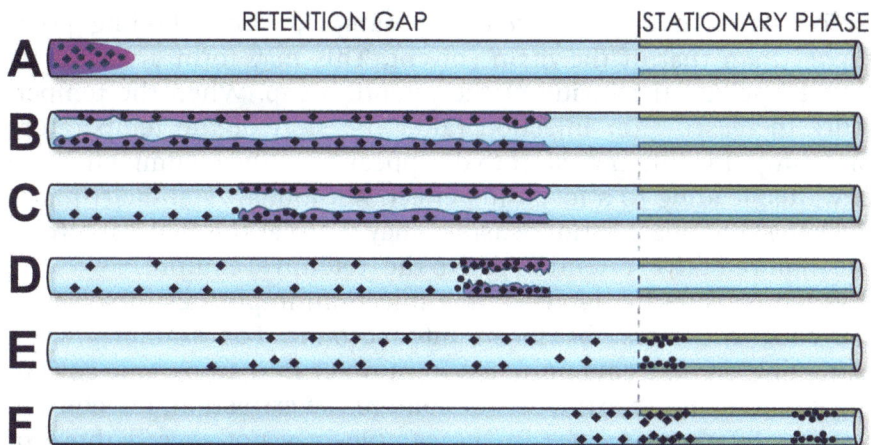

Figure 2.11 Analyte focussing on a retention gap. (A) The solvent plug enters the column and (B) forms a film on the wall of the retention gap. (C) The solvent plug evaporates, and as a result (D), the more volatile analytes will focus in the solvent. Ultimately, the more volatile analytes will reach the stationary phase where they continue to be focussed (E), followed (F) by the less volatile analytes.

retention gap to the entrance of the column, where a stationary-phase film appears. This effect may be enhanced if "phase soaking" (swelling) of the stationary-phase film occurs, following the injection of a relatively large amount of solvent into the column. It is tantamount that the stationary-phase polarity is adapted to that of the solvent to avoid droplet formation (and the resulting "Christmas-tree peaks") and to enhance phase soaking. Finally, the solvent (or a possible co-solvent that is accidentally present or purposefully added to the sample) may temporarily form a film that acts as a dynamic stationary phase in the retention gap. The polarity (deactivation layer) of the latter should also match that of the sample solvent. When the solvent film gradually evaporates from the entrance of the retention gap onwards, less volatile analytes will be increasingly concentrated in a progressively shorter flooded zone.

2.4.2.1 Large-volume Injection

When injecting large sample volumes into a retention gap with an on-column injector, conditions are typically chosen such that (part of) the solvent evaporates during the process. In such a (partially) concurrent solvent evaporation process, the syringe plunger may be moved slowly to introduce the sample for a relatively long time.

A co-solvent may be added to the sample, with a boiling point higher than that of the main solvent. This may form a temporary liquid layer on the inside of the retention gap. When the temperature increases after injection (after the start of the temperature program), this film may start to disappear from the column entrance onwards, leading to a focussing effect. Additional "stationary-phase focusing" or "phase-ratio focusing" may occur at the transition from the (non-retentive) retention gap to the (retentive) column.

When considering large-volume injection, impurities in solvents, reagents, *etc.*, must be taken into account. For example, when extracting organic contaminants from water into an organic solvent using liquid–liquid extraction or solid-phase extraction, the concentrations of trace analytes in the sample may not exceed those of impurities in the solvent. Using the purest solvents available and performing blank experiments for every batch (bottle) of solvent are important measures to avoid erroneous conclusions.

2.4.3 Programmed-temperature Vaporizer Injector

The programmed-temperature vaporizer **(PTV) injector** (Figure 2.8, right) is the most flexible (and most complex) of the devices designed for introducing liquid samples into a GC. Various parameters can be programmed, including the temperature of the injector, the split (or "solvent vent") flow and the flow into the column. The evaporation chamber of the PTV is similar to that of a split/splitless injector, except that the temperature can be increased very rapidly (and also decreased rapidly). Rapid heating is achieved through resistive heating, heating cartridges, or both. Rapid cooling is achieved by compressed air, liquid carbon dioxide, liquid nitrogen, or a Peltier element.

In principle, it is possible to use the PTV at a constant high temperature, but this does not serve many purposes (except, perhaps, for comparing the results of different injection techniques). Almost always, liquid samples are injected in a cold PTV injector, so that needle discrimination is avoided. It is then possible to evaporate the solvent (split flow on and column flow off) and direct the less volatile compounds to the column in splitless mode in the next stage.

The number of parameters that can be varied and the dependence between these (for example, the temperature program of the injector and the vent-flow program must depend on each other) imply that optimizing a PTV injection for a specific type of sample requires time, effort and, ideally, expertise. Various groups have used experimental design strategies for this purpose. However, **large-volume injections** tend to be much easier with a PTV injector than with an on-column injector. The (often packed) liner should accommodate at least the volume of the liquid sample, but this does not evaporate instantaneously as in hot (split/splitless) injectors. The speed of injection (*i.e.* the movement of the syringe plunger) can be much higher in the case of a PTV than in the case of on-column injection, where partially concurrent solvent evaporation is a requirement. For the same reason, the injection temperature is much less critical (and can be lower) in the case of PTV injection.

2.4.4 Headspace Injection

In headspace injection, the gaseous phase above the sample ("gas cap") is analyzed. The term **headspace** can be used for the gas cap above solid or, more commonly, liquid samples. In the case of solid samples, the term "**thermal desorption**" (TD) (see Section 2.4.5) is also commonly used. There is no fundamental distinction between headspace and thermal-desorption techniques, but the typical applications and instrumentation tend to be different. In all cases, the intention is to bring volatile analytes from the sample to the gas phase prior to injection. If the sample is purposefully degraded to form volatile products, we speak of pyrolysis, which is described in Section 8.3.5.

There are two different modes of headspace analysis, *i.e.* static and dynamic headspace. These are illustrated in Figure 2.12A and B. In **static headspace**, equilibrium between the sample and the gas phase is sought. The composition of the gas phase, as measured by GC, is then related to the composition of the sample. A "purge-closed loop" system, in which an inert gas is circulated through or across a sample until equilibrium is reached, is a form of accelerated static headspace sampling.

In **dynamic headspace**, a gas is continuously flushed over a solid sample or bubbled through a liquid sample with the aim to collect and focus all analytes. The latter process is also commonly known as **purge-and-trap**. Typically, the analytes stripped from the sample are focussed in a cold spot (or a very cold spot known as a cryo-trap) or at the top of the column. Headspace analyses are typically slow (say, 30 min or 1 h) to reach equilibrium or deplete the sample. On the other hand, sample preparation steps may be avoided, and non-volatile components are kept away from the column. Volatile

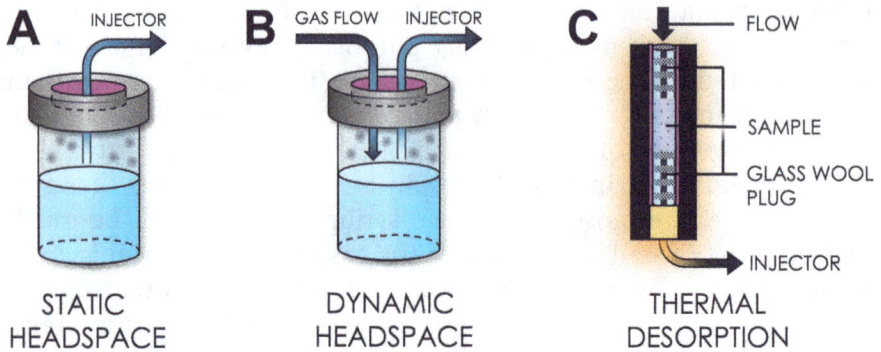

Figure 2.12 Schematic overviews of different alternative sample interfaces for GC. (A) Static headspace, (B) dynamic headspace, and (C) thermal desorption.

solvents present in high concentrations (such as water) may form an obstacle for successful headspace analysis. For example, in the case of aqueous samples, highly volatile analytes with boiling points (far) below 100 °C may be analyzed directly. The evaporation of low-polarity analytes, such as hydrocarbons, from the sample may be aided by high activity coefficients (see Module 2.2). Analytes that are much less volatile than the solvent may be analyzed after evaporating the latter. This is more easily achieved using a PTV injector than with a headspace setup.

Repeat injections may, in principle, be used to determine absolute concentrations or amounts of analytes through static experiments. The total amount of the analyte present $(Q_{i,\text{tot}})$ follows from

$$Q_{i,\text{tot}} = \frac{V_{\text{inj}} \cdot c_{i,\text{hs1}}{}^2}{c_{i,\text{hs1}} - c_{i,\text{hs2}}} = Q_{i,1} \cdot \frac{1}{1 - c_{i,\text{hs2}}/c_{i,\text{hs1}}} \tag{2.13}$$

where V_{inj} is the volume from the headspace injected into the GC, $c_{i,\text{hs1}}$ and $c_{i,\text{hs2}}$ are the measured concentrations of the analyte in the first and second analyses, respectively, and $Q_{i,1}$ is the amount of the analyte detected in the first experiment. In practice, repeat experiments may not be ideal for several reasons. For example, not all of the gas volume extracted from the headspace is introduced into the GC or if a pierced septum leads to losses of headspace gases after the first experiment.

2.4.5 Thermal Desorption

Thermal desorption (TD) is very similar to dynamic headspace. A TD unit (Figure 2.12C) allows specific samples to be introduced, such as tubes containing adsorbents with analytes immobilized on their surface. For example, pollutants in the air can be sampled (statically or dynamically, see Section 2.4.4) on such adsorbents. Sampling is then performed independently from the analysis in a different place and at a different time. In case very few samples need to be analyzed, headspace units or PTV injectors can be used with some improvization (*e.g.* by using a liner as an adsorption tube). If such analyses are performed more frequently, a dedicated TD unit may be used, which allows easy – and possibly automatic – change of adsorption tubes prior to each analysis. Apart from such a sample changer, the TD units work very similarly to a PTV injector. Stages may be programmed (and optimized) that allow evaporation of water or other solvents. Focussing steps at or before the top of the column are usually in order.

Thermogravimetric analysis (TGA) is different from TD analysis in that the main interest is in the amount of material that evaporates or otherwise emerges from the sample. This is monitored by measuring the weight of the sample during analysis. The emerging products can be characterized, but this is more often done by direct coupling to a mass spectrometer (TGA-MS) than with GC or GC-MS.

In pyrolysis-GC (Py-GC), the sample is rapidly heated (in the absence of oxygen) to the point where the molecules disintegrate. The volatile products of the pyrolysis process are analyzed by GC, often followed by MS. This reveals qualitative and sometimes quantitative information on the sample. Py-GC is discussed in Module 8.3.5.

2.5 Detectors for Gas Chromatography (B)

Detection is one of the most important aspects of any chromatographic method. After all, chromatography is "just" a separation method, and a good detection system is needed to turn it into a tool for chemical analysis. In this section, we focus on detectors for capillary (open-tubular) GC. Detectors need to be rather small (volumes of the order of microlitres; see Table 1.3) and extremely sensitive. Typical amounts of analytes injected into the column are in the nanogram range. Fortunately, there are several detectors (and detection principles) that are eminently suited for combination with

GC. The operation of GC detectors is aided considerably by the inertness of the mobile-phase gases. In many cases, the carrier gas is not detected, creating a perfect blank signal when no peaks are eluted.

An ideal (GC) detector should (i) be sensitive; (ii) exhibit little noise, resulting in a low detection limit; (iii) universally respond to all compounds (with equal response factors) or respond specifically to certain elements or certain compound classes; (iv) show a fast, linear response across a very large range; (v) be precise (repeatable), robust, and reliable; (vi) be non-destructive; (vii) be safe; (viii) be simple; and (ix) be cheap.

We distinguish between specific and selective detection. Detection is **specific** if certain compounds are detected, whereas the signal for other compounds is zero. It is called **selective** if the signal for one type of compound is significantly higher than that for another type of compound. There is a degree of selectivity, whereas specificity is absolute. However, when following the above definition rigorously, specificity tends to never be achieved. A detector that we call specific may, in fact, be extremely selective.

2.5.1 Flame-ionization Detector

The FID (see Figure 2.13A) is the most popular and the most celebrated GC detector. It possesses many of the properties identified above for an ideal detector, as summarized in Table 2.5. The FID is based on a hydrogen flame, sustained by flows of hydrogen and compressed air (or oxygen). The combustion of organic compounds in the flame generates ions, which in turn create a current when collected by a (typically cylindrical) negative electrode ("collector") above the flame, with the nozzle head acting as the positive electrode.

An FID responds to nearly all organic compounds. The response is proportional to the mass flow of the analyte (weight time^{-1} units) but is recorded as an electronic signal (pA s^{-1}). Hydrocarbons show fairly similar response factors, although aromatic compounds yield a somewhat lower response than aliphatic hydrocarbons. Electronegative heteroatoms, such as oxygen, sulfur, or halogens reduce the response. Very few compounds, including carbon dioxide (CO_2), carbon monoxide (CO), carbon disulfide (CS_2), and carbon tetrachloride (CCl_4) are virtually non-detectable. An important workaround used, for example, in element-analysis systems based on the combustion of the sample, is a so-called **methanizer**. This is a

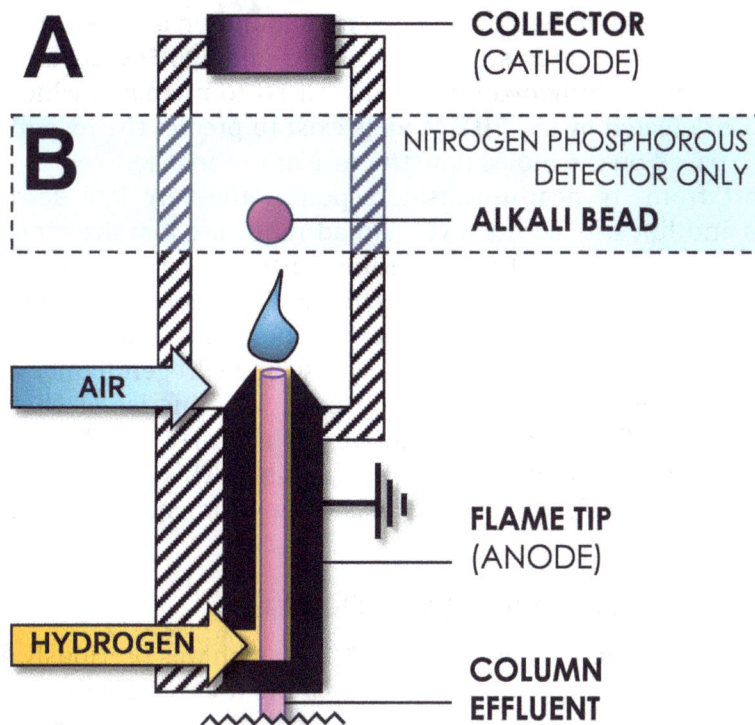

Figure 2.13 Schematic picture of (A) an FID and (B) a nitrogen–phosphorus detector (NPD). The latter is discussed in Section 2.5.4.1.

Table 2.5 Characteristics of the flame-ionization detector (FID).

Property	Value	Comments
Type of response	Mass flow (mass time^{-1})	Linear response
Response time	*ca.* 5 ms	Limited by electronics (trade-off against noise)
Responds to	All organic compounds	Response is fairly constant for hydrocarbons; response is reduced by (electronegative) heteroatoms
Detection limit	1 pg s^{-1}	Typical value for hydrocarbons
Dynamic range	10^7	Possibly more (not explored yet)
Precision	High	—
Complexity	Simple and robust	Does require hydrogen and compressed-air (or oxygen) flows
Destructive	Yes	Typically operated at high temperatures (*e.g.* 250 °C)
Safety	Moderate	Requires hydrogen flow; flame
Cost	Low	In terms of both investment and consumables

catalytic reactor containing a nickel compound on a carrier material catalyst. At temperatures between 350 °C and 400 °C, CO and CO_2 can be quantitatively hydrogenated with H_2 to methane, which can then be detected by the FID. Models exist to predict the response of an FID based on the molecular structure of the analyte.[21]

Apart from its near-universal response, the very low detection limits and linearity across a very broad range are massive strengths of the FID. It is perfectly compatible with capillary (open-tubular) GC, and it is simple, cheap, and reliable. FIDs require little maintenance. All of these traits make the FID the go-to detector for many applications of GC. One disadvantage is the hydrogen flame, which cannot be used everywhere. For example, in refineries or chemical plants, where flammable or explosive gas mixtures may be encountered, the use of flames is strictly prohibited. This necessitates the use of alternative detectors in process-GC instruments.

2.5.2   Thermal Conductivity Detector

The thermal conductivity detector (TCD), which was previously often referred to as a katharometer, measures the thermal conductivity of the effluent of a GC column through thermally induced changes in the resistance of a filament. The typical configuration is a Wheatstone bridge, with a second filament flushed with carrier gas acting as a reference (see Figure 2.14). In principle, any compound (except the carrier gas) can be detected with a TCD. The response is linear with concentration, but the detection limits (in comparable units of the total mass of the analyte) and the linear working range are both several orders of magnitude worse than those of the FID. The detector is extremely simple. It is robust and reliable, provided that oxygen (or air) is kept away from hot filaments. Typical characteristics are listed in Table 2.6.

2.5.3   Photoionization Detector

In a photoionization detector (PID), high-energy radiation is used to ionize analyte molecules. Different energies, corresponding to different wavelengths in the (vacuum) UV region, can be used to ionize different classes of compounds so that the selectivity of the PID detector can be varied. In principle, all analytes with an

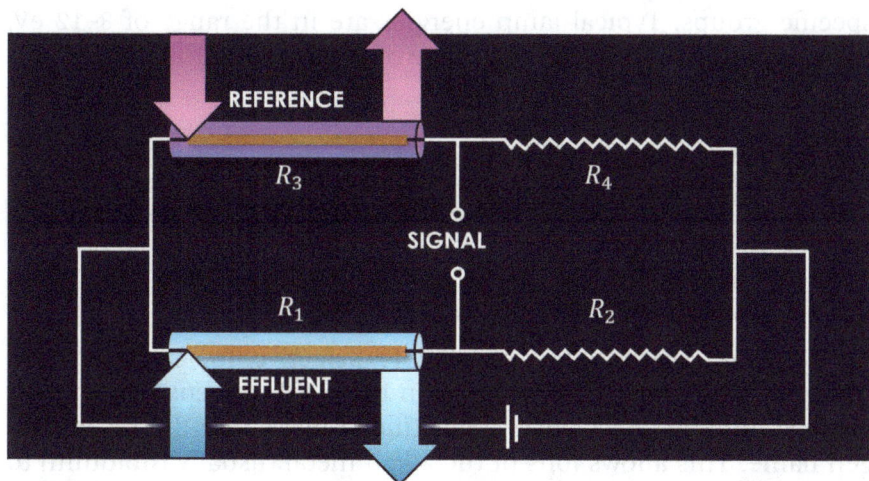

Figure 2.14 Schematic picture of a thermal conductivity detector (TCD).

Table 2.6 Characteristics of the thermal conductivity detector (TCD).

Property	Value	Comments
Type of response	Concentration (weight volume^{-1})	Linear response
Response time	100 ms	Much slower than FID
Responds to	Any volatile compound (except the carrier gas)	Hydrogen and helium exhibit exceptionally high thermal conductivity, making all other compounds detectable
Detection limit	High ng to µg	Total amount of the analyte injected
Dynamic range	>10^3	Much narrower than FID
Precision	High	Stable, low noise
Complexity	Simple and robust	No gas other than the carrier gas required
Destructive	No	Typically operated at high temperatures (*e.g.* 250 °C)
Safety	Excellent	No known safety issues
Cost	Low	In terms of both investment and consumables

ionization threshold below the energy of the light can be detected. For example, at a light energy of 10 eV (124 nm), aromatics show a high response, while alkanes are not ionized. At 12 eV (103 nm), both alkanes and aromatics are detected (the latter with slightly higher molar response factors),[22] and the PID is a near-universal detector. Thus, a range of analyte groups can be targeted rather than specific groups. Typical lamp energies are in the range of 8–12 eV, corresponding to a wavelength range of about 150 to 100 nm.

2.5.4 Element-specific Detectors

Erika Cremer (Box 2.2) was one of the first scientists to investigate selective GC detectors.

2.5.4.1 Nitrogen–Phosphorus Detector

An NPD, or **thermionic detector,** is selective for nitrogen and phosphorus (Figure 2.13B). It is akin to an FID, except that a small alkali-containing glass bead is positioned just above the hot hydrogen flame. This allows ions of the alkali metal (usually rubidium) to enter the plasma. A significant disadvantage is the gradual depletion of the bead, which causes a decrease in response over time. Heating the rubidium bead above the operating temperature provides a temporary increase in response, but not to the original level.

In an alkali-flame detector (AFD), the alkali ions are added as a salt aerosol to the hydrogen flow, which may result in a more constant response over time.

Box 2.2 Hero of analytical separation science: Erika Cremer. Reproduced from ref. 23 with permission from Elsevier, Copyright 1979.

Erika Cremer (1900–1990, Germany and Austria)
Erika Cremer was born in München, Germany, but she spent most of her scientific life at the University of Innsbruck, Austria. She was one of the true pioneers of gas chromatography, starting her studies on the subject as early as 1944. She contributed to the underlying theory by performing physical experiments and to the development of selective GC detectors. She was the daughter of a university professor, the granddaughter of a professor, and the great-granddaughter of a professor. She also became a full professor, but this was by no means logical. In 1940, she was first appointed lecturer in Innsbruck on a temporary basis, until "the men" would return from the war.

2.5.4.2 Flame-photometric Detector

A **flame-photometric detector** (FPD) is also akin to an FID, except that light emitted from sulfur- or phosphorus-containing analytes is observed. Halogen-containing analytes also emit light, but a specific response for sulfur or phosphorus can be obtained by using suitable optical filters. The signal for sulfur is not proportional to the mass flow of sulfur atoms. Instead, it is proportional to the square of the mass flow. An electronic correction is commonly applied, but this is a potential source of error, for example, due to a background signal.

A very interesting extension of the FPD has been developed by Jing and Amirav.[24] This pulsed **flame-photometric detector** (PFPD) makes elegant use of the delayed emission of compounds such as S, HPO, and HNO to increase the selectivity of the detector towards heteroatoms and to lower detection limits. A time gate of 5 ms effectively eliminates any signal due to carbon. It is more difficult to differentiate between the various heteroatoms based on the time delay. Filters can be used to focus on different heteroatoms based on emission wavelengths. Sulfur invariably yields a quadratic response.

2.5.4.3 Electron-capture Detector

In an **electron-capture detector** (ECD), the effluent from the GC column is bombarded with a constant flux of electrons supplied by a radioactive foil (beta emitter). Most commonly, ^{63}Ni isotopes are used for this purpose. The electrons ionize the carrier gas, which may be nitrogen, but often is argon containing 5% methane, creating a current. When electrons are captured by analyte molecules, this current decreases. Thus, a form of indirect detection is used, where the presence of an analyte is detected as a negative peak against a constant high background (although, of course, the signal is easily converted to display positive peaks). The ECD is a concentration-sensitive detector, and it is selective towards dense electron clouds in analyte molecules, which implies a high response (and very low detection limits, down to the sub-pg range of the total amount of the analyte injected) for halogen-containing analytes. Electron-withdrawing functional groups such as nitro ($-NO_2$) and nitrile ($-C\equiv N$) also show significant responses with ECDs. Working with radioisotopes requires special training and qualifications. This, in combination with the ever-increasing popularity of mass spectrometers, has led to a decline in the use of ECDs.

2.5.4.4 *Chemiluminescence Detectors*

Chemiluminescence is a phenomenon in which light is produced through a chemical reaction. **Sulfur-chemiluminescence** detectors (SCDs) rely on a reaction of sulfur monoxide with ozone (O_3), while **nitrogen-chemiluminescence** detectors (NCDs) are based on the reaction of nitrogen monoxide (NO) with O_3. Both detectors are specific for the respective elements, which implies that there is virtually no response to other compounds. However, large amounts of carbon will affect the detector response. They give rise to an equimolar and linear response and very low detection limits.[25]

In an SCD, analytes leaving the GC column are first combusted with oxygen at a temperature up to 1000 °C to yield SO_2 (plus CO_2 and H_2O). In the next stage, hydrogen is added to reduce SO_2 to SO at about 850 °C. Finally, ozone (O_3) is added to convert SO to sulfur dioxide in an excited state (SO_2^*), the decay of which results in the emission of light in the UV/visible region (180–460 nm), which is detected using a photomultiplier tube (PMT). The principle of the detector is schematically illustrated in Figure 2.15B.

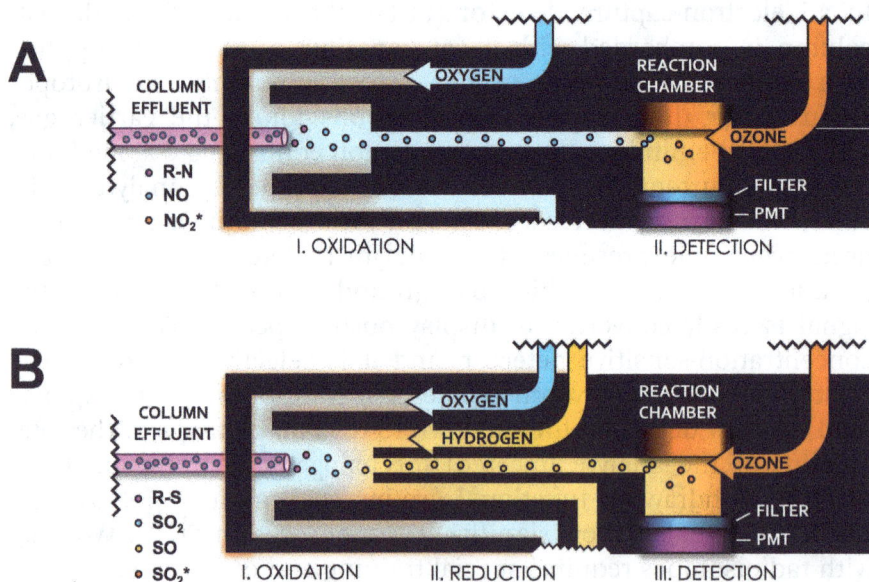

Figure 2.15   Schematic illustration of the functioning of a nitrogen-chemiluminescence detector (NCD, panel A) and a sulfur-chemiluminescence detector (SCD, panel B) for GC. PMT = photomultiplier tube. R-N and R-S depict nitrogen- and sulfur-containing analytes, respectively.

An NCD works along similar principles (Figure 2.15A), but the reduction stage with hydrogen is obsolete because the combustion stage produces almost exclusively NO as the nitrogen-containing compound. Upon reaction with O_3, the NO_2^* formed produces chemiluminescence in the near-infrared region around 1200 nm.

Excellent results may be obtained with SCD and NCD detectors, but a lack of robustness has been their main disadvantage. With the increased availability of chemiluminescence detectors from multiple sources, this issue may gradually be resolved.

2.5.5 Atomic-emission Detector

A plasma is a state of matter in which charged particles (ions and electrons) dominate. If a sample for analysis is brought into a plasma, atoms of specific elements can be excited and made to emit light at very specific wavelengths when the atoms fall back to lower energy states. We have seen this previously for flame-photometric detection, which uses light emission of sulfur and phosphorus atoms in a hydrogen flame. Some metals, such as sodium, potassium, barium and calcium, also emit specific coloured light when subjected to a flame. The effect can be used more generally if (i) higher-energy plasmas are used, (ii) both visible and UV ranges are covered, and (iii) monochromators (or "polychromators", to simultaneously monitor several emission lines) are used, instead of optical filters.

In elemental analysis, the inductively coupled plasma (ICP), which is based on an argon plasma, is extremely popular for the sensitive and specific detection of almost all elements in the periodic system. Apart from light emission, MS is frequently used to monitor ions formed in the plasma. In the context of GC detection, the key word above is the "almost" all elements. ICP optical-emission spectrometry (ICP-OES) and ICP-MS are not suitable for analyzing the lighter elements (CHNO) and the halogens, which make up the vast majority of organic analytes encountered in GC.

For GC, a microwave-induced plasma (MIP) based on excited helium is highly attractive. It is a tuneable selective detector for almost all the elements one may encounter. It may, for example, be tuned to detect sulfur compounds, chlorine, or even carbon or hydrogen. This can make it instantly clear which elements are present in which peaks (or analytes). The resulting instrument, known as an **atomic-emission detector** (AED), has been amply demonstrated as a useful tool for such purposes. However, the commercial implementation uses a diode array to monitor a selected part of the spectra, which means that only a group of elements with emission lines at similar wavelengths can be monitored simultaneously.[26] While the signal (amount of light emitted) is, in principle, proportional to the mass flow of the element monitored, there are some caveats associated with absolute quantification. The detection limits tend to be higher than those encountered with element-specific detectors, such as the ECD for halogens and the SCD for sulfur.

2.5.6 Infrared Spectroscopy

Infrared (IR) spectroscopy is a well-established method to study all kinds of materials and molecules. Light in the mid-infrared range (wavelength *ca.* 2.5–25 µm, corresponding to 4000 to 400 wavelength per centimetre on the cm^{-1} scale, which is still favoured by IR spectroscopists) resonates with specific vibrations in molecules, such as C-H stretch or HOH bend vibrations. As a result, absorption bands in the mid-infrared spectrum are indicative of specific functional groups present in analyte molecules. Since they are related to well-defined physical processes, the position of such bands is highly reproducible and so can be IR spectra. This has led to large collections (libraries) of IR spectra that are available commercially. Infrared spectra reveal chemical information that may be of aid in identifying analytes or add to the confidence of their (tentative) identification with MS. In addition, IR spectra may be used to distinguish between isomers that yield (near-)identical mass spectra.

There are basically two ways in which IR spectroscopy can be coupled to gas chromatography. The simplest of these is to use a flow cell. Relatively long path lengths are desirable due to the limited sensitivity of IR spectroscopy and, especially, the low volumetric concentrations of the analytes in the gaseous sample. This has led IR practitioners to use the term "light pipe" instead of flow cell.[27] The light path is often coated with gold, as this is an excellent reflector for IR light, and multiple reflections of the light from the walls of the light pipe occur before the IR light reaches the detector. The light pipe and the connecting tubing (transfer lines) must be heated above the oven temperature to avoid the adsorption of analytes. Such a light-pipe GC-IR system is robust, but the detection limits are orders of magnitude higher than those routinely obtained with GC-MS. This discrepancy and the band broadening caused by the large light pipe (typically 15 mm length × 1 mm i.d.; $V_{det} \approx 120$ μL) and the transfer lines make it unrealistic to couple the two detectors in series (even though the IR detector is non-destructive).

The second approach is to deposit the analyte on a suitable surface and to record the spectrum of the immobilized material. The fundamental advantage of this approach is that more time can be devoted to the collection of a spectrum, which may increase the signal-to-noise ratio of Fourier-transform (FT) IR spectra. In a previously commercialized matrix-isolation (MI) instrument, analytes were frozen in an inert matrix (typically 2% argon added to the carrier gas) on a rotating gold-plated drum kept at cryogenic temperatures (10–15 K) using a closed-loop helium refrigeration system. The low temperature (eliminating rotational motion of the analyte molecules) and the inert matrix resulted in very narrow IR absorption bands.[28] Although the technique was demonstrated on a number of different types of compounds and yielded low detection limits (*e.g.* spectra of 50 pg of fluorene and pyrene showed excellent signal-to-noise ratios),[28] it was arguably too complex and possibly too expensive to be successful. Ironically, the higher quality of the IR spectra obtained made the system less compatible with existing spectral libraries.

In the related but simpler direct-deposition (DD) approach, the analyte bands are frozen in small (100 μm diameter) spots on an IR-transparent (*e.g.* ZnSe) disc at liquid nitrogen temperatures (77 K) under vacuum.[29] Transmission spectra can then be obtained using IR microscopy optics. Because the spot size was three times smaller than that obtained with the MM interface, a higher mass sensitivity could be obtained, and the spectra obtained could be compared with those obtained from samples analyzed pressed in KBr tablets. Many such spectra are collected in libraries.

2.5.7 Ultraviolet Spectroscopy

The combination of GC with UV detection is not new, but it has recently been spurred by the emergence of good commercial instrumentation that allows for the recording of high-quality **vacuum-UV** (VUV) spectra in the wavelength range of 115–240 nm.[30] The lower part of this wavelength region (115 nm corresponds with 10.8 eV) overlaps with the energies used in photoionization detectors (see Section 2.5.3). This means that some analytes (*e.g.* aromatics) will be ionized. However, this does not impair their detection with

GC-VUV spectroscopy. The system uses a deuterium lamp and a heated flow cell (10 mm length × 1 mm i.d.; $V_{det} \approx 80$ µL) with a highly UV-reflective coating on the inner wall. The principle is similar to the GC-IR light pipe described above (Section 2.5.6), but the materials used are vastly different. Cell windows need to be constructed from VUV-transparent materials (*e.g.* magnesium fluoride, MgF_2, which is transparent in the specified wavelength range). A make-up carrier-gas flow is used to control the residence time of the analytes in the flow cell and to minimize extra-column band broadening. Like GC-IR, GC-UV adds information on compound identity and allows differentiation between isomers and other molecules with very similar mass spectra. After background correction, the UV spectra are very reproducible, but the construction of large libraries of gas-phase UV spectra is still in its infancy. The linearity of the UV response allows automatic deconvolution of the spectra.[31]

2.5.8 Olfactometry

GC is a non-destructive separation method, and in an ideal separation, pure-component vapours are eluted from the GC column. Olfactometry is the use of human smell for scientific or analytical purposes, and it can be combined with GC in a very elegant and productive manner. GC olfactometry is the direct coupling of a physical separation method with a biological-function detector. It may be of great help in identifying chemical compounds that contribute positively to the aroma of food or beverages or the fragrance of consumer products, or contribute negatively to the smell of products.

Dedicated sniffer masks, colour classification codes and speech recognition systems exist to facilitate olfactometry. The most powerful systems combine olfactometry with parallel detection with MS so that odour information can be combined directly with mass-spectral information. Restrictions may be imposed on human volunteers. For example, they must typically be non-smokers, and they may have to postpone their coffee and/or lunch until after they have completed their laboratory duty.

2.6 Gas Chromatography-Mass Spectrometry (M)

GC-MS was a near-instant success, with the first devices already being reported in the 1950s.[32,33] The reasons for this were the obvious and immense value of the combination of the two techniques and their compatibility, which implied that their online coupling was relatively straightforward. The components of a GC-MS system are schematically depicted in Figure 2.16.

The characteristic parameter of an ion in MS is not so much its mass but the ratio of its mass m and its charge z. Thus, an ion with mass 100 and charge +1 will behave identically to an ion with mass 200 and charge +2, as for both ions $m/z = 100$. Some authors use the notation m/e instead of m/z for the mass-to-charge ratio. The two are equivalent.

The mass analyzer either separates ions according to their m/z values, or the ions are measured within the analyzer. Different types of mass analyzers will be discussed in Section 3.10.1.2.

The volumetric flow rate in open-tubular GC (about 1 mL min^{-1} at room temperature and atmospheric pressure) is such that small turbopumps suffice to maintain a low enough pressure in the MS. A critical point was the realization of a heated transfer line between the two instruments, as any cold spots could jeopardize the entire process. Modern GC-MS instruments tend to be small and compact with very short transfer lines. If this is not the case, transfer lines remain critical.

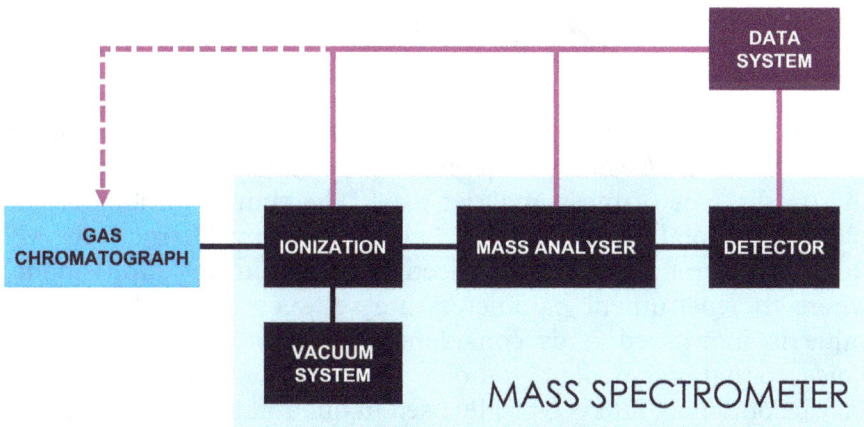

Figure 2.16   Schematic outline of a gas chromatography–mass spectrometry system.

Mass spectrometers are extremely sensitive. Femtogram amounts of analytes can typically be detected. Arguably, the greatest obstacle to the development and proliferation of GC-MS instruments was the lack of suitable and affordable computers early on. Computers are no longer a limiting factor. Good software and spectral libraries (collections of MS spectra) are amply available.

The resolution or **resolving power** of MS instruments is defined differently from that in chromatography. Whereas chromatographers define resolution as the distance between two peaks relative to their (average) width, mass spectrometrists always look at two peaks that overlap at the 10% level and then define resolving power (RP_{MS}) as

$$RP_{MS} = \frac{M}{\Delta M} \qquad (2.14)$$

where the two peaks have masses (or, rather, mass-to-charge ratios) of M and $M + \Delta M$. Alternatively (more conveniently and a bit more favourably), the width of a peak in the mass spectrum (at half height) may be substituted for ΔM. The resolving power typically depends on the mass at which it is established. For example, it may be defined as $RP_{MS} = 10\,000$ at $m/z = 1000$. Mass accuracy (ΔM^*) can be defined as an absolute number ($\Delta M^* = M_{exp} - M_{calc}$, in Da), where M_{exp} is the experimentally obtained mass and M_{calc} is the mass calculated from the molecular formula (based on the most-common isotopes for each element), or as a relative number (in ppm)

$$\Delta M^* = \frac{M_{exp} - M_{calc}}{M_{calc}} \times 10^6 \qquad (2.15)$$

Mass spectrometry may be performed in (full-)scan mode, looking at all ions (*i.e.* m/z values) sequentially or simultaneously – depending on the type of mass analyzer used. The sum of all ions is displayed in a **total-ion-current** (TIC) chromatogram. Alternatively, MS instruments may be used in **selective-ion-monitoring** (SIM) mode, where the instrument parameters are set such that only one m/z value is monitored. This considerably reduces the measurement noise, so that higher signal-to-noise ratios and lower detection limits may be obtained. SIM is typically used in quantitative target analysis, while non-target analysis requires using the full-scan mode. Different m/z values may be selected during different time slots to quantify a number of peaks in the chromatogram.

2.6.1 Ionization Methods

2.6.1.1 Electron Ionization

Electron ionization (previously known as electron-impact ionization; both conveniently abbreviated EI) is based on bombarding the analyte molecules with electrons emitted from a filament (often made of tungsten – as in lightbulbs – or rhenium) in a high vacuum (*ca.* 10^{-4} Pa). The energy of the electrons can be varied by changing the potential difference between the filament and the ionization chamber but is usually chosen to be high (typically 70 eV). The word "bombarding" is actually rather deceptive. The high-energy electrons are much more likely to "drag" electrons away when skimming past analyte molecules (forming $M^{+\cdot}$ radical ions) rather than engage in a collision. Thus, the EI process resembles a "hurricane" of electrons more closely than a "bombardment" with electrons. The internal energy of the $M^{+\cdot}$ ion is so great that it easily splits into a number of characteristic fragments. When this happens, we speak of a "hard" ionization method.

There is considerable variation in the intensities of the various fragments, characterized by their mass-to-charge ratio (m/z), and there is some variation in the relative intensities, but the overall pattern is sufficiently reproducible to render EI mass spectra extremely useful for the (tentative) identification of analytes. To distinguish between homologues, GC retention indices are invaluable. Many geometrical isomers can be distinguished based on their mass spectra, but at some point, the differences become insignificant, and GC selectivity becomes indispensable. This is, for example, true for stereoisomers.

The interpretation of mass spectra threatens to become a skill of the past, thanks to the availability of spectral libraries and due to a rapidly increasing gap between the number of MS instruments employed and the number of analysts who are skilled in the art of interpreting spectra. Yet, understanding some of the principles of mass spectra is extremely useful to critically assess and validate automatic assignments.

2.6.1.2   Chemical Ionization

In chemical ionization, ion–molecule reactions are used to produce ions from analyte molecules. A reaction gas is added at pressures much higher than those encountered in EI (*ca.* 10–20 Pa). The reagent gas is "bombarded" with high-energy electrons (100–200 keV) to form an excess of reagent ions, which then collide and react with analyte ions. Common reagent gases are methane (yielding predominantly $[M + H]^+$ ions) and ammonia (yielding $[M + H]^+$ and $[M + NH_4]^+$ ions). The $[M + H]^+$ ions are usually the most abundant peaks in CI-MS spectra, and they are commonly referred to as "quasi-molecular ions" or "pseudo-molecular ions". These ions degrade much less than the radicals formed in EI. Therefore, CI is particularly useful to establish the molecular weight (and, in the case of high-resolution MS, the molecular formula) of analytes.

Recently, **atmospheric-pressure chemical ionization** (APCI) has been advocated as an attractive alternative to conventional CI.[34] In APCI, ionization takes place at atmospheric pressure (or at least pressures much higher than those used for conventional EI or CI). A corona discharge is used to produce electrons, which charge a reagent gas, such as nitrogen. Charged nitrogen (*e.g.* N^{2+}) ions will then ionize analyte molecules (forming M^+ ions). If water molecules are added, MH^+ ions may be mainly formed. The much higher concentration of molecules at atmospheric pressure results in a much higher fraction of analyte molecules ionized and, as a result, a much higher sensitivity. Detection limits were found to be 10 times lower for APCI-GC/GC than for EI-GC/GC for the analysis of polychlorinated biphenyls (PCBs) and polybrominated diphenyl ether (PBDEs).[35]

Note that mass spectrometrists tend to refer to all CI ionization systems as APCI.[36] For them, the pressure encountered in the ionization chamber is very high (although "atmospheric" is a significant exaggeration). For gas chromatographers, there is nothing special about the pressures encountered in GC-CI-MS interfaces or in the photoionization interfaces discussed below, which are referred to by mass spectrometrists as atmospheric-pressure photoionization (APPI).

2.6.1.3   Photoionization

Like chemical ionization (CI), photoionization (PI) is a soft-ionization technique capable of preserving molecular ions, including those of relatively unstable, polar analytes.[37] There is a trade-off between the applicability or universality of the ionization and the extent of fragmentation. Higher light energies produce a more universal response, but increased fragmentation. Krypton discharge lamps, which emit photons with energies of 10.0 and 10.6 eV, are commonly used.[36] A gas flow containing a highly ionizable "dopant", such as acetone or toluene, may be added to the GC effluent to enhance the ionization efficiency.[38] PI is used much less often than CI in (industrial) practice.

2.6.2 Mass Analyzers

The most common type of mass spectrometer applied in combination with GC is a (single) **quadrupole MS** ("single quad" or QMS). A quadrupole mass analyzer consists of four parallel metal rods. Radiofrequency voltages with DC offsets are applied to the rods, identically for the diagonal pairs but differently for any pair of adjacent rods. At a given voltage, only ions with specific mass-to-charge ratios (m/z) have a stable axial path through the quadrupole, so that specific ions can be selected or a spectrum can be recorded through a voltage scan within a second or less.

Quadrupole mass spectrometers are relatively simple instruments, which may be miniaturized, and therefore require a relatively small vacuum capacity. These kinds of MS instruments, together with user-friendly software, are readily available from several manufacturers. They are fully compatible with routine applications of GC-MS, employing EI and spectral libraries. Thus, even small GC-MS instruments are very powerful tools in the laboratory. The disadvantages of such small mass spectrometers are their limited speed and limited resolution.

The speed of the instrument becomes an issue when very fast GC is performed. This is commonly encountered in comprehensive two-dimensional gas chromatography (GC×GC), a technique that is discussed in detail in Module 7.2. In GC×GC, a long series of very narrow peaks enter the detector, with standard deviations down to 100 ms or less. For such very fast peaks, a scanning QMS spectrometer is inappropriate because the concentration of the analyte may vary significantly during the recording of one spectrum, resulting in a highly biased mass spectrum. For very fast GC, a time-of-flight (ToF) mass spectrometer is more appropriate. In such an instrument, a sample of ions produced in the source is accelerated into a flight tube at regular intervals. Heavier ions (higher m/z values) move slower through the flight tube, and the detected ion current as a function of time can be converted to a mass spectrum. ToF-MS instruments can also provide a higher resolution than QMS instruments, but there is a trade-off between speed and resolution.

2.6.2.1 Gas Chromatography-High-resolution Mass Spectrometry

High-resolution MS (HRMS) instruments have the ability to provide exact (*i.e.* accurate) m/z values within several decimals, as opposed to QMS instruments, which are typically regarded as unit-mass-resolution instruments. HRMS systems are not typically associated with GC-MS, but the advent of HRMS instruments in recent decades has also led to increased interest in GC-HRMS.[39] Such systems offer the advantage of producing unique molecular formulae to considerably reduce the number of possible candidates in detecting unknowns. In addition, HRMS allows the selection of narrower m/z windows for quantification, which may reduce (chemical) background and noise, resulting in lower detection limits.

GC-HRMS first started to emerge for the analysis of **dioxins** (polychlorinated dibenzo-*p*-dioxins or PCDDs). Out of 75 PCDD congeners with different numbers and locations of chlorine atoms, seven are known to be extremely toxic. Hence, it is not just important to measure the total amount of dioxins in a sample, but also the concentrations of these individual congeners. GC-MS experiments were performed on what can rightly be considered old-fashioned magnetic-sector instruments.[40] These are an ingenious type of MS instrument, which were long the benchmark of HRMS. They show a broad dynamic range and offer low detection limits (in the femtogram or even attogram range). However, such instruments are relatively slow (few spectra per second), large, complicated, and expensive (in terms of both purchase and maintenance). As a result, newer types of HRMS instruments are progressively displacing magnetic-sector systems.

Modern ToF-MS systems offer a resolution approaching that of sector instruments while allowing the acquisition of a spectrum every few milliseconds. This renders them the instrument of choice for GC×GC MS experiments. One of the weaknesses of ToF instruments is their limited dynamic range. GC–Orbitrap-MS systems do offer low detection limits and a broad dynamic range. Such systems provide a better resolution than magnetic-sector instruments, but the trade-off between resolution and speed renders these systems less attractive for coupling with GC×GC.

2.6.2.2 Gas Chromatography-Tandem Mass Spectrometry

In tandem MS or MS/MS, specific ions can be selected and subjected to (additional) fragmentation. The resulting fragments may be characterized as a complete mass spectrum (full-scan mode) or specific ions may be selected for the quantification of target analytes (selected-ion-monitoring or SIM mode). Even more than in the case of HRMS, soft ionization methods, such as CI, are preferred in MS/MS. This allows the selection of the molecular ion in the first MS stage.

Triple-quadrupole (QqQ) instruments are popular in combination with LC for quantification of target compounds because (i) the initial (electrospray) ionization tends to be soft and (ii) retention information is much less useful than in the case of GC. However, GC-QqQMS systems have seen increasing application in the past decade. The attainable detection limits of ToF can be lowered by adding a quadrupole mass filter and operating in SIM or multiple-reaction-monitoring (MRM) modes. Such hybrid quadrupole–time-of-flight (QToF) instruments provide higher mass resolution than QqQ systems and allow rapid operation in full scan mode.[41]

2.7 Controlling Flow and Temperature (A)

2.7.1 Flow Control

During the course of this century, reducing valves and manually adjusted needle valves have gradually been replaced by electronics (although the former are still abundant to meet the inlet requirements of modern controllers). **Electric-pneumatic controllers** (EPCs) are the core devices to control flows and pressures in chromatographic systems. An EPC typically contains several variable metering valves that serve to reach set values of flow or pressure. For example, the pressure in the injector (top of the column) and the flow through a split line may be controlled in closed-loop systems. In the case of a split injector (or PTV), the flow through the column is more difficult to control, as it will be a small difference between the total flow of carrier gas, the split flow and possibly a septum-purge flow.

The EPC is able to calculate the required column pressure if the column dimensions (length and internal diameter), the temperature, and the type of carrier gas are correctly specified. If the calculation is correct, a t_0 marker should elute at the expected time. The flow can also be checked using an electronic flow meter. When the system is changed, for example, when a different column has been installed, this is always a good idea.

An immediate complication is that GC analyses are rarely performed isothermally. In the case of a temperature program, an EPC may allow constant-pressure or constant-flow operation. In the former mode, the column flow rate and the linear velocity decrease during the run because the viscosity of gases increases with increasing temperature. In the latter case, the pressure must increase during the run following a set profile calculated by the EPC. Constant-flow operation tends to result in narrower peaks, as the bands leave the column more quickly.

The viscosity of carrier gases can be accurately described by the following equation:

$$\eta = \eta_0 \left(\frac{T}{273.15} \right)^{\varepsilon} \tag{2.16}$$

where η is the viscosity of the gas in μPa s, η_0 is its viscosity at 0 °C (273.15 K) and ε is an empirical coefficient. Values for η_0 for common carrier gases are 18.6, 8.35 and 16.59 μPa s for helium, hydrogen and nitrogen, respectively. The respective values for ε are 0.646, 0.68, and 0.725, respectively.

The much lower viscosity of hydrogen directly implies lower column pressure drops, which is especially relevant for performing fast GC analysis in narrow-bore (*e.g.* 50 μm i.d.) open-tubular columns. Nitrogen is less viscous than helium but much less attractive in terms of diffusion coefficients. With increasing analyte molecular weight, the diffusion coefficients in helium quickly approach 2.6 times those in nitrogen, while the values in hydrogen approach 3.6 times higher values than those in nitrogen – and thus 40% higher than those in helium.[42]

2.7.2 Temperature Control

When using the typical coiled capillary GC column in an air circulation oven, it is tantamount that the temperature is identical everywhere in the oven. Possible small variations close to the injector and the detector (which may both be at different temperatures in the oven) will not immediately jeopardize the separation. However, if parts of the coiled column are experiencing different temperatures, the analyte zones will be accelerated and decelerated multiple times, leading to excessive band broadening and deformed peaks. There are basically two ways to maintain constant temperatures in GC ovens. The first one is an oven with a high thermal mass. Once a constant temperature is reached, it is robust. A significant amount of energy is required to heat the oven, but programming the temperature can be achieved with sufficient heating power and an adequate control loop. Cooling the oven may take considerable time, especially if the initial oven temperature (T_{init}) is close to room temperature because the heat flux is proportional to the temperature difference between the oven (T_{oven}) and the outside ($T_{ambient}$). Cooling can be expedited using evaporating nitrogen as a coolant. This requires a control unit for the GC and a supply of liquid nitrogen. Forced cooling allows low initial temperatures, including values below ambient ($T_{init} < T_{ambient}$). This may create additional selectivity and alleviate the need for thick film or even packed columns for separation of highly volatile analytes or permanent gases. Therefore, forced cooling of the oven is an attractive, albeit costly, option.

The second approach involves a **low-thermal-mass GC** oven with a fast control loop to correct for deviations from the set temperature.[43] This is potentially attractive because it requires much less energy and may lead to much shorter cooling times, without the need for forced cooling, but it is more difficult to realize. A low-thermal-mass GC is less robust, and achieving adequate control is challenging. However, it should require much less energy. Low-thermal-mass GC appears to still be in its infancy. A low-thermal-mass unit as an accessory to a conventional (high-thermal-mass) GC system seems to be defeating the purpose.[43] Direct resistive heating of "Silco-steel" columns (metal columns with an internal silica coating as an alternative to fused silica) has yielded very promising results,[44] but it is probably still insufficiently robust for routine application.

2.7.2.1 Temperature Programming

The vast majority of GC analyses are carried out under programmed-temperature conditions, and in the vast majority of these, a single-segment linear temperature program is employed, in which the temperature varies linearly from an initial temperature T_{init} to a final temperature T_{final} during a time t_G at a rate γ_T that follows from

$$\gamma_T = \frac{T_{final} - T_{init}}{t_G} = \frac{\Delta T}{t_G} \qquad (2.17)$$

As in LC, the temperature program scales with the hold-up time (t_0) of the column. Comparable results are obtained if T_{init}, T_{final}, and t_G/t_0 are kept constant. However, the assumptions underlying this conclusion include constant pressure operation,[45] while constant flow operation is generally preferred for temperature-programmed GC. The elution times of analytes can be kept constant under temperature-programmed conditions if t_0 is kept constant. This is known as "retention-time locking".[45]

2.7.2.2 Spatial Temperature Gradients

Arguments in favour of a **negative temperature gradient** along the length of the column (*i.e.* a higher temperature towards the column inlet and a lower temperature towards the column outlet) have been around for decades. One argument is that the tail of the peak is always at a higher temperature than the front so that the peak is progressively compressed. However, the distance between peaks may also be reduced, which could have a negative impact on the resolution. Blumberg demonstrated that a negative temperature gradient (or "focussing" gradient) mainly served to compensate for aberrations due to imperfect injections, variations in the stationary-phase film thickness, *etc.*[46] With the current quality of columns and the various post-injection-focussing techniques available, this is likely to be less useful now than it was at the time. More importantly, a specific negative "static thermal gradient" may only improve the separation of a few analytes, akin to an isothermal separation.

Better results may be obtained by imposing a negative axial temperature gradient while (linearly) increasing the temperature at each point along the column. Simulations suggest that such a "dynamic thermal gradient" may improve the separation of a range of analytes.[47] It is predicted to yield narrower peaks (*i.e.* higher sensitivities), lower elution temperatures and higher resolution between peaks. However, the predicted improvements (gains in resolution up to 13%) are modest in comparison with the added complexity of the system.

It is especially daunting to realize a dynamic thermal gradient along a many-metre-long open-tubular GC column, but in recent years, an ingenious system has been described and commercialized, in which a longer column is wound around a cone, along the axis of which a thermal gradient is created.[48,49] This system allows fast and efficient analysis, but it is yet unclear whether this is due to the use of a relatively short and narrow column (10 m × 100 μm i.d.) or to the negative thermal gradient.

2.8 Column Characterization (A)

A convenient way to characterize GC stationary phases was pioneered by **Rohrschneider**[50] and later modified by **McReynolds**.[51] There is some injustice in the first name being largely forgotten by chromatographers, while the latter lives on. The Rohrschneider approach to stationary-phase characterization bears some resemblance to linear-free-energy models (see Module 3.8), although there is no straightforward relationship between a free-energy difference (ΔG) and the retention-index increment that is at the heart of the Rohrschneider model.[50] His assumption was that the retention-index increment ($\Delta I_{i,sp}$), *i.e.* the difference between the retention index of a polar (analyte) probe (denoted by i) on a given stationary phase ($I_{i,sp}$) and that of the same probe on a reference, non-polar stationary phase ($I_{i,squalane}$), both measured at 100 °C, was indicative of the polarity of the phase. The non-polar reference phase was chosen to be squalane, a branched alkane (full name: 2,6,10,15,19,23-hexa-methyltetracosan, $C_{30}H_{62}$). Rohrschneider then postulated a linear model with five terms, containing five coefficients that characterized the stationary phase (x_{sp}, y_{sp}, z_{sp}, u_{sp}, and s_{sp}) and five analyte parameters (a_i through e_i),

$$\Delta I_{i,sp} = I_{i,sp} - I_{i,squalane} = a_i x_{sp} + b_i y_{sp} + c_i z_{sp} + d_i u_{sp} + e_i s_{sp} \qquad (2.18)$$

Rohrschneider then selected five diverse analytes that were loosely connected with five types of analyte–stationary-phase interactions and assigned these arbitrary values for a_i through e_i (see Table 2.7). With the computers currently available, a set of five linear equations with five known (measured) values of $\Delta I_{i,sp}$ and five sets of known (postulated) analyte parameters a_i through e_i would be easy to solve for the five stationary-phase parameters x_{sp}, y_{sp}, z_{sp}, u_{sp}, and s_{sp}. However, Rohrschneider's choice to exclusively assign one term to each probe rendered such a calculation obsolete. The stationary-phase parameters simply follow from

$$x_{sp} = \frac{\Delta I_{benzene,sp}}{100} \tag{2.19a}$$

$$y_{sp} = \frac{\Delta I_{ethanol,sp}}{100} \tag{2.19b}$$

$$z_{sp} = \frac{\Delta I_{butanone,sp}}{100} \tag{2.19c}$$

$$u_{sp} = \frac{\Delta I_{nitromethane,sp}}{100} \tag{2.19d}$$

$$s_{sp} = \frac{\Delta I_{pyridine,sp}}{100} \tag{2.19e}$$

Characterization of a stationary phase just requires measuring the retention indices of the five probes, which involves measuring the retention times of the probes, a mixture of n-alkanes and a t_0 marker, such as methane, all at 100 °C. The retention indices on a squalane column are known (see Table 2.7), which is helpful because such columns are no longer in use. Squalane was typically used in packed columns – adsorbed on a solid carrier – and it would be difficult to coat it on the inner wall of a fused-silica open-tubular column. Squalane columns had low maximum operating temperatures of about 130 °C, showed high bleeding, and were prone to oxidation. They have been replaced by poly(dimethyl siloxane) columns, which exhibit a marginally higher polarity (*cf.* Table 2.2; the corresponding polarity index of squalane would be 0).

The impracticality of squalane columns is one of the reasons why the characterization of analytes using the Rohrschneider (or McReynolds) scheme is cumbersome and, as a result, not very meaningful. To characterize an analyte, its retention index must be determined on five

Table 2.7 Probes for the Rohrschneider[50] stationary-phase characterization protocol.

Rohrschneider (100 °C)	$I_{i,squalane}^{50}$	a_i	b_i	c_i	d_i	e_i
Benzene	649	100	0	0	0	0
Ethanol	384	0	100	0	0	0
2-Butanone	531	0	0	100	0	0
Nitromethane	457	0	0	0	100	0
Pyridine	695	0	0	0	0	100

previously characterized columns (so that their x_{sp}, y_{sp}, z_{sp}, u_{sp}, and s_{sp} values are known) and on squalane. An inconvenience that is minor in comparison is that a set of five linear equations with five unknowns (a_i through e_i) would need to be solved.

McReynolds has made some relevant and some not-so-relevant changes to the Rohrschneider scheme.[51] He has replaced some of the probes with some larger homologues. This is useful because the retention times for ethanol and nitromethane (and to a lesser extent 2-butanone) on non-polar columns are very low.

The McReynolds probes are shown in Table 2.8. Retention indices below 500 imply that analytes elute before *n*-pentane, which is the smallest *n*-alkane that can conveniently be stored and injected as a liquid. Butanol and nitropropane elute between *n*-pentane and *n*-heptane on non-polar columns, which is a more convenient range. The increase in temperature from 100 °C to 120 °C is in line with this somewhat higher range of *n*-alkanes. Extending the number of probes from five to ten has received limited interest. Only the first five McReynolds probes are being used. Changing the analyte parameters by a factor of 100 is irrelevant. It just causes McReynolds parameters to be 100 times larger than the corresponding Rohrschneider parameters. McReynolds did characterize an amazing number of stationary phases that were available at the time (1970) with the purpose of demonstrating the similarity of many of these to reduce the number of columns thought to be needed.[51] One common way to assign a single number to the polarity of a stationary phase is to take the average of the first five McReynolds coefficients. This is the **polarity indicator** listed in Table 2.2.

Table 2.8 Probes for the McReynolds stationary-phase characterization protocol.

McReynolds (120 °C)	$I_{i,squalane}$[51]	a_i	b_i	c_i	d_i	e_i	H	J	K	L	M
Benzene	653	1	0	0	0	0	0	0	0	0	0
Butanol	590	0	1	0	0	0	0	0	0	0	0
2-Pentanone	627	0	0	1	0	0	0	0	0	0	0
Nitropropane	652	0	0	0	1	0	0	0	0	0	0
Pyridine	699	0	0	0	0	1	0	0	0	0	0
2-Methyl-2-pentanol	690	0	0	0	0	0	1	0	0	0	0
1-Iodobutane	818	0	0	0	0	0	0	1	0	0	0
2-Octyne	841	0	0	0	0	0	0	0	1	0	0
1,4-Dioxane	654	0	0	0	0	0	0	0	0	1	0
cis-Hydrindane	1006	0	0	0	0	0	0	0	0	0	1

> See the website for references to lists of stationary-phase parameters: ass-ets.org

2.9 Correcting for Mobile-phase Compressibility (A)

So far, we have ignored the compressibility of the carrier gas in our description of dispersion (*e.g.* in the van Deemter and Golay plate-height equations; see Module 1.7). We used the following simple form of the **Golay equation** for open-tubular chromatography (eqn (1.131)):

$$h = \frac{2}{\nu} + f(k) \cdot \nu + g(k) \cdot \delta_f^2 \cdot \nu \tag{2.20}$$

No one else than J. Calvin Giddings derived already in 1960[52] that this equation can be corrected for the expansion of an ideal carrier gas as follows:

$$h = \left\{ \frac{2}{\nu_{out}} + f(k) \cdot \nu_{out} \right\} j_1 + g(k) \cdot \delta_f^2 \cdot \nu_{out} \cdot j_2 \tag{2.21}$$

where ν_{out} is the reduced linear velocity at the column outlet and j_1 and j_2 are the correction factors that depend on the pressure ratio across the column ($P_{ratio} = P_{inlet}/P_{outlet}$) as follows:

$$j_1 = \frac{9}{8} \frac{\left(P_{ratio}^4 - 1\right)\left(P_{ratio}^2 - 1\right)}{\left(P_{ratio}^3 - 1\right)^2} \tag{2.22}$$

and

$$j_2 = \frac{3}{2} \frac{\left(P_{ratio}^2 - 1\right)}{\left(P_{ratio}^3 - 1\right)} \tag{2.23}$$

The values of the two pressure correction factors are illustrated as a function of P_{ratio} in Figure 2.17.

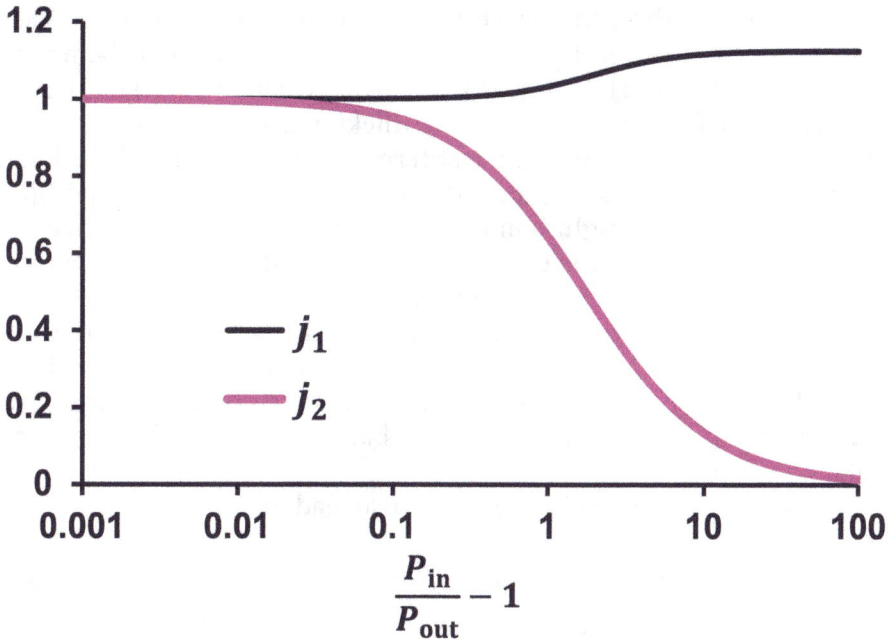

Figure 2.17 Pressure correction factors j_1 (eqn (2.22)) and j_2 (eqn (2.23)) as a function of the relative pressure drop across the column.

It is seen in Figure 2.17 that the j_1 pressure correction factor remains close to unity for a pressure drop up to about 100% (*i.e.* $P_{ratio} - 1 = 1$), after which it increases only slightly. At very high-pressure drops, j_1 increases by 12.5% (approaching the limiting value of 9/8), which implies that the first term in eqn (2.21) increases by 12.5% and the efficiency (plate count may decrease by such a percentage). In contrast, however, the j_2 pressure correction factor remains close to unity only up to a pressure drop of about 10% (*i.e.* $P_{ratio} - 1 = 0.1$), after which it starts to decrease rapidly. Therefore, the effect of the last (non-equilibrium, or C_s) term diminishes at high-pressure drops. As a result, at high-pressure drops, somewhat thicker stationary-phase films than the conventional δ_f 0.3 or $d_f < d_c/1000$ may be used without punishment in terms of efficiency loss. As a modified rule of thumb, the chromatographer should adhere to

$$\delta_f \leq 0.3/\sqrt{j_2} \tag{2.24}$$

The pressure drop in conventional (one-dimensional) open-tubular GC tends to be low (typically $P_{ratio} - 1 < 0.2$), but it may be much higher in GC-MS. This suggests that somewhat thicker stationary-phase films (higher reduced film thickness, δ_f) may be used in GC-MS while still keeping the last term in eqn (2.21) small. As long as the last ($g(k) \cdot \delta_f^2$ or C_s) term is much lower than the second ($f(k)$ or C_m) term on the right-hand side of eqn (2.20), we may conclude from Figure 2.17 that the effect of the pressure drop on the efficiency is typically small. At high-pressure drops, for example, in GC-MS or when using very narrow (100 µm or 50 µm i.d.) columns, the effect of the pressure drop is at worst a loss in efficiency of 12.5%, but likely less. This largely justifies why we have ignored the compressibility factors in our simplified treatment of band broadening in open-tubular chromatography.

Giddings' compressibility factors[52] also lead to

$$\bar{u} = u_{out} \cdot j_2 \tag{2.25}$$

and

$$t_0 = \frac{L}{\bar{u}} = \frac{L}{u_{out} j_2} \tag{2.26}$$

Box 2.3 Hero of analytical separation science: Karel Cramers. Photo courtesy of Prof. Hans-Gerd Janssen (Unilever, The Netherlands).

Karel Cramers (1935–2024, The Netherlands)
Karel Cramers built an amazing analytical group at the Technical University of Eindhoven (The Netherlands) that excelled in electromigration methods (Frans Everaerts) and Karel's own speciality: capillary gas chromatography. Cramers had a great understanding of the theoretical foundations of open-tubular chromatography, and he used this basis to keep pushing GC, SFC and LC forward. The limits of speed and resolution were continuously tested in Eindhoven. Karel was a great teacher of the fundamentals of capillary chromatography, and numerous separation scientists benefitted from courses he taught all over the world.

This implies that the residence time in the column is longer than suggested by the outlet velocity. Karel Cramers (Box 2.3) contributed significantly to the theory of GC.

2.10 Practical Tips and Troubleshooting for GC (M)

Troubleshooting is both a massive and a massively important subject. It may be the subject of an entire book. In the context of the present textbook, we will limit ourselves to some practical guidelines and some general advice for troubleshooting. A general advice is as follows: when confronted with an issue that you have not yet identified, it always helps to narrow down the possible causes.

2.10.1 Carrier Gases

We have seen in Module 2.3 that hydrogen is fundamentally the best carrier gas, with helium a good second choice. It is much more expensive than hydrogen, but it is also safer and, therefore, used most often. Using nitrogen results in much slower analyses (lower optimum velocity due to lower analyte diffusion coefficients) or lower efficiencies and higher pressure drops if the velocity is not reduced. It is important that the carrier gas is very pure to avoid degradation of stationary phases (*e.g.* due to traces of oxygen) or detector noise (*e.g.* due to traces of methane). Very pure gases are available commercially, and it is essential not to contaminate these before they reach the GC. Laboratory gas lines should ideally be welded and not plumbed between the gas source and the connection in the lab. When installing filters in line before the GC, this should be done prudently and correctly so as not to defeat the purpose. If the carrier gas and the supply lines are sufficiently clean, filters may not be needed.

2.10.2 GC Columns

There is such a thing as a default column for GC. The default stationary phase is a non-polar poly(dimethyl siloxane) or PDMS. However, for polar analytes or sample matrices, more polar columns are frequently required. In most cases, you can follow a good experience. Check the literature carefully, and unless you are an academic researcher, go for established (mainly polysiloxane) phases. The most sensible rule is to "Use non-polar (or next medium polar) columns if you can, use polar columns if you must".

A 250 µm i.d. column can be routinely used on most contemporary instruments, and 320 µm i.d. columns are fine. Rarely, a case can be made for using 530 µm i.d. columns. They offer a somewhat higher sample loadability but considerably less separation power. Such wide-bore columns used to be a stepping stone for chromatographers used to – and instruments designed for – packed-GC columns, but both are increasingly rare. On-column injection is traditionally thought to be more practical with slightly wider columns, but 320 µm i.d. and even 250 µm i.d. columns are considered practical with contemporary on-column injectors, using 235 µm outer diameter (o.d.) and 170 µm o.d. tapered needles, respectively. Some manufacturers still specify needle diameters in the "gauge" system (an old vocational system for specifying wire diameters), but this is totally unacceptable in this day and age.

2.10.3 Poor Peak Shapes

There are many reasons why peaks may not be as narrow or as symmetrical as you may reasonably expect. In most cases, you (fortunately) do not need to replace your column. If you just installed the column, it may be positioned imperfectly at the detector or – more likely – the injector side. There is a rather precise length of column that should be inserted in the device at the end of the column and the end ("cut") should be straight and clean.

When working with dirty samples, a (coated) guard column (similar to the analytical column, except much shorter) or an uncoated retention gap may be used. Even better is the use of a PTV injector, using which any unwanted (non-volatile) material can be contained in the liner. Dirt may build up in a liner (or in a guard column or pre-column), but these are much easier and cheaper to replace than the analytical column. Injection problems are a common cause of peak broadening or deformation.

2.10.4 Flow Control

The ability to measure gas flows is invaluable. Electronic flow meters are very attractive, but conventional bubble flow meters often still suffice. If you want to measure a split flow and a column flow (and possibly a septum-purge flow), you are looking at different magnitudes. Indirectly, a t_0 signal can be used as an indicator for the column flow. Methane (natural gas) – if available in the lab – or propane from a gas lighter, both injected using a gas syringe, are good t_0 markers (but injecting these together with significant amounts of air is not ideal for your column).

When using an EPC, you should check whether your settings are consistent with your observations. One common mistake is incorrect input data (especially column length and diameter). If these are incorrect, your system may work, but not how you think it does. Your pressure and flow (t_0) data should be consistent. When using an EPC, you may need to reset the zero-pressure point occasionally.

2.10.5 Bleeding Septa

With the exception of on-column injectors, which tend to be equipped with a syringe-guiding valve, and gas-injection valves, GC injectors tend to be equipped with a septum, which is an inert rubber disc (Figure 2.8, purple disc at the top) that can withstand the temperature of the injector without giving rise to significant bleed and can be pierced many times by a syringe before it starts to leak. Because zero bleed cannot be achieved, many modern injectors have a separate (small) flow of carrier gas ("septum purge") that flushes the surface of the septum and ends up at waste. Because a septum cannot be pierced indefinitely, it is typically very easy to replace (with the possible exception of some unfortunate auto-injector configurations). You should be able to replace the septum easily yourself and you should be able to diagnose a leaking septum, as it affects peak areas (recovery) and retention times (the latter especially in the case of constant flow operation).

> See the website for chromatograms and more troubleshooting tips: ass-ets.org

Acknowledgements

Prof. Hans-Gerd Janssen is acknowledged for his review of this chapter.

References

1. Z. Ji, R. E. Majors and E. J. Guthrie, *J. Chromatogr. A*, 1999, **842**, 115–142.
2. M. H. Abraham, C. F. Poole and S. K. Poole, *J. Chromatogr. A*, 1999, **842**, 79–114.
3. C. Liang, J.-Q. Qiao and H.-Z. Lian, *J. Chromatogr. A*, 2017, **1528**, 25–34.
4. S. F. Donovan and M. C. Pescatore, *J. Chromatogr. A*, 2002, 47–61.
5. L. Blomberg, *J. High Resolut. Chromatogr.*, 1982, **5**, 520–533.
6. M. Roth and J. Novák, *J. Chromatogr. A*, 1982, **234**, 337–345.
7. G. Betzenbichler, L. Huber, S. Kräh, M. L. K. Morkos, A. F. Siegle and O. Trapp, *Chirality*, 2022, **34**, 732–759.
8. Z. Witkiewicz, J. Oszczudłowski and M. Repelewicz, *J. Chromatogr. A*, 2005, **1062**, 155–174.
9. C. F. Poole and S. K. Poole, *J. Sep. Sci.*, 2011, **34**, 888–900.
10. N. Etxebarria, O. Zuloaga, M. Olivares, L. J. Bartolomé and P. Navarro, *J. Chromatogr. A*, 2009, **1216**, 1624–1629.
11. M. L. Lee and B. W. Wright, *J. Chromatogr. A*, 1980, **184**, 235–312.
12. J. Buijten, L. Blomberg, S. Hoffmann, K. Markides and T. Wännman, *J. Chromatogr. A*, 1984, **289**, 143–156.
13. R. D. Dandeneau and E. H. Zerenner, *J. High Resolut. Chromatogr.*, 1979, **2**, 351–356.
14. G. Boczkaj, A. Przyjazny and M. Kamiński, *Anal. Bioanal. Chem.*, 2011, **399**, 3253–3260.
15. S. R. Lipsky and M. L. Duffy, *J. High Resolut. Chromatogr.*, 1986, **9**, 376–382.
16. S. R. Lipsky and M. L. Duffy, *J. High Resolut. Chromatogr.*, 1986, **9**, 725–730.
17. Y. Takayama and T. Takeichi, *J. Chromatogr. A*, 1994, **685**, 61–78.
18. S. E. Stein, *J. Am. Soc. Mass Spectrom.*, 1999, **10**, 770–781.
19. E. N. Fuller, P. D. Schettler and J. C. Giddings, *Ind. Eng. Chem.*, 1966, **58**, 18–27.
20. J. Luong, R. Gras and R. Tymko, *J. Chromatogr. Sci.*, 2003, **41**, 550–559.
21. J. T. Scanlon and D. E. Willis, *J. Chromatogr. Sci.*, 1985, **23**, 333–340.
22. J. N. Driscoll, J. Ford, L. F. Jaramillo and E. T. Gruber, *J. Chromatogr. A*, 1978, **158**, 171–180.
23. L. S. Ettre and A. Zlatkis, in *Journal of Chromatography Library: 75 years of chromatography - a historical dialogue*, 1979, vol. **17**, pp. 21–30.
24. H. Jing and A. Amirav, *J. Chromatogr. A*, 1998, **805**, 177–215.
25. X. Yan, *J. Sep. Sci.*, 2006, **29**, 1931–1945.
26. U. A. T. Brinkman, *J. Chromatogr. A*, 2008, **1186**, 109–122.
27. D. E. Henry, A. Giorgetti, A. M. Haefner, P. R. Griffiths and D. F. Gurka, *Anal. Chem.*, 1987, **59**, 2356–2361.
28. G. T. Reedy, D. G. Ettinger, J. F. Schneider and S. Bourne, *Anal. Chem.*, 1985, **57**, 1602–1609.
29. S. Bourne, A. M. Haefner, K. L. Norton and P. R. Griffiths, *Anal. Chem.*, 1990, **62**, 2448–2452.
30. K. A. Schug, I. Sawicki, D. D. Carlton, H. Fan, H. M. McNair, J. P. Nimmo, P. Kroll, J. Smuts, P. Walsh and D. Harrison, *Anal. Chem.*, 2014, **86**, 8329–8335.
31. P. Walsh, M. Garbalena and K. A. Schug, *Anal. Chem.*, 2016, **88**, 11130–11138.
32. J. C. Holmes and F. A. Morrell, *Appl. Spectrosc.*, 1957, **11**, 86–87.
33. R. S. Gohlke, *Anal. Chem.*, 1959, **31**, 535–541.
34. Y. Niu, J. Liu, R. Yang, J. Zhang and B. Shao, *TrAC, Trends Anal. Chem.*, 2020, **132**, 116053.
35. J. Fang, H. Zhao, Y. Zhang, M. Lu and Z. Cai, *Trends Environ. Anal. Chem.*, 2020, **25**, e00076.
36. D. X. Li, L. Gan, A. Bronja and O. J. Schmitz, *Anal. Chim. Acta*, 2015, **891**, 43–61.

37. C. Bressan, J. F. Ayala-Cabrera, F. J. Santos, S. Cuadras, L. Garrostas, N. Monfort, É. Alechaga, E. Moyano and R. Ventura, *Anal. Bioanal. Chem.*, 2020, **412**, 7837–7850.
38. D. B. Robb, T. R. Covey and A. P. Bruins, *Anal. Chem.*, 2000, **72**, 3653–3659.
39. I. Špánik and A. Machyňáková, *J. Sep. Sci.*, 2018, **41**, 163–179.
40. W. A. Traag, W. Kulik and L. G. M. T. Tuinstra, *J. Chromatogr. A*, 1992, **595**, 289–299.
41. A. Fontana, I. Rodríguez and R. Cela, *J. Chromatogr. A*, 2017, **1515**, 30–36.
42. E. N. Fuller, K. Ensley and J. C. Giddings, *J. Phys. Chem.*, 1969, **73**, 3679–3685.
43. J. Luong, R. Gras, R. Mustacich and H. Cortes, *J. Chromatogr. Sci.*, 2006, **44**, 253–261.
44. V. R. Reid, A. D. Mcbrady and R. E. Synovec, *J. Chromatogr. A*, 2007, **1148**, 236–243.
45. L. M. Blumberg and M. S. Klee, *Proceedings Ninet. Int. Symp. Capill. Chromatogr. Electrophor.*, 1992, **405**, 17.
46. L. M. Blumberg, *Chromatographia*, 1994, **39**, 719–728.
47. S. Avila, H. D. Tolley, B. D. Iverson, A. R. Hawkins, S. L. Johnson and M. L. Lee, *Anal. Chem.*, 2021, **93**, 11785–11791.
48. J. Leppert, P. J. Müller, M. D. Chopra, L. M. Blumberg and P. Boeker, *J. Chromatogr. A*, 2020, 460985.
49. J. Leppert, L. M. Blumberg, M. Wüst and P. Boeker, *J. Chromatogr. A*, 2021, 461943.
50. L. Rohrschneider, *J. Chromatogr. A*, 1966, **22**, 6–22.
51. W. O. McReynolds, *J. Chromatogr. Sci.*, 1970, **8**, 685–691.
52. J. C. Giddings, S. L. Seager, L. R. Stucki and G. H. Stewart, *Anal. Chem.*, 1960, **32**, 867–870.

3 Liquid Chromatography

There is much to discuss in what is justifiably the largest chapter of this book. The basics of LC are reviewed, followed by more detailed discussions on stationary phases and instrumentation. Gradient elution is a very important technique, the essence of which is described early in the chapter, with an advanced closer look reserved for the end of the chapter. The most important LC techniques, such as reversed-phase LC, normal-phase LC, including hydrophilic-interaction liquid chromatography (HILIC), and separation methods for ions, receive ample attention in three dedicated modules. Retention models for these techniques are discussed coherently in a dedicated module. Detection and hyphenated techniques, such as LC-MS, are discussed in some detail, followed by a brief overview of troubleshooting and advanced modules on preparative LC, chiral recognition, and protein separations. Finally, kinetic plots are discussed as a good tool for optimizing LC columns and conditions (Figure 3.1).

3.1 Basics of Liquid Chromatography (B)

3.1.1 Introduction

In this chapter, we will discuss column liquid chromatography in detail (see Section 1.1.2 for a brief overview of thin-layer chromatography). Gas chromatography (Chapter 2) is a fantastic analytical technique, but it is only applicable to analytes that are sufficiently volatile and stable to be eluted intact within a reasonable time at the maximum operating temperature of the GC instrument and column. This implies that only a small fraction of all known chemical compounds are amenable to GC. The vast majority of compounds

Analytical Separation Science
By Bob W. J. Pirok and Peter J. Schoenmakers
© Bob W. J. Pirok and Peter J. Schoenmakers 2025
Published by the Royal Society of Chemistry, www.rsc.org

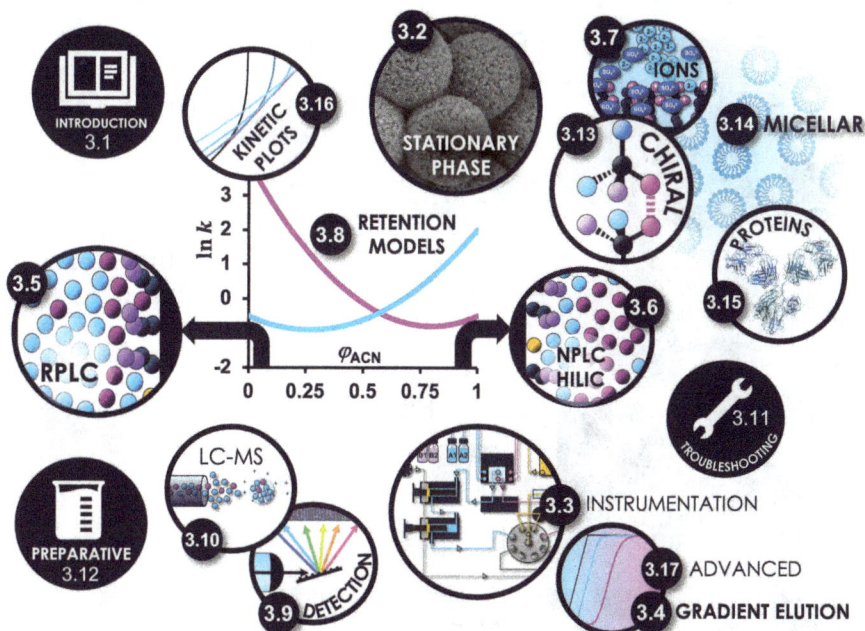

Figure 3.1 Graphical overview of the modules in this chapter.

would need to be either chemically modified ("derivatized" or "pyrolysed", see Sections 8.3.4 and 8.3.5) or analysed with a different separation technique. For analytes that are not sufficiently volatile, a dense mobile phase is required that can "extract" the analytes from the stationary phase. By far, the most common technique for achieving this is liquid chromatography, where the mobile phase is a liquid. Other possible techniques will be discussed in Module 4.5 (field-flow fractionation), Chapter 5 (electrophoresis), and Chapter 6 (supercritical-fluid chromatography).

3.1.2   Basic Instrumentation and Operation

A basic LC instrument is schematically illustrated in Figure 3.2. Similar to GC, an LC instrument is equipped with an injector, to introduce a sample mixture, and one or more detectors, to monitor the effluent containing the separated analytes. Injection is a simpler process in LC than in GC because the phase change that is usually required in GC (from a liquid sample into a GC mobile phase) is not necessary. However, detection in LC (Module 3.9) is not as straightforward as in GC. This is due to the liquid mobile phase

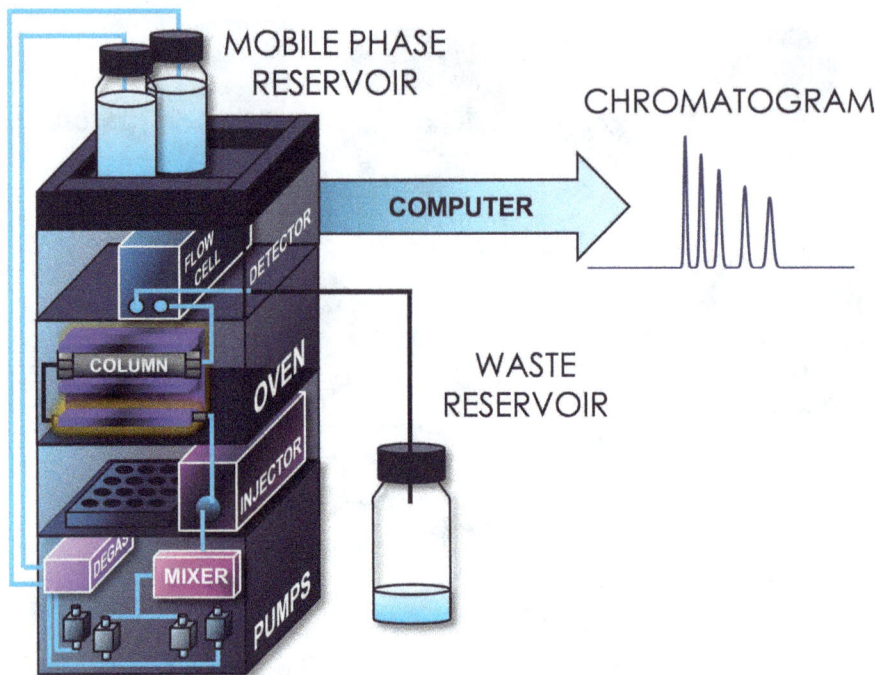

Figure 3.2 Schematic illustration of a basic LC setup. Contemporary LC systems are modular, and the different modules are stacked on top of each other.

being much denser and less inert than a gaseous mobile phase. The nearly ideal flame-ionization GC detector (FID, see Section 2.5.1) is incompatible with liquid chromatography, as the mobile phase swamps the signal of the analytes. Detection in LC usually requires a selective response to analyte molecules against a baseline of mobile phase (*e.g.* ultraviolet or fluorescence spectroscopy) or the removal of the mobile phase molecules (*e.g.* evaporative light-scattering or mass-spectrometric detection). The many different detection methods in LC derive from a continuous effort to achieve sufficient sensitivity (combined with sufficiently low noise), universality, and linearity. In practice, not all analytes feature properties that allow selective detection (*e.g.* chromophore moieties for UV detection), and not all mobile phases allow easy removal (*e.g.* non-volatile buffers). The presence of a dense mobile phase also renders the hyphenation of LC with MS (*i.e.* LC-MS, Section 3.10.1) less straightforward than that of GC with MS (*i.e.* GC-MS, Module 2.6).

The lower diffusion coefficients of liquids and high-molecular-weight analytes also introduce new challenges in limiting dispersion. Very high plate counts (*e.g.* 100 000) are more difficult to achieve in LC than in GC. The low diffusion coefficients in LC make open-tubular columns an unrealistic option (see Section 1.7.8). Instead, packed columns are the norm in LC. The lower permeability of such columns, as well as the high viscosity of the mobile phase, limits the lengths of the columns that can be used, and the performance of LC separations is ultimately limited by the available pressure. This, in itself, justifies the continued use of the abbreviation HPLC, which denotes high-pressure liquid chromatography.

Despite all the above challenges, LC has become an enormously successful analytical separation method. In large part, this is due to its unparalleled versatility. While GC only offers temperature and stationary-phase structure and polarity as options to manipulate retention and selectivity, LC offers a diverse range of retention mechanisms that can be exploited by chromatographers to resolve complex mixtures of non-volatile analytes.

In the most basic LC separation system (Figure 3.2), a pump is used to transport a pre-mixed mobile phase through the injector, column and detector. Common injectors feature a valve-based system with a sample loop that is (temporarily) placed in line with the column so that the pump can displace the injection plug towards the column. By pumping a viscous liquid instead of a gas, a larger force is required to transport the mobile phase through the column. The **backpressure** experienced at the head of the column is thus high. The pressure decreases (approximately) linearly along the length of the column, down to atmospheric pressure at the detector outlet. The difference between the pressures at the column inlet and its outlet is often referred to as the **pressure drop.**

In early LC experiments, glass tubes were used as columns and gravitational forces were relied upon to transport the mobile phase through the column. It was long clear that the key to faster and more efficient separations was the use of smaller particles and that this required higher pressures. However, several hurdles had to be taken, including the development of safe high-pressure equipment, the preparation of small particles (with a sufficiently narrow size distribution), and finding a way to pack such particles in columns. At the time (the 1960s), these were tough challenges, but they were effectively overcome by several groups of people. In the United States, Csaba Horváth at Yale University (New Haven, CT) and Jack Kirkland at DuPont (Wilmington, DE) were the first to realize HPLC experiments. In Europe, Jozef Huber (Box 3.1) and Johan Kraak at

Box 3.1 Hero of analytical separation science: J. F. K. Huber. Photo courtesy of the University of Amsterdam.

J. F. K. (Jozef) Huber (1925–2000, Austria)
J. F. K. (Jozef) Huber was born in Salzburg (Austria). He started his academic career in Innsbruck (Austria, 1958–1960), followed by several years (1960–1963) at the TU Eindhoven (The Netherlands). He then became a full professor at the University of Amsterdam before returning to Austria in 1974 as a professor at the University of Vienna. Jozef Huber contributed massively to the theory and practice of chromatography. He pioneered heart-cut two-dimensional LC. Together with Johan Kraak, he was the first to realize HPLC in Europe in the 1960s.

the University of Amsterdam (The Netherlands) performed genuine HPLC experiments. These successes and the great need for fast, efficient separations of non-volatile analytes, especially from the pharmaceutical industry, led to rapid developments in technology and methodology. Within ten years, reliable instrumentation and good columns became available from several manufacturers. New and better stationary phases with smaller, spherical particles were developed, and gradient elution (Module 3.4) became commonplace. It did not take long for HPLC to become the number one analytical technique in terms of market size and the number of scientific publications. As most samples in practice are complex mixtures containing non-volatile analytes or matrices, the development of LC has never stopped.

3.1.3 Retention in LC

We have seen in Section 1.6.1 that retention in LC depends on the interactions of the analyte with the mobile phase (expressed as the activity coefficient at infinite dilution, $\gamma_{i,\mathrm{m}}^{\infty}$) and with the stationary phase ($\gamma_{i,\mathrm{s}}^{\infty}$), culminating in a distribution coefficient ($K_{\mathrm{d},i}^{x,\infty}$) and ultimately a **retention factor** (k_i), which can be expressed as

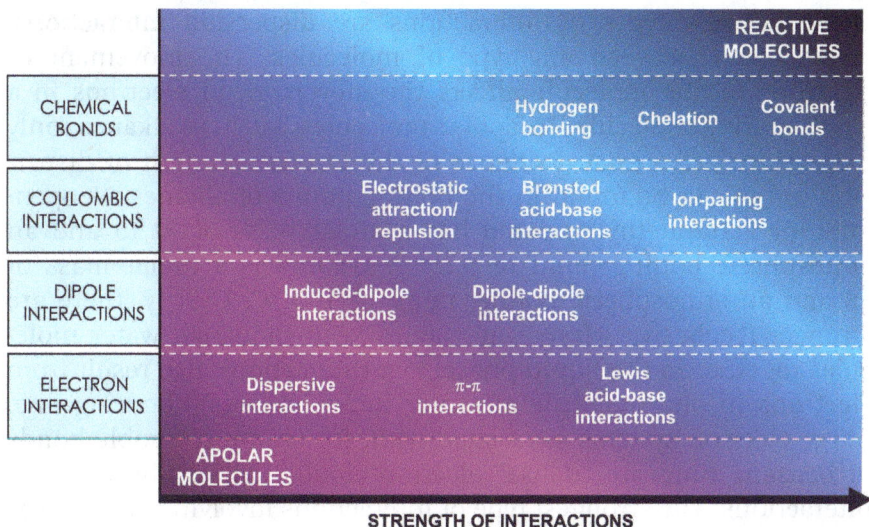

Figure 3.3 Overview of the types and strengths of interaction between molecules and ions encountered in LC.

$$k_i = K_{d,i}^{x,\infty} \frac{n_s}{n_m} = \frac{\gamma_{i,m}^{\infty}}{\gamma_{i,s}^{\infty}} \frac{n_s}{n_m} \tag{3.1}$$

where n_s and n_m are the numbers of moles of the respective phases in the column. It is important to realize that the mobile phase and any additives also interact with the stationary phase. The nature and the composition of the stationary phase often change if the mobile-phase composition is altered. This may, for example, take the form of (ion) exchange or competitive adsorption on active sites or a dynamically generated stationary layer. Thus, the stationary phase and the mobile phase are interdependent. We will discuss this in more detail for different forms of LC in later modules in this chapter (especially Modules 3.5, 3.6, and 3.7).

Figure 3.3 provides an overview of the types of **interactions** between molecules, ions, or ligands encountered in liquid chromatography. The colour shading is a rough indication of the "polarity" of molecules, with apolar molecules, such as alkanes, shown in pink (bottom-left corner) and polar molecules or ions in blue. At the extreme end, molecules that react, rather than interact, are located in the top right corner (dark blue). The strength of the interaction roughly increases from left to right.

The weakest types of interactions are **dispersion interactions**, which occur between any type of molecules. The movement of electrons in one molecule affects the movement of electrons in a neighbouring molecule. In apolar molecules, such as alkanes, only dispersion forces exist. Evidence of the limited strength of dispersion forces can be found in the boiling points of alkanes. Neopentane (or 2,2-dimethylpropane) has a molar mass of 72.15 and an atmospheric boiling point of 9.5 °C. Water has a molar mass of 18 and an atmospheric boiling point of 100 °C. Clearly, there are stronger ("cohesive") forces at play to keep the small water molecules together in the liquid phase. π–π Interactions also result from electrons in one molecule affecting those in a neighbouring one, but π–electron systems associated with, for example, double bonds or benzene rings are "dense" electron clouds that exhibit stronger interactions. The strongest type of interactions involving electrons is Lewis acid–base interactions, in which a Lewis base (a molecule with a "free" pair of electrons that is not involved in chemical bonding) interacts with a Lewis acid (a molecule with an incompletely filled outer shell). A typical example of a Lewis base (electron donor) is ammonia (NH_3), whereas boron trifluoride (BF_3), a molecule that is often used as a catalyst to form fatty acid methyl esters from triglycerides with methanol prior to their analysis, is a typical example of a Lewis acid (electron acceptor).

A second category of interactions involves **permanent dipoles**, *i.e.* molecules in which the electrons are distributed asymmetrically. In dipole-induction interactions, a molecule with a permanent dipole moment induces a temporary dipole in a neighbouring molecule, which itself does not need to have a permanent dipole moment. Stronger, so-called dipole–dipole, interactions occur when two molecules with permanent dipole moments align.

Coulombic or **electrostatic interactions** occur between ions. These are strong interactions that can occur across longer ranges than the other interactions shown in Figure 3.3, such as between a negative surface of a packing material and analyte ions in solution. When oppositely charged ions approach each other more closely, they may engage in ion-exchange or ion-pairing interactions (see Module 3.7). The transfer of a proton from a conventional (Brønsted–Lowry) acid, such as acetic acid, to a conventional base, such as triethylamine, can be portrayed as a form of coulombic interaction.

Covalent chemical bonds are the ultimate results of "interaction" or, in that case, "reaction" between molecules. Hydrogen bonding and the formation of chelates do not lead to the formation of covalent bonds and are sometimes explicitly identified as "non-convalent" bonds. However, they are very strong interactions that play major roles in LC. For example, hydrogen bonds are of crucial importance in all separations that involve water as a mobile-phase component, including reversed-phase liquid chromatography (RPLC, see Module 3.5) and HILIC (Section 3.6.2). Chelation plays a crucial role in the separation of amino-acid enantiomers (see Module 3.13).

Not all effects can be captured in Figure 3.3. For example, steric effects may amplify or diminish other interactions.

3.1.4 Determining the Hold-up Volume in Liquid Chromatography

Eqn (3.1) shows the retention factor (k_i), which can be determined from an isocratic measurement of the retention time $(t_{R,i})$ of analyte i and the hold-up time t_0, based on (eqn (1.13)), *i.e.* $k_i = (t_{R,i} - t_0)/t_0$. This implies that an accurate value of t_0 is needed to precisely determine k_i or the selectivity $(\alpha_{j,i} = k_j/k_i)$, to study relationships of retention *vs.* temperature or retention *vs.* competition, and to link LC retention with thermodynamic properties (eqn (3.1)). Unfortunately, determining accurate values for t_0 is a significant challenge in LC. The main cause of this problem is the diffuse boundary between the stationary and the mobile phase. The strong mobile-phase component tends to adsorb on or absorb in the stationary phase, causing an excess concentration of strong solvent near the surface. It is hard to determine what should be considered the mobile or stationary phase. In addition, the small pores in LC packing materials may be more or less accessible for different analytes and at different mobile-phase compositions. However, working with such a variable t_0 value is very complex in LC, as it would dramatically complicate the description of gradient elution (Module 3.4) and retention models (Module 3.8).

No generally accepted method exists for determining t_0. It is easy to verify that a sensible value is obtained. The unretained volume $(V_0 = Ft_0$, where F is the volumetric flow rate) should be in the range of 60% to 80% of the empty column volume $(V_0 = \pi d_c^2 L/4$, where L is the length and d_c is the internal diameter of the column). Table 3.1 provides an overview of suggested methods with associated comments. A good starting point for interested readers are the review articles in ref. 1 and 4.

Table 3.1 Overview of suggested methods for determining t_0 in liquid chromatography.

Method	Description	Comments
"Unretained" compound	Measuring the elution time of a supposedly unretained compound (*e.g.* acetone or uracil for RPLC[a] or dodecylbenzene or nonadecane-2-one for HILIC[1,b])	Depending on the phase system (mobile and stationary phase), such compounds may actually be retained
Minor disturbance	Injecting a solvent mixture with a slightly different composition from the mobile phase or injecting a minute amount of one of the mobile-phase components	Yields complex, hard-to-interpret signals that are a combination of analyte peaks and "system peaks" (or "vacancy peaks")
Deuterated solvent	Measuring the elution time of a deuterated analogue of a mobile-phase component	Different components may elute after (for the preferentially absorbed/adsorbed component) or before t_0
ACN method[c]	Measuring the elution time of deuterated acetonitrile (CD_3CN) with a pure acetonitrile mobile phase[2]	Yields the total volume of mobile phase in the column
Salt method	Measuring the elution time of a salt (*e.g.* UV-active KBr)[3]	Injection of small volume at low ionic strength yields the exclusion volume (volume outside the particles); high injection concentration and/or high ionic strength causes the elution volume to approach t_0
Homologues	Linearize plot of ln k *vs.* the number of carbon atoms in a homologous series by varying t_0	Resulting t_0 value is found to vary with mobile-phase composition[3]
Pycnometry	Measure the difference in weight (Δw) of the column when fully equilibrated with two pure solvents with different densities ($\Delta \rho$)	Yields the total volume of mobile phase in the column from $V_0 = |\Delta w| / |\Delta \rho|$; stationary phase may behave ("wet") differently in different solvents

[a]Reversed-phase liquid chromatography (see Module 3.5).
[b]Hydrophilic-interaction liquid chromatography (see Section 3.6.2).
[c]Applicable to both RPLC and HILIC.

3.2 Stationary-phase Materials (B)

3.2.1 Column Packing

The requirements for (ideal) packing materials for high-pressure (and ultra-high-pressure) LC are summarized in Table 3.2. The vast majority of LC columns contain a **packed bed** of particles with diameters between 1.5 and 10 μm. As explained in Chapter 1, the smallest particles (smaller than 2 μm) allow the fastest analyses, but they require high pressures, and the maximum attainable number of plates is limited (see Section 3.3.2). The somewhat larger particles require longer analysis times for a similar efficiency, but they allow longer columns, so that higher numbers of plates can be obtained for the separation of very complex samples, such as protein digests (see also Module 3.16).

Developing the best possible packing particles is one challenge for optimizing LC separations. Another significant challenge is to pack these particles into a column effectively. Because the particles are very small, high pressures are required during the analysis, but also to pack the particles tightly in a column. Particles may aggregate and such interactions then need to be broken. The common packing procedure is based on preparing a high-concentration "**slurry**", which is a suspension of stationary-phase particles. The end of the column is closed off using a frit, which is permeable for mobile-phase molecules but not for packing particles. The **packing process** (Figure 3.4) is in essence a filtration of the slurry, ideally under constant-flow (rather than constant-pressure) conditions.[5] After the bed is formed, it is consolidated at a pressure two or three times higher than that expected during the use of the column.[6] The inner wall of a steel column should be polished to reach a mirror finish. LC is such a high-performance technique that even the quality of the steel will be reflected in the eventual efficiency and peak shape.

The slurry should be as homogeneous as possible, *i.e.* there should be no aggregation or agglomeration of particles (except when packing capillary columns, where this may be advantageous).[6] Preparing a homogeneous slurry is most commonly achieved using an ultrasonic bath. A narrow **particle-size distribution** (no particles that are much smaller than the mean, known as "fines", or larger particles present) is as important in packing the column as it is during its use for LC separations. A slurry solvent with a density similar to that of the particles has been advocated to promote slurry stability. Other scientists have promoted high-viscosity solvents, as

Table 3.2 Summary of the requirements for ideal packing materials in LC.

Property	Comments
Micrometre-sized particles ($1 \geq d_\mathrm{p} \geq 10$ µm)	Particles of 3 or 5 µm are most commonly used in HPLC; sub-2 µm particles are used in UHPLC (see Section 3.3.2); particles larger than 10 µm are sometimes used in size-exclusion chromatography (see Module 4.2) and preparative LC (see Module 3.12)
Spherical particles	Spherical particles have been found to provide a higher efficiency (lower reduced plate height) and a higher column permeability (lower pressure drop) than irregular particles
High surface area	A high surface area (usually >100 m² g⁻¹) ensures sufficient retention and sample capacity (loadability)
Narrow particle-size distribution (particle SD)	A narrow particle-size distribution is essential for preparing ("packing") efficient LC columns to obtain high efficiencies and high permeabilities (low pressure drops).
Fully porous or superficially porous	Fully porous particles are still most commonly used; superficially porous (core–shell) particles show higher efficiencies (lower reduced plate heights) than fully porous particles; this is mainly attributed to a narrower particle-size distribution; superficially porous particles mostly have diameters ($2 \geq d_\mathrm{p} \geq 3$ µm), but sub-2 µm superficially porous particles have emerged
Well-defined pore-size distribution	Small pores (≤100 Å) correspond to the highest surface area and may result in lower efficiency and slower column equilibration for high-molecular-weight analytes; the latter typically require wide-pore (≥200 Å) columns; in size-exclusion chromatography (see Module 4.2), the pore-size distribution determines the selectivity; it is typically not narrow
Homogeneous	Heterogeneous surfaces lead to non-linear adsorption isotherms and asymmetrical peaks (see Module 1.8)
Stable	Ideally, particles and their surfaces are not irreversibly affected by changes in mobile-phase composition, temperature, pH, *etc.*; particles should ideally not swell, as this would greatly reduce the column permeability
Chemically inert	The particles should not engage in any chemical reactions and show no catalytic activity
Adaptable	If the surface can be chemically modified with ligands or coated with a stable stationary-phase film, this increases the applicability of a packing material
High mechanical strength	Particles should withstand shear caused by high flow rates during analyses and (slurry) packing; the packed bed should allow a high pressure drop and should be robust to (sudden) changes in pressure or flow rate

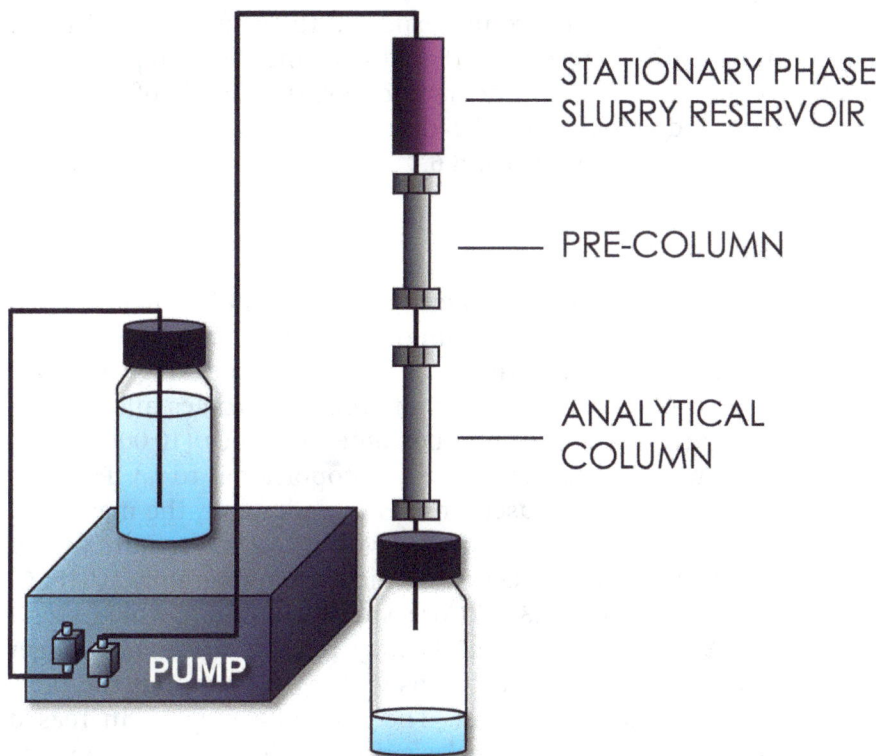

STATIONARY PHASE
SLURRY RESERVOIR

PRE-COLUMN

ANALYTICAL
COLUMN

PUMP

Figure 3.4 Schematic illustration of a setup used to pack LC columns. Pumps used for packing LC columns should allow high pressures (up to 200 MPa). Pneumatic pumps are frequently used for this purpose.

these would exert a greater drag force on the particles. Packing at very high pressure, packing columns vertically (from the bottom), and providing pressure pulses during packing have all been advocated.

Packing under supercritical conditions (see Chapter 6) has been promoted, as the lower viscosity of high-density carbon dioxide-based solvents allows using very high packing flow rates at moderate pressures. Electro-osmotic flow (EOF, see Chapter 5) may be used, which relies on a high voltage rather than a high pressure to push the slurry into the column. In addition, centripetal forces have been used to pack HPLC columns. For the latter purpose, the rotating parts of a centrifuge were modified to rotate within the confines of a tractor tyre, allowing a number of capillary columns to be packed simultaneously.[7] For an overview of packing techniques, see ref. 6.

Column manufacturers are not expected to tread on any of these adventurous paths. They have the best – and most repeatable – slurry-packing procedures. Understandably, the details of these are proprietary. A detailed scientific discussion on the slurry-packing process can be found in ref. 5 and 6.

3.2.2 Silica

Silica (SiO_2) is the material that best meets the set of demands listed in Table 3.2. It is the most commonly used packing material by far. The important characteristics of silica-based packing materials are listed in Table 3.3. It is seen that smaller particles allow much shorter columns to be used for the same efficiency (10 000 plates) while providing much faster analyses (proportional to t_0). Particles with $3 < d_p < 5$ μm can be used on HPLC equipment, the maximum pressure of which is 40 MPa. Particles with $d_p < 2$ μm require higher pressures and ultra-high-pressure liquid chromatography (UHPLC) equipment (see Section 3.3.2). Core–shell columns (particle size 2 μm $< d_p <$ 3 μm; Section 3.2.2.1) may require somewhat longer analysis times than UHPLC columns with $d_p < 2$ μm, but the pressure drop is considerably lower. The flow rate may be increased above $u = 2u_{opt}$ (reducing the analysis time) within the limitations of conventional HPLC equipment. For a more detailed discussion on column performance, see Module 3.16 (kinetic plots).

Silica particles are usually made from sodium silicate or tetra-ethoxy silane in a **sol–gel process**. The latter reagent may yield a silica with a somewhat better pH stability, as the presence of residual sodium increases the rate of base-induced hydrolysis. Other metal impurities (Fe and Ni) contribute to peak broadening and tailing, emphasizing the need for **high-purity silica** for application in LC. A distinction used to be made between (relatively) low-purity "Type-A" silica and high-purity "Type-B" silica, but such a classification has become largely obsolete, since all mainstream silica materials for HPLC are now of high purity. If the pH of the solution is decreased, the monomers start polymerizing, forming a colloidal "sol".

Depending on various conditions, including the rate of acidification, ionic strength, and presence of an organic solvent, a colloidal solution of spherical droplets is formed, ideally with a desired diameter and a narrow droplet-size distribution. As the polymerization continues, a cross-linked "gel" starts forming in the droplets. This is the reason why **"silica gel"** is still a common name for

Table 3.3 Typical characteristics of silica-based packing materials for LC (reduced parameters defined as $h = H/d_p$ and $v_0 = u_0 d_p/D_m$).

Designation	Particle size (μm)	Pore size (Å)	Surface area (m² g⁻¹)	Flow resistance factor (ψ)[a]	Optimum reduced velocity ($\nu_{0,opt}$)	Minimum reduced plate height (h_{min})	Column length (mm for 10 000 plates)	Optimal linear velocity ($u_{0,opt}$; mm s⁻¹)[b]	Hold-up time (t_0) at $u = 2u_{opt}$; (s)	Pressure drop (MPa)[c]
HPLC particles (fully porous)	10	60–1000	100–400	800	10	2	200	1.0	100	3
	5						100	2.0	25	13
	3						60	3.3	9.0	36
UHPLC particles (fully porous)	1.7	90–300	80–400	800	10	2	34	5.9	2.9	111
Core-shell particles (superficially porous)	2.7	80–500	100–200	800	7	1.7	46	2.6	8.9	16
	1.7					5	1.4	24	2.9	4.0	24

[a]See Section 1.6.3.
[b]Using $D_m = 10^{-9}$ m² s⁻¹.
[c]Using the Darcy equation (eqn (1.69)) with a viscosity (η) of 10^{-3} Pa s.

the final particles. Drying the particles hardens the silica and – under appropriate conditions – gives rise to pore formation. More recently, two-step processes have been developed, in which first a water-insoluble siloxane polymer is formed, which is then emulsified to form spherical silica particles in a second step, with porogens added to control the porosity of the resulting particles.

In another approach, microparticles are created from aggregates of very small silica particles within droplets of polymerizable organic materials. After the particles are bound together, the polymer is burned off and the microparticles are sintered. The diameter of the pores in the resulting microparticles can be controlled by the size of the original nanoparticles.

Narrow-pore (60–120 Å) particles are typically applied for the LC separation of low-molecular-weight analytes, whereas materials with larger pores (200–1000 Å) are preferred for separations of macromolecules. The analyte molecules need to penetrate the pores in order to have a sufficiently large stationary-phase surface available for interaction. In size-exclusion chromatography (SEC), the pore-size distribution plays a different role. The exclusion of molecules of different sizes from the pores to different extents determines SEC selectivity. This will be discussed in Module 4.2. It may safely be assumed (and experimentally confirmed by comparing LC with capillary electro-chromatography; see Chapter 5) that there is no flow through the particles, even if pores of up to 500 Å extend throughout. This follows from Darcy's law (eqn (1.69)) and the realization that any pores that extend through the particle are much narrower than the interstitial channels and are unlikely to align with the direction of the pressure gradient (*i.e.* with the column axis). The idea of "perfusion particles" is an unlikely proposition from earlier days of HPLC.

The chemical process of preparing silica particles is intricate and involves many parameters. Manufacturers of silica particles and LC packing materials have each perfected their own proprietary recipes. Chromatographers have a habit of complaining that different batches of silica behave slightly differently, but given the complexity of the synthesis and the extreme effect of even marginal changes in surface interactions on selectivity in LC (see Section 1.6.1.1), it is actually an enormous achievement that nearly identical batches and chromatographic columns can be produced.

The resulting silica gel is an amorphous material, the surface of which is hydrated to yield **silanol** ($\equiv$Si–OH) groups. Silanol groups can be isolated, vicinal (*i.e.* two silanol groups on adjacent silicon atoms), or geminal (*i.e.* two silanol groups on the same silicon atom). Not only the number of silanols (*i.e.* chemically bonded water) but also the amount of physically adsorbed water considerably affects the chromatographic and reactive properties of the silica material. Extensive heat treatments (*e.g.* twelve hours at 200 °C under vacuum) can be used to remove the physically absorbed water. At higher temperatures, silanols start disappearing, but very high temperatures (exceeding 1000 °C) are needed to fully dehydroxylate the surface, turning it hydrophobic and unsuitable for LC. Dry surfaces without physically adsorbed water may be needed to perform reactions with silanols (see below) or to determine the number of silanol groups, but such a dry surface is not stable under LC conditions, as the surface will collect traces of water from the mobile phase or from samples. The presence of surface silanols turns the silica material acidic. The pK_a of silica is said to be 6.8 ± 0.5,[8] but the transition from protonated $\equiv$SiOH groups at high pH to deprotonated negatively charged $\equiv$SiO$^-$ groups occurs more gradually than is the case for a pure acid, due to the variety of silanols and their surroundings. Silica or, more precisely, (residual) silanol groups at the surface interact strongly with basic analytes resulting in the so-called **secondary interactions** in *e.g.* RPLC. At a pH above about 5, silica exhibits cation-exchange properties. Metal ions, such as Fe^{3+}, present in trace amounts in the eluent or leached from the instrumentation, will exchange with protons to form anion-exchange sites. At lower pH, silanol groups may also be protonated to $\equiv$Si–OH$_2^+$ groups. The iso-electric point of silica is around 2.5.[8]

The number of silanols at the surface can be determined by titration, for example, with methyl lithium (measuring the amount of methane released using GC) or with trimethylchlorosilane (TMCS, measuring the amount of HCl released). Both methods provide a total number of silanol groups, along with physically absorbed water. Both methods also measure only accessible silanol groups. The size of, especially, the TMCS molecule implies that not all silanols can react. Spectroscopic methods allow a distinction between different groups at the dried surface. Fourier-transform-infrared (FTIR) spectroscopy can distinguish between isolated silanols, internal silanols, and vicinal hydrogen-bonded silanols (bands at 3747, 3680 and 3535 cm^{-1}, respectively). Solid-state nuclear magnetic resonance (NMR) spectroscopy of the ^{29}Si nucleus can distinguish between geminal silanols, isolated silanols, and silicon atoms that are connected only with siloxane (Si–O–Si) bridges (signals at 91, 100 and 109 ppm, respectively).

The ability of silica to **absorb** significant amounts of water is used in HILIC (see Section 3.6.2), a technique in which a dynamically generated aqueous stationary phase is combined with a moderately polar mobile phase, often more than 90% of acetonitrile (ACN), for the separation of highly polar analytes, such as carbohydrates (sugars). Silica can also be used in combination with low-polarity mobile phases in traditional normal-phase liquid chromatography (NPLC; see Section 3.6.1), a technique that is more suitable for analytes of low to medium polarity. Silica can also be chemically modified to yield a wide range of additional LC materials (see Section 3.2.2.2).

Silica dissolves very slowly under mild LC conditions, but its dissolution rate increases rapidly with increasing temperature and, especially, with increasing pH (above 7). If the surface is well shielded by the stationary phase (see Section 3.2.2.2), a higher maximum pH may be specified. One function of a guard column (installed between the injector and the column) is to protect the column from dirty samples. Another function of the guard column can be to saturate the mobile phase with silicates. In this case, the guard column may also be installed before the injector, where it does not contribute to the dead volume (V_0) or the extra-column dispersion (but does contribute to the dwell volume; see Module 3.4).

3.2.2.1 *Superficially Porous Particles*

Most particles used are still fully porous, but **superficially porous** or **core–shell particles** have proven very successful in recent years. The idea of superficially porous particles is not new. LC pioneers Jack Kirkland (Box 3.2) and Csaba Horváth (Box 3.3) already experimented with what were then called pellicular particles in the first decades of HPLC. The pellicular particles at the time were relatively large (up to 50 μm in diameter) and mostly consisted of a large non-porous core, with a thin porous shell. The idea was to minimize the stationary-phase contribution to the slowness of **mass transfer** (the C_S term in the van Deemter equation; see Section 1.7.4.2) so that diffusion in and out of the pores would not be a limiting factor for efficiency. The early pellicular particles did not prove successful because they were too large (leading to increased contributions to the plate height, such as eddy diffusion and the mobile-phase contribution to mass transfer) and the porous layer was too thin (giving rise to a low mass loadability).

Contemporary core–shell particles start with a very small, solid **core** (*e.g.* 1.7 μm) and feature a thick porous **shell** (*e.g.* 0.5 μm), giving rise to small particles (in the present example, $d_c = 1.7 + 2 \times 0.5 = 2.7$ μm). The thick shell implies that overloading is barely an issue. The volume of a fully porous particle with a diameter of 2.7 μm is about 10 fL. The volume of a solid core with a diameter of 2.7 μm is about 2.5 fL. This implies that about 75% of the particle volume is porous and only 25% is non-porous. Thus, there is no high price to pay in terms of pore volume and surface area when moving from fully porous particles to contemporary core–shell particles.

An example of a process for preparing core–shell particles is schematically illustrated in Figure 3.5. It starts with a slurry of non-porous silica particles, at a pH where its surface is negatively charged. The charge allows coating the particles with a positively charged (*i.e.* cationic) polymer. The excess polymer can be removed by filtration and rinsing. In the next step, negatively charged silica nanoparticles (typically $10 < d_p < 16$ nm) are coated onto the now positive surface. The two steps (positive polymer and negative nanoparticles) are repeated a number of times. Up to 50 cycles have been reported. In an alternative method, the so-called one-step

---

Box 3.2 Hero of analytical separation science: J. J. Kirkland. Photo courtesy of The Analytical Scientist.

J. J. Kirkland (1925–2016, United States of America)

Jack Kirkland was one of the pioneers who fulfilled the promise of small particles and high pressures by realizing HPLC in practice. He made numerous contributions to HPLC packing materials, including silica particles, chemically bonded phases and core–shell particles. He also contributed significantly to the development and proliferation of size exclusion chromatography and field-flow fractionation. Jack was also an outstanding lecturer and teacher. Scores of scientists learned the basics of HPLC from Jack Kirkland and Lloyd Snyder during the 25 years (1971–1996) in which they taught short courses for the American Chemical Society.

coacervation process, a number of layers of particles are added following each single polymer-coating step, so as to reduce the total number of cycles required. Finally, the polymer is burned off and the nanoparticles are sintered together at a temperature that leaves the pores between the particles intact.

Both the processes to synthesize the non-porous cores and to create the shell are tightly controlled, resulting in narrow particle-size distributions. This (rather than a low C_s term) is now generally believed to explain the very high efficiencies (opt $\ll 2$) that can be obtained with core–shell particles. A contributing factor may be the higher density of core-size particles, which could make them easier to pack. However, the difference in density of the entire particles is not very large, as most of the volume of core-size particles is porous (see the estimate of 75% porous volume above).

The pore-size distributions are not very narrow. This property, together with a total pore volume that is not much smaller than that of totally porous particles, implies that core–shell particles can be used successfully in size-exclusion chromatography[9] (see Module 4.2).

Box 3.3 Hero of analytical separation science: Csaba Horváth. Courtesy of Yale University (New Haven, CT, USA).

Csaba Horváth (1930–2004, United States of America)

Csaba Horváth was born in Hungary. He is generally credited as one of the very first to realize HPLC in the 1960s. Working together with Wayne Melander at Yale University (New Haven, CT, USA), he did much to demonstrate the high performance of the technique. He developed elaborate theories for dispersion and retention in LC, such as the solvophobic theory for RPLC. Csaba Horváth was also a master of languages and linguistics. Among his many great findings was the notion that Mikhail Tsvet wanted to call chromatography "Tsvetography", describing the increasingly uncommon normal-phase LC as reversed-reversed-phase LC and the "fornicated" (*i.e.* porous) surface.

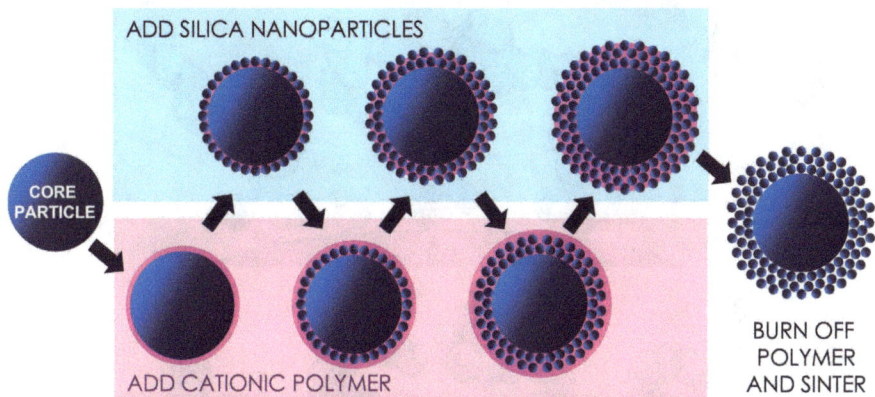

Figure 3.5 Schematic illustration of a process to prepare core–shell particles. In each step, the negatively charged particle is coated with a positively charged polymer. Three coating cycles are shown, but more cycles are performed in practice. The nanoparticles are also smaller in practice relative to the core than suggested in the figure. In the final step, the organic polymer is burned off and the nanoparticles are sintered together in a porous shell.

3.2.2.2 Chemically Bonded (Silica-based) Phases

Silica particles are the dominant LC packing material, but the so-called "bare silica" (the hydrated SiO_2 surface) is not often used. In the vast majority of cases, the surface is modified to yield a **chemically bonded phase** (CBP). The most common CBPs feature *n*-octadecyl chains, as illustrated in Figure 3.6A. The dried surface (physically bound water removed) reacts with, for example, *n*-octadecyl-dimethyl-methoxy silane

$$\equiv Si\text{–}OH + MeO\text{–}Si(CH_3)_2\text{–}C_{18}H_{37} \rightarrow\ \equiv Si\text{–}O\text{–}Si(CH_3)_2\text{–}C_{18}H_{37} + MeOH \quad (3.2)$$

The reaction takes place from a liquid solution, for example, under refluxing conditions with toluene (boiling point 111 °C) as a solvent. Chlorosilanes (such as *n*-octadecyl-dimethyl-chlorosilane) are also suitable reagents, but methanol (MeOH) and ethanol (EtOH, in the case of ethoxysilanes) are more attractive reaction products than is hydrochloric acid. The latter may undergo side reactions with other functional groups, so that chlorosilanes cannot be used to bind polar ligands, such as γ-aminopropyl-dimethylsilyl groups, to the surface.

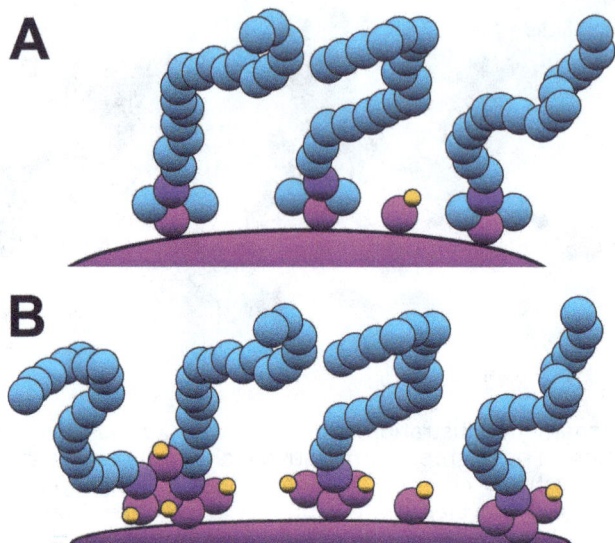

Figure 3.6 Schematic illustration of "monomeric" (A) and "polymeric" (B) octadecyl-silica (ODS) stationary phases for RPLC. Blue spheres represent CH_2 or CH_3 groups, purple spheres are silicon atoms, pink spheres denote oxygen atoms, and yellow spheres denote hydrogen atoms.

Another alternative reaction is with a trifunctional reagent, such as *n*-octadecyl-trimethoxy-silane. This leaves methoxy groups on the bonded groups, but during or after the synthesis, these are rapidly hydrolysed to yield silanol groups. Traces of water also will cause trifunctional reagents to react with each other. The result is a so-called polymeric C18 phase, as illustrated in Figure 3.6B. One way to characterize octadecyl-silica (ODS) phases (and CBPs in general) is the total (weight) percentage of carbon (%C) sometimes referred to as the **carbon load**. Polymeric ODS phases typically show a higher carbon percentage than monomeric phases (prepared using the same starting silica). A better idea of the bonding density is provided by the surface coverage (SC) in μmol m^{-2}, which can be found from

$$SC = \frac{\%C \times 10^6}{S_A(100 \cdot MW_C - \%C \cdot MW_L)} \tag{3.3}$$

where S_A is the specific surface area of silica in m^2 g^{-1}, MW_C is the carbon fraction of the molecular weight of the ligand, and MW_L is the molecular weight of the entire ligand. In the case of a monomeric ODS phase, the formula for the *n*-octadecyl-dimethyl-silyl ligand is $C_{20}H_{43}Si$, so that $MW_C = 240$ and $MW_L = 311$. This implies that a carbon percentage of $\%C = 10$ and a specific surface area of the silica of $S_A = 100$ m^2 g^{-1} correspond to a surface coverage of about 4.8 µmol m^{-2}.

When reacting with a monofunctional reagent (Figure 3.6A), the number of silanol groups on the surface is seen to decrease. When reacting with a trifunctional reagent (Figure 3.6B), the total number of silanol groups present increases. Steric hindrance prohibits the reaction of all silanols. Of the 9 µmol m^{-2} of silanols typically present on the silica surface, only about 50% can be derivatized, resulting in a bonding density of about 4.5 µmol m^{-2}. To avoid strong interactions of the remaining silanol groups with, especially, basic analytes, an **end-capping** step is required with a smaller reagent, such as TMCS or trimethyl-methoxy-silane (TMMS). Not all silanols will react, but the remaining ("residual") ones are less accessible. The result of end-capping barely shows in the total carbon percentage, but it can be observed in LC practice, as it leads to more linear distribution isotherms and, thus, more symmetrical peaks (see Module 1.8). "Polymeric" ODS columns have been found to be more stable against acid hydrolysis (at pH < 2), but less efficient (lower plate counts) than "monomeric" ODS columns and less reproducible due to a greater variation in carbon content.

Examples of different types of CBPs are given in Table 3.4. There have been many attempts to improve the performance and the pH stability of ODS-type materials. Replacing the methyl groups in *n*-octadecyl-trimethoxy-silane (MeO-Si(CH$_3$)$_2$–C$_{18}$H$_{37}$) with iso-propyl (MeO-Si(C$_3$H$_7$)$_2$–C$_{18}$H$_{37}$) or iso-butyl (MeO-Si(C$_4$H$_9$)$_2$–C$_{18}$H$_{37}$) groups may protect the silica surface through steric hindrance although the same effect may reduce the coverage of the surface (in µmol m^{-2}). Such steric hindrance proved to yield a material that was more stable against hydrolysis of the Si–O–Si–C bonds on the surface at low pH. Replacing the methyl group with, for example, urea or carbamate groups yielded phases with embedded polar groups, showing better symmetry for polar analytes.

Silica hydrides are a special kind of stationary-phase material.[10] They can be formed by binding a trifunctional R$_3$SiH to the surface to form a monolayer with hydride groups instead of the common silanol groups. The hydride groups can react with alkenes or alkynes to produce a variety of CBPs. Alkynes allow "bidentate" attachment of the ligand to two surface hydride groups, resulting in the most stable stationary phase. Due to their hydrophobic nature, silica-hydride surfaces retain much less water than silica

Table 3.4 Examples of different chemically bonded phases (CBPs) used in LC. Only the main ligand (R) is mentioned. Small groups may be immersed in the ligands, represented as $\equiv$Si-O-SiXY-R. For example, X = Y = methyl or X = hydroxy and Y = trimethylsilyl.

Ligand (R)	Formula (example)	Used in
Octadecyl Octyl	$-C_{18}H_{37}$ $-C_8H_{17}$	Reversed-phase liquid chromatography (RPLC, Module 3.5)
Butyl	$-C_4H_9$	Hydrophobic-interaction chromatography (HIC, Section 3.15.1)
Phenyl Biphenyl	$-C_2H_4-(C_6H_5)$ $-(C_6H_4)-(C_6H_5)$	RPLC, with additional π–π interactions
Perfluoro Perfluorophenyl	$-C_2H_4-C_6F_{13}$ $-C_2H_4-(C_6F_5)$	RPLC, selectivity differs from alkyl silicas
γ-aminopropyl ("amino") γ-aminocyano ("cyano") 2,3-dihydroxypropyl ("diol")	$-C_3H_6-NH_2$ $-C_3H_6-CN$ $-CH_2-CHOH-CH_2OH$	Normal-phase liquid chromatography (NPLC; Section 3.6.1), Hydrophilic-interaction liquid chromatography (HILIC; Section 3.6.2)
β-alkylsulfobetaine (zwitterionic)	$-C_6H_{12}-NH_2{}^+-C_2H_4-SO_3{}^-$	Hydrophilic-interaction liquid chromatography (HILIC)
Strong anion exchange (SAX)	$-C_2H_4-(C_6H_4)-SO_3{}^-$	Separation of anions by ion-exchange chromatography (IEC; Section 3.7.1)
Weak anion exchange (WAX)	$-C_2H_4-(C_6H_4)-COO^-$	Separation of anions by ion-exchange chromatography (IEC; Section 3.7.1)
Strong cation exchange (SCX)	$-C_2H_4-(C_6H_4)-NH_3{}^+$	Separation of cations by ion-exchange chromatography
Weak cation exchange (WCX)	$-C_2H_4-(C_6H_4)-NH_2$	Separation of cations by ion-exchange chromatography
Silica hydride	$-H$	NPLC/HILIC of polar analytes Somewhat hydrophobic, but not chemically inert[a]

[a]Can react with alkenes or alkynes to form relatively (pH) stable CBPs.

surfaces. Nevertheless, they exhibit HILIC-like retention behaviour (see Section 3.6.2) for polar compounds in high-organic (>90% acetonitrile in water) mobile phases for very polar analytes, presumably due to a negative surface charge caused by the auto-dissociation of water molecules.

3.2.3 Alternatives to Silica

The greatest weakness of silica-based packing materials is their vulnerability at basic pH. The upper pH limit is often specified as pH = 8 or pH = 9, but for conventional packings, pH = 7 may be a more sensible upper limit.[11] It is higher for some specific bonded phases designed for increased stability (see Section 3.2.2) and for hybrid materials. The best known of the latter incorporate methyl groups or ethyl bridges in the synthesis of silica particles. Methyl groups can, for example, be introduced by adding triethoxymethylsilane as a comonomer to tetraethoxysilane in a sol–gel process. In the case of ethyl bridges, a suitable comonomer would be 1,2-bis(triethoxysilyl)ethane. When particles with ethyl bridges are chemically bonded to a trifunctional silane, and thereafter subjected to exhaustive end-capping, columns can be prepared that are stable up to pH = 12, which makes them suitable for the separation of basic drugs as neutral molecules.

A number of other alternatives have been explored. **Alumina** (Al_2O_3) was commonly used in early applications of normal-phase liquid chromatography (see Section 3.6.1) in combination with low-polarity mobile phases. It offers much greater pH stability than silica, but this will mainly be relevant in combination with aqueous mobile phases. As the alumina surface does not feature the equivalence of silanol groups, it cannot be chemically modified using the processes described in Section 3.2.2. As a result, alumina-based materials are not used in reversed-phase liquid chromatography (Module 3.5), which is the dominant LC method. **Zirconia** and **titania** also provide greater stability at high pH than silica but do not produce equally efficient CBPs. Attempts to modify the surfaces of such materials to enhance their applicability in LC have mainly resulted in phases with an immobilized polymeric coating, which have found limited application. The main success of titanium oxide and immobilized titanium ions is for the enrichment of phosphopeptides from protein digests. Such metal-ion "affinity" chromatography can be used as a separation column for gradient-elution LC-mass spectrometry (LC-MS) or as a sample-preparation (enrichment) step prior to high-resolution reversed-phase LC-MS. The use of the term affinity chromatography in this context is different from the antibody–antigen affinity chromatography discussed in Section 3.15.2.

Copolymers of styrene and divinyl benzene (PSDVB) can be prepared in the form of porous particles of suitable size for application in HPLC. Such materials show excellent pH stability across a very broad range (1 < pH < 14). However, the efficiency of such columns is inferior to that of ODS columns, and overloading occurs at low analyte concentrations. PSDVB phases are mostly used in size-exclusion chromatography, where interactions between the stationary-phase surface and the analyte molecules are avoided (see Module 4.2). The most common application of chemically modified PSDVB is for ion-exchange chromatography (IEC; Section 3.7.1), with ionic (strong ion exchangers) or ionizable (weak ion exchangers) groups present on the surface.

Porous **graphitic carbon** (PGC, also known as "porous glassy carbon", or, affectionately, as "black magic") is an interesting material. In a process developed by Mary Gilbert in the group of John Knox in Edinburgh (Scotland),[12] who, incidentally, is the only chromatographer known to have decorated an official banknote (Figure 3.7), an organic polymer (phenol-formaldehyde resin) was formed in the pores of a silica template. The polymer was then pyrolysed under an inert (nitrogen) atmosphere to obtain a carbonaceous material. After cooling, the silica template was washed away with a basic solution, and the carbonaceous material was finally heated ("fired") to a temperature in the range of 2000 to 2800 °C to anneal the surface and remove micropores. A mechanically strong, partly graphitized carbonaceous product was produced.

PGC is a highly stable, retentive material. It typically requires stronger eluents than the CBPs described in Section 3.2.2. For example, pure methanol was used in the original work to elute methyl benzenes. When used in GC, the material yielded significantly negative McReynolds indices (see Module 2.8), indicating that it behaved as much less polar than squalane (the branched alkane used as a reference).[12] Numerous published applications of PGC columns for analytical separations or as highly retentive trap columns have not resulted in a great proliferation of this stationary-phase material. The possible interconnected reasons for this are poor repeatability and long equilibration times. However, there remains a steady interest in the material.[13] Particularly promising is the application of PGC columns for the separation of glycan isomers.[14]

Figure 3.7 Reproduction of a Scottish twenty-pound note, featuring Mary Gilbert from John Knox' group performing an HPLC experiment. It is no longer a valid currency.

An alternative way to produce stationary phases with carbonaceous surfaces is to treat zirconia particles with organic vapours at very high temperatures (700 °C). The resulting carbon-clad zirconia columns exhibit low polarity and high stability across a broad range of pH values and temperatures. However, the widths and shapes of chromatographic peaks have been disappointing so far.

3.2.4 Monolithic Columns

So far, we have almost exclusively discussed particulate stationary phases that had to be packed into LC columns. In an alternative approach, the stationary phase can be synthesized in the column or in any other confined space. Such column materials are known by the Greek name **monolith** (English: one stone; German: ein Stein). To be useful for chromatography, it is essential that the monolith is sufficiently porous to allow the mobile phase (and analytes) to be transported through the column. Through-pores of sufficient size and number are required for this purpose. This can be realized by adding one or more non-reacting solvents – so-called porogens – to the polymerization mixture and by controlling the number of nuclei of polymerization (amount of catalyst), the rate of polymerization (temperature and intensity of irradiation), and, ultimately, the phase separation occurring between the formed polymer and the remaining (porogen-rich) reaction mixture. The phase separation is greatly affected by the nature of the polymer (monolith) and the nature and composition of the porogen. This description already makes it clear that many parameters affect the formation of porous monoliths, making it hard to control and reproduce the process.

3.2.4.1 Silica Monoliths

Silica is an obvious target material, given its dominance as a packing material in LC. Silica can be formed from alkoxysilane solutions ($SiOR_4$, where R is a small alkyl group, typically methyl or ethyl) through a sol–gel process. In producing monoliths, the polymerization is induced by acid hydrolysis of the Si–OR bonds and the formation of Si–O–Si bonds. Conditions should be such that at the gelation stage, a continuous silicon skeleton is formed, rather than individual particles. The simultaneous formation of a continuous silica phase and a continuous solvent (porogen) phase is known as spinodal decomposition.[15] Poly(ethylene glycol) (PEG) is a common additive in the formation of silica monoliths. The pertaining idea is that the phase separation and consequently the average size of pores and the skeleton thickness are mainly determined by the ratio of PEG and alkoxysilane in the polymerization mixture. The total volume of pores is mainly determined by the volume fraction of porogenic solvent in the mixture. In principle, this would allow independent control of the diameter of the through pores (*i.e.* the permeability of the monolith) and the porosity of the monolith.[15] The silica skeleton features micropores (smaller than 50 Å in diameter). A treatment with ammonia allows enlarging these to mesopores (diameters of 100 Å or more). This produces a stationary phase with a suitable pore structure and surface area for LC.

A disadvantage of the formation of silica by the above polycondensation process is the significant shrinkage of the monolithic structure. If the monoliths were to be prepared *in situ* in, say, a 4.6 mm i.d. column, it would end up leaving large open spaces between the monolith and the walls. Large-diameter silica monoliths can be prepared in a mould and then encapsulated in a piece of heat-shrinking tubing, typically poly(tetrafluoroethane) (PTFE). Such "silica-rod" columns have become commercially available since the turn of the century. Due to their high porosity and high permeability, they allowed the use of high volumetric flow rates at moderate pressures, thus decreasing the analysis time. With the advent of UHPLC (Section 3.3.2) and core–shell columns (Section 3.2.2.1), other options for fast analyses have emerged. The plate heights of the original silica-rod columns were about 10 μm,[16] which is 2.5 to 3 times higher than what can be achieved with sub-2 μm fully porous or sub-3 μm superficially porous particles. Thus, the latter solutions allow much shorter columns for equal efficiencies. As UHPLC columns are necessarily also narrower (see Section 3.3.2), the volume of solvent required for a fast analysis decreased by more than an order of magnitude in comparison with silica-rod columns.

Silica monoliths can be made *in situ* in fused-silica capillary columns (50 to 530 μm i.d.). Silica rods are limited in length to approximately 250 mm because shrinking also occurs in the axial direction during both preparation and use. In contrast, capillary monolithic-silica columns can be several metres long, without requiring excessive pressures to work at or above the optimum linear velocity. The latter is due to the high permeability coefficients (K_p; eqn (1.69)) of high-efficiency silica monoliths, which are 2 to 4 times higher than those of columns packed with 5 μm particles. By activating the tube wall with sodium hydroxide, the monolith can be anchored to the wall of the column through covalent bonds. To limit the extent of cross-linking and thereby reduce shrinkage of the silica gel during polymerization, methyl-trimethoxy-silane may be added to tetramethoxysilane as a comonomer. By also including urea in the mixture, the process of post-polymerization mesopore formation can be simplified to just heating the capillary to 120 °C. Residual organic material, including PEG, can be removed by pyrolysis at 330 °C. Optimizing the process has led to so-called second-generation silica monoliths that may outperform packed columns for difficult separations (see also Module 3.16).

The surface of silica monoliths can be modified to obtain CBPs in very much the same way as silica particles. The most common reagent to prepare ODS phases *in situ* in capillary columns is octadecyldimethyl-*N*,*N*-diethylaminosilane. End-capping can be performed with hexamethyldisilazane. Many other modification reactions have been reported for the application of silica monoliths in other forms of LC.

3.2.4.2 *Organic-polymer Monoliths*

A great deal of effort has been devoted to the preparation, study and application of organic-polymer monoliths (OPMs). Many scientists have contributed to this effort, but no one more than František Švec, who was long affiliated with the Berkeley National Laboratory (CA, USA). OPMs are easier to prepare than silica-based monoliths.

Free-radical polymerization (*e.g.* of acrylates or of mixtures of styrene and divinylbenzene) involves less shrinkage than is experienced in the preparation of silica through a sol–gel polycondensation process. Also, the resulting polymer is more elastic. Thus, detachment from the column wall or rupture of the monolith are less likely to occur with OPMs. Free-radical polymerization can be initiated with UV light (photopolymerization) or by heating (thermal polymerization). As with silica monoliths, the porosity and permeability can be influenced by the nature, composition and total concentration of porogens present. For all these reasons, there are numerous reports on the *in situ* preparation of organic monoliths in columns and other confined (separation) spaces.

The ease of preparing organic-polymer monoliths stands in contrast with the difficulty of preparing high-efficiency columns. Frustratingly, OPMs tend to show poor performance (*i.e.* large plate heights) for small molecules and for isocratic separations. OPMs are competitive only for gradient separations of high-molecular-weight analytes. The best example is the high-resolution separation of protein digests on long monolithic polystyrene-*co*-divinylbenzene (PSDVB) columns. In such separations, full use is made of the high permeability of monolithic columns. The long columns (≥ 1 m) result in high values of t_0 and, consequently, long analysis times (often ≥ 10 h). OPMs have also been applied successfully for fast gradient-elution separations of high-molecular-weight analytes.[17]

Organic polymers, and thus OPMs, tend to swell (and shrink) based on the composition of a solvent in which they are immersed. For example, PSDVB polymers swell significantly in tetrahydrofuran (THF), which is a strong solvent for polystyrene. This "free volume" within the polymer may give rise to slow equilibration in the case of water–organic gradients (see Modules 3.4 and 3.5) and may be accessed to some extent by low-molecular-weight analytes, contributing to slow mass transfer.

The heterogeneity of the porous structure is thought to be a main factor contributing to the limited efficiency of OPMs. Free-radical polymerization and phase separation (gelation) are uncontrolled processes. In photopolymerization, the light intensity is likely to vary with the location, if only because part of the light is absorbed by the reaction mixture, so that less light penetrates to greater distances from the UV-transparent wall. Likewise, in the case of thermal polymerization, radial differences in temperature will lead to local differences in the number of the initial polymerization nuclei. Exothermic polymerization reactions complicate temperature control during the formation of monoliths. Heterogeneities in terms of variations in the diameter of through pores will lead to band broadening (see Sections 1.4.4 and 1.7.4.3).

There have been many attempts to improve the performance of OPMs (for example, see ref. 18 and 19). These include systematic experimental design to optimize the preparation conditions, early termination of the polymerization process, performing "hyper-cross-linking" after the initial phase separation, and incorporating nanoparticles in or at the surface of the polymer. More controlled, "living" polymerization methods have been used in attempts to obtain pore-homogeneous OPMs. The results of all these efforts have not made OBPs truly competitive with packed columns, except in specific cases (long columns for complex separations of high-molecular-weight analytes; geometries other than columns). While the stability of OBPs in various mobile phases (including high pH eluents) is very good, the batch-to-batch reproducibility of OBPs is believed not to be as good as that currently achieved by manufacturers of packed columns.

The wide choice of monomers available allows straightforward preparation of a variety of monoliths, including polar and non-polar surfaces; anionic, cationic and zwitterionic stationary phases; and chiral selectors. Alternatively, post-polymerization surface-modification or "grafting" reactions can be used to obtain columns with desired selectivities.

3.2.4.3 Hybrid Monoliths

Hybrid monoliths attempt to combine the efficiency of silica monoliths with the enormous variety of OPMs. The most common way to prepare them is by polymerizing mixtures of tetramethoxysilane with trifunctional (*e.g.* trimethoxy) silanes that feature a different organic group, *i.e.* $(CH_3O)_3$–Si–R monomers, where R is, for example, an alkyl, phenyl, γ-aminopropyl, or 3-glycidyl-oxypropyl group. This implies that the skeleton of such monoliths is a functionalized silica network. The preparation of the monolith from a mixture of tri- and tetrafunctional silanes is akin to the formation of a silica monolith. Trifunctional reagents are still network formers, so that both monomers contribute to the formation of polymeric (silica) networks. The extent of shrinkage and the structure (domain size, pore-size distribution) of the resulting monolith depend on the nature and concentration of the comonomer so that the polymerization conditions must be reoptimized.

Incorporation of a reactive comonomer, with R equal to 3-chloro-propyl, vinyl or allyl, produces a monolith open to further chemical modification. Combinations of (early) polycondensation, followed by later radical addition, can also be realized in a "one-pot" process. For example, a mixture of tetramethoxysilane, vinyl-trimethoxysilane and 3-sulfopropyl-methacrylate (SPMA; as potassium salt) yielded a hybrid monolith with a strong cation-exchange functionality.

Many applications of hybrid monoliths have focussed on sample preparation (solid-phase extraction and protein digestion; see Chapter 8), rather than on analytical separations. However, highly promising results have also been obtained in chromatographic separations. For example, the group of Hanfa Zou (Box 3.4) used polyhedral oligomeric silsesquioxane (POSS) in combination with epoxy–amine ring-opening polymerization and with thiol-based click polymerization to obtain hybrid monoliths with excellent performance for small molecules.[20] One of their columns yielded more than 180 000 plates m^{-1} for n-butylbenzene at a linear velocity of 0.75 mm s^{-1}, which would allow achieving 20 000 plates with $t_0 \approx 150$ s.

Box 3.4 Hero of analytical separation science: Hanfa Zou. Photo courtesy of *The Analytical Scientist*.

Hanfa Zou (1961–2016, People's Republic of China)

Hanfa Zou was an inspirational separation scientist. From his PhD research onwards, he was affiliated with the Dalian Institute of Chemical Physics of the Chinese Academy of Sciences. Hanfa made great contributions to the development of new column technologies, including high-performance monoliths, and separation methods, such as capillary LC and capillary electro-chromatography. These were all dedicated to the separation of very complex samples, especially in the field of proteomics. He also possessed great communication skills, which he used extensively to explain his ideas to the separation-science community.

3.2.5 New Column Technologies

Heterogeneities in columns, for example, in monoliths (Section 3.2.4), give rise to poor chromatographic performance. In contrast, one of the reasons for the high efficiencies obtained with core–shell particles (Section 3.2.2.1) is thought to be the homogeneity of the packed bed, resulting from a narrow particle-size distribution. Based on these observations, modern technologies have been introduced in attempts to create highly structured separation channels for LC. These include micromachined micro-pillar-array columns (Section 3.2.5.1) and 3D-printed columns (Section 3.2.5.2).

3.2.5.1 Micro-pillar-array Columns

Micro-pillar-array "columns" (µPACs) use silicon-machining technology to generate flat separation channels, in which the flow is guided along perfectly ordered obstacles ("pillars") (Figure 3.8). They have been pioneered by Fred Regnier (Purdue, IN, USA) and brought to fruition by Wim de Malsche and Gert Desmet (VUB, Brussels, Belgium). Other than in a packed column, the pillars should not touch to allow liquid to pass through the channel, and the ratio between the volume occupied by pillars and by the liquid mobile phase can be varied. In addition, unlike in packed columns, wall effects can be minimized by implementing "half pillars" in the wall of the channel.

Special machining techniques, such as deep reactive-ion etching, are required to obtain near-vertical pillars of significant height (*e.g.* 100 µm). The shape of the pillars can be designed at will, within the constraints of the fabrication process. For example, in comparison with cylindrical pillars, pillars with an intersection in the form of a diamond extended in the flow direction yielded very much lower pressure drops across the channel, whereas extension of the diamond in a direction perpendicular to the flow yielded a much higher number of plates per unit channel length.[22] To generate sufficient retention, a porous surface needs to be created, for example, by growing a gel layer on the surface by electrochemical anodization.[21] Very impressive separation performance can be obtained from µPACs. For example, a number of µPACs have been developed on a 4″ (about 100 mm) silicon chip and connected in series in a serpentine manner to achieve a separation channel with a total length of 3 m. Pillars that were 5 µm in diameter and positioned 2.5 µm apart in this channel yielded about one million plates

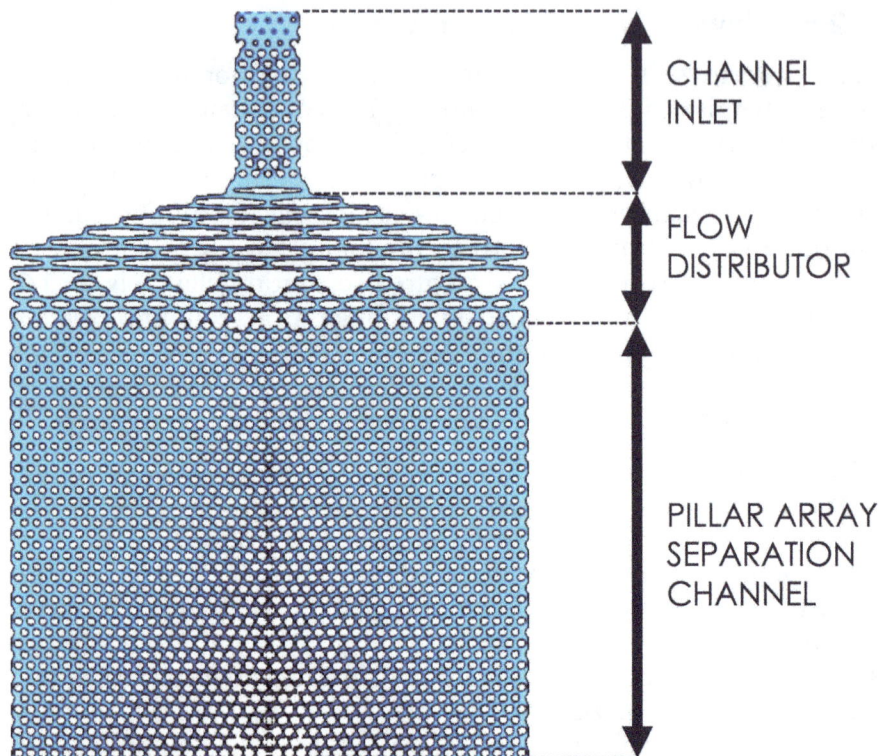

Figure 3.8 Impression of a micro-pillar-array column for chromatographic separations. The coloured channels are etched out from the silicon chip (white areas) and contain liquid during the operation of the chip in LC. Adapted from ref. 21 with permission from the Royal Society of Chemistry.

in 20 min at a pressure of 35 MPa.[23] µPACs for routine use in a miniaturized LC system are commercially available.

3.2.5.2 3D-printed Columns

3D printing or additive manufacturing is another potential method to produce highly ordered structures. While pillar-array columns approach perfect order in two dimensions, 3D printing can potentially achieve this in three dimensions. 3D printing has progressed spectacularly in the last decade. The choice of materials that can be used and the resolution of the printers have increased dramatically. We are not at a point where LC columns can be printed from suitable materials with adequate resolution, but it seems only a matter of time before such is the case.[24] 3D printing allows designing an ideal separation channel, as with pillar-array columns, but now in three dimensions. A homogeneous packed bed can be created, but a wide range of other structures can also be printed and studied. Structures with two continuous phases have been proposed that may be seen as perfect – and perfectly ordered – monoliths. The column walls (or other housings) can be printed in conjunction with the separation channel. Flow distributors at the column inlet and collectors at the column outlet can be designed based on theoretical considerations and optimized experimentally through rapid prototyping using 3D printing.

Figure 3.9 shows a classification of the most important current options for 3D-printing HPLC columns. Fused deposition modelling (FDM) printers, which rely on melting and extruding a polymer filament, are the cheapest and most accessible instruments, but the spatial resolution and the quality of the product do not suffice for HPLC. Various types of digital light processing (DLP) printers provide more resolution, especially in the hybrid stereolithography mode.[25] The highest resolution is provided by two-photon polymerization (2PP), but this process is terribly slow. Matheuse *et al.*[26] demonstrated that highly structured tetrahedral-skeleton monoliths with very high porosity ($\varepsilon_{tot} = 0.8$) could be printed with through pores as narrow as 800 or 500 nm. However, printing nanocolumns (150 mm length × 75 μm i.d.) with such a high resolution took about two weeks (330 h) and three weeks (470 h), respectively, for the two different pore sizes.

DLP, stereolithography, and 2PP printers all rely on photopolymerization. The resulting resins are imperfect and subject to deformation ("warping") and ageing. Adding, for example, carbon nanofibres to the photopolymerization mixture may strengthen the printed objects. When a high concentration of glass nanoparticles is present in the polymerization mixture, the resin can be burned off and the glass sintered, resulting in 3D-printed glass objects. Such processes have been demonstrated but are not yet robust. Selective laser melting (SLM) printers allow printing metals, by locally melting powders to form designed metal objects. A metal such as titanium holds promise for LC because of its strength and the possibility of modifying the surface after oxidation to TiO_2. However, the spatial resolution of SLM does not suffice for printing HPLC columns. In all cases, the smoothness of printed channel walls is currently much inferior to that of the polished stainless steel used to produce high-quality LC columns.

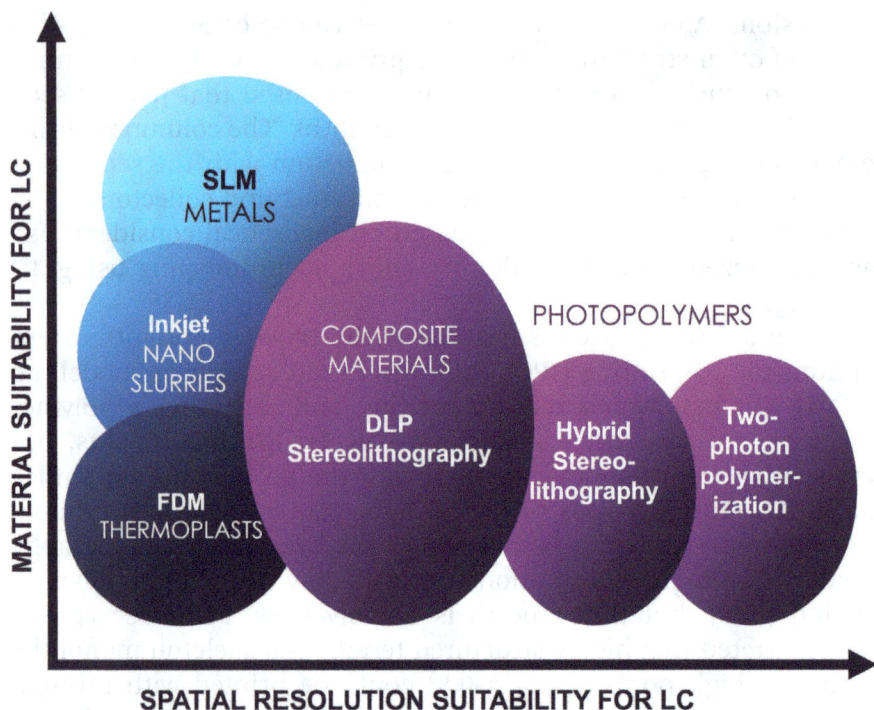

Figure 3.9 Classification of options for 3D-printing LC columns. Material classes are indicated in capitals. SLM = selective laser melting, FDM = fused deposition modelling, DLP = digital light processing.

3.2.5.3 Slip Flow

Experiences with superficially porous particles with narrow particle-size distributions (Section 3.2.2.1) strongly suggest that significant gains in efficiency may also still be achieved in slurry-packed columns. Non-porous silica particles can be prepared with very narrow particle-size distributions, but Darcy's law (eqn (1.69)) and the localized frictional heating resulting from very high pressure drops across very short columns (see Section 3.3.2) indicate that the use of such particles is extremely challenging. If we extrapolate the data given in Table 3.3 towards smaller particles, we find that 1 μm fully porous particles require an unrealistic 320 MPa to yield 10 000 plates (with a column length of 20 mm and a t_0 of 1 s). At the highest pressure commercially available (150 MPa), such particles would yield less than 5000 plates (with a column length of less than 10 mm and a t_0 of less than 0.5 s). Hypothetical particles with a diameter of 0.5 μm will yield about 1000 plates at 150 MPa (with $L = 1$ mm and $t_0 = 30$ ms). The diffusion coefficient shown in Table 3.3 was that for a low-molecular-weight analyte ($D_m = 10^{-9}$ m^2 s^{-1}). For a high-molecular-weight analyte with $D_m = 2 \times 10^{-10}$ m^2 s^{-1}, the optimal linear velocity becomes smaller. Particles with a diameter of 1 μm will deliver 10 000 plates (with $L = 20$ mm and $t_0 = 5$ s) at 64 MPa; 0.5 μm particles will deliver 5000 plates (with $L = 5$ mm and a $t_0 = 0.65$ s) at about 125 MPa. None of these numbers take frictional heating into account. It is clear that very small particles allow fast analysis, but also that the high pressures required pose serious obstacles.

Slip flow[27] potentially provides a way to bypass Darcy's law. Moreover, it reduces flow differences in the channels between sub-μm particles. The principle of slip flow is schematically illustrated in Figure 3.10. It has been pioneered in LC by Mary Wirth (Box 3.5). The big difference between standard Poiseuille flow and slip flow is that in the former case the velocity at the wall is zero, whereas in the latter case, it is not. As a result, the differences in velocities are much smaller in the case of slip flow. Slip flow is thought to occur if the interaction between the mobile-phase molecules and the wall of the channel is weak, as is expected, for example, in reversed-phase liquid chromatography, where the mobile phase is aqueous (hydrophilic) and the stationary phase is non-polar (hydrophobic).

Slip flow was found to become significant for particles smaller than 1 μm in diameter. For larger particles, the velocity in the middle of the channel is so high as to obscure the slip-flow effect. Slip flow causes the flow experienced with 0.5 μm particles at constant pressure to be several times higher than expected – or at the constant flow rate, the pressure to be several times lower. In addition, sub-μm particles were found to self-assemble in near-perfect crystalline structures, so as to drastically reduce the eddy dispersion in the column. Reduced plate heights as low as 0.7 were reported.[27] These results seemed extremely promising, but they have not been followed up by experiments from other groups.

3.3 Instrumentation (B)

In the previous module, we addressed the chromatographic column as the key element for achieving a separation. In this module, we will discuss the other hardware components and describe the configuration of a complete setup for liquid chromatography. We will also discuss some theoretical concepts that guide instrument

LAMINAR FLOW

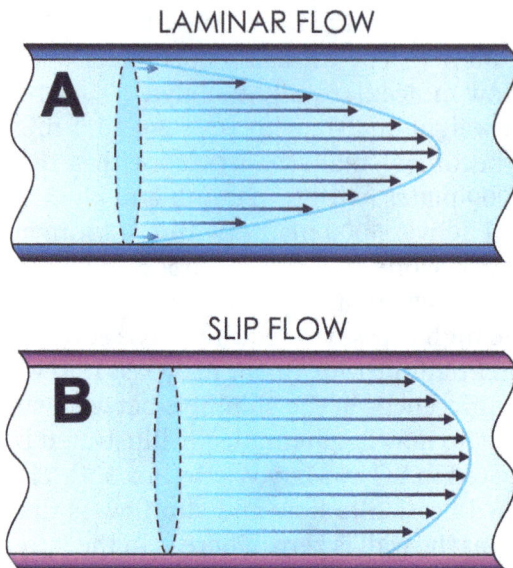

SLIP FLOW

Figure 3.10 Schematic flow profiles for (A) Poiseuille flow and (B) slip flow. The channel walls are different in the two cases. For example, if water is the eluent, the blue walls in (A) are hydrophilic, whereas the pink walls in (B) are hydrophobic.

Box 3.5 Hero of analytical separation science: Mary Wirth. Courtesy of Purdue University (West Lafayette, IN, United States).

Mary Wirth (United States of America)

Before her retirement, Mary Wirth was the W. Brooks Fortune Distinguished Professor at the Department of Chemistry, Purdue University (IN, USA), the institute from which she obtained her PhD. In between, she worked at the University of Arizona in Tucson. Mary worked on spectroscopy during the first part of her career but developed into a serious separation scientist, focusing on new materials for separating proteins. Her most spectacular research concentrated on slip flow, which, unlike Poiseuille flow, implies a non-zero velocity at the surface of packing particles. It allowed her to perform protein separations on columns packed with sub-micrometre particles. It would be fantastic if we could exploit slip flow routinely.

development and focus on the most important principles that are at the basis of modern instruments.

3.3.1 Hardware

3.3.1.1 Degasser

We start the coverage of a modern LC instrument upstream of the mobile-phase flow. Contemporary instruments will almost always feature a degasser unit that prepares the mobile phase by removing air. Air (nitrogen and oxygen) dissolved in the mobile phase (in the case of isocratic operation with a pure or pre-mixed solvent) or the mobile-phase components (in the case of online mixing) can frustrate the operation of LC pumps and in some cases detectors (fluorescence and electrochemical detection). Both nitrogen and oxygen dissolve to significant extents in liquid solvents typically used as LC mobile phases. The gases can be removed by ultrasonic vibration for, say, a minute, or, more effectively, by sparging with helium. Helium itself exhibits very low solubilities in liquid solvents. Unfortunately, when exposed to air, the solvent takes up air again until equilibrium is reached. Continuous sparging with helium may lead to a drift in the mobile-phase composition or the buffer concentration and is also expensive. The solution applied in most modern LC instrument is the incorporation of an online vacuum degasser (Figure 3.11), which makes use of a semi-permeable membrane or polymeric tubing that allows oxygen and nitrogen to pass but is non-permeable to solvents. By applying vacuum on the other side of the membrane or on the outside of the tube, gases are continuously withdrawn from the solvents.

3.3.1.2 Pumps

Contemporary solvent delivery systems used in LC are almost exclusively **reciprocating piston pumps.** Such pumps are capable of operating up to at least 40 or 100 MPa (400 or 1000 bar) and are thus suitable for high-pressure LC (HPLC) and UHPLC (see Section 3.3.2). An example of a reciprocating piston pump is shown schematically in Figure 3.12A. During the filling cycle, the **piston** retracts, thus drawing liquid from the mobile-phase reservoir to fill the pump chamber. Once the piston has sufficiently retracted and the pump chamber is filled, the pump enters the delivery cycle where the piston pushes the liquid out of the pump chamber.

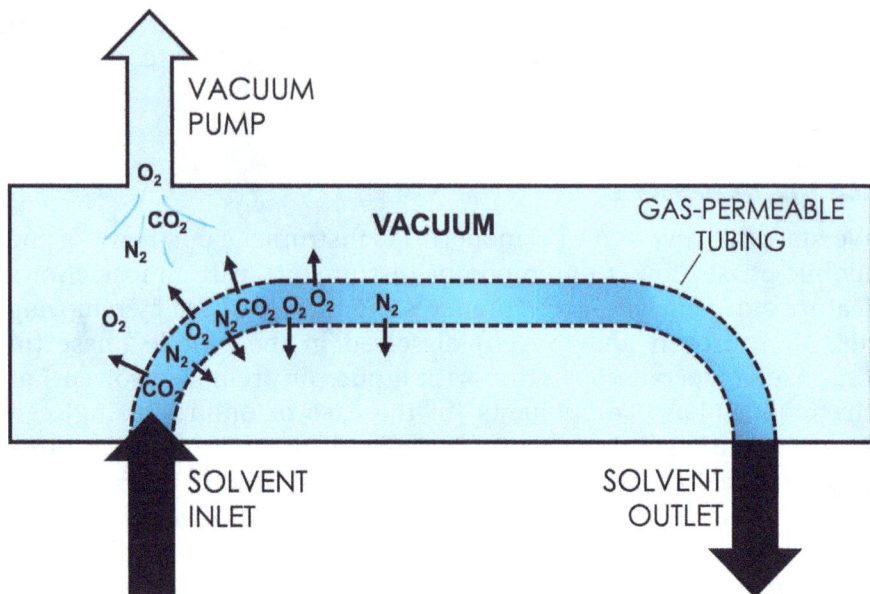

Figure 3.11 Schematic illustration of the principle of a vacuum-degasser unit.

Check valves are used to restrict the flow direction. The inlet check valve prevents that the mobile phase is pushed back into the reservoir during the delivery cycle, whereas the outlet check valve ensures that the mobile phase is not drawn back into the pump chamber during the filling cycle.

The most common check valve employs a ball that is pushed into a slot by the pressure differential. For example, during the delivery cycle, the relatively high pressure inside the pump chamber will push the ball of the inlet check valve into its slot, thus closing off the flow towards the mobile-phase reservoir. Ball-based check valves can feature a spring (such as depicted in Figure 3.12) to aid the ball in sealing the flow direction.

To prevent leakage from the pump head, the pump chamber is sealed off by a **seal** (depicted in yellow in Figure 3.12), which contains a hole through which the piston can perpetually move back and forth. Modern pumps also feature a **seal wash** unit, which serves to wash contaminants (and, in case aqueous buffers are used, salt deposits) off the seal and piston, so as to enhance the lifetime of these parts. High pressures, elevated temperatures, and corrosive mobile-phase additives all promote wear of the seals and check valves.

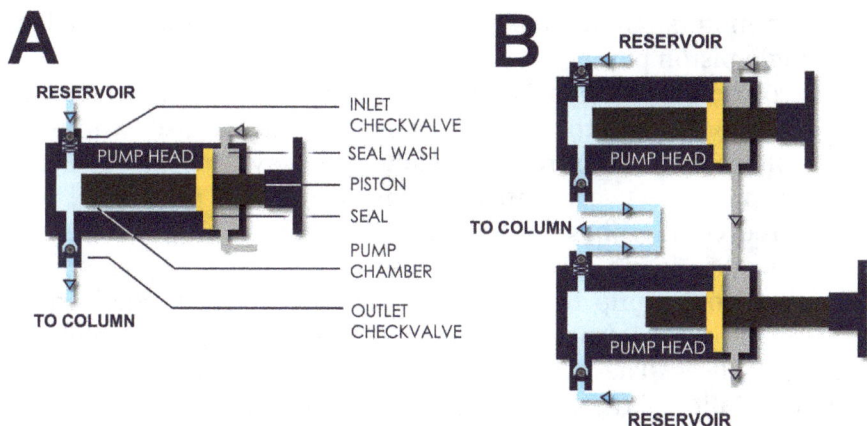

Figure 3.12 (A) Schematic illustration of a reciprocating piston pump used for mobile-phase delivery in LC, with various components denoted. (B) Dual-piston reciprocating pump designed to provide continuous flow. The top piston is delivering the mobile phase, while the bottom piston is being filled.

During its delivery cycle, the piston will experience high back-pressure as the piston pushes the mobile phase out of the pump chamber into the column. We have seen in Section 1.6.4 that the backpressure depends on several variables, including the viscosity of the solvent, the flow rate, and also the particle size of the column packing. Higher backpressures give rise to larger pressure spikes, which are undesirable from the perspectives of both the lifetime of the column and the accuracy of the flow. Different tools have been developed to mitigate this. For example, the piston is moved using a motor that propels a driving cam (not shown in the figures). The shape of this driving cam can be optimized to allow a single-piston or dual-piston pump to deliver flow more smoothly. In addition, the piston stroke speed can be automatically adjusted during a delivery cycle. The piston will then move fast at the start of the delivery cycle and then slow down once the backpressure of the previous delivery cycle is attained. Nevertheless, it is fundamentally not possible for a single-piston pump to deliver a constant flow rate across several cycles. The best way to use a single-piston pump successfully in LC is to ensure that an entire analysis can be performed without the need for a refill cycle. This can be achieved by using a large piston or a very small column volume and by triggering a refill cycle before each analysis. Larger pistons may require greater force (increasing with the square of the diameter of the piston chamber) and, thus, stronger motors. Gradient elution cannot be performed with just one single-piston pump.

To provide a stable continuous flow irrespective of the volume of the piston or that of the column, it is necessary to use two pump heads alternatingly. Figure 3.12B shows the very common **dual-piston pump**, where two pump heads are connected in parallel and configured counterphase to provide a continuous flow. While one pump head is in its filling cycle (bottom piston in Figure 3.12B), the other (top) pump head is in its delivery cycle. The delivery of flow by the two alternating pump heads leads to greatly reduced pressure pulses. The main remaining cause of such pulses is the compressibility of the mobile phase. Accelerated movement of the piston at the start of its delivery cycle serves to compress the mobile phase, reducing its volume until the column pressure is exceeded and the outlet check valve opens.

Another setup is the accumulator-piston pump, as shown in Figure 3.13. Here, two pump heads are coupled in series rather than in parallel, as was the case for the dual-piston setup. During its delivery cycle, the primary pump head supplies the accumulator pump head, which is in its filling cycle, with the mobile phase, while at the same time also delivering the mobile phase to the column (Figure 3.13A). To accomplish this, the primary unit can be envisaged as to pump twice as much volume as the accumulator. For example, if the pump is configured to pump 0.5 mL min^{-1}, this means that the primary pump will deliver 1.0 mL min^{-1}, while the accumulator will pump at 0.5 mL min^{-1}. During its filling cycle (Figure 3.13B), the primary pump will fill its pump chamber, while the accumulator will deliver the mobile phase to the column. The accumulator-piston pump offers high precision at high flow rates. Compared to the dual-piston setup, the accumulator-piston pump has the advantage that fewer check valves are required (two rather than four).

Modern pumps feature sophisticated algorithms and sensors to compensate for changes in compressibility and thermal expansion. This is especially complicated for the accumulator-piston pump, at the point where the two pump heads switch cycle and a stable pressure and flow must be maintained with minimum pulsation. For such systems, the instrument-control software will typically allow the user to specify the identity of the solvent (mixture) that is being pumped. The compressibility should not be underestimated. The compressibility of water is about 0.5% per 10 MPa. For common organic modifiers, this is about 1% (ACN and THF) to 1.2% (methanol). This means that when a pump unit displaces THF at a flow rate

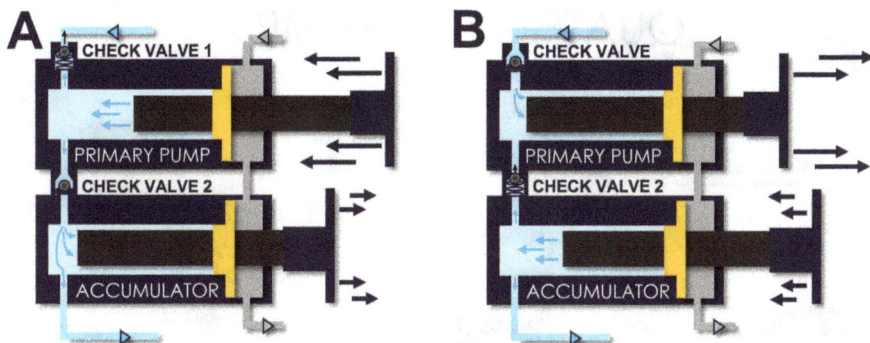

Figure 3.13 Schematic of an accumulator-piston pump with the primary pump (A) in the delivery phase and (B) in the filling phase.

of 1 mL min^{-1} at 1000 bar, the solvent will gradually expand as it passes through the column. When the pressure reaches atmospheric levels at the end of the column, the exit flow will be 1.1 mL min^{-1}. Compensating for thermal changes and compressibility becomes particularly important when mobile-phase composition programs are used (*i.e.* gradients, see Module 3.4).

The setups described thus far represent what can be referred to as **isocratic pumps**, *i.e.* mobile-phase delivery units that deliver a constant mobile-phase composition. However, many contemporary LC modes use mixtures of different solvent systems (*e.g.* reversed-phase LC, see Module 3.5). Pumps that deliver mixtures of mobile-phase solvents are thus more common.

A **quaternary pump** is a low-pressure pump that can prepare and deliver mixtures of four solvents. An example is shown in Figure 3.14A. A **solvent selector**, which is a proportioning manifold, allows different solvents to be blended by opening the valve for each selected solvent for a fraction of the time needed to achieve the set concentration. The total volume is displaced by a single-pump unit (typically a dual-piston or accumulator-piston setup). The same principle can be applied for binary or ternary solvent-delivery systems, with different manifolds. To prepare, for example, a mixture of 25% solvent B and 75% solvent A, one may open the solvent-A line for 0.75 s and the solvent-B line for 0.25 s, provided that the aspiration flow of the pump is constant. However, a major complication arises from the operation of common HPLC pumps, which are designed to deliver a constant flow. To achieve this, such pumps typically have very non-constant inlet ("suction" or "aspiration") flows, an effect that can easily be observed if an air bubble is

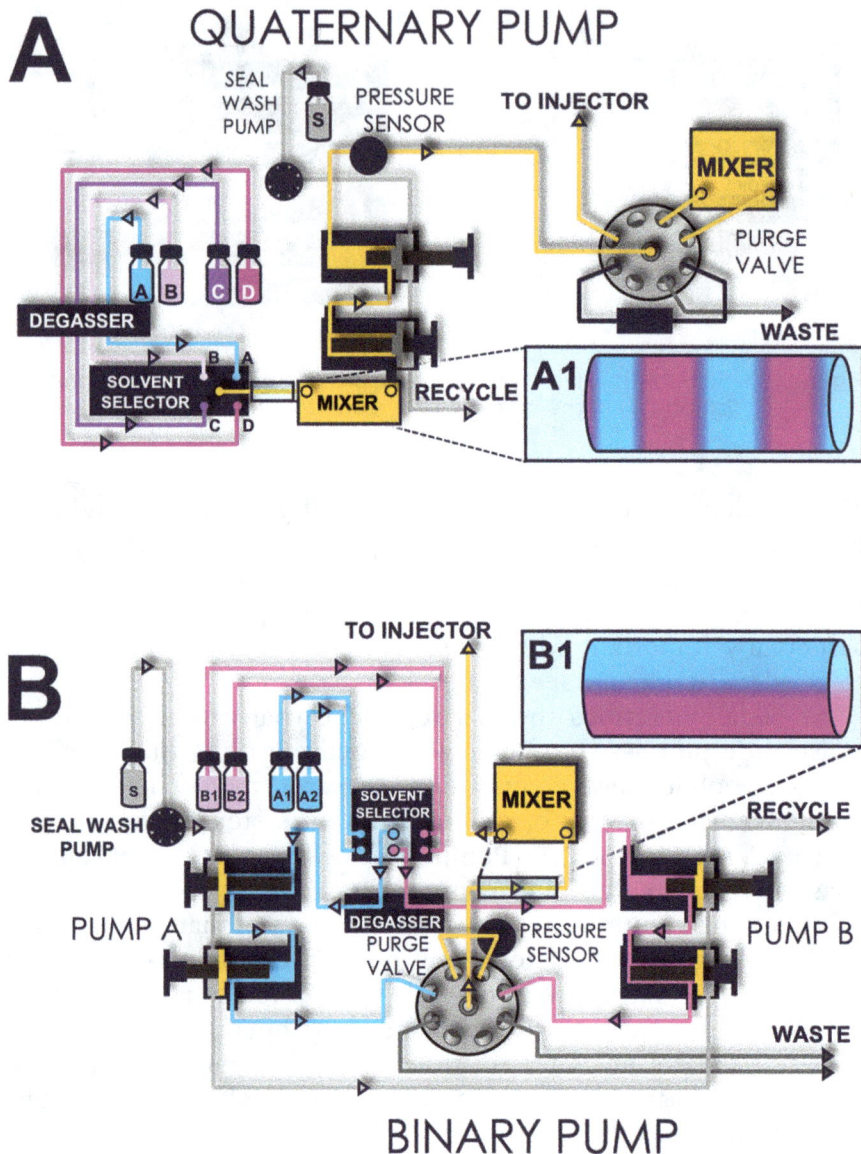

Figure 3.14 Schematic illustration of a typical (A) quaternary pump and (B) binary pump. The insets A1 and B1 display the resulting pre-mixing solvent composition. This figure also displays an example of low-pressure mixing (A) and high-pressure mixing (B). Note that for (B1), the effect of pump strokes has been ignored, which would in reality induce a variation over time in composition. Yellow flow paths contribute to the dwell volume. Details may differ for different manufacturers.

present in the plastic solvent line connecting the solvent reservoir to an HPLC system. As a result of the non-constant suction flow, switching the proportioning manifold at exactly the right times to achieve a desired mobile-phase composition is a significant challenge.

Modern low-pressure **mixing** systems synchronize valve switching with the suction-flow profile. The resulting solvent pattern is illustrated for a binary mixture in Figure 3.14A1. To ensure that the mobile-phase system is homogeneous, significant mixing is required. Quaternary pumps may use one or two mixers and also the pump chambers contribute to the mixing to achieve homogenization. All of these elements contribute to the large dwell volumes of low-pressure mixing systems (indicated by the yellow flow path in Figure 3.14A). The dwell volume (V_D) is the total volume between the initial point where the solvents are brought together (the proportioning manifold) and the top of the column. It is of high importance in gradient-elution LC and is discussed in more detail in Module 3.4. A two-pump solvent delivery system or, colloquially, a "binary pump" features two pump units (each often two-piston units) to deliver mixtures of two solvents (Figure 3.14B). As a result, the mobile-phase composition in the tubing after merging the two streams features continuous volumes of both solvents (see Figure 3.14B1). Contrary to common belief, these two solvents – even when similar – will not blend instantly, except for the volume near the interface. Mixing is thus also required for binary pumps.

Using a different high-pressure pump for each solvent implies more expensive hardware in comparison with the low-pressure mixing approach that leads to an affordable quaternary system. However, the individual pumps make it more straightforward to control the eventual composition of the eluent. The accuracy of the delivered composition depends on the accuracy of the pumps. The latter may need to be calibrated (depending on the compressibility of the solvents), and they may be challenged at the extreme ends of the scale (*e.g.* 1% A or 1% B). As illustrated by the yellow flow paths in Figure 3.14, high-pressure mixing systems have inherently smaller dwell volumes than low-pressure mixing systems. This does not disqualify the latter, but it must be borne in mind when developing gradient-elution LC methods that are to be used on multiple instruments. When developing such a method on a high-pressure mixing system, it is prudent to program a significant initial hold volume (V_{init}) that is larger than the largest dwell volume (V_D) that can possibly be expected. This initial hold volume can be translated into an initial hold time using the flow rate $t_{init} = V_{init}/F$. The advantages and disadvantages of low-pressure and high-pressure mixing systems for (gradient-elution) LC are summarized in Table 3.5.

3.3.1.3 Mixers

In the insets in Figure 3.14, we have seen that solvent-delivery systems produce two-phase systems (or, in the case of a quaternary pump, three-phase or four-phase systems). In the case of the quaternary pump, we have to mix consecutive bands, and the mixing volume should necessarily be larger than a train of AB (or ABC or ABCD) solvent segments. The initial mixing takes place before the pump (low-pressure mixing). Additional mixing takes place in the pump heads and even this may not suffice, so that a final mixer may be installed between the pump and the injector (top right in Figure 3.14A). All these mixing volumes add to the dwell volume (V_D). If a step change in composition is induced by the user (through the instrument-control software), it takes a time $t_D = V_D/F$ before the change in composition effectively reaches the top of the column. Mixing also induces a deformation of the step change. An ideal mixer is defined such that the composition is the same anywhere in the mixer but varies with time depending on what comes in (which the user can control) and what goes out. The latter is a convolution of the incoming profile (in our case, the step gradient)

Table 3.5 Advantages and disadvantages of low-pressure and high-pressure mixing systems for (gradient-elution) LC.

Low-pressure mixing	High-pressure mixing
Advantages	**Advantages**
• Relatively low-cost hardware (just one high-pressure pump) • Conceptually easy expansion from binary to ternary or quaternary systems	• Inherently simple control of resulting mobile-phase composition • Inherently small contributions to the dwell volume
Disadvantages	**Disadvantages**
• Difficult to accurately control mobile-phase composition • Significant mixing volume required • Inherently larger dwell time and more serious gradient deformation	• Expensive (one high-pressure pump for each mobile-phase component) • Requires high pump accuracy • Requires an efficient high-pressure mixer

and the **response function** of the mixer, which, for an ideal mixture, is an exponential decay, the time constant (τ_{mix}) of which equals the volume of the mixer (V_{mix}) divided by the flow rate, *i.e.* $\tau_{mix} = V_{mix}/F$ (see Figure 3.15). Every mixing step contributes to the deformation, but for the same total mixing volume, a series of small mixers outperform a single large one.[28] In high-pressure mixing, as with the binary pump shown in Figure 3.14B, the required mixing volumes are smaller, as the average composition is more correct at every time point (see Figure 3.14B1). In reality, the pump piston strokes will also induce composition variations for such pumps, and the need for mixing should not be underestimated. Two fully miscible solvents can coexist for a long time without some form of stirring or agitation, so as to greatly enhance the contact area. This is the goal of smart, low-volume mixers used in LC.

Numerous such mixers have been developed and described in the literature using microfluidic and 3D-printing technologies. However, few of the proposed designs meet all the requirements of HPLC, which include (i) small volume (microlitre range), (ii) low pressure drop, (iii) robust operation without clogging, (iv) compatible with common LC solvents, and, ideally, (v) resistant to high solvent pressures. A few examples are shown in Figure 3.16. The "herringbone" mixer[31] (Figure 3.16A) is an example of a chaotic mixer, which relies on counter-rotating vortices, created by the relief structure on the (bottom) surface of a micromachine channel. Figure 3.16B shows a mixer with modified Tesla structures,[30] which features an

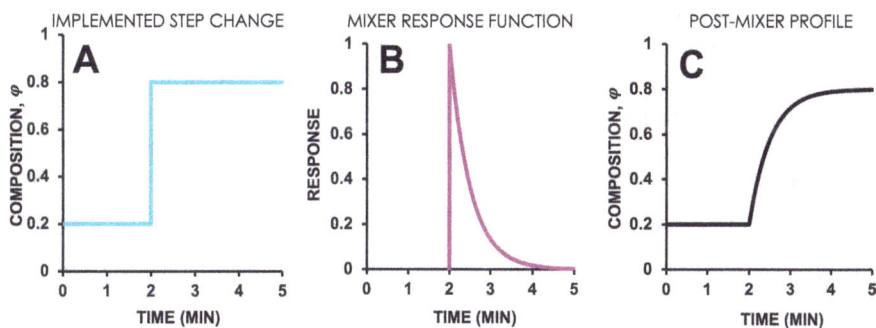

Figure 3.15 (A) Implemented step change program of the mobile-phase composition. (B) Example of a response function of a mixer (exponential decay with a time constant of 0.5 min). (C) Resulting post-mixer profile as a result of the step change (A) and mixer response function (B).

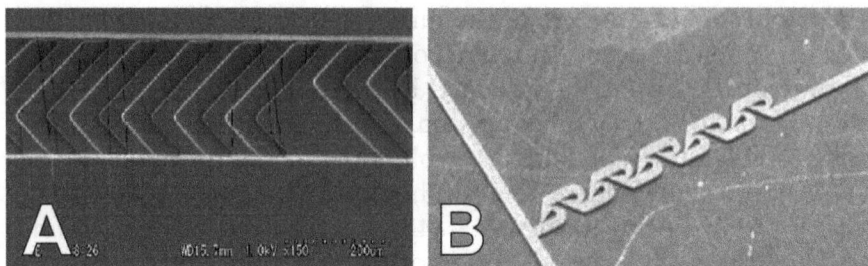

Figure 3.16 Examples of different mixing strategies: (A) herringbone mixer, (B) mixer with modified Tesla structures. Panel A reproduced from ref. 29, https://doi.org/10.1371/journal.pone.0039057, under the terms of the CC BY 4.0 license, https://creativecommons.org/licenses/by/4.0/. Panel B reproduced from ref. 30 with permission of the Royal Society of Chemistry.

ingenious flow path. Static mixtures from various other manufacturers are offered especially for LC, but their exact design is proprietary.

3.3.1.4 Injection

Injection is typically performed using a six-port **valve**, as illustrated in Figure 3.17. Panel A displays a side-view schematic of a six-port two-position valve (only three ports visible). The different LC tubings are connected to the static ("**stator**") component. These are interconnected through the motor-powered moving component ("**rotor**"). Both the rotor and stator are – similar to virtually all components in LC – made of an inert material. To withstand high pressures, the stator is usually made of stainless steel. The rotor is typically made of Vespel, a polyimide-based polymer, or a ceramic material. Many different types of valves exist for different applications, such as the standard two-position six-port injection valves (with or without a syringe port), two-position 8- or 10-port valves for comprehensive two-dimensional LC (LC×LC, see Section 7.3.4), and a remarkable 7-position 14-port valve used in heart-cut 2D-LC (see Section 7.3.3). Manual and pneumatically activated valves exist, but most modern valves are electronically powered.

Figure 3.17B shows how a two-position six-port valve can be used for sample injection in LC. In the **load** position, the mobile phase bypasses the injector and flows directly to the column. A syringe (pump) or metering device can be used to fill the sample loop. In the inject position, the pump flow is directed through the sample loop

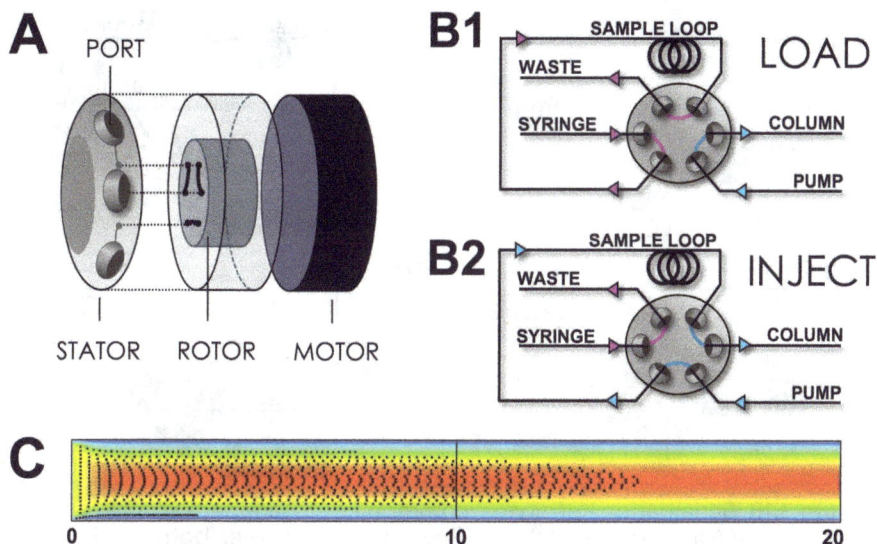

Figure 3.17 (A) Side-view schematic illustration of a valve assembly comprising a static component ("stator") to the ports of which the different LC tubings are connected, and the motor-powered rotating component ("rotor") that connects different ports together depending on its position. (B) Front-view schematic illustration of a two-position six-port valve used for injection in LC in the load (B1) and inject (B2) positions. (C) Computational fluid dynamics (CFD) simulation of the flushing of a 20 µL sample loop at a flow of 10 µL min^{-1} during 1 min. Black dots represent analyte molecules. All analyte molecules (black dots) to the right of the vertical line would be lost if the sample volume would be equal to the loop volume. Panel C reproduced from ref. 32, https://doi.org/10.1002/jssc.201700863, under the terms of the CC BY 4.0 license, https://creativecommons.org/licenses/by/4.0/.

prior to entering the column, transporting the sample to the column inlet.

The flow in the sample loop is laminar (Poiseuille flow, see Section 1.6.3), which implies a parabolic flow profile with zero flow at the wall (light blue in the computational fluid dynamics simulation[32] shown in Figure 3.17C) and twice the average velocity at the column centre (red region). In the simulation, the loop is filled to half its volume. All analyte molecules (black dots) that pass the vertical line in the middle would have exited a loop filled to its full volume. The exact loss depends on the analyte diffusion coefficient (greater losses for lower diffusion coefficients, *i.e.* for high-molecular-weight analytes) and on the diameter of the loop (smaller losses for narrower loops – with greater lengths to maintain the same volume).

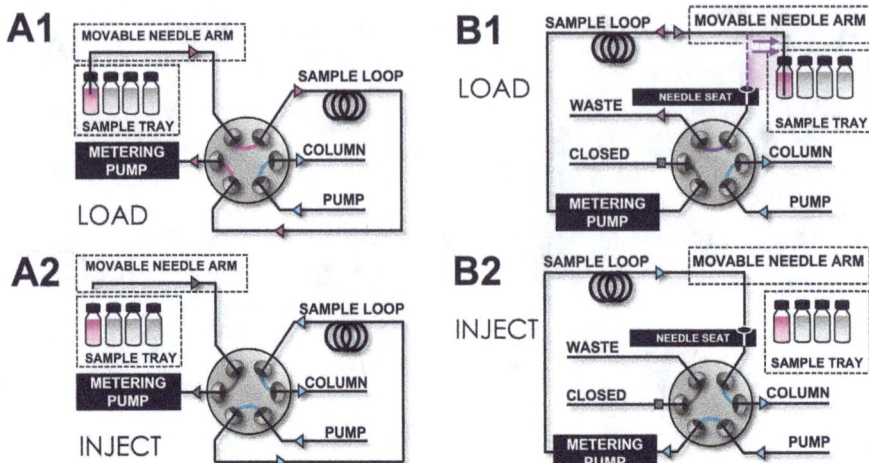

Figure 3.18 Schematics of two common autosampler configurations. (A)Pull-to-fill and (B) flow-through-needle, both in the load (A1/B1) and inject (A2/B2) positions In B1, the metering device first pushes out liquid to waste (purple path) and then needle moves (purple arrows) towards the sample to load (pink path). Note that different manufacturers may use different variations of these strategies.

Modern analytical laboratories typically equip their LC instruments with **autosamplers**,[33] especially when they process hundreds or thousands of samples per day. Such an automatic injector typically features one or several sample trays that can optionally be thermostatted if this is required to ensure the integrity of the samples. A moving robot arm equipped with a needle can be directed to take a sample from a specific vial. The **push-to-fill configuration** is analogous to conventional injection. The sample is aspired by the syringe, which is subsequently inserted in a needle port and emptied. Possibly, a loop is installed between the syringe and the sample, and the need for moving parts can be reduced by using a low-pressure valve to switch between aspiration and emptying of the syringe.

The **pull-to-fill configuration** shown in Figure 3.18A works similarly, but the sample is drawn directly into the loop, without the use of a sample port. The system draws liquid from the specified sample vial in the **load** position. The contents in the sample loop can then be transported to the column in the **inject** position.

One issue with both the pull-to-fill and push-to-fill configurations is that the tubing connecting the needle to the valve will still contain the sample once the sample is injected. To clean the needle, modern systems will feature a **wash** option. This can be either a dedicated washing line or simply a sample vial containing a wash solution. If the needle is not properly cleaned, the sample from the previous analysis may contaminate the next run. Such so-called **carry-over** must be avoided at all costs. Nevertheless, even if the needle is properly rinsed, the pull-to-fill setup still wastes significant quantities of the sample.

The **flow-through needle configuration** (or "split-loop") shown in Figure 3.18B prevents waste of the sample. Here, the metering pump is in line with the sample loop. The load position works similarly to the pull-to-fill configuration, but in the inject mode, the needle is pressed into a high-pressure **needle seat**. This special holder allows the needle to be placed directly leak-tight into the LC tubing. Upon injection, the mobile phase flows directly through the sample loop, which also helps prevent carry-over.

The loop installed on an injection valve, as shown in Figure 3.18A, can have different volumes, depending to a large extent on the LC system used (*e.g.* 10 μL on a conventional LC system or 1 μL on a modern ultra-high-pressure LC system; see Section 3.3.2). Injection may involve a fully filled loop (which requires flushing it with a sample volume exceeding the loop size by at least a factor of 2) or a partially filled loop. The former approach is more precise; the second one is more flexible. For injection of very small volumes, injection valves exist in which the external loop is replaced by a small channel inside the valve rotor. For injecting large volumes (relative to the volume of the column) of a sample dissolved in a weak eluent, a loading pump may be used to bring the sample onto the column, where the analytes are then focussed until a stronger eluent arrives.

A point of concern is a potential difference in viscosity between a sample zone in the injector or at the top of the column. If this becomes large, the flow profile may become unstable and break up in a phenomenon known as **viscous fingering**. Different parts of the sample zone may move at different velocities, and this may result in significant injection band broadening. In a detailed study,[34] Samuelsson *et al.* found that such dramatic effects did not occur in analytical LC below viscosity differences of about a factor of three. Such large differences are uncommon, but they may, for example, be encountered in polymer separations (Chapter 4) because polymer (sample) solutions may be much more viscous than mobile phases. At much lower differences in viscosity, as low as about 20%, the sample zone may be elongated by "pre-viscous fingering effects", leading to injection band broadening, especially in the case of isocratic elution.

> See the website for more on viscous fingering: ass-ets.org

3.3.1.5 Oven

Columns are typically installed inside a **thermostatted** column compartment (often referred to as a column oven), with either still air or forced air circulation. Working at an elevated temperature may be desirable, for example, to lower the viscosity of the mobile phase, which would lower the pressure drop or allow an increase in the flow rate (see also Section 3.3.4, high-temperature LC). In other cases, it may be advantageous to work at sub-ambient temperatures, for example, to maintain the integrity of thermally unstable analytes or to benefit from a possible gain in selectivity at lower temperatures. However, a column oven is usually of limited use in controlling the temperature inside the column due to the poor thermal conductivity of the packing material and the mobile phase and the low heat-transfer coefficients between the various materials of the column and its contents. The wider the column and the higher the temperature, the more difficult it becomes to control the temperature. In the case of UHPLC), columns are narrower than in HPLC, and it is advisable to thermally insulate the column rather than to try and control its temperature (see Section 3.3.2). To control the temperature inside the column, it is essential to pre-heat (or pre-cool) the mobile phase before it arrives at the column. A pre-column heat exchanger is an important feature of a modern HPLC (or UHPLC) system.

3.3.1.6 Fraction Collection

One of the advantages of a liquid mobile phase is the relative ease with which fractions of the column effluent (ideally representing single peaks) can be collected. This is typically done automatically using a **fraction collector**. Specific fractions can be collected based on their elution times, in which case maintaining a constant flow rate is essential, or based on the detector signal, with collection triggered by the onset of a (specific) peak. In all cases, the dwell volume between the detector and the fraction collector is critical. The post-column dwell volume and the flow rate must be known exactly to avoid errors in peak collection, and dispersion must be kept minimal, so as to maintain the integrity of the separation.

3.3.2 Ultra-high-pressure Liquid Chromatography

For more than 30 years, HPLC was confined to pressures up to about 40 MPa, while the typical diameter of stationary-phase particles gradually decreased from 10 to 5 and then 3 μm. However, in 2005, Waters ignited a step change by introducing systems operating at pressures up to 100 MPa. Quickly thereafter all major manufacturers followed suit, and this led to a class of instruments generically known under the name ultra-high-pressure liquid chromatography or UHPLC. Such instruments have three main characteristics: (i) capability to operate at maximum pressures in the range of 100 to 150 MPa, (ii) use of particles with diameters smaller than 2 μm (so-called sub-2 μm particles), and (iii) greatly reduced extra-column volumes and dispersion. UHPLC pumps are not fundamentally different from HPLC pumps, but there is more strain on seals and check valves. Sub-2 μm particles (with diameters of often 1.7 μm and sometimes as small as 1.5 μm) are amply available from several manufacturers. As explained in Section 1.7.6, working at very high pressures causes a dramatic increase in the amount of frictional heat generated in the column, and there is a greatly increased risk of radial temperature gradients that jeopardize the chromatographic efficiency. To reduce such gradients, UHPLC columns tend to be narrower than HPLC columns, with typical internal diameters of 2.1 mm and 4.6 mm, respectively. The small column volumes and high efficiencies put severe restrictions on the maximum permissible extra-column dispersion. This will be discussed in Section 3.3.3.

3.3.3 Extra-column Band Broadening

In LC, everything is small because of the low diffusion coefficients in liquids. We use very small particles and short columns. In the previous section, we have seen that contemporary UHPLC columns are not only shorter but also narrower than common HPLC columns. This forces us to meticulously consider everything that happens outside the column. For a band emerging from the column, we can write

Table 3.6 Band broadening in LC columns. Calculated from eqn (3.4) using $\varepsilon = 0.625$ and $h = 2$.

	L, mm	d_c, mm	d_p, μm	N	$\sigma_{V,col}$ $(k = 0)$, μL	$\sigma_{V,col}$ $(k = 3)$, μL
HPLC	300	4.6	10	15 000	25.4	102
	150	4.6	5	15 000	12.7	51
	100	4.6	3	16 700	8.0	32
UHPLC	100	2.1	1.7	29 400	1.26	5.0
	50	2.1	1.7	14 700	0.89	3.6
	50	1	1.7	14 700	0.20	0.8

$$\sigma_{V,col} = \frac{V_0(1 + k)}{\sqrt{N}} = \frac{(\pi/4)\varepsilon \cdot d_c^2 \cdot L(1 + k)}{\sqrt{L/(h \cdot d_p)}}$$

$$= (\pi/4)\varepsilon \cdot d_c^2 \sqrt{L \cdot h \cdot d_p} \cdot (1 + k) \tag{3.4}$$

Table 3.6 lists the band dispersion calculated for typical HPLC and UHPLC columns, both for an unretained peak ($k = 0$) and for a peak eluting with $k = 3$. It can be seen from Table 3.6 that the band dispersion decreased gradually during the first three decades of HPLC, while the field progressed from 10 μm particles to 5 μm and then 3 μm particles. However, the progression to UHPLC with 1.7 μm particles has heralded a new era in LC instrumentation, not

Box 3.6 Hero of analytical separation science: Monika Dittmann. Photo courtesy of *The Analytical Scientist*.

Monika Dittman (Germany)
Monika spent a career at Agilent Technologies (previously Hewlett Packard), focusing largely on HPLC, capillary-electro chromatography, and microfluidics. She stood out among the many scientists associated with instrument manufacturers, thanks to her relentless focus on a sound understanding of the fundamentals of our science. Monika studied the separation performance of (electro-) chromatographic systems and the contributions to dispersion from within and outside the column. She represented the honest science behind the glamorous LC technology.

just in terms of pressure but also in terms of band dispersion. The volume of a band eluting from a UHPLC column is about an order of magnitude smaller than that eluting from an HPLC column. This poses significant challenges to the equipment. One of the scientists who has been addressing this is Monika Dittmann (Box 3.6).

In Chapter 1 (Section 1.7.5), we have assumed that the various contributions to the observed (subscript "obs") band broadening from the column, from outside the column ("ec"), the injector ("inj"), detector ("det"), and connections ("con"), are independent, allowing variances to be added as follows:

$$\sigma_{obs}^2 = \sigma_{col}^2 + \sigma_{ec}^2 = \sigma_{col}^2 + \sigma_{inj}^2 + \sigma_{det}^2 + \sigma_{con}^2 \tag{3.5}$$

The different contributions can in turn be expanded if the contributions are independent. For example, σ_{con}^2 may be the sum of the variances due to each piece of tubing and each connector. The variance produced by the column includes contributions from the packed bed and the frits. Although the latter are part of the column (as are the column connectors), their contribution to the overall band dispersion should be seen as part of the extra-column variance (σ_{ec}^2). If we want to make sure that the observed peak width is for at least 90% due to the chromatographic process, we have

$$\sigma_{col} \geq 0.9\,\sigma_{obs} \tag{3.6}$$

$$\sigma_{col}^2 \geq 0.81\sigma_{obs}^2 = 0.81\left(\sigma_{col}^2 + \sigma_{ec}^2\right) \tag{3.7}$$

$$\sigma_{ec}^2 \leq \left(\frac{0.19}{0.81}\right)\sigma_{col}^2 = 0.235\,\sigma_{col}^2 \tag{3.8}$$

$$\sigma_{ec} \leq 0.484\,\sigma_{col} \tag{3.9}$$

Thus, roughly, to observe real chromatographic peaks, we must keep the extra-column variance less than a quarter of the column variance, and the standard deviation due to extra-column effects must be less than half that due to the column. In a 2019 survey of extra-column band-broadening studies on commercial instrumentation, Desmet and Broeckhoven found values of σ_{ec}^2 for typical HPLC systems between 20 and 100 μL^2, while values between 5 and 10 μL^2 were found for UHPLC systems.[35] A comparison with the numbers given in Table 3.6 suggests that contemporary HPLC systems are adequate at least for retained peaks, while the performance of UHPLC systems is borderline in relation to that of contemporary UHPLC columns. Clearly, controlling the extra-column band broadening while maintaining the performance in terms of other aspects of HPLC (*e.g.* mixing of the sample and mobile phase and temperature control) is a challenge, even for expert engineers.

The observed extra-column band broadening will depend significantly on the diffusion coefficient of the analyte in the mobile phase and thus on the analyte molecular weight and the eluent viscosity. This means that the observed variance ($\sigma_{obs,i}^2$) for one analyte (*i*) cannot be corrected by subtracting the extra-column variance ($\sigma_{ec,j}^2$) measured for another analyte (*j*). In gradient-elution experiments, the pre-column band broadening is mitigated by on-column focussing for analytes for which the retention factor at the initial mobile-phase composition is high. The column band broadening is determined by the retention factor at the moment of elution (k_e; see Module 3.4). A value of $k_e = 3$ is a reasonable estimate for typical gradient-elution separations of low-molecular-weight analytes so that the numbers in the last column of Table 3.6 provide an indication of the peak width.

A common method to estimate the extra-column variance is to perform measurements of peak profiles with the column replaced by a zero dead volume (zdv) union. This method has significant practical issues and fundamental flaws because the dispersion contributions before and after the zdv union cannot be assumed independent, given that the flow profile is barely disturbed.[35] The extra-column variance can also be estimated by extrapolating the variance measured for analytes with different retention factors to the point where $(1 + k)^2 = 0$ or by measuring the variance for columns of different lengths and extrapolating to $L = 0$. In the first method, it is assumed that N is invariable despite variations in k and D_m. The influence of these parameters on the observed value for σ_{ec}^2 is reduced when working at a velocity above u_{opt} and at relatively low k values (below 3). When extrapolating the variance observed for a single analyte on columns with varying lengths, the assumption made (H independent of L) is more correct, although packing quality, pressure, and other factors may have minor effects. Varying the length by connecting small columns in series is more problematic, as it adds more connectors and tubing.

3.3.3.1 Connection Capillaries

Connection capillaries may contribute significantly to extra-column band broadening. The dispersion in long straight capillaries can be described by the **Taylor–Aris equation** (eqn (1.90)). However, for short capillaries (with diameter d_{cap} and length L_{cap}), a transition is required as follows[35]

$$\sigma_{V,cap}^2 = \frac{\pi \cdot F \cdot L_{cap} \cdot d_{cap}^4}{384\, D_m}\left(1 - \frac{1 - e^{-a \cdot L_{cap}}}{a \cdot L_{cap}}\right) \tag{3.10}$$

where F is the volumetric flow rate and $a = 15.04 \cdot \pi D_m/F$. The (bracketed) correction factor is seen to depend on the length of the connection capillary and the flow rate but not on the diameter of the capillary. For a 200 mm-long capillary at a flow rate of 1 mL min^{-1}, the correction factor is 0.24, indicating that the Taylor–Aris approximation (eqn (1.90)) greatly overestimates the dispersion. Still, the dispersion due to LC connections is a critical factor, and the internal diameter of the capillary has a dramatic fourth-power effect on the Taylor–Aris factor. If the above 200 mm capillary has an internal diameter of 250 μm, then it gives rise to a variance (at 1 mL min^{-1}) of about 25 μL^2 (or a standard deviation of about 5 μL). When comparing this value with the data given in Table 3.6, while bearing in mind eqn (3.9), the connector is seen to be adequate for HPLC, but not for UHPLC. For UHPLC, a 200 mm capillary should be as narrow as 100 μm i.d., in which case $\sigma_{V,cap} = 1.2$ μL. The flow rate has only a small effect. When $\sigma_{V,cap}$ is reduced to 0.5 mL min^{-1}, $\sigma_{V,cap} = 1.1$ μL. In addition, if the length of the connector can be reduced to 100 mm, we obtain $\sigma_{V,cap}= 0.6$ μL. These short and narrow connectors need to be cut straight and installed properly to avoid losing a significant fraction of the column efficiency. In the case of "dirty" (inhomogeneous) samples or in situations where analyte solubility may be a limiting factor (*e.g.* in polymer separations), such narrow capillaries are susceptible to clogging.

Special attention is needed for the coupling of LC to MS. Modern MS instruments have been optimized so as to contribute little to extra-column band broadening when connected properly with LC instruments.[36] However, the connection is often still the responsibility of the analyst, and connection capillaries that are too long and/or too wide are still all too common.

3.3.3.2 *Injection Band Broadening*

Injection band broadening can be ascribed to the volume of injection (V_{inj}) and to all dispersion processes taking place in and around the injector (flow path, connectors, *etc.*). If we group all dispersion that takes place before the column under injection band broadening, this can be summarized as

$$\sigma_{V,inj}{}^2 = \frac{V_{inj}{}^2}{\theta_{inj}} + \Sigma\sigma_{V,\text{pre-column}}{}^2 \tag{3.11}$$

where θ_{inj} is a factor related to the injection profile and the sum covers all band-broadening processes from the injector to the column, which are assumed to be independent. If a perfect plug of sample were to be injected ("block injection"), θ_{inj} would be equal to 12, but in practice, much lower values are found, down to $\theta_{inj} \leq 1$ for small volume injections. Due to the enormous range of θ_{inj} values reported, it is impossible to recommend any value.

3.3.3.3 *Detection Band Broadening*

For detection band broadening, an extra term may be introduced to account for the time constant τ_{det}. If we group all post-column dispersion under detection band broadening, we have

$$\sigma_{V,det}{}^2 = \frac{V_{det}{}^2}{\theta_{det}} + \frac{\tau_{det}{}^2 \cdot F^2}{12} + \Sigma\sigma_{V,\text{post-column}}{}^2 \tag{3.12}$$

A value of $\theta_{det} = 1$ is often used, but as in the case of injection band broadening, a broad range of values can be found in the literature, in part because some of the band broadening in capillary tubing and connectors has been accumulated in the first term on the right-hand side of eqn (3.12). The time constant must be low in UHPLC to minimize the contribution of the second term on the right-hand side of eqn (3.12). At a flow rate of 0.5 mL min^{-1}, a time constant of 1 s is unacceptable, as it yields a variance of $\sigma_{V,det}{}^2 = 5.8~\mu L^2$ ($\sigma_{V,det} = 2.4~\mu L$), which is much larger than half the values for the standard deviations under UHPLC conditions listed in Table 3.6. Values of $\tau_{det} = 0.2~s$ ($\sigma_{V,det}{}^2 = 0.23~\mu L^2$; $\sigma_{V,det} = 0.48~\mu L$) or $\tau_{det} = 0.1~s$ ($\sigma_{V,det}{}^2 = 0.06~\mu L^2$; $\sigma_{V,det} = 0.24~\mu L$) do suffice for UHPLC detectors operated at 0.5 mL min^{-1}. A small time constant does imply more detector noise, as does a small detector cell. In contrast, peak dilution (see Section 1.4.5, eqn (1.34)) is limited, thanks to the high number of plates produced in a small column volume.

3.3.3.4 *Pre-column vs. Post-column Band Broadening*

If an analyte is injected in an injection solvent that is weaker than the mobile phase, it will be focussed at the top of the column. In contrast, when an analyte band leaves the column, its volume will increase because the analyte needs to be "extracted" from the stationary phase. As a result, we may write[35]

$$\sigma_{col}^2 = \sigma_{V,inj}^2 \left(\frac{1 + k_{i,e}}{1 + k_{i,inj}} \right)^2 + \sigma_{V,det}^2 \tag{3.13}$$

where $k_{i,inj}$ is the retention factor of analyte i at the moment of injection and $k_{i,e}$ is its retention factor at the moment of elution. $\sigma_{V,inj}^2$ is given by eqn (3.11) and $\sigma_{V,det}^2$ by eqn (3.12). If $k_{i,inj} \gg k_{i,e}$, all dispersion of the analyte band before the column is effectively overcome by on-column focussing and the first term on the right-hand side of eqn (3.13) can be neglected. The easiest way to achieve focussing is to apply gradient elution starting with a weak solvent. However, large volumes of sample dissolved in a solvent that is a strong eluent may give rise to a breakthrough (see also Section 3.5.5). Among the strategies to achieve analyte focussing are some of the active-modulation approaches discussed in Section 7.3.4 (active solvent modulation and at-column dilution).

> See the website for a calculation tool on extra-column band broadening: ass-ets.org

3.3.4 High-temperature Liquid Chromatography

Working at elevated temperatures brings some advantages (see Table 3.7). Diffusion coefficients increase with increasing temperature, causing the optimum linear velocity ($u_{opt} = v_{opt} \cdot D_m/d_p$) to increase. When working above the optimum velocity, the efficiency may increase with increasing temperature, but it does not when working at u_{opt} (at all temperatures), as $H_{min} = h_{min} \cdot d_p$ is independent of D_m. The viscosity of liquids decreases with increasing temperature, which implies that at a constant linear velocity, the column pressure will decrease. According to the Darcy equation (eqn (1.69)), $\Delta P \propto \eta \cdot$

u_0. However, if we work at the optimal velocity and apply the Stokes–Einstein equation (eqn (1.89)), which predicts that $\eta \cdot D_m \propto T$, we obtain

$$\Delta P_{opt} \propto \eta \cdot u_{opt} \propto \eta \cdot D_m \propto T \qquad (3.14)$$

This implies that the required pressure drop will increase somewhat with increasing temperature. For example, when moving from 25 °C (298 K) to 80 °C (353 K), the pressure drop is expected to increase by about 20%.

When the temperature increases, retention factors usually decrease. To avoid this leading to a decrease in selectivity and resolution, it may be compensated by a decrease in the eluent strength, usually by increasing the concentration of the weaker solvent in the mobile phase. The effect of temperature tends to be much smaller than that of the mobile-phase composition. For example, in reversed-phase LC, the above increase in temperature from 25 °C to 80 °C may be compensated by increasing the water content of the mobile phase by 5% to 10%. One optimistic scenario for HTLC is to develop into a "green" separation method, working with pure water as the eluent and the highly sensitive flame-ionization detector (see Section 2.5.1). Unfortunately, even at high temperatures, only a very limited number of analytes yield optimal retention factors with 100% water as the eluent. Moreover, high-temperature water is very corrosive, and FIDs can easily be damaged. High-temperature mobile phases also tend to reduce the lifetime of columns, especially those containing silica-based stationary phases. The solubility of silica rapidly increases with increasing temperature so that alternative materials are needed (see Section 3.2.3).

Controlling the temperature in LC is not trivial. The temperature of the mobile phase entering the analytical column is essential because it cannot be changed rapidly once inside the column due to the relatively small surface area of the column and the poor heat-conducting properties of the packing. A heat exchanger can be positioned before the injector so that it does not contribute to extra-column band broadening (but does to the dwell time; see Module 3.4). Inside the column, frictional heating may occur, which also affects the mobile-phase temperature (see Section 3.3.2). Detectors such as mass spectrometry or an evaporative light-scattering detector may benefit from a high temperature mobile phase, but some other detectors may require prior cooling of the mobile phase.

Table 3.7 Advantages and disadvantages of high-temperature LC.

Advantages	Disadvantages
• Higher diffusion coefficients → higher optimum linear velocities	• Adequate temperature control is difficult
• Lower viscosity → lower pressure drop (at the same linear velocity)	• Mobile phase may become corrosive (especially aqueous solvents) • Column lifetime decreases
• High temperatures may be needed to dissolve analytes (*e.g.* high-temperature size-exclusion chromatography; see Module 4.2)	• Cold spots or power failure may cause analyte precipitation
• Reduced amounts of organic modifier needed ("greener" methods)	• Risk of analyte degradation or chemical reactions
• May be favourable for coupling with MS (see Section 3.10.1) or with an evaporative light-scattering detector (ELSD, see Section 3.9.4)	• Cooling before or pressurization after the detector may be required • HTLC with pure water as eluent, and flame-ionization detection is largely an illusion

A heat exchanger between the column and the detector does contribute to the extra-column band broadening. In case the temperature approaches the boiling point of the solvent, it may be necessary to have a restrictor in line after the detector. This restrictor (*e.g.* a piece of capillary tubing) should only slightly increase the pressure in the detector (by 0.1 to 0.3 MPa) because most LC detectors cannot withstand high pressures.

Analyte stability is a trade-off between the rate of degradation, which is expected to increase with increasing temperature, and the time spent in the column, which may decrease considerably.[37] The risk of analyte degradation may not increase much at moderate temperatures, but this will need to be verified for individual analytes. The first indication that analyte degradation takes place may be a peak that is broader than expected based, for example, on a comparison with a chromatogram recorded at a lower temperature or with other peaks in the chromatogram. Reactions at the time-scale of the separation may give rise to distorted or bimodal peaks (including so-called "Batman peaks").[38] Such peak shapes may be used to deduce reaction-rate constants.

3.3.5 Miniaturization

Packed capillaries have no fundamental advantages in comparison with larger packed columns, but there are some practical advantages, such as easier ways to make long columns, better temperature control, and good possibilities for interfacing with MS. The first two advantages are due to the properties of fused silica, which allow rapid heating and cooling. Nano-LC has been advocated for high-resolution LC-MS separations of protein isoforms using long (e.g. 1 m) packed capillaries at very low flow rates.[39] Such separations can be efficient due to the very low diffusion coefficients of proteins, resulting in low optimal linear velocities (see also Module 3.16). Long analysis times result, but these pale in comparison to the time required to interpret the resulting MS data.

Chip-based LC has had limited success. The main reason for this is the need for high pressures. This means that for a microchip to be applicable in HPLC, both the chip itself and its connections to the outside world, especially to the pump, need to be resistant to pressure. If large instruments (such as pumps and mass spectrometers) need to be connected to a chip, the (ultra-)miniaturization seems to be defeating the purpose, unless the chip format brings genuine advantages when compared to column-based systems. One example is the micro-pillar-array columns discussed in Section 3.2.5.1, which have the potential to outperform packed columns. The group of Detlev Belder (Universität Leipzig, Germany), who have been pioneering high-pressure LC chips, have identified the low thermal mass and the resulting potential for very fast temperature control as another advantage of chip-based LC. They demonstrated high-temperature and temperature-programmed LC-MS separations with short analysis times (20 to 200 s) and very short cooling down times. However, the chromatographic performance was limited.[40] Although good separation power per column length was reported (in excess of 50 000 plates per metre), the short column length (35 mm) implied an available plate count below 2000. Spatial multidimensional LC is another possible unique application of microchip LC.[41] Comprehensive two-dimensional spatial liquid chromatography is akin to two-dimensional thin-layer chromatography but performed at high pressure in a miniaturized system. Comprehensive three-dimensional spatial liquid chromatography would add a third dimension, which would provide unprecedented possibilities in terms of selectivity and separation power (peak capacities of the order of one million at modest pressures of 5 MPa.[42] However, such high-performance spatial separations have proven hard to realize in practice.

3.4 Gradient Elution (M)

3.4.1 Introduction to Gradient Elution

Due to the dramatic effect of the mobile-phase composition on the retention of analytes, changing the former during the run is a very powerful tool in LC. In that sense, mobile-phase composition in LC plays the same role as temperature in GC, *i.e.* the main parameter by which to tune retention and encompass a wide range of analytes in a single chromatogram. In GC, programming the temperature ("temperature programming") accommodates a wide range of boiling points of analytes that can be eluted. In broad terms, programming the mobile-phase composition ("solvent programming") accommodates a wide range of analyte properties (*e.g.* polarities). There are two main goals of solvent programming: (i) to provide conditions such that all relevant analytes can be eluted in a single run under near-optimal conditions and (ii) to quickly scan the sample to establish the "polarity" (or, better, the "retentivity") of the relevant analytes.

LC separations can basically be conducted in two ways. In **isocratic elution**, the mobile-phase composition does not change during the analysis; in **gradient elution**, it does. The word "gradient" suggests a gradual change in composition, but sudden changes in composition are included in the gradient category as "step gradients" (even if this seems an oxymoron). Thus, there are linear, convex, concave, step and multi-segment gradients. This is shown in Figure 3.19, in which φ is used as a universal parameter to indicate composition (*e.g.* the volume fraction of modifier in RPLC or the ionic strength in IEC). Linear gradients, in which the mobile-phase composition changes linearly with time, are by far the most common. As long as a single column is considered, gradients are always programmed such that the elution strength increases, so that gradually more components start migrating through the column. The opposite situation, where components gradually get more stuck in the column, does not make sense. The one exception to this rule, gradient size-exclusion chromatography, will be described in Module 4.3.

When performing isocratic analysis, the mobile phase can be premixed. This allows maintaining a constant composition,

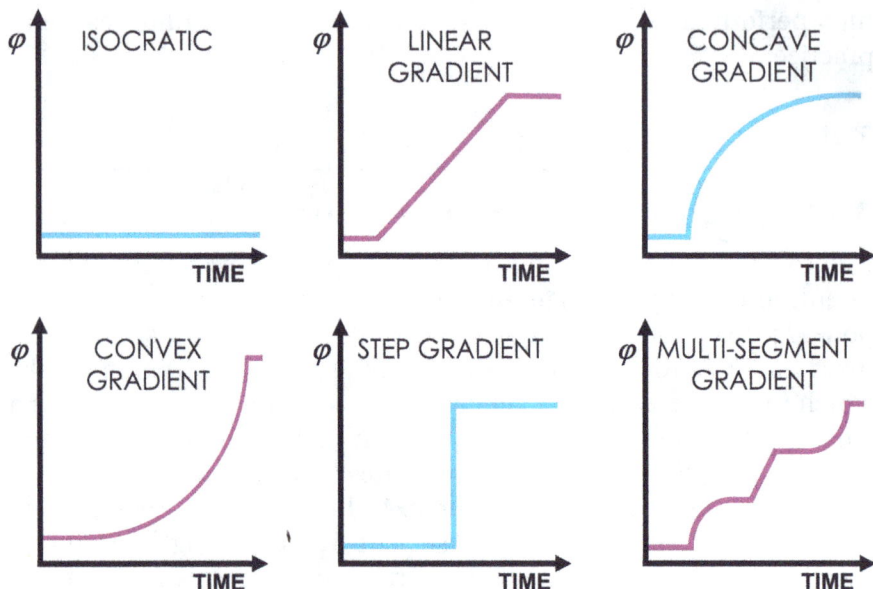

Figure 3.19 Different types of LC gradients.

provided that the composition of the solvent in the reservoir does not change over time. The latter may occur, for example, because of selective evaporation of more-volatile components, which may be accelerated if helium purging is used for degassing. Isocratic LC requires the simplest LC equipment, with just a single high-pressure pump. Flushing with a strong solvent to clean the system and the column may be achieved with a solvent-selection valve, installed before the pump. Gradients are almost invariably produced by dedicated instruments (see Section 3.3.1.2), which can be programmed to produce mobile phases with specified compositions, by mixing two or more solvents or solvent mixtures (the mobile-phase components) from different reservoirs. Using such instruments, the composition may be constant (isocratic elution) or a function of time, as indicated in Figure 3.19. There are basically two ways in which solvents can be mixed online in LC systems, *viz.* high-pressure and low-pressure mixing, as illustrated for binary systems in Figure 3.14 (see Section 3.3.1.2).

Gradient elution has numerous significant advantages in comparison with isocratic elution. Most of all, a greater range of analytes can be analysed in one run. In some situations, isocratic elution is still dominant. For example, the purity of an active pharmaceutical ingredient (API) with known contaminants may be monitored by isocratic LC. However, purity screening for unknown contaminants benefits greatly from gradient elution (and from universal detection methods). As we often deal with complex samples in LC, and because good equipment has become amply available, gradient elution has become the dominant mode.

The advantages and disadvantages of gradient-elution LC are summarized in Table 3.8. It is seen that gradient elution comes at a price. Universal detectors cannot be used because the large changes in solvent composition will obscure the presence of analytes, invariably at low concentrations. However, the advantages of gradient elution tend to outweigh the disadvantages in most cases.

3.4.2 Retention Times Under Gradient-elution Conditions

Two main factors determine the behaviour of analytes under gradient conditions: (i) the gradient program (*i.e.* the relationship between the mobile-phase composition and time) and (ii) the relationship between retention and the mobile-phase composition (see Figure 3.20). It is important to scale the first relationship to the hold-up time (t_0) of the column. For example, if we change the mobile-phase composition from 0 to 1 (*e.g.* 0% to 100% of ACN in RPLC) in 10 min and the conditions (column and flow rate) are such that t_0 equals 1 min, then the gradient encompasses ten column volumes $(t_G/t_0 = 10)$, which may be considered a typical gradient. However, if such a gradient is applied on a longer column at a slower flow rate with $t_0 = 10$ min, the entire gradient takes place within a single column volume and it is considered a fast gradient. Hence, the effective (dimensionless) slope of the gradient is related to $d\varphi/d(t/t_0)$ or for a linear (segment of a) gradient $(\Delta\varphi \cdot t_0)/t_G$, where $\Delta\varphi$ is the change in composition $(\varphi_{final} - \varphi_{init})$ and t_G is the duration of the linear segment. The second factor (ii above) is called the retention model. In Module 3.8, we will discuss several retention models in detail. In Module 3.17, we will take a close look at the relation between the gradient program and the observed retention behaviour of the analytes. In this module, we will only discuss the simplest (but most important) case briefly.

The simplest form is the log-linear model (eqn (3.46); see Section 3.8.3.1), in which the logarithm of the retention factor varies linearly with φ and the extent of variation (slope) is defined by $S' = \partial(\ln k)/\partial\varphi = 2.303 \cdot \partial(\log k)/\partial\varphi = 2.303 \cdot S$. If both the retention model ($\ln k$ or $\log k$ *vs.* φ) and the gradient (φ *vs.* t/t_0) are linear, a gradient results in which $\ln k$ (or $\log k$) varies linearly with time (t/t_0). Such a gradient was defined by Snyder *et al.* as a linear solvent strength (LSS) gradient and stood at the basis of a detailed description of gradient-elution LC.[43-45] The log-linear model is often used to describe retention in RPLC. In other cases (*e.g.* IEC), a log–log retention model is more appropriate, and a convex gradient is required to meet the conditions of the LSS theory.

We should consider three different situations for a simple (one-segment) linear gradient that starts with an initial hold time t_{init} at the initial composition φ_{init}, followed by a linear increase in the organic-modifier content during a time t_G to a final composition φ_{final}, where it is held for a time t_{final} before returning to the initial composition for re-equilibration. Early eluting analytes may migrate through the column without experiencing the gradient at all. Their retention time will be determined by their retention factor at the initial composition (k_{init}), *viz.*

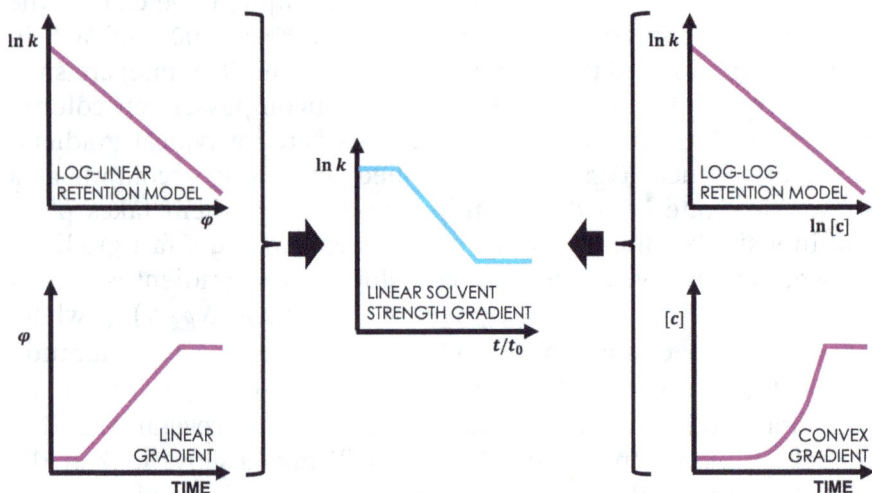

Figure 3.20 Factors contributing to the generation of a linear solvent strength (LSS) gradient, in which by definition the (natural) logarithm of the analyte retention factor ($\ln k$) varies linearly with the (normalized) time (t/t_0). The log-linear model (left) is frequently used to describe retention in reversed-phase LC (ϕ is the volume fraction of organic modifier). The log–log model describes ion-exchange chromatography ($[c]$ is the dimensionless counterion concentration, *i.e.* the actual concentration divided by unit concentration).

Table 3.8 Advantages and disadvantages of isocratic and gradient-elution LC.

Isocratic separations	Gradient elution
Strengths	**Strengths**
• No time lost between analyses for column re-equilibration • Best possible separation for any specified pair of analytes • Simplest, cheapest, and most reliable mode of LC	• Broad range of analytes covered • Optimum elution conditions for most peaks • Sharper (late-eluting) peaks • Increased sensitivity (for late-eluting peaks)
Weaknesses	**Weaknesses**
• Limited range of analytes in one run • Possible build-up of (non-eluted) contaminants	• Time needed for column re-equilibration • Limited choice of detectors • Higher noise (due to incomplete mixing) • Higher background (baseline drift) • Spur peaks (contaminants in weak eluent) • Increased (instrument) complexity and cost

$$t_{R,\text{before}} = t_0(1 + k_{\text{init}}) \tag{3.15}$$

The onset of the gradient, which is programmed to occur at $t = t_{\text{init}}$ will arrive at the end of the column (length L) at $t = t_{\text{init}} + t_0 + t_D$, where t_D is the dwell time, which is the time elapsed between a software instruction to change the composition and the time this change arrives at the column inlet (see also Section 3.3.1.2). Hence, eqn (3.15) is applicable if

$$t_{R,\text{before}} = t_0(1 + k_{\text{init}}) \le t_{\text{init}} + t_0 + t_D \tag{3.16}$$

or

$$k_{\text{init}} \le \frac{t_{\text{init}} + t_D}{t_0} \tag{3.17}$$

If this is not the case, the gradient will catch up with the analyte. For analytes eluting during the linear part of the gradient, the following equation can be derived (see Module 3.17)

$$t_{\text{R,gradient}} = \frac{t_G}{S'\Delta\varphi}\ln\left\{1 + S'\frac{\Delta\varphi}{t_G}k_{\text{init}}\left[t_0 - \frac{t_D + t_{\text{init}}}{k_{\text{init}}}\right]\right\} + t_0 + t_D + t_{\text{init}} \qquad (3.18)$$

This equation applies if eqn (3.16) does not hold (*i.e.* if the analyte does not elute before the onset of the gradient) and

$$t_{\text{R,gradient}} \leq t_{\text{init}} + t_0 + t_D + t_G \qquad (3.19)$$

where $t_{\text{R,gradient}}$ is obtained from eqn (3.18) and the sum on the right-hand side corresponds to the time at which the end of the (linear segment of the) gradient arrives at the end of the column. If eqn (3.19) does not hold, we have

$$t_{\text{R,after}} = k_{\text{final}}\left(t_0 - \frac{t_D + t_{\text{init}}}{k_{\text{init}}}\right) + \frac{t_0}{b}\cdot\frac{k_{\text{final}} - 1}{k_{\text{init}}} + t_0 + t_D + t_G \qquad (3.20)$$

provided that

$$t_{\text{R,after}} \leq t_0 + t_D + t_G + t_{\text{final}} \qquad (3.21)$$

If the last condition is not met, the analyte will not be eluted before the eluent weakens again, so that it will remain on the column. If this happens during a series of runs, the analyte may accumulate and contaminate or ultimately block the column.

The different regions are illustrated in Figure 3.21. In this case, $t_D = 2$ min, $t_{\text{init}} = 7$ min, and $t_0 = 3$ min, so that eqn (3.15) applies during the first 12 min of the chromatogram. The linear segment of the gradient is 30 min long, so that eqn (3.18) applies in the range $12 \leq t_R \leq 42$ min. The final hold time (t_{final}) equals 28 min, so that eqn (3.20) applies when $42 \leq t_R \leq 70$ min. No peaks will elute later than 70 min because the eluent gets weaker after this point. Peaks eluting during the gradient are approximately equally broad (barring co-elution), but peaks eluting after 42 min elute under isocratic conditions, causing them to broaden significantly.

Let us now focus on analytes that elute where we want them to, *i.e.* in the linear part of the gradient, where eqn (3.16) does not hold, but eqn (3.19) does. If we introduce $b = S' \cdot B \cdot t_0$ as the effective gradient steepness in eqn (3.18) and if conditions are such that the analytes are highly retained under the starting conditions (large k_{init}), we obtain

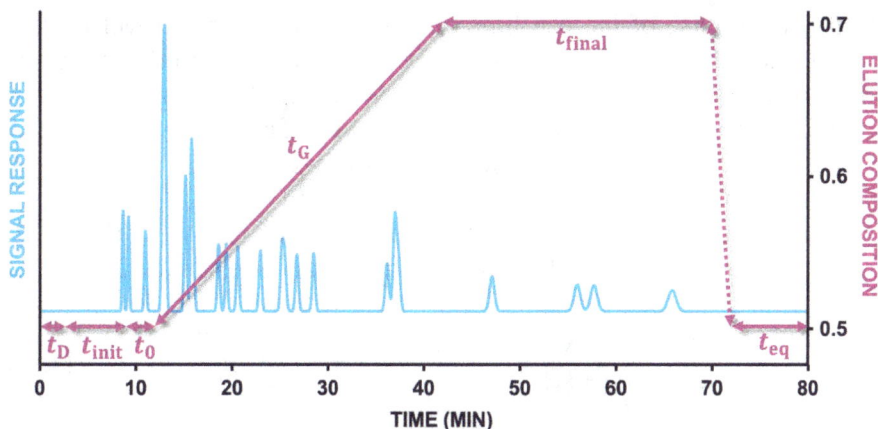

Figure 3.21 Gradient-elution chromatogram of a mixture of 32 small-molecule aromatic compounds. Some of these co-elute, while some do not elute in this program. Dwell time (t_D) 2 min; initial hold time (t_{init}) 7 min; hold-up time (t_0) 3 min; gradient duration (t_G) 30 min; final hold time (t_{final}) 28 min; return to the initial composition in 2 min; and equilibration time (t_{eq}) 8 min.

$$t_{R,gradient} \approx \frac{t_0}{b}\ln(b \cdot k_{init}) + t_0 + t_D + t_{init} \tag{3.22}$$

At this point, the composition at the end of the column (the elution composition, φ_e) is

$$\varphi_e \approx \varphi_{init} + \frac{\Delta\varphi}{t_G} \cdot \frac{t_0}{b}\ln(b \cdot k_{init}) = \varphi_{init} + \frac{1}{S'}\ln(b \cdot k_{init}) \tag{3.23}$$

and the retention factor of the analyte at the moment of elution (k_e), which is the all-important parameter determining the peak width observed in the chromatogram, follows from

$$\ln k_e \approx \ln k_0 - \varphi_{init} \cdot S' - \ln(b \cdot k_{init}) = \ln k_{init} - \ln(b \cdot k_{init}) = -\ln b \tag{3.24}$$

or simply

$$k_e \approx \frac{1}{b} = \frac{t_G}{S' \cdot \Delta\varphi \cdot t_0} \tag{3.25}$$

When b is large and k_e is very small. It can also be demonstrated that $k_e \approx 2/b$ at the halfway point through the column, implying that (in the case of large values of b) the components travel largely unretained through the column. This gives rise to limited separation power (and typically little resolution) when applying fast gradients to high-molecular-weight analytes (see Module 4.3). Values of 0.3 < b < 1 (*i.e.* < k_e < 3) are typically considered optimal for efficient gradient-elution separations.

If we assume equal S' values for a group of analytes, the peak capacity under gradient-elution conditions can be estimated from

$$n_p \approx \frac{t_G}{4\sigma_t} \approx \frac{t_G \cdot \sqrt{N}}{4t_0(1 + k_e)} \approx \frac{t_G \cdot \sqrt{N}}{4t_0} \cdot \frac{b}{1 + b} = \frac{S'\Delta \cdot \varphi \cdot \sqrt{N}}{4(1 + b)} \qquad (3.26)$$

This implies that slow gradients (*i.e.* low b values) can result in high peak capacities, especially for high-molecular-weight analytes (high S' values; proteins/peptides and polymers). In case all S' values are large and the gradient is relatively fast, the limiting peak capacity is

$$n_p \approx \frac{t_G \cdot \sqrt{N}}{4t_0} \qquad (3.27)$$

This shows that long gradients are favourable, while longer columns may not be. Lengthening the column, while keeping all other factors constant, increases t_0 and N proportionally, so that the net result is a decrease in the maximum attainable peak capacity $\left(n_p \propto 1/\sqrt{L}\right)$. The best way to maximize the factor $\sqrt{N}/t_0$ is to use kinetic plots, as described in Module 3.16.

3.5   Reversed-phase Liquid Chromatography (B)

3.5.1   Introduction to Reversed-phase Liquid Chromatography

By definition, in reversed-phase liquid chromatography (RPLC), the mobile phase is more polar than the stationary phase. RPLC conquered the analytical world for good as early as the 1970s, *i.e.* in the second decade of HPLC, when good chemically bonded octyl and octadecyl columns started to emerge. The possibilities of RPLC are enormous. The vast majority of known (organic) chemicals have a polarity somewhere in between that of saturated hydrocarbons and that of water. For all these compounds, it is likely possible to create an RPLC system in which they are distributed (approximately) evenly between a stationary phase (that may be as apolar as saturated hydrocarbons) and a mobile phase (that may be as polar as water). Moreover, in a gradient-elution experiment, all kinds of molecules from such a diverse range may be eluted under near-optimal conditions (*e.g.* a retention factor at the moment of elution $k_e \approx 3$; see Module 3.4) in a single run.

For compounds that are very non-polar, we may have to resort to normal-phase liquid chromatography (Section 3.6.1) or super-critical-fluid chromatography (Chapter 6). For very polar or ionic analytes, we may need hydrophilic-interaction liquid chromatography (Section 3.6.2) or ion-separation methods (Module 3.7 for LC; Chapter 5 for capillary electrophoresis) to generate sufficient retention, but ionization may be suppressed by adapting the pH or by adding an ion-pair reagent to render many ionogenic analytes compatible with RPLC. For all other types of analytes, RPLC can be applied, and it comes with several attractive attributes.

- RPLC is immensely flexible because the polarity range spanned by water and the low-polar THF (two fully miscible solvents) at the extremes is enormous.
- RPLC provides a high selectivity, thanks to the large polarity differences possible between a water-based mobile phase and a hydrocarbon-based stationary phase.
- RPLC allows the optimization of retention and selectivity for a very broad range of samples on a single (type of) column, by just varying the mobile phase.
- RPLC can deal with aqueous samples, of which there are very many.
- RPLC allows fast equilibration, making (fast) gradient-elution LC possible.
- RPLC allows the addition of salts, buffers and ion-pair reagents to the mobile phase to further control retention and enhance selectivity.

Thanks to all these advantages – and more – RPLC has by far become the most widely applied LC technique, prompting Csaba Horváth to suggest that the "increasingly abnormal" normal-phase liquid chromatography (NPLC) be renamed "reversed-reversed-phase" LC.[46]

The **octadecyl silica (ODS)** phases described in Section 3.2.2.2 are the most used stationary phases in RPLC. Other stationary phases (see Table 3.4) are used less often. Columns can be classified in terms of (more or less) retention, selectivity (differences in retention), peak width and peak shape. Usually, a specific mixture is used to characterize a column. Examples include the Neue test and the Tanaka test (see Section 3.5.6). More advanced stationary-phase characterization and classification schemes are discussed in Module 3.8.

The first place to look for optimizing retention and, to a lesser extent, selectivity is the mobile phase, which typically is a mixture of water with an organic modifier. In the rare case of non-aqueous RPLC (used, for example, for separating triglycerides), the mobile phase is still miscible with water and the separation system is not distorted by the presence of water in the sample. Methanol (MeOH) and ACN are used most often as organic modifiers. THF is the strongest common modifier in RPLC, but it is less attractive than MeOH and ACN. Some of the strengths and weaknesses of possible modifiers are summarized in Table 3.9.

Water is the **weak solvent** in RPLC. For most compounds, retention factors in highly aqueous mobile phases are very high. Operation of ODS columns with pure water as the mobile phase is not recommended, as this leads to reduced retention and poorly shaped peaks, presumably because the layer of alkyl chains is not solvated by the mobile phase ("non-wetting" conditions). A highly aqueous mobile phase may also be excluded from the hydrophobic pores.[47] Generally, it is not advisable to operate alkyl-modified silica columns with mobile phases containing more than 95% water because equilibration times may increase and chromatographic precision (repeatability of retention times and peak shapes) may decrease. Stationary phases with embedded polar groups may alleviate the "dewetting" problem to some extent.

Retention decreases exponentially upon adding increasing concentrations of organic modifier to the mobile phase (see Module 3.8). Columns with (somewhat) more polar stationary phases may exhibit fewer problems with dewetting, but these may defeat the purpose because they are less retentive and the very reason to explore highly aqueous mobile phases is the failure to achieve sufficient retention otherwise. For highly polar analytes that do not

Table 3.9 Some of the strengths and weaknesses of the most common organic modifiers used in RPLC.

Modifier	Strengths	Weaknesses
Methanol (MeOH)	• Relatively cheap • Ecologically friendly ("green")	• High viscosity when mixed with water • Relatively high UV cut-off (205 nm)
Ethanol	• Less toxic ("greener") alternative to methanol	• Higher viscosity than methanol (in mixtures with water)
2-Propanol	• Strong modifier (lower percentages required) • May be used to clean LC systems and columns, thanks to high drag forces	• Highly viscous
Acetonitrile (ACN)	• Relatively low viscosity • Low UV cut-off (190 nm)	• Toxic • Expensive
Acetone	• Relatively cheap • Relatively low viscosity	• Not transparent in UV
Tetrahydrofuran (THF)	• Strongest solvent (for lipids and polymers)	• High UV cut-off (approximately 215 nm)[a] • Unstable

[a]This applies to "fresh" (recently opened bottle), non-stabilized THF. The cut-off wavelength is often higher (220–230 nm) in practice.

exhibit sufficient retention in RPLC, HILIC has been developed (see Section 3.6.2).

The ODS phase is not akin to a film of a liquid alkane. In contact with the mobile phase, an equilibrium state is established between the stationary phase and the mobile phase. Under isocratic conditions, the mobile-phase composition inside the column reaches a steady state where it equals the composition delivered to the column, as it keeps being replenished while equilibrium is established. The stationary layer features alkyl chains solvated by organic

SOLVENT	η (mPa s)
Water	0.89
Methanol	0.54
Acetonitrile	0.37
Tetrahydrofuran	0.46
50% Methanol/Water	1.55
50% Acetonitrile/Water	0.80

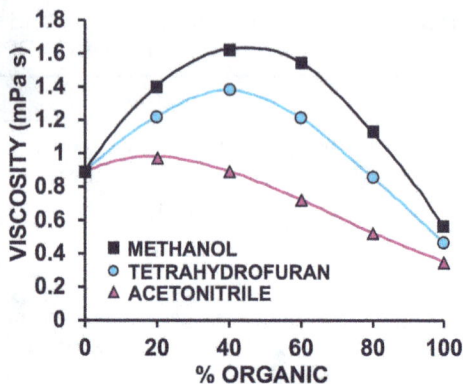

Figure 3.22 Viscosity of pure mobile-phase components and mobile-phase mixtures at 25 °C. Squares: methanol, circles: THF, triangles: acetonitrile.

modifiers but may also contain some water molecules, for example, in the vicinity of residual silanols at the surface. Water is more likely to accumulate at the surface at high concentrations of organic modifier (HILIC range; see Section 3.6.2). The apparent polarity of the stationary phase will be somewhere in between that of an alkane (ODS) and the modifier.

This dynamic equilibrium makes it difficult to establish exactly which molecules belong to the stationary phase and the mobile phase. In practical terms, it is difficult to establish an exact value for t_0 in RPLC (see Section 3.1.4). The value of t_0 may vary with the mobile-phase composition in isocratic RPLC and will even vary during the run in gradient RPLC. In describing gradient-elution LC (Module 3.4), a pragmatic assumption is usually made on a constant value for t_0, but one should be aware that this is a simplification.

3.5.2 Mobile-phase Viscosity

The viscosity of hydro-organic solvent mixtures is illustrated in Figure 3.22. It is seen that the **viscosity** of mixtures can greatly exceed that of the pure components. This is especially true for mixtures of methanol and water (squares) and THF and water (circles). In the case of gradient elution, the mobile-phase viscosity varies drastically during the run.

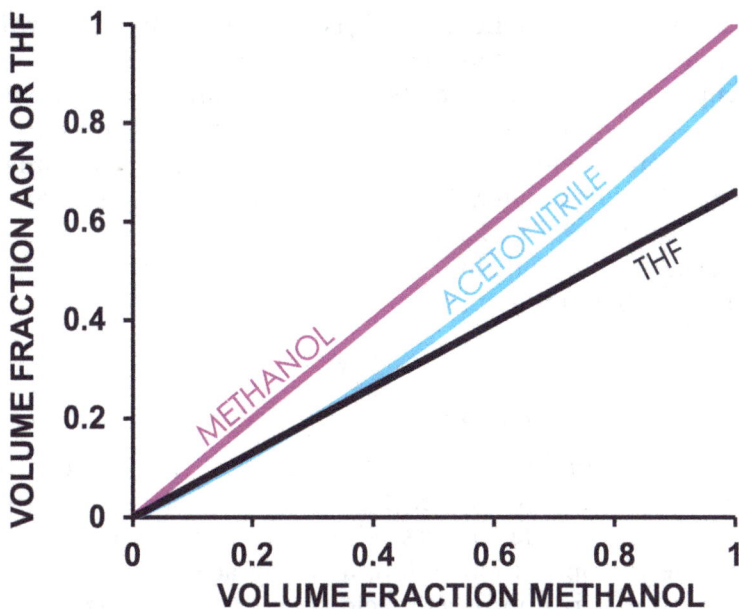

Figure 3.23 Iso-eluotropic binary mixtures in RPLC. Based on data from ref. 48.

3.5.3 Iso-eluotropic Mixtures

We mentioned earlier that the strength of modifiers increases in the order MeOH < ACN < THF. This has been quantified by measuring the retention of a series of low-molecular-weight compounds in mixtures of water with the three different modifiers and establishing which compositions yielded the same retention on average.[48] For such mixtures, the term **iso-eluotropic** was introduced. The result of the comparison of the three modifiers is illustrated in Figure 3.23 and can be expressed by the following simple relationships

$$\overline{\varphi_{ACN}} = 0.32\, \varphi_{MeOH}^{2} + 0.57\, \varphi_{MeOH} \tag{3.28}$$

$$\overline{\varphi_{THF}} = 0.66\, \varphi_{MeOH} \tag{3.29}$$

where $\overline{\varphi_{ACN}}$ and $\overline{\varphi_{THF}}$ are the average volume fractions of ACN and THF, respectively, which yield the same retention as a volume fraction φ_{MeOH} of MeOH (in all cases with water being the other component in the binary mobile-phase mixture).

For individual components, such iso-eluotropic mixtures will not provide identical retention factors. Different modifiers give rise to different selectivities and may be used to optimize separations (see Chapter 10). Any mixture of two iso-eluotropic mixtures is again approximately iso-eluotropic. For example, mixtures containing 50% methanol and 50% water, 36.5% ACN and 63.5% water, 33% THF and 67% water, and 18.25% ACN, 16.5% THF, and 65.25% water are all expected to yield similar retention times for the analytes in a given sample but different selectivities.

3.5.4 Mobile-phase pH

The separation of ionizable analytes usually requires control of the pH of the mobile phase. The ionization of weak acids and surface silanols can be suppressed by acidifying the mobile phase. To effectively suppress ionization, a strong acid can be added. For example, 10 mM trifluoroacetic acid (pK_a = 0.23) results in a pH of about 2. However, this is not a buffer, and the acidity of the mobile phase may be affected, for example, by the sample. Better control of the mobile-phase pH requires the addition of a buffer. In Figure 3.24, the buffer capacity of some common RPLC buffers *in water* is indicated (as discussed later in this section, the pK_a and pH values change when an organic modifier is added to the mobile phase). A buffer forms when significant amounts of a conjugated acid–base pair are present in the solution. Adding a small amount of an acid or a base (at a much lower concentration than the buffer, *e.g.* in the form of a sample) will shift the equilibrium slightly but will barely affect the pH. Buffers with an accurate pH (again, *in water*) can be prepared by adding a weighted amount of an acid and a conjugated base.

As seen in Figure 3.24, there are several options for low-pH buffers. Phosphate buffers can be used around pH = 2 or 7, but phosphate buffers with 4 < pH < 5.5 do not exist. Citric acid buffers are attractive for LC because they cover most of the common working range on ODS-type stationary phases. Citrate salts are also better soluble in aqueous–organic mixtures than phosphate salts, and they are more compatible with MS detection, although ammonium acetate or ammonium formate is generally preferred in the latter case. Several buffers are shown in Figure 3.24 that allow buffering at high pH, where the ionization of weak bases may be suppressed. This is challenging in the case of silica-based columns. Although some

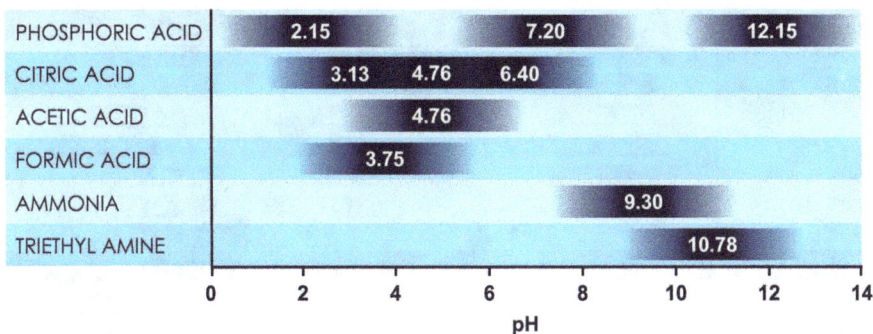

Figure 3.24 Ranges of some common RPLC buffers. Darker shading indicates higher buffer capacity.

modern silica-based or hybrid phases allow operation at pH 9 or 10, the deprotonation of silanol groups creates the potential for long-range ionic interactions with (partially) charged analytes.

The fact that RPLC separations almost always use mixed aqueous–organic mobile phases has enormous consequences when dealing with ionizable analytes. There are basically three ways to measure and deal with pH in RPLC, as documented in detail by Martí Rosés from Barcelona (Spain):[49]

- measuring or specifying the pH of the aqueous component of the mobile phase before adding an organic modifier (^W_wpH method);
- measuring the pH in the prepared aqueous–organic mobile phase after the calibration of the pH meter using aqueous standard buffers (^s_wpH method); and
- measuring the pH in the prepared aqueous–organic mobile phase after the calibration of the pH meter using standard buffers in the same mobile-phase mixture (^s_spH method).

The results can be dramatically different, as the actual pH (*i.e.* the negative logarithm of the hydrogen-ion activity) and the pK_a values of buffers and analytes vary substantially with the composition of the mobile phase. As an illustration, the approximate variations in pK_a values for phosphate and ammonia buffers are summarized in Figure 3.25. Preparing a buffer from a 1:1 (molar) ratio of mono-hydrogen phosphate (*e.g.* K_2HPO_4) and di-hydrogen phosphate (*e.g.* KH_2PO_4) in water yields a pH of 7.2, while in 60% ACN, the pH becomes about 8.7, and in 80% methanol, it becomes about 9.7. The

PHOSPHORIC ACID			
IN WATER	2.15	7.20	12.15
20% ACN	ca. 2.25	ca. 7.64	ca. 12.52
40% ACN	ca. 2.49	ca. 8.08	ca. 12.95
60% ACN	ca. 2.85	ca. 8.66	ca. 13.76
40% MeOH	ca. 2.64	ca. 8.19	ca. 12.29
60% MeOH	ca. 2.95	ca. 8.94	ca. 12.36
80% MeOH	ca. 3.76	ca. 9.65	ca. 12.15
AMMONIA			
IN WATER		9.3	
40% MeOH		ca. 8.89	
60% MeOH		ca. 8.62	
80% MeOH		ca. 8.59	

0   2   4   6   8   10   12   14

pH

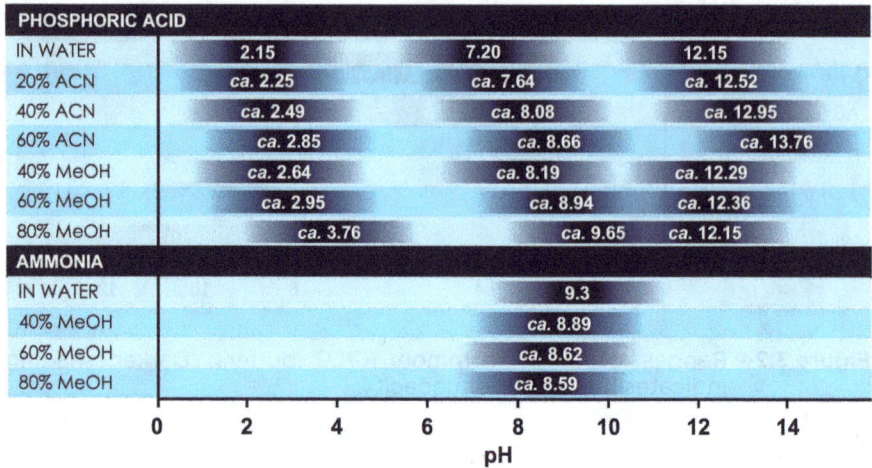

Figure 3.25 Approximate shifts in buffer pK_a values in aqueous–organic solvents. Based on data from ref. 49.

latter two numbers are $_w^s$pH values. If the $_w^w$pH method had been used, a pH of 7.2 would have been assumed in all cases, which clearly does not relate to the true hydrogen-ion activity.

The pK_a values of phosphoric acid and ammonia are seen to shift in opposite directions in Figure 3.25. The same differences in the direction of shifts are found for weakly acidic analytes that deprotonate at high pH (increasing pK_a values when adding modifier) and for weakly basic analytes that protonate at low pH (decreasing pK_a values). Hence, the chromatographic selectivity for different classes of analytes may be dramatically affected by changes in the pH of the mobile phase.

The $_w^w$pH method is easiest to use in practice, but fundamentally least correct. It may work well for a routine method, in which the preparation of the buffer and/or the pH measurement are tightly controlled and well described. Describing the amounts (ideally weights) of buffer components that must be added (*e.g.* 2.72 g KH_2PO_4 plus 3.48 g K_2HPO_4 in 1 L water to prepare a 20 mM phosphate buffer of pH 7.2) is a way to avoid errors in pH measurements. However, measuring the pH of aqueous solutions using one or (preferably) more standard buffers should also be reasonably possible. If a different buffer is prepared and subsequently mixed with an organic modifier, the resulting hydrogen-ion activity, and, thus, chromatographic retention factors and selectivities, will likely be different. In the case of gradient-elution RPLC, it is unrealistic to measure the pH as a function of composition and the ^w_wpH method is commonly used, but it is possible to correct for pH changes as a function of mobile-phase composition (ammonium acetate buffers in MeOH–water and ACN–water gradients; see ref. 49 and references cited therein).

A limited number of standard buffers with known pH in a given aqueous–organic mixture are available, but such measurements using the ^s_spH method are recommended for measuring physical data (*e.g.* pK_a values) or for situations in which different buffers may be used (*e.g.* in the case of a method in which only pH and not the type and concentration of buffer are specified). When the pH meter is calibrated with aqueous buffers (^s_wpH), a correction may be applied.[49]

3.5.5 Injection Solvents

As retention varies exponentially with the mobile-phase composition, the composition of the injection solvent can have dramatic effects in RPLC. When injecting a sample in a stronger eluent than the mobile-phase composition (*i.e.* higher in organic modifier) at the time of injection, the local conditions in the sample plug will be such that the analyte is moving faster through the column, until it is sufficiently diluted. This may cause the analyte zone to be extended and the peaks appearing upon elution of the analyte to be broadened. This situation is described in eqn (3.13) with $k_{i,\mathrm{inj}} <$ $k_{i,\mathrm{e}}$. Oppositely, when the sample solvent is weaker, peak focussing may occur and the injection band broadening will be eliminated (eqn (3.13) with $k_{i,\mathrm{inj}} \gg k_{i,\mathrm{e}}$). Thus, an injection solvent weaker than the (starting) eluent composition is strongly recommended.

In the case of gradient elution, injection band broadening may be eradicated further down the column, provided that the sample plug is diluted strongly enough and soon enough to retain the analytes. It is not evident that this will occur because the chromatographic column is designed for the opposite purpose, *i.e.* to yield minimal axial dispersion. If the sample plug (injection volume) is large and the injection solvent is very strong, (part of) the sample may be eluted together with an intact sample plug. This phenomenon is called **breakthrough**. The risk of breakthrough is especially high for large molecules that are (partly) excluded from the pores (see Module 4.3). In a zone of strong solvents, such analytes may experience exclusion conditions, causing them to move through the column at greater speed than the strong solvent. As a result, these analytes are focussed into narrow peaks at the front of the solvent plug, with only a small fraction (from the tail of the initial injection band) being properly retained on the column.[50]

3.5.6 Column Testing

Several different tests are in use for RPLC columns. Important parameters to test are the column efficiency (typically measured using a non-polar analyte), the peak symmetry for basic compounds, the hydrophobic selectivity (*e.g.* the methylene-group increment) and the selectivity for polar analytes. Two of the most established tests are summarized in Tables 3.10 and 3.11.

The **Neue test**[51] (summarized in Table 3.10) is quite simple to perform, as it involves a single mobile phase, *viz.* 35% (by volume) of a 20 mM, pH = 7 phosphate buffer and 65% methanol. A flow rate of 1.4 mL min^{-1} has been specified for a 4.6 mm i.d. column packed with 5 μm particles, but this should be adapted for different columns ($F \propto d_c^2$ and $F \propto d_p^{-1}$). The test provides measures for column hydrophobicity, silanol activity and other polar activity. The latter aspect, based on butylparaben and dipropyl phthalate as analytes, allows discrimination between columns with and without embedded polar groups. Propranolol and amitriptyline allow the characterization of the effect of electron-rich stationary-phase ligands, such as those encountered with perfluorinated RPLC columns.

Using the **Tanaka test**[52,53] (summarized in Table 3.11), the extent of (hydrophobic) retention and the methylene-group increment are determined from the ratio of the retention factors of *n*-pentylbenzene and *n*-butylbenzene using an 80 : 20 (by volume) methanol : water mobile phase. The absolute values of the retention factors correlate with the surface area of the packing material. The α_{CH_2} parameter is found to correlate strongly with the carbon content of the stationary phase. The selectivity of the stationary phase for triphenylene and *ortho*-terphenyl (α_{TO}) using the same mobile phase is indicative of the shape selectivity and relates to the surface coverage and the type of CBP (monomeric or polymeric). The selectivity between caffeine and phenol (α_{CP}), measured using a 30 : 70 (by volume) methanol : water mobile phase, provides an indication of the number of (accessible) silanols. This parameter allows a distinction between phases that have and those that have not been end-capped. The selectivity between procaine amide and benzyl alcohol ($\alpha_{PB,2.7}$) using a 30 : 70 (by volume) methanol : buffer mobile phase (20 mM phosphate at pH = 2.7)

Table 3.10 Test procedure for RPLC columns according to Neue.[51]

Analytes	Function
Uracil	t_0 marker
Naphthalene Acenaphthene	Hydrophobic compounds used to measure column retentivity and hydrophobic selectivity
Propranolol Amitriptyline	Basic probes used to characterize silanol activity
Butylparaben Dipropyl phthalate	Polar probes used to characterize polar selectivity (*e.g.* from embedded polar groups)

Mobile phase (isocratic)

35% 20 mM K_2HPO_4/KH_2PO_4 buffer, pH 7.00 in water, 65% methanol

Table 3.11 Test procedure for RPLC columns according to Tanaka.[52]

Analytes	Function
n-Butylbenzene *n*-Pentylbenzene	Used to measure the retentivity of a stationary phase and the methylene-group increment on retention (αCH_2) using mobile phase (1)
Triphenylene *ortho*-Terphenyl	Used to characterize the shape selectivity using mobile phase (1)
Caffeine Phenol	Used to characterize silanol activity using mobile phase (2)
Benzyl amide Phenol	Used to measure ion-exchange effects using mobile phases (3 and 4)

Mobile phase (isocratic)

(1) 80% methanol in water
(2) 30% methanol in water
(3) 30% methanol in 20 mM aqueous phosphate buffer pH = 2.7
(4) 30% methanol in 20 mM aqueous phosphate buffer pH = 7.6

provides an indication of the cation-exchange capacity of the mobile phase, which has proven different from the concentration of accessible silanols, almost all of which are protonated at pH = 2.7. The combined effects of silanol groups on retention and peak shape are characterized by measuring the same selectivity factor at pH = 7.6 ($\alpha_{PB,7.6}$). The various measurements on apolar and polar analytes also provide information about chromatographic efficiency and peak symmetry. The Tanaka test is very thorough. It provides good insights into stationary phases and their behaviour. On the downside, it is slightly laborious because it requires experiments in four different mobile phases.

In any kind of test, it is crucial to minimize extra-column contributions to band broadening and to correct for extra-column residence times, as described in Section 3.3.3.

3.6 Normal-phase Liquid Chromatography (B)

3.6.1 Classical Normal-phase Liquid Chromatography

Many of the early applications of LC have involved polar adsorbents and non-polar mobile phases. Separations on solid stationary phases, such as silica, are known as liquid–solid chromatography (LSC), but more generally, LC separations in which the stationary phase is more polar than the mobile phase are referred to as **normal-phase liquid chromatography** (NPLC). Because most polar analytes prefer the polar stationary phase, they elute last in NPLC, while non-polar analytes prefer the mobile phase and elute first. In classical NPLC, bare silica is the most common stationary phase and an alkane (*e.g. n*-heptane) is used as the weak solvent, using which few other compounds than saturated hydrocarbons can be eluted. Somewhat more polar solvents that are miscible with the alkane, such as ethyl acetate or dichloromethane, are used as strong solvents (modifiers). Small concentrations of strong solvents decrease retention drastically and re-equilibration (*i.e.* removing the strong solvent from the column by washing with the weak solvent) is very slow, frustrating the use of gradient elution. Small amounts of water, which are hard to avoid in solvents such as dichloromethane, greatly affect retention and are even more difficult to remove from the column. Pre-saturation of the column with water using water-saturated pre-columns is cumbersome.

As explained in Module 3.5, RPLC has become by far the dominant LC method, leaving few types of applications for other methods. Three types of applications of NPLC are worth mentioning here. First, these include the separation of very non-polar samples. Certain classes of lipids, such as triglycerides, may be separated by non-aqueous RPLC, with mixtures of ACN and THF as mobile phases. However, because of the small differences in polarity between the mobile phase and the stationary phase, the selectivity of such systems is limited. For the least polar analyte mixtures, such as mineral oil products, NPLC is still an attractive option. Saturated alkanes tend to be essentially unretained, even with the weakest (*i.e.* alkanes, such as *n*-heptane) NPLC mobile phases. Unsaturated hydrocarbons are successively more retained, and a change in the mobile phase is typically required to elute aromatics. Methods for such group-type separations of hydrocarbons by NPLC are well established, but supercritical-fluid chromatography (SFC) has emerged as an attractive alternative in some cases (see Chapter 6).

The latter is even truer for a second type of application of NPLC, which relies on strong polar interactions between (functional groups of) analyte molecules and the stationary surface. A typical example is the separation of enantiomers (see Module 3.13). Strong interactions that are different for different stereoisomers may provide sufficient selectivity for separating these. Other types of isomers may also exhibit greater selectivity in NPLC (or in SFC) than in RPLC.

A third type of separation for which RPLC is inadequate concerns very polar analytes, which show little retention even when applying (nearly) pure water as the mobile phase. Clearly, such analytes can be retained more strongly on polar stationary phases. However, very polar analytes are often present in aqueous samples and some water is otherwise likely present. This implies that classical NPLC, with non-polar mobile phases that are not miscible with water and a strong effect of water on the stability of the phase system, is not an attractive option. For such highly polar samples, HILIC (Section 3.6.2), with a polar stationary phase and a water-containing mobile phase, has become the most common approach.

3.6.1.1 Competitive Adsorption

The prevailing model for retention in liquid–solid NPLC (*i.e.* NPLC with a solid adsorbent, such as silica, as the stationary phase) is **competitive adsorption**. This model was developed by Lloyd Snyder (see Module 3.4) and Edward Socsewiński (from the Medical Academy in Lublin, Poland). The underlying idea is that the active sites on the stationary surface are all occupied by the strong component of the mobile phase (or the modifier, M) and that these need to be replaced by analyte molecules (A) to realize adsorption. This can be described as an adsorption equilibrium.

$$A_{mob} + \frac{n_A}{n_M}M_{ads} \rightleftarrows A_{ads} + \frac{n_A}{n_M}M_{mob} \tag{3.30}$$

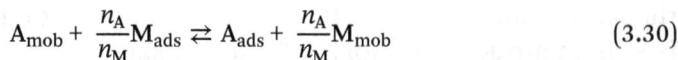

where the subscripts "ads" and "mob" denote molecules adsorbed on the stationary surface and present in the mobile phase, respectively. n_A denotes the number of adsorption sites occupied by an analyte molecule and n_M denotes the number of adsorption sites occupied by a modifier molecule. For example, if an analyte molecule occupies three sites and a modifier molecule one site, then the coefficient in the equilibrium equation would be $n_A/n_M = 3/1 = 3$. The coefficient does not need to be an integer number.

The corresponding equilibrium coefficient is

$$K_{AM,ads} = \frac{[A_{ads}]}{[A_{mob}]} \cdot \frac{[M_{mob}]^{n_A/n_M}}{[M_{ads}]^{n_A/n_M}} \tag{3.31}$$

Here, the ratio $[A_{ads}]/[A_{mob}]$ is the adsorption coefficient of analyte A $(K_{A,ads})$, which is proportional to its retention factor k_A with a (phase ratio) proportionality factor ϕ^{-1}. If we work at the low analyte concentrations required for good LC with reasonably symmetrical peaks, $[A_{ads}] \ll [M_{ads}]$ and the latter concentration can be considered constant. We then obtain

$$\log k_A = \log\left(K_{A,ads}\phi\right) = \log\left(K_{AM,ads}\phi[M_{ads}]^{n_A/n_M}\right) - \frac{n_A}{n_m}\log[M_{mob}] \tag{3.32}$$

The first term on the right-hand side of this equation is a (temperature-dependent and phase-ratio-dependent) constant that is numerically equal to the logarithm of the retention factor at the unit concentration of the modifier $(\log k_{A,1})$.

$$\log k_A = \log k_{A,1} - \frac{n_A}{n_m}\log[M_{mob}] \tag{3.33}$$

This equation shows that in LSC with competitive adsorption, a log–log model describes retention as a function of modifier concentration. Different units can be used for this concentration (*e.g.* mol L^{-1}, g L^{-1}, or volume fraction), with the conversion factor accounted for by different values for $\log k_{A,1}$. Note that eqn (3.33) does not hold at very low concentrations of modifier, where $-\log[M_{mob}]$ becomes infinitely large. This is because our assumption that the stationary surface is fully covered with modifier molecules is no longer correct.

The retention *vs.* composition lines will be steeper for larger analyte molecules that occupy a greater number of adsorption sites (larger values of n_A). The strength of the modifier is reflected in the adsorption coefficient $K_{AM,ads}$. If M is a strong adsorber, the equilibrium will be towards the left in eqn (3.30), and $[A_{ads}]$ and $K_{AM,ads}$ will be low. This will result in a low value for $\log k_{A,1}$ and a low value for $\log k_A$. The **strength of the modifier** roughly increases in the order chloroform < dichloromethane < THF < ethyl acetate < MTBE < 2-propanol < methanol. Methyl-*t*-butyl ether (MTBE) is amply available and transparent for UV light, but it smells strongly and may be toxic. 2-Propanol has a very high viscosity.

Polar CBPs, such as cyano, diol, or amino columns, may be considered for application in NPLC, but retention (k values) and selectivity (α values) tend to be much lower on such phases. Operation of such columns may be easier (faster equilibration and better compatibility with gradient elution), but one of the few remaining strong points of NPLC, its high selectivity for molecules with different polar groups or for isomeric analytes, diminishes. Retention in LSC tends to decrease with increasing temperature, usually following the van 't Hoff equation (linear variation of $\log k_A$ with $1/T$; see Section 1.6.1.1).

Table 3.12 summarizes the strengths and weaknesses of conventional NPLC, using a non-polar base solvent (usually an alkane) and moderately polar modifiers. The strengths of NPLC, and the possible applications derived from these, are all confronted with alternative methods that have gained popularity in the 21st century. As a result, classical normal-phase liquid chromatography is increasingly abnormal. This shift away from classical NPLC is motivated by the inherent weaknesses of the technique. When working with silica columns, eliminating water or controlling its concentration is inconvenient at best. Similar sensitivity to other strongly adsorbing compounds suggests gradient elution as a method to wash the column at the end of each analysis, but this is not helped by the slow equilibration of silica columns. Efficiency (plate count) is typically lower in NPLC than in RPLC. This may be due to slow adsorption/desorption kinetics (see Section 1.7.4.4), as well as to an inhomogeneous surface that causes non-linear distribution isotherms (see Module 1.8). Strongly adsorbing additives (*e.g.*

Table 3.12 Summary of some of the strengths (and possible applications) and weaknesses of NPLC.

Strengths of normal-phase LC

Strength	Possible application(s)	Possible alternative(s)
Separation of non-polar samples (incompatible with RPLC)	Group-type separation of hydrocarbons (group-type) separation of lipids	Supercritical-fluid chromatography (SFC, see Chapter 6)
High selectivity for isomers	Separation of (non-chiral) isomers on silica columns Separation of enantiomers using chiral columns (see Module 3.13)	Supercritical-fluid chromatography (SFC, see Chapter 6)
Separation of extremely polar analytes (insufficiently retained in RPLC)	Carbohydrates, peptides	Hydrophilic-interaction liquid chromatography (HILIC, see Section 3.6.2)[a]

Weaknesses of normal-phase LC

Stationary phase	Weaknesses
Silica	Slow equilibration Very sensitive to traces of water Subject to fouling by other strongly adsorbing compounds Poor (day-to-day and column-to-column) repeatability Moderate efficiency and a strong tendency towards peak tailing
Chemically bonded phases (cyano, diol, and amino)	Low retention and low selectivity

[a]HILIC is a form of NPLC.

amines) may be added to the mobile phase to mask the most active adsorption sites and suppress peak tailing. LSC columns have limited mass loadability when considering peak shapes, but when poor peak shapes are accepted, their high selectivity may be useful for (semi-)preparative separations (see Module 3.12).

3.6.2 Hydrophilic-interaction Liquid Chromatography

HILIC is a name introduced by Andrew Alpert (PolyLC, Columbia, MD, USA) for a form of LC designed for the separation of very polar analytes that cannot be sufficiently retained in RPLC – not even with (nearly) 100% water in the mobile phase. In HILIC, the mobile phase is quite polar, consisting largely of an organic solvent (commonly ACN) with a low concentration (typically <10%) of water and possibly other additives. The stationary phase is polar, for example, silica or a polar bonded phase (*e.g.* amino or diol, see Table 3.4). The essence of HILIC is that a significant amount of water covers the stationary surface, forming a dynamically generated stationary phase with a very high polarity. ACN is an aprotic solvent and it does not participate in hydrogen-bonding interactions, which is thought to be favourable in the formation of a stable water layer that contrasts with the mobile phase. With the stationary-phase polarity approaching that of water and the mobile-phase polarity approaching that of the organic solvent (usually ACN), the stationary phase is more polar than the mobile phase, making HILIC a special form of NPLC. Any analytes with a polarity similar to ACN or lower are likely to prefer the mobile phase ($K_{d,i} = C_{i,s}/C_{i,m} \approx 0$; see (eqn (1.2))) and will elute unretained. Due to the water layer formed, non-polar analytes may actually be excluded from some of the pores. Only very polar analytes may distribute evenly across the two phases. These include some important compound classes, such as carbohydrates (sugars and glycans), amino acids, peptides, (oligo-)nucleotides, and polar drugs.

HILIC and RPLC behaviour on a moderately polar (*e.g.* diol) column is schematically illustrated in Figure 3.26. When the mobile phase consists almost exclusively of ACN (right-hand side, HILIC), water molecules are distributed towards the polar stationary phase and a dynamic stationary **water layer** is formed. Under such conditions, highly polar analytes are retained, as indicated by the light-blue curve. When the volume fraction of water in the mobile phase increases from, say, 5% to 10% or 20%, the retention of highly polar compounds decreases. When the mobile phase consists almost exclusively of water (left-hand side, RPLC), ACN molecules are "expelled" from the extremely polar water phase and distributed towards the stationary phase, forming a dynamic stationary ACN layer. Under such conditions, moderately and low-polar analytes are retained, as indicated by the dark-pink curve. When the volume fraction of ACN in the mobile phase increases, the retention of

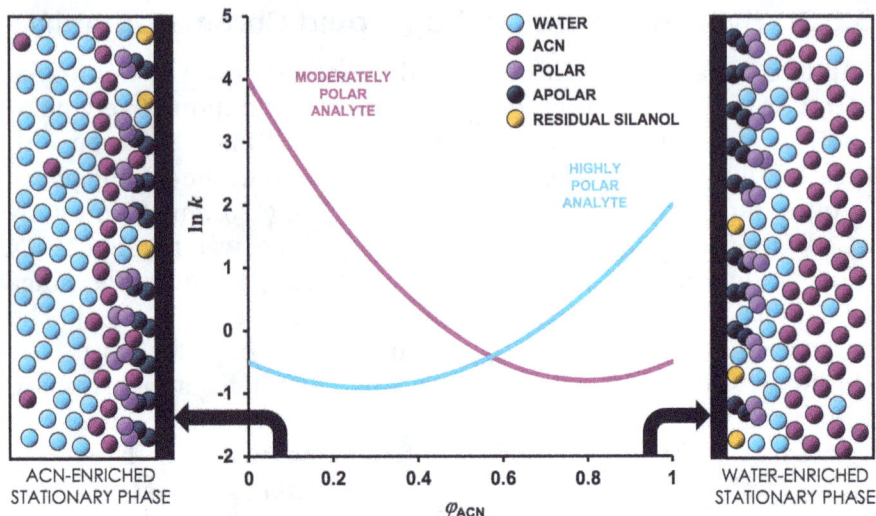

Figure 3.26 Schematic illustration of RPLC (left) and HILIC (right) retention on a moderately polar (*e.g.* diol) column (polar groups
denoted in purple; apolar spacer in dark blue; residual silanol
groups in yellow). When an RPLC mobile phase consists
almost exclusively of water (light blue), the stationary phase
becomes enriched in ACN (maroon) as shown on the left, and
a moderately polar compound (dark-pink curve) becomes
significantly retained. When a HILIC mobile phase consists
almost exclusively of ACN (right-hand side), the stationary
phase becomes enriched in water and a highly polar compound (light-blue curve) becomes significantly retained.

moderately polar compounds decreases, until at high ACN concentrations retention may go up slightly. However, moderately polar
analytes are not significantly retained under HILIC conditions.

The **retention** process in HILIC is complex. Polar analytes will
interact with the water layer and with the stationary-phase surface. The latter may show different kinds of interactions, involving the chemically bonded groups (if present), as well as possible
ion-exchange effects with bonded ion-exchange groups or the silica
substrate. The main challenge is to obtain sufficient retention, and
this is achieved with stationary phases that absorb thick water layers.
Silica with amine, amide or diol groups are commonly used, as are
zwitterionic phases and, less frequently, hydrophilic (*e.g.* carbohydrate) polymers.

In comparison with NPLC, HILIC is more convenient. It is much
easier to control the composition of the mobile phase and the
activity of the stationary-phase surface because the amount of water

in the eluent can be easily controlled. The stationary phase behaves more homogeneously in HILIC, as the mechanism does not rely on competitive adsorption on a restricted number of sites. This results in more symmetrical peaks, without a need for tailing suppressants in the mobile phase. Another advantage of HILIC is greater compatibility with electrospray ionization (ESI) MS (see Section 3.10.1). However, chromatographers are unlikely to base their choice of methods on these advantages because the application domains of HILIC (highly polar analytes) and that of classical NPLC (low-to-medium polarity analytes; see Section 3.6.1) do not really overlap. The combination of HILIC and RPLC is thought to be equipment friendly because vulnerable parts, such as seals, will be exposed to the same solvent mixture (water–ACN) at all times.

The effective **diffusion** coefficient (see Section 1.7.4.1) tends to be lower in HILIC than in RPLC due to slower interparticle diffusion. This is thought to be due to strong interaction with the surface and slow movement in the viscous water layer. Slower effective diffusion results in a lower B-term and a higher C-term in a plate-height equation.[54] This implies that the optimal linear velocity is lower in HILIC than in RPLC. **Re-equilibration** of HILIC columns after a gradient separation may also be slower in comparison with RPLC, but it may be argued that complete re-equilibration is not always necessary, provided that the process (composition, times, and flow rates) is identical before each run.

The **injection solvent** can have a dramatic effect on the peak shape and efficiency in HILIC. Injecting samples with a high water content (higher than that in the mobile phase) is similar to injections in RPLC of samples with high concentrations of strong organic modifiers (see Section 3.5.3). However, the effects are thought to be more severe in HILIC because (i) the amount of water in HILIC eluents is often just a few per cent, (ii) the variation of retention with mobile-phase composition is less steep, making it more challenging to (re-)focus analyte bands, and (iii) equilibration is thought to be slower. For these reasons, diluting aqueous samples with acetonitrile to mimic the mobile-phase composition is highly recommended in HILIC.

RPLC (Module 3.5) is an excellent technique for separating aqueous samples. However, one of its main weaknesses is its inability to achieve sufficient retention for highly polar and ionogenic compounds. For example, in the field of metabolomics, scientists aim to detect, identify, and eventually quantify all compounds with molecular weights of 1 kDa or lower in biological samples. As in many other fields, RPLC–MS is the most important analytical technique. However, the metabolome (*i.e.* the collection of all metabolites) may feature many compounds that are too polar to be retained by RPLC, including carbohydrates, amino acids, organic acids, and nucleosides or nucleotides.[55] Depending on the separation conditions, some of these classes of compounds may be charged. HILIC cannot replace RPLC for all compounds that the latter technique does separate well, but it can be complementary. As common mobile phases used in both RPLC and HILIC have water and acetonitrile as major components, the techniques may be combined more easily, with a pre-separation by HILIC followed by a separation of the fractions by RPLC, or *vice versa* (see Module 7.3). As the compounds that are unretained in RPLC are either very polar or charged, a complementary method should ideally cover these classes, which is why HILIC columns that combine hydrophilic (partitioning) and ion-exchange selectivities are increasingly used. Zwitterionic HILIC phases have the potential to cover all component classes that are insufficiently retained in RPLC.

3.7   Separation of Ions (M)

In most cases, RPLC is not a good technique for separating charged analytes (ions). Small ions tend not to be retained and any effect of residual (ionized) silanol groups on the chromatography of (partially) charged basic analytes tends to be detrimental in that it leads to band broadening and peak tailing, rather than retention and selectivity. Sometimes charges are specifically induced on small analytes because their separation as ions is preferred to their separation as neutral molecules. For example, carbohydrates are normally neutral, and their separation by HILIC (Section 3.6.2) is well established. However, they can also be ionized at high pH (*i.e.* pH > 12) for their separation as anions. This may be advantageous in combination with highly sensitive pulsed-amperometric detection (PAD; see Section 3.9.5.1).

Large **ions**, which possess charged groups and hydrophobic parts, such as fatty acids, may be retained in their charged forms because of hydrophobic interactions between the alkyl chains and stationary-phase ligands. Likewise, very large molecules, such as peptides or proteins, may be retained in RPLC if they feature significant non-charged segments. In HILIC (Section 3.6.2), electrostatic forces and ion-exchange mechanisms play a role in the retention of charged analytes.

Reversed-phase columns can be used for the separation of small, ionogenic molecules if their ionization can be suppressed with (solvated) protons by varying the pH or by a "pairing-ion" with a stronger suppression effect. Such ion-pairing chromatography (IPC) will be described in Section 3.7.3. Small ions can be retained on ion exchangers, leading to the well-established IEC technique described in Section 3.7.1. Ion chromatography (IC), which is a special form of IEC, is discussed briefly in Section 3.7.2. For small anions, ion-exclusion chromatography is a niche technique that will be briefly described in Section 3.7.4.

Mixed-mode columns do not fall under any of these headings. In HILIC (Section 3.6.2), often both partitioning and ion exchange play a role in determining retention and selectivity. Hydrophobic ion exchangers provide a combination of RPLC and IEC mechanisms. Mixed-mode columns may be useful for separating samples that contain both charged and neutral analytes, such as the complex mixtures of dyes studied by Pirok *et al.*[56] Mixtures of (weak) acids and bases may be turned into mixtures of charged and neutral analytes if the ionization of one class of molecules can be suppressed.

3.7.1 Ion-exchange Chromatography

In ion-exchange chromatography (IEC), the exchanging ions have like charges, *i.e.* they are all positively (cations) or all negatively (anions) charged. Groups on the stationary phase have the opposite charge. For example, negatively charged groups on the stationary surface engage in cation-exchange processes. Initially, all of the charged groups on the surface are paired with counterions that are present in the mobile phase. For example, the surface of a cation exchanger may be covered with "protons" (H_3O^+ ions) or with sodium ions, *etc.*, depending on the composition of the mobile phase. This association may be loose, creating an electrical double layer, while the surface retains a negative charge. This would be the case, for example, for a stationary surface covered with sulfonate ($-SO_3^-$) groups, which are always charged, irrespective of the pH. Such a stationary phase is called a strong cation exchanger (SCX). In contrast, stationary surfaces covered with carboxylate ($-COO^-$) groups are negatively charged at high pH but become neutral at low pH. Such stationary phases are called weak cation exchangers

(WCXs). Likewise, there are strong anion exchangers (SAXs), for example, those based on quaternary ammonium groups ($-NR_3^+$) and weak anion exchangers (WAXs), for example, based on ternary or secondary amine groups ($-NR_2H^+$ or $-NRH_2^+$). Stationary phases that contain both anion-exchange and cation-exchange functionalities are called mixed phases if the functionalities are on separate sites or **zwitterionic** if both functionalities are present on the same ligand. Alkylsulfobetaine phases are an example of the latter class of ion exchangers.

Weak ion exchangers allow moderation of the ion-exchange process. For example, a multi-valent polymeric cation (a "polyelectrolyte") may be held on a cation-exchange column. If the latter is an SCX column, these strong coulombic forces retaining the analyte polymers will not be affected by the pH. If the column contains a WCX phase, the forces between the stationary surface and the analyte polymer may be greatly reduced when the pH is lowered. In other cases, the pH may be used to moderate the charge on analyte ions. The crucial role of the pH in ion-exchange processes makes it desirable for stationary phases to be compatible with a broad range of pH values. Therefore, IEC packing materials are often based on organic polymers (*e.g.* styrene-divinylbenzene or methacrylate) rather than on silica.

Figure 3.27 illustrates a gradient-elution IEC separation using an SAX column. Figure 3.27A shows the obtained chromatogram (with time running from left to right), and Figure 3.27B illustrates the processes taking place in the column (from the column entrance on the left to the exit on the right). The latter figure illustrates the processes taking place in the column. In stage I (shown on the right-hand side), neutral molecules migrate with the mobile phase, as they do not engage in electrostatic interactions. Next (stage II), monovalent anions elute at a low concentration of counterions. Divalent counterions are eluted at progressively higher sulfate concentrations (stage III). Finally (stage IV), the column is regenerated with a large concentration of chloride ions, so as to remove the strongly interacting sulfate ions from the stationary phase. The ionic strength of the mobile phase is lowered prior to the next injection of a sample. Ions of equal charge are separated by the size of their electron cloud (charge density) and other contributions to retention (*e.g.* hydrophobic interaction).

The mental image of an ion exchanging with another ion at a specific site on the surface is helpful in the sense that it provides a model for the effects of the counterion concentration and the pH on

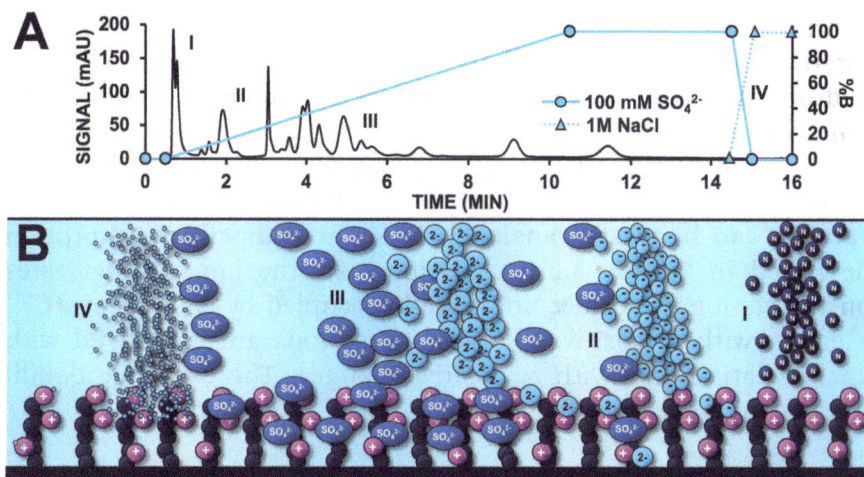

Figure 3.27 (A) Strong anion-exchange separation of small analyte ions with the mobile-phase composition program highlighted. (B) Schematic depiction of strong anion exchange with (I) elution of the neutral analytes at the dead time, (II) elution of the single-charged anions at a low concentration of sulfate ions, (III) elution of the double-charged anions with significantly higher concentrations of sulfate, and (IV) regeneration of the stationary phase using a high concentration of a weak salt (NaCl). The different stages are also depicted in the chromatogram in panel A. Differences between different ions of the same charge are not indicated in panel B. Chromatogram of panel A reproduced from ref. 57 with permission from Elsevier, Copyright 2016.

retention. However, the model is quite specific and rather incorrect. A more realistic model is that of an electrical double layer, which we will return to when we discuss ion exclusion in Section 3.7.4. In the **exchange model**, counterions are stoichiometrically connected with exchange sites. However, this is an entropically unfavourable situation in comparison with the counterions moving freely in a diffuse layer. There will be an excess of counterions in the layer close to the surface, and the electrostatic potential changes gradually away from the surface. A thinner layer results when the ionic strength of the mobile phase is higher. Describing the ion-exchange mechanism quantitatively through an accurate model of the double layer is extremely difficult. A compromise may be to consider a fixed stationary-phase layer with a constant potential, known as the **Donnan potential**. In the Donnan layer, there will be an excess of counterions equal to the number of surface charges, so as to obtain electric neutrality overall. An excellent discussion on realistic mechanisms for the separation of ions has been provided by Ståhlberg.[58] For educational purposes, we return to the stoichiometric model in the following paragraphs.

In discussing the IEC retention process in stoichiometric terms, we describe the case for cations. The treatment of anions is identical. IEC can be treated analogously to the competitive adsorption described in Section 3.6.1.1. In this case, the ion-exchange sites on the stationary surface are initially occupied by counterions $C^{z_c^+}$, a cation with z_c positive charges. These ions are exchanged with analyte cations $A^{z_a^+}$ with z_a positive charges. The exchange equilibrium can be described by

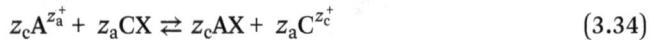

$$z_c A^{z_a^+} + z_a CX \rightleftarrows z_c AX + z_a C^{z_c^+} \tag{3.34}$$

where X denotes the ion-exchange stationary phase. In the simple case of monovalent analyte ions and monovalent counterions, this equation reduces to

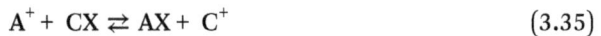

$$A^+ + CX \rightleftarrows AX + C^+ \tag{3.35}$$

The corresponding equilibrium coefficient is

$$K_{AC,iec} = \frac{[AX]^{z_c} \left[C^{z_c^+}\right]^{z_a}}{\left[A^{z_a^+}\right]^{z_c} [CX]^{z_a}} \tag{3.36}$$

The ratio $[AX]/\left[A^{z_a^+}\right]$ is proportional to the retention factor k_A of analyte A $\left(k_A = \phi[AX]/\left[A^{z_a^+}\right]\right)$, where ϕ is a phase ratio. We may consider nearly all exchange sites to be occupied with counterions, so that $[CX]$ is a constant and

$$\log k_A = \frac{1}{z_c}\log(K_{AC,iec}\phi[CX]^{z_a}) - \frac{z_A}{z_c}\log\left[C^{z_c^+}\right] = \log k_{A,1} - \frac{z_A}{z_c}\log\left[C^{z_c^+}\right] \tag{3.37}$$

On a given column and at a given temperature, $\log k_{A,1}$ is a constant that is numerically equal to the logarithm of the retention factor at unit concentration of counterion. Eqn (3.37) shows that a log–log model describes retention as a function of counterion concentration in IEC. $\left[C^{z_c^+}\right]$ can be expressed in one of the various concentration units (*e.g.* mol L^{-1} or g L^{-1}). The conversion factor will be reflected in the value of log $k_{A,1}$. Straight lines are predicted for retention *vs.* counterion concentration, with a slope proportional to the charge of the analyte ions. Lines will be steeper for larger analyte ions that occupy a greater number of adsorption sites (larger values of n_A). Eqn (3.37) cannot be used at very low concentrations of counterions because in this range our assumption that the stationary surface is always fully covered with counterions does not hold. One consequence of the log–log behaviour is that an absolute change (*e.g.* by 1 mM) in counterion concentration has a much larger effect on log k at low concentrations than at high concentrations. The difference induced by increasing the counterion concentration from 1 mM to 2 mM is large, while the effect of an increase from 0.01 to 0.011 M is small and that of a change from 0.1 to 0.101 M is negligible. As a result **convex gradients** are more appropriate than linear gradients (see Figure 3.20).

Ion-exchange retention will be strongly affected by the mobile-phase pH for weakly acidic analytes (WAAs) or weakly basic analytes (WBAs) or whenever weak ion exchangers (WAX or WCX) are used. The ion-exchange retention can be corrected for all these effects by assuming that it is proportional to the fraction of analyte molecules that are charged, so as to make them ions, and the fraction of ion-exchange ligands on the stationary phase that are charged. For example, in the case of monovalent weak acids that lose charge through protonation at low pH (*e.g.* carboxylic acids) and weak bases that lose charge through deprotonation at high pH (*e.g.* amines), we can write

$$\ln k_i = \ln k_{i,\text{IEC}} - \log\left(1 + 10^{(pK_{a,\text{WAA}}-\text{pH})}\right) - \log\left(1 + 10^{(pK_{a,\text{WCX}}-\text{pH})}\right)$$
$$- \log\left(1 + 10^{(\text{pH}-pK_{a,\text{WBA}})}\right) - \log\left(1 + 10^{(\text{pH}-pK_{a,\text{WAX}})}\right) \tag{3.38}$$

The effects of pH on retention are summarized in Figure 3.28. If either a weakly acidic analyte is considered or a WCX column is used, retention diminishes below the pK_a value due to protonation. For weakly basic analytes or WAX columns, retention diminishes above the pK_a value due to deprotonation.

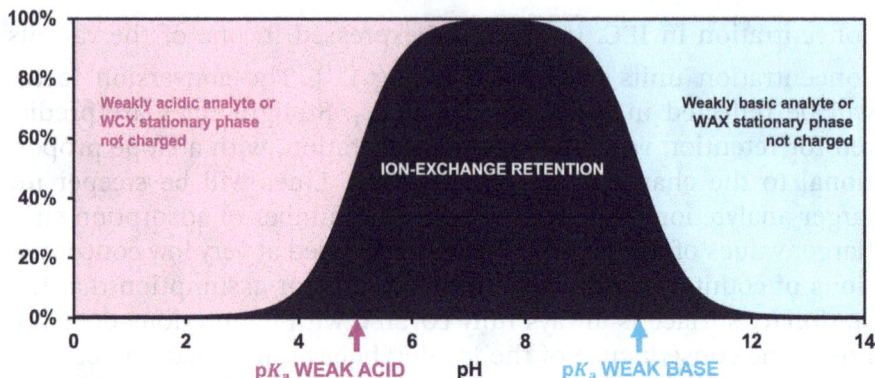

Figure 3.28 Reduction in ion-exchange retention when considering weakly acidic or weakly basic analytes or when using WCX or WAX columns. 100% on the vertical axis indicates the full ion-exchange retention that would be obtained if the analytes and stationary phases were fully charged.

3.7.2 Ion Chromatography

IC is a special form of IEC. The name was coined by Hamish Small, then from Dow Chemical (Midland, MI, USA), for specific separations of small **inorganic ions** with suppressed conductivity detection.[59] Prior to the emergence of this method, the quantitative determination of even one or two of the anions, using, for example, titrimetric or spectrophotometric methods, could be time consuming and difficult, especially at low concentration levels. IC suddenly allowed a whole series of anions to be determined with low detection limits, starting with fluoride, chloride, bromide, iodide, sulfate, nitrite, nitrate, and phosphate, with many other analyte anions soon to follow.[60] For inorganic cations, IC represented less of a revolution because atomic-spectroscopy techniques were already well established. However, IC allowed non-metallic ions, such as ammonium, to be included. Moreover, IC allowed the speciation of metal ions, such as Cu^+ and Cu^{2+}.

Highly sensitive detection of small anions or cations was made possible by selecting ion exchangers with a relatively low capacity, so that relatively low-ionic-strength mobile phases could be used. This in turn allowed suppressing the conductivity of the eluent using a "stripper" column. Initially, this was an ion-exchange column, such as a cation exchanger in the H^+ form to exchange Na^+ ions for solvated protons. For example, the conductivity of a sodium-bicarbonate (Na^+ and HCO_3^-) buffer could be suppressed by forming the neutral carbonic acid (or $H_2O + CO_2$). Interestingly, in the commercial version of Small's system that soon followed, the "stripper" column was renamed to "suppressor" column. Later **suppressors** used the principle of dialysis ("membrane suppressors") rather than that of ion exchange to achieve even lower detection limits. Hydroxide eluents can be generated electrolytically online, which avoids contamination with carbonate due to the absorption of CO_2 from the air.[61] IC has become the benchmark method for the separation of inorganic (and some organic) anions, as well as for many small cations.

3.7.3 Ion-pair Chromatography

In ion-pair (or ion-pairing) chromatography (IPC), an ion-pair reagent (IPR) with a charge opposite to that of the analyte ions is added to the mobile phase. This may result in the formation of a neutral ion pair, which can be retained on an RPLC column. A significant boost to the development of HPLC in the 1970s and 1980s came from the realization by scientists, such as Phyllis Brown (Box 3.7), that it could be instrumental in addressing questions from biology and biochemistry. Separations of peptides with additives such as trifluoroacetic acid, and separations of oligonucleotides with alkyl-amine additives can be classified as IPC.

The IPC process is significantly more complex than that of IEC, as is schematically illustrated in Figure 3.29 (large frame on the right). Four different **equilibria** are indicated in the figure.

1. The absorbed IPR acts as a dynamic ion exchanger, analogously to the IEC equilibrium illustrated in the small frame on the left. Analyte ions A^+ exchange with counterions C^+, which is indicated as equilibrium #1.
2. The IPR, denoted with the dark-blue hydrophobic groups and a light-blue anionic group, in this case (left-hand side to the left of the number 2) associated with a counterion C^+, engages in equilibrium #2 between the RPLC (typically octyl or octadecyl) stationary phase and the mobile phase to form a dynamic cation exchanger.
3. IPR in the mobile phase engages in equilibrium #3, where it exchanges its coordination with a counterion C^+ for an analyte ion A^+, forming a neutral ion pair (IPR-A).

Box 3.7 Hero of analytical separation science: Phyllis Brown. Photo courtesy of *LC-GC International*.

Phyllis R. Brown (1924 – 2015, United States of America)

Phyllis Brown (1924–2015) obtained her PhD from Brown University (Providence, RI, USA; not named after her). She later became a professor at the University of Rhode Island in Kingston. Phyllis was among the first to realize the enormous potential of liquid-phase separations for elucidating biological structures and biochemical processes. She stood at the basis of liquid-chromatographic and electrophoretic techniques for characterizing nucleic acids (DNA and RNA) and the constituting nucleotides and nucleosides. In doing so, she helped lay the foundation for major development in the life sciences, such as the Human Genome Project, and forensic science.

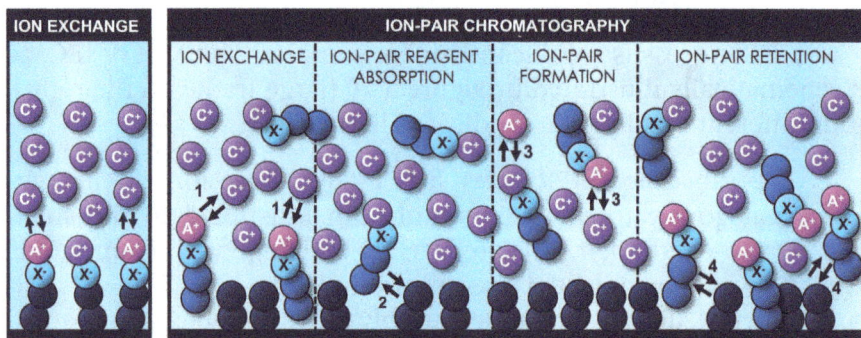

Figure 3.29 Schematic illustration of the equilibria underlying ion-exchange chromatography (IEC, left) and ion-pairing chromatography (IPC, right). All four equilibria that underlie IPC occur simultaneously and at the same location in the column. The different equilibria are explained in the main text. A^+ denotes analyte cations, C^+ denotes counterions, and X^- denotes functional groups on the cation exchanger or ion-pair reagent. Dark grey or dark blue "blobs" denote hydrophobic anchor groups or chains.

4. The neutral ion pair IPR-A engages in equilibrium #4 between the mobile phase and the stationary phase.

The IPC retention mechanism is a combination of all four equilibria, and they all occur simultaneously and at the same place in the chromatographic column. As a result, many factors influence retention in IPC. For example, retention can be increased by (i) decreasing the concentration of the modifier in the mobile phase, with the logarithm of the retention factor typically varying linearly with the volume fraction of the modifier as in regular RPLC; (ii) increasing the concentration of the **ion-pairing reagent**, with (if both the analyte and the IPR are singly charged) $\log k$ typically varying linearly with $\log[\text{IPR}]$; (iii) increasing the size of the IPR (in the case of an alkyl chain with n_C carbon atoms on the IPR, this implies increasing the length of the chain) with $\log k$ typically varying linearly with n_C; and (iv) if any of the ions (analyte, IPR, or stationary phase) is weak, pH will have a large effect on retention and potentially on the selectivity. The different effects may also be correlated, especially in case pH plays a role. All these effects combined create a complex picture for retention and selectivity in IPC.

Figure 3.30 shows an example of a gradient-elution IPC separation of low-molecular-weight anions. With a sodium-hydroxide mobile phase (bottom trace), ion-pairing mechanisms barely play a role, and other characteristics of the analytes (*e.g.* hydrophobic moieties) determine retention. Ion-pairing effects are introduced by adding 10 mM tetramethylammonium hydroxide instead of NaOH to the mobile phase (middle trace), and IPC retention is greatly enhanced by increasing the size of the IPR cation to tetrabutylammonium (top trace).

3.7.4 Ion-exclusion Chromatography

Earlier in this module, we mentioned the electrical double layer that separates a charged surface from a bulk mobile phase. This layer contains an excess of ions with opposite charge to the surface, so as to create overall neutrality. Ions of equal charge to the surface are depleted in this layer. The thickness of the double layer depends on the ionic strength of the mobile phase. At high ionic strength, there is an abundance of ions and the double layer is thin. At low ionic strength, the double layer is thick. If the double-layer thickness exceeds the diameter of pores in the packing material, double-layer overlap implies that a potential extends throughout the pore,

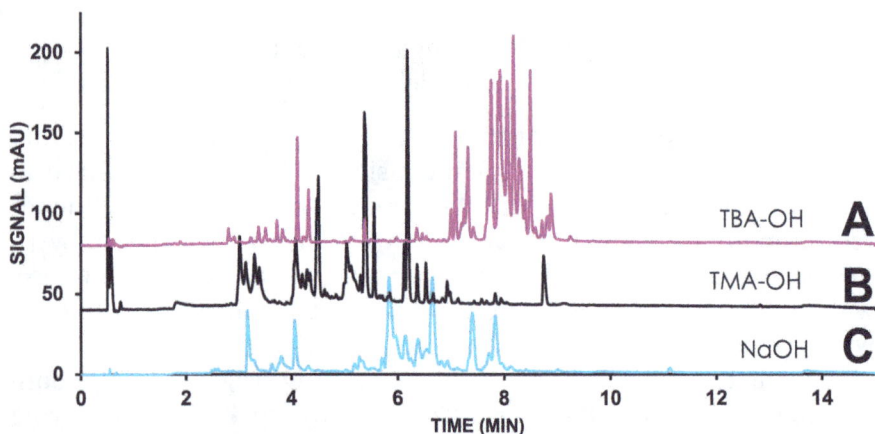

Figure 3.30 Separation of small molecules by ion-pair reversed-phase LC using 10 mM (A) tetrabutylammonium hydroxide, (B) tetramethylammonium hydroxide, and (C) sodium hydroxide as ion pair in the mobile phase. Chromatograms reproduced from ref. 57 with permission from Elsevier, Copyright 2016.

causing depletion or exclusion of like-charged ions from the pores. We speak of ion-exclusion chromatography, which is commonly abbreviated ICE (because IEC was already taken).

In the most common case, a high-capacity SCX resin, such as sulfonated polystyrene-divinylbenzene (PSDVB), is used in combination with a dilute strong acid (*e.g.* sulfuric or perchloric acid) and anions will be excluded from the pores. Less commonly, a high-capacity SAX resin (*e.g.* quaternary ammonium) may be used in combination with a dilute strong base (*e.g.* sodium hydroxide) for IEC of cations. ICE has also been shown to work with weak ion exchangers and weak-acid or weak-base mobile phases, but such situations are much less common.[62] Exclusion can be turned into separation between different analytes if (i) it is partial and (ii) the extent of exclusion is dependent on the analyte. This implies that ICE can be used for partially charged species.

In ideal ICE, neutral compounds can penetrate all pores freely and elute with the mobile-phase volume ($V_m = V_{excl} + V_{pores}$), whereas ions with the same charge as the surface will be excluded from all pores and elute at the exclusion volume (V_{excl}). In fact, injecting small samples of KBr, which possesses a UV-active anion, at low ionic strength can be used to estimate V_{excl} of silica-based columns, whereas injecting larger amounts or using a mobile phase with a high ionic strength may provide an indication of V_m (see also Section 3.1.4).

For the elution of partially ionized acids of the form HA $\leftrightarrows$ A$^-$ + H$^+$ from an SCX column, a simple approximation is

$$V_e = V_{excl} + \frac{1}{\frac{[A^-]}{[HA]}+1} V_{pores} = V_{excl} + \frac{1}{10^{(pH-pK_a)}+1} V_{pores} \tag{3.39}$$

For eluting partially ionized bases of the form HB$^+$ $\leftrightarrows$ B + H$^+$ from an SAX column, this becomes

$$V_e = V_{excl} + \frac{1}{\frac{[A^-]}{[HA]}+1} V_{pores} = V_{excl} + \frac{1}{10^{(pK_a-pH)}+1} V_{pores} \tag{3.40}$$

Figure 3.31 illustrates how ICE separations can be achieved according to this simple model. With a dilute acid of pH = 4 as the mobile phase (dark-blue sinusoidal curve), weak acids can be separated on an SCX column based on their acidity approximately in the range $3 < pK_a < 5$. At pH = 6 (light-blue curve), the selective range is approximately $5 < pK_a < 7$. More generally, if the mobile-phase pH is pH$_{mp}$, the selective range is approximately pH$_{mp}$ − 1 < pK_a < pH$_{mp}$ + 1. Weak bases can be separated on an SAX column at a high pH (for example, pH = 10, pink line, selective range $9 < pK_a < 11$).

Large-diameter columns (*e.g.* 7.8 mm i.d.) are conventionally used for ICE, but there is no fundamental reason for this. The pragmatic reason is that a wider column makes it easier to deal with extra-column band broadening (see Section 3.3.3). In addition, ICE requires sample volumes that are relatively small in comparison to the column volume (V_C), and this is easier to achieve when V_C is large. Depending on the application, ideal ICE conditions may or may not be approached. For example, in the separation of aromatic acids on PSDVB-SCX columns, energetic interactions of the analytes with the surface are hard to avoid. Reversely, ICE effects may be observed in other forms of LC. ICE suffers from a limited elution window and a modest efficiency. A long column may be needed to achieve sufficient separation. Gradient elution is not an option, limiting the use of ICE for complex samples. The technique is most interesting for situations in which high selectivity can be obtained for a group of analytes with similar, but different pK_a values. The best example

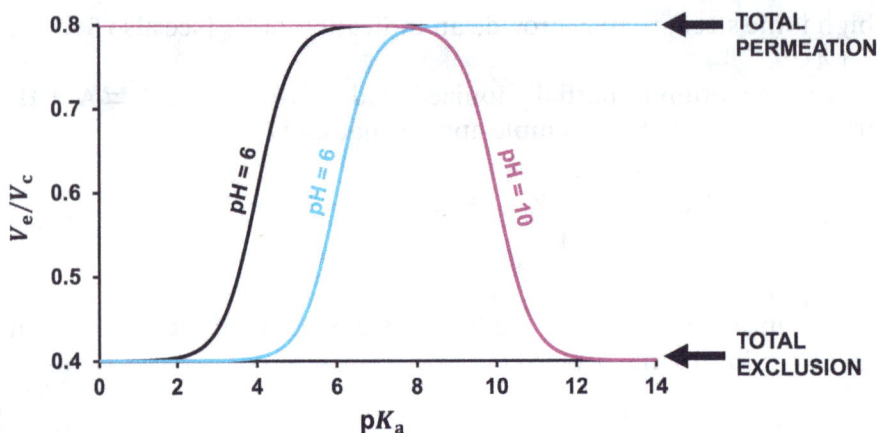

Figure 3.31 Illustration of fractional elution volumes as a function of analyte pK_a. Calculated using eqn (3.39) and eqn (3.40), using $V_{excl} = 0.4 \ V_C$ and $V_{pores} = 0.4 \ V_C$. Blue lines: separation of weak acids on an SCX column; pink line: separation of weak bases on an SAX column.

is carboxylic acids, the pK_a values of which typically range between 4 and 5.

3.8 Retention Models (M)

3.8.1 Usefulness of Retention Models

As mentioned in Section 3.4.2 (see Figure 3.20), a relationship between retention and one or more operating variables is called a retention model. Retention models serve several purposes. Retention models

- serve to discuss different retention mechanisms;
- may help predict physical–chemical parameters (*e.g.* octanol–water partition coefficients);
- provide an understanding of the effects of different parameters; for example, it is good to know (also in practice) whether an effect is linear or exponential;
- aid in the characterization and comparison of different stationary phases;
- help predict changes resulting from (accidental or purposeful) changes in the operating conditions;

- are essential for describing the behaviour of analytes under gradient-elution conditions;
- play a crucial role in systematic method development and optimization;
- facilitate a successful transfer of methods to different LC systems; and
- help establish the robustness of a method and help establish sensible system-suitability parameters.

Thus, there are plenty of reasons to discuss retention models for various forms of LC in this module.

3.8.2 Quantitative Structure–Retention Relationships

Quantitative structure–retention relationships (QSRRs) aim to describe LC retention as a function of many parameters related to the chemical structure of analyte molecules.[63] A schematic illustration of QSSRs is provided in Figure 3.32. The starting point consists of numerous molecular descriptors for an analyte molecule. There are numerous tools that produce large numbers of such descriptors based on a digital (one-, two-, or three-dimensional) coding for the molecule. Based on a set of compounds for which retention is known, a model can be built, for example, using multiple linear regression (MLR) or partial least squares (PLS) techniques. The spectacular progress made in the field of artificial intelligence in recent years has generated many new options for this step of the process, with more progress expected. The number of molecular descriptions is much too large in relation to the training sets available, so that a careful selection of variables ("feature selection") is an important aspect of the model-building process. Local models, built from data on a limited number of compounds that resemble the target analytes, have been more accurate for retention prediction in LC than global models, built using large numbers of diverse compounds in the training set.

A weak point in the application of QSSRs in LC is the lack of a standardized way to record and share retention data. In GC, retention indices (see Section 2.1.3) provide a reasonable option, and such data are increasingly added to GC–MS spectral libraries. In LC, there is no generally accepted retention-index scheme, and retention factors, let alone retention times, are hard to reproduce because they

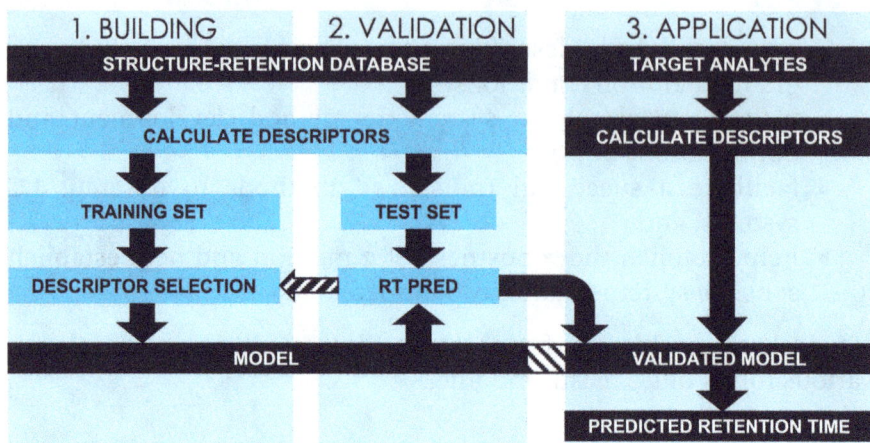

Figure 3.32 Schematic illustration of Quantitative Structure–Retention Relationships (QSRRs).

are influenced by a whole array of experimental conditions. Readers interested in retention indices for LC may consult a recent review.[64]

3.8.2.1 Octanol–Water Partition Coefficients

Octanol–water partition coefficients (K_{ow} or P_{ow}) are used as a measure for lipophilicity (hydrophobicity) in many fields, for example, in pharmaceutical sciences as a measure for possible drug uptake and in toxicology as a measure for possible bioaccumulation. A substantial number of studies have been devoted to the correlation between RPLC retention factors and log P_{ow}, mainly with the goal of avoiding laborious "manual" measurements of the distribution coefficients by liquid–liquid extraction using two solvents that do not phase separate rapidly. Conversely, the large numbers of log P_{ow} values tabulated, as well as any method aimed at predicting log P_{ow} values from molecular structure, are potentially valuable for liquid chromatographers.

LC can also be used to study biological partition/distribution more generally than just measuring log P_{ow} values. Other partition coefficients may be accessible. "Biomimetic stationary phases", such as artificial membranes, and immobilized-protein stationary phases are examples of the potential of LC.[65,66]

3.8.2.2 Linear Free-energy Relationships

Linear free-energy relationships (LFERs) express a free-energy difference (ΔG) in terms of several linear terms, each representing specific interactions. An example of such a model that has been particularly well tested in LC is the **Solvation-Parameter Model** (SPM) developed by Mike Abraham and expanded by Marti Rosés.[67] The basic **Abraham model** for neutral analytes reads

$$\ln k_i = c_{sys} + e_{sys}E_i + s_{sys}S_i + a_{sys}A_i + b_{sys}B_i + v_{sys}V_i \tag{3.41}$$

The various terms and coefficients are described in Table 3.13. The analyte coefficients (E_i through V_i) can either be determined from chromatographic experiments or calculated based on molecular characteristics. Parameter values have been tabulated for at least 9000 chemical compounds (see ref. 67 and references cited therein). The model seems to be consistent in the sense that identical probe values are obtained from experiments on different chromatographic (mobile phase/stationary phase) systems and *vice versa*. It has been applied for neutral (*i.e.* non-charged) analytes in various modes of LC (RPLC, NPLC, and HILIC).

The Abraham model gets considerably more complex when ionic or ionogenic analytes are considered. Various expansions of the model have been investigated. One of the most promising expansions includes two additional terms as follows

$$\ln k_i = C_{sys} + e_{sys}E_i + S_{sys}S_i + a_{sys}A_i \\ + b_{sys}B_i + v_{sys}V_i + j^+_{sys}D^+J^+_i + j^-_{sys}D^-J^-_i \tag{3.42}$$

where D^+ is the molar fraction of the cationic (*e.g.* protonated) form of the analyte and D^- is the molar fraction of the anionic (*e.g.* dissociated) form. The molar fraction of the neutral form of the analyte is denoted as D^0, and the various analyte parameters become a weighted sum of the values for the different forms, for example

$$E_i = D^+ E_i^+ + D^0 E_i^0 + D^- E_i^- \tag{3.43}$$

To obtain correct values for the molar fractions – and implicitly for the other parameters in the model – it is imperative that both the actual value of the pH and that of the analyte dissociation constant

Table 3.13 Explanation of the coefficients in the Abraham Solvation-Parameter Model (SPM).

Symbol(s)	Description
C_{sys}	System constant (includes the phase ratio)
$e_{sys}E_i$	$\Delta G/RT$ contribution from n– and π–**electron pairs**
e_{sys}	System (mobile phase/stationary phase) electron-interaction coefficient
E_i	Analyte excess molar refractivity
$s_{sys}S_i$	$\Delta G/RT$ contribution from **dipole interactions** (orientation and induction)
s_{sys}	System (mobile phase/stationary phase) dipole-interaction coefficient
S_i	Analyte dipolarity/polarizability
$a_{sys}A_i$	$\Delta G/RT$ contribution from **hydrogen-bond donation** from the analyte to the mobile phase and stationary phase
a_{sys}	System (mobile phase/stationary phase) hydrogen-bond-basicity coefficient
A_i	Analyte overall hydrogen-bond acidity
$b_{sys}B_i$	$\Delta G/RT$ contribution from **hydrogen-bond donation** from the mobile phase and stationary phase to the analyte
b_{sys}	System (mobile phase/stationary phase) hydrogen-bond-acidity coefficient
B_i	Analyte overall hydrogen-bond basicity
$v_{sys}V_i$	$\Delta G/RT$ contribution from the difference in **cavity formation** in the mobile phase and stationary phase and residual solute–solvent dispersion interactions
v_{sys}	System (mobile phase/stationary phase) cavity-formation coefficient
V_i	Analyte McGowan molar volume
$j^+_{sys}J^+_i$	$\Delta G/RT$ contribution from cation-exchange interaction
j^+_{sys}	System (mobile phase/stationary phase) cation-exchange coefficient
$J^+_iJ^+_i$	Positive for analyte cations and zero for anions and neutral molecules
$j^-_{sys}J^-_i$	$\Delta G/RT$ contribution from anion-exchange interaction

(continued)

Table 3.13 (*continued*)

Symbol(s)	Description
j_{sys}^-	System (mobile phase/stationary phase) anion-exchange coefficient
J_i^-	Positive for analyte anions and zero for cations and neutral molecules

(*i.e.* $pK_{a,i}$) are established in the aqueous–organic mixture, as these values differ considerably from those in a purely aqueous solvent.[68]

The Abraham model and its various modifications provide a consistent description of retention in various forms of chromatography. It provides insights into which interactions contribute most to the retention of various compounds. For neutral compounds, it is relatively simple to obtain parameters for chromatographic systems based on experiments performed with a small set of probe solutes. Using the model is more demanding for ionizable analytes due to the need to establish accurate values for the molar factions of the various charged and non-charged forms.

Other LFER models have also been developed for LC, although none as elaborately as the Abraham model. One example is the **hydrophobic-subtraction model (HSM)**, which describes the selectivity (relative retention, α) of analytes relative to ethylbenzene (EB). The assumption is that hydrophobic interactions are reflected in the retention of the reference compound, revealing more specific information in the LFER model, which reads

$$\log\left(\frac{k_i}{k_{EB}}\right) = \eta H - \sigma S + \beta A + \alpha B + \kappa C \tag{3.44}$$

where the lower-case Greek letters refer to properties of the analyte, *viz.* hydrophobicity (η), bulkiness (σ), hydrogen-bond basicity (β) and acidity (α), and effective ionic charge (κ), and the capital letters refer to the corresponding column parameters, *viz.* hydrophobicity (H), "steric interaction" (S), hydrogen-bond acidity (A) and basicity (B), and ion exchange (C).

The most striking application of the hydrophobic-subtraction model is the characterization of a large number of different columns (see Supplementary Material of ref. 69). A "column-comparison function" ($F_{s,12}$) has been developed to describe the extent to which different columns behave differently (or identically). It is based on the squared differences of the stationary-phase parameters, with weighting factors for each term,[70] *i.e.*

$$
F_{s,12}^2 = 156(H_1 - H_2)^2 + 10\ 000(S_1 - S_2)^2 + 900(A_1 - A_2)^2 \\
+ 20\ 450(B_1 - B_2)^2 + 6900(C_1 - C_2)^2
$$

(3.45)

For non-ionic analytes, it is realistic to omit the last (C) term from this equation. The HSM model was recently improved to capture isomer selectivity[71].

Comparing columns pairwise is rather unattractive, and various groups of researchers have tried to analyse data sets of LC column characteristics. Lesellier and West[72] have suggested spider plots to visualize the differences between different groups of columns. Another approach, which is easier to visualize, but more difficult to interpret, is to use the multivariate techniques described in Chapter 9 (Sections 9.9.1, Principal Component Analysis, and 9.9.2, Hierarchical Cluster Analysis).

3.8.3 Single-variable Models

Single-variable models describe the effect of one variable (*e.g.* mobile-phase composition, temperature, salt concentration, or pH) under the assumption that all other variables, including the stationary phase, remain constant. The latter usually implies that the model is developed on one given column. When using different columns (*i.e.* columns of different dimensions, packed with identical stationary particles) to establish coefficients in retention models or when using retention models to predict retention on columns of different dimensions, very careful attention should be paid to the effects of (pre-injection) dwell volumes and dead volumes in the chromatographic system, including connections, connecting capillaries and frits.

The most important parameter in LC is arguably the mobile-phase composition. It typically allows changing the retention factor by orders of magnitude. This is much greater than, for example, the variation that can be achieved by changing the temperature. While temperature is the most important parameter in GC and temperature programming is the common *modus operandi*, LC centres around the mobile-phase composition and gradient elution is commonly used. Therefore, we will first explore models that describe the effect of the mobile-phase composition on retention in LC.

3.8.3.1 Log-linear Model

The log-linear model is the best-known and most used model, especially in RPLC. It is an empirical model, which is usually described as

$$\log k = \log k_0 - S\varphi \tag{3.46a}$$

or

$$\ln k = \ln k_0 - S'\varphi \tag{3.46b}$$

where φ is the volume fraction of strong solvent, k_0 is the extrapolated retention factor at $\varphi = 0$ (100% weak solvent; in RPLC 100% water), and S or S' (with $S' = 2.303\ S$) is a (positive) slope parameter. The two equations are equivalent. More people are familiar with the conventional 10-based logarithm ($\log x = {}^{10}\log x = \log_{10} x$), but the natural logarithm ($\ln x = {}^{e}\log x$) can be more readily integrated. We will describe the following models in terms of natural logarithms (ln). The 10-based-log versions can be found in Table 3.14.

The log-linear model is a useful approximation across a limited range of retention factors (*e.g.* $1 < k < 10$) in RPLC. Retention factors below 1 ($\ln k < 0$) are known to deviate significantly from the log-linear model, and use of the model is not recommended in this range. The log-linear model is often erroneously called the LSS model (see Section 3.4.2). The main advantage of the log-linear model is its simplicity. It only features two coefficients, and it is linear in terms of these (*i.e.* the derivative of $\ln k$ towards one of the coefficients – k_0, S or S' – is independent of this coefficient). To obtain estimates of the two coefficients, two experimental data points suffice. These can either be obtained from isocratic or gradient-elution experiments (see Section 3.17.2).

The retention factor generally increases exponentially with increasing molecular weight for chemically similar compounds, such as homologues (Martin rule; see Section 2.1.3). The slope (S or S') increases with molecular weight in a similar manner, and linear correlations have been observed between $\log k_0$ or (or $\ln k_0$) and S (or S'). For high-molecular-weight analytes, the retention lines become

Table 3.14 Summary of the single-variable models discussed in this module to describe the effect of the composition or salt concentration on retention.[a]

Model	RPLC	NPLC	HILIC	IEC
Log-linear $\log k = \log k_0 - S\varphi$ $\ln k = \ln k_0 - S'\varphi$ $S' = 2.303 \cdot S$	~ Limited range of composition • High-MW analytes	~ High-MW analytes	—	—
Quadratic $\log k = A\varphi2 + B\varphi + C$ $\ln k = A'\varphi2 + B'\varphi + C'$ $[A', B', C'] = 2.303 \cdot [A, B, C]$	• Broader range of composition • Extrapolation dangerous	• Good fit across limited ranges of composition	~ Also used for HILIC ~ Qualitatively describes effects at low and high $m' = 2.303 \cdot m$	—
Log-log $\log k = \log k_1 - m\log \varphi$ $\ln k = \ln k_1 - m'\ln \varphi$ $m' = 2.303 \cdot m$	~ Not intended for RPLC, but sometimes used across limited ranges of composition	• First proposed for NPLC • Grasps dramatic effects of low concentrations	~ Also used for HILIC ~ No flexibility in describing the curvature	—
$\log k = \log k_1 - m_c\log c_{ci}$ $\ln k = \ln k_1 - m_c'\ln c_{ci}$ $m_c' = 2.303 \cdot m_c$	—	—	—	• Describes the diminishing effect of increasing concentrations • m_c relates to the charge of the ion

(continued)

Table 3.14 (*continued*)

Model	RPLC	NPLC	HILIC	IEC
Mixed-mode $\log k = \log k_0 - S'_{mm,1}\varphi - S_{mm,2}\ln\varphi$ $\ln k = \ln k_0 - S'_{mm,1}\varphi - S'_{mm,2}\ln\varphi$ $[S'_{mm,1}, S'_{mm,2}] = 2.303 \cdot [S_{mm,1}, S_{mm,2}]$	—	—	• Largely applied to HILIC • Limited flexibility to describe curvature	—
Neue–Kuss $\ln k = \ln k_0 - \dfrac{S_{nk,1}\varphi}{1 + S_{nk,2}\varphi} + 2\ln$ $(1 + S_{nk,2}\varphi)$	• Can be integrated to describe gradients • Extrapolation dangerous	~ Also used for NPLC	~ Also used for HILIC	—

[a]The symbol ~ indicates approximate validity of the model. φ denotes the volume fraction of the strong solvent; c is the concentration of the counter ion.

extremely steep, and their exact shape becomes impossible to discern and practically irrelevant (see Module 4.3).

Many other models have been proposed. All of these have in common that they try to grasp the experimentally observed non-linearity of the log k or (or ln k) vs. φ curve.

3.8.3.2 Log–log Model

The greatest advantages of the log-linear model are its simplicity and the possibility to estimate the coefficients from only two experiments. The only other model that can match this is the log–log model, which takes the following form

$$\ln k = \ln k_1 - m' \ln \varphi \qquad (3.47)$$

where k_1 is the extrapolated retention factor at $\varphi = 1$ (100% strong solvent) and m' is a (positive) slope parameter. Clearly, the log–log model yields a linear plot of ln k or vs. ln φ (or log k or vs. log φ). However, in a plot of ln k vs. φ, as shown in Figure 3.33, the log–log model tends to be very steep at low concentrations of strong solvent. This is in line with traditional NPLC using solid adsorbents (such as silica) as stationary phase. Such surfaces tend to have a finite number of strongly adsorbing groups. The strong solvent, even in small amounts, may engage in competitive adsorption (see Section 3.6.1.1). Trace amounts of water in the weak solvent may have dramatic effects on retention in such NPLC systems. Such effects are very difficult to control, and equilibration times tend to be long. Such complications have contributed to the advent of HILIC in recent years.

Note that the log–log model is invalid for $\varphi \downarrow 0$ (because the model suggests $k \uparrow \infty$).

A log–log model is also appropriate for IEC, with the concentration of counterion (c_{ci}) taking the place of the volume fraction of strong solvent and the slope (m_c') in principle being proportional to the charge of the analyte ion.

$$\ln k = \ln k_1 - m_c' \ln c_{ci} \qquad (3.48)$$

The mechanism of ion exchange can be compared with that of competitive adsorption. Ions and counterions compete for a finite number of ionic groups on the stationary-phase surface.

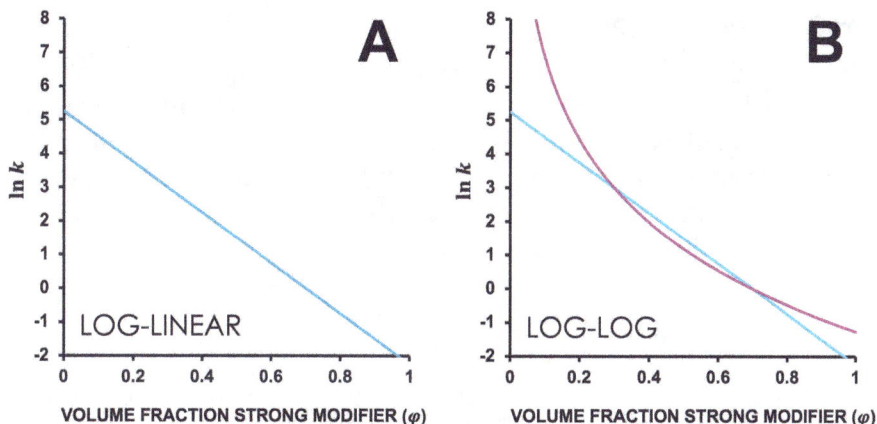

Figure 3.33 Illustration of the log-linear (A, blue) and log–log (B, pink) retention models. Lines are constructed through presumptive data points at ln k = 3 for φ = 0.3 and ln k = 0 for φ = 0.7. The log-linear model is drawn as a blue line in the plot of the log-log model (B) to allow comparison.

Therefore, similar to small fractions of polar solvents in NPLC, low concentrations of counterions may have dramatic effects on retention in IEC.

3.8.3.3 Quadratic Model

Although it has been derived from solubility-parameter theory[73] (see Section 2.1.2), the quadratic model is essentially used as an empirical model of the form

$$\ln k = A'\varphi^2 + B'\varphi + C' \tag{3.49}$$

This equation is linear in its three coefficients, A', C' (both usually positive) and B' (usually negative). Eqn (3.49) is valid across broader ranges of retention or composition than eqn (3.46a) and eqn (3.46b). Three data points suffice to estimate the coefficients, but it is dangerous to extrapolate outside the range covered experimentally.

One of the disadvantages of the quadratic model is that it is difficult to integrate when inserted in the migration equation for gradient elution (see Module 3.17). Integration[73] yields a so-called error function, erf(x), which cannot easily be computed, but is available on most computation platforms and in spreadsheets, such as Excel.

3.8.3.4 Mixed-mode Model

The mixed-mode model has emerged from the school of the famous Chinese professor Lu Peichang (Dalian Institute of Chemical Physics). It is a combination of the log-linear and the log–log models, arising from the idea that both RP-like partition and NP-like competitive adsorption may contribute to retention in methods such as HILIC. It takes the form

$$\ln k = \ln k_0 - S'_{mm,1}\varphi - S'_{mm,2}\ln \varphi \qquad (3.50)$$

where $S'_{mm,1}$ is the slope parameter and $S'_{mm,2}$ is a curvature parameter (both are usually positive). The logarithmic term may account for some curvature in the $\ln k$ vs. φ curve but only to a limited extent. If the curvature is too strong, there is no solution for the model passing through all three data points.

3.8.3.5 Neue–Kuss Model

The Neue–Kuss model, proposed by Uwe Neue (Box 3.8) and Hans-Joachim Kuss, is a slightly more complex variant of the mixed-mode model, i.e.

Box 3.8 Hero of analytical separation science: U. D. Neue. Photo courtesy of HPLC2013 Amsterdam.

Uwe Dieter Neue (1948–2010, Germany)
Uwe Dieter Neue obtained his PhD from the University of Saarbrücken (Germany), working with Heinz Engelhardt and in the footsteps of István Halász, another contributor to the development of HPLC. For more than 30 years Neue played a key role in advancing liquid chromatography within the Waters company, but beyond that, he published important studies on the characterization of stationary phases, retention mechanisms and selectivity, and he could explain all these subjects with great skill and enthusiasm. The Neue–Kuss retention model is just one of the reasons we should remember him.

$$\ln k = \ln k_0 - \frac{S'_{nk,1}\varphi}{1 + S'_{nk,2}\varphi} + 2\ln(1 + S'_{nk,2}\varphi) \tag{3.51}$$

where $S'_{nk,1}$ is the slope parameter and $S'_{nk,2}$ is a curvature parameter (both are usually positive). The model can be flexibly fit through three data points, and it can – in principle – be used in RPLC, NPLC and HILIC. An important property of the Neue–Kuss model is that it can be integrated when used in the gradient-elution equation (see Module 3.17).

3.8.3.6 Summary of Models Describing the Effect of Mobile-phase Composition

All three single-variable (φ) models with three coefficients, *i.e.* the quadratic, mixed-mode, and Neue–Kuss models, can accommodate some curvature in an $\ln k$ *vs.* φ plot. This is illustrated in Figure 3.34. The models have been forced through two data points, *viz.* $\ln k = 3$ at $\varphi = 0.3$ and $\ln k = 0$ at $\varphi = 0.7$. Lines are drawn for different values of the curvature parameters (A' for the quadratic model, $S'_{mm,2}$ for the mixed-mode model, and $S'_{nk,2}$ for the Neue–Kuss model), as indicated in the figure. In all cases, a curvature coefficient of 0 yields a straight line passing through the two fixed points, *i.e.* the log-linear model. If we imagine a third data point at $\varphi = 0.5$, then it is clear from Figure 3.34 that small experimental errors may lead to very large changes in the extrapolated values below $\varphi = 0.3$ (above $\ln k = 3$) and above $\varphi = 0.7$ (below $\ln k = 0$). Extrapolation is dangerous in all three cases. The quadratic and Neue–Kuss models provide sufficient flexibility to fit the curve through three data points with more or less curvature. The mixed-mode model allows little flexibility in its present form. The curvature cannot be increased further than shown in the figure ($S'_{mm,2} = 2.5$).

Table 3.14 provides a summary of the retention models for LC that have been discussed in this module.

3.8.3.7 Effect of Temperature on Retention

The retention factor k is proportional to a partition coefficient K, which in turn can be expressed as some difference in partial molar free energies or chemical potentials (μ; see Section 1.6.1). Therefore, we have

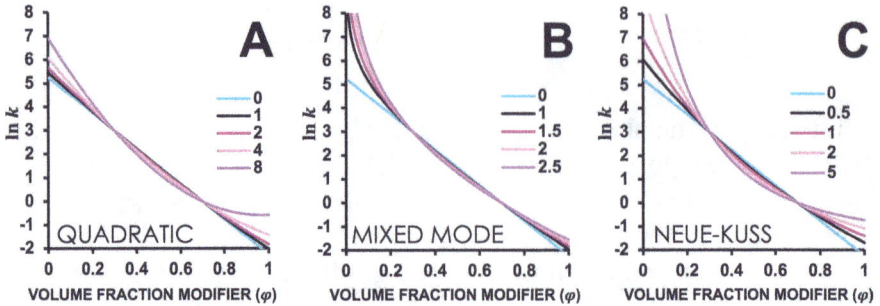

Figure 3.34 Illustration of extrapolation using the quadratic (A), mixed-mode (B), and Neue–Kuss (C) retention models. Values of the curvature parameters (A' for the quadratic model, $S'_{mm,2}$ for the mixed-mode model, and $S'_{nk,2}$ for the Neue–Kuss model) are indicated in the figure.

$$\ln k = \ln K + \ln \beta = \frac{\Delta\mu}{RT} + \ln \beta = \frac{\Delta H}{RT} - \frac{\Delta S}{R} + \ln \beta \qquad (3.52)$$

where β is a phase ratio, equal to V_s/V_m in liquid–liquid chromatography, but less well defined in other forms of LC. Eqn (3.52) is known as the **van 't Hoff equation,** named after University of Amsterdam professor Jacobus Henricus van 't Hoff, the first-ever Nobel Prize winner in chemistry. In practice, a plot of k *vs.* $1/T$ is expected to yield a straight line, with the slope related to ΔH and the intercept related to a combination of ΔS and the phase ratio.

Usually, retention decreases with increasing temperature, which implies that ΔH in eqn (3.52) is positive. However, this is not fundamental. If we perform liquid–liquid chromatography, we could, in principle, swap the mobile and stationary phases (bar trivial factors such as viscosity). If we do so, ΔH would keep the same value and just change the sign. Such an experiment is usually not practical, but it demonstrates that ΔH may feasibly be negative. However, practical LC systems in which retention increases with increasing temperature are rare.

It can be argued that the ΔH values obtained from the **van 't Hoff plot** can be deceptive if more than one retention mechanism is in play (*i.e.* if different adsorption or absorption sites in the column contribute to the retention).[74] However, the relationship between k and $1/T$ is definitely more linear than that between k and T. Assuming the latter kind of linear relationship is rather lackadaisical.

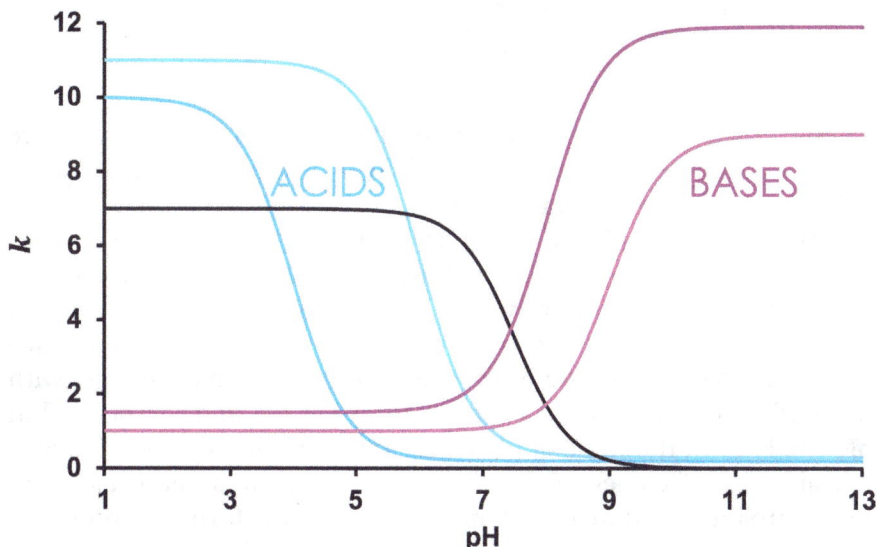

Figure 3.35 Schematic illustration of the retention behaviour of some weak acids (such as carboxylic acids; black and blue lines) and weak bases (such as amines; pink lines) as a function of pH.

3.8.3.8 Sigmoidal Model for Effect of pH

The retention behaviour of monovalent weak acids and bases follows a sigmoidal pattern, reminiscent of the titration curve (see Figure 3.35). For a weak monoprotic acid in RPLC, we may write

$$k = \frac{k_{\text{HA}} + k_{\text{A}^-} \times 10^{(\text{pH}-\text{p}K_a)}}{1 + 10^{(\text{pH}-\text{p}K_a)}} \tag{3.53}$$

where k_{HA} is the retention factor of the intact (protonated) acid, k_{A^-} is the retention factor of the dissociated acid (conjugated base), and $\text{p}K_a$ is the negative logarithm of the acid dissociation constant. For a weak monoprotic base, we have analogously

$$k = \frac{k_{\text{B}} + k_{\text{HB}^+} \times 10^{(\text{p}K_a-\text{pH})}}{1 + 10^{(\text{p}K_a-\text{pH})}} \tag{3.54}$$

where k_{B} is the retention factor of the intact acid, k_{HB^+} is the retention factor of the protonated base, and $\text{p}K_a$ is the negative logarithm of the acid dissociation constant, which is equal to

$$K_a = \frac{K_w}{K_b} \tag{3.55}$$

where K_w is the dissociation constant of water. It is clear from Figure 3.35 that retention, as well as selectivity, can vary very strongly with pH for weak acids and weak bases.

Varying the pH in RPLC is not as clear-cut as the retention curves in Figure 3.35 suggest. The pH in aqueous–organic mixtures is different from that in pure water, and pK_a of the analytes will also be affected by the type and volume fraction of organic solvents (see Section 3.5.4). The nature of the stationary phase may change, with most silanol groups being protonated at pH < 4 and dissociated at pH > 6. Finally, the value of pH tends to affect not only the selectivity but also the efficiency of the chromatographic systems. Peaks tend to be broader around pK_a. Apparently, the two forms (protonated and deprotonated) of the analyte add to the band broadening.

3.8.4 Multi-variable Models

The effects of the different variables (mobile-phase composition, temperature, pH, *etc.*) on retention tend to be intertwined. For example, the optimum mobile-phase composition at one temperature may be different from that at another temperature. The typical way to construct models that describe the effect of two variables simultaneously is to embed one single-variable model into another. For example, the coefficients in one of the models describing the effect of composition can be made to depend on temperature. The simplest case is the log-linear model

$$\ln k = \ln k_0(T) - S'(T)\varphi = \ln {}^0k_0 + \frac{\ln {}^1k_0}{T} - \left({}^0S' + \frac{{}^1S'}{T} \right)\varphi \tag{3.56}$$

In the case of pH optimization, it is almost unavoidable to consider the mobile-phase composition simultaneously because the effects of pH on selectivity tend to be accompanied by large effects on retention. The latter can be accounted for by adapting φ. A simple example of one of many possible models[75] is obtained by combining the log-linear model with eqn (3.53)

$$k = \frac{k_{HA} + k_{A^-} \times 10^{(pH-pK_a)}}{1 + 10^{(pH-pK_a)}} = \frac{10^{(\log k_{HA,0} - S'_{HA}\varphi)} + 10^{(\log k_{A^-,0} - S'_{A^-}\cdot\varphi + pH + \log K_{a,0} - S'_{K_a}\varphi)}}{1 + 10^{(pH + \log K_{a,0} - S'_{K_a}\varphi)}} \quad (3.57)$$

3.9 LC Detectors (B)

A wide range of detectors exist for LC. A few of these are in common use, while others are eminently suited for specific niche applications. Unfortunately, LC does not have an equivalent to the FID in GC (see Section 2.5.1), which is extremely sensitive and linear across many decades and responds nearly universally to all compounds with predictable response factors. The simple (and largely correct) explanation for this is that the detection after GC separation takes place against a background of an inert gas, whereas the detection after LC involves measurements of dilute solutions in a dense liquid phase. Attempts have been made to combine an FID with LC, after evaporating the solvent. A moving-wire interface for the purpose was commercially available from Pye Unicam (Cambridge, UK)[76] for a period of time. Even years after they were taken out of production, these systems continued to generate interest among liquid chromatographers hoping to create the ultimate LC detector. LC–FID has not really been made to work, not even with (super-)heated water as the mobile phase, an eluent that exhibits a low background level, but significant corrosion or "warping" of detector parts. FID can be used in combination with supercritical fluid chromatography (SFC, see Chapter 6), but only with pure carbon dioxide as the mobile phase, in which case the number of possible applications is extremely limited.

The ideal LC detector exhibits

- a high sensitivity (*i.e.* a high slope of signal-*vs.*-concentration or signal-*vs.*-mass-flow calibration curves),
- linear calibration curves (across a broad range),
- low noise (together with a high sensitivity, this results in low detection limits),
- uniform or otherwise predictable response factors or, alternatively, highly specific response to certain (classes of) compounds, with no response to all other (classes of) compounds,
- no response to the mobile phase, including in the case of gradient elution,

- no response to changes in temperature, flow rate or pressure (on the column or in the detector cell),
- no contribution to extra-column band broadening,
- robustness with regard to pressure build-up (*e.g.* due to a blocked exit capillary) and
- no destruction of the sample, so as to allow the collection of separated fractions after the detector.

In this chapter, we will first discuss the most common LC detectors and then briefly describe some detectors used for niche applications. In Module 3.10, we will discuss the hyphenation of LC with larger detection systems, such as mass spectrometers.

3.9.1 Refractive Index Detection

Differential refractive index (DRI) detectors are the most universal among LC detectors because every liquid has a refractive index. The refractive index is a measure of the extent to which light is bent when crossing an interface (Figure 3.36A). Snell's law provides a relation between the angle of incidence (relative to the "normal", which is a line perpendicular to the interface), θ_{inc}, and the angle of the refracted light (θ_{ref}) when passing from a medium with refractive index n_1 to a medium with refractive index n_2. It reads

$$n_1 \sin \theta_{inc} = n_2 \sin \theta_{ref} \qquad (3.58)$$

   In DRI detectors, an interface is created between a flow cell containing the LC effluent and a reference cell containing the mobile phase (two diffractions on a window between these cells effectively cancel). The reference cell and the *differential* measurement are needed because the change in refractive index is small compared to its absolute value. The angle of diffraction, *i.e.* the extent to which the light beam is shifted from the centre, is measured by a dual photodiode or with a diode array. More sensitive DRI instruments based on interferometry have been developed and commercialized. However, operating such systems is challenging because the refractive index is sensitive to variations in temperature, flow rate, and pressure. Usually, the detector-cell housing has a high thermal mass, and a heat exchanger is installed between the column and the detector cell. However, the latter contributes to the

A

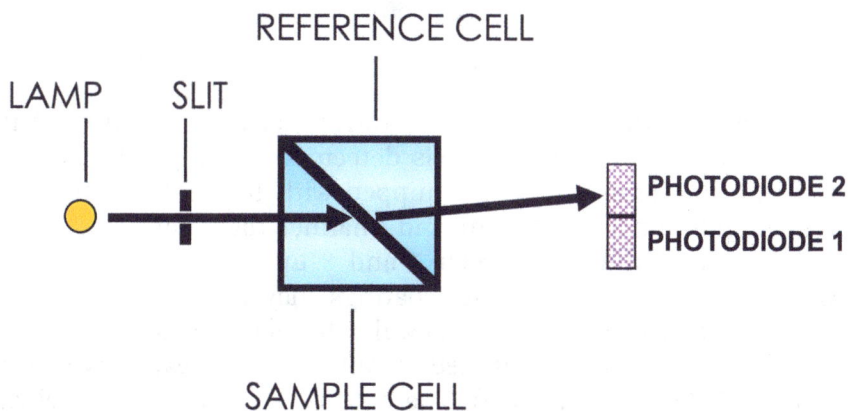

B

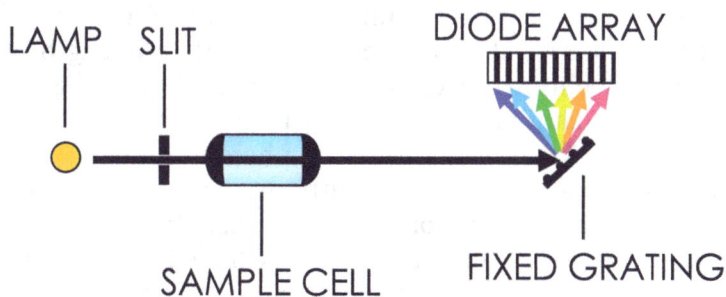

Figure 3.36 Schematic illustration of (A) differential refractive index (DRI) detector and (B) UV–vis diode-array detector (DAD).

extra-column residence time (see Module 1.2) and the extra-column band broadening (see Section 3.3.3). DRI detectors tend to require a warm-up period before a stationary temperature is achieved in the detector cell.

DRI detectors are not compatible with gradient elution, which greatly limits the number of possible applications. One isocratic form of LC where DRI detectors still find frequent use is SEC (see Module 4.2). SEC is mostly applied for the characterization of polymers, and lack of UV absorbance is an asset for synthetic polymers because it tends to imply lightfast materials. In addition, natural polysaccharides (starch and cellulose) lack UV absorbance.

3.9.2 UV–Vis Detection

Spectrometers that operate in the ultraviolet and (sometimes) the visible wavelength region (UV–vis detectors) are arguably the most used LC detectors. They are equipped with flow cells with typical path lengths of 5 to 10 mm and channel diameters of 1 or 0.5 mm for total volumes between 1 and 8 µm. Tapered cells with a widening flow path or special coatings may reduce the effects of light reflecting from the channel walls. Special Z-cells or U-cells are used for applications of (packed) capillary LC and capillary electrophoresis (CE; see Chapter 5). The detection cell is typically closed at both ends with high-purity quartz windows that are transparent for UV light down to 190 nm. Depending on the light source, UV detectors allow selection from a number of narrow emission lines (*e.g.* low-pressure mercury lamp) or from a continuous spectrum (*e.g.* deuterium lamp). The former allows for single-wavelength detection (usually with a filter to select a line, *e.g.* 254 nm for mercury). A continuous light source allows selecting a narrow wavelength range with simple monochromator optics (typically a diffraction grating). Detection of a transmitted light beam can best (lowest noise and highest sensitivity) be performed with a photomultiplier tube, but photodiodes are also used for the purpose. Figure 3.36B shows a photodiode array in combination with a grating. This allows recording complete UV spectra on the fly with a high frequency (up to about 200 Hz). Unless special provisions are taken to construct high-pressure cells, UV detectors cannot withstand high pressure. In a different setup, optical fibres are used to bring the light to a cell in the flow path and to record the transmission. This provides significant flexibility and allows the minimization of extra-column band broadening.

The response of UV–vis detectors is proportional to the concentration of the analyte in the cell. It is accurately described by the **Lambert–Beer law**, which describes the absorbance A_i of analyte i as

$$A_i = -\log T_m = -\log \frac{I}{I_0} = a_{is}(\lambda)\, l\, c_i \qquad (3.59)$$

where T_m is the transmission, I is the intensity of the transmitted light, I_0 is the intensity of the incident light, $a_{is}(\lambda)$ is the wavelength (λ)-dependent extinction coefficient (or absorption coefficient or absorptivity) of analyte i in solvent s, l is the path length of the cell, and c_i is the concentration of the analyte in the cell. Up to an absorbance of about 2 (1% of the light transmitted), the absorbance varies linearly with the concentration. Thereafter, the calibration curve flattens and noise starts to prevail.

If the extinction coefficient is known, the amount of analyte can be quantified from the area of the chromatographic peak. More commonly, (a) calibration standard(s) are used for each analyte, as $a_{is}(\lambda)$ values are not usually known and are difficult to predict. Unfortunately, $a_{is}(\lambda)$ values vary widely for different analytes, with many classes of analytes exhibiting no or negligible absorption. This limits the usefulness of the convenient and affordable UV detector for some applications.

Some examples of molar extinction coefficients are listed in Table 3.15, together with minimum wavelengths for operation with some common LC solvents. There is some variation in the specified UV cut-off, which depends on the solvent purity. HPLC-quality or UV-labelled solvents are usually high-purity (and high-price) options. In the case of a cell path length of 5 mm, a concentration of $1/a_i(\lambda)$, with l in units of m, results in an absorbance of 0.005. Thus, $1/a_i(\lambda)$ is an indication of the detection limit in mole L^{-1}. Note that this refers to the concentration in the detector cell (see Section 1.4.5 for conversion into the injection concentration). Saturated hydrocarbons, aliphatic alcohols, ethers and esters show very low absorption in the UV–vis range.

3.9.3 Fluorescence Detection

Fluorescence is a process in which (analyte) molecules absorb light of a certain wavelength, resulting in an increase in the energy level from the ground state by a very specific amount (corresponding to the excitation wavelength). They then fall back to an energy level higher than the ground state, emitting a photon with lower energy (corresponding to the emission wavelength). Thanks to the very well-defined wavelengths, fluorescence is very specific. Therefore, signal background and noise can be very low, which results in very low detection limits. In a fluorescence detector (FLD), the

Table 3.15 Examples of molar extinction coefficients of some chemical compounds and UV cut-off (minimum wavelength) for application of different LC solvents.

Example compounds			LC solvents	
Name	$a_i(\lambda)$ (L mol^{-1} m^{-1})	λ (nm)	Name	λ_{min} (nm)
Benzene	180	255	Water and acetonitrile	190[a]
Toluene	300	262	n-Hexane and n-heptane	195–200
Phenol	2000	273	Methanol, ethanol, 2-propanol, and c-hexane	205–210
Nitrobenzene	6000	262	1,4-Dioxane	215
Styrene	15 000	245	Tetrahydrofuran	220
Acetophenone	17 000	231	Dichloromethane	230–235
Acetone	18 000	265	Chloroform	245
1,3,5-Trinitrobenzene	25 000	222	Ethyl acetate	255
Vitamin-A	51 000	328	Toluene	286
β-Carotene	140 000	450	Acetone and methyl-ethyl ketone	330

[a]190 nm is typically the minimum operation wavelength of a detector with quartz elements (lamp and cell windows).

emitted light is usually measured at a 90° angle. At such a high angle relative to the transmitted light, straylight – and thus noise – is minimized. The fluorescence signal increases linearly with the concentration in dilute samples, but at high concentrations of the analyte (or co-eluting UV-active compounds), significant absorption of both the excitation light and the emitted light occurs, resulting in the non-linearity of the calibration. Due to the sensitivity of fluorescence detection and because of the possibility to direct laser light at very precise, small locations, laser-induced fluorescence detection (LI-FLD) is often used in combination with miniaturized systems, such as (packed) capillary LC and capillary electrophoresis (see Chapter 5). Nel Velthorst (Box 3.9) was a pioneer of high-resolution FLD.

Box 3.9 Hero of analytical separation science: Nel Velthorst. Photo courtesy of the Vrije Universiteit Amsterdam.

Nel Velthorst (1932–2023, The Netherlands)

Nel Velthorst found her way from a farmer's daughter to a remarkable chemistry professor. Throughout her career, she remained loyal to her alma mater, the Vrije Universiteit Amsterdam. She was a great advocate of chemistry, and as a spectroscopist, she appealed to (prospective) students with a "lightshow" about fluorescence, phosphorescence and chemiluminescence. In her research, she developed new spectroscopic techniques, such as high-resolution fluorescence and cryogenic laser spectroscopy and coupled these with liquid chromatography and capillary electrophoresis to achieve extremely low detection limits.

Natural fluorescence is limited to a small fraction of compounds. These are obvious targets for FLD. To use sensitive and specific FLD for other (classes of) compounds, derivatization reactions are used (see Module 8.3). A prime example is the derivatization of primary amines with fluorescamine. The resulting derivatives allow the detection at very low levels by fluorescence with an excitation wavelength of 400 nm and an emission wavelength of 480 nm.

3.9.4 Evaporative Light-scattering Detection

Various light-scattering detectors are used in conjunction with LC. Light scattering in the condensed phase is discussed in Module 4.6, as its use is restricted to high-molecular-weight analytes. In this section, we discuss light scattering in the gas phase, which is an increasingly common detection principle for LC.

In an evaporative light-scattering detector (ELSD), the effluent of an LC column is nebulized in a stream of heated nitrogen or air (Figure 3.37), after which the mobile phase is evaporated from the droplets in a heated drift tube. Non-volatile analyte particles remain, which **scatter** light from a continuous source (*e.g.* a xenon lamp) in all directions. Part of the scattered light is detected by, typically, a photomultiplier tube. The principle is akin to the light scattered from very small airborne particles in a cinema. They scatter light from the projector and can be detected from a distance, even by the

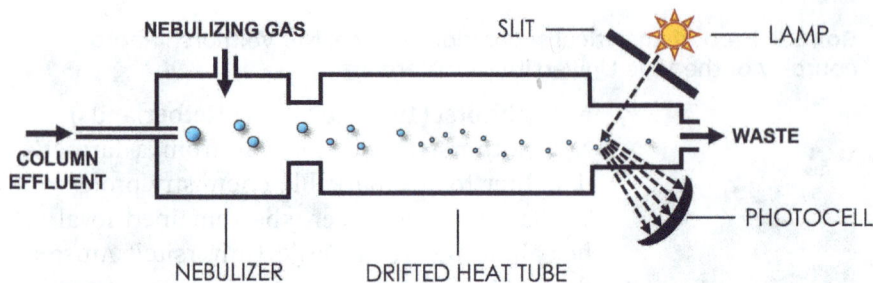

Figure 3.37 Schematic of an evaporative light-scattering detector (ELSD).

human eye. ELSD detection is quite universal. Almost all non-volatile compounds can be detected. However, the response depends on the structure and, in the case of polymers, the molecular weight of the analyte. The response also depends on the mobile-phase composition and the temperature inside the detector, but the background is limited if no non-volatile components (such as inorganic buffer salts) are present in the mobile phase. This property makes the ELSD compatible with gradient elution. A serious disadvantage is that the response is distinctly non-linear, with higher response factors obtained for higher analyte concentrations. One consequence of this is that chromatographic peaks will appear sharper than their true concentration profiles, as the low signals near the edges of the peak are attenuated, while the peak top is amplified.

ELSD detection is the most sensitive option for detecting low concentrations of non-UV-active polymers after separation by SEC (see Module 4.2), but converting SEC–ELSD profiles into molecular-weight distributions is quite complicated because of the concentration dependence (*i.e.* non-linearity) and molecular-weight dependence of the response.

Manufacturers mainly focus their latest developments on improving the linear range and sensitivity of the ELSD using, for example, a laser as the light source. Other improvements include increasing the temperature range to better adjust the ELSD to the solvent so that effects by the latter can be minimized. Computer algorithms are used to digitally improve the universality of the detector.

3.9.4.1 Charged Aerosol Detector

The charged aerosol detector (CAD) is a variant of the ELSD (Figure 3.38). After nebulization and eluent evaporation, the stream of particles in drift gas is mixed with a stream of ions (typically nitrogen ionized through a corona discharge). The ions charge the

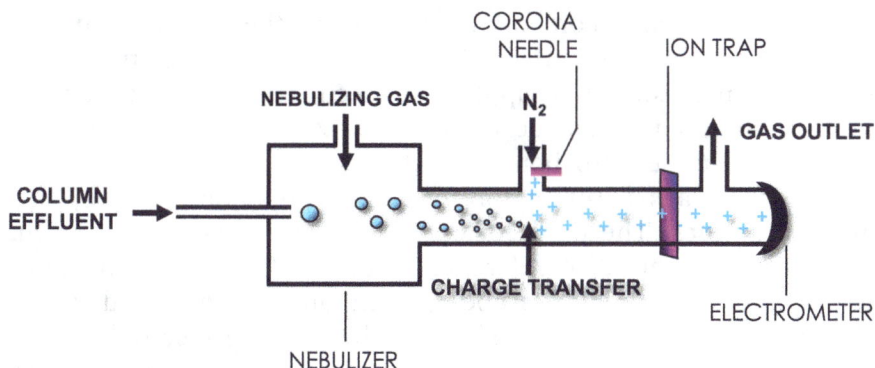

Figure 3.38 Schematic of a charged aerosol detector.

particles, and after the removal of excess gas ions, the charge on the particles is recorded. Lower detection limits have been reported for CAD relative to ELSD, but the response is not truly linear, with a curvature opposite to that shown by the ELSD (*i.e.* lower response for higher concentrations, resulting in seemingly broader peaks).

3.9.5   Electrical Detection

In this section, we will discuss detection principles that are based on electrodes and voltages or currents. We call it "electrical detection" to cover both detection principles that involve electrochemical reactions (electrochemical detection, Section 3.9.5.1) and detection that relies on physical principles, such as conductivity detection (Section 3.9.5.2).

3.9.5.1   Electrochemical Detection

Several electrochemical principles are traditionally used in analytical chemistry. In **amperometry**, a voltage is applied between a working electrode and a reference electrode, and the current is measured (in amperes, A; hence amperometry). In **coulometry**, the working electrode is designed such that the electrochemical reaction is carried to completion, and the total electrical charge supplied to the system during the process (in coulombs, C) is used to determine the amount of analyte present. In chromatographic terminology, a detector based on amperometry is concentration sensitive, while a coulometric detector is mass-flow sensitive. In **voltammetry**, the current is varied, while the voltage (in volts, V) is measured, providing information on both the nature and the amount of analyte present. Because the LC effluent varies in time, amperometry is most easily and most commonly applied for electrochemical detection.

Many compounds can be reduced or oxidized and can thus be detected using electrochemical detection, including many compounds that lack chromophores, such as aliphatic alcohols, sugars and amines. The mode of operation can be varied. If current runs from the negative reference electrode (anode) to the positive working electrode (cathode), then analytes are oxidized at the working electrode. Oppositely, if the working electrode is the anode, the reduction of analytes is studied. Amperometric detection is, in principle, simple to implement. The power requirements are low, and sensitive, low-noise electronics are readily available. Electrochemical detection lends itself to miniaturization because the concentration sensitivity is increased if the ratio of the surface of the working electrode to the volume of the detector cell increases. Electrochemical detection is an obvious choice in combination with chip-based (microfluidic) separations. The limited number of electrochemically active analytes and a lack of robustness, especially due to electrode fouling, are the main weaknesses of electrochemical detection.

Pulsed amperometric detection (PAD) is one of the most successful answers to the problem of electron fouling, at the expense of increased complexity. Short amperometric measurement periods are alternated with cleaning periods, with a complete cycle taking, for example, half a second. By alternating oxidation and reduction cycles, the build-up of reaction products on the (often gold) working electrode is avoided. One striking example is the detection of carbohydrates after anion-exchange chromatography at high pH. Very low detection limits can be achieved with PAD in such analyses.

Coulometric detection also suffers less from electrode fouling than amperometry. The former makes use of electrodes with large surface areas and short diffusion distances, such as a flow-through porous graphite electrode. In contrast to the small electrodes used in amperometric detection, the electrodes used in coulometry are stable over longer periods, after which they can be regenerated by programming an inverted voltage period.

Gradient elution is reasonably possible in combination with electrochemical detection, provided that the ionic strengths of the mobile-phase components (solvents A and B) are equal.

3.9.5.2 Conductivity Detection

In conductivity detection, a constant voltage is applied between a pair of electrodes and the current is measured. In contrast to the case of amperometric detection, no chemical reaction is meant to occur. The **conductivity** of the LC effluent (ρ_{effl}, in S m^{-1}) is found by measuring the resistance across the cell (R_{cell}, in ohm)

$$\rho_{effl} = \frac{d_{cell}}{A_{elec}R_{cell}} \tag{3.60}$$

where d_{cell} is the thickness of the cell (distance between the electrodes, in m) and A_{elec} is the surface area of each of the electrodes (assumed to be the same, in m^2). If ions are present in the effluent, they are mainly responsible for the conductivity. The conductivity depends on the type (specifically the effective charge and mobility) and concentration of ions present. When the mobile phase contains ions (*e.g.* a buffer or a salt), the background conductivity is high, and it is not possible to measure low concentrations of analytes against such a background. Suppressing the conductivity of the mobile phase, so as to reduce the detection background, is one of the main principles of ion chromatography (see Section 3.7.2). This has led to very low detection limits.

When using a capacitively coupled contactless conductivity detector (C4D), the electrodes are not in touch with the effluent. Instead, the detector unit is positioned around a non-conducting (*e.g.* fused silica or PTFE) capillary. A typical C4D consists of two ring electrodes around the capillary, with grounding (*e.g.* using copper foil) between these.[77] An AC voltage is applied on the first electrode. From the signal measured using the second electrode, the conductivity can be derived. Using C4D, the conductivity can be measured at the outlet of a packed capillary column, with minimal extra-column residence time and band broadening. It can also be positioned before or at the top of the column to measure gradient profiles or the shapes of injection pulses. As such, it has become a very useful and flexible tool for liquid chromatographers.

3.9.6 Comparison of Main Detection Methods

A qualitative comparison of the main detectors used in LC is provided in Figure 3.39. Five criteria are used in the comparison. Good

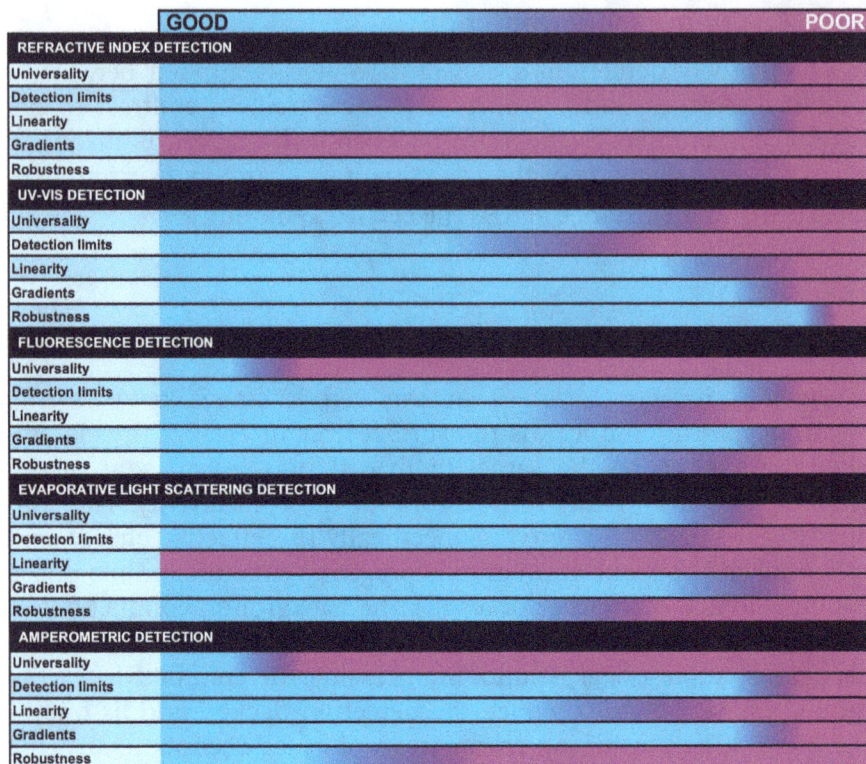

Figure 3.39 Qualitative comparison of the main types of LC detectors.

performance is indicated in light blue and poor performance in pink. "Universality" describes the range of applicability of a detection method, which is perfect if it can be applied to all analytes and poor if it applies to a limited number of specific ones. In the case of "Detection limits", good means low and poor means high. "Linearity" includes the dynamic range of a method. For example, fluorescence is linear at low concentrations, but not at higher levels. "Gradients" indicate the compatibility of the detector with gradient elution, and "Robustness" includes the ease of use of a detector. Clearly, there is no perfect LC detector, and the choice depends on the requirements of a specific application.

3.9.7 Other LC Detectors

Element-specific detectors, which are fairly common in GC (see Module 2.5), are not frequently encountered in LC. One exception is the hyphenation of LC with an inductively coupled plasma (ICP) spectrometer (see Section 3.10.2). Another exception is the use of **chemiluminescence** detectors designed for GC (see Section 2.5.4.4) in LC. Element-specific sulfur-chemiluminescence detectors (SCDs) and nitrogen chemiluminescence detectors (NCDs) have sporadically been used in combination with LC, but such systems have not caught on, possibly because of a lack of robustness. Chemiluminescence in the liquid phase is more commonly used for trace analysis by LC in combination with post-column derivatization (see Section 8.3.3).

Dedicated detectors may be used in conjunction with chiral (enantiomer) separations (see Module 3.13). **Polarimeters** are conventionally used to measure the rotation of polarized light by enantiomer solutions. For a given compound, the extent of rotation is related to its concentration. Some of the LC detectors based on this principle are known as optical-rotary-dispersion (ORD) detectors. For UV-active chiral compounds, circular-dichroism detectors may be used based on the principle that different enantiomers show different absorption of left and right circularly polarized light.

Dedicated detectors are also commonly used in conjunction with the separation of macromolecules and nanoparticles, as will be described in Module 4.6.

3.10 Hyphenation and LC-MS (M)

3.10.1 LC-MS

Increasingly, LC systems are equipped with mass spectrometric (MS) detectors or, if you want, mass spectrometers are equipped with LC inlets. There are two groups of reasons for this. The first is demand. MS is one of the most sensitive and selective LC detectors, and it provides a wealth of structural information that is not provided by LC itself. The other reason is supply. An ever-increasing range of MS instruments, ionization interfaces, and data-interpretation tools are available. MS systems are increasingly robust, and they range from complex, ultra-high-resolution instruments to affordable, easy-to-use ones.

Some characteristics of mass spectrometry, such as the resolving power (RP_{MS}) and mass accuracy (ΔM^*), were defined in Module 2.6 (eqn (2.14) and (2.15)). Mass spectrometrists speak of the duty cycle as the time needed to conduct one scan across the spectrum and the sampling rate (in Hz) as the number of scans that can be produced within 1 s. In some cases, the scan rate is specified in Da s^{-1}, but, as pointed out by Holčapek *et al.*,[78] the scan rate in kDa s^{-1} is equal to that in Hz for a typical scan range of 1000 *m/z* units.

LC-MS has not always been as robust and accessible as it is today. It was not the instant success that GC-MS (Module 2.6) was as early as the 1970s. Numerous obstacles had to be overcome as summarized in Table 3.16. The number of different interfaces developed is a testament to the struggles faced by numerous groups of researchers. The great breakthrough came with the introduction of atmospheric pressure ionization techniques, in which the ionization precedes the vacuum stage. The most important of these is ESI (see Section 3.10.1.1), for which John Bennett Fenn received the Nobel Prize in Chemistry in 2002.

Clearly, the presence of liquid solvents in LC is unattractive from the perspective of MS. Large amounts of solvent (but not the analytes) must be removed by the vacuum system. In Chapter 2 (Module 2.4, Table 2.3), we saw that the introduction of 1 µL of water into a hot GC injector gives rise to more than 1 mL of water vapour in the GC injector at 120 kPa. When we introduce 1 mL min^{-1} of an aqueous (RPLC) mobile phase into an MS ionization chamber under vacuum ($\ll$1 Pa), we are dealing with hundreds of litres of vapour. Turbo-molecular pumps have emerged for this purpose, but narrow-bore LC columns have long been preferred by mass spectrometrists. A 1 mL min^{-1} flow rate on a standard 4.6 mm column yields the same linear velocity as a flow of 0.2 mL min^{-1} on a 2.1 mm i.d. column. For a 1 mm i.d. column, the corresponding flow rate is about 50 µL min^{-1}. The MS instrument may benefit from a lower flow rate, resulting in higher sensitivities. However, 1 mm i.d. columns are thought to be much less robust from an LC perspective. Such columns put high demands on the solvent delivery system, especially with an eye on generating accurate mobile-phase compositions and accurate gradients. Hence, 2.1 mm i.d. columns have been found a good compromise. When using UHPLC, the flow tends to be higher than 0.2 mL min^{-1} because the optimum linear velocity increases in inverse proportion to the particle size. In contrast, lower optimum flow rates exist for high-molecular-weight analytes due to their low diffusion coefficients (see Section 1.7.3). This justifies the use of low flow rates in the analysis of intact proteins. Restrictions on mobile phases and possible contaminations (from eluents or samples) are – and are likely to remain – a weakness of MS. However,

Table 3.16 Overview of the obstacles encountered in the development of LC-MS.

Obstacle	Description	Status
Vacuum	• Large amounts of solvent • Relatively high source pressures	• Highly efficient (but expensive) turbo pumps • Many LC-MS applications involve low flow rates, but analytical columns (*e.g.* 2.1 mm i.d.) can be used
Mobile phase	• Only volatile solvents • No salts • No non-volatile buffers	• No change
Analytes	• Difficult to apply to high-molecular-weight analytes • Volatile analytes can be lost through the vacuum system • Low ionization and transmission efficiencies for "heavier" (higher-molecular-weight), non-volatile analytes	• LC-MS used successfully for some classes of macromolecules (especially proteins)
Source pollution	• Non-volatile compounds (in samples or mobile phases) contaminate the source • Frequent, laborious cleaning of the source required	• Contamination is still a major issue, but atmospheric-ionization sources (*e.g.* electrospray) are easier to clean
Reproducibility	• Poor quantitative reproducibility (ionization and transmission efficiencies) • Poor reproducibility of the extent of fragmentation • Not possible to create spectral libraries	• Repeatability (for quantitation within a series of experiments) has improved, but reproducibility has not • MS/MS transitions are tabulated including approximate (relative) intensities

(*continued*)

Table 3.16 (*continued*)

Obstacle	Description	Status
Gradient elution	• Changing the composition of the LC effluent during time	• Not a limitation for contemporary MS

the range of analytes that are amenable to MS has expanded greatly. Proteins are the most dramatic example.

Whereas electron ionization (EI), which is the most common approach in GC-MS, yields extensive fragmentation, so-called **soft ionization methods** that lead to little or no fragmentation are dominant in LC-MS (see Section 3.10.1.1). HRMS may yield unambiguous molecular formulae for undissociated ("molecular") ions, but dual-stage or tandem MS (MS/MS) or even multi-stage MS (MS^n) are needed to obtain more information on the structure of molecules. In tandem MS, a specific **precursor ion** (*i.e.* a specific *m/z* value) is physically selected from the population using one of the mass analysers described in Section 3.10.1.2. These precursor ions are then purposefully fragmented in a high-energy process, such as **collision-induced dissociation** (CID) or **electron-transfer dissociation** (ETD). The **product ions** are characterized in the same mass analyser as for the first selection or in a separate analyser. In MS/MS, a scan can be performed in the first stage (precursor-ion scan) and/or in the second stage (product-ion scan). Scanning modes are typically associated with non-target analysis. In target analysis, selectivity and sensitivity can be increased, and noise can be decreased by operating in the **selected reaction monitoring** (SRM) mode, which is analogous to the selective ion monitoring (SIM) mode in single-stage MS (see Module 2.6). Different precursor ions and product ions can be defined in the instrument software for each chromatographic peak. By selecting several product ions to monitor simultaneously (**multiple reaction monitoring** or MRM mode), the signal-to-noise ratio may be further improved.

Unfortunately, MS/MS measurements are more difficult in practice than they appear on paper. Other ions and neutral molecules may pass the first stage mass analyser and may impair the fragmentation. Hence, severe matrix effects may be observed, the spectral information obtained after the second stage may be of poor quality, and no useful product ions are obtained for a large number of analytes.

MS/MS operation is one of the aspects of MS that is still open to significant improvements.

3.10.1.1 Ionization Methods

The initial LC-MS interfaces involved bringing the LC effluent into the vacuum, where ionization was attempted in a similar manner as in GC. Among these low-pressure-ionization techniques were capillary–inlet interfaces, which suffered from unstable source pressures and premature evaporation of the effluent in the capillary, and offered a very narrow working range of up to about 400 Da. This meant that few of the compounds that could be analysed by LC, but not with GC, were amenable to such LC-MS systems. Moreover, many small analytes were lost through the vacuum system.

The next generation of direct-liquid-introduction systems involved the use of a diaphragm (with an opening of 2 to 5 μm) or a restrictor, to avoid early evaporation of the effluent, and a desolvation chamber. Such interfaces could accommodate LC flow rates between 10 and 50 μL min^{-1}, which were usually achieved by post-column splitting of the LC effluent. Aqueous (RP-)LC mobile phases allowed chemical ionization (CI) with water as a reagent, but jet stability was poor and the interfaces were susceptible to clogging, which led to poor repeatability and robustness. In nebulizer–jet interfaces, the effluent from 100 μm i.d. fused-silica-capillary LC columns was nebulized using a stream of helium.

The particle-beam interface (PBI), also known as the monodisperse-aerosol-generator interface (MAGIC), operated under near-ambient conditions. The interface allowed performing electron ionization (Section 2.6.1.1) and chemical ionization (Section 2.6.1.2), and spectral searches in common libraries were conducted successfully. PBIs have mainly been applied for relatively low-polar analytes of low molecular weight.

In contrast, continuous fast-atom bombardment (CFAB) interfaces were specifically developed and used for characterizing peptides and small proteins. In CFAB, a small flow of effluent (1 to 10 μL min^{-1}) was mixed with a viscous solvent, such as glycerol, which in itself is challenging, and then flowed down a slanted surface, where it was hit with accelerated argon atoms. Because singly charged ions were almost exclusively obtained, the upper mass limit was determined by the capabilities of the mass analyser.

Moving-belt interfaces follow a very different approach. They have some attractive features, which cause them to reappear from time to time in research papers. The effluent of the LC is deposited on a slowly moving "belt", from which the solvents can be evaporated, followed by EI or CI ionization. Initially, the belt was envisaged as an interface between LC and flame-ionization detection (*i.e.* LC-FID), but currently, it is of more interest as a potential interface between LC and **matrix-assisted laser-desorption/ionization** (MALDI) MS, a technique for which Koichi Tanaka from Shimadzu received his share of the 2002 Nobel Prize in Chemistry.

In MALDI, the sample is mixed with a dye that absorbs the light of laser pulses directed at the dried sample. Salt ions present in the sample solution (LC effluent) or admixed with the matrix help create adduct ions with multiple charges, without breaking down analyte molecules. This makes MALDI arguably the best ionization technique for characterizing large macromolecules, such as proteins or polymers, with MS. Unfortunately, MALDI-MS cannot easily be interfaced with LC because the liquid effluent must be dried, usually after adding a matrix (alternatively, the matrix can be pre-deposited before the analysis). Typical LC-MALDI-MS interfaces operate off-line, with droplets of effluent being deposited on a MALDI plate at a number of predefined spots. The plate is transported to the MALDI-MS, where it can be analysed without time constraints from the LC separation, allowing analysts to focus on peaks (spots) of interest. The off-line coupling of LC and MALDI-MS is illustrated schematically in Figure 3.40.

The repeatability of spectra obtained from each laser pulse is extremely poor. MALDI spectra are always the average of a number of spectra resulting from multiple pulses, aimed at different locations of the sample spot. Even when considering such averaged spectra, repeatability is poor and MALDI can hardly be used for quantitative analysis, nor for proving the absence of analytes in the spot. Proving their presence is a more realistic proposition. MALDI may provide excellent qualitative information, provided that the eluent and the added matrix (*i.e.* the dye) do not lead to interferences, which usually are confined to the low-molecular-weight range, and the analytes of interest are sufficiently separated. The latter aspect makes LC-MALDI-MS infinitely more useful than MS alone. When aiming to apply MALDI-MS to synthetic polymers, the sample spot must have a very narrow molecular-weight distribution (very low PDI; see Module 4.1). If not, only the smallest molecules will appear in the MALDI spectrum, and it will by no means represent the actual molecular weight distribution. This is because there are simply many more small molecules than big ones if their weight fractions are identical

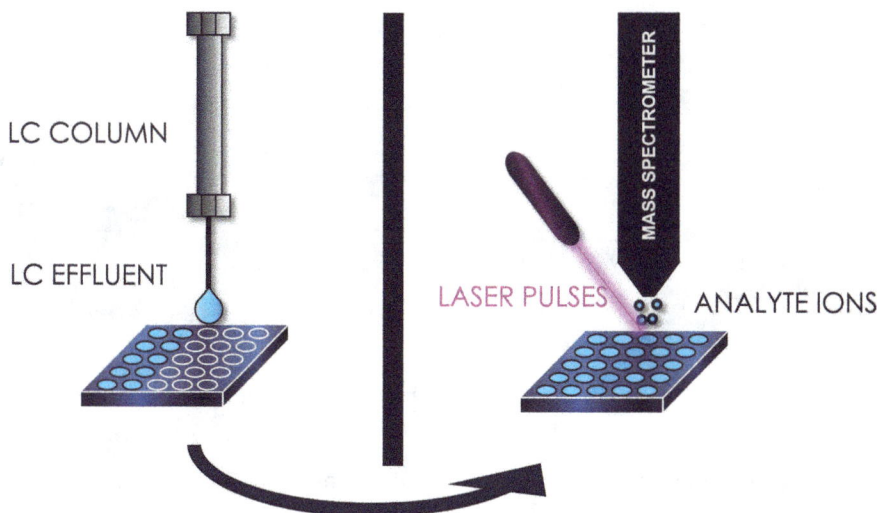

Figure 3.40 Illustration of the off-line coupling of LC with matrix-assisted laser-desorption/ionization (MALDI)-MS.

and because the MS sensitivity (efficiency of ionization and transmission) is much higher for the small molecules than it is for the larger ones.

The obvious choice of a mass analyser to combine with MALDI is a time-of-flight (ToF) instrument (see Section 3.10.1.2) because this works with pulses of ions entering the spectrometer, allowing synchronization with the laser. It provides high-resolution data on the analyte molecular weight, which can be obtained from extrapolation of the charge distribution to the single-charge ion. Further characterization would require fragmentation of the selected molecular ion, for example by collision-induced dissociation (CID) or electron-transfer dissociation (ETD), and measuring the spectrum of product ions in a second MS stage. ToF analysers do not allow such MS/MS experiments, but ToF/ToF spectrometers have emerged for this very purpose.

The last stepping stone to modern online LC-MS interfaces was the **thermospray** (TSP) interface, which was the first technique that allowed the routine use of analytical LC systems directly coupled with MS. The spray emerging from a heated probe (hence "thermospray") was thought to induce solvent-induced chemical ionization. It was concentrated in a desolvation chamber, and a skimmer prevented all but the centre of the spray from entering the MS. Because TSP was especially useful for polar mobile phases (such as RPLC

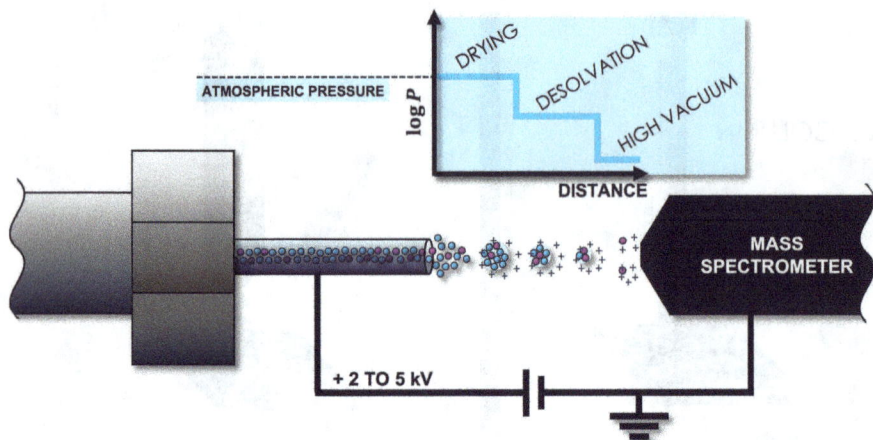

Figure 3.41 Schematic illustration of the online coupling of LC and MS through electrospray ionization (ESI).

effluents) and polar analytes, it opened the field of LC-MS to the analysis of biomolecules. In the case of less-polar mobile phases, filaments and discharge electrodes were used in conjunction with TSP. Although a high temperature usually resulted in a stable spray, TSP interfaces were not easy to use. In addition, repeatability was poor.

Atmospheric ionization methods, particularly **electrospray ionization (ESI)**, heralded the coming of age of LC-MS. For the first time, reasonably robust interfaces were obtained that were routinely applicable to a wide range of analytes and analyte classes. Atmospheric-ionization interfaces stand out by having ionization start before the sample enters the MS. This is schematically illustrated in Figure 3.41. Ionization occurs when the sample (LC effluent) emerges from a narrow tip due to a very high potential difference (typically between 2 and 5 kV) applied between the tip and the MS (the latter is usually grounded). The high voltage is thought to produce an ionized spray (aerosol). In Figure 3.41, the spray tip is at a positive voltage, producing positive ions, but ESI can equally well be applied in negative-ion mode, just by inverting the voltage.

Charged microdroplets are formed, which are reduced to nano-droplets upon drying (at atmospheric pressure and optionally with hot nitrogen gas) and desolvation (at intermediate vacuum). At some point, known as the Rayleigh limit, these contain too much charge to remain stable, producing multiply charged analyte ions with little or no fragmentation ("**soft ionization**"). Most of the (uncharged) solvent does not enter the MS or is removed at the desolvation stage. The latter effect can be enhanced by directing the electrospray along the inlet of the MS (at a perpendicular angle) rather than at the MS inlet (other than what is schematically shown in Figure 3.41).

LC-ESI-MS has revolutionized research in the life sciences. Due to the multiple charges on analyte ions, and the algorithms embedded in instruments to deduce the molecular weight from the charge pattern, very high analyte masses can be accurately determined. The upper limit of conventional MS is thought to be around 20 MDa.[79] Charge-detection MS[80] has (re-)emerged as a characterization technique for "heavy" analytes, such as viruses or nanoparticles. At the low-molecular-weight end, polar analytes that needed derivatization to be amenable to GC-MS, along with the compatibility of aqueous samples with RPLC, meant that metabolite studies moved from GC-MS to LC-MS. The same reasoning applies to most environmental samples and many food or food-related samples.

ESI interfaces can accommodate a wide range of flow rates. Regular ESI interfaces are capable of handling flows in the mL min^{-1} range and these are routinely combined with HPLC and UHPLC separations on analytical-size columns (4.6 mm and 2.1 mm i.d., respectively). Micro-ESI systems typically operate in the range of 50 to 500 µL min^{-1}, making them compatible with 1 or 2 mm i.d. HPLC columns. Nano-ESI interfaces work best at flow rates between 50 and 500 nL min^{-1}, which are used for studying proteins or peptides in (very) high-resolution MS systems, using very narrow (100 µm i.d. or less) and often long LC columns. The very low flows and small amounts of volatile acids or buffers result in a high MS sensitivity and very small amounts of sample (and very small volumes of eluents) suffice. Very low flow rates on long columns may yield efficient separations for high-molecular-weight analytes such as proteins because their low diffusion coefficients lead to low optimal linear velocities (see also Module 3.16).

Low-polarity analytes tend to show low response in ESI-MS. **Atmospheric-pressure chemical ionization** (APCI) is a popular technique for low-molecular-weight, low-polarity analytes. An ionizing gas is used to produce analyte ions, analogous to GC-CI-MS (see Section 2.6.1.2). A corona discharge may be added to initiate the ion–molecule reactions, using solvent ions as intermediates. Similar to ESI-MS, the interface can be configured such that only charged particles are drawn into the MS vacuum. Also analogous to ESI is the fact that the interface can be operated in both positive-ion and negative-ion modes. However, other than with ESI-MS, the process yields almost exclusively single-charged

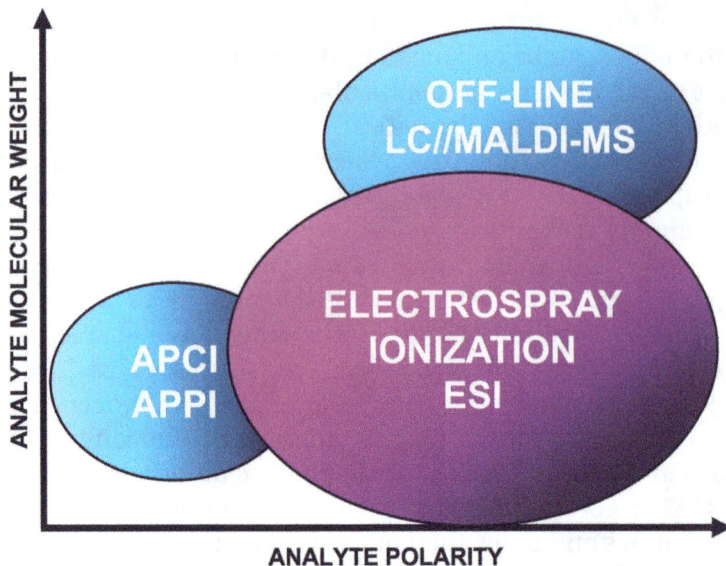

Figure 3.42 Schematic overview of the application fields of the ionization interfaces most commonly used in LC-MS. MALDI = matrix-assisted laser-desorption/ionization; APCI = atmospheric-pressure chemical ionization; APPI = atmospheric-pressure photoionization. "//" indicates "offline". Adapted from ref. 81 with permission from the Author, Copyright 2019.

ions. Current APCI interfaces can easily be used in conjunction with analytical HPLC and UHPLC columns at flow rates up to 2 mL min^{-1}. Interfaces also exist that can be used for both ESI and APCI.

While APCI is typically a technique used for low-polarity analytes, it does not work well in conjunction with low-polarity organic mobile phases, such as those encountered in NPLC or SEC. On such occasions, an excess of an ionizable dopant may be combined with the vapour formed after evaporating the LC effluent and subjected to radiation with a vacuum UV lamp. The resulting atmospheric-pressure photo-ionization (APPI) improves the sensitivity of APCI detection for hard-to-ionize compounds.

A schematic picture of the ionization methods currently used most often in LC-MS is shown in Figure 3.42. ESI is the preferred technique for most applications. For non-polar macromolecules, such as polyolefins, a good LC-MS interface still does not exist. This is one reason for the continued interest in LC-FTIR (Section 3.10.2.1) and LC-NMR (Section 3.10.2.3) systems.

Table 3.17 Typical characteristics of different types of mass spectrometers. Adapted and updated from ref. 78 with permission from Elsevier. Copyright 2012.

Analyser	Resolving power[a]	Mass accuracy (ppm)	Speed (Hz)[b]	Maximum m/z (kDa)	Linear dynamic range (decades)	MS/MS	Cost
Quadrupole (Q)	3000–5000	>5	2–10	2–3	5–6	No	Low
Triple quadrupole ("triple-quad") or QqQ	3000–5000	>5	2–10	4–6	5–6	Yes	Moderate
Ion trap (IT)	5000–20 000	>5	2–10	4–6	4–5	Yes	Moderate
Time-of-flight (ToF)	20 000–100 000	1–5	10–50	10–20	4–5	No	Moderate
Quadrupole–time-of-flight (QToF)	20 000–100 000	1–5	10–50	10–20	4–5	Yes	Moderate to high
Orbitrap	100 000–1 000 000	1–3	1–5	4	3.5	Yes	High to very high
Fourier-transform ion-cyclo-tron resonance (FT-ICR)	750 000–2 500 000	0.3–1	0.5–2	4–10	4	Yes	Very high to enormous[c]

[a]At low m/z (e.g. 400).
[b]May depend on the mass range scanned and resolving power.
[c]Depends on magnet strength.

3.10.1.2 Mass Analysers

Several different types of mass analysers exist. Indicative properties of the most common types used in LC-MS are summarized in Table 3.17. Quadrupole (Q) analysers are the most affordable and most common mass spectrometers in analytical laboratories. They are formed by a characteristic set of four parallel metal rods, arranged in a square in the perpendicular direction. A radiofrequency field (with a DC offset) is applied to each pair of opposing rods. The result is that there is only a stable direction through the quadrupole in the length direction for ions with a particular m/z value. The quadrupole is a low-end mass analyser with modest resolving power. However, by combining three quadrupoles in a QqQ ("triple quad") configuration, one of the best systems for targeted quantitative analysis is obtained. In a QqQ system, the first quadrupole is used to select the precursor ion. The second quadrupole (q) is the collision cell, and the third quadrupole is used to separate the product ions. The resolving power of each analyser is limited, but the selectivity of the QqQ setup suffices for many applications. Quadrupoles are relatively slow, and the upper m/z limit is often only a few thousand (10 000 only for specific instruments).

In an ion-trap MS, dynamic electric fields are used to trap ions. Ion traps were developed by Wolfgang Paul, and this achievement earned him the Nobel Prize in Physics in 1989. The electric fields have to switch at radiofrequencies to keep the ions trapped. Ion traps have a characteristic symmetrical djembe shape, and they can be miniaturized to create small (potentially portable) mass spectrometers. Speed and resolving power are limited, but MS^n experiments can be performed with relatively simple instrumentation.

In time-of-flight (ToF) MS, ions are periodically injected into a field-free flight tube, and the time it takes for them to pass the length of the tube is measured. The flight time increases with the length of the tube and the square root of m/z, which implies that resolution decreases with increasing analyte molecular weight. However, the upper m/z limit is the highest among the common mass spectrometers. Among these, ToF instruments are also the fastest. With scan rates of 50 Hz or more, they are compatible with fast UHPLC separations. ToF analysers are more stable and more sensitive than Fourier-transform ion-cyclotron resonance (FT-ICR) spectrometers at high sampling rates.[82] ToF analysers do not allow MS/MS measurements, but hybrid quadrupole–time-of-flight (QToF) instruments have become popular for this purpose. QToF instruments are slower when scanning a range of masses (as in untargeted analysis), but they can be fast in targeted analysis, with the quadrupole used as a filter for a specific ion (m/z value) at a specific elution time.

The Orbitrap analyser is based on harmonic ion oscillations in electrostatic fields and uses Fourier transformation to convert the measured data into a mass spectrum. High-end Orbitrap spectrometers show very high resolving power (greater than one million at m/z = 300–400).[83] They may compete with FT-ICR spectrometers (see below), especially for high-molecular-weight analytes. Because Orbitrap spectrometers are much smaller and less expensive to purchase and operate than FT-ICR machines, they have become the go-to HRMS instruments in many laboratories.

In an FT-ICR spectrometer, ions are brought to oscillate circularly in a homogeneous magnetic field. Both mass resolution and scan rate are proportional to the strength of the magnet, so that very large magnets yield amazing performance. A resolving power of three million (at m/z = 200) has been reported for an FT-ICR instrument with a 21 T magnet.[84] However, FT-ICR spectrometers are relatively slow. Off-line couplings have been described (similar to those discussed for MALDI in Section 3.10.1.1, except that the fractions are collected in micro-vials), but this is not attractive if multiple samples are to be analysed.

3.10.1.3 Ion Mobility

Ion-mobility spectrometers (IMS instruments) measure the drift time of ions in a dilute "buffer" gas (usually helium). The migration velocity of ions is determined by their m/z value and by their **collision cross section** (CCS). The latter can be seen as a hydrodynamic radius under IMS conditions. CCS values have been measured for a great number of analytes, and there have been several efforts to predict CCS values from the molecular structure.

IMS systems are used in airport security for detecting drugs or explosives because of their extreme speed. This is also their great strength as an analytical instrument. MS cannot distinguish between isobaric compounds (*i.e.* compounds with the same molecular formula) without resorting to fragmentation. For compounds that only differ in their spatial configurations, such as optical isomers, even the most powerful mass analysers are useless. Potentially, analytes with different spatial configurations exhibit different CCS values, and this point has been demonstrated by numerous researchers. Thus, IMS-MS combinations promise extremely fast analysis of pairs of isomers.

A weak point of IMS-MS is that all the biases of the ionization stage (matrix effects, ion suppression, and discrimination) and all kinds of interferences at the IMS and MS stages bear on the process. For complex samples, it is crucial to first clean up the sample, so as to introduce greatly simplified fractions – ideally just pairs of analytes – to the IMS-MS system at any one time. Thus, LC-IMS-MS is a suitable system for analysing complex samples. However, in many cases, LC can also distinguish between isomers, reducing the value of IMS as an intermediate stage.

3.10.2   Spectroscopy Hyphenated with LC

LC-MS has seen tremendous progress during the last decades, and it is pervasively used in laboratories worldwide. However, LC-MS is still imperfect. It is not applicable to all (classes of) analytes and does not provide all possible information. Therefore, the hyphenation of LC with spectroscopic techniques is still of interest. In this section, we will briefly discuss two interesting options: LC-FTIR and LC-NMR. Other systems in which LC is hyphenated with detection systems exist. For example, it is not difficult to couple an LC system with an ICP spectrometer, with minimizing post-column band broadening the main technical challenge. The LC effluent instantly decomposes to the level of atoms and single-atom ions in excited states. The element composition can be determined using optical-emission spectroscopy (ICP-OES) or mass spectrometry (ICP-MS). The LC-ICP combination is useful for specific detection of any kind of metal, for example, in catalyst research. Proteins sometimes contain metals and often contain sulfur, which allows their quantitative detection in the LC effluent with an ICP detector.

However, not all LC mobile phases are perfectly compatible with LC-ICP-MS. High concentrations of organic solvents (*e.g.* in RPLC or HILIC) may lead to plasma instability and soot formation at the inlet of the torch and possibly that of the MS. This leads to signal drift, reduced sensitivity, and increasing noise levels. In addition, high concentrations of non-volatile salts (*e.g.* in IEC) may cause salt deposits.

3.10.2.1 LC-FTIR

Fourier-transform infrared spectroscopy readily provides a good deal of information on molecular structure. For example, FTIR spectra often signal the presence or absence of specific functional groups in a molecule. Moreover, IR spectra are potentially unique, reproducible fingerprints of chemical compounds. This makes LC-FTIR a very interesting companion for LC-MS. An exact mass and, from that, a molecular formula obtained by MS still correspond to a large number of structures. An IR spectrum may allow quick pruning of this collection. FTIR is quite sensitive, and modern (microscopy) optics allow measurements in small volumes, compatible with analytical LC. FTIR can also be selective and potentially universal. However, this last aspect is also its main weakness. LC mobile phases invariably absorb much of the IR light, with water being particularly detrimental. Dealing with a high and often variable background (*e.g.* in the case of gradient elution) is further complicated by the fact that FTIR is almost always a single-channel measurement, in the sense that signal and background are not mentioned simultaneously.

There have been significant attempts to improve flow-cell LC-FTIR interfaces, for example, with quantum-cascade lasers as light sources and advanced chemometric baseline-correction strategies.[85,86] However, most LC-FTIR interfaces rely on the deposition of the sample on an IR-compatible substrate and evaporation of the eluent prior to the spectroscopic measurement. This kind of coupling is not very different from that outlined for LC//MALDI-MS in Section 3.10.1.1. Suitable substrates are gold, which is a very good reflector of IR light, zinc selenide or silicon. These latter two materials are transparent in the mid-IR region. Such deposition interfaces are not particularly attractive because they are complex, lack robustness, and do not allow good quantitative measurements.

Dispersive IR spectroscopy, *i.e.* wavelength-resolved rather than time-resolved acquisition, has its niche if a wavelength can be found where target analytes absorb the light, while the mobile phase does not. The most striking example is the analysis of polyolefins, such as poly(ethene) or poly(propene) in an aromatic mobile phase, such as 1,3,5-trichlorobenzene. The aromatic C–H stretch vibration is found above 3000 cm^{-1}, whereas the aliphatic C–H stretch vibrations are found below this threshold. IR spectroscopy may be used as a quantitative detector, but it does not provide spectral information.

3.10.2.2 *LC-Raman Spectroscopy*

Raman spectroscopy is related to mid-IR spectroscopy in that there is some similarity between the spectra obtained, but the measurement principle and the instrumentation are quite different. Raman is an inelastic scattering technique. Scattering implies that light is measured that is scattered by molecules. Inelastic refers to the fact that energy changes, so that the scattered light has a different wavelength from the incident light. Raman spectra can be recorded by using a specific wavelength produced by a laser as the incident radiation and filtering out this wavelength from the scattered light using a notch filter. The laser light excites analyte molecules to a virtual energy level, from which it decays to a well-defined energy state. Thus, while the excited level is virtual, the starting point and the end point of the process are well defined, so that the difference in wavelength between the incident light and the inelastically scattered light is the difference between two energy levels in the molecule. This energy level corresponds to that observed in FTIR spectra. However, the selection rules are different. IR absorption only occurs if the change in energy levels produces a change in dipole moment, whereas Raman bands require a change in polarizability. If a transition changes both, it corresponds to a band that is visible in both (FT)IR and Raman spectra.

Instrumentally, Raman spectroscopy is somewhat similar to laser-induced fluorescence spectroscopy, but the two do not pair well together. Fluorescence of the sample tends to thwart Raman spectroscopy because it yields light across a broad range of wavelengths. This results in a very high background in the Raman spectrum. Unlike fluorescence, Raman spectroscopy also produces light of higher energy (lower wavelength) than the incident light, but measuring such anti-Stokes Raman signals reduces the already low sensitivity of Raman spectroscopy. Choosing a low energy for the incident light, such as a near-infrared (NIR) laser, is the most practical way to reduce fluorescence. A very attractive aspect of Raman spectroscopy is that it is – unlike FTIR – insensitive to the presence of water. This makes Raman spectroscopy an interesting option in combination with LC, provided that sufficient sensitivity can be achieved.

3.10.2.3 *LC-NMR*

Without a doubt, NMR spectroscopy is the best technique for detailed characterization and structure elucidation of organic compounds. NMR spectra can be obtained for different nuclei, with ^1H and ^{13}C being best known, while numerous other one- and two-dimensional NMR techniques may provide a wealth of information on connectivity within a molecule, spatial configuration, *etc.* Obtaining NMR spectra "on the fly" after an LC separation – or any other kind of NMR information – is an exciting prospect. There are, however, a few problems. One is the inevitable presence of solvents in the LC effluent. Proton NMR spectra, which are the fastest and most sensitive way to obtain spectral information, are usually recorded in deuterated solvents, such as deuterated chloroform ($CDCL_3$) or deuterated acetone (CD_6O), to suppress the background signals and reveal the analyte spectra. The most common RPLC modifiers, ACN and MeOH, are relatively attractive because both show single (singlet) peaks in the ^1H-NMR spectrum (except for the 1.1% of the protons connected to a ^{13}C nucleus). It is technically preferred but generally not a realistic proposition to perform LC with deuterated solvents, so that suppression techniques become much more critical. The greatest obstacle, however, is the low sensitivity of NMR spectroscopy. Spectra can be recorded of small amounts of analyte, if a sufficiently large number of scans ("pulses") are taken. To obtain quantitative information, these also need to be sufficiently far apart to allow for relaxation.

The most exciting results on the hyphenation of LC with NMR and several other detectors (UV, MS, or FTIR, resulting in "hypernation") have been reported by Ian Wilson from AstraZeneca (Macclesfield, UK).[87] FTIR was performed using an attenuated total reflection (ATR) flow-cell interface, and MS and NMR were necessarily configured in parallel, with about 5% of the effluent directed to the ToF–MS and 95% to the NMR. Analyte amounts down to 50 μg yielded good ^1H-NMR spectra on a 500 or 600 MHz NMR spectrometer, but the best results were obtained in stop-flow mode. Such analyte amounts were appropriate for the online ATR–FTIR detection but resulted in massive overloading of the UV detector. In addition, such amounts of analyte are at least 100 times above the common amounts of analytes present in bands eluting from an analytical HPLC system. Developments in NMR (*e.g.* "cryoprobes") have led to lower detection limits. In addition, LC peaks may be concentrated on solid-phase-extraction (SPE) cartridges or trap columns in a manner similar to stationary-phase-assisted modulation (SPAM) used in comprehensive two-dimensional liquid chromatography (LC×LC), a process described in Section 7.3.

When performing LC-NMR experiments online, neither of the techniques is used under optimal conditions. In addition, a high-field NMR instrument is kept occupied during the entire time. Therefore, most laboratories have found offline coupling between LC and NMR preferable. Fractions of the LC effluent can be collected in NMR tubes if desired, and it is possible to change the solvent after collection by evaporation and reconstitution. Fractions can then be put in the autosampler queue of the NMR instrument, and the desired experiments can be programmed for each individual fraction.

The strongest arguments for online LC-NMR can be made for samples that cannot be separated into pure-component peaks but show a continuous envelope of multiple chemical structures. This is the case, for example, with synthetic polymers, and in this domain, online LC-NMR is still of interest.[88]

3.11   LC Troubleshooting (M)

3.11.1   Introduction

When all goes well, the presence, retention times, and areas of peaks obtained from LC separations provide information to answer analytical questions. Unfortunately, all does not always go well, and it is of paramount importance that chromatographers are able to certify their results. At a minimum, this requires that they can (i) realize that a problem occurs. Ideally, chromatographers are also able to (ii) identify the problem, (iii) establish the cause, and (iv) find a solution. Together, these four stages describe the competence called **troubleshooting**. Just realizing that a problem has occurred is not trivial. Either the analyst who performs the analysis or the one responsible for it should ensure that the results are correct. This requires knowledge and experience, especially when the problem is not instantly apparent from the chromatogram. This is even more true to diagnose the problem and establish what caused it. If this can be done, it will point to a solution, which may involve making a fresh mobile phase, washing or replacing the column, repairing the pump, *etc.* In this module, we discuss some aspects of troubleshooting and present some tools to use in the process. Extensive literature exists on the subject. John Dolan from LC Resources deserves credit for producing a substantial part of this.[89]

> See the website for more on LC troubleshooting: ass-ets.org

3.11.2 Test Mixture

The first good advice is to have a test mixture available. This could be a general mixture, like the one used to create Figure 3.43 for a mixture containing compounds associated with your own application. The mixture is there to help you solve problems, not to create new ones. Thus, your test mixture should be stable and available (*e.g.* on a standard position in your autosampler). Figure 3.43 shows three RPLC chromatograms of a standard test mixture consisting of uracil (approximately unretained), 2-hydroxynaphthalene (polar and low retention factor), and toluene (apolar and high retention factor). The chromatograms are recorded under identical conditions on nominally identical columns with different histories. Such chromatograms may be recorded when a problem is suspected, but in a routine situation, it is prudent to run the test mixture much more often to monitor the performance of the LC system and the column. **Shewart charts** may be used to monitor the performance, plotting the retention times and peak width of the three peaks against the run number. Warning limits and action limits can be indicated in such a plot, determined from the variations observed in the initial repeat runs. See Figure 3.44 for an example. In the present case, control charts may be maintained for the retention time and for the peak width (plate number) of one of the peaks or ideally automatically for all three peaks. Different control systems also exist, such as Youden charts that allow both precision and accuracy to be evaluated.

> See the website for examples of control charts, such as Shewart and Youden charts: ass-ets.org

Some shifts in retention times may occur as the column gets older (light-blue line *vs.* dark-blue line in Figure 3.43). Some shifts may be acceptable, and the changes may be accounted for using retention-time-alignment tools (see Section 9.11.4). The pink chromatogram

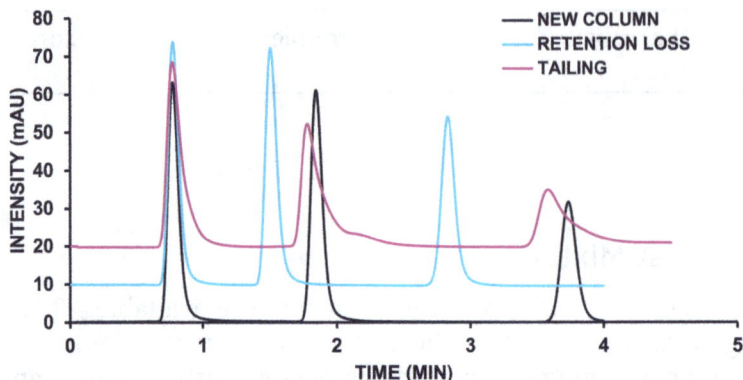

Figure 3.43 Overlay of chromatograms obtained with columns of the same selectivity and physical parameters, but different histories, 50 × 2.1 mm, 1.8 µm. Separation of three analytes: uracil (no retention), 2-hydroxynaphthalene (moderate retention), and toluene (high retention). Courtesy of Gerben van Henten and Rick van den Hurk (University of Amsterdam, The Netherlands).

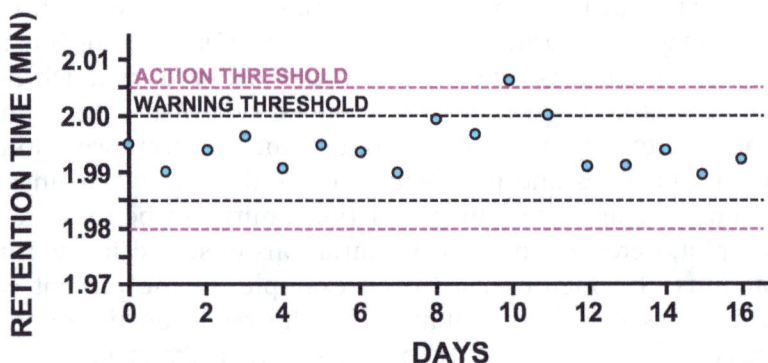

Figure 3.44 Example of a Shewart chart where the retention time is monitored over time. Warning and action thresholds can be defined based on the variations observed in the initial repeat runs (repeatability test, see Module 10.5).

indicates a situation where the column may no longer be acceptable. It may be washed with a strong solvent and re-equilibrated. As a last resort, the flow direction may be reversed, but a new column may be needed. Such desperate measures, to try and salvage a column against the odds, are more common in an academic setting than in an industrial setting.

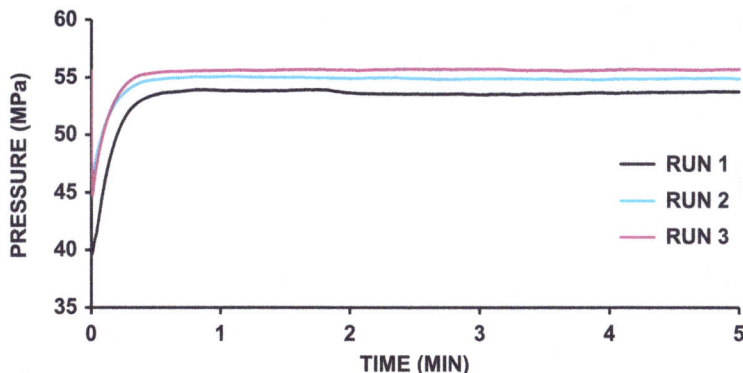

Figure 3.45 Overlay of pressure curves obtained for the same column during three selected isocratic runs. Note that the selected runs were not consecutive. The initial increase in the pressure follows the injection of the sample. Courtesy of Gerben van Henten and Rick van den Hurk (University of Amsterdam, The Netherlands).

3.11.3 Pressure Traces

One very valuable tool, which is freely available with most data stations, is to record the pressure during the run. If the pressure trace is stored together with the chromatogram, it provides a great deal of information on the chromatographic performance and the status of the system. Again, in a routine situation, it is prudent to (automatically) maintain a Shewart chart that shows the system backpressure as a function of the run number. Figure 3.45 shows pressure curves recorded during three isocratic runs at different times. It is not uncommon that the pressure drop across the column increases slightly during usage. One possible reason for this is the blockage of narrow flow channels in the column or the frit. If part of the frit or the top of the column gets blocked, **channelling** may occur, with different fractions of the eluent – and with it the sample – following different paths through the column. This may lead to band broadening or systematic double peaks. Replacing the inlet frit is a last resort in such a situation.

Apart from the observed increase in backpressure during column usage, the traces in Figure 3.45 are smooth, instilling confidence in the correct functioning of the system. An example where the **pressure traces** indicate malfunctioning of the system is shown in Figure 3.46. Panel A shows the pressure trace during the actual analysis. Trace B shows a reference trace from an earlier run. The

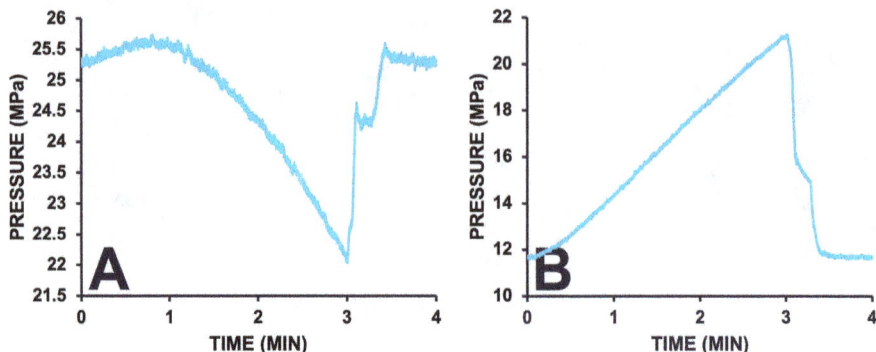

Figure 3.46 Pressure curves from the HILIC analysis linear gradient from 10 to 50% aqueous buffer in ACN. (A) is the actual pressure trace; (B) is a stored reference trace. Figure reproduced from ref. 90 with permission from *LC-GC International*. Courtesy of Dwight Stoll (Gustavus Adolphus College, United States of America).

linear segment of the gradient is expected to yield an approximately straight line for a linear ACN–water gradient (see Section 3.5.2, Figure 3.22). The pressure trace is very revealing. It clearly shows that there is a problem and provides some indication as to where to look. There is a high pressure, indicating that the pumps are functioning. This may indicate a blockage in the capillary between the ACN pump (of the binary system, see Section 3.3.1.2) and the high-pressure mixer or in the mixer itself. However, the reason for this problem was much simpler. The A and B bottles had been swapped, or the gradient program was entered wrongly so that the gradient ran from 50 to 10% water (RPLC gradient) rather than from 10 to 50% water (HILIC gradient).[90] Of course, in this case, the chromatograms also showed that something was going dramatically wrong.

Short-term variations in the pressure are also extremely valuable diagnostic tools, as indicated in Figure 3.47. Such pressure pulses should be kept as small as possible. In a well-functioning system, they are mainly due to the compressibility of the mobile phase, and this may be corrected for in the pump software. However. If there is an **air bubble** in one of the pump heads or if a check valve is malfunctioning, much stronger pressure pulses may be obtained, as shown in Figure 3.47. Air bubbles can usually be removed by purging the pump, and check valves can be cleaned in an ultrasonic bath or replaced. We recommend that analysts learn how to replace inlet

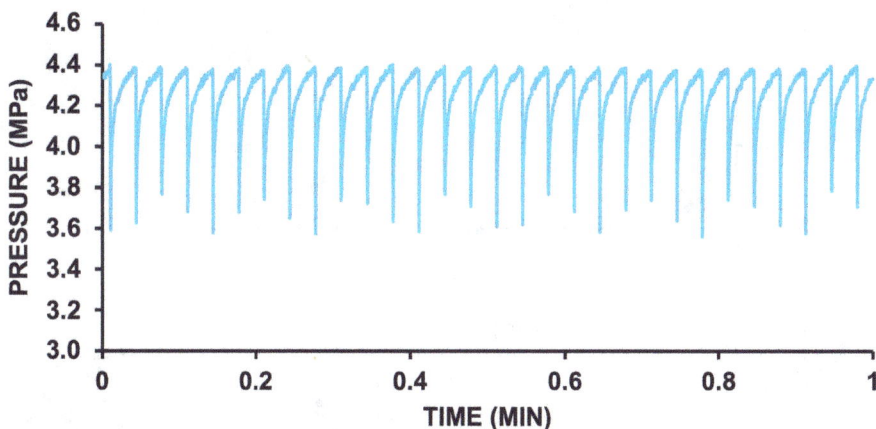

Figure 3.47 Fluctuations in pressure due to an air bubble or a malfunctioning check valve. Courtesy of Gerben van Henten and Rick van den Hurk (University of Amsterdam, The Netherlands).

filters, check valves and pump **seals** to keep their systems running. Instruments often feature diagnostic tools, such as pressure ramps, to identify sources of malfunctioning.

3.11.4   Peak Shapes

Peak shapes and peak widths are highly informative about the performance of the column and the status of the chromatographic system. A distinction may be made between setting up a method and applying it. When setting up a method, a system-suitability test (see Module 10.5) may be appropriate. Extra-column band broadening should be minimized. (Near-)symmetrical peaks should be obtained, and this may require adapting the mobile phase. The injection solvent is especially important, as explained for RPLC in Section 3.5.5 and demonstrated in Figure 3.48. The analyte (2-hydroxynaphthalene) is dissolved in 100% strong solvent (ACN), which is not generally advised. It is seen that for all but the very small injection volumes, split peaks are observed both in isocratic (A and B) and in gradient-elution (C) experiments. Until the sample zone is sufficiently diluted, the analyte moves unretained. To inject larger volumes, a weaker sample solvent (*e.g.* 50% ACN or 50% water) is required.

If the peak shape deteriorates during the application of a method, troubleshooting may be required. Running a test mixture (see

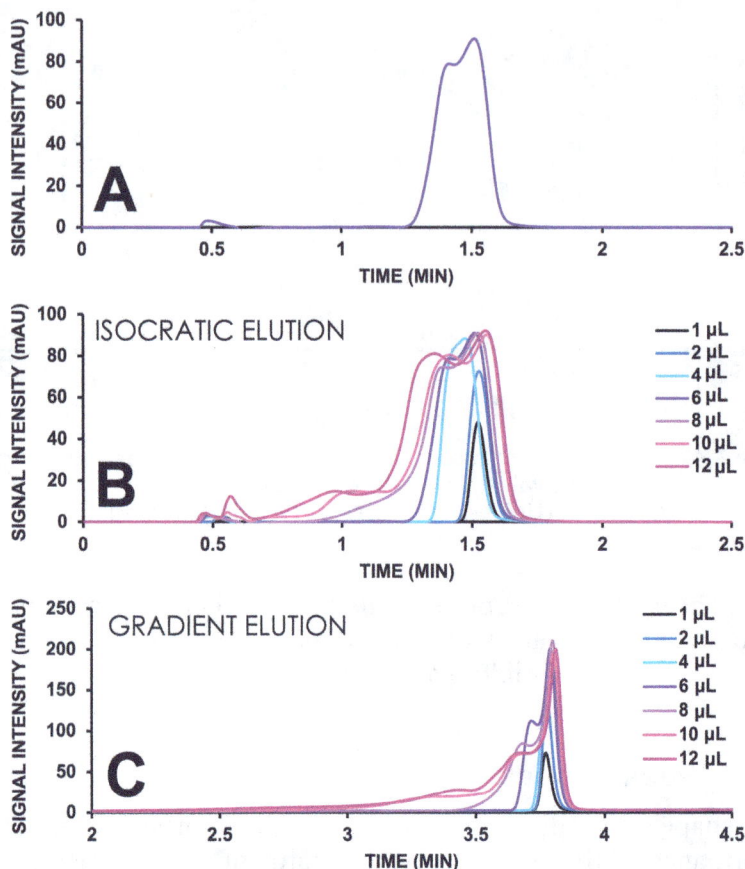

Figure 3.48 (A) Example of a split peak for 2-hydroxynaphthalene in RPLC (injection volume 6 µL). (B) Chromatograms of 2-hydroxy-naphthalene injected (50 mg L⁻¹ 2-naphthol in 100% ACN) in acetonitrile with various injection volumes; isocratic analysis with a mobile phase of water and acetonitrile (50/50, v/v). (C) Chromatograms of toluene injected in acetonitrile at various injection volumes; gradient program from 100% water to 100% acetonitrile. Courtesy of Gerben van Henten and Rick van den Hurk (University of Amsterdam, The Netherlands).

Section 3.11.2) is an important first step. Ageing of the column is a possible explanation (see Figure 3.43), but injector malfunctioning may also lead to broadened and distorted peaks. This may be revealed by trying another column (or another instrument, if available).

3.11.5 Other Issues

There are many other issues that may be encountered in LC practice (see *e.g.* ref. 89). Some **baseline drift** is always observed in the case of gradient-elution LC. This may be minimized by using pure solvents, but the drift may increase with ageing of the solvents. THF, which is mainly used for SEC and other polymer separations, is especially prone to oxidation (peroxide formation) and polymerization. Increased **noise** may, for example, be due to inadequate degassing or to an ageing UV lamp. **Incompatibility** of mobile-phase components and additives, such as buffer salts, is a common problem. Inorganic salts may deposit in the pump or in a mixer. Organic (volatile) buffers are not only desirable for coupling LC with MS, but they may also reduce the risk of salt deposits.

The actual **flow rate** is not usually measured in LC, but it is relatively straightforward to measure it at the system outlet using a stopwatch and a volumetric flask. In the case of low flow rates, the amount of solvent collected in a given time may be determined by weighing. When the flow rate is lower than expected and the pressure is normal, there may be a leak at the downstream side of the column. If the pressure is low, a leak upstream or a pump malfunction may be the cause. If the **pressure** is too high, immediately disconnect the detector, so as to avoid blowing up a detector cell. If this lowers the pressure, you know that there is a blockage in the detector or one of the connection capillaries. Working backwards from the detector, the source of the excess pressure can easily be established.

3.12 Preparative LC (A)

3.12.1 Introduction to Preparative LC

This textbook is about analytical separation science. This implies that large-scale preparative LC is outside the scope of this book. However, we do want to briefly touch on the subject because of its great importance and because many of the concepts described for analytical-scale LC are equally relevant for preparative and even industrial-scale LC. On a smaller scale, preparative LC has analytical goals. It is relatively straightforward to collect fractions for further characterization. The combination of LC and NMR is a case in point, as discussed in Section 3.10.2.3. When collecting fractions for offline characterization, time is no longer critical, and the sample can be

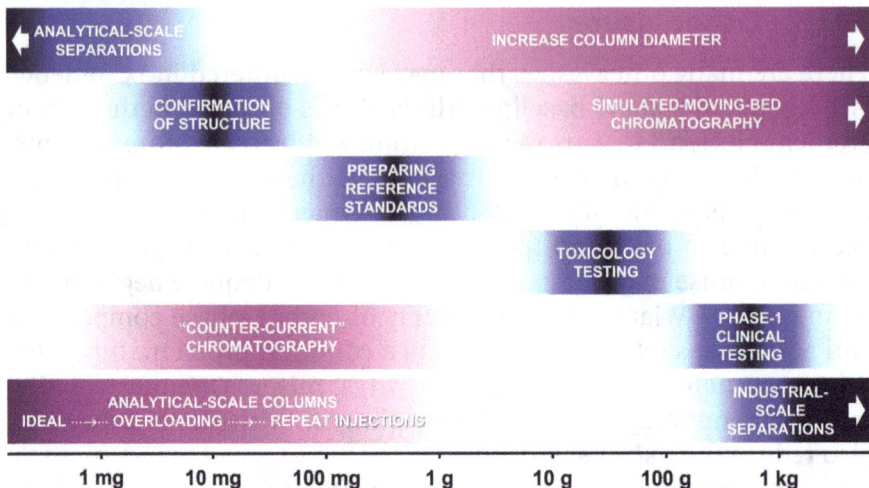

Figure 3.49 Overview of the different sample ranges for (preparative) LC separations (purple) and some of the main strategies (dark pink).

manipulated prior to NMR measurements (*e.g.* the solvent can be replaced with a deuterated one). Figure 3.49 gives an overview of many different ranges from a pharmaceutical perspective. It is seen that analytical chemists can be largely self-sufficient by stretching the possibilities of analytical-scale columns. For analytical purposes, the strategies shown in the pink box at the bottom left of the figure usually suffice. Analytical goals may include the confirmation of molecular structure (mainly by NMR techniques) or the preparation of reference standards (for quantitative analysis or, for example, SEC calibration). Overloading the column up to the point that separation is just sufficient and, if necessary, performing several repeat injections can be undertaken on the analytical column. When overloading the column, we leave the linear part of the calibration curve, which usually leads to broad and highly asymmetrical peaks, as explained in Module 1.8 (see *e.g.* Figure 1.30). Usually, adsorption isotherms become concave and peaks become tailing. Therefore, target peaks, such as APIs, can be separated more easily from minor contaminants that elute earlier than from contaminants eluting after the target peak. Repeat injections can be performed automatically on conventional LC equipment if it is equipped with a fraction collector.

An elegant way to perform multiple repeat injections, which the authors described as multicolumn ("counter-current") solvent gradient purification, was described by Müller-Späih *et al.*[91] and later by De Luca *et al.*[92] They used two columns. While one of these (column 1) was used to perform a gradient-elution separation of a mixture including a target compound, the other column (column 2) was regenerated. The first irrelevant part of the effluent of column 1 was sent to waste. Next, the impure leading edge of the target compound was sent to column 2, the pure central part of the target peak was collected, and the impure tailing edge was again sent to column 2. The final irrelevant part of the gradient was used to clean the column with the effluent sent to waste. After adding a fresh sample to column 2 to replenish the amount of target compound collected, the columns switched roles. Column 1 was regenerated, while the gradient-elution program was applied to column 2, *etc.*

For large-scale separations, it is no longer realistic to perform numerous repeat injections and is unavoidable to enlarge the column diameter. When the column length and particle diameter are kept constant, the sample volume and the flow rate scale with the square of the column diameter, provided that the sample is distributed evenly across the entire column upon injection, and there are no radial differences in flow rate in the column. Starting from an analytical separation on an analytical column with 4.6 mm i.d., an injection volume of 10 µL, and a flow rate of 1 mL min^{-1}, the latter two values become about 30 µL and 3 mL min^{-1} on a 7.8 mm i.d. column, about 50 µL and 5 mL min^{-1} on a 10 mm i.d. column, 400 µL and 50 mL min^{-1} on a 30 mm i.d. column, and about 1.2 mL and 120 mL min^{-1} on a 50 mm i.d. column. Clearly, large volumes of samples and large amounts collected require large volumes of solvents, which, of course, is also the case when performing repeat injections on an analytical column. Salts and non-volatile buffers in the eluent are undesirable if they need to be removed after fraction collection. NPLC (Section 3.6.1) and, especially, SFC (see Chapter 6) are more attractive for preparative separations than RPLC because aqueous solvents require much energy and high temperatures to remove. Together with the polar (reactive) nature of RPLC eluents, this increases the risk of analyte degradation. High flow rates require different pumps (or larger pump heads) and larger diameters of the connecting capillaries to avoid excessive pressure drops across these. Injectors need to allow larger injection volumes. This may, for example, require the installation of larger sample loops and larger syringes in autosamplers. Very large volumes may be injected using a dedicated sample pump, which operates similarly to a modifier pump. Dilution of the sample prior to injection can be beneficial, provided that the initial mobile phase is a weak eluent. Step gradients are often used for preparative separations. In the case of isocratic elution, the mobile-phase pump should be switched off temporarily, while the second pump injects the sample. Using a single-pump gradient system (see Figure 3.14A) is not recommended because of the large dwell volumes and mixing volumes associated with such systems. UV photometers are by far the most common detectors used in preparative LC. For non-chromophores, DRI detectors may be needed, but much care should be taken to avoid pressure build-up in the detector cell, as it tends to be vulnerable. For high-value analytes, such as biopharmaceuticals, mass spectrometers are sometimes used to establish the best possible cut-off points for the collected fraction.

Larger detector cells may also be used, but the path length of a UV detector cell may need to be shortened when overloading conditions are applied. Commercial cells for (semi-)preparative LC experiments are available from many manufacturers, as are semi-preparative LC columns. Wider columns have lower pressure limits because the total force exerted on the wall is proportional to the wall area, which in turn is proportional to the column diameter. Therefore, larger particles tend to be used in wider columns. Longer columns and lower flow rates may be required to maintain the original separation (obtained on an analytical column) on a large-i.d. column.

3.12.2   Counter-current Chromatography

We have not been discussing liquid–liquid chromatography (LLC) in this textbook so far. The main reason for this is that LLC systems proved unstable in the early days of LC and were poorly compatible with gradient elution. When CBPs (see Section 3.2.2.2) appeared, LLC became obsolete. However, LLC systems have some advantages. Columns can be flexible. The stationary liquid can be exchanged for a different one or it may be renewed (replaced by a clean liquid) after extensive use. The phase ratio may be varied (within certain limits), and the sample loadability is usually much higher than for liquid–solid chromatography or for LC with CBPs. In the modern implementations of LLE described in this section, less stress is exerted on analyte molecules than in columns packed with microparticles, so that the conformation, aggregation, and activity of biomacromolecules can be maintained. Finally, contrary to the case of silica-based CBPs, there are no restrictions on the pH, provided that both phases and the analytes remain intact.

Counter-current chromatography (CCC) aims to capitalize on these advantages. The term has become synonymous with a form of LLC in which the stationary liquid is held in place by centrifugal forces.[93] The name is slightly misleading, as it suggests that the mobile and stationary phases are moving in opposite directions, which is not the case. In CCC, the phase ratio can (and should) be measured, and some of the LC conventions described in Chapter 1 and in this chapter are slightly modified in CCC. Eqn (1.38) can be adapted to obtain an equation that is explicit in the distribution coefficients K_i and K_j of the first- and last-eluting analytes, respectively, *i.e.*

$$R_{S,j,i} = \frac{\sqrt{N}}{2}\left(\frac{k_j - k_i}{2 + k_i + k_j}\right) = \frac{\sqrt{N}}{2}\left(\frac{K_j - K_i}{\frac{2V_m}{V_s} + K_i + K_j}\right) = \frac{\sqrt{N}}{2}\left(\frac{K_j - K_i}{2\left(\frac{1}{S_F} - 1\right) + K_i + K_j}\right) \quad (3.61)$$

where S_F is the stationary-phase fraction, defined as $S_F = V_s/V_c = V_s/(V_s + V_m)$, and V_c is the volume of the column. The last part of eqn (3.61) suggests that the resolution is penalized for small values of S_F. If 50% of the column volume is occupied by the stationary phase, the denominator is equal to $2 + K_i + K_j$. When the stationary phase occupies only 10%, this becomes $18 + K_i + K_j$. This implies that with low stationary-phase fractions, similar compounds $(K_i \approx K_j)$ will be hard to resolve, especially when they are weakly retained (low K values). In CCC, the stationary phase may occupy up to 90% of the column volume.

Two implementations of CCC are commonly used and commercially available. These are illustrated in Figure 3.50. In panel A, an implementation of hydrostatic CCC is depicted that is more commonly (and more correctly) known as **centrifugal partition chromatography** (CPC). There is one rotary axis in this system (identified as ROTOR) that creates a constant centrifugal field (denoted by G), and four partition chambers (or "loculi") are drawn out of a larger number. Each chamber is thought to represent one theoretical plate, so that achieving a plate count of 1000 will be a major challenge. The connecting tubes fill themselves with the mobile phase (dark pink), which in this case has a lower density than the stationary phase.

The implementation shown in Figure 3.50B is an example of **hydrodynamic CCC**, which is also referred to as high-speed CCC. It is

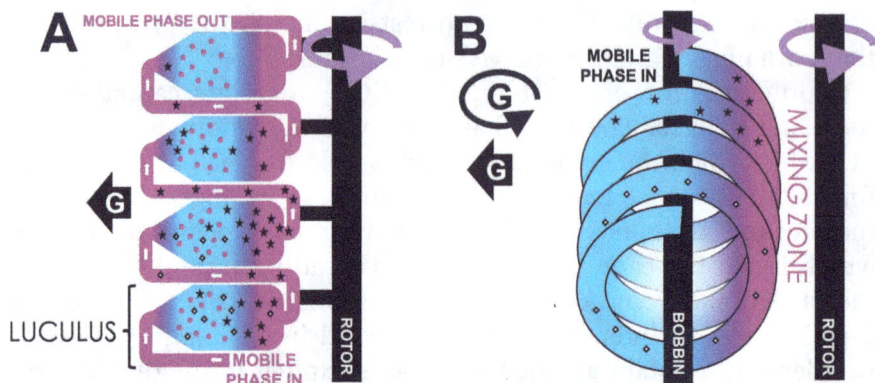

Figure 3.50 Two implementations of counter-current chromatography. (A) Centrifugal partition chromatography (CPC), a form of hydrostatic CCC with a single rotor and a constant centrifugal field. The less-dense solvent (pink) becomes the mobile phase, and analytes that distribute towards this phase (stars) move quickly through the system. The denser solvent (blue) stays confined to the partition chambers, and analytes that distribute towards this phase (diamonds) migrate slowly. (B) Hydrodynamic CCC or high = speed CCC. This system has two axes (identified as ROTOR and BOBBIN), which cause a variable centrifugal field. In this case, the dense (blue) solvent becomes the mobile phase and the less-dense (pink) solvent becomes the stationary phase. Consequently, the light-blue-diamond analytes now migrate more quickly than the light-pink-star analytes.

characterized by two axes: the main axis, identified as ROTOR, and a second axis labelled BOBBIN, which is a "planetary" axis causing the helical channel to rotate around its own axis, while the entire helix rotates around the main axis. This results in a constant centrifugal field with a superimposed variable centrifugal field, indicated by the two arrows denoted "G". There are no chambers and connecting tubes, the mobile phase and stationary phase are continuously in contact, and the pressure drop across the channel is much lower than in the case of a CPC system. Each turn of the spiral column is thought to represent up to four theoretical plates, which means that at least 250 turns are needed to achieve a plate count of 1000.

In CCC, changing the composition of the mobile phase may affect the composition of the stationary phase, making gradient elution more complicated than in conventional RPLC. However, we should be aware that the composition of a stationary-phase layer anchored by C18 chains also changes during a gradient, as does the composition (and volume) of the water-rich stationary phase during a gradient in HILIC.

CCC is a way to measure octanol–water partition coefficients directly for many compounds, provided that these distribute to some extent (*e.g.* −1.3 < log K_{ow} < 1.3). log K_{ow} values outside this range will lead to peaks that seem unretained or "indefinitely" retained and require adaptation of the phase system, making the measurement of log K_{ow} less direct. One way that has been used to measure very high log K_{ow} is "concurrent chromatography". In this case, two pumps are used, with the stationary phase moving very slowly in the same direction as the mobile phase. Another way to elute highly retained compounds from a CCC column is to change the mobile phase by stopping the flow of the mobile phase and starting to pump the stationary phase in the opposite direction. The stationary phase then becomes the mobile phase and *vice versa*. Unlike in the case of backflushing LC columns, the resolution of the analytes is not jeopardized in the process.

CCC can deal with complex samples, such as plant extracts, without a risk of fouling the column. On the other hand, it offers modest efficiency (1000 plates is a high number) and modest speeds (at best). The high loadability is a serious advantage, especially for compounds that have a tendency to adsorb strongly on (residual silanols on) silica-based phases. This makes basic compounds from plant extract, such as alkaloids, a prime target for CCC. The spur in interest in cannabis and its constituents has also renewed interest in CCC.

3.12.3 Simulated Moving Bed Chromatography

Simulated moving bed (SMB) chromatography is an ingenious way to continuously separate a mixture into two fractions: an early eluting one (to become the extract) and a late-eluting one (to become the raffinate). Two such processes may be needed to separate a target compound (*e.g.* a biopharmaceutical product) from a complex matrix. Developing an SMB process may be a challenge and may require hiring specialized help, but once the process is going, it can run continuously. Eight identical columns are required in the setup illustrated in Figure 3.51, but the amount of stationary phase required per amount of product obtained is very low. In addition, the process is highly efficient in terms of energy and amounts of solvent needed. Despite the rather elaborate hardware, with valves and capillaries needed to allow four different functions (enter feed, F, or desorbent, D; tapping off extract, E, or raffinate, R) to alternate around eight different connection points, the approach can be quite cost effective.

The SMB process is illustrated in Figure 3.51. For didactical purposes, we show how the process could start with introducing the (continuous) feed into the empty system. Flow through the system is in a clockwise direction, as indicated by the arrows in the centre. The feed fills the first column (step 1), with the early eluting fraction (pink) running to the end of the column and the late-eluting fraction not further than halfway. After this first step, the feed is switched to the next position (step 2). Moving the feed forward is equivalent to making the bed move backwards, which is the core idea of the SMB approach. Nothing happens to the contents of the first column during step 2 because there is no flow through this column. In step 2, the second column is filled with fresh feed; in step 3, column 3 is filled; *etc.* In step 4, there is flow again through the first column, and the early eluting fraction is collected in the extract (E, in this step located at the top). From this point on, the extract is obtained continuously, but in each step, it is obtained from a subsequent connection point. The late-eluting fraction is moved towards the column exit in steps 4 and 5, but it is not eluted to the raffinate until step 8 (there is no flow through the first column during steps 6 and 7). From this point onwards, there is a continuous flow of raffinate through the moving connection that is marked R in each step. From step 9 onwards, the process enters a steady cycle, with step $8n +1$ in Figure 3.51 representing steps 9, 17, 25, *etc.* All four streams (feed and desorbing solvent in; extract and raffinate out) are continuous as long as the process is running.

3.13   Chiral Recognition (A)

Chirality is a massively important concept, and chromatography – particularly LC – is massively important in dealing with chirality in many different fields, but most dramatically in pharmaceutical science and medicinal chemistry. The word chirality is derived from the Greek word for hand ("χέιρ"). The logic behind this is that our hands are notoriously chiral. Our left hand and right hand are mirror images of each other, but they cannot be superimposed. This makes them chiral objects by definition.

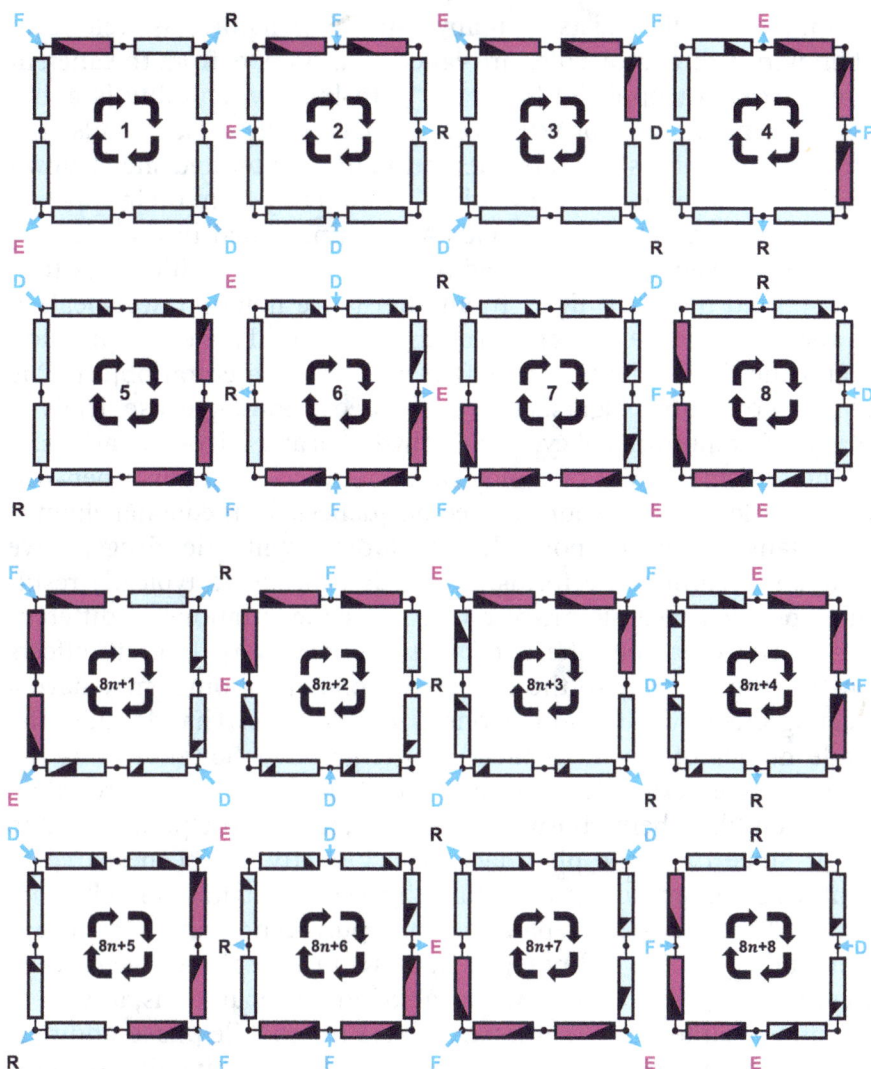

Figure 3.51 Explanation of the simulated moving bed approach to preparative chromatography. The first eight steps illustrate how the process develops. The fast-eluting fraction of the sample feed (F) that ends up in the extract (E) from step 4 onwards is indicated in pink. The slow-moving fraction that ends up in the raffinate (R) from step 8 onwards is indicated in dark blue. D is the desorbing eluent, assumed to be identical to the solvent of the sample feed. After the first eight steps, the process repeats itself every eight steps (steady cycle). The arrows in the middle indicate the (clockwise) direction of flow.

Similar chirality exists in many aspects of nature, as well as in chemistry. Chiral molecules are called enantiomers from the ancient Greek word for opposite ("ἐναντίος"). In biology, chirality is abundant, from the level of complete animals, such as sea snails that almost exclusively show one orientation in their twisted shells, down to chiral molecules that act distinctly differently from their opposite forms. For example, amino acids are distinguished in their left (L) and right (D) enantiomers based on the direction in which they turn transmitted polarized light. In an alternative notation, we speak of *R*- and *S*-isomers. Only L-amino acids are involved in gene expression – and the double helix of our DNA is in itself a chiral object. The chirality of simple sugars, such as D-glucose and L-glucose, leads to many different forms of cyclic carbohydrates and oligosaccharides.

The extreme ability of biological systems to distinguish between chiral molecules has enormous consequences for medicinal chemistry. Many bioactive molecules, including synthetic drugs, have different enantiomeric forms. Synthetic procedures typically result in a so-called **racemic mixture** of equal concentrations of different forms. However, the different forms may have very different effects on the human body to the extent that one enantiomer may have a healing effect, while another one is highly toxic. Thalidomide, the trade names of which included Contergan and Softenon, was the most tragic example. Racemic medicine led to more than 10 000 children being born with severe defects in the period 1957 to 1961. Since then, the pharmaceutical industry has been geared to synthesizing, producing, and clinically testing enantiomerically pure drugs. The analytical separation and purification of chiral drugs were indispensable technologies in this transition. When we succeed in separating enantiomers, we speak of chiral separations, as well as enantio-separations. A stationary phase (or mobile-phase additive) that enables us to achieve such feats is said to exhibit chiral selectivity or to be enantioselective. Apryll Stalcup (Box 3.10) is one of the scientists who have contributed to the science of chiral separations.

Box 3.10 Hero of analytical separation science: Apryll Stalcup. Photo courtesy of *The Analytical Scientist*.

Apryll Stalcup (Ireland)

Apryll Stalcup is brat. Military brat, that is (in her own words). Her father served in the military, which meant she moved around a lot as a child. She stayed in a moving mode for much of her time as a separation scientist, working in Washington, DC, Cincinnati (OH), and Hawaii, until she settled as a Professor of Chemical Sciences at Dublin City University (Ireland) and Director of the Irish Separation-Science Cluster. Apryll also worked on very many subjects, within and outside analytical separation science, including LC, CE, and briefly SFC (with Tom Chester in Cincinnati). Her strongest focus has been on chiral separations and, more recently, on ionic-liquid statiory phases.

A measure for the chiral purity of a compound (in *R/S* notation) is the percentage enantiomeric excess (ee$_R$) of the dominant form, here assumed to be the *R*-form. The enantiomeric excess is defined as

$$ee_R = \frac{C_R - C_S}{C_R + C_S} \cdot 100 \tag{3.62}$$

where C_R is the molar concentration of the *R*-form and C_S that of the *S*-form. For the pure *R*-enantiomer, ee$_R$ approaches 100%. Chromatographic methods are best suited to separate enantiomers and accurately determine ee$_R$, especially when the enantiomeric excess is high (ee$_R > 90\%$). Ideally, the low-concentration enantiomer is eluted before the high-concentration one, as tailing peaks are more commonly encountered in LC than fronting peaks.

In Section 1.6.1, we have demonstrated that in a typical LC system (with 20 000 plates), a mere difference of 50 J mol^{-1} in the way two enantiomers interact with the phase system (stationary phase and mobile phase) can make a difference between the complete overlap and (near-) baseline resolution. This fabulous attribute of liquid chromatography explains why enantiomers can almost always be separated. It also makes chiral separations hard to predict, as extreme accuracy of any computational approach is required.

Without useful computational models, analysts must rely on trial-and-error screening of columns, as explained later in this section.

Figure 3.52 shows a schematic overview of analytical separation methods applied to chiral analytes. Some of these analytes are volatile, allowing them to be analysed by GC (see Section 2.1.4). Because GC has attractive features, chiral analytes can be chemically modified ("derivatized") as a sample-preparation strategy to increase their volatility (see Section 8.3.4). Analytes may also be derivatized with other chiral molecules [chiral derivatizing agents (CDRs)] to form diastereomers, *i.e.* molecules with two chiral centres, which are more different than molecules with one chiral centre, and may feasibly be separated on a conventional achiral separation system (left-hand side in Figure 3.52). This derivatization route is sometimes referred to as the **indirect approach** to enantiomer separation. Essential requirements for the derivatization approach are that the CDR is enantiomerically pure, the reaction goes to completion, and the reaction product (diastereomer) is stable. The analyst should also be aware that different diastereomers may exhibit different detector response factors. A way to establish relative response factors is to subject a racemic mixture to the analysis method. Having both enantiomers available in their pure form allows changing the elution order, so that a minor enantiomeric contaminant can always be made to elute before the large peak of the other, nearly pure enantiomer.

However, as shown by the large pink box in the background of the figure, LC and "supercritical"-fluid chromatography (or, almost always, "subcritical"-fluid chromatography, SFC) are by far the most important techniques for separating chiral compounds. As explained in Chapter 6, common implementations of packed-column SFC are a form of LC, in which carbon dioxide is a major component of the mobile phases. SFC has some attractive features, including fast analyses and easy removal of the solvent, and offers high chiral selectivity.

Occasionally (purple arrow in Figure 3.52), enantiomers are derivatized and then separated by LC or SFC. In such a case, derivatization is performed either to enhance the enantioselectivity of the phase system, or to enhance the detectability of the analytes, or both. A famous example is the derivatization of difficult-to-detect chiral amino acids with a mixture of o-phthalic-dialdehyde (OPA) and a chiral derivatization reagent, such as N-acetylcysteine. The former reagent facilitates highly sensitive fluorescence detection,

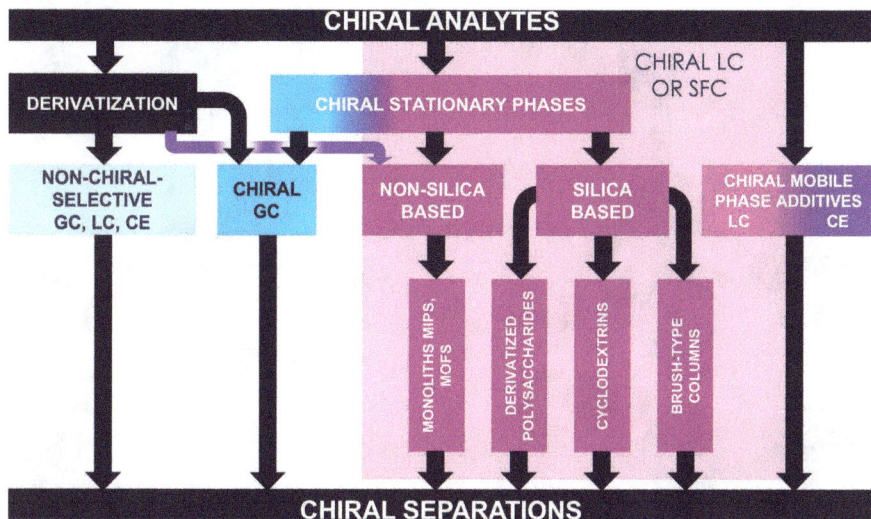

Figure 3.52 Overview of the various ways to achieve chiral separations using analytical separation methods. The pink-shaded box in the background indicates the importance of chiral LC (and chiral SFC; see Chapter 6) in this field.

while the latter produces a bulky substituent, which makes it easier to separate the enantiomers.

Generally, it is attractive if an intricate and time-consuming derivatization step can be avoided, in what is known as the **direct approach** to chiral separation. Chiral stationary phases (CSPs) are by far the most common way to achieve chiral selectivity. One possible type of chiral selectivity, based on three-point interaction, is illustrated in Figure 3.53. In this schematic example, it is assumed that atoms or groups of equal colour can interact strongly with each other. The two molecules at the top are enantiomers. The two chiral ligands that are immobilized on the stationary phase are identical. It is clear that the options for the two different enantiomers to interact are different. Therefore, three-point interaction is a potential mechanism to achieve chiral separations, and immobilizing chiral centres on a stationary surface can lead to chiral selectivity.

In LC, it is, in principle, also possible to add a chiral selector to the mobile phase. The formation of transient complexes between the analyte and a chiral additive is analogous to the IPC technique discussed in Section 3.7.3. The strength of the interaction (*i.e.* the association coefficient) should be different for the enantiomeric analytes that must be separated. Like an ion-pairing reagent, the free (non-complexed) chiral selector may be distributed across the

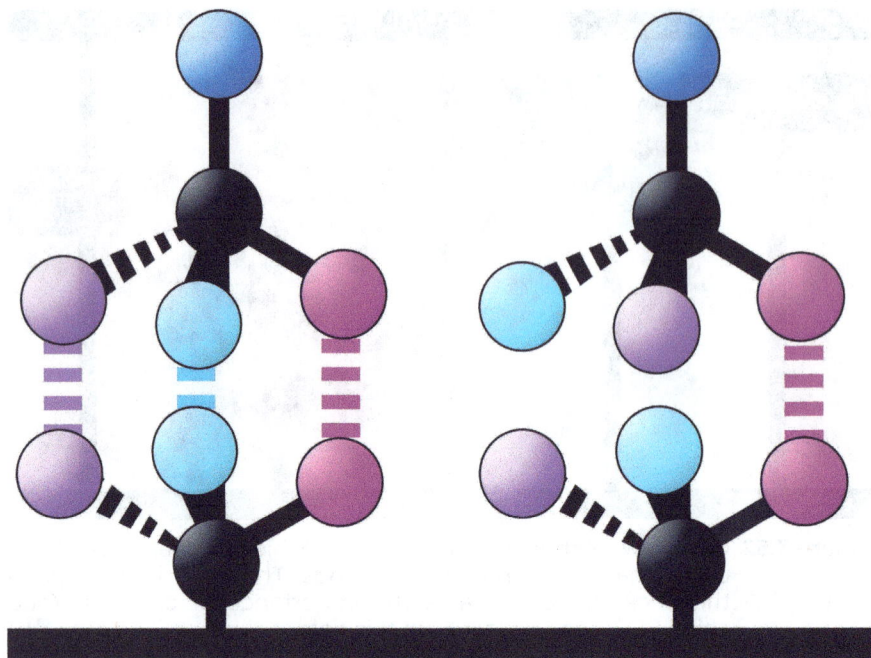

Figure 3.53 Illustration of chiral selectivity based on three-point interaction between two enantiomeric analytes and a chiral selector immobilized on a stationary surface. In this example, the enantiomer on the left can engage in strong interactions with the stationary phase simultaneously, while this is not possible for the enantiomer on the right.

mobile and the stationary phases, as will be the transient complexes. Common examples of chiral mobile-phase additives include cyclodextrin (cyclic oligosaccharides; described later) and amino acids – the latter in combination with metal ions.

The addition of a chiral additive has several disadvantages. Some of these are similar to those of IPC, such as the complexity of method development and the sensitivity of the method to changes in many different parameters (*e.g.* temperature and pH). Interfering system peaks may appear in the chromatogram, as in IPC. Chiral additives may complicate detection. For example, the selector may be incompatible with MS. The (pure)chiral selector needs to keep being supplied to the LC system, and this may be expensive. For all these reasons, CSPs are the dominant approach to achieving chiral separations in LC. In CE (see Chapter 5), where no stationary phase is present, chiral additives to the buffer are commonly used. This is illustrated by the purple shading at the right of Figure 3.52, outside the pink box that represents LC and SFC.

The tetrahedral chirality illustrated in Figure 3.53 was first described by J. H. van 't Hoff back in 1874. It occurs if four different chemical structures (atoms, functional groups or larger structures) are attached to a tetravalent atom (such as carbon or silicon) or if three different chemical structures plus a lone pair of electrons are attached to a trivalent atom (such as nitrogen or phosphorus). The three-point-interaction principle was, if not created, at least made popular by William (Bill) Pirkle from the University of Illinois Urbana-Champaign (IL, USA), who stated that "chiral recognition requires a minimum of three simultaneous interactions between the CSP and at least one of the enantiomers, with at least one of these interactions being stereochemically dependent".[94] CSPs that rely on the three-point-interaction principle have become known as "Pirkle-type columns" or π–π interaction columns. Interactions on such columns are not necessarily confined to π–π interactions of electron clouds. For example, if we imagine that in the two enantiomers depicted in Figure 3.53, the light blue functional group is acidic and the purple group is basic, then the enantiomer on the right would show much stronger interactions with the stationary phase than the one on the left, and chiral recognition (*i.e.* chiral separation) can be achieved. Ionic (attractive or repellent) interactions tend to be detrimental to chiral selectivity. They operate across longer distances than other interactions, such as hydrogen bonding, and they are less localized and directed. The main method for suppressing the ionization of analytes and stationary surfaces is control of the pH. If necessary, the addition of an ion pair may be considered. Operating at high ionic strength also reduces the effect of ionic interactions, but this approach is generally not attractive.

There are more types of chirality than the tetrahedral geometry described earlier, and there are more types of CSPs than the three-point-interaction (Pirkle-type) phases illustrated in Figure 3.53. A conceptually simple way to create a CSP is to cover the surface of a silica substrate with a protein. The retention mechanism and the chiral recognition mechanism cannot be easily explained because of the multitude of chiral and achiral interaction sites on the protein. Polysaccharides also possess many chiral centres, and they have long been used as chiral selectors. In early LC applications, polysaccharides, such as micro-crystalline cellulose, were used as the stationary phase, but the common approach has become to physically or covalently cover silica surfaces with polysaccharides, such as (incompletely) derivatized cellulose polymers. Rather than cellulose itself, its derivatives, such as cellulose ethers, tend to be well soluble, allowing the preparation of chromatographically efficient CSPs. Arguably, the most popular and most broadly applicable CSPs consist of silica particles coated with another polysaccharide, amylose, which is one of the main components of starch. Derivatization of amylose with, for example, carbamate groups has resulted in columns with significant chiral selectivity for a variety of analytes. Cyclodextrins are cyclic oligosaccharides that form cavities of a very specific size and structure. They consist of six, seven, or eight saccharide units (α-, β-, and γ-cyclodextrin, respectively). The cyclodextrins feature exactly defined cavities, the size of which, together with the positioning of hydrogen-bonding OH groups, determines the extent of interaction with different enantiomers. Chemically bonded phases with small chiral ligands are known as "brush-type" CSPs, and they usually aim to create the type of three-point interaction illustrated in Figure 3.53. Other silica-based CSPs contain macrocyclic antibiotics or crown ethers.

Polymer monoliths, with chiral centres incorporated during the synthesis or grafted on the surface later, and molecularly imprinted polymers (MIPs) are among the non-silica-based CSPs. An MIP is created by forming a polymer, for example, a polymeric monolith, in the presence of a pure enantiomer. After the synthesis, the probe enantiomer is washed out, and the idea is that a perfect adsorption cavity remains for that specific enantiomer, so that it will be retained longer than its counterpart. Results obtained with MIPs are erratic in terms of the selectivity achieved and rather disappointing in terms of the chromatographic efficiency. Metal–organic frameworks have been explored, but their value in practice has yet to be proven.

The CSP is the most important factor determining the selectivity (and eventual separation) of enantiomers using the direct method, but the efficacy of any column for separating a new pair of enantiomers, for example, a newly synthesized (candidate) API, cannot be predicted with confidence. Therefore, automatic column-screening procedures are in place in industrial laboratories, in which many columns are tested. Based on the chromatograms obtained, the best (or the most promising) column is selected to start a more thorough method-development process. There are some challenges associated with such **high-throughput screening**. When selecting columns based on single chromatograms, obtained using one specific mobile phase, a mediocre column may be selected, for which the test conditions are near the optimum. A column for which the test conditions are sub-optimal, but the performance under optimal conditions is superior, may be discarded. High-throughput screening may be performed under LC conditions or under SFC conditions. The screening process can be conducted more rapidly in SFC, and the chances of success may be higher. Selectivity screening is described in more detail in Module 10.2.

3.14   Micellar Liquid Chromatography (A)

Micellar LC (MLC) is a form of LC that uses little or no organic solvent in the mobile phase. In daily life (*e.g.* taking showers or washing dishes), we often use the effect that apolar (fatty) compounds that do not dissolve in water can be dissolved in aqueous soap solutions. This is due to the self-structuring of surface-active molecules (or surfactants) to form micellar nanostructures, above a certain critical micelle concentration (cmc) of the surfactant in the mobile phase. Most commonly, in aqueous solutions, the apolar parts (usually hydrocarbon chains) aggregate in the centre of a spherical micelle, whereas the polar or ionic groups form the outside. Sodium dodecyl sulfate (SDS) is a typical example of surfactant molecules used in MLC. The opposite situation, with so-called reverse micelles present in an apolar solvent, is quite rare and not used in LC. The stationary phase, which is usually an RPLC column with octadecyl or octyl chains, is saturated with surfactant molecules, forming a type of bilayer. Analyte molecules are distributed between the stationary phase and the micellar mobile phase. Celia García Álvarez-Coque (Box 3.11) has been one of the champions of MLC for many years.

Potentially, MLC is a form of "green" chromatography that helps reduce the ecological footprint of analytical separations. However, the chromatographic efficiency of MLC is disappointing due to slow (stationary-phase) mass transfer.[96] A small amount of organic solvent is usually added to the mobile phase because this has been found to enhance the efficiency. The amount of modifier used in MLC is typically about a factor 3 lower in MLC than in conventional RPLC with hydro-organic (*e.g.* water–ACN) mobile phases. Liquid chromatographers have been exploiting several other ways to reduce the amounts of organic solvents used (smaller particles, narrower columns, and higher temperatures), and MLC is not commonly used. In electrophoresis, micellar buffers are the only known way to achieve separations of uncharged molecules. This technique, known

> **Box 3.11** Hero of analytical separation science: María Celia García
> Álvarez-Coque. Photograph reproduced from ref. 95 with permission
> from Elsevier, Copyright 2016.
>
>
>
> **María Celia García Álvarez-Coque (Spain)**
> María Celia García Álvarez-Coque is a professor
> of analytical chemistry at the University of
> Valencia (Spain), where she has built a very
> successful group of researchers. She has made
> a long series of important contributions to
> analytical separation science and chemomet-
> rics, with an eye on fundamental aspects of LC
> stationary and mobile phases. She has investi-
> gated numerous retention mechanisms and
> different modes of operation of LC, such as
> isocratic *vs.* gradient elution. Her work on retention modelling and
> optimization in LC is particularly outstanding. Celia is an expert on
> micellar LC and, more recently, micro-emulsion LC.

as micellar electrokinetic chromatography (MEKC), will be discussed
in Section 5.1.4.6 and Module 5.4.

3.15   Specific Protein-separation Methods (A)

Proteins are a very important class of analytes. They play crucial
roles in biological systems and are exciting new biopharmaceutical
drugs. They are also structurally immensely complex, as a copolymer
of some 20 comonomers, with a uniquely defined primary structure
(the main chain, its length and its sequence), secondary structure
(the spatial conformation of specific parts of the molecule, such as
alpha helices or beta sheets) and tertiary structure (the three-dimensional structure of the protein as a whole). On top of that, there
are supramolecular aggregates and numerous (post-)translational
modifications. The latter are changes to the chemical structure,
such as glycosylation, oxidation, and deamidation. The importance
of proteins, on the one hand, and their structural complexity, on
the other, make proteins one of the most prominent targets for
developing analytical separation methods. Many of the techniques
discussed in this chapter and in the next chapters (Chapter 4,
Separation of Large Molecules; Chapter 5, Capillary Electrophoresis)

can be applied to separate proteins. In this module, we will discuss a few methods that specifically relate to proteins.

3.15.1 Hydrophobic-interaction Chromatography

In many cases, including the all important preparative separations, it is imperative to maintain the activity (and thus the conformation) of proteins. For this to be the case, organic solvents, such as methanol or acetonitrile, must be excluded from the mobile phase. This can be achieved with IEC (Section 3.7.1) or with aqueous SEC (Module 4.2), but certainly not with RPLC (Module 3.5). Although the latter is frequently used for separating proteins, it invariably requires significant amounts of organic modifier in the mobile phase, causing denaturing of analyte proteins. Hydrophobic-interaction chromatography (HIC) is akin to RPLC in that the mobile phase is (much) more polar than the stationary phase and that analytes are retained by hydrophobic interactions. However, these interactions are much weaker in HIC than in RPLC, which is achieved by a much less-apolar stationary surface. Traditionally, this was achieved by using silica particles modified with short alkyl chains, such as butyl chains (C_4H_9), instead of the octadecyl chains ($C_{18}H_{37}$) commonly used in RPLC. More recently, dedicated stationary phases have been developed for HIC that show quite weak hydrophobic interactions, so as to create "gentle" conditions for polymer separations. These include alkylamide stationary phases (based on wide pore, 1000 Å silica) and non-porous hydrophilic polymer resins, functionalized with butyl groups.[97] With such stationary phases, no organic modifier needs to be present in the mobile phases used for HIC. Instead, a negative salt gradient is used in HIC, with high salt concentrations resulting in the initial high retention conditions and increasingly low salt concentrations leading to the elution of increasingly hydrophobic proteins. Retention factors are found to decrease exponentially with decreasing (molar) salt concentration, *i.e.*

$$\ln k_{i,\text{HIC}} = a_{i,\text{HIC}}[\text{Salt}] + b_{i,\text{HIC}} \tag{3.63}$$

where $k_{i,\text{HIC}}$ is the retention factor of protein i under HIC conditions, and the coefficients $a_{i,\text{HIC}}$ and $b_{i,\text{HIC}}$ depend on the analyte protein, the stationary phase, and the type of salt used. Interestingly, retention in HIC tends to increase with increasing temperature. This may be due to the (partial)unfolding of the protein, resulting in more sites becoming available for hydrophobic interaction. Working at elevated temperatures increases the risk of denaturing the analyte proteins.

Typical salts used are sulfates, such as ammonium sulfate ($(NH_4)_2SO_4$), or sodium sulfate (Na_2SO_4). Other salts, such as sodium chloride (NaCl), have also been used. Different salts have been found to yield different selectivities.[97] Salt concentrations encountered in HIC are high, often exceeding 1 mol L^{-1}, which implies that HIC cannot be coupled directly with mass spectrometry. Occasionally (*e.g.* for separating membrane proteins), surfactants or small amounts of organic solvents are added to the mobile phase to reduce retention.

3.15.2   Affinity Chromatography

Affinity chromatography (AFC) is based on very specific interactions, usually involving at least one protein. The specific interaction can result in high selectivity, but chromatographic efficiency may be impaired by the strength of the interaction (see Section 1.7.4.4). Very strong interactions may be desirable if affinity chromatography is used for protein purification or in case the specific interaction is exploited for sample preparation. A protein can be immobilized on a stationary phase to separate small molecules or study the strength of the interactions (*e.g.* drug–protein binding). Usually, the protein undergoes some modification to ensure stable bonding to the surface. Proteins may also be the target analytes. For example, an antigen may be immobilized on the surface to separate or purify antibodies from immune serum. A substrate (or substrate analogue) may be immobilized to separate enzymes. A polysaccharide surface may be used to separate lectin proteins. Reversely, immobilized lectins may be used to separate glycoforms. Oligonucleotides can be purified with AFC by immobilizing complementary base sequences. In some cases, both the immobilized molecules and the targets are proteins, such as for the purification of immunoglobulins using Protein A or Protein G.

3.15.3   Metal-ion Affinity Chromatography

Metal-ion affinity chromatography [MAC or immobilized-metal-ion affinity chromatography (IMAC)] bears little resemblance to what is traditionally known as affinity chromatography, except that, indeed, there is specific affinity of analytes with the stationary phase and, yes, there is chromatography. In MAC, the stationary phase is chemically modified with a chelating group, such as 2,2'-azanediyldiacetic acid or tricarboxy methylethylenediamine. Metal ions are then immobilized on the stationary phase through chelation forces. In principle, the metal can be removed by washing with a high concentration of a chelating agent, such as ethylenediaminetetraacetic acid (EDTA). Metal ions may interact specifically with certain amino acids, such as histidine, which has an imidazole group.

MAC is a denaturing separation process, and protein recovery can be high under relatively mild conditions. Therefore, it is finding use in protein purification or, for analytical scientists, in sample preparation. A good example of the latter is the selective concentration of phosphopeptides using a titanium-loaded MAC column.[98] Protein phosphorylation is an important post-translational modification, for example, for the activation or deactivation of enzymes. However, the analysis of phosphoproteins or – after digestion – phosphopeptides is complicated by their low concentrations and limited stability. MAC with Ti^{4+} ions can be used to strongly enrich phosphopeptides in proteomic samples. Ti^{4+} is much more effective for this purpose than other ions, such as Fe^{3+} or Zr^{4+}.[98]

3.15.4 Chromatofocussing

In chromatofocussing, analytes are eluted from a (strong) ion-exchange column using a pH gradient. Typically an anion-exchange column is used and the separation starts at a high pH, where (almost) all proteins are negatively charged and strongly retained on the column. The pH gradient is generated by a mixture of ampholytes. Proteins position themselves within the gradient around the position of their pI values. Chromatofocussing offers a high sample loadability and, due to a band-sharpening (focussing) effect, a high resolving power. Therefore, it finds use as a tool for protein purification. A disadvantage is that the solubility of proteins is affected by the pH, so that precipitation may occur at some point in the column. This can be counteracted by adding a zwitterion, such as glycine, to the mobile-phase buffer.

3.16   Kinetic Plots (A)

A brilliant starting point for a discussion on the kinetic performance of packed LC columns is the study by Knox and Saleem[99] augmented by the reflections of Hans Poppe.[100] In Section 1.5.1, we have outlined the best way to achieve the separation of a given pair of compounds. Once sufficient retention and the best possible selectivity have been attained (and secondary effects, such as mass overloading or peak tailing, have been overcome), the required number of plates for the separation is found from eqn (1.41) and the required column length can be found from

$$L_{\text{req}} = N_{\text{req}} \cdot H = N_{\text{req}} \cdot h \cdot d_{\text{p}} \tag{3.64}$$

If we now recall the Darcy equation ($\Delta P = \psi \cdot \eta \cdot L \cdot u_0/d_{\text{p}}^2$; eqn (1.69)), we find

$$t_{0,\text{req}} = \frac{\psi \cdot h^2 \cdot N_{\text{req}}^2 \cdot \eta}{\Delta P_{\text{max}}} = \frac{E \cdot N_{\text{req}}^2 \cdot \eta}{\Delta P_{\text{max}}} \tag{3.65}$$

where $E = \psi h^2$ is defined as the "separation impedance" by Knox. Eqn (3.65) is quite revealing. It tells us that the required analysis time depends strongly on the reduced plate height (h) and not on the reduced velocity (v). This implies that the shortest possible analysis time (in isocratic LC, $t_{R,\text{req}} = t_{0,\text{req}} (1 + k_w)$, where k_w is the retention factor of the last eluting analyte) always corresponds to the lowest h, i.e. the minimum in the h vs. v curve. The required analysis time also depends on the column permeability factor ψ, the mobile-phase viscosity (η; low viscosity is favourable), and the all-important maximum available pressure drop (ΔP_{max}). Clearly, the progression from HPLC to UHPLC has enabled faster analyses. The occurrence of N_{req}^2 in eqn (3.65) indicates a very strong dependence and underlines that sufficient retention and selectivity must absolutely be realized first (see eqn (1.41)).

Eqn (3.65) specifies the price we have to pay for the number of plates we need for our separation, but the study by Knox and Saleem[99] teaches us more. We can also rearrange the Darcy equation to find the optimum particle size, i.e.

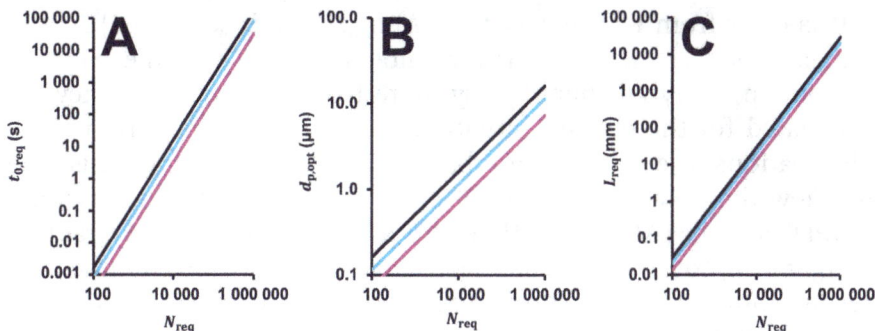

Figure 3.54 Required analysis time (A), optimum particle size (B) and required column length (C) as a function of the required number of plates. Calculations based on eqn (3.64–3.66) with parameter values specified in the text. Dark-blue lines: $\Delta P = 20$ MPa; light-blue lines: $\Delta P = 40$ MPa; and pink lines: $\Delta P = 100$ MPa.

$$d_{p,opt} = \sqrt{\frac{\psi \cdot D_m \cdot h_{min} \cdot \nu_{opt} \cdot N_{req} \cdot \eta}{\Delta P_{max}}} \qquad (3.66)$$

Eqn (3.66) shows explicitly what we have been hinting at before (see, for example, the beginning of Module 3.2), *i.e.* that we may need to choose larger particles if we need more plates. If we can operate at a higher pressure (or if the viscosity of the mobile phase is lower), we can use smaller particles. Interestingly, for smaller molecules (high D_m), larger particles are optimal compared to macromolecules. This is because the latter diffuse so slowly that we are less likely to end up working in the *B*-branch of the van Deemter curve (*i.e.* below $\nu = \nu_{opt}$).

Thus, if we know how many plates we need, we do not only know how much time this will take (eqn (3.65)) but also what size particles we should use (eqn (3.66)), and, once we put the latter information into eqn (3.64), we can determine how long our column should be. The results are shown in Figure 3.54 for three situations, *i.e.* HPLC with $\Delta P_{max} = 20$ MPa (the number used by Poppe) and with a typical $\Delta P_{max} = 40$ MPa, and UHPLC with $\Delta P_{max} = 100$ MPa. In all cases, we have assumed $\psi = 1000$, $D_m = 10^{-9}$ m^2 s^{-1}, and $\eta = 0.001$ Pa s. We used the same reduced plate height equation as Poppe did,[100] *i.e.* $h = A\nu^{1/3} + \dfrac{B}{\nu} + C\nu$, with $A = 0.8$, $B = 1.5$ and $C = 0.05$. These coefficients imply a value for ν_{opt} of 2.87 and h_{min} of 1.80.

It is clear from Figure 3.54 that (A) $t_{0,\text{req}}$, (B) $d_{p,\text{opt}}$ and (C) L_{req} all increase as N_{req} increases. The numbers are in the same ballpark as contemporary LC, but slightly more favourable because they are calculated for the optimal situation, *i.e.* $h = h_{\text{min}}$. Apart from these observations, the numbers are barely useful. For example, it is nice to know that if we want 100 000 plates, we can achieve this with a UHPLC system ($\Delta P = 100$ MPa) with a t_0 value of 325 s. A gradient of $t_G = 10t_0$ duration would (accounting for the dwell time and re-equilibration time) require about an hour. However, where can you buy a 410 mm-long column packed with 2.28 µm particles? The columns that are available to us are quantized. We can choose between a limited selection of particle sizes and column lengths. When deviating from Figure 3.54, it is clear in which direction one should move. Choosing smaller particles implies a lower velocity than ν_{opt}. We learnt in Chapter 1 (Section 1.4.4) that working in this B-branch of the van Deemter curve is a lose–lose situation, as it yields fewer plates in a longer time. Thus, Figure 3.54 provides a lower limit for the particle size. When it suggests a particle size of 2.28 µm, it is better to choose 3 µm particles rather than 1.7 µm particles. However, we need better advice than just a direction in which to move. This can be found in **kinetic plots**.

In the type of kinetic plot proposed by Poppe (Figure 3.55), the "time equivalent of a theoretical plate" (TETP, equal to H/u) is plotted against the required number of plates for a pressure drop of 100 MPa (UHPLC conditions). The curved lines each represent a given particle size. Each such line is seen to touch the pink line, which is the **Knox-and-Saleem limit** illustrated in Figure 3.54. For each value of N_{req}, this line corresponds to the opttimum particle size. For example, the darkest line, representing 1.6 µm particles, corresponds to the optimum for $N_{\text{req}} \approx 50\,000$. If the required number of plates is higher, this line quickly goes through the roof and larger particles should be selected. On the left-hand side of the figure, we see that smaller particles are favourable. Even particles smaller than 1.6 µm in diameter are attractive for fast separations, as long as we are able to avoid radial temperature gradients in the column due to frictional heating (see Section 3.3.2) and extra-column band broadening (Section 3.3.3). These factors have not been taken into account in constructing Figure 3.54 and Figure 3.55.

Figure 3.55, which is affectionately known as a **Poppe plot,** contains a lot of information on LC systems, but it is not the clearest presentation. For example, the analysis time is the product of the

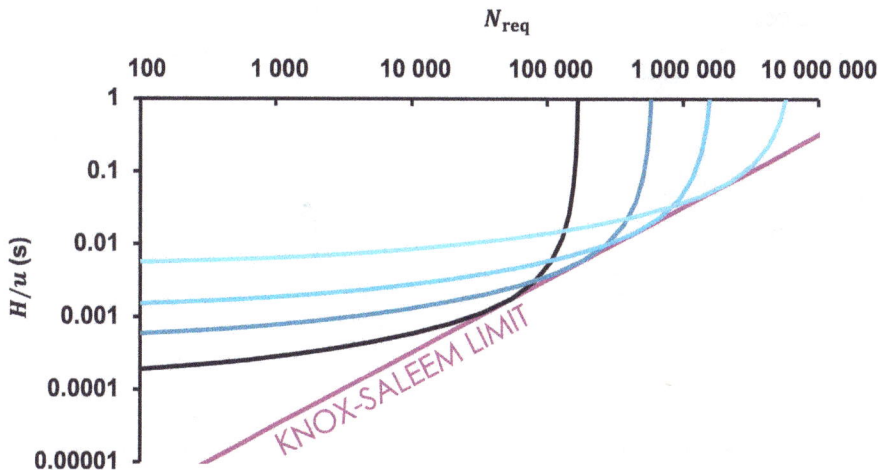

Figure 3.55 Classical Poppe plot,[100] in which the time per plate (H/u) is plotted against the difficulty of the separation (N_{req}). The pink line is the Knox-and-Saleem limit that corresponds to Figure 3.54. $\psi = 1000$, $D_m = 10^{-9}$ m^2 s^{-1}, $\eta = 0.001$ Pa s, and $\Delta P = 100$ MPa. The blue lines (from light to dark) denote particle diameters of 10, 5, 3, and 1.6 μm.

number of plates required and the time per plate. A line representing a constant time is a downward diagonal in the figure. More recently, researchers have proposed other types of kinetic plots, and they have been used for a great variety of chromatographic systems. Ken Broeckhoven and Gert Desmet from the VUB (Brussels, Belgium) are arguably the most prolific experts in this field.[101,102]

As a first example, we may replot the data of Figure 3.55 in a more practical way by relating the required number of plates (separation complexity) directly to the required analysis time, which is given by $t_{0,req} = N_{req} \cdot H/u$. This result is shown in Figure 3.56.

The information from Figures 3.55 and 3.56 is the same, but the latter figure is more accessible. Again, it is clear that fast analysis benefits from smaller particles (darker curves), whereas high plate counts require larger particles. It should be noted that the figure still applies to good columns (with $h_{min} = 1.8$) and high pressure ($\Delta P = 100$ MPa). Under such conditions, 1.6 μm particles may be attractive until about 50 000 plates, above which 3 μm particles become more time efficient. The 3 μm particles are attractive until about $N_{req} = 200\,000$.

Figure 3.56 features so-called kinetic-performance limits (KPLs), valid for columns of variable length operated at the maximum

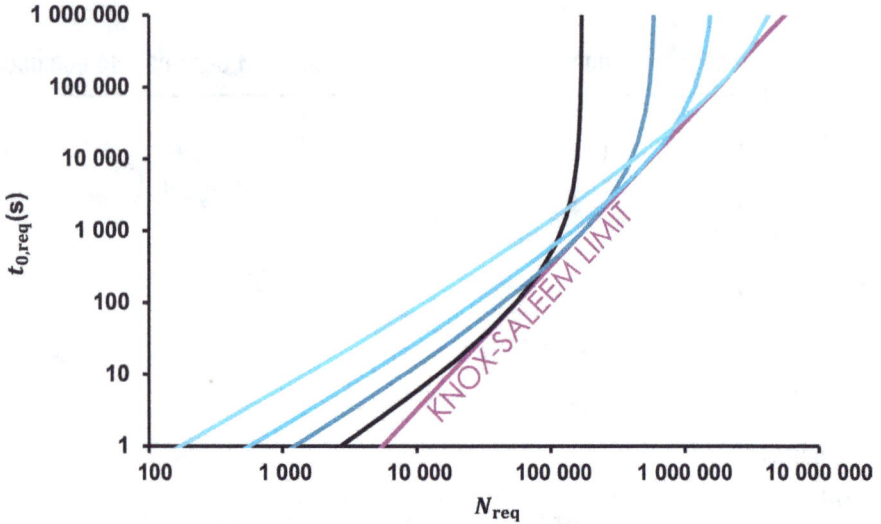

Figure 3.56 Kinetic plot in which the required analysis time (t_{req}) is plotted against the required number of plates (N_{req}) at maximum UHPLC pressure ($\Delta P = 100$ MPa). The blue lines (from light to dark) represent particle diameters of 10, 5, 3, and 1.6 μm. The pink line is the Knox-and-Saleem limit (eqn (3.66)). Parameter values are as given in Figure 3.55.

pressure drop. In fact, this is how the lines are easily established for a given particle size and ΔP_{max}. Assume a length N; calculate u_0 from the Darcy equation; calculate H from the plate height equation; calculate the plate count from $N = L/H$; and calculate the dead time from $t_0 = L/u_0$. A similar procedure can be followed for any column for which the permeability factor (ψ) and H vs. u data have been established. This may be a packed column, a monolithic column, a pillar-array column, *etc.*

Broeckhoven and Desmet[102] indicate a simple method to obtain points on the KPL line from experimental data points obtained at a lower pressure, based on the ratio λ of the maximum pressure and the experimental pressure.

$$\lambda = \frac{\Delta P_{max}}{\Delta P_{exp}} \tag{3.67}$$

If the linear velocity is assumed constant, while the length is assumed to increase to reach the maximum pressure drop across the column, the plate count increases proportionally with λ.

$$N_{KPL} = \lambda N_{exp} \tag{3.68}$$

This is also true for the hold-up time

$$t_{0,KPL} = \lambda \cdot t_{0,exp} \tag{3.69}$$

and for the isocratic retention time.

$$t_{R,KPL} = \lambda \cdot t_{R,exp} \tag{3.70}$$

Eqn (3.70) is also valid under gradient conditions, provided that the duration of the gradient is adapted accordingly ($t_{G,KPL} = \lambda \cdot t_{G,exp}$) and the dwell time and the initial hold-up time are accounted for. If this is the case, the peak capacity can also be found from

$$n_{p,KPL} = \sqrt{\lambda}\left(n_{p,exp} - 1\right) + 1 \tag{3.71}$$

Plotting the peak capacity $n_{p,KPL}$ against the time needed ($t_{G,KPL}$) is a good way to display the kinetic performance under gradient conditions.

The above transformations allow the comparison of different systems based on a limited number of experiments. To accentuate the difference between different systems, the Desmet group often show plots of t_0/N^2 as shown in Figure 3.57. In such a plot, the Knox-and-Saleem limit (pink line) becomes horizontal. Low values are favourable and it is very clear which particle diameter is most attractive for which plate count (using the plate height curve, permeability factor, and mobile-phase viscosity as used by Poppe[100] and a maximum pressure of 100 MPa).

See the website for a calculation tool: ass-ets.org

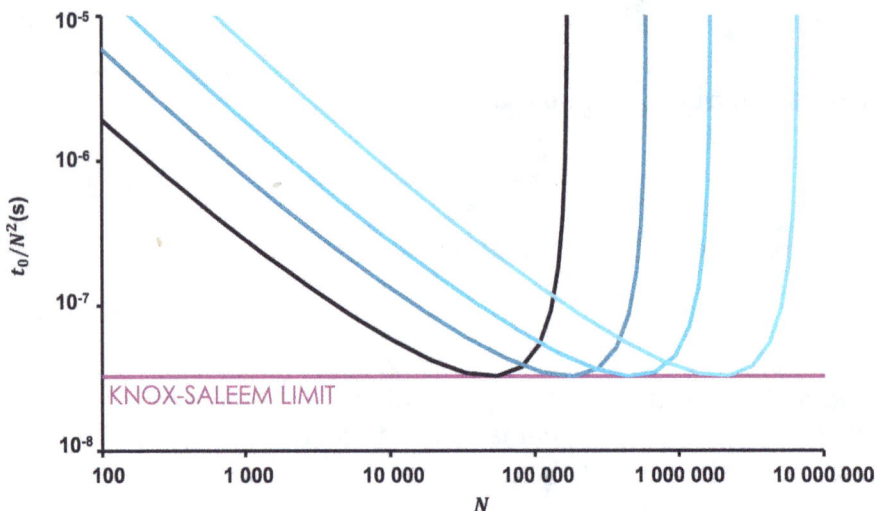

Figure 3.57 Kinetic plot in which the analysis time divided by the square of the plate count (t_0/N^2) is plotted against the number of plates (N) at maximum UHPLC pressure (ΔP = 100 MPa). The blue lines (from light to dark) represent particle diameters of 10, 5, 3, and 1.6 µm. The pink line is the Knox-and-Saleem limit (eqn (3.66)). Parameter values are as given in Figure 3.55.

3.17   Gradient Elution – A Closer Look (A)

The basic principles of gradient-elution LC are introduced in Module 3.4. In this module, we will discuss some aspects of gradient-elution LC in more detail.

3.17.1   Deriving Retention Equations

As explained in Section 3.4.2 (Figure 3.19), two factors determine the gradient-elution process. These are (i) the gradient program as it is entered by an analyst in the control software of the instrument, and (ii) the retention model, which describes retention as a function of the programmed variable(s), such as mobile-phase composition, temperature, and pH. We will consider a single programmed variable, the composition φ, but the other variables (or combinations thereof) can be treated in an analogous manner. The gradient program can be described as a function

$$\varphi(t) = \begin{cases} \varphi_{\text{init}} & \text{for } t < 0 \\ f(t) & \text{for } 0 \le t \le t_{\text{G}} \\ \varphi_{\text{final}} & \text{for } t > t_{\text{G}} \end{cases} \qquad (3.72)$$

where, as before, φ_{init} is the initial composition, φ_{final} is the final composition, and t_{G} is the duration of the gradient program. The time is (mathematically) negative before the gradient arrives at the column inlet (see below). The actual gradient arrives at the column inlet after a dwell time t_{D} and an initial hold time t_{init} programmed by the analyst. Moreover, the composition at a specific distance z into the column lags behind by a further time z/u that it takes for the mobile phase to migrate through the column with a velocity u. This yields

$$\varphi(z,t) = f\left(t - t_{\text{D}} - t_{\text{init}} - \frac{z}{u}\right) \qquad (3.73)$$

To express t in terms of φ, we need to introduce the inverse function

$$t = t_{\text{D}} + t_{\text{init}} + \frac{z}{u} + f^{-1}(\varphi) \qquad (3.74)$$

the derivative of which is

$$dt = \frac{dz}{u} + d\{f^{-1}(\varphi)\} \qquad (3.75)$$

Now we turn our attention to the analyte i, the retention of which is described by a model $k_i(\varphi)$ and the migration velocity of which through the column is

$$\frac{dz}{dt} = \frac{u}{1 + k_i(\varphi)} \qquad (3.76)$$

Eliminating dt from eqn (3.75 and 3.76) yields

$$\frac{d\{f^{-1}(\varphi)\}}{k_i(\varphi)} = \frac{dz}{u} \qquad (3.77)$$

To integrate this equation from the time the analyte enters the column (at $z = 0$ and $t = 0$) until it leaves the column (at $z = L$, $z/u = t_0$ and $t = t_{R,i}$), we find from eqn (3.74) that $f^{-1} = \varphi$ runs from $(-t_D - t_{init})$ to $(t_{R,i} - t_D - t_{init} - t_0)$. Between $t = -t_D - t_{init}$ and $t = 0$, we have $\varphi = \varphi_{init}$ (see eqn (3.72)) and thus $k(\varphi) = k_{i,init}$. If we introduce a corrected time $t''_{R,i} = t_{R,i} - t_D - t_{init} - t_0$, which is the time the analyte has been exposed to the gradient when it leaves the column, we obtain

$$\int_{-t_D-t_{init}}^{0} \frac{d[f^{-1}(\varphi)]}{k_{i,init}} + \int_{0}^{t''_{R,i}} \frac{d[f^{-1}(\varphi)]}{k_i(\varphi)} = \int_{0}^{L} \frac{dz}{u} = t_0 \tag{3.78}$$

As $k_{i,init}$ is a constant, the first integral yields a constant, and we obtain

$$\int_{0}^{t''_{R,i}} \frac{d[f^{-1}(\varphi)]}{k_i(\varphi)} = t_0 - \frac{t_D + t_{init}}{k_{i,init}} \tag{3.79}$$

Eqn (3.79) is the fundamental gradient-elution formula. It can be solved (sometimes) analytically or (always) numerically if the gradient program, $\varphi(t)$, and the retention model for the analyte, $k_i(\varphi)$, are known.

If the analyte elutes before the onset of the gradient, the second integral in eqn (3.78) does not come into play, and we have

$$\int_{-t_D-t_{init}}^{t''_{R,i}} \left[\frac{d[f^{-1}(\varphi)]}{k_{i,init}}\right] = t_0 \tag{3.80}$$

which yields

$$t_{R,i} = t_0(1 + k_{i,init}) \tag{3.81}$$

This is the same result as obtained for isocratic elution at $\varphi = \varphi_{init}$, which is correct, because the migration of the analyte is not affected by the impending gradient.

If the analyte elutes after the completion of the gradient, we must split the integral of eqn (3.79), and we obtain

$$t_{R,i} = k_{i,\text{final}} \cdot \left\{ t_0 - \int_0^{t_G''} \frac{d[f^{-1}(\varphi)]}{k_i(\varphi)} - \frac{t_D + t_{\text{init}}}{k_{i,\text{init}}} \right\} + t_G \tag{3.82}$$

where $t_G'' = t_G - t_D - t_{\text{init}} - t_0$ and $k_{i,\text{final}}$ is the retention factor of the analyte at the final composition (φ_{final}).

We will now limit ourselves to linear gradients of the form

$$\varphi(t) = \varphi_{\text{init}} + \frac{\varphi_{\text{final}} - \varphi_{\text{init}}}{t_G} t = \varphi_{\text{init}} + \frac{\Delta\varphi}{t_G} t = \varphi_{\text{init}} + Bt \tag{3.83}$$

This yields with eqn (3.73)

$$\varphi(z,t) = f\left(t - t_D - t_{\text{init}} - \frac{z}{u}\right) = \varphi_{\text{init}} + B\left(t - t_D - t_{\text{init}} - \frac{z}{u}\right) \tag{3.84}$$

from which we obtain

$$t - t_D - t_{\text{init}} - \frac{z}{u} = f^{-1}(\varphi) = \frac{\varphi - \varphi_{\text{init}}}{B} \tag{3.85}$$

and thus

$$f^{-1}(\varphi) = \frac{\varphi - \varphi_{\text{init}}}{B} \tag{3.86}$$

with as its derivative

$$d[f^{-1}(\varphi)] = \frac{1}{B}d\varphi \tag{3.87}$$

Inserting this in the fundamental eqn (3.79), we find for the specific case of a linear gradient

$$\int_0^{t_{R,i}''} \left[\frac{d[f^{-1}(\varphi)]}{k_i(\varphi)}\right] = \frac{1}{B} \int_{\varphi_{\text{init}}}^{\varphi_{\text{init}}+Bt_{R,i}''} \left[\frac{d\varphi}{k_i(\varphi)}\right] = t_0 - \frac{t_D + t_{\text{init}}}{k_{i,\text{init}}} \tag{3.88}$$

Eqn (3.88) is valid for a simple linear gradient for any kind of retention model, *i.e.* any function $k\varphi$. The log-linear model $(\ln k = \ln k_0 - S'\varphi)$ can be rewritten as

$$k = k_0 e^{-S'\varphi} \tag{3.89}$$

so that

$$\frac{1}{B} \int_{\varphi_{init}}^{\varphi_{init}+Bt''_{R,i}} \left[\frac{d\varphi}{k_i(\varphi)}\right] = \frac{1}{Bk_0} \int_{\varphi_{init}}^{\varphi_{init}+Bt''_{R,i}} [e^{S'\varphi}d\varphi]$$

$$= \frac{1}{Bk_0 S'} [e^{S'\varphi}]_{\varphi_{init}}^{\varphi_{init}+Bt''_{R,i}} = t_0 - \frac{t_D + t_{init}}{k_{i,init}} \tag{3.90}$$

after substitution of the integration limits and rearranging, this yields

$$t_{R,i} = \frac{1}{S'B} \ln\left\{1 + S'Bk_{init}\left[t_0 - \frac{t_D + t_{init}}{k_{init}}\right]\right\} + t_0 + t_D + t_{init} \tag{3.91}$$

or

$$t_{R,i} = \frac{t_G}{S'\Delta\varphi} \ln\left\{1 + S'\frac{\Delta\varphi}{t_G}k_{init}\left[t_0 - \frac{t_D + t_{init}}{k_{init}}\right]\right\} + t_0 + t_D + t_{init} \tag{3.92}$$

which is identical to eqn (3.18).

3.17.2   Gradient Scanning

In the previous section, we have seen that, if the solvent program is accurately known (and accurately delivered by the instrument (see Section 3.17.4), the retention time of an analyte can be calculated, whether it is eluted before, during or after the actual gradient, provided that a retention model, $k_i(\varphi)$, is available. The reverse is also true. If the solvent program is known and retention times are available, then the parameters of the retention model can be computed. This is known as **gradient scanning** or the use of scouting gradients to establish retention parameters. As illustrated in Figure 3.58, retention-time prediction (left) and gradient scanning (right) are each other's complement. In both cases, gradient-elution

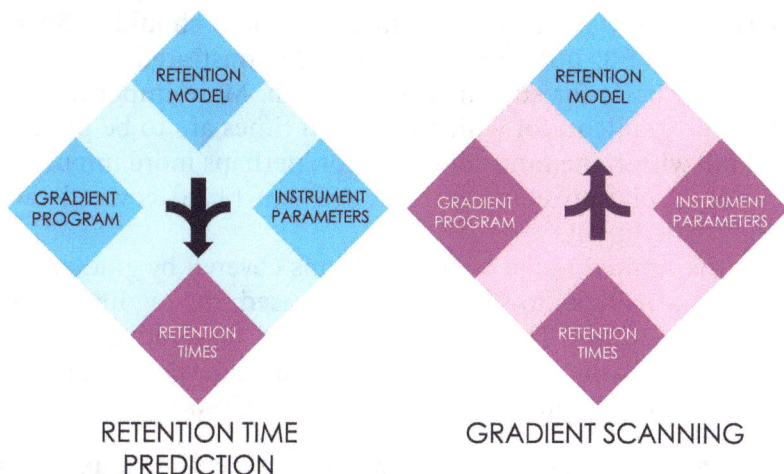

RETENTION TIME   GRADIENT SCANNING
PREDICTION

Figure 3.58 Figure illustrating the complementarity of retention-time prediction and scanning experiments with gradient-elution LC.

theory (Section 3.17.1) is required, and the solvent program needs to be accurately known, as well as instrument parameters, such as the dwell time and the flow rate. In principle, two retention times suffice for determining the parameters of a linear or linearized retention model (such as the log-linear or log–log models, see Sections 3.8.3.1 and 3.8.3.2, respectively), and three retention times suffice for determining the model parameters of the quadratic (Section 3.8.3.3), mixed-mode (Section 3.8.3.4), or Neue–Kuss (Section 3.8.3.5) models. According to a study by Vivó-Truyols *et al.*,[103] gradient scans only provide information across a narrow range of compositions. As a consequence, they found considerable uncertainty in determining three-parameter retention models from gradient-elution data. Conversely (see Figure 3.58), this suggests that two-parameter models (linear or linearized) will usually suffice for predicting gradient-elution retention data. The input data should be recorded using gradients with different effective slopes ($b = SBt_0$). Exact equations to obtain the model parameters from the retention data are typically not available, but numerical solutions can easily be found using widely available spreadsheet tools.

See the website for example calculations: ass-ets.org

Conventional wisdom suggests that the slopes should differ by at least a factor of 3, but den Uijl et al.[104] demonstrated that there is no scientific basis for such a rule of thumb. More importantly, the slope of the gradient, for which retention times are to be predicted, should fall within the range of slopes (or, perhaps more importantly, the range of elution compositions, see eqn (3.25) covered by the scanning experiments.

Given the limited range of compositions covered by gradients and the limited sensitivity to the exact model used, the log-linear model may usually suffice. The recommendations for gradient-scanning experiments with the goal of establishing the model parameters ($\ln k_{0,i}$ and S'_i) for analyte i (and other analytes) are as follows.

- Make sure that you have suitably accurate values of the instrument parameters available (V_D, V_0; by expressing these in terms of volumes, you can find the values at any flow rate from $t_D = V_D/F$ and $t_0 = V_0/F$). If not, measure them.
- Ensure that the analyte is eluted within the linear range of the gradient, i.e.

$$t_0 + t_D + t_{init} \leq t_{R,i} \leq t_0 + t_D + t_{init} + t_G \qquad (3.93)$$

- This is best done by running a full-scope gradient, from 0% (or 5%) to 100% B in the first gradient. Based on the results of this first experiment, you may narrow down the composition range in the second gradient.
- You may vary the effective gradient slope ($b = S'Bt_0 = S'\Delta\varphi t_0/t_G$) in several ways, i.e. by varying the duration of the linear part of the gradient (t_G), while keeping the composition span ($\Delta\varphi$) constant, or vice versa. You may also change the flow rate (F), but be aware that this changes the values of t_D and t_0 in your calculations (see below).
- Make sure you select the corresponding peaks for the same analyte ("peak matching" or "peak tracking", see also Module 10.6).
- Use the following two equations for the two gradients denoted with the prescripts 1 and 2. Be aware that t_D and t_0 may be different for the two gradients if you vary the flow rate and that $^1k_{init}$ will vary if you vary the initial composition (φ_{init}).

$$^1t_{R,i} = \frac{^1t_G}{S_i'\,^1\Delta\varphi}\ln\left\{1 + S_i'\frac{^1\Delta\varphi}{^1t_G}k_0 e^{-S_i'^1\varphi_{init}}\left[^1t_0 - \frac{^1t_D + ^1t_{init}}{k_0 e^{-S_i'^1\varphi_{init}}}\right]\right\} + ^1t_0 + ^1t_D + ^1t_{init} \quad (3.94)$$

$$^2t_{R,i} = \frac{^2t_G}{S_i'\,^2\Delta\varphi}\ln\left\{1 + S_i'\frac{^2\Delta\varphi}{^2t_G}k_0 e^{-S_i'^2\varphi_{init}}\left[^2t_0 - \frac{^2t_D + ^2t_{init}}{k_0 e^{-S_i'^2\varphi_{init}}}\right]\right\} + ^2t_0 + ^2t_D + ^2t_{init} \quad (3.95)$$

- Find the two unknowns (S_i' and $k_{0,i}$) through a numerical tool (*e.g.* Excel-Solver).

3.17.3 Multi-segment Gradients

Gradients are not limited to simple, one-segment linear gradients as discussed so far in this module. Numerous different gradient profiles are shown in Figure 3.19 (Section 3.4.1). For example, in IEC, retention varies linearly with the logarithm of the counterion concentration, and a linear-solvent strength gradient, in which by definition ln k decreases linearly with time, is a convex gradient in which the counterion concentration increases exponentially (see Figure 3.20). Modern LC systems do not always allow gradient segments of different (convex or concave) shapes, but they do allow gradients composed of multiple linear segments. In this way, a convex gradient can be mimicked by, for example, a slow segment, followed by a faster segment, and completed with a fast segment.

Bos *et al.*[105] have developed retention equations for gradients composed of multiple linear segments. The result for the retention time $[t_{R,i}]_n$ of a peak eluting during the nth segment is

$$\begin{aligned}
[t_{R,i}]_n = {} & \frac{t_{G,n}}{\Delta\varphi_n S_i'}\ln\left[1 + \frac{\Delta\varphi_n}{t_{G,n}}S_i' k_{init,n}\right. \\
& \times \left(t_0 - \frac{t_{init,1} + t_D}{k_{init,i,1}} - \frac{t_{init,2}}{k_{init,i,2}} - \cdots - \frac{t_{init,n}}{k_{init,i,n}}\right. \\
& + \frac{t_{G,1}}{\Delta\varphi_1 S_i'}\left(\frac{1}{k_{init,i,1}} - \frac{1}{k_{init,i,2}}\right) + \frac{t_{G,2}}{\Delta\varphi_2 S_i'}\left(\frac{1}{k_{init,i,2}} - \frac{1}{k_{init,i,3}}\right) \\
& + \cdots + \left.\left.\frac{t_{G,n-1}}{\Delta\varphi_{n-1}S_i'}\left(\frac{1}{k_{init,i,n-1}} - \frac{1}{k_{init,i,n}}\right)\right)\right] \\
& + t_0 + t_D + t_{init,1} + t_{init,2} + \cdots + t_{init,n-1} \\
& + t_{init,n} + t_{G,1} + t_{G,2} + \cdots + t_{G,n-1}
\end{aligned} \quad (3.96)$$

where $t_{G,n}$ is the duration of the nth gradient segment and $\Delta\varphi_n = \varphi_{final,n} - \varphi_{init,n}$ is the composition span of the nth segment. $t_{init,n}$ is the initial time programmed at the start of segment n, which in practice is often equal to zero. $k_{init,i,n} = k_i(\varphi_{init,n})$ is the retention factor of analyte i at the start of the nth segment.

3.17.4   Gradient Deformation

So far, we have assumed that a gradient program is delayed prior to its arrival at the column inlet by a dwell time t_D, at any position z in the column by a time equal to $t_D + z/u$, and ultimately at the end of the column by a time $t_D + t_0$. We have ignored any deformation of the gradient, although we already noted in Section 3.3.1.3 that the indispensable mixers in gradient-elution LC systems cause deformation of the gradient (see Figure 3.15). The behaviour of a mixture can be characterized by an (exponential-decay) response function. In a similar manner, a response function can be determined to describe how an imposed gradient is deformed in an LC system. This is illustrated in Figure 3.59, which shows the observed gradients and the calculated response functions for a high-pressure-mixing ("binary pump") system with a small dwell volume (light-blue lines) and a low-pressure-mixing ("quaternary pump") system characterized by a large dwell volume (pink lines).

Gradient profiles, like those shown on the left-hand side of Figure 3.59, can easily be measured using a UV detector by removing the column from the system and using water as solvent A and water containing 0.1% acetone as solvent B.[105] Niezen *et al.*[106] measured the actual profiles at high pressures using a capacity-coupled contactless conductivity detector (C4D) and demonstrated that the observed profiles varied, depending on the solvent combinations used (involving water, methanol, ACN, and THF), as well as on the flow rates and gradient durations. The C4D, which was operated around a narrow piece of fused-silica tubing (75 μm i.d.; 0.05 μL detector-cell volume), yielded a nearly linear response to changes in the mobile-phase composition, without adding any additives (tracers).

Bos *et al.*[105] found that a so-called stable function adequately described the response function for a water-to-water gradient, but Niezen *et al.*[106] described the response functions for hydro-organic gradients as a weighted sum of two such stable functions. Generally, the greatest distortion occurs with very fast gradients. The

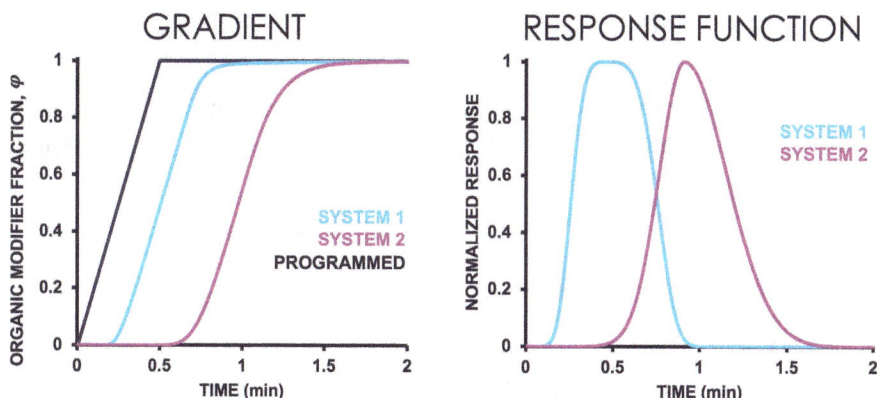

Figure 3.59 Illustration of how the imposed gradient (left panel, black line) is deformed on two different instruments. The deformation can be characterized by a response function (right panel). Light-blue lines (System 1): high-pressure-mixing ("binary pump") system with a small dwell volume; pink lines (System 2): low-pressure-mixing ("quaternary pump") system with a large dwell volume. Measurements are described in ref. 106.

ratio between the dwell volume ($V_D = t_D \cdot F$), which is mainly determined by the solvent-delivery system, and the gradient volume ($V_G = t_G \cdot F$) is critical for the extent of gradient deformation observed. Accurate prediction of retention times using simple equations for linear gradients is still possible if the distorted profile is seen as the combination of many linear segments (500 in ref. 106) and eqn (3.96) is applied. The response–function approach does so far not take into account deformations induced by the column. In principle, it is possible to study such deformations by subtracting the gradient profiles measured using the C4D signals before and after the column, for the same gradient (on the same instrument). A column with a substantial number of theoretical plates is not expected to give rise to significant axial dispersion that would lead to distortion of the gradient. However, one of the eluent components (*e.g.* an organic modifier in water) may adsorb (or absorb) onto the column to a significant extent. This may require a different type of correction, as the gradient in eqn (3.79) would be affected by composition changes along the length of the column.

Acknowledgements

Prof. Sebastiaan Eeltink, Prof. James Grinias, and Prof. Ken Broeckhoven are acknowledged for their extensive review of this chapter and for fruitful discussions.

Recommended Reading

For more about contemporary liquid chromatography see L. Nováková, M. Douša, P. Česla and J. Urban, *Modern HPLC separations in theory and practice*, 2024, ISBN: 9788011044725.

References

1. L. Redón, X. Subirats and M. Rosés, *Molecules*, 2023, **28**, 1372.
2. F. Gritti, Y. Kazakevich and G. Guiochon, *J. Chromatogr. A*, 2007, **1161**, 157–169.
3. G. E. Berendsen, P. J. Schoenmakers, L. de Galan, G. Vigh, Z. Varga-puchony and J. Inczédy, *J. Liq. Chromatogr.*, 1980, **3**, 1669–1686.
4. C. A. Rimmer, C. R. Simmons and J. G. Dorsey, *J. Chromatogr. A*, 2002, **965**, 219–232.
5. F. Gritti and M. F. Wahab, *LC-GC Eur.*, 2018, **31**, 90–101.
6. M. F. Wahab, D. C. Patel, R. M. Wimalasinghe and D. W. Armstrong, *Anal. Chem.*, 2017, **89**, 8177–8191.
7. A. M. Fermier and L. A. Colón, *J. Microcolumn Sep.*, 1998, **10**, 439–447.
8. A. Berthod, *J. Chromatogr. A*, 1991, **549**, 1–28.
9. B. W. J. Pirok, S. J. M. Hoppe, M. Chitty, E. Welch, T. Farkas, S. van der Wal, R. Peters and P. J. Schoenmakers, *J. Chromatogr. A*, 2017, **1486**, 96–102.
10. M. Matyska and J. Pesek, *Separations*, 2019, **6**, 27.
11. K. D. Wyndham, J. E. O'Gara, T. H. Walter, K. H. Glose, N. L. Lawrence, B. A. Alden, G. S. Izzo, C. J. Hudalla and P. C. Iraneta, *Anal. Chem.*, 2003, **75**, 6781–6788.
12. M. T. Gilbert, J. H. Knox and B. Kaur, *Chromatographia*, 1982, **16**, 138–146.
13. M. Russo, M. R. T. Camillo, F. Rigano, P. Donato, L. Mondello and P. Dugo, *J. Chromatogr. A*, 2024, 464728.
14. T. Zhang, W. Wang and M. Wuhrer, *Anal. Chem.*, 2024, **96**, 8942–8948.
15. O. Núñez, K. Nakanishi and N. Tanaka, *J. Chromatogr. A*, 2008, **1191**, 231–252.
16. T. L. Chester, *Anal. Chem.*, 2013, **85**, 579–589.
17. S. Eeltink, B. Wouters, G. Desmet, M. Ursem, D. Blinco, G. D. Kemp and A. Treumann, *J. Chromatogr. A*, 2011, **1218**, 5504–5511.
18. I. Nischang and T. J. Causon, *TrAC, Trends Anal. Chem.*, 2016, **75**, 108–117.
19. J. Urban, *J. Sep. Sci.*, 2016, **39**, 51–68.
20. H. Lin, L. Chen, J. Ou, Z. Liu, H. Wang, J. Dong and H. Zou, *J. Chromatogr. A*, 2015, **1416**, 74–82.

21. M. Callewaert, J. O. De Beeck, K. Maeno, S. Sukas, H. Thienpont, H. Ottevaere, H. Gardeniers, G. Desmet and W. De Malsche, *Analyst.*, 2014, **139**, 618–625.
22. M. Callewaert, G. Desmet, H. Ottevaere and W. De Malsche, *J. Chromatogr. A*, 2016, **1433**, 75–84.
23. W. De Malsche, J. Op De Beeck, S. De Bruyne, H. Gardeniers and G. Desmet, *Anal. Chem.*, 2012, **84**, 1214–1219.
24. C. Salmean and S. Dimartino, *Chromatographia*, 2019, **82**, 443–463.
25. F. Gritti and S. Nawada, *J. Sep. Sci.*, 2022, **45**, 3232–3240.
26. F. Matheuse, K. Vanmol, J. Van Erps, W. De Malsche, H. Ottevaere and G. Desmet, *J. Chromatogr. A*, 2022, **1663**, 462763.
27. B. A. Rogers, Z. Wu, B. Wei, X. Zhang, X. Cao, O. Alabi and M. J. Wirth, *Anal. Chem.*, 2015, **87**, 2520–2526.
28. S. Nawada and F. Gritti, *J. Chromatogr. A*, 2021, **1653**, 462357.
29. K. Kitazoe, Y.-S. Park, N. Kaji, Y. Okamoto, M. Tokeshi, K. Kogure, H. Harashima and Y. Baba, *PLoS ONE*, 2012, **7**, e39057.
30. C.-C. Hong, J.-W. Choi and C. H. Ahn, Lab Chip, 2004, **4**, 109.
31. M. A. Ianovska, P. P. M. F. A. Mulder and E. Verpoorte, *RSC Adv.*, 2017, **7**, 9090–9099.
32. B. W. J. Pirok, A. F. G. Gargano and P. J. Schoenmakers, *J. Sep. Sci.*, 2018, **41**, 68–98.
33. C. Paul, F. Steiner and M. W. Dong, *LCGC North Am.*, 2019, **37**, 514–521.
34. J. Samuelsson, R. A. Shalliker and T. Fornstedt, *Microchem. J.*, 2017, **130**, 102–107.
35. G. Desmet and K. Broeckhoven, *TrAC, Trends Anal. Chem.*, 2019, **119**, 115619.
36. S. Buckenmaier, C. A. Miller, T. Van De Goor and M. M. Dittmann, *J. Chromatogr. A*, 2015, **1377**, 64–74.
37. J. D. Thompson and P. W. Carr, *Anal. Chem.*, 2002, **74**, 4150–4159.
38. A. Sepsey, D. R. Németh, G. Németh and A. Felinger, *J. Chromatogr. A*, 2018, **1564**, 155–162.
39. Z. Zhai, D. Mavridou, M. Damian, F. G. Mutti, P. J. Schoenmakers and A. F. G. Gargano, *Anal. Chem.*, 2024, **96**, 8880–8885.
40. C. Weise, H. Westphal, R. Warias and D. Belder, *Anal. Bioanal. Chem.*, 2024, **416**, 1023–1031.
41. B. Wouters, E. Davydova, S. Wouters, G. Vivo-Truyols, P. J. Schoenmakers and S. Eeltink, *Lab. Chip.*, 2015, **15**, 4415–4422.
42. E. Davydova, P. J. Schoenmakers and G. Vivó-Truyols, *J. Chromatogr. A*, 2013, **1271**, 137–43.
43. L. R. Snyder and J. W. Dolan, in *High-Performance Gradient Elution: The Practical Application of the Linear-Solvent-Strength Model*, Wiley Blackwell, 2006.
44. J. W. Dolan, J. R. Gant, L. R. Snyder, J. W. Dolan and J. R. Gant, *J. Chromatogr. A*, 1979, **165**, 3–30.
45. J. W. Dolan, J. R. Gant and L. R. Snyder, *J. Chromatogr. A*, 1979, **165**, 31–58.
46. C. Horváth and W. R. Melander, in *High-performance liquid chromatography; Advances and perspectives*, ed. C. Horváth, Academic Press, New York, 1980, vol. **2**, pp. 114–319.
47. T. H. Walter, P. Iraneta and M. Capparella, *J. Chromatogr. A*, 2005, **1075**, 177–183.
48. P. J. Schoenmakers, H. A. H. Billiet and L. de Galan, *J. Chromatogr. A*, 1981, **205**, 13–30.
49. M. Rosés, *J. Chromatogr. A*, 2004, **1037**, 283–298.
50. X. Jiang and P. J. Schoenmakers, *J. Chromatogr. A*, 2002, **982**, 55–68.
51. U. D. Neue, K. Van Tran, P. C. Iraneta and B. A. Alden, *J. Sep. Sci.*, 2003, **26**, 174–186.

52. K. Kimata, K. Iwaguchi, S. Onishi, K. Jinno, R. Eksteen, K. Hosoya, M. Araki and N. Tanaka, *J. Chromatogr. Sci.*, 1989, **27**, 721–728.

53. E. Cruz, M. R. Euerby, C. M. Johnson and C. A. Hackett, *Chromatographia*, 1997, **44**, 151–161.

54. F. Gritti, B. A. Alden, J. McLaughlin and T. H. Walter, *J. Chromatogr. A*, 2023, **1692**, 463828.

55. I. Kohler, M. Verhoeven, R. Haselberg and A. F. G. Gargano, *Microchem. J.*, 2022, **175**, 106986.

56. B. W. J. Pirok, M. J. den Uijl, G. Moro, S. V. J. Berbers, C. J. M. Croes, M. R. van Bommel and P. J. Schoenmakers, *Anal. Chem.*, 2019, **91**, 3062–3069.

57. B. W. J. Pirok, J. Knip, M. R. van Bommel and P. J. Schoenmakers, *J. Chromatogr. A*, 2016, **1436**, 141–146.

58. J. Ståhlberg, *J. Chromatogr. A*, 1999, **855**, 3–55.

59. H. Small, T. S. Stevens and W. C. Bauman, *Anal. Chem.*, 1975, **47**, 1801–1809.

60. P. R. Haddad and P. E. Jackson, in *Ion Chromatography, Principles and Applications*, Elsevier, New York, 1990.

61. Z. Lu, Y. Liu, V. Barreto, C. Pohl, N. Avdalovic, R. Joyce and B. Newton, *J. Chromatogr. A*, 2002, **956**, 129–138.

62. B. K. Głód and M. Baumann, *J. Sep. Sci.*, 2003, **26**, 1547–1553.

63. Y. Wen, R. I. J. Amos, M. Talebi, R. Szucs, J. W. Dolan, C. A. Pohl and P. R. Haddad, *Electrophoresis*, 2019, **40**, 2415–2419.

64. F. Rigano, A. Arigò, M. Oteri, P. Dugo and L. Mondello, *J. Chromatogr. A*, 2021, 461963.

65. K. Valkó, *J. Chromatogr. A*, 2004, **1037**, 299–310.

66. C. Liang and H.-Z. Lian, *TrAC, Trends Anal. Chem.*, 2015, **68**, 28–36.

67. S. Soriano-Meseguer, E. Fuguet, M. H. Abraham, A. Port and M. Rosés, *J. Chromatogr. A*, 2021, 461720.

68. S. Espinosa, E. Bosch and M. Rosés, *J. Chromatogr. A*, 2002, **945**, 83–96.

69. D. R. Stoll, T. A. Dahlseid, S. C. Rutan, T. Taylor and J. M. Serret, *J. Chromatogr. A*, 2021, **1636**, 461682.

70. L. R. Snyder, J. W. Dolan and P. W. Carr, *J. Chromatogr. A*, 2004, **1060**, 77–116.

71. S. C. Rutan, T. Kempen, T. Dahlseid, Z. Kruger, B. Pirok, J. G. Shackman, Y. Zhou, Q. Wang and D. R. Stoll, *J. Chromatogr. A*, 2024, **1731**, 465127.

72. E. Lesellier and C. West, *J. Chromatogr. A*, 2018, **1574**, 71–81.

73. P. J. Schoenmakers, H. A. H. Billiet and R. Tussen, *J. Chromatogr. A*, 1978, **149**, 519–537.

74. A. Sepsey, É. Horváth, M. Catani and A. Felinger, *J. Chromatogr. A*, 2020, 460594.

75. R. M. L. Marques, P. J. Schoenmakers, C. B. Lucasius and L. Buydens, *Chromatographia*, 1993, **36**, 83–95.

76. R. J. Maggs, *Chromatographia*, 1968, **1**, 43–48.

77. K. A. Oudhoff, M. Macka, P. R. Haddad, P. J. Schoenmakers and W. T. Kok, *J. Chromatogr. A*, 2005, **1068**, 183–187.

78. M. Holčapek, R. Jirásko and M. Lísa, *J. Chromatogr. A*, 2012, **1259**, 3–15.

79. J. Snijder, R. J. Rose, D. Veesler, J. E. Johnson and A. J. R. Heck, *Angew. Chem. Int. Ed.*, 2013, **52**, 4020–4023.

80. D. Z. Keifer, E. E. Pierson and M. F. Jarrold, *Analyst*, 2017, **142**, 1654–1671.

81. C. G. De Koster and P. J. Schoenmakers, in *Hyphenations of Capillary Chromatography with Mass Spectrometry*, 2019, pp. 279–295.

82. S. Forcisi, F. Moritz, B. Kanawati, D. Tziotis, R. Lehmann and P. Schmitt-Kopplin, *J. Chromatogr. A*, 2013, **1292**, 51–65.

83. E. Denisov, E. Damoc, O. Lange and A. Makarov, *Int. J. Mass Spectrom.*, 2012, **325**, 80–85.

84. W. Bahureksa, T. Borch, R. B. Young, C. R. Weisbrod, G. T. Blakney and A. M. McKenna, *Anal. Chem.*, 2022, **94**, 11382–11389.
85. T. F. Beskers, M. Brandstetter, J. Kuligowski, G. Quintás, M. Wilhelm and B. Lendl, *Analyst.*, 2014, **139**, 2057.
86. J. Kuligowski, G. Quintás, R. Tauler, B. Lendl and M. De La Guardia, *Anal. Chem.*, 2011, **83**, 4855–4862.
87. I. D. Wilson, in *NMR Spectroscopy in Pharmaceutical Analysis*, Elsevier, 2008, pp. 449–469.
88. W. Hiller, P. Sinha, M. Hehn and H. Pasch, *Prog. Polym. Sci.*, 2014, **39**, 979–1016.
89. J. W. Dolan and L. R. Snyder, in *Troubleshooting LC Systems: A Comprehensive Approach to Troubleshooting LC Equipment and Separations*, Springer Science and Business Media Deutschland GmbH, 1989.
90. D. R. Stoll, *LCGC North Am.*, 2018, **36**, 860–866.
91. T. Müller-Späih, G. Ströhlein, O. Lyngberg and D. Maclean, *Chim. Oggi - Chem. Today*, 2013, **31**, 56–60.
92. C. De Luca, S. Felletti, G. Lievore, A. Buratti, S. Vogg, M. Morbidelli, A. Cavazzini, M. Catani, M. Macis, A. Ricci and W. Cabri, *J. Chromatogr. A*, 2020, **1625**, 461304.
93. A. Berthod, T. Maryutina, B. Spivakov, O. Shpigun and I. A. Sutherland, *Pure Appl. Chem.*, 2009, **81**, 355–387.
94. W. H. Pirkle and T. C. Pochapsky, *Chem. Rev.*, 1989, **89**, 347–362.
95. T. Alvarez-Segura, J. R. Torres-Lapasió, C. Ortiz-Bolsico and M. C. García-Alvarez-Coque, *Anal. Chim. Acta*, 2016, **923**, 1–23.
96. M. J. Ruiz-Ángel, S. Carda-Broch and M. C. García-Álvarez-Coque, *Sep. Purif. Rev.*, 2013, **42**, 1–27.
97. R. E. Ewonde, K. Böttinger, J. De Vos, N. Lingg, A. Jungbauer, C. A. Pohl, C. G. Huber, G. Desmet and S. Eeltink, *Anal. Chem.*, 2024, **96**, 1121–1128.
98. H. Zhou, M. Ye, J. Dong, G. Han, X. Jiang, R. Wu and H. Zou, *J. Proteome Res.*, 2008, **7**, 3957–3967.
99. J. H. Knox and M. Saleem, *J. Chromatogr. Sci.*, 1969, **7**, 614–622.
100. H. Poppe, *J. Chromatogr. A*, 1997, **778**, 3–21.
101. K. Broeckhoven, D. Cabooter, S. Eeltink and G. Desmet, *J. Chromatogr. A*, 2012, **1228**, 20–30.
102. K. Broeckhoven and G. Desmet, *J. Sep. Sci.*, 2021, **44**, 323–339.
103. G. Vivó-Truyols, J. R. Torres-Lapasió and M. C. García-Alvarez-Coque, *J. Chromatogr. A*, 2003, **1018**, 169–181.
104. M. J. den Uijl, P. J. Schoenmakers, G. K. Schulte, D. R. Stoll, M. R. van Bommel and B. W. J. Pirok, *J. Chromatogr. A*, 2021, **1636**, 461780.
105. T. S. Bos, L. E. Niezen, M. J. den Uijl, S. R. A. Molenaar, S. Lege, P. J. Schoenmakers, G. W. Somsen and B. W. J. Pirok, *J. Chromatogr. A*, 2021, **1635**, 461714.
106. L. E. Niezen, T. S. Bos, P. J. Schoenmakers, G.W. Somsen and B. W. J. Pirok, *Anal. Chim. Acta*, 2023, **1271**, 341466.

4 Separation of Large Molecules

This chapter deals with the separation of large (*i.e.* high-molecular-weight) molecules, which include synthetic polymers and bio-macromolecules. The stage is set with a brief introduction to "large molecules" and the various distributions that are encountered. Size-exclusion chromatography (SEC) is discussed as the most important chromatographic technique in this field, but interaction-based ("interactive") separations of (synthetic) polymers are also shown to be relevant. Hydrodynamic chromatography (HDC) and field-flow fractionation (FFF) are specific techniques for separating large molecules and nano-(or sometimes micro-)particles. Finally, liquid-phase light scattering and viscometric detection techniques are discussed (Figure 4.1).

4.1 Introduction (B)

In Chapters 1–3, we have seen how both liquid and gas chromatography can be used to establish the separation of an analyte mixture. In all cases discussed so far, chromatographic separation was based on (differences in) the chemical interaction of analytes. In packed-column liquid chromatography (LC), retention is determined by the balance between interactions of the analyte with the stationary phase and the mobile phase. Analytes enter the pores of the stationary-phase particles and are retained depending on the strength and type of chemical interactions. Elution of analytes is often facilitated by gradually increasing the strength of the mobile phase (gradient elution). In gas chromatography (GC), interactions

Analytical Separation Science
By Bob W. J. Pirok and Peter J. Schoenmakers
© Bob W. J. Pirok and Peter J. Schoenmakers 2025
Published by the Royal Society of Chemistry, www.rsc.org

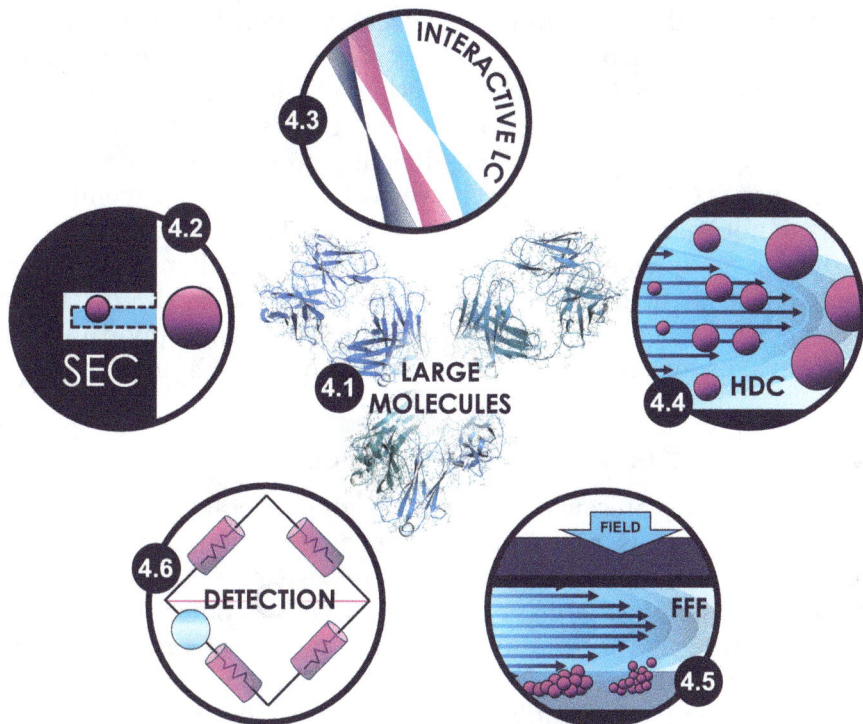

Figure 4.1 Graphical overview of modules in Chapter 4.

with the mobile phase are negligible, and retention is determined by interactions between the analyte molecules and the stationary phase and between analyte molecules mutually (pure-component vapour pressure). These separation mechanisms can therefore collectively be referred to as **interaction chromatography**. However, in our treatment of packed-column LC thus far, we have ignored the possibility that analyte molecules may not be able to (fully) enter the pores of the particles. At low ionic strength, it is possible that ions are excluded from the pores due to electrostatic interactions. Such an ion-exclusion mechanism has been discussed in Section 3.7.4. Molecules may be larger than the diameter of (some of) the pores, causing them to be excluded due to their size. This is the basis for what is known as **size-exclusion chromatography** (SEC, or, under older names, **gel-permeation chromatography** (GPC); and gel-filtration chromatography, GFC).

In this chapter, we will focus on the separation of "large" molecules. We realize, of course, that all molecules are small to the human eye, but the apparent oxymoron "large molecules" is commonly used to refer to high-molecular-weight molecules. These include many types of highly complex analytes that continue to pose serious challenges for scientists. Although interaction LC is often used for the analysis of large molecules, their size results in strong chemical interactions, which significantly affect their behaviour in interaction LC. In this chapter, we will also describe techniques that are not based on molecular interactions. These include chromatographic methods and separation methods other than chromatography, such as **field-flow fractionation** (FFF). Finally, the detection of large molecules also brings unique problems and opportunities, which will be discussed in the final module of this chapter.

4.1.1 Characteristics of Large Molecules

Recognizing the presence of "large" (high-molecular-weight) molecules around us is not difficult. One field of study, proteomics, focuses on all the proteins that form the basis of a biosystem, such as our own body. One class of **proteins** is antibodies, which play a central role in the immune response system of organisms. An example of the latter is immunoglobulin-G antibodies, which are the most common type of antibodies found in blood. These antibodies protect the organism from infections by binding antigens on the surface of pathogens, such as bacteria, viruses and fungi. Figure 4.2 illustrates the structure of an IgG1 monoclonal antibody (immunoglobulin). With a molecular weight of about 150 000 g mol^{-1} (or 150 kDa) and a width of approximately 14.5 nm, these molecules easily fit our description of a large molecule or **macromolecule.**

Such a complex molecule exhibits a great number of properties, and it is difficult to capture its chemistry in a single value or chemical property that can be targeted by a well-defined separation mechanism. The analysis of large proteins is, therefore, not straightforward. For example, the macromolecular representation contains primary, secondary, tertiary and quaternary protein structures. The higher-order structures (not shown in Figure 4.2) may change depending on the surroundings of the protein. In Section 3.15.1, we have discussed hydrophobic interaction chromatography (HIC) as a useful method to separate proteins under so-called native conditions, in which their functional morphology is maintained.

It is hard – but not impossible – to imagine separation mechanisms with the resolving power to achieve selectivity for proteins with very minor differences, such as the presence of small carbohydrate groups ("glycans") added to the macromolecular structure.

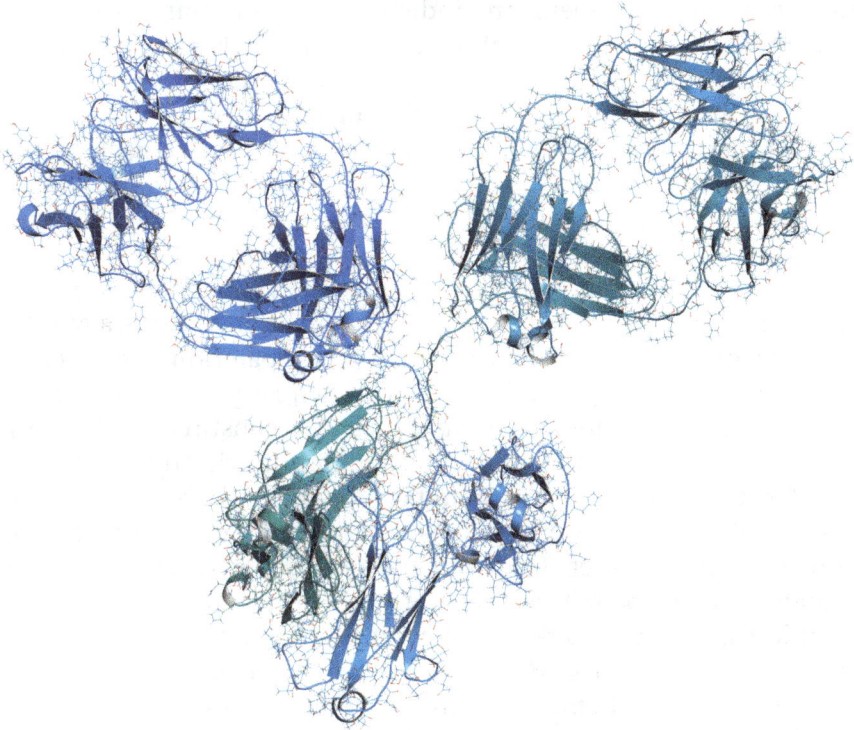

Figure 4.2 Schematic depiction of an IgG1 monoclonal antibody (immuno-globulin).

Because intact proteins are difficult to separate and characterize, scientists often resort to breaking down proteins with enzymes to obtain a number of smaller peptides. The resulting "digests" form a fingerprint of the parent protein. Individual peptides can be identified and used to deduce the presence of specific proteins in the original sample.

There are many other natural polymers. Deoxyribonucleic acid (DNA) is a **polynucleotide** formed by two spirally intertwined chains built up of four different nucleotides. The main building blocks of plants are lignin and cellulose, both organic polymers that contain – next to the inevitable carbon and hydrogen – substantial amounts of oxygen. Cellulose and starch are chain polymers with repeating carbohydrate (D-glucose) units. Cellulose typically consists of very long chains (10 000 units or more). The regular structure and hydrogen bonding groups cause cellulose to be essentially insoluble.

Repetitive polymers, created by connecting large numbers of monomeric units, are also synthesized by people in large quantities.

Such **synthetic polymers** are indispensable for many aspects of contemporary life. Our telephones and computers, and our cars – or, if you prefer, our trains – are full of synthetic polymers. We paint our houses (and coat many other objects, such as solar panels) with synthetic polymers. Synthetic polymers are used to package and conserve our food, *etc.* We cannot do without polymers, but one of the great challenges of the 21st century is that we need to reuse our cups and bottles and recycle all polymeric products that we cannot reuse to achieve a sustainable society. The characterization of synthetic polymers and the necessary transition towards a circular economy pose great challenges to analytical separation scientists.

Hermann Staudinger (Nobel Prize in Chemistry 1953) proved that polymers are large molecules in which the constituent units, the monomers, are all covalently attached. Although "dynamic polymers" exist, in which the subunits are connected by bonds other than covalent bonds, we will consider covalently bonded macromolecules in this chapter. However, the techniques discussed in this chapter can also be relevant for the study of non-covalently bonded complexes, such as drug–protein complexes or aggregates of several large molecules (see *e.g.* Module 4.5, Field-Flow Fractionation).

A polymer is a chemical compound that is composed of one or multiple types of monomers. A synthetic polymer does not consist of molecules with a single, well-defined molecular structure. Instead, distributions of different molecules exist, as illustrated schematically in Figure 4.3. If a single type of monomer is present, it is referred to as a **homopolymer**. Any larger number of types of monomers (indicated by different shapes and colours in Figure 4.3) gives rise to a **copolymer**. Sometimes the word "terpolymer" is used for a polymer that contains three different types of monomers, but this type of counting is not typically used in describing polymers. Copolymers with more than two types of monomers may be called complex copolymers, but the term "complex" is typically used for all kinds of polymers that exhibit more than one **distribution of structures**, such as those illustrated in Figure 4.3.

Polymers may be formed in different ways. Step-growth polymers grow one step at a time. Important examples are polycondensation reactions that lead to the formation of polyesters and polyamides. It is challenging to produce high-molecular-weight polymers in this manner. This can only be achieved by pushing the reaction to full conversion and – in the case of a reaction of a diacid with a dialcohol or diamine – ensuring the exact stoichiometric amounts of the two monomers in the reaction mixture. Even then, relatively

END-GROUP DISTRIBUTION   CHEMICAL COMPOSITION & SEQUENCE DISTRIBUTION

RANDOM COPOLYMER

ALTERNATING COPOLYMER

BLOCK COPOLYMER

MOLAR-MASS DISTRIBUTION   BRANCHING DISTRIBUTION   BLOCK-LENGTH DISTRIBUTION

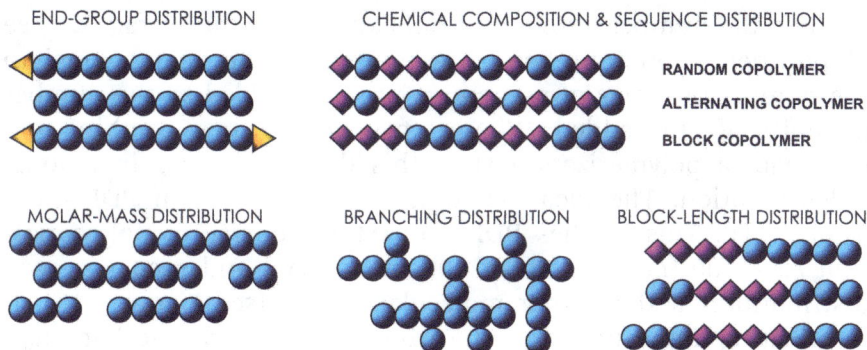

Figure 4.3 Examples of different polymer structure distributions. Elements of different colours and shapes represent different monomers or end groups. Actual synthetic polymers contain many more units than suggested in this schematic picture.

low-molecular-weight polymers tend to be obtained with broad distributions (PDI = $1 + r_{conv}$, where $0 < r_{conv} < 1$ is the degree of conversion), as shown in Figure 4.4A. The **polydispersity** index (PDI; defined below, eqn (4.3)) indicates the width of a distribution, with a lower limit of 1 (extremely narrow). Ring-opening polymerization reactions yield similar widths of distributions.

Free-radical polymerization is a process in which rapid chain growth is started by an *initiation* step, effected by an "initiator" molecule that is usually activated by light (UV polymerization) or by raising the temperature (heat-induced polymerization). While the radical is active on a polymer chain, rapid *propagation* occurs until the radical is deactivated by a *termination* reaction. The resulting polymers again only reach high molecular weight at full conversion, and broad molecular-weight distributions are again obtained PDI = 1 + r_{conv}/2 (see Figure 4.4B). All kinds of monomers with double bonds, such as alkenes, vinyl compounds (including vinyl chloride and styrene), acrylates, *etc.*, polymerize through chain-growth free-radical polymerization.

In contrast to polycondensation and free-radical polymerizations, which tend to approach polymers with PDI = 2 and PDI = 1.5, respectively, at full conversion, "living-polymerization" methods may directly result in narrowly distributed polymers (see Figure 4.4C). In living polymerizations, all chains start growing at the same time, and there are no termination or chain transfer reactions. This is well demonstrated in practice by anionic polymerization. The polymerization is initiated by introducing an anion through, for example, butyl lithium (BuLi), with the number of chains formed equal to the number of moles of BuLi added and the average degree of polymerization (number of monomers in the chain) equal to the number of moles of monomers present divided by the number of moles of BuLi added. Whereas two radicals may combine and terminate a polymerization step, this does not occur in anionic polymerization. The ideal outcome is a polymer with PDI = 1 + $n_p/(n_p + 1)^2$. This implies PDI ≈ 1.01 for a degree of polymerization n_p = 100 (*i.e.* a chain of 100 monomers) and even narrower distributions at the higher molecular weight (see Figure 4.4C). If terminating impurities (oxygen, carbon dioxide, and water) are kept out of the system, the polymers remain active, and a new monomer can be added. In this way, anionic polymerization is used to produce highly controlled diblock or triblock copolymers, such as styrene–butadiene–styrene rubbers.

Distributions are essential properties of most polymers. Some types of polymers, such as DNA and proteins, have a well-defined chain sequence (primary structure). The molecular weight of such polymers is basically well-defined, although it may be altered by attachments to the chain (*e.g.* methylation of DNA or glycation of proteins). DNA and proteins can be seen as high-definition polymers that are essentially monodisperse. This is quite different for natural macromolecules, such as cellulose, and for all synthetic polymers. All of these are polydisperse. For homopolymers, the chain sequence may be well-defined since there is only one type of monomer, but the chain length is not. It typically varies across a broad range. The size

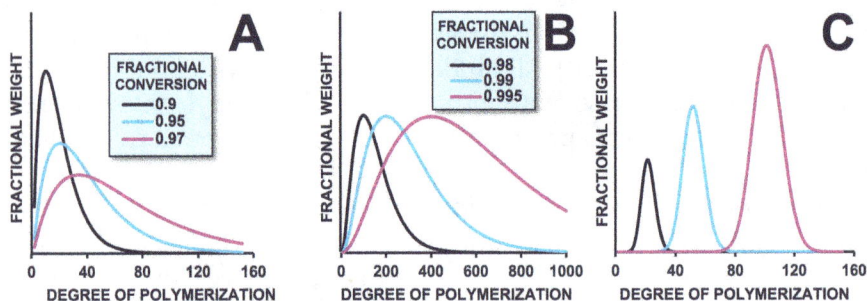

Figure 4.4 Examples of polymer molecular-weight distributions arising from different polymerization methods. (A) Step-growth (polycondensation, ring-opening) polymerization with PDI approaching a value of 2 at full conversion. (B) Chain-growth (addition, free radical) polymerization with PDI approaching a value of 1.5 at full conversion. (C) Anionic polymerization with PDI decreasing with increasing molecular weight.

distribution of polymers is usually characterized by the **molecular-weight distribution** (or, equivalently, the **molar-mass distribution**), and this distribution is in turn characterized by several molecular-weight averages. Molecular weight is typically expressed in atomic mass units or **Daltons** (Da). One Dalton is defined as one-twelfth of the mass of a ^{12}C carbon atom. Molar mass is expressed in g mol^{-1}. For unknown reasons, the International Union of Pure and Applied Chemistry (IUPAC) recommends against the use of "molecular mass". Following this logic, we should speak of weight spectrometry instead of mass spectrometry.

If we divide the molecular-weight distribution into n_f fractions, the number of molecules in fraction i is n_i and the (average) molecular weight in the fraction is M_i. We can then define the **number-average molecular weight** (M_n) as

$$M_n = \frac{\sum_i^{n_f} (n_i M_i)}{\sum_i^{n_f} n_i} \tag{4.1}$$

and the **weight-average molecular weight** (M_w) as

$$M_w = \frac{\sum_i^{n_f} (w_i M_i)}{\sum_i^{n_f} w_i} = \frac{\sum_i^{n_f} (n_i M_i^2)}{\sum_i^{n_f} (n_i M_i)} \tag{4.2}$$

where w_i is the weight of fraction i. M_w is always larger than M_n, and the ratio between the two parameters

$$PDI = \frac{M_w}{M_n} \tag{4.3}$$

is known as the PDI or the dispersity. Although it is the IUPAC-suggested terminology, we avoid the latter and its common symbol D because, in this book, D usually denotes a diffusion coefficient. The PDI is an indication of the width of the distribution. To convert it to the relative standard deviation (rsd) of the distribution, we have

$$rsd = \frac{\sigma_p}{M_n} = \sqrt{PDI - 1} \tag{4.4}$$

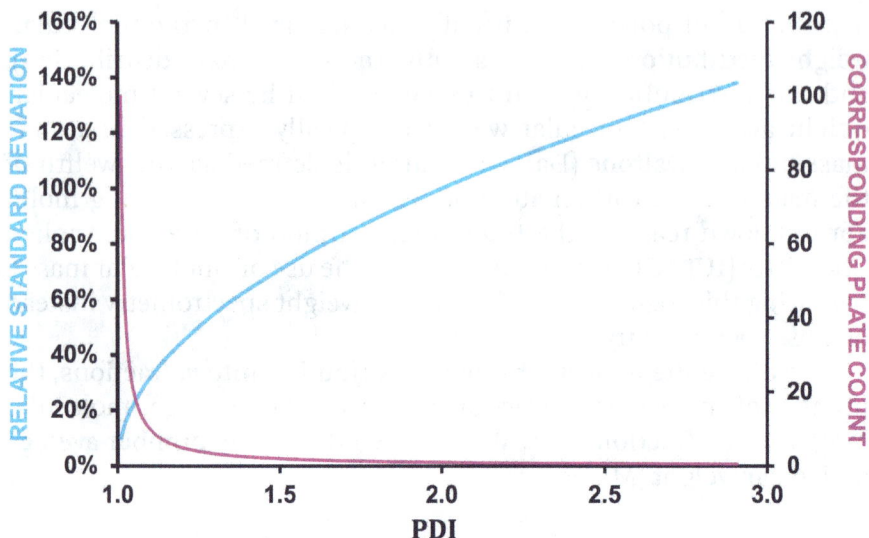

Figure 4.5 Relationship between the relative standard deviation of a distribution and the polydispersity index (PDI) of a polymer. On the right-hand side is shown the number of chromatographic plates corresponding to the distribution.

where σ_p is the standard deviation of the polymer distribution. The relationship is illustrated in Figure 4.5, with an apparent number of plates indicated on the right-hand side to show how chromatographers would perceive a peak that represents the distribution. Narrow polymer standards, used for calibrating SEC experiments (see Module 4.2), have PDI values of 1.05 or 1.1. The distributions of such standards, with rsd values of about 20 and 30%, respectively, appear very broad to the eyes of a chromatographer, as the widths correspond to 20 and 10 plates, respectively.

Polymer chemists refer to narrow distributions with a PDI of 1.2, while most chromatographers would be embarrassed to show a chromatogram with just a handful of plates. We should just be aware that a narrow standard of a polymer with a specified molecular weight of 10 000 and a PDI of 1.05 will show all masses between 6000 and 14 000 in significant concentrations. Calling such standards monodisperse is an oversimplification, which may thwart our understanding of polymer separations.

Polymers may feature many distributions other than the molecular-weight distribution (MWD); some important examples are listed in Table 4.1 (and are schematically illustrated in Figure 4.3). Functional groups or end groups greatly affect polymer properties.

Table 4.1 Overview of the most important distributions that may be encountered in polymers.

Property	Distribution	Abbreviation
Molecular weight (or molar mass)	Molecular-weight distribution (molar-mass distribution)	MWD (MMD)
Functional groups/end groups	Functionality-type distribution	FTD
Chemical composition (different monomers)	Chemical-composition distribution	CCD
Length of (homopolymeric) blocks (sequence distribution)	Block-length distribution (average length of "mini-blocks")	BLD
Branching	Degree-of-branching distribution	DBD
Tacticity (stereoregularity)	Tacticity distribution	TD

Reactive groups allow polymers to be formed or to change properties *in situ*, for example, in glues or construction materials. The chemical composition determines the behaviour of polymers in coatings. The block lengths affect, for example, the morphology and thus the elasticity, hardness, *etc.*, of polymeric materials, while the degree of branching greatly affects the processability. In many polymers, tacticity plays a role. For example, stereoregular (isotactic) polypropene is a strong, solid material, whereas irregular (atactic) polypropene is rubbery and soft. To understand and control the properties of polymers and to ultimately be able to develop and produce tailor-made, sustainable properties, it is essential that the various interdependent distributions can be accurately measured. Polymer separations are indispensable for this goal.

A general concern in any liquid-phase separation is the solubility and continued dissolution of the analytes. Analytes need to be dissolved to an extent appropriate for the analysis. For example, if we need to determine an MWD, there should be no aggregates of molecules in the sample solution. In many cases, we want to avoid analyte precipitation (*e.g.* in capillary connectors). However, in gradient-elution LC methods, precipitation of analytes at the top of the column may be desirable, provided that redissolution is fast when the eluent strength increases. Dissolving high-molecular-weight analytes may be a very slow process, and it may be the most important factor in determining the minimum analysis time. Dissolution may be accelerated by increasing the temperature, at the risk of analyte degradation (*e.g.* oxidation) and solvent evaporation.

4.2 Size-exclusion Chromatography (B)

4.2.1 Fundamental Concept

4.2.1.1 Mechanism

The MWD (or molar-mass distribution (MMD)) is the most important polymer distribution, and SEC is the most important technique to characterize it. This makes SEC the dominant polymer separation technique.[1] It is used ubiquitously in both research and development, as well as in production environments. The basic principle of SEC is simple. The technique employs chromatographic columns packed with porous particles. The chemistry of these particles and the nature of the mobile phase are chosen such that there are no enthalpic interactions between the analyte and the surface (see Section 4.2.2). The mobile phase is a good solvent for the analytes, which implies that the latter migrate with the solvent – with one important caveat. The pores of the packing material are so small relative to the size of the analyte molecules in the solution that the analytes cannot penetrate all the pores, resulting in a smaller volume of the mobile phase available (see Figure 4.6). The mental picture here is that a polymer molecule – depicted as a sphere rather than as a random coil – cannot approach the particle surface or the inner wall of a pore more closely than its radius. For a relatively small polymer molecule and a relatively large pore (Figure 4.6A1), most of the pore volume can still be sampled by (the centre of gravity of) the polymer molecules (darkish blue area), while the polymer molecule is only excluded from a narrow zone (the light blue region). In the case of a narrow pore, the excluded region becomes quite substantial in comparison with the entire pore volume (Figure 4.6B1). A larger molecule can only sample a small fraction of a medium-sized pore (Figure 4.6A2) and may be totally excluded from a narrow pore (Figure 4.6B2). Finally, a very large molecule may be totally excluded from all pores (Figure 4.6A3 and B3).

Each analyte elutes at a time corresponding to its available mobile-phase volume. This means that very large analyte molecules, which are excluded from all the pores because of their size, elute first, followed by molecules that can penetrate part of the pore volume and, finally, small analytes, which can permeate the entire mobile-phase volume and elute at the conventional t_0. Another way to picture this is from the time spent in different zones. In the interstitial space, the mobile phase moves through the column (with

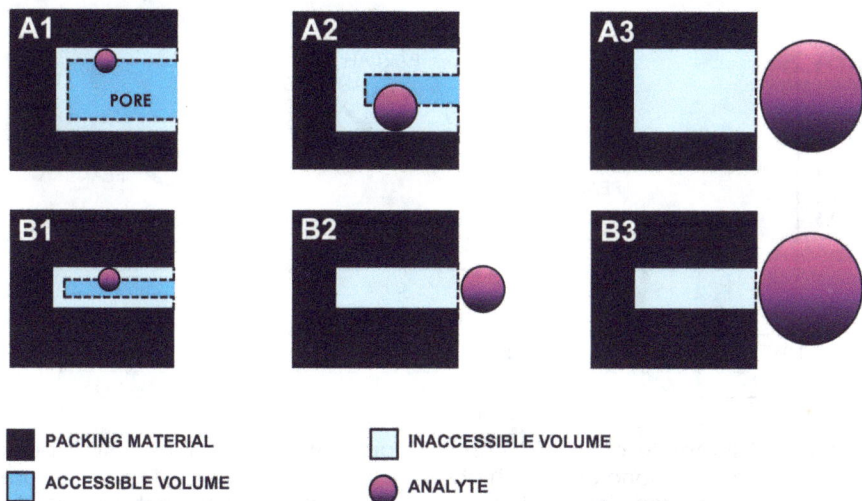

Figure 4.6 Schematic depictions of pores (A: large; B: small) with analytes of different sizes (1: small, 2: medium, 3: large). The small analyte can access a significant fraction of both pore volumes (blue). For the medium analyte, this is only the case for the larger pore (A2), and it is fully excluded from the smaller pore (B2). In both cases, the largest analyte cannot fit in any of the pores and is fully excluded. In SEC, the separation is achieved using a column, for example, with the pores depicted in (A) because the three different analytes can sample different fractions of the pore volume and thus reside there for different times.

average velocity u_{int}), whereas the pores contain a "stagnant mobile phase" (seemingly an oxymoron).

From this description, it follows that all analytes are eluted within a narrow range. At the lower end, this range is confined by the point of **total exclusion** from all the pores. In the case of uniformly sized spherical packing particles, this interstitial volume (V_{excl}) is about 40% of the empty column volume. At the upper end is the sum of the interstitial (exclusion) volume and the total pore volume (V_{pore}) in the column, together known as the **total permeation** volume ($V_{perm} = V_{excl} + V_{pore}$). All molecules larger than any of the pores (and well dissolved and not interacting with the surface) elute at the exclusion limit. If a significant fraction of the analyte molecules fall into this category, this may give rise to a sharp peak at the exclusion volume, superimposed on the apparent size distribution (see Figure 4.7). Solvent molecules, residual monomers and possibly (depending on the pore-size distribution of the column) oligomers and some additives all elute at the permeation limit, as long as interactions

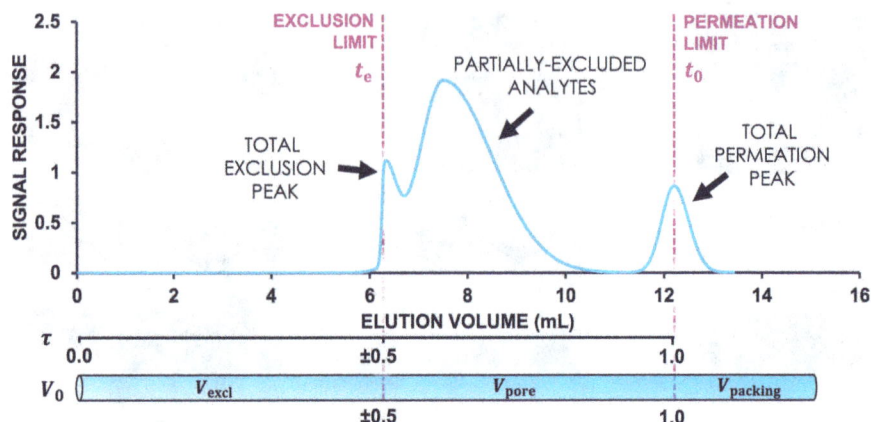

Figure 4.7 Example of a SEC chromatogram with a total-exclusion peak corresponding to high-molecular-weight analytes that are excluded from all pores, a main peak representing analyte molecules that are partially excluded from the pore volume and separated based on their size in solution (hydrodynamic diameter), and a total-permeation peak representing all non-adsorbing low-molecular-weight analytes. The second *x*-axis denotes the τ-scale, which is defined as $\tau = V_e/V_0$.

with the surface are absent for all of these small molecules. Between the exclusion limit and the total permeation limit, there is SEC selectivity, and analyte molecules may be separated based on their size. Synthetic polymers and repetitive natural polymers show elution profiles that can be converted into an MWD. Proteins can also be separated, although the many different structural elements make it difficult to suppress all enthalpic interactions. If the latter is achieved successfully, proteins elute as narrow peaks from SEC columns because of their very narrow MWDs. For a recent review on the SEC of proteins and other biopharmaceutical products, see ref. 2.

Note that in Figure 4.7, a second *x*-axis is depicted on the τ-scale. This is defined as $\tau = V_e/V_0$ so that the permeation limit is equal to $\Delta\mu_{SEC,i}^e = 1$, while the exclusion limit is approximately $\tau = 0.5$. The τ-scale is typically used to compare the SEC separations and calibration curves from different columns with different dimensions and particles.

The equilibrium coefficient described in Chapter 1 (eqn (1.2)), $K_{d,i} = c_{i,s}/c_{i,m}$ is equal to 0 for all analytes and thus not meaningful in SEC. Instead, it is possible to define a **SEC equilibrium coefficient** as the ratio of the average concentration of the analyte in the stagnant mobile phase inside the pores and its average concentration in the moving mobile phase outside the particles,

$$K_{SEC,i} = \frac{c_{i,pore}}{c_{i,int}} \qquad (4.5)$$

In the absence of enthalpic interactions, the **excess chemical potential** $\Delta\mu^e_{SEC,i}$ is determined solely by entropy effects, and we obtain

$$\ln K_{SEC,i} = \frac{\Delta\mu^e_{SEC,i}}{RT} = -\frac{\Delta S^e_{SEC,i}}{R} \qquad (4.6)$$

where $\Delta S^e_{SEC,i}$ is the partial molar excess entropy of transferring analyte i from the interstitial volume to the pore volume, and the corresponding excess enthalpy ($\Delta H^e_{SEC,i}$) is equal to 0. Differences in entropy can be related to the confined spaces available for the analyte molecules. Analyte permeation into the pores of a SEC column is associated with a loss of **conformational entropy** of the solution, and SEC is correctly designated as an entropic process. Elution times should not be affected by changes in temperature (assuming that, across a limited temperature range, $\Delta S^e_{SEC,i}$ is constant). Actually, changing the temperature is a good method to check whether interaction effects play a (usually undesired) role in SEC separations.

4.2.1.2 Calibration and Resolution

On a given column, with a given pore-size distribution, the elution volume (V_e) in SEC is indicative of the size of analyte molecules in the solution. This is illustrated in Figure 4.8. Note that we refer to an elution volume and not to a retention volume. There is no retention in SEC, so this term should be avoided. Note also that we refer to the size of molecules, not their molecular weight. As the name implies correctly, size-exclusion chromatography is a technique that yields an MSD. To convert such an MSD or, more practically, a SEC chromatogram, to an MWD, a conversion is required. This is commonly referred to as **calibration**. Calibration in SEC implies the transformation of the horizontal (time axis) from elution volume to molecular weight, and it does not involve a relationship between signal intensity and the amount of polymer present (the vertical axis). A SEC

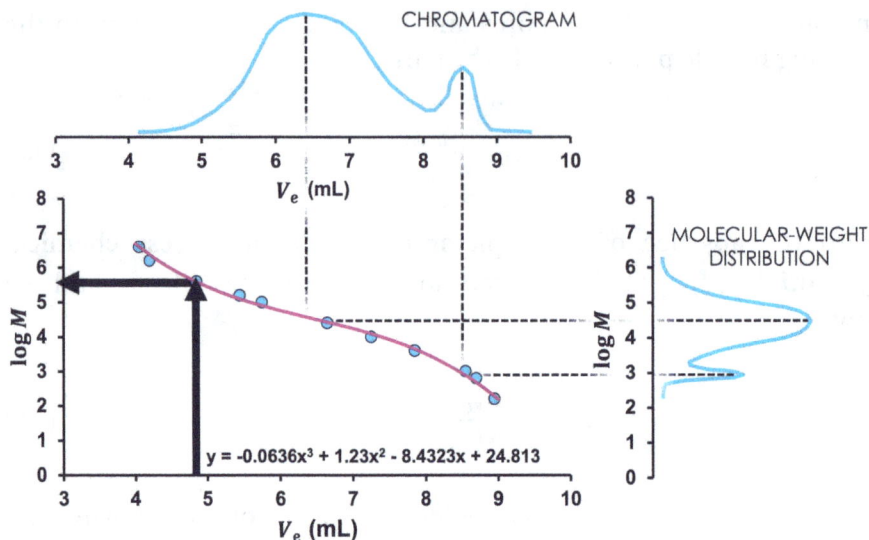

Figure 4.8 Example of a SEC calibration curve. Dots denote calibration points obtained by injecting narrow standards. The calibration curve is often described by a third-order polynomial that is fitted to the experimental data points. The drawn line represents such an equation, which is shown in the figure (here, $y =$ log M and $x = V_e$).

calibration curve, such as the one illustrated in Figure 4.8, converts the elution volume to the molecular weight of the analyte. For example, in Figure 4.8, an elution volume of 4.85 mL corresponds with a polymer molecular weight of about 400 000 Da (black arrows). It is important to realize that the calibration curve is different for each different column (determined by its pore-size distribution) and for each chemically or architecturally different type of polymer. The size of polymer molecules – and thus the SEC calibration curve – is also affected by the solvent (mobile phase).

Resolution in SEC is defined analogously to other forms of chromatography, but a SEC chromatogram may show a single (envelope) peak rather than multiple peaks. Therefore, we write

$$R_{S,j,i} \approx \frac{V_{e,j} - V_{e,i}}{4\sigma_v} = \frac{\log M_i - \log M_j}{4b\sigma_v} = \frac{\log(M_i/M_j)}{4b\sigma_v} \tag{4.7}$$

where $V_{e,i}$ is the elution volume of the first peak and $\log M_i$ the corresponding molecular weight, while $V_{e,j}$ and $\log M_j$ illustrate these same properties for the second peak. $b = \log (M_i/M_j)/(V_{e,j} - V_{e,i})$ is the absolute slope of the calibration curve. It is assumed that both analytes (i and j) elute within the shallow, (approximately) linear part of the calibration curve. An arbitrary M_i/M_j range of one order of magnitude can be used to define a **specific resolution** of

$$R_{sp} = \frac{0.25}{b\sigma_v} \tag{4.8}$$

that allows comparison between different SEC columns, and this can be normalized with the column length (L) to obtain a **packing resolution factor**

$$R_{sp}{}^* = \frac{0.25}{b\sigma_v\sqrt{L}} \tag{4.9}$$

(with units of length$^{-0.5}$) that allows comparison between different SEC packing materials.[1,3] Columns with higher R_{sp} (in a certain molecular-weight range) and particles with higher $R_{sp}{}^*$ values provide greater SEC resolution.

It can easily be understood that the relationship between size in solution and molecular weight is different for different types of polymers. For example, a polyethene (also known as polyethylene, PE) molecule, which is a long chain of methylene (CH_2) groups, must have a lower weight than a poly(vinyl chloride) (PVC) molecule of the same size due to the heavy chlorine atoms present in the monomeric vinyl chloride (CH_2–$CHCl$) units.

By determining the calibration curve as a function of the elution volume rather than the elution time, it is, in principle, independent of the flow rate. It may, however, be influenced by other instrumental conditions. To minimize the influence of the instrument, all elution volumes should be corrected for the **extra-column volume** between the injector and the detector (V_{ec}). While such a correction does not affect the calibration on a single column (or set of columns connected in series), it will facilitate the transfer of SEC methods and a more fundamental interpretation of SEC data (see, for example, Section 4.6.4, Universal Calibration).

Most commonly, SEC calibration curves are constructed by injecting a number of **narrow standards** (typically PDI ≤ 1.1) and plotting the logarithm of their specific (peak) molecular weights against the recorded elution volumes. To obtain accurate (*i.e.* correct) molecular weight information, it is essential that these standards are chemically and architecturally identical to the analyte polymer. Thus, narrow polystyrene (PS) standards can be used to record a calibration curve that is valid for obtaining MWDs of PS samples, while narrow PVC standards are required for PVC samples, *etc.* The required sets of standards are available for many different types of homopolymers – at a price. In practice, accurate MWD data are often not required. Precise (repeatable) data often suffice. In a production environment, the MWD of a product polymer (PX) may, for instance, be measured using a calibration curve recorded using (linear) PS standards, and the obtained molecular weight data may be specified for the polymer (*e.g.* $M_w \geq 25\,000$ and PDI ≤ 2). The actual molecular weight limit may be different (*e.g.* PS molecules of $M_{w,PS} = 25\,000$ may have the same size in solution as PX molecules of $M_{w,PX} = 20\,000$), but the specification may be precisely controlled in the production environment. Comparison between batches and collection of historical data are totally legitimate based on precise measurements, but it is good to be aware of the limitations and to specify the background of the data (in the present example, data are relative to linear PS).

4.2.2 SEC Columns

4.2.2.1 Suppression of Interactions

The most important characteristic of a SEC column is something it should not have. SEC columns should not exhibit any enthalpic interactions with the analyte macromolecules. In some cases, this requires special attention to the mobile phase and/or the (surface of the) packing material. Common packing materials for SEC are polystyrene–divinylbenzene and conventional silica gel. Both types of columns can be used with organic solvents such as tetrahydrofuran (THF) for characterizing many non-polar or medium-polar synthetic polymers. More polar polymers, such as polyamides or polyurethanes, usually require more polar surfaces, such as polyacrylate or polyester particles or diol-modified silica, in combination with more polar solvents, such as dimethylformamide (DMF), *N*-methylpyrrolidone (NMP), dimethylacetamide

(DMAc), or hexafluoroisopropanol (HFIP or HFIPA). In some cases, polar additives, such as butylamine or triethylamine, are added to the mobile phase. Such additives compete with polar groups on the polymer for active sites on the surface. Based on the (usually higher) concentration of the additive in comparison with the analyte polymers and the stronger interactions of the former with the stationary phase, enthalpic interactions can be effectively suppressed. However, such additives tend to irreversibly alter the stationary phase and may even be hard to remove from the instrument. In addition, they may have detrimental effects on detection. Thus, if it is possible to select a combination of a packing material and a neat solvent that provides genuine SEC conditions, the use of strongly adsorbing additives should be avoided. To dissolve biological carbohydrate polymers (starch and cellulose), significant amounts of inorganic salts, such as lithium chloride (LiCl), are often added to the mobile phase.

Special columns are typically used for SEC of proteins, in combination with aqueous mobile phases. Chemically modified silicas, with surfaces similar to those used in hydrophilic interaction liquid chromatography (HILIC; see Section 3.6.2) but with larger pores, can be used for SEC of proteins. However, such columns have a limited pH range (often 1 or 2 < pH < 8 or 9). Polar organic surfaces, such as OH-functional polyacrylates, offer greater pH stability. Even on dedicated columns, it is difficult to suppress all interactions of proteins with the surface, especially for samples containing diverse analytes with different hydrophobicities and isoelectric points (pI).

4.2.2.2 Selectivity

In most forms of LC, selectivity can be manipulated (or tuned; see Module 10.1) by varying the mobile phase. This is not the case in SEC, where the mobile phase is chosen to avoid any interactions between the analytes and the stationary phase. To change the selectivity in SEC, we must change the column. More specifically, the selectivity is determined by the **pore-size distribution** of the packing material. While this is sometimes abbreviated as PSD, we will not abbreviate it in this chapter to avoid confusion with the particle-size distribution.

The highest selectivity corresponds to the lowest slopes in the calibration curves shown in Figure 4.9, which are observed in columns with a relatively narrow pore-size distribution. The vertical (molecular-weight) axis reveals that such columns can

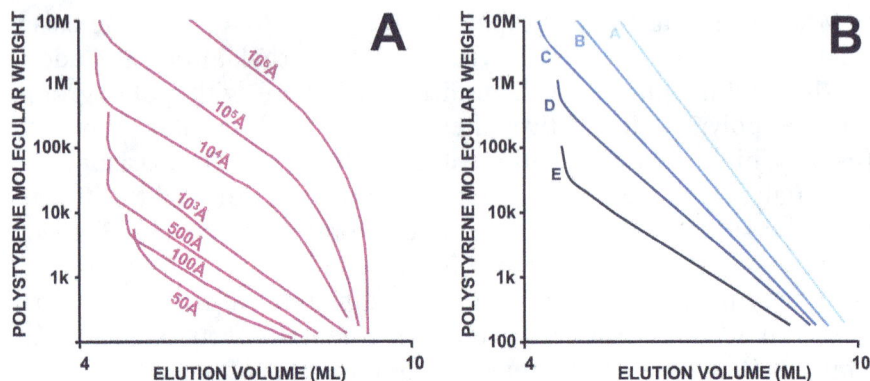

Figure 4.9 SEC calibration curves for different types of columns, specifically aimed at narrow size ranges (left) or "mixed" or "linear" columns, aimed at covering a broad range of molecular weights (right). Curves were constructed based on public data pertaining to PLgel columns from PolymerLabs (now Agilent Technologies).

exhibit large elution-volume differences for polymers that differ in molecular weight across a very narrow range. It is worth noting that the designation of these columns does not correspond to the **pore diameter** in Ångstroms (equal to 10^{-10} m). If this were the case, the largest pore diameter (10^6 Å) would equal 10^{-4} m = 100 µm. Since the particle diameter used to pack these columns is 20 µm, such pores are physically impossible. Moreover, to be even partially excluded from pores in this range, the analyte particles would need to be many micrometres in diameter, much larger than the macromolecules or (sometimes) nanoparticles subject to separation by SEC. The column designation of 10^6 Å is said to refer to the length of a fully extended polymer, the diameter of which, in a random-coiled configuration, corresponds roughly to the mean pore size in 10^4 Å columns.

Columns that cover limited molecular-weight ranges are highly useful for polymer samples that fall in these specific ranges, but all larger molecules will elute around the exclusion volume, and all smaller molecules around the void volume. This makes such a dedicated, specific column highly inflexible – and often impractical. To gain sufficient flexibility and selectivity for the samples that may be encountered in a practical situation, it has been common practice to couple several such dedicated columns in series, starting with a large-pore column, followed by columns with stepwise smaller pores. Ideally, such a column set exhibits a smooth calibration curve across a broader range, but this is not guaranteed, as it depends on how the different pore size distributions complement each other. A **pore-size mismatch** between columns results in irregularities ("kinks") in calibration curves and apparent molecular-size distributions. To remedy this problem, Yau *et al.* introduced the concept of **bimodal pores.**[4] They demonstrated that a broad linear calibration range can be obtained by coupling just two columns with narrow pore-size distributions, mean values that differ by about an order of magnitude, and approximately equal pore volumes. In terms of Figure 4.9, this implies that the calibration curves are essentially non-overlapping, and the linear parts are roughly parallel. Alternatively, a column manufacturer may create a packing material with a broad pore-size distribution or blend different packing materials (with equal particle sizes, to avoid a large increase in pressure drop) to create single columns with smooth calibration curves across a broad range. Such columns are designed to have a broad linear range in a log M *vs.* V_e calibration curve (Figure 4.9) and are, therefore, known as **linear columns**. Because of the packing materials, **mixed** or **mixed-bed columns** are alternative designations.

4.2.2.3 Particle Size

It is often proclaimed that resolution in SEC is determined by the pore volume, and this is one reason why SEC columns are typically wider (7 or 8 mm i.d.) than traditional (4.6 mm i.d.) or modern (2.1 mm i.d.) high-performance liquid chromatography (HPLC) columns. Like in all other forms of LC, the number of plates, and thus the column length in relation to the size of the packing particles, determines the efficiency. Together with the calibration curve (column selectivity), this determines the resolution. There is no fundamental reason for SEC columns to be wider and, therefore, more expensive and to require much larger amounts of eluents (since the flow rate scales with the square of the column internal diameter).

There are practical reasons why very wide columns have traditionally been preferred for SEC, and the main one of these is **extra-column band broadening**. To avoid extra-column band broadening, liquid chromatographers use very narrow capillaries to connect the various system components (injector, column, detector). Such narrow capillaries clog more easily than wider ones if (high molecular weight) polymers precipitate, for instance, due to variations in temperature. In addition, solutions of high-molecular-weight polymers tend to be quite viscous, even at low concentrations. Both these factors make it tempting to choose slightly wider connection capillaries. However, the **Taylor–Aris equation** discussed in Chapter 1 (eqn (1.90)), which states that

$$\sigma_{V,ec}^{2} = \frac{\pi d_c^4 LF}{384 D_m}$$

(4.10)

shows that increasing the diameter of the capillaries can have dramatic effects because it occurs to the fourth power. The extra-column band broadening is especially threatening for polymers because they exhibit values for D_m that decrease with increasing molecular weight. SEC is also unforgiving for injection band broadening. This is in contrast with gradient-elution LC, where the analytes can be focussed at the column inlet and analysts can get away with broad injection plugs, provided that the injection solvent is a weak eluent. Typical detectors used in SEC, such as multi-angle light scattering (MALS) detectors, tend to have relatively large (detection) cell volumes, appropriate for large (wide-bore) SEC columns. In recent years, 4.6 mm i.d. SEC columns have become increasingly popular. In comparison with 7.8 mm i.d. columns, the eluent consumption can be reduced by about a factor of 3.

4.2.2.4 Stress Exerted on Analyte Molecules

The general principles of LC are valid in SEC. Thus, much faster separations can be performed if the particle size is reduced. The very small particles used in ultra-high-performance liquid chromatography (UHPLC) have sufficiently large pore sizes to allow very fast and efficient separations of polymers up to at least 50 000 in molecular weight.[5] Wide-pore (300 Å) UHPLC particles have also become available for large-molecule (protein) separations. One disadvantage of UHPLC is that greater stress is exerted on the analyte molecules. This can be expressed in the **Deborah number** (De),

$$\mathrm{De} = k_{pac} \frac{u_{int}}{d_p} \frac{6.12 \Phi \eta r_g^3}{RT}$$

(4.11)

where k_{pac} is a constant that depends on the packing structure, with typical values ranging between 6 and 9, depending on the kind of packing[1]; u_{int} is the interstitial linear velocity; d_p is the particle size; Φ is the Flory–Fox parameter (equal to 2.5×10^{23} mol^{-1} under typical

SEC conditions); η is the viscosity of the mobile phase; r_g is the radius of gyration of the analyte molecules; R is the gas constant; and T is the absolute temperature.

Comparable conditions in LC imply a constant reduced linear velocity ($v_{int} = u_{int}d_p/D_m$ = constant), so that $u_{int,opt}$ is inversely proportional to d_p and De is inversely proportional to d_p^2. As a consequence, the Deborah number increases by about a factor of 35 when progressing from a traditional SEC column packed with 10 μm particles to an "ultra-performance" SEC column packed with 1.7 μm particles. Thus, much greater stress is exerted on the analyte molecules when smaller particles (and very fast analyses) are used. In the study of sensitive analytes, such as protein aggregates, it is necessary to use larger particles. For determining the MWD of synthetic polymers, the increased stress has been demonstrated not to break molecular bonds, except for very large molecules ($\gtrsim 1$ MDa) and at very high flow rates.[6] A practical consideration is that ultra-performance SEC columns packed with sub-2 μm particles are more liable to clogging than the large, robust SEC columns. Because extra-column band broadening is a major limitation in miniaturized SEC, installing filters or pre-columns may subvert fast and efficient analysis.

Figure 4.10 illustrates the Deborah numbers that may be encountered in SEC analysis of polystyrene with a THF mobile phase. High stress and possible deformation of macromolecules occur in the pink zone shown in the figure (De ≥ 0.5). Such conditions only occur for high-molecular-weight polymers in combination with small packing particles. For such analytes, columns packed with 10 μm or 20 μm particles may be used, not just to reduce the exerted stress but also to mainly accommodate the very large pores required to allow partial penetration of the pores.

The approximations used to construct Figure 4.10 may be useful as general estimates for SEC. The radius of gyration of PS molecules in THF in nanometres was approximated as[6]

$$r_g = 0.018M^{0.6} \tag{4.12}$$

The analyte diffusion coefficient in m² s⁻¹ at 25 °C is approximated as[7]

$$D_m = 3.45 \times 10^{-8}M^{0.564} \tag{4.13}$$

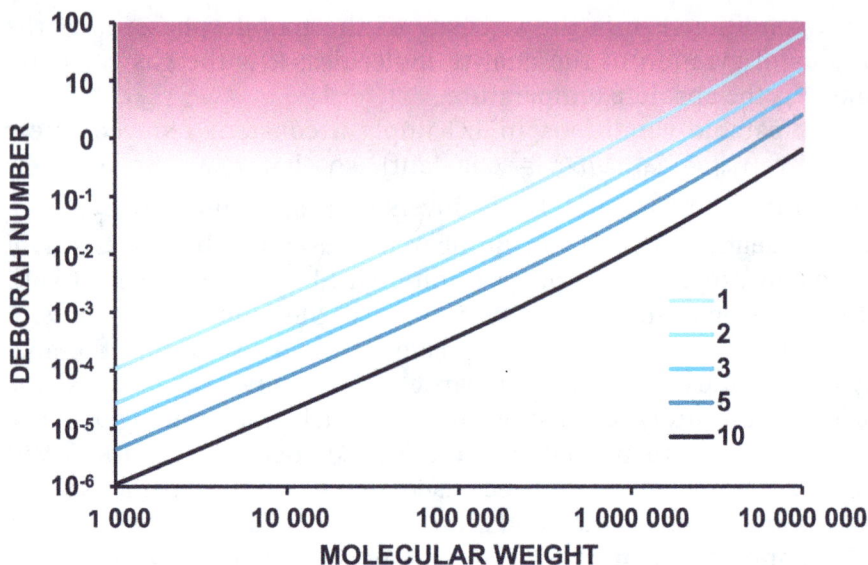

Figure 4.10 Calculated SEC Deborah numbers for polystyrene in THF. Numbers in the legend indicate the particle size of the packing material in micrometres. A reduced interstitial velocity of 100 is assumed, which implies lower flow rates for larger molecules. The pink zone indicates the danger area, where significant stress is exerted on the analyte molecules.

The intrinsic viscosity (see Section 4.6.3) in mL g^{-1} at 25 °C is approximated as[8]

$$[\eta] = 0.0141\, M^{0.7} \tag{4.14}$$

while assuming a concentration of 5 mg mL^{-1} in converting this to an actual viscosity.

To reduce the stress exerted on analyte molecules (or molecular aggregates), it helps to choose larger particles. These are routinely used in SEC columns intended for characterizing (ultra-)high-molecular-weight polymers. However, smaller particles may be encountered when connecting columns with different pore-size distributions in series. Lowering the concentration of the analyte also helps (as it leads to lower viscosities). By positioning the columns with the largest pores (and particles) first in a set of columns, the analytes are diluted by the time they reach the column packed with smaller particles. Finally, reducing the flow rate always reduces the stress exerted on the analytes.

4.2.2.5 Injection Volume and Concentration

When the polymer concentration in the solution exceeds a certain "critical overlap concentration", a so-called **molecular crowding effect** causes analyte molecules in the solution to become more compact, resulting in longer elution times. This will cause errors in the conversion from elution volume to molecular weight using a calibration curve (Figure 4.8). It also invalidates many of the assumptions made for other ways of molecular-weight calibration, such as light scattering and viscometry (see Module 4.6). As a rule of thumb, the critical concentration can be taken as $c_{i,max} = 1/[\eta]$. This implies that the maximum analyte concentration decreases with the polymer molecular weight. For example, using eqn (4.14) for PS in THF leads to $c_{i,max} \approx 70/M^{0.7}$, which suggests that for PS with $M = 10\,000$, an injection concentration of 100 mg mL^{-1} may be acceptable, while for $M = 100\,000$, this becomes 20 mg mL^{-1}. For $M = 1\,000\,000$, a concentration of 5 mg mL^{-1} is already pushing the limit, as is a concentration of 1 mg mL^{-1} for $M = 10\,000\,000$. For polymers up to $M = 1\,000\,000$, concentrations of 1 to 2 mg mL^{-1} are commonly used.

As to the maximum injectable volume, Striegel (ref. 1, p. 87) uses one-third of the peak volume of the total-permeation (solvent) peak as the upper limit. This leads to

$$V_{inj,max} \approx 1.5\sigma_v = \frac{1.5 V_0}{\sqrt{N}} = 1.5 \cdot \frac{\pi}{4} d_c^2 \cdot L \cdot \varepsilon_{tot} \cdot \sqrt{\frac{h \cdot d_p}{L}} \qquad (4.15)$$

Assuming a reduced plate height of $h = 2.5$, we obtain as a rule of thumb

$$V_{inj,max} \approx 1.4 \cdot d_c^2 \cdot \sqrt{L \cdot d_p} \qquad (4.16)$$

This implies that on a conventional SEC column with $L = 600$ mm, $d_c = 7.8$ mm, and $d_p = 10$ μm (abbreviated as a 600 × 7.8 mm; $d_p = 10$ μm column), we may inject volumes up to approximately 275 μL. Injecting volumes of 100 or 200 μL is indeed not uncommon in conventional SEC using such large columns. On a 150 × 4.6 mm, $d_p = 5$ μm column, $V_{inj,max}$ would be approximately 35 μL, whereas on a modern 50 × 4.6 mm, $d_p = 1.7$ μm UHP-SEC column, the $V_{inj,max} \approx 8$ μL.

4.2.2.6 Band Broadening

Band broadening in SEC obviously suffers from the low diffusion coefficients of the analyte polymers in solution, but this is aggravated by a slow diffusion in and out of pores that are barely larger than the analyte pores. Ironically, this is exactly the range of pore diameter to analyte hydrodynamic diameter where SEC selectivity is greatest. This is illustrated in Figure 4.11. The efficiency is lower (*i.e.* plate heights are greater) for all polymers in comparison with a low-molecular-weight analyte (toluene), but much lower for the partially excluded 30 000 Da standard (estimated r_h = 93 Å) than for the almost fully excluded 52 000 Da standard (estimated r_h = 126 Å).

The observed width of SEC peaks (σ_{obs}^2) is not solely determined by chromatographic (σ_{chrom}^2) and extra-column (σ_{ec}^2) band broadening but also by the MWD of the analyte polymer (σ_{PDI}^2),

$$\sigma_{obs}^2 = \sigma_{PDI}^2 + \sigma_{chrom}^2 + \sigma_{ec}^2 \tag{4.17}$$

The challenge is to ensure conditions such that $\sigma_{chrom}^2 + \sigma_{ec}^2 \approx 0$ and, therefore, $\sigma_{obs}^2 \approx \sigma_{PDI}^2$. Long columns are favourable because σ_{PDI} (SEC selectivity, *i.e.* differences in retention volume) increases linearly with the column length L, whereas, as usual, σ_{chrom} increases

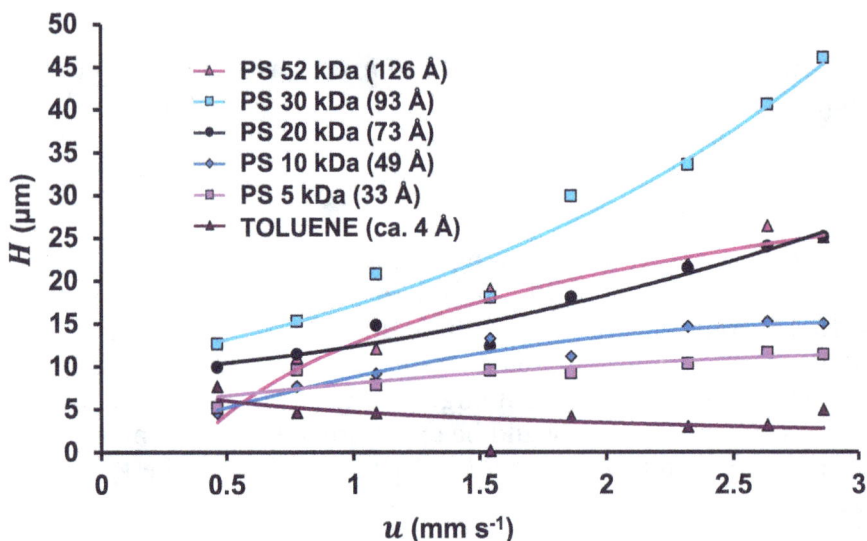

Figure 4.11 Observed plate height for various polystyrene standards on a SEC column with an average pore diameter of 130 Å.

in proportion with $\sqrt{L}$ (or $\sqrt{N}$). Wide-bore columns may help minimize the relative contribution of σ_{ec}.[6] However, longer columns increase the analysis time, and wider columns significantly increase the required volume of solvent (mobile phase). The task of chromatographers is to find a good balance based on eqn (4.17).

4.2.2.7 SEC with Core–Shell Particles

The core–shell or superficially porous columns described in Section 3.2.2.1 that have become popular in LC can also be used in SEC. Columns packed with core–shell particles offer high efficiencies, presumably because of the narrow particle-size distribution of the packing material. A typical particle size is 2.7 µm, which is significantly less than the particles currently used for SEC (3–20 µm) but significantly larger than the 1.7 µm particles typically used in UHPLC. For SEC, this means less stress on the analyte polymers and potentially fewer problems due to clogging of the system. Intuitively, core–shell columns seem unattractive for SEC because the pore volume is decreased by the presence of a solid core in the particles. However, a simple calculation shows this effect to be small; for example, a solid particle that is to become the core with a diameter of 1.5 µm has a volume of 1.77 fL. Adding a shell with a thickness of 0.6 µm yields a particle with a diameter of 2.7 µm and a volume of 10.3 fL. This means that 83% of the resulting particle is porous and only 17% of its volume consists of the solid core.

Pirok *et al.* have studied the application of solid-core particles for SEC.[3] Figure 4.12 shows fitted calibration curves for three conventional SEC columns packed with fully porous particles (pink lines) and three core–shell columns (blue lines). In the figure, the logarithm of the molecular weight is plotted against the fractional elution volume ($\tau = V_e/V_0$), where V_e is the elution volume and V_0 is the total volume of the mobile phase in the column. Indeed, some selectivity is sacrificed, as is apparent from the somewhat steeper blue lines and the narrower range of τ values spanned by the core–shell columns. However, it was shown that this can be more than compensated for by the higher chromatographic efficiency offered by core–shell columns.

The packing resolution factor defined by eqn (4.9) can be used to compare the SEC resolution offered by fully porous and core–shell particles. Core–shell particles were found to yield comparable resolution in a much shorter time.[3]

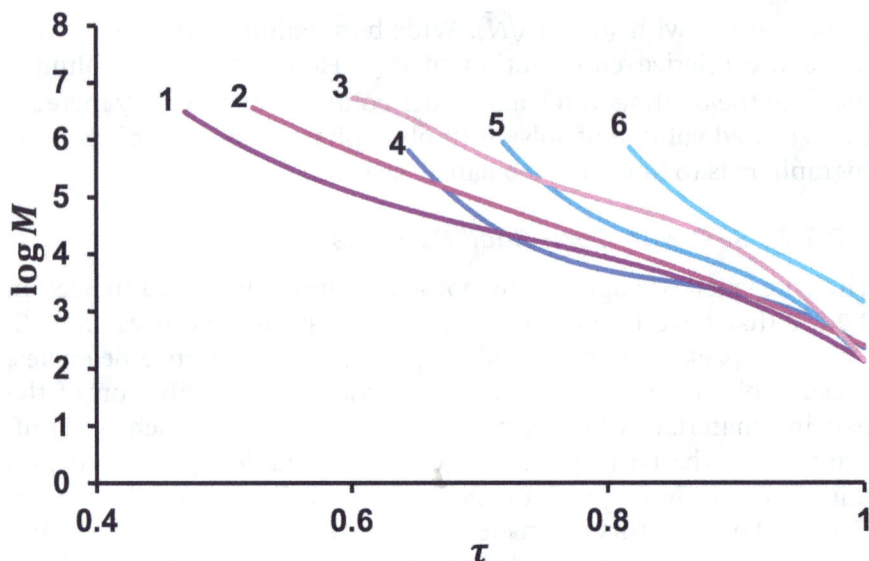

Figure 4.12 Overlay of fitted calibration curves for columns packed with fully porous particles (pink) and core–shell particles (blue). Columns shown are as follows: (1) Agilent Mixed D (300 × 7.5 mm, d_p = 5 μm, pore size: MIXED); (2) Agilent Mixed C (100 × 4.6 mm, d_p = 5 μm, pore size: MIXED); (3) Agilent PLgel 10^5 (300 × 7.5 mm, d_p = 5 μm, pore size: 10^5 Å); (4) 150 × 4.6 mm, d_p = 2.9 μm, pore size: 98 Å; (5) Phenomenex (150 × 4.6 mm, d_p= 3.6 μm, 175 Å); (6) Phenomenex (150 × 4.6 mm, d_p = 3.6 μm, pore size: 477 Å). Reproduced from ref. 3 with permission from Elsevier, Copyright 2017.

4.3 Interactive Liquid Chromatography (M)

4.3.1 Isocratic (Critical) Chromatography of Polymers

In Chapter 1, we have seen that the partition coefficient and thus also the retention factor are thermodynamic properties. We may recall eqn (1.54),

$$\ln k_i = \ln K_{d,i}^{x,\infty} + \ln \frac{n_s}{n_m} = \frac{-\Delta H_i^e}{RT} + \frac{\Delta S_i^e}{R} + \ln \frac{n_s}{n_m} = \frac{-\Delta G_i^e}{RT} + \ln \frac{n_s}{n_m} \qquad (4.18)$$

which implies that retention is determined by three factors, appearing in the logarithmic equation as an enthalpy term, an entropy term, and a phase ratio term. In SEC (Module 4.2), we do all we can to avoid energetic (enthalpic) interactions, and, if the SEC mechanism prevails, there is no effect of temperature on retention. This justifies considering SEC an entropic process, so the thought *SEC is entropy* is not necessarily incorrect. However, the opposite, *entropy is SEC,* is a gross misunderstanding. Entropy plays a role in all equilibria. We measure compounds that are fairly evenly distributed across the mobile and the stationary phases because unretained and endlessly retained analytes are of little interest. This means that, in practice, conditions are often such that significant "enthalpy–entropy compensation" is observed.

In this module, we will first consider **repetitive polymers** that consist of one or a few building blocks (monomers). As a plausible first guess, we may consider ΔG_i^e to be built up from contributions of the structural elements in the molecule. For a linear homopolymer, we obtain

$$\Delta G_i^e = n_{mon}\Delta G_{mon}^e + \Delta G_{end\text{-}groups}^e \qquad (4.19)$$

where we distinguish between the contributions of the monomeric unit and the end groups to ΔG_i^e and n_{mon} is the number of monomer units in the chain, commonly referred to as the **degree of polymerization**. Such an assumption is in line with the Martin rule (see Section 2.1.3), considering a homopolymer as an extended version of a homologous series.

The magnitude of ΔG_{mon}^e between successive members of the polymeric series, *i.e.*

$$\ln \alpha_{i+1,i} = \frac{\Delta G_{mon}^e}{RT} \qquad (4.20)$$

If ΔG_{mon}^e is substantial (*e.g.* $\Delta G_{mon}^e = 450$ J mol^{-1}), the selectivity is large ($\alpha_{i+1,i} = 1.2$), and individual members of the series ("oligomers") can be separated. An isocratic chromatogram could look like the one shown in Figure 4.13A. If the selectivity is lower (*e.g.* $\Delta G_{mon}^e = 80$ J mol^{-1}, $\alpha_{i+1,i} = 1.03$), individual members can no longer be separated and low-molecular-weight polymers appear as a single peak with dramatic tailing (Figure 4.13B). If the selectivity between successive members becomes negligible ($\Delta G_{mon}^e = 4$ J mol^{-1}, $\alpha_{i+1,i} = 1.0015$), the polymer may elute as a single peak (Figure 4.13C) at a retention time determined by the free-energy contribution of the end-groups and the phase ratio.

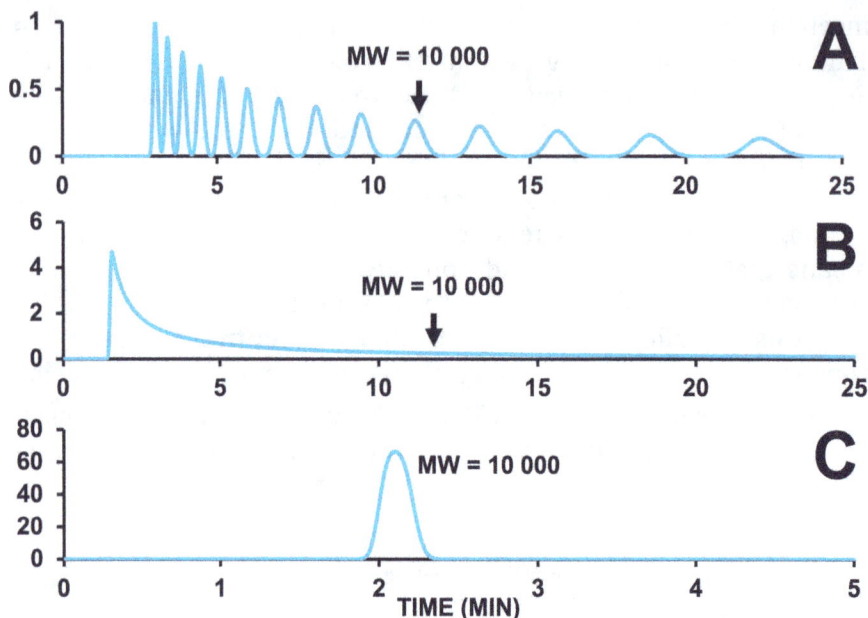

Figure 4.13 (A) Separation of oligomers with $\alpha_{i+1,i}$ = 1.2 and N = 2000. (B) Separation of low-molecular-weight polymers ($\alpha_{i+1,i}$ = 1.03). (C) Near-critical elution of a polymer ($\alpha_{i+1,i}$ = 1.0015). For explanation, see text.

This type of molecular-weight-independent elution of polymers is known as liquid chromatography at the critical point (of elution), liquid chromatography at the critical composition, or, shortly, **critical chromatography** of polymers. It may be possible to conduct SEC (using a very strong eluent), adsorption chromatography (using a weak eluent), and critical chromatography (using an intermediate eluent with a very specific, critical composition) all on the same column by just varying the eluent or eluent composition (see Figure 4.14). It is correct to state that the enthalpic contribution of the monomeric unit is cancelled by the entropic contribution. It is not correct to say that exclusion (SEC behaviour) is cancelled by adsorption behaviour. Indeed, it is possible to realize critical chromatography on a column packed with non-porous particles.

Together, the critical and adsorption/absorption domains are often referred to as "**interactive**" LC of polymers. These methods rely on differences in interactions of the analytes with (mainly) the stationary phase, as opposed to SEC, where such interactions are meant to be non-existent.

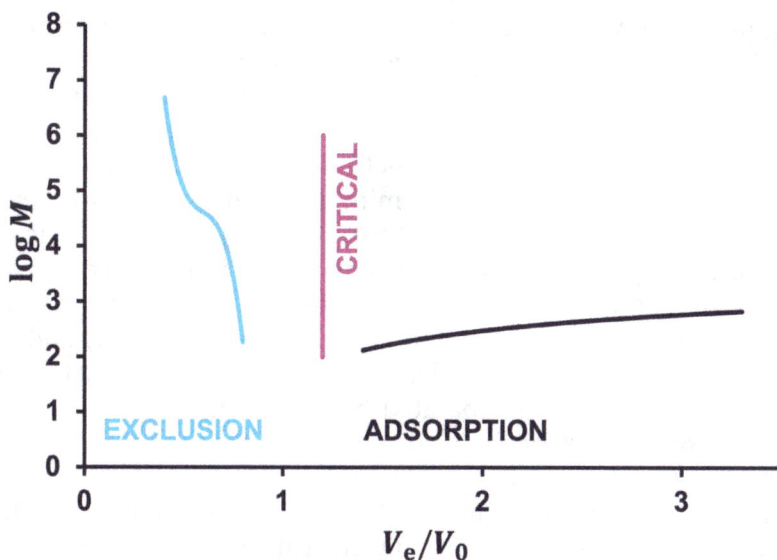

Figure 4.14 Illustration of the SEC, critical, and adsorption or absorption domains that may all be encountered on a single LC column, depending on the mobile-phase composition.

As shown in Figure 4.14, SEC separations span a narrow range of elution volumes, as explained in Module 4.2. To overcome this limited selectivity, SEC is often performed on long columns or a series of columns with a considerable total length. Isocratic ad- or absorption chromatography provides much more selectivity but is highly impractical and unattractive for polymers (see Figure 4.13).

If we combine eqn (4.18) and (4.19), while realizing that at the critical point $\Delta G_{\text{mon}}^{\text{e}} = 0$, we find

$$\ln k_i = -\frac{n_{\text{mon}}\Delta G_{\text{mon}}^{\text{e}} + \Delta G_{\text{end-groups}}^{\text{e}}}{RT} + \ln\frac{n_{\text{s}}}{n_{\text{m}}} = \frac{-\Delta G_{\text{end-groups}}^{\text{e}}}{RT} + \ln\frac{n_{\text{s}}}{n_{\text{m}}} \quad (4.21)$$

This shows that retention at the critical point is largely determined by the end-groups, while selectivity (*i.e.* differences in retention) is only determined by differences in (the free energy of transfer of) the end-groups. The greatest differences can be created when dealing with polar end-groups (*e.g.* hydroxyl groups) and polar surfaces (*e.g.* silica). Indeed, such NPLC-like surfaces are most successfully applied for critical separations. However, other types of interactions can also be used advantageously.

Figure 4.15 shows an example of a critical separation of benzal-dehyde-derivatized poly(ethylene glycols) based on π–π interactions between carbonyl end-groups and a phenyl stationary phase. It can be seen in the figure that PEGs of different molecular weights with two hydroxyl end-groups (PEG 2000, 4000, and 6000) elute at the same retention time around 6.5 min, as do PEGs with two benzal-dehyde end-groups (PEGDCHO 2000, 4000, and 6000), confirming the critical separation. Excellent separation is obtained between these series and other series with different end-groups, *i.e.* with one hydroxy and one benzaldehyde end-group (PEGCHO) and with one methyl and one benzaldehyde end-group (MPEGCHO). The latter two series are not separated because neither the methyl nor the hydroxyl groups interact strongly with the phenyl groups on the stationary phase, whereas the benzaldehyde group does. The separation depicted in Figure 4.15 concerns relatively low-molecu-lar-weight polymers. Realizing truly critical separations tends to get more difficult when the molecular weight increases. This can be understood from eqn (4.21). The larger the degree of polymeriza-tion (n_{mon}), the more a very small contribution to the free energy ($-\Delta G^e_{mon}$) will contribute to the observed retention.

Critical chromatography is not just "critical". It is extremely critical. Critical conditions involve a very specific combination of composition and temperature on a very specific column. A nomi-nally identical but slightly different column may give rise to a change in the critical composition. In addition, the history of the column may have an effect, as the surface may have been altered or contami-nated. A slight deviation from the critical composition will cause the separation to drift towards the SEC side (*i.e.* higher-molecular-weight molecules starting to elute earlier) or towards the adsorption or absorption side (*i.e.* higher-molecular-weight molecules starting to elute later). Moreover, retention is a function of temperature. A specific critical mobile-phase composition is only valid at a specific temperature. Finally, as described in Section 1.6.1.2, the effects of pressure on retention are strongest for high-molecular-weight analytes, such as polymers. This causes the critical composition to depend on the pressure. The result may be that if critical conditions prevail at the top of an LC column, they may be lost later in the column, where the pressure is lower. This would suggest that critical separations are best performed with relatively large particles (*e.g.* 10 μm diameter) and at relatively low linear velocities (1 mm s^{-1} or lower) to minimize the pressure drop across the column.

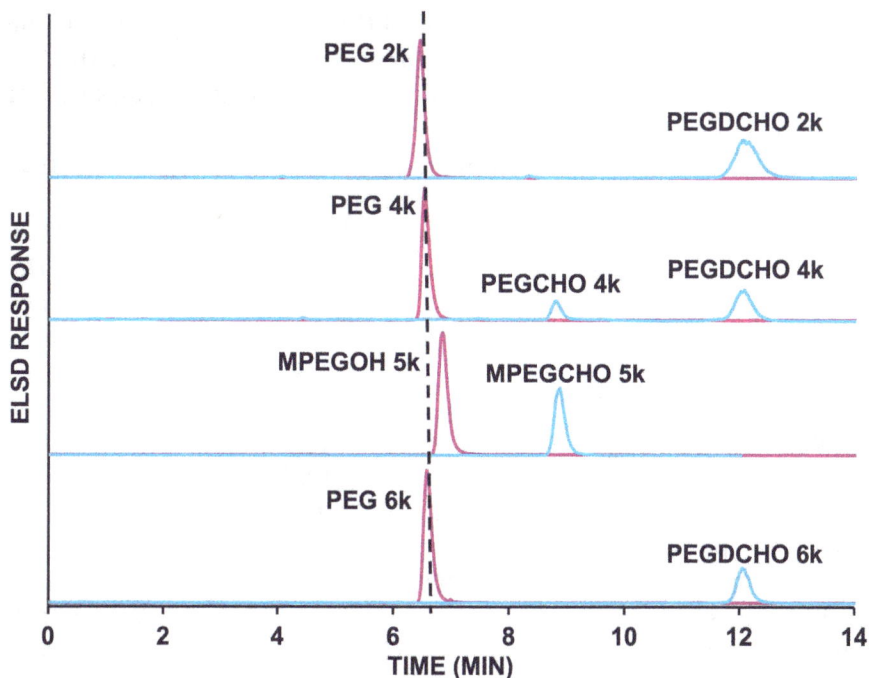

Figure 4.15 Example of a critical separation of polyethylene glycols (PEGs) with different end-groups on an XB-phenyl column at 30 °C. End-groups: PEG, –OH and –OH; PEGCHO, –OH and –BA; MPEGOH, –CH$_3$ and –OH; MPEGCHO, –CH$_3$ and –BA; PEGD-CHO, –BA and –BA (BA = benzaldehyde). Mobile phase: acetonitrile : water 45:55 by volume. Reproduced from ref. 9 with permission from the Royal Society of Chemistry.

More robust (near-)critical separations of polymers can be performed by comprehensive two-dimensional liquid chromatography.[10]

4.3.2 Gradient Liquid Chromatography of Polymers

Isocratic, critical LC is potentially relevant for separating (synthetic) polymers based on the nature or number of the end groups. However, such separations are notoriously non-robust. Much more commonly, gradients are applied to study the chemical composition of polymers with LC in the adsorption/absorption domain. This allows analytes that differ greatly in molecular weight and composition to be eluted under (near-)optimal conditions within a single run. As long as the final solvent (mobile phase) is a strong eluent for all analytes, they will all elute when the gradient has passed the end of the column. If the initial solvent is a weak eluent, the analytes will be focus at the top of the column, negating any pre-column band broadening arising, for example, from imperfect injections.

Only analytes that are unretained in the initial mobile phase may experience size-exclusion effects. If a molecule is migrating during the gradient, it cannot move ahead of its position because it would then experience a weaker solvent and be retained until it is back in the band where it belongs. Likewise, molecules that fall behind experience a stronger eluent, allowing them to catch up with identical analyte molecules. Any SEC effects here would reduce band broadening but not affect the migration speed (and eventual retention time) of the peak. The natural focussing effect in gradient-elution LC that has been described in Module 3.4 is especially strong for high-molecular-weight analytes, for which retention varies more strongly with mobile-phase composition than for small molecules.

Retention *vs.* composition lines are extremely steep for high-molecular-weight analytes. For such lines, it does not really matter whether they are linear or curved. Because of this, and because size-exclusion effects can be ignored, we can describe polymer retention using a simple retention model, such as the log-linear one (Section 3.8.3.1). Empirically, a correlation between slope and intercept has been observed,

$$S = p + q \ln k_0 \tag{4.22}$$

This equation holds especially well for a homologous series or a specific type of polymer. Substitution into eqn (3.46b) yields

$$\ln k = \ln k_0 - S \cdot \varphi = \ln k_0 - (p + q \ln k_0)\varphi = (1 - q \cdot \varphi)\ln k_0 - p \cdot \varphi \tag{4.23}$$

This equation shows that retention is independent of $\ln k_0$ and thus the same for all members of the series when $\varphi_{\text{crit}} = 1/q$. Hence, this φ_{crit} is the critical composition. Retention at this point follows from $\ln k_{\text{crit}} = -p/q$. The model is schematically illustrated in Figure 4.16. Different homopolymeric polymeric series, each with different monomeric units and end-groups, yield three different bundles of lines, each with clearly discernible critical compositions. A combination of such series cannot possibly be eluted under isocratic conditions because, at any composition, many analytes would either be unretained or retained for an infinitely long time.

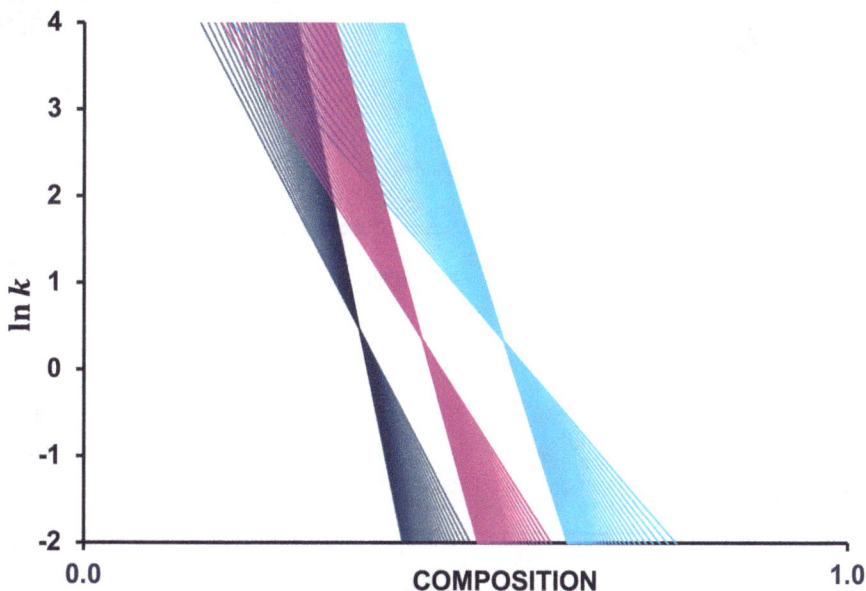

Figure 4.16 Schematic illustration of the retention model of eqn (4.23) for three chemically different polymeric series. The respective p, q, M_n and PDI values are as follows: -1.3, 2.77, 50000, and 1.20 for series 1; 0.8, 2.26, 200000, and 1.04 for series 2; 0.6, 1.82, 2000, and 1.04 for series 3. Different members of the series follow the Martin rule, with $\ln k_0$ increasing at regular intervals.

Gradient elution is much more appropriate for such samples. By varying the composition during the run, all analytes can be eluted. This is illustrated in Figure 4.17. When performing a fast gradient (Figure 4.17A), analytes elute with low retention factors, in line with the theory outlined in Module 3.4. Under such conditions, as shown in Figure 4.17A, retention is independent of molecular weight. This is known as **pseudo-critical** (gradient) elution. In contrast, a slow gradient results in higher retention factors at the moment of elution and a significant influence of molecular weight on retention, as is evident from peaks that are broader and fronting (Figure 4.17B).

Domain experts may spend much time discussing the mechanism of gradient-elution LC of polymers. It could be similar to that of other forms of LC (RPLC, NPLC), although the mechanisms involved in such separations are also still the subject of discussion. In the case of polymers, some scientists distinguish between an interactive LC mechanism and a precipitation/redissolution mechanism, in which the analyte polymers are thought to be insoluble under the starting conditions. The two are hard to distinguish in practice.

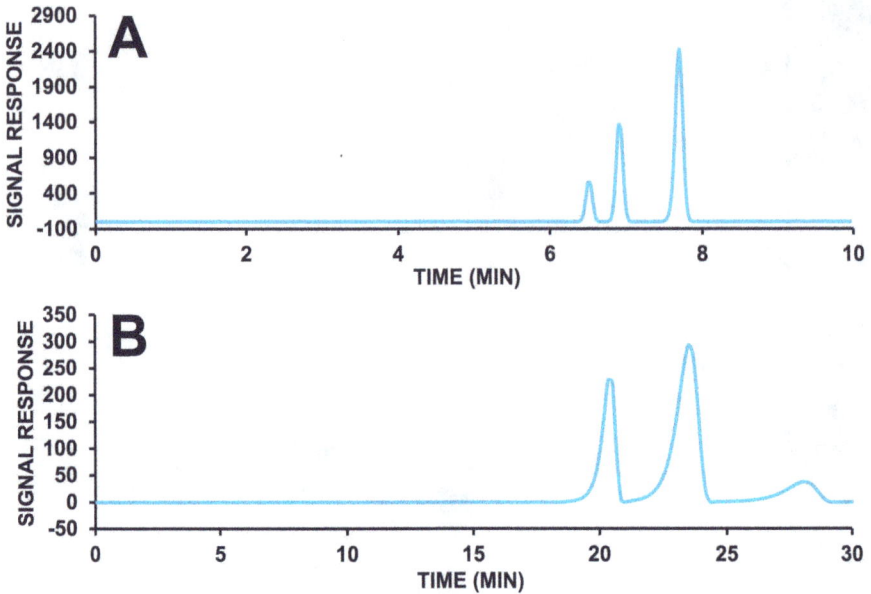

Figure 4.17 Calculated chromatograms based on the retention models in Figure 4.16. Hold-up time: t_0 = 60 s; dwell time: t_D = 120 s; initial composition: 0% strong solvent; final composition: 100% strong solvent. Top: fast gradient (t_G = 10 min = 10 t_0), showing near-critical elution independent of molecular weight for each series. Bottom: slow gradient (t_G = 48 min = 48 t_0), showing a significant effect of molecular weight on retention within each series.

The latter mechanism may depend more strongly on the injection amount. Some polymers may also dilute slowly, for example, due to high crystallinity, giving rise to band broadening and delayed elution. If the separation is good, the exact mechanism is usually irrelevant.

A highly relevant problem is **breakthrough.** If a sample is dissolved in a strong solvent and the injection volume is large enough for a more-or-less intact solvent plug to travel through the column, most of the analyte polymers may elute with the solvent peak. This problem is especially serious for high-molecular-weight analytes because they may move to the front of the solvent plug due to exclusion effects.[11]

The effect is schematically illustrated in Figure 4.18. In the top part of the figure, a plug of strong, non-polar solvent (purple) resulting from a large-volume injection is depicted at four different positions in the column (and four different corresponding time points). Although the mobile phase is polar (blue) and high retention factors are anticipated, the analytes do not come into contact with the mobile phase as long as they are in the solvent plug. At the tailing end of the plug, analyte molecules get retained as soon as the mobile phase becomes sufficiently weak. In the case of gradient elution, they may then be refocussed into a reasonably narrow peak at a specific mobile-phase composition. For large molecules, which are big enough to be excluded from a significant fraction of the pores, the situation is different. These molecules move quicker than the solvent when they are in the strong-solvent plug. However, when they are about to leave the plug, the mobile phase becomes weaker, and they get retained. Thus, such large analyte molecules are focussed on the leading edge of the solvent peak, as shown in Figure 4.18A (A3 and A4). When a smaller volume of solvent is injected or an injection solvent can be chosen that is weaker, analyte breakthrough can be circumvented (Figure 4.18B).

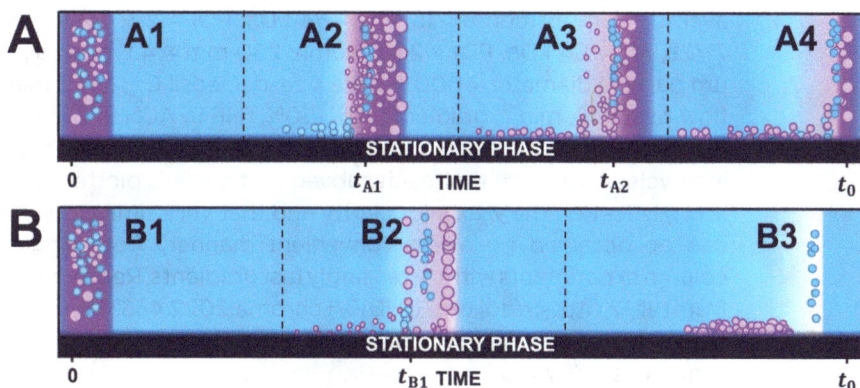

Figure 4.18 (A) Schematic explanation of the breakthrough phenomenon. Blue depicts polar, and purple depicts non-polar solvents. At the moment of injection (A1, $t = 0$), the strong injection solvent (purple) inhibits the adsorption of analytes. Within the strong-solvent plug, large molecules experience exclusion, causing them to move to the front of the plug (A2), which results in the focussing of the larger hydrophobic analytes (A3). An intact strong-solvent plug exits the column (A4, t_0), with large hydrophobic molecules at its leading edge, while a fraction of smaller analyte molecules elutes unretained. This entire effect can be avoided when the injection plug is (i) sufficiently weak or (ii) sufficiently small so that the solvent band dissolves sufficiently (B2) by the time that it reaches the exit of the column (B3).

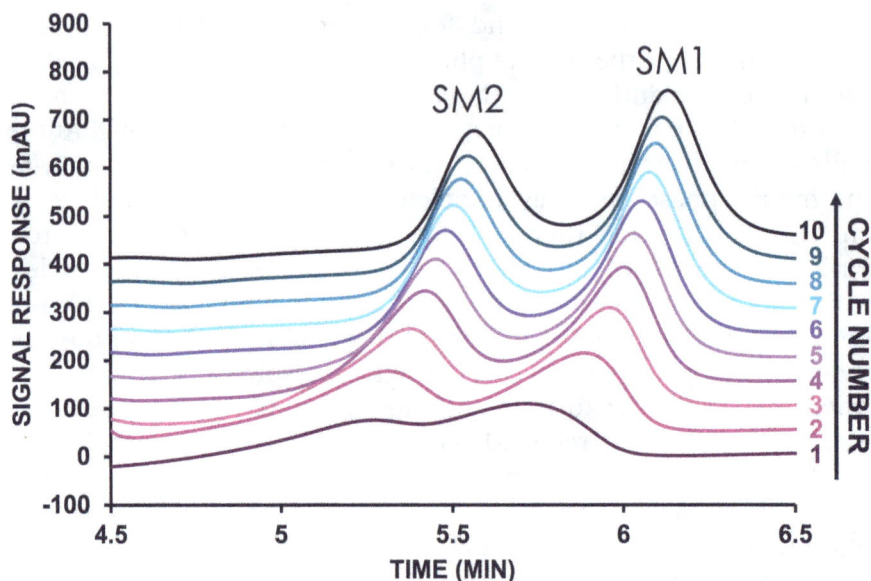

Figure 4.19 Recycling LC (LC◌LC) experiments that illustrate the effect of the gradient duration (t_G/t_0) on the elution profile. Styrene–methyl methacrylate statistical copolymers with SM1: S/M average composition 84/16, M_w = 54 kDa, PDI = 2.3; SM2: S/M 71/29, M_w = 64 kDa, PDI = 2.1. Column: 250 mm × 4.6 mm i.d., 5 μm particle diameter, 4000 Å pore size, nucleosil C_{18}; $t_0 \approx 3$ min; flow rate 1 mL min^{-1}; gradient: 30%–50% THF in ACN in 2.5 min. Gradient duration ranges from $t_G/t_0 \approx 0.8$ to $t_G/t_0 \approx 0.08$ following cycles 1–10. Note that each subsequent cycle is plotted with a +50 offset on the y-axis for clarity and that very similar results can be obtained (in a less-convenient manner) from single-column experiments with increasingly fast gradients. Reproduced from ref. 12, https://doi.org/10.1016/j.chroma.2022.463386, under the terms of the CC BY 4.0 license https://creativecommons.org/licenses/by/4.0/.

Recycling LC (LC◌LC) is a very elegant way to study the effect of gradient duration (Figure 4.19). A gradient spanning a total volume well within a single column volume (*e.g.* t_G = 150 when t_0 = 180 s) is applied to the sample, and the entire effluent from the gradient is eluted from the column into a second identical column. The effluent from this second column is then guided back to the first column, *etc.* During every cycle, the effective t_G/t_0 decreases from 0.83 after the first cycle to 0.42 after the second cycle, 0.28 after the third cycle, *etc.* The recycling process can be repeated until gradient deformation starts to interfere.

The most important application of gradient-elution LC for polymers is not for separating "blends" of different polymers but for studying copolymers. When analyzing random (statistical) copolymers of sufficiently high molecular weight and with sufficiently fast gradients, retention will solely depend on the chemical composition of the polymers, and a chemical-composition distribution can be obtained from a single (one-dimensional) LC experiment.

For statistical copolymers (monomeric units AABABBBABB in some "random" order), retention varies continuously between that of the respective homopolymers. For block copolymers (with di-block, AAAAABBBBBB, or tri-block, AAAAABBBBBBBBAAAA structures), the elution profile presents a less clear picture of the chemical structure. Retention tends to be dominated by the retained block and its size ("block length"), with the unretained block becoming largely "invisible", *i.e.* not contributing to retention.

For good reviews on interactive LC of polymers, see ref. 13 and 14.

4.4   Hydrodynamic Chromatography (A)

SEC is the bread-and-butter technique for determining molecular weight distributions. It is ubiquitously applied for this purpose, and a large variety of SEC columns are commercially available. SEC also has its weaknesses, which mostly relate to the porous packing material. The diffusion of large molecules in and out of the pores reduces the chromatographic efficiency of SEC. This latter factor does not play a role in **hydrodynamic chromatography** (HDC), a separation method that does not rely on pores to achieve partial exclusion. The principle of HDC is illustrated in Figure 4.20, which is a schematic depiction of a flow channel, the diameter of which ($d_{channel}$) is not much greater than the hydrodynamic diameter ((d_h = $2r_h$) of the largest analyte particles (for illustration purposes). In such a case, these large analytes cannot approach the wall closely enough to sample the slow flow lines in this region. They are confined to a narrower central range of the channel, where the flow lines are faster.

The all-important parameter in HDC is the aspect ratio, defined as

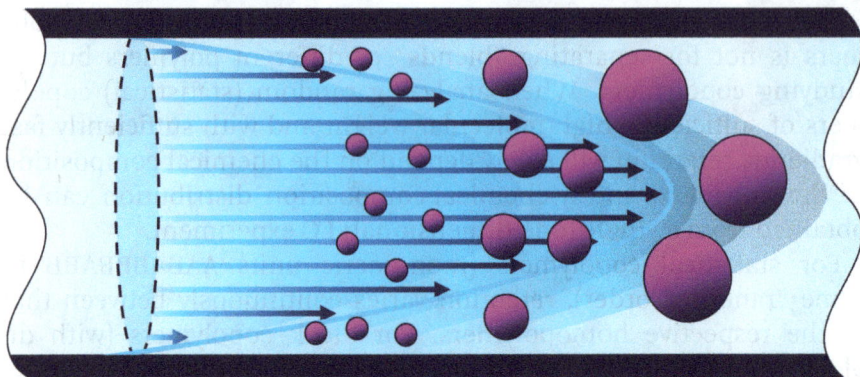

Figure 4.20 Schematic illustration of hydrodynamic chromatography. Particles that are relatively large in comparison with the diameter of the channel are sterically excluded from the slow flow lines close to the wall and thus elute earlier than small molecules that can sample all flow lines in the channel.

$$\lambda = \frac{d_h}{d_{channel}} \tag{4.24}$$

Approximate theories for HDC have led to a simple relationship between λ and a dimensionless retention parameter $\tau_i = V_{e,i}/V_0 = \bar{u}_0/\bar{u}_i$, where $V_{e,i}$ is the elution volume of analyte i and $\bar{u}_i$ is its average linear velocity, and V_0 and $\bar{u}_0$ are the corresponding properties of the mobile phase (*i.e.* the hold-up volume and the average mobile-phase velocity, respectively). Retention follows

$$\tau = \frac{1}{1 + 2\lambda - C\lambda^2} \tag{4.25}$$

where the coefficient C is said to account for phenomena that are not captured by the theory, such as the rotation of analyte molecules or particles and perfusion of the analyte molecules. Discrete values of C have been proposed for different types of polymers or particles (*e.g.* idealized linear exclusion, for which $C = 0$; impermeable hard spheres, for which $C = 4.9$, *etc.*). For a dilute polymer solution under thermodynamically good solvent and temperature conditions, a value of $C = 2.7$ has been derived.[15]

It is clear from Figure 4.21 that HDC, like SEC, has a limited elution window. If anything, the window is narrower in HDC, ranging from about $0.7 < \tau < 1$, *i.e.* 30% of the mobile-phase volume, whereas the range in SEC encompasses up to 50% of the mobile-phase volume. The narrowest channels show an inversion of the elution order above a certain molecular weight. This is suggested by the curve for 1000 nm (1 μm) channels and is very clear from 500 nm (0.5 μm) channels. At this point, the size of the molecules approaches the diameter of the channel. For a molecular weight of 10 000 000 and a 500 nm channel, $\lambda = 0.93$. In this region, molecules are likely to deform (elongate) when passing through the channel. This is the domain of **slalom chromatography**[16] (Figure 4.22) or molecular-topology fractionation.[17] It may be possible to distinguish between different conformations of the polymer molecules (*e.g.* linear *vs.* branched), but such separations will be confounded, as very different (large and small) molecules may show the same elution times. Practical applications are, therefore, likely to require comprehensive two-dimensional separations (see Chapter 7).

HDC may be performed in open-tubular columns, but Figure 4.21 indicates that selectivity will be quite limited unless the column diameter is as small as 2 μm or less. This is extremely challenging. As is the case with SEC and, unlike, for example, gradient-elution LC, there is no focussing mechanism at the column inlet. Unlike

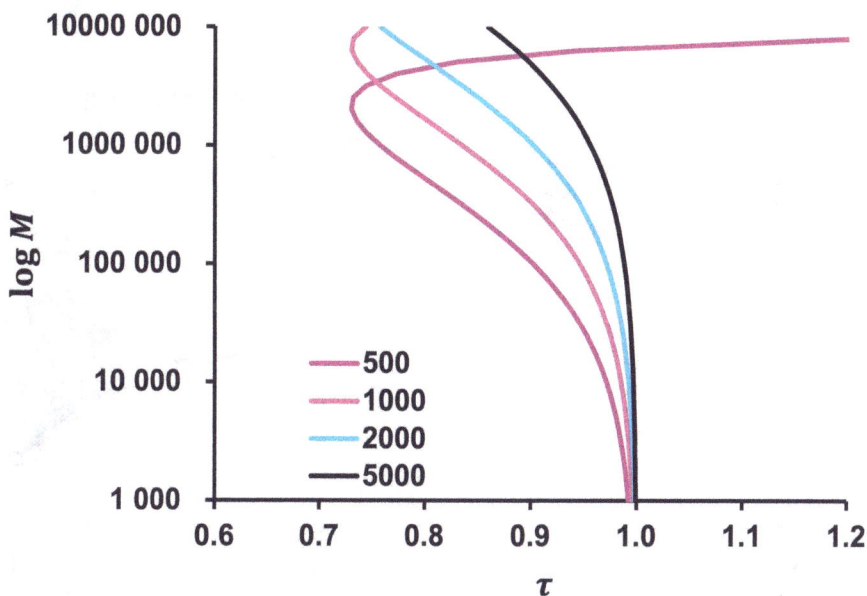

Figure 4.21 Schematic HDC calibration curves plotted using eqn (4.25) with $C = 2.7$. The legend indicates the channel diameter in nanometres. eqn (4.12) was used to calculate r_g, and $r_g/r_h = 0.8$ was assumed.

SEC, where slow mass transport in and out of the pores is a limiting factor, open-tubular HDC can be a high-resolution technique. This is aided by the short diffusion distances across narrow channels. The flip side of the high efficiency is that there is very little tolerance for injection-band broadening and other extra-column dispersion. The same is true for overloading effects, which are readily encountered for the high-molecular-weight analytes for which HDC seems promising. Overloading may be avoided by injecting low concentrations, but this, in combination with the very small tolerated injection and detection volumes, renders detection extremely challenging. The (linear) working range between the detection limit and the point where significant overloading occurs may well be negative. The few demonstrations of open-tubular HDC that have been reported describe amazing experimental achievements but otherwise underline the impracticality of the technique.

HDC may also be performed in packed columns, with steric exclusion occurring in the irregular channels between very small particles.[18] This is a far more practical proposition than open-tubular HDC because the column diameter is not limited. Indeed, packed-column HDC may be performed on commercial LC or SEC

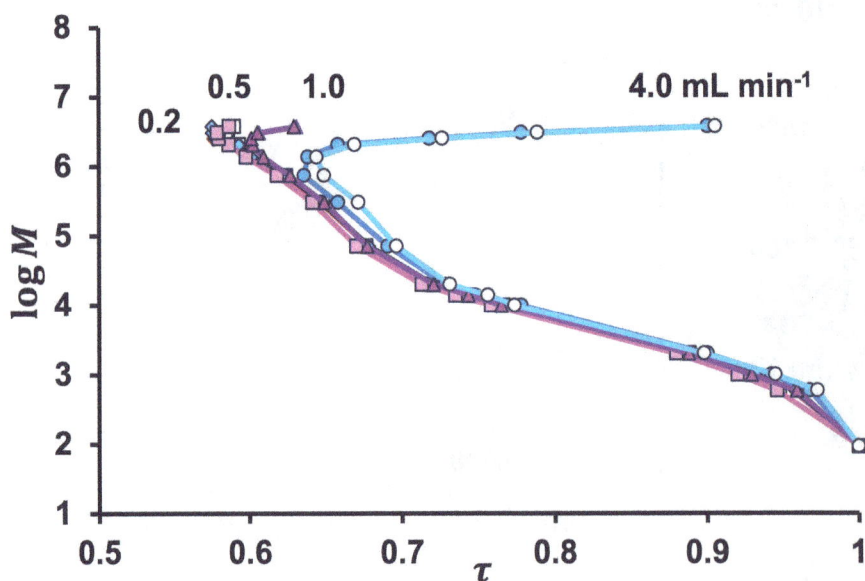

Figure 4.22 Overlay of calibration curves obtained for a core–shell column (150 × 4.6 mm, d_p = 2.6 μm) at four different flow rates. Adapted from ref. 3 with permission from Elsevier, Copyright 2017.

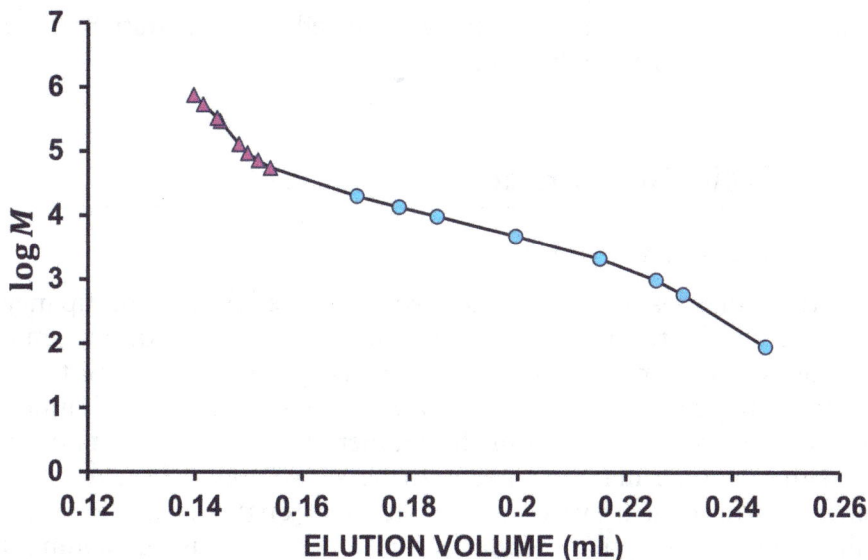

Figure 4.23 Calibration curve and fast SEC and HDC separations obtained on a commercial UHPLC-RPLC column. Data from ref. 6

equipment. The advent of UHPLC with sub-2 μm particles has brought packed-column HDC within the realm of readily available commercial instrumentation. Instead of V_0 and $\bar{u}_0$, the interstitial properties V_{int} and $\bar{u}_{int}$ can be used, but HDC can be performed on commercial columns packed with porous particles. This is because high-molecular-weight analytes that are excluded from the surface of the particles, in principle, do not enter the vicinity of the pores. SEC and HDC can be shown to operate consecutively on commercial UHPLC columns.[6] This is illustrated in Figure 4.23, where the SEC calibration curve (blue dots) is seen to connect with an HDC branch (pink triangles). At a flow rate of 1 mL min⁻¹, the elution window for SEC for polystyrenes up to about 50 kDa is 20 s. The elution window for HDC for standards of about 50 kDa to about 1 MDa is a mere 6 s.

A dedicated commercial solution (described in ref. 19) exists that is specifically dedicated to the separation of nanoparticles. The spherical silica packing particles (15 μm in diameter) and the column (800 mm in length × 7.5 mm i.d.) are both relatively large. The mobile phase is a proprietary, concentrated detergent solution, which conditions the column in such a way that good selectivity can be obtained. However, considerable band broadening is observed, which needs to be computationally corrected for if accurate particle size distributions are to be obtained.[20]

4.5   Field-flow Fractionation (A)

4.5.1   Concept and Modes

In HDC, analytes are separated based on differences in occupancy of different (faster and slower) flow lines. The size (hydrodynamic radius) of the large analyte molecules or particles restricts them from occupying slow flow lines close to the wall of a flow channel. For this to work, the ratio of the diameters of the analyte particle and the channel diameter needs to be substantial. Therefore, the channels need to be very narrow, either through the use of unrealistically narrow open capillaries or through the use of columns packed with very small particles. In **field-flow fractionation (FFF)**, this situation is fundamentally changed by applying a force, *i.e.* the **field**, that pushes analyte molecules (or particles) towards (one of) the wall(s) of an open flat or cylindrical channel. In this way, the analyte will be forced to occupy more of the slow streamlines.

The basic principle of FFF is illustrated in Figure 4.24. The entire family of FFF techniques was developed by Calvin Giddings (see Box 4.1). The field presses the particles towards the accumulation wall, and the largest particles stay closest to the wall because the counteraction of diffusion is slowest. This is the opposite situation from the one encountered in HDC, where the largest particles are

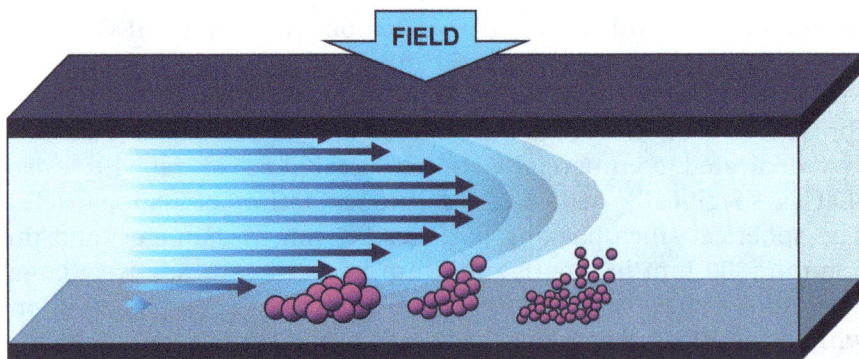

Figure 4.24 Schematic illustration of the principle of FFF in a flat open channel. The wall at the bottom is known as the accumulation wall.

excluded from the wall region due to their size. Thus, in HDC, the largest molecules or particles elute first (just as they do in SEC), whereas in most forms of FFF, they elute last. The extent to which analytes are pressed to the wall and retained in the FFF channel can be influenced by changing the type and strength of the field. Various FFF techniques are classified based on the type of field (see Table 4.2). The different FFF techniques rely on different forces to push analyte molecules or particles towards the accumulation wall. The asymmetric implementation of flow-FFF (AF4) is by far the most prevalent.

The field applied in FFF causes a transverse drift of the analytes towards the accumulation wall. This causes a concentration gradient, with a higher concentration of the analytes closer to the wall, which in turn causes a diffusional flux of analytes away from the wall. When the two processes reach equilibrium, an exponential concentration profile remains, with the highest concentration at the wall. The counterforce is always based on molecular diffusion (the term used for molecules) or Brownian motion (the equivalent term, used mainly for particles).

In **sedimental (centrifugal) FFF** (SdFFF; Figure 4.25A), a curved channel is mounted in a centrifuge that rotates at speeds of up

Box 4.1 Hero of analytical separation science: J. C. Giddings. Photo courtesy of the University of Utah.

J. Calvin Giddings (1930–1996, United States of America)
Calvin Giddings, a professor at the University of Utah (Salt Lake City, UT, USA), was one of the great minds, if not the greatest, in the history of separation science. In his 1965 book *Dynamics of Chromatography* (Vol. 1), he laid the theoretical foundations for much of our field. The book had just one major flaw: there was never a Vol. 2. Apart from his many contributions to various forms of chromatography, Cal Giddings created field-flow fractionation, an entire range of separation and fractionation methods for higher-molecular-weight analytes and small (micro-/nano-)particles. Cal was ahead of his time in caring for nature, environment and our planet.

Table 4.2 Classification of FFF techniques.

Name	Abbreviation	Field	Analyte property	Commercial
Flow-FFF	FFF	Cross-flow	—	—
Asymmetric flow-FFF	AF4	Cross-flow	Size	Yes
Hollow-fibre flow-FFF	HF5	Cross-flow	Size	Yes
Sedimentation FFF	SdFFF	—	—	—
Centrifugal FFF	CfFFF	Centrifugal	Density[a]	Yes
Thermal FFF	T(h)FFF	Temperature gradient	Thermal diffusion[a]	Yes
Electrical FFF	ElFFF	Voltage gradient	Charge[a]	Yes
Gravitational	GrFFF	Gravity	Density[a]	Yes
Magnetic	MFFF	Magnetic	Susceptibility	No
Acoustic	AcFFF	Sound waves	Compressibility[b]	No
Dielectrophoretic	DEP-FFF	Electric (non-uniform)	Charge[a]	No

[a]And size.
[b]And density.

to 2500 rpm. Intricate roll bearings at the centre of the circle are needed to transport the carrier fluid in and out of the channel and create the necessary longitudinal flow. SdFFF is especially suitable for the separation of particles in the nanometre – and sometimes micrometre – range. Separation is based on differences in density. If the particle densities exceed the density of the carrier liquid, centrifugal forces push them towards the accumulation wall, and the most dense analytes are retained longest in the channel. The separation based on density is accompanied by separation based on size, as the centrifugal field is counteracted by Brownian motion. This can be an advantage when fractionating a homogeneous sample (*e.g.* gold nanoparticles), but a disadvantage when the

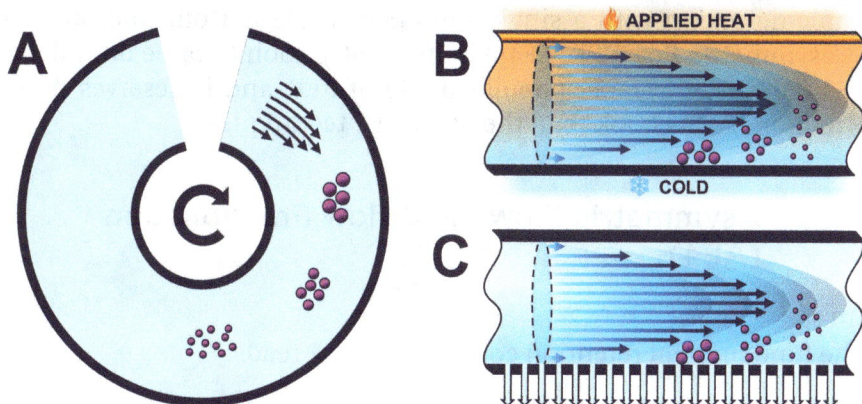

Figure 4.25 Schematic illustration of three implementations of FFF. (A) centrifugal sedimentation FFF (SdFFF); (B) thermal FFF (ThFFF); (C) asymmetric flow FFF (AF4).

sample is more complicated, containing (many) different kinds of nanoparticles.

In the case of **thermal FFF** (ThFFF; Figure 4.25B), thermal diffusion is the force that drives analyte molecules away from the "hot" wall towards the "cold" accumulation wall. This effect is mainly observed for synthetic polymers, and ThFFF has occasionally been used to study these. The temperature of the "hot" wall is limited by the boiling point of the carrier fluid and the temperature of the "cold" wall by its freezing point.

In commercial instrumentation for ThFFF, the maximum temperature of the hot wall is 190 °C, while the cold wall is typically kept at 15 or 20 °C. Whether or not thermal diffusion occurs, let alone its magnitude, is hard to predict. It basically depends on the nature of the analyte molecules and that of the carrier liquid. Thermal diffusion is typically absent in water but often occurs in organic solvents. Retention depends on the chemical composition of the analytes, suggesting that polymer blends (*i.e.* mixtures of polymers) may be separated or chemical composition distributions may be obtained for copolymers. However, the counteraction of molecular diffusion depends on the size (hydrodynamic radius) of the analyte molecules, so any separation according to composition is confounded with a separation according to molecular size (or weight). This greatly reduces the value of the ThFFF method.

Flow FFF was initially envisioned with two parallel semi-permeable walls at the top and bottom of the channel. However, the implementation with a single semi-permeable bottom wall, known as **asymmetric flow FFF** (AF4, Figure 4.25C), soon became dominant. AF4 is by far the most popular mode of FFF, and it deserves more attention in this book than the other FFF techniques.

4.5.2 Asymmetric Flow Field-flow Fractionation (AF4)

4.5.2.1 Concept

The fundamental retention equation for FFF reads

$$R_{L} = \frac{t_{R}}{t_{0}} = \frac{1}{6\lambda\left\{\coth\left(\frac{1}{2\lambda}\right) - 2\lambda\right\}} \approx \frac{1}{6\lambda} \tag{4.26}$$

where R_{L} is the relative retention time or "retention level",[21] and the dimensionless retention parameter λ is defined as

$$\lambda = \frac{l}{w} \tag{4.27}$$

with l being the mean distance of the analyte molecules (or particles) from the accumulation wall, also known as the **characteristic layer thickness**, and w being the channel thickness. The approximation $R_{L} \approx 1/6\lambda$ is quite accurate for low values of λ ($\lambda \leq 0.04$, or $t_{R} \gtrsim 4t_{0}$), but it underestimates the retention of less-retained analytes. eqn (4.26) implies that retention increases if l and, consequently, λ decreases. In other words, retention increases if the analyte molecules are compressed closer to the wall. The equation predicts that if the analyte molecules are, on average, 3% of the channel thickness away from the wall (*i.e.* $\lambda = 0.03$), the retention time is about six times the unretained time ($R_{L} = 5.91$).

In flow FFF, the balance between diffusional forces and the flow field results in

$$l = \frac{D_{m}}{u_{cross}} \tag{4.28}$$

where D_{m} is, as always, the molecular diffusion coefficient of the analyte in the mobile phase and u_{cross} is the linear cross-flow velocity.

It is important to realize that in flat-channel flow FFF, the thickness (height) of the channel is smaller than the spacer thickness if soft, compressible membranes are used. When assembling the AF4 channel, they are compressed in areas covered by the spacer and are relatively thicker in the fractionation channel. Additionally, the spacer itself is compressible to some extent. Therefore, an effective channel thickness (w_{eff}) is introduced in flow FFF. The surface area (A_{mem}) of the membrane in the channel follows from $A_{mem} = V_0/w_{eff}$, which allows us to convert the cross-flow velocity to the volumetric cross-flow through $u_{cross} = F_{cross}/A_{mem} = F_{cross}w_{eff}/V_0$ so that eqn (4.28) yields an equation that relates the diffusion coefficient to flow-FFF retention data,

$$D_m = l \cdot u_{cross} = \lambda w_{eff} \cdot \frac{F_{cross} \cdot w_{eff}}{V_0} = \frac{\lambda \cdot F_{cross} \cdot w_{eff}^2}{V_0} \approx \frac{t_0 \cdot F_{cross} \cdot w_{eff}^2}{6V_0 \cdot t_R} \qquad (4.29)$$

In AF4 practice, it is difficult to determine t_0 accurately. A "void peak" observed in the fractogram does not necessarily provide a good estimate of t_0. Instead, t_0 may be calculated from[22]

$$t_0 = \frac{V_0}{F_{cross}} \ln\left(1 + \frac{F_{cross}}{F_{out}}\right) \qquad (4.30)$$

This yields for the analyte diffusion coefficient

$$D_m \approx \frac{w_{eff}^2}{6t_R} \ln\left(1 + \frac{F_{cross}}{F_{out}}\right) \qquad (4.31)$$

Finally, we can use the **Stokes–Einstein equation** to also arrive at an equation for the hydrodynamic diameter of the analyte particles (d_h),

$$\begin{aligned}
d_h &= \frac{k \cdot T}{3\pi \cdot \eta \cdot D_m} \approx \frac{2k \cdot T \cdot V_0 \cdot t_R}{\pi \cdot \eta \cdot t_0 \cdot F_{cross} \cdot w_{eff}^2} \\
&= \frac{2k \cdot T \cdot t_R}{\pi \cdot \eta \cdot w_{eff}^2} \cdot \frac{1}{\ln\left(1 + \dfrac{F_{cross}}{F_{out}}\right)}
\end{aligned} \qquad (4.32)$$

Thus, flow FFF can be used to estimate diffusion coefficients and particle diameters from retention data, provided that retention is sufficiently high for the approximation $R_L \approx 1/6\lambda$ to be valid. To estimate D_m or d_h, w_{eff} must first be estimated through "calibration" with an analyte for which the diffusion coefficient in the mobile phase is known (*e.g.* bovine serum albumin, BSA) or with reference particles with a known hydrodynamic diameter (*e.g.* "monodisperse" polystyrene latex particles). A risk in dealing with long-chain macromolecules is that they may get entangled, even at moderate concentrations, causing the apparent diffusion coefficients to be lower and the hydrodynamic diameters to be larger than their molecular values. Calibration with analytes with known D_m can be avoided if D_m (or d_h) can be measured online. This can be achieved with dynamic light-scattering detection (see Section 4.6.2).

Rewriting eqn (4.29) and (4.31) in terms of retention, we have

$$t_R \approx \frac{t_0 \cdot F_{cross} \cdot w_{eff}^2}{6V_0 \cdot D_m} \approx \frac{\pi \cdot \eta \cdot t_0 \cdot F_{cross} \cdot w_{eff}^2 \cdot d_h}{2k \cdot T \cdot V_0} \tag{4.33}$$

or

$$t_R \approx \frac{w_{eff}^2}{6D_m} \ln\!\left(1 + \frac{F_{cross}}{F_{out}}\right) \approx \frac{\pi \cdot \eta \cdot w_{eff}^2 \cdot d_h}{2k \cdot T} \ln\!\left(1 + \frac{F_{cross}}{F_{out}}\right) \tag{4.34}$$

From eqn (4.33) and (4.34), we can see that (i) retention increases with the increasing size of the analyte molecules or particles (*i.e.* with decreasing D_m and increasing d_h); (ii) retention increases with increasing cross-flow (F_{cross}); (iii) the cross-flow should be adapted to the channel volume (it is the ratio F_{cross}/V_0 that matters); and (iv) a thinner channel (smaller w_{eff}) results in (much) shorter retention times, as, at the same distance from the wall, the flow lines are faster in a narrow channel.

It should be noted that the limit at $F_{cross} = 0$ according to eqn (4.34) is $t_R = 0$. This is, of course, not realistic. eqn (4.29), (4.31), (4.33) and (4.34) are all approximate equations, and their validity cannot be stretched to the impractical region where F_{cross} approaches 0.

The efficiency of AF4 separations, *i.e.* the widths of observed peaks and the processes leading to band broadening, is not studied nearly as intensively as in LC. The focus is on selectivity rather than efficiency. The apparent plate height in AF4 (H_{AF4}) can be approximated as[21]

$$H_{AF4} = \frac{w_{eff}}{9D_m}\left(\frac{t_0}{t_R}\right)^3 \overline{u_0} \qquad (4.35)$$

where $\overline{u_0}$ is the average linear (longitudinal) velocity. Typically, peaks in AF4 are not observed to broaden at higher retention times, while resolution increases. It is favourable to operate AF4 systems at relatively high channel flow rates (resulting in low t_0). High cross-flows (which would result in high t_R/t_0 ratios) are not always recommended, as they may lead to various undesirable effects (*e.g.* aggregation of analyte molecules, interaction with or adsorption on the membrane).

4.5.2.2 Strengths and Weaknesses of AF4

AF4 is a very useful method for separating analytes of high molecular weight, non-covalently bonded molecular aggregates, and nanoparticles. Unlike HDC (Module 4.4) or SEC (Module 4.2), it does not require a packed column, and it can be performed in open channels. This is because the analytes are pressed towards the accumulation wall (*i.e.* the membrane) by the field (*i.e.* the cross-flow) so that they are sampling the slow flow lines near the wall much more often. HDC can be seen as "fieldless" FFF. The absence of particles implies that the stress exerted on the analyte units (molecules, aggregates, or particles), as expressed by the Deborah number (eqn (4.11)), is greatly reduced. However, when calculating the Deborah number for FFF, it may be more appropriate to use the mean distance from the wall (l, eqn (4.27)) as the characteristic length parameter instead of the effective channel width (w_{eff}).

The absence of particles also implies that the surface area available for undesirable energetic interactions is drastically reduced. In AF4, there is a possibility of analytes interacting with the membrane, but this constitutes a much smaller surface area, and inert materials can be sought. However, the possibility of interactions is increased by the fact that the cross-flow pushes the analytes towards the membrane, increasing their local concentration. The low stress exerted on the analytes renders FFF eminently suitable for studying higher-order structures of non-covalently bonded molecules, such as micelles and liposomes, or protein aggregates. Many contemporary drug-delivery systems feature nano-sized particles, and AF4 has become an important technique to characterize these. Likewise, AF4 can be used to characterize viruses. AF4 can also be used as a purification technique, for example, to remove low-molecular-weight additives from protein formulations. Table 4.3 provides an overview of the strengths and weaknesses of the AF4 technique.

4.5.2.3 AF4 Practice

The FFF channel has a typical **trapezoidal** shape, as shown in Figure 4.26. This shape is intended to keep the linear velocity more or less constant, despite the loss of longitudinal flow due to the cross-flow through the semi-permeable membrane that forms the bottom of the channel. The linear velocity is not exactly constant. This would require a more intricate shape and, unrealistically, a constant ratio

Table 4.3 Summary of the strengths and weaknesses of asymmetric flow field-flow fractionation (AF4).

Strengths	Weaknesses
No packing particles, resulting in	Presence of membrane implies
• Low stress exerted on analytes (including non-covalent aggregates) • Low-pressure drop across the channel • Much reduced surface area, reducing the possibility of interactions	• Possible interactions of analytes with the surface of the membrane • Pore-size distribution (and molecular-weight cut-off) not well-characterized • Permeability of the membrane may change over time • Membrane may eventually get clogged
Potential to analyze very large molecules, higher-order structures or nanoparticles (provided that a homogeneous solution, emulsion or suspension can be obtained)	Low-molecular-weight analytes are not observed
AF4 separates strictly on size (unlike other FFF techniques, which yield confounded distributions; SdFFF: size/density; ThFFF composition/molecular weight)	Chemically or architecturally different structures with identical sizes may co-elute
Physical properties (d_h, D_m) can be obtained from AF4 measurements	To obtain size parameters (d_h) or diffusion coefficients (D_m), the AF4 channel must be calibrated (*i.e.* weff must be determined) using reference standards
Obtained fractions may be characterized with a variety of methods (online or offline)	Online coupling is complicated by need to maintain constant channel flow (*i.e.* channel outlet pressure)

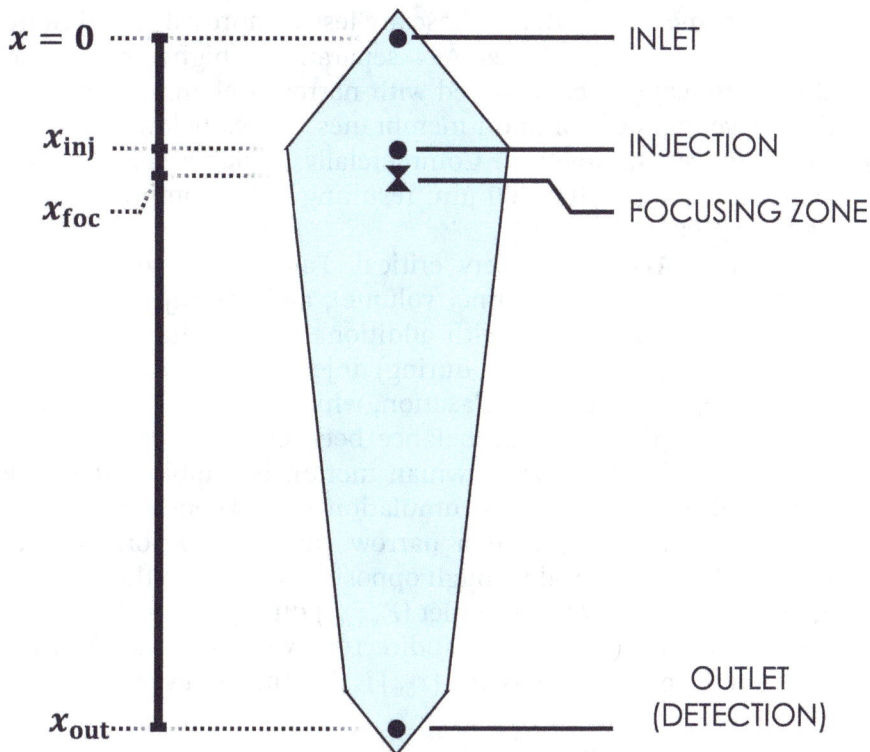

Figure 4.26 Schematic top view of a channel for asymmetric flow field-flow fractionation (AF4). The location of the focussing zone is determined by the magnitudes of opposing flows from the inlet and outlet ports during the focussing step (see eqn (4.36)). Instead of reversing the flow through the outlet port, a separate port may be used for the focussing flow.

between the channel (outlet) flow and the cross-flow. In an alternative configuration, the channel is a semi-permeable fibre (hollow-fibre flow FFF or HF5). In this latter case, the longitudinal flow decreases with the distance travelled through the fibre.

Typical membranes are prepared from organic microfibres, such as regenerated cellulose or polyether sulfone. Such membranes are inherently inhomogeneous, implying that there is no exact cutoff in terms of analyte size or molecular weight. A cutoff specified as 10 kDa (a common value) should be interpreted as an indicative number. Organic membranes are also compressible, creating a good seal between the flow channel and the areas compressed by the spacer and causing the effective channel diameter (w_{eff}) to be smaller than the thickness of the spacer. Better-defined membranes exist (*e.g.* ceramic membranes), but these are less compressible and more difficult to use in AF4. Better AF4 separations (higher resolution in shorter times) can be achieved with narrower channels, but due to the unevenness of common membranes, values below $w_{eff} \approx 100$ μm have proven unpractical. Commercially available spacers have thicknesses of about 200–500 μm, resulting (after compression) in $100 \lesssim w_{eff} \lesssim 300$ μm.

Injection in AF4 is not very critical. Fairly large volumes are injected (relative to the channel volume), and the injection loops and tubing may be washed with additional carrier fluid to ensure complete injection. After (or during) injection, a relaxation and focussing step is included. Relaxation, which is critical but usually quite rapid, implies that the balance between the cross-flow and the molecular diffusion or Brownian motion is established in the transverse direction to the accumulation wall. Focussing implies concentrating the sample in a narrow zone in the longitudinal direction. This is achieved through opposing flows from the channel inlet ($F_{in,foc}$) and the channel outlet ($F_{out,foc}$) during this stage. If x is the coordinate in the longitudinal direction, with $x_{in} = 0$ at the inlet point, then the point of focussing (x_{foc}) is determined by

$$\frac{x_{foc}}{x_{out}} = \frac{F_{in,foc}}{F_{in,foc} + F_{out,foc}} \tag{4.36}$$

Usually, the focussing point is chosen slightly further into the channel than the injection point ($x_{foc} > x_{inj}$). The focussing step is less critical than the relaxation step, especially when cross-flow programming is applied during the run. Too much focussing (*i.e.* too high an (initial) cross-flow, too high an (initial) analyte concentration, or too high a flow ($F_{in,foc} + F_{out,foc}$) during focussing) increases the risk of analyte adsorption to the membrane, leading to a decrease of recovery, and analyte–analyte interactions, resulting in band broadening and incorrect assessment of analyte properties.

Cross-flow programming is the counterpart in AF4 to temperature programming in GC or gradient elution in (interactive) LC. It is the obvious approach if the target analytes differ greatly in molecular weight or particle size. A common program is a (time-delayed) exponential decay, which implies that the cross-flow decreases exponentially after an initial hold period.

When analyzing relatively large particles (say, 500 nm in diameter or larger), there is a possibility for a reversal of the elution order, known as the steric mode of FFF. This situation is akin to that encountered in HDC, despite the presence of the field. Under conditions in which all analyte particles are strongly compressed near the accumulation wall (the membrane), the largest particles still cannot occupy the slowest flow lines because they are sterically excluded from the wall. If their mean position (l) concurs with a significant longitudinal migration velocity, such large particles will elute from the channel. Elution in steric mode may feasibly be used to fractionate relatively large particles, but a more common consequence is a confounding effect on the particle-size distribution, with particles of different sizes eluting simultaneously (smaller particles in normal mode and larger ones in steric mode). The best way to diagnose whether the observed PSD is obscured by steric-mode effects is to check the effect of the cross-flow using eqn (4.31) (or eqn (4.29)) and eqn (4.32) to estimate the analyte diffusion coefficient and hydrodynamic diameter, respectively. In normal mode, the observed D_m and d_h should be independent of the cross-flow. If the estimated value of D_m increases and that of d_h decreases with increasing F_{cross}, there is a strong indication that steric effects are at play.

4.6 Detection Techniques for Large Molecules (A)

The same detectors commonly used in the LC of small molecules can be used for polymers. There are some differences in the application or applicability of certain common LC detectors that have been described in Module 3.9. There are also some dedicated detectors, which have not yet been discussed and are particularly suited for macromolecules. These include (liquid-phase) light-scattering detectors (Sections 4.6.1 and 4.6.2) and viscometers (Section 4.6.3).

Differential-refractive-index (DRI) detectors are more commonly used in SEC than in other forms of LC for several reasons. One reason for the unpopularity of DRI detectors in LC is their incompatibility with gradient elution. This is not a disqualifier in SEC, HDC, or AF4 because these are isocratic methods. The relatively high detection limits are also not a main obstacle for these size-based separation methods in most cases. Finally, many synthetic polymers are not UV-active, as this is a favourable property when aiming to create lightfast materials.

Evaporative light-scattering detectors (ELSDs) are also frequently used for detecting polymers, either in combination with size-based separation methods or interactive (gradient-elution) LC. However, the sensitivity of ELSD varies with analyte concentration (*i.e.* the response is non-linear) and analyte molecular weight. In the case of gradient elution, the mobile-phase composition may also affect the ELSD sensitivity. As a result, quantitative analysis of polymers, including the quantitative characterization of MWDs, is very complicated when using ELSD.

Mass spectrometry (MS) is often used in the analysis of proteins and peptides (see Module 3.10). It is used (much) less often for the characterization of synthetic polymers. One reason is that extensive separation is required prior to MS detection to avoid bringing a wide variety of molecules into the ionization chamber at the same time and, as a result, a great bias in the data. Even after extensive liquid-phase separation, MS tends to be difficult for high-molecular-weight polymers and virtually impossible for very non-polar or very polar polymers, such as polyolefins (polyalkenes) or polysaccharides, respectively.

For a more detailed overview of the challenges faced in the quantitative detection of polymers after LC separations, see ref. 23.

4.6.1 Static Light Scattering Detection

Static light scattering (SLS) detectors are often used in combination with isocratic separation methods, such as SEC, HDC, and AF4, in research environments. They may provide measurements of the molecular weight and size in solution of eluting molecules. Such online measurements provide an instant validation of the separation mechanism, for example, to confirm the absence of adsorption effects in SEC or to verify operation in normal mode (rather than steric mode) in AF4. In some cases (mainly for homopolymers), SLS detection may alleviate the need for calibration standards.

Light is scattered by a solution of a polymer (or a suspension or emulsion of particles) in all directions. In static light scattering, the intensity of the scattered light is assumed to be constant, as long as the sample does not change. In contrast, dynamic light scattering (Section 4.6.2) is based on variations in the intensity of the scattered light. The amount of light $I(\theta)$ detected at an angle θ from the incident light (with intensity I_0) and at a distance r_{det} from the sample is given by the excess **Rayleigh ratio**

$$R(\theta) = \left[\frac{I(\theta)r_{det}^2}{I_0}\right]_{solution} - \left[\frac{I(\theta)r_{det}^2}{I_0}\right]_{solvent} \tag{4.37}$$

The distance r_{det} is frequently omitted in eqn (4.36), but in that case, it should be added to the calibration factor K^*,

$$K^* = \frac{4\pi^2 n_0^2}{\lambda_0^4 N_A} \cdot \left(\frac{\partial n}{\partial c}\right)^2 \tag{4.38}$$

which contains both factors related to the instrumental setup, such as the vacuum wavelength of the incident radiation, λ_0, and possibly r_{det} (see above), as well as factors related to the solvent (its refractive index, n_0) and the analyte. The specific refractive-index increment $\partial n/\partial c$ is a critically important parameter that varies with the analyte polymer and the solvent, as well as the temperature and wavelength of the experiment. Knowledge of $\partial n/\partial c$ is required to draw quantitative inferences from SLS data. N_A in eqn (4.38) is Avogadro's number.

Both the excess Rayleigh ratio and the calibration factor K^* appear in the all-important Rayleigh–Gans–Debye equation for static light scattering,

$$\frac{K^* c}{R(\theta)} = \frac{1}{P(\theta)}\left(\frac{1}{M_w} + 2A_2 c + 3A_3 c^2 + \ldots\right) \tag{4.39}$$

where c is the concentration of the analyte polymer and A_2, A_3, etc., are virial coefficients. In the case of a good solvent, as is a necessity in SEC, A_2 is positive, but the low concentrations that are typically encountered in online SEC-SLS systems (after dilution of the analytes due to the chromatographic process) imply that the virial coefficient terms can usually be neglected. The angular dependence of the SLS process is described by the form factor $P(\theta)$ (more officially known as the particle-scattering factor), which is described by

$$\frac{1}{P(\theta)} = 1 + \frac{2}{3!}\left[\frac{4\pi n_0}{\lambda_0} \cdot r_{rms} \sin\frac{\theta}{2}\right]^2 + \frac{2}{5!}\left[\frac{4\pi n_0}{\lambda_0} \cdot r_{rms} \sin\frac{\theta}{2}\right]^4 + \ldots \tag{4.40}$$

Usually, the root-mean-square radius, r_{rms}, which is similar but not equal to the radius of gyration, is small enough for the fourth-order term to be neglected. For small molecules, there is hardly any angular dependence of the light scattering and $P(\theta) \approx 1$. As a rule of thumb, $M_w \lesssim 50\,000$ is often stated, but the molecular weight above which angular dependence becomes significant depends strongly on the (refractive index of the) solvent and the wavelength of the SLS light source. In the absence of angular dependence and if the concentration is also low, eqn (4.39) and (4.40) condense to

$$M_w \approx \frac{R(\theta)}{K^*c} \qquad (4.41)$$

where the excess Rayleigh ratio can be measured at any angle (*e.g.* at 90°, as in a commercial right-angle light-scattering (RALS) detector. More commonly, **multi-angle light scattering (MALS) detectors** are used, which contain several photosensors at fixed positions that allow measuring the scattered light at multiple angles simultaneously. Such detectors allow measuring M_w and r_{rms} simultaneously for each fraction of polymer solution eluting from a SEC column or an FFF channel, provided that the concentration of the analyte can be measured simultaneously. For this purpose, an SLS detector is usually combined with a concentration detector, such as a UV spectrometer or a differential refractive index detector. Care must be taken to correct for the time delay between the two detectors so that data from corresponding points in the elution profile (*i.e.* the same fraction) are used in the calculations.

SLS detectors are often called "absolute" detectors because they determine molecular weight without the use of a calibration curve so that the determined molecular weight is not relative to calibrants (*e.g.* relative to PS or to dextran). However, (i) the response of the SLS unit must be calibrated. This is usually done using a solvent with a known (and large) Rayleigh ratio (*e.g.* toluene), for which the refractive index at the experimental conditions is known. This calibration does not have to be repeated when changing solvents, temperature, *etc.*, although it is prudent to check the response occasionally (*e.g.* quarterly); (ii) the response of the various photodiodes must be normalized relative to an arbitrary one (most conveniently, that placed at 90°), and (iii) the specific refractive-index increment of the analyte polymer (in the eluent, at the detection temperature and wavelength) should be accurately known. The latter condition is the main bottleneck. It may already be a challenge to measure $\partial n/\partial c$ for a homopolymer because this must be done using the

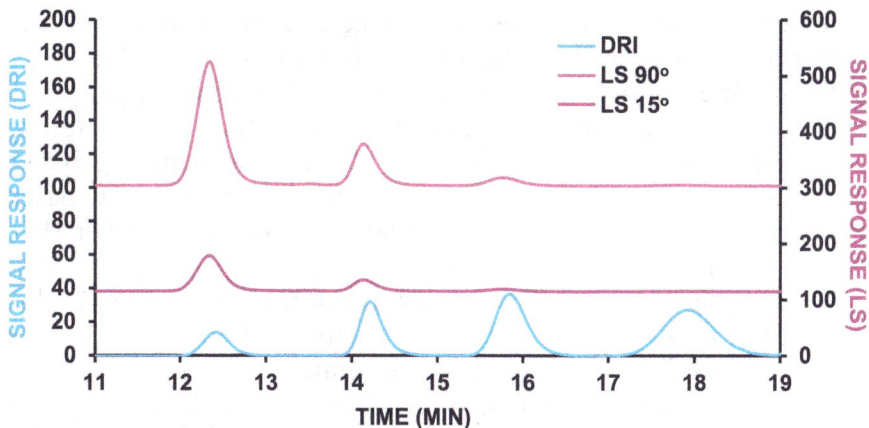

Figure 4.27 Chromatograms of polystyrene standards recorded with a differential-refractive-index (DRI) detector (pink trace, top) and with a right-angle (90°) static-light-scattering (RALS) detection (blue trace, bottom). Added in the middle is the trace from static light scattering at a 15° angle (purple trace). Data courtesy of Dr. Ton Brooijmans (Envalior).

same wavelength as that used in the MALS detector. For copolymers, the situation is dire. Any composition drift, *i.e.* any variation of the chemical composition with the molecular weight, causes $\partial n/\partial c$ to vary along the SEC profile and a variety of different molecules with different $\partial n/\partial c$ values to elute simultaneously. For copolymers, accurate information may be obtained on r_{rms} from the angular dependence of the excess Rayleigh ratio, but accurate information on M_w remains elusive.

The intensity of the scattered light increases with the increasing molecular weight of the analyte. This follows directly from rewriting eqn (4.41) as $R(\theta) = K^* \cdot c \cdot M_w$. Thus, high-molecular-weight analytes yield much more intense signals on light-scattering traces than in the corresponding chromatograms recorded with a concentration-sensitive (DRI or UV) detector (see Figure 4.27).

If both M_w and r_{rms} are obtained from light-scattering experiments, important information can be deduced on the conformation of analyte molecules or the morphology of analyte particles. This is typically done by plotting $\log r_{rms}$ against $\log M_w$ to obtain a so-called conformation plot. Particles with different structures exhibit different slopes in such plots. For rod-like particles or hollow spheres, the slope may approach unity. For solid spheres, the slope should be closer to 0.3. Randomly coiled linear polymers exhibit slopes between 0.5 and 0.6, but the branching of polymers causes the molecules to be more compact in solution so that a smaller value of r_{rms} results in a given value of M_w. In the likely event of an increasing number of branching points with increasing molecular weight, the conformation plot will be flattening towards higher values of M_w. The slope of the conformation plot also depends on the thermodynamic quality of the solvent, but in size-based separations, "good" solvents are usually encountered, rather than "poor" solvents or "theta" solvents.

The analysis of **branched polymers** illustrates both the power and the limitations of the SEC–MALS combination. At any time point (or elution volume) in the SEC trace, molecules elute with equal size (hydrodynamic radius, r_h) in solution. These may be linear polymers of a certain molecular weight or more compact branched molecules of a higher molecular weight. Thus, at any slice of the chromatogram, a distribution of molecules is eluted. Online MALS detection will yield (weight) average values of M_w and r_{rms}. The ratio between the hydrodynamic radius and the average root-mean-square radius is indicative of the average degree of branching of the molecules.

As with copolymers, a one-dimensional (SEC) separation falls short of comprehensively characterizing a polymer sample that features two essential distributions (MWD and CCD in the case of copolymers; MWD and DBD in the case of branched polymers), even with highly informative detection systems. To achieve such goals, two-dimensional separations, as described in Chapter 7, are a prerequisite.

4.6.2 Dynamic Light Scattering Detection

Dynamic light scattering (DLS) has long been available for characterizing nanoparticles, such as latexes. It is also known as quasi-elastic light scattering (QELS) or photon-correlation spectroscopy (PCS). Because particles in a solvent are continuously subjected to Brownian motion, the intensity of scattered light will vary. The rate at which such variations in the intensity of the light occur is indicative of the size of the scattering particles. The larger they are, the slower they move, and the slower the variations in the scattered light. DLS yields the molecular (or "translational" diffusion coefficient) (D_m) of the particles, which can be connected to the hydrodynamic diameter (d_h) of the particles using the Stokes–Einstein relation (see eqn (4.32)). AF4 measurements also provide a measure for d_h, but this assumes the validity of the FFF retention equations. Such assumptions do not need to be made when using DLS. In theory, DLS may provide a particle-size distribution of a polydisperse sample, but online DLS detectors in combination with SEC or, more commonly, FFF instruments provide a less ambiguous picture of the particle-size distribution.

4.6.3 Viscometric Detection

In Chapter 1, we discussed laminar (Poiseuille) flow. (Eqn (1.64)) can be rewritten as an explicit equation for the viscosity (η) of a fluid,

$$\eta = \frac{\Delta P \cdot \pi r_c^4}{8LF} = \frac{\Delta P \cdot \pi d_c^4}{128LF} \tag{4.42}$$

where, as before, ΔP is the pressure drop across a capillary tube, L is the length of the capillary, r_c is its inner radius, d_c is its inner diameter, and F is the volumetric flow rate. Thus, if we know the dimensions of the capillary and the flow rate, we can determine the viscosity by measuring the pressure. This is precisely what is done in the simplest (single capillary) viscometric detector. Measuring the viscosity $\eta(t)$ of a polymer solution as a function of time and the baseline viscosity of the solvent (η_0) yields the specific viscosity $\eta_{sp}(t)$

$$\eta_{sp}(t) = \frac{\eta(t) - \eta_0}{\eta_0} \tag{4.43}$$

However, the specific viscosity is not the most meaningful characteristic of the polymer solution, and a single-capillary viscometer is sensitive to fluctuations in flow rate and temperature. Therefore, viscometers with more than one capillary have been developed. The most elegant of these designs is akin to an electrical **Wheatstone bridge** and the thermal-conductivity detector for GC described in Section 2.5.2. Such a design is illustrated in Figure 4.28. It allows direct measurement of $\eta_{sp}(t)$ through

$$\eta_{sp}(t) = \frac{\eta(t) - \eta_0}{\eta_0} = \frac{4\Delta P_1(t)}{\Delta P_2(t) - 2\Delta P_1(t)} \tag{4.44}$$

where ΔP_1 is the pressure difference across the centre of the bridge (see Figure 4.28) and ΔP_2 is the pressure difference between the inlet and the outlet of the bridge. The presence of the reservoir before

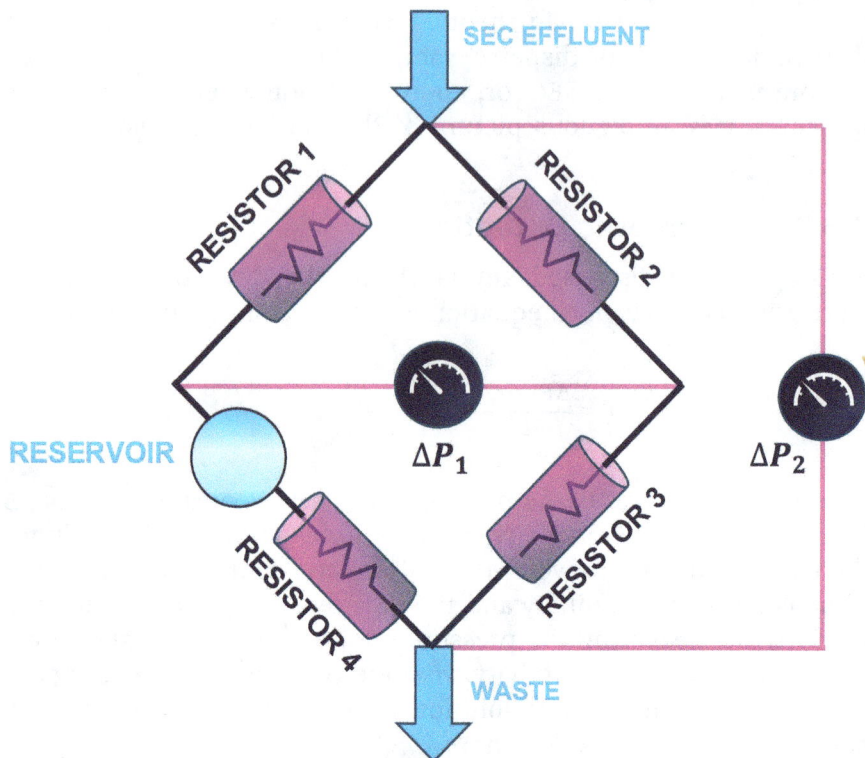

Figure 4.28 Schematic illustration of a viscometer design analogous to a "Wheatstone bridge" circuit. The four resistors are capillaries with known lengths and diameters. The differential pressure transducers measure differences across the bridge in two directions.

Resistor #4 implies that this resistor contains the solvent rather than the polymer solution at the point of measurement. The volume of the reservoir (and its design) should be such that no sample solution enters Resistor #4 during the analysis. Between analyses, the polymer solution has to be washed from the reservoir to ensure a good baseline during the next run.

Fundamentally, the specific viscosity that is measured directly by the viscometer is not as meaningful as the intrinsic viscosity, which is defined as

$$[\eta] = \lim_{c_i \downarrow 0} \frac{\eta_{sp}}{c_i} \qquad (4.45)$$

where c_i is the concentration of the analyte polymer. In practice, c_i is measured with a concentration detector (DRI or UV), and it is supposed to be small enough to assume $[\eta] \approx \eta_{sp}/c_i$. As with light-scattering detectors, the inter-detector delay must be corrected when calculating the intrinsic viscosity. The intrinsic viscosity relates to the molecular weight through the **Mark–Houwink** (or Mark–Houwink–Sakurada) equation

$$[\eta] = K_{MH}M^{a_{MH}} \tag{4.46}$$

where K_{MH} and a_{MH} are the so-called Mark–Houwink coefficients, which elate to the intercept and the slope, respectively, in a plot of $\log [\eta]$ *vs.* $\log M$. Both coefficients are generally positive so that the intrinsic viscosity increases with increasing molecular weight. Negative values of a_{MH} have been encountered for dendrimers and hyperbranched polymers. A viscometer is also a molecular weight-sensitive detector, be it less so than static light scattering detectors. This is because a_{MH} is typically smaller than 1, whereas the excess Rayleigh ratio measured in light scattering detectors is approximately proportional to M_w (eqn (4.41)). The Mark–Houwink coefficients depend on the polymer, the solvent and the temperature. Many values have been reported in the literature, but their reproducibility and validity across a broad range of molecular weights are questionable. If at all possible, the coefficients should be determined experimentally for a specific polymer. An online viscometer in combination with a (calibrated) concentration detector offers a very elegant tool for this purpose.

Using a viscometric detector, an **intrinsic viscosity distribution** (IVD) can be obtained directly.[24] It may be argued that the IVD is a better and more robust indication of polymer properties than the MWD, but the latter is generally accepted, while the former is not.

4.6.4 Universal Calibration

The fundamental principle of SEC, *i.e.* that molecules of equal size elute at the same elution time or elution volume, can be phrased in more exact terms by defining the hydrodynamic volume (v_h) of a polymer in solution, for which

$$v_h \propto [\eta]M \tag{4.47}$$

The intrinsic viscosity has inverse concentration dimensions, *i.e.* volume/mass, and the molar mass has dimensions of mass/mole, so that v_h is a molar volume (volume/mole). The principle of universal calibration states that when plotting $\log v_h$ against the elution volume, all types of polymers fall on the same line. This has largely proven correct over the years, with hyperbranched or dendrimeric polymers as the most notable exceptions. A combination of a light-scattering detector, a viscometer, and a concentration detector – a so-called **triple-detector** or triple-SEC setup – allows measuring both $[\eta]$ and M_w at each time point, providing all necessary data for a universal-calibration plot, although the actual plot does not need to be constructed to determine an absolute MWD. Knowing the Mark–Houwink coefficients of different polymers, in principle, allows the construction of universal calibration plots, but such a procedure is sensitive to inaccuracies in the Mark–Houwink coefficients.

The accurate molecular weight of the analyte polymer at an elution volume V_e may be obtained from

$$\log M_i(V_e) = \frac{1}{1 + a_{MH,i}} \log \frac{K_{MH,ref}}{K_{MH,i}} + \frac{1 + a_{MH,ref}}{1 + a_{MH,i}} \log M_{ref}(V_e) \qquad (4.48)$$

If the following conditions are met: (i) a calibration curve is constructed for a reference polymer under reference conditions (mobile phase and temperature); (ii) the Mark–Houwink coefficients for this polymer under these conditions ($K_{MH,ref}$ and $a_{MH,ref}$, *e.g.* for polystyrene in THF at 25 °C, see eqn (4.14)) are known; and (iii) the coefficients for the analyte polymer $K_{MH,i}$ and $a_{MH,i}$) are established at its own measurement conditions (possibly using a different mobile phase and/or temperature, but using the same column or column set, instrument, and connection capillaries).

> 🔘 See the website for additional literature and resources: assets.org

Acknowledgements

Dr. André Striegel, Mr. Ab Buijtenhuijs and Rick van den Hurk are acknowledged for their reviews of this chapter.

Recommended Reading

A. M. Striegel, W. W. Yau, J. J. Kirkland and D. D. Bly, in *Modern Size-Exclusion Liquid Chromatography*, John Wiley & Sons, Inc, 2nd edn, 2009.

A. M. Striegel and A. K. Brewer, Hydrodynamic Chromatography, *Annu. Rev. Anal. Chem.*, 2012, 5, 15–34.

References

1. A. M. Striegel, W. W. Yau, J. J. Kirkland and D. D. Bly, in *Modern Size-Exclusion Liquid Chromatography*, John Wiley & Sons, Inc, Hoboken, NJ, USA, 2nd edn, 2009.
2. V. D'Atri, M. Imiołek, C. Quinn, A. Finny, M. Lauber, S. Fekete and D. Guillarme, *J. Chromatogr. A*, 2024, **1722**, 464862.
3. B. W. J. Pirok, P. Breuer, S. J. M. Hoppe, M. Chitty, E. Welch, T. Farkas, S. van der Wal, R. Peters and P. J. Schoenmakers, *J. Chromatogr. A*, 2017, **1486**, 96–102.
4. W. W. Yau, C. R. Ginnard and J. J. Kirkland, *J. Chromatogr. A*, 1978, **149**, 465–487.
5. E. Uliyanchenko, P. J. Schoenmakersand S. van der Wal, *J. Chromatogr. A*, 2011, **1218**, 1509–1518.
6. K. Terao and J. W. Mays, *Eur. Polym. J.*, 2004, **40**, 1623–1627.
7. W. Mandema and H. Zeldenrust, *Polymer*, 1977, **18**, 835–839.
8. H. L. Wagner, *J. Phys. Chem. Ref. Data*, 1985, **14**, 1101–1106.
9. Y.-Z. Wei, Y.-F. Chu, E. Uliyanchenko, P. J. Schoenmakers, R.-X. Zhuo and X.-L. Jiang, *Polym. Chem.*, 2016, **7**, 7506–7513.
10. X. Jiang, A. van der Horst, V. Lima and P. J. Schoenmakers, *J. Chromatogr. A*, 2005, **1076**(1–2), 51–61.
11. X. Jiang and P. J. Schoenmakers, *J. Chromatogr. A*, 2002, **982**, 55–68.
12. L. E. Niezen, B. B. P. Staal, C. Lang, H. J. A. Philipsen, B. W. J. Pirok, G. W. Somsen and P. J. Schoenmakers, *J. Chromatogr. A*, 2022, **1679**, 463386.
13. T. Chang, *Adv. Polym. Sci.*, 2003, **163**, 1–60.
14. H. J. A. Philipsen, *J. Chromatogr. A*, 2004, **1037**, 329–350.
15. A. M. Striegel and A. K. Brewer, *Annu. Rev. Anal. Chem.*, 2012, **5**, 15–34.
16. Y. Liu, W. Radke and H. Pasch, *Macromolecules*, 2006, **39**, 2004–2006.
17. R. Edam, E. P. C. Mes, D. M. Meunier and P. J. Schoenmakers, *J. Chromatogr. A*, 2014, **1366**, 54–64.
18. G. Stegeman, J. C. Kraak, H. Poppe and R. Tijssen, *J. Chromatogr. A*, 1993, **657**, 283–303.
19. A. Williams, E. Varela, E. Meehan and K. Tribe, *Int. J. Pharm.*, 2002, **242**, 295–299.
20. B. W. J. Pirok, N. Abdulhussain, T. Aalbers, B. Wouters, R. A. H. Peters and P. J. Schoenmakers, *Anal. Chem*, 2017, **89**(17).
21. K. G. Wahlund, *J. Chromatogr. A*, 2013, **1287**, 97–112.
22. K. G. Wahlund and J. C. Giddings, *Anal. Chem.*, 1987, **59**, 1332–1339.
23. W. C. Knol, B. W. J. Pirok and R. A. H. Peters, *J. Sep. Sci.*, 2021, **44**, 63–87.
24. T. H. Mourey, *Int. J. Polym. Anal. Charact.*, 2004, **9**, 97–135.

5 Capillary Electrophoresis

This chapter is intended to provide the reader with a sound under-standing of the principles of capillary electrophoresis (CE) and other electromigration techniques and a good sense of where and how such techniques can be applied. The basic principles are outlined and band-broadening mechanisms are discussed, followed by a brief discussion of a range of techniques. Apart from capillary electro-phoresis, these include capillary gel electrophoresis (CGE), capil-lary isotachophoresis (cITP), capillary isoelectric focussing (cIEF), capillary electrochromatography (CEC) and micellar electrokinetic chromatography (MEKC). The latter technique will also be discussed in greater detail at the end of the chapter, following a discussion of instrumentation for CE and its coupling with mass spectrometry. The chapter is summarized in Figure 5.1.

5.1 Principles of Capillary Electrophoresis (M)

Electrophoresis is a family of techniques based on the separation of analytes in solution under the influence of a strong electric field. Electrophoresis is not a new technique. One name that should not to be forgotten is that of Arne Tiselius (see Box 5.1), who developed electrophoretic techniques to fractionate samples containing nanopar-ticles or proteins. The detection methods that are used today did not exist at that time, and most observations of "moving boundaries" were made visually, possibly after tagging. The biggest problem, however, was convection. Regions of the solutions in electrophoresis cells have different densities, either due to differences in composition or heating from a non-negligible electrical current, which creates convective movements in the cell.

Analytical Separation Science
By Bob W. J. Pirok and Peter J. Schoenmakers
© Bob W. J. Pirok and Peter J. Schoenmakers 2025
Published by the Royal Society of Chemistry, www.rsc.org

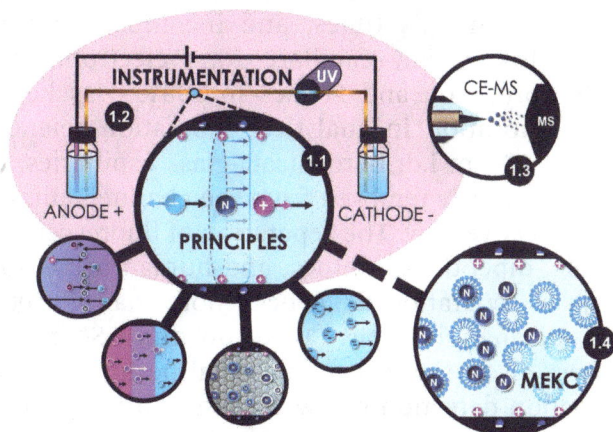

Figure 5.1 Graphical overview of the Modules in Chapter 5.

Box 5.1 Hero of analytical separation science: Arne Tiselius. Figure in Public Domain.[1]

Arne Wilhelm Kaurin Tiselius (1902–1971, Sweden)

Arne Wilhelm Kaurin Tiselius performed his doctoral studies on the "moving-boundary" method, which later became known as zone electrophoresis. He was fascinated by the study of proteins in biological samples, including human serum. He was awarded the Nobel Prize in Chemistry in 1948, at the age of 46. The following quote is attributed to Tiselius: "We live in a world where, unfortunately, the distinction between true and false appears to become increasingly blurred by manipulation of facts, by exploitation of uncritical minds, and by the pollution of the language." Arguably, this is even more true today than it was in Tiselius' era.

One way to mitigate convection was by stabilizing the separation medium. Initially, this was done with paper strips and, later, with agar (a natural "jelly" consisting of polysaccharides and pectin) or polyacrylamide gels. "Slab-gel" electrophoresis is still used for separating protein samples in biochemical and clinical studies. However, it comes with some challenges, including tedious sample preparation, long analysis times, and incompatibility with mass spectrometry. This is why capillary zone electrophoresis (CZE), introduced by Jorgenson and Lukacs in 1981,[2] was heralded as a revolutionary innovation in analytical separation science. By performing CZE in 75 μm i.d. borosilicate-glass capillaries, dissipated heat could be effectively removed from the channel, eliminating the problem of convective flow. The separation of fluorescent derivatives of amino acids, dipeptides, amines, and human urine was remarkable in terms of separation efficiency (more than 400 000 theoretical plates) and analysis times (10 to 30 min). Soon, fused-silica capillaries were also used, providing equal performance but allowing UV absorption detection below 300 nm. Although some restrictions apply, capillary electrophoresis is definitely a highly efficient analytical separation method. Analysis times are typically 5–15 min, but much faster separations are feasible using short capillaries. CZE is a capillary method with usually low requirements for consumables and low operating costs. Small amounts of sample suffice but – on the downside – the associated small amounts of analyte need to be detected in small volumes. CE soon earned a place for itself in the field of analytical separation science, playing a major role in elucidating the sequence of the human genome. Today, CE is an indispensable analytical technique in the biopharmaceutical industry for quality assessment of therapeutic proteins (including monoclonal antibodies), peptides and oligonucleotides. It definitely deserves a chapter in this textbook.

5.1.1 Basic Theory

Electrophoresis is based on the migration ("**electromigration**") of charged analytes in solution under the influence of an electric field. The phrase "in solution" is relevant here. The same principle in the gas phase yields ion mobility spectrometry, which could also be called gas-phase electrophoresis. Electrophoretic separation occurs because different ions or charged nanoparticles exhibit different migration velocities due to differences in their size and charge or, generally, their charge-to-size ratio.

An analyte cation is drawn towards the anode by an electrostatic force (F_{el})

$$F_{el} = q_i \cdot E = q_i \cdot \frac{\Delta V}{L} \tag{5.1}$$

where q_i is the charge of the ion, E is the field strength, which is the ratio of the voltage difference (ΔV) applied on the cathode and anode, and L is the distance between these two electrodes. The movement of the charged ions through the solvent induces a viscous counterforce, to which we can apply **Stokes' law**

$$F_{\text{visc}} = 6\pi \cdot \eta \cdot r_i \cdot v_i \qquad (5.2)$$

where η is the viscosity of the solution, r_i is the radius of the ion and v_i is its migration velocity. After the electric field is applied, ions in solution quickly reach a steady state, where the opposing forces are equal and the velocity is constant ($F_{\text{el}} = F_{\text{visc}}$; see Figure 5.2). By combining eqn (5.1) and (5.2), we derive the migration velocity

$$v_i = \frac{q_i \cdot E}{6\pi \cdot \eta \cdot r_i} = \frac{q_i \cdot \Delta V}{6\pi \cdot \eta \cdot r_i \cdot L} \qquad (5.3)$$

The **electrophoretic mobility** of the analyte ion is defined as the ratio of the migration velocity and field strength, *i.e.*

$$\mu_i = \frac{v_i}{E} = \frac{q_i}{6\pi \eta r_i} \qquad (5.4)$$

The electrophoretic mobility depends on the charge-to-size ratio of the analyte particle (q_i/r_i) and the viscosity of the solution, which means that, for common ions (*e.g.* small inorganic anions or metal cations), the mobilities can be tabulated in a specific solvent. In water, the hydration shell adds to the ion radius, and the electrophoretic mobilities of small inorganic ions typically are in the 10^{-8}–10^{-7} m^2 V^{-1} s^{-1} range. Molecular ions are mostly not spherical, but generally, charge-to-molecular mass dependency of their electrophoretic mobility still prevails.

Eqn (5.4) is, strictly speaking, only valid for analyte ions in pure solvent. However, CE is commonly carried out in a salt or buffer solution, the **background electrolyte** (BGE), which, to a certain extent, reduces the analyte ion mobility by retardation and relaxation effects. The **retardation effect**, indicated by the small light-blue arrows in Figure 5.2, is an extra viscous drag caused by the diffuse cloud of BGE ions that surround the analyte cation. To compensate for the positive charge of the analyte ion, the diffuse cloud has a net

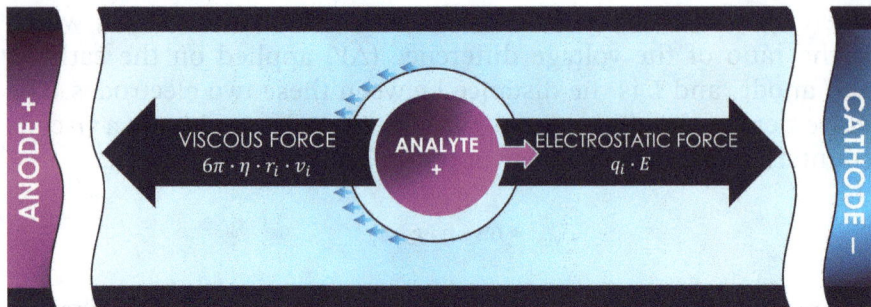

Figure 5.2 Forces (large, dark-blue arrows) applied to an analyte cation as it migrates towards the negative cathode in an electrophoretic system, as indicated by the pink arrow. The small light-blue arrows pointing to the left indicate a retardation effect caused by the net negative charge of the diffuse ion cloud (light blue) surrounding the analyte cation.

negative charge, implying that it is attracted to the anode. Moreover, when the analyte ion and diffuse cloud move in counter directions, it will take time for the ion cloud to rearrange. During this relaxation time, the analyte ion will be slowed down by the overall oppositely charged ion cloud. The diffuse ion cloud becomes more compact when the ionic strength of the BGE (I) increases (just like the electric double layer described in Section 3.7.1). The decrease in the electrophoretic mobility caused by retardation and relaxation is roughly proportional to $\sqrt{I}$. For example, the electrophoretic mobility of a series of aromatic sulfonic acids could be empirically described by

$$\mu_i = \mu_i^0 \, e^{-0.77\sqrt{|q_i| \cdot I}} \tag{5.5}$$

where μ_i^0 is the electrophoretic mobility of analyte i in pure water.

The effect of the pH of the BGE (see Figure 5.3) is often important in CZE, as it determines the effective or average charge of analyte ions. In Module 3.5, we described that the retention of weak acids or weak bases in reversed-phase liquid chromatography (RPLC) was determined by the fraction of the analyte present as non-charged molecules, assuming the ionic forms to be unretained. In CZE, the opposite is true. The electrophoretic mobility is determined by the fraction of the species that is charged, as $\mu = 0$ for neutral species. This yields effective mobility for a weak acid (HA) that loses a proton upon ionization (*e.g.* a carboxylic acid)

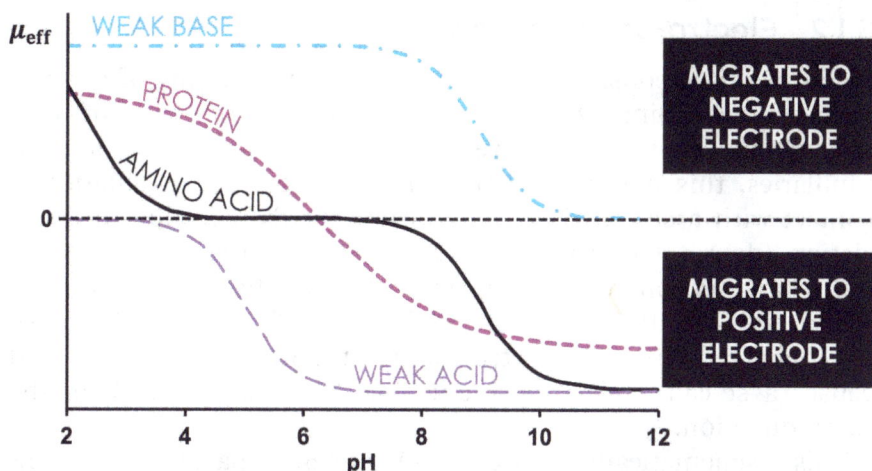

Figure 5.3 Schematic illustration of the change in mobility with pH for acidic, basic, amphoteric and proteinaceous compounds.

$$\mu_{eff,HA} = \frac{K_{a,HA}}{K_{a,HA} + [H^+]} \mu_{A^-} \tag{5.6a}$$

and for a weak base that is protonated at low pH (*e.g.* an amine)

$$\mu_{eff,B} = \frac{[H^+]}{K_{a,B} + [H^+]} \mu_{BH^+} \tag{5.6b}$$

For multivalent and amphoteric compounds, there are several steps in how mobility varies with pH (namely one at every K_a value). For proteins, which contain many acidic and basic groups, there is a gradual change in mobility. At low pH, proteins are net positively charged (due to the association of acid groups and the protonation of basic groups), and the electrophoretic mobility causes them to migrate towards the negative electrode. At high pH, proteins are net negatively charged (due to the dissociation of acid groups), and their electrophoretic mobility is towards the positive electrode. At its isoelectric point, a protein has a net charge of zero and hence has no electrophoretic mobility. As a constant pH is obviously essential to obtain consistent analyte electrophoretic mobilities, BGEs most commonly are or comprise buffering systems, such as phosphate, acetate or borate buffers.

5.1.2 Electro-osmotic Flow

So far, we have considered the electrophoretic mobility of charged particles, assuming that there is no volumetric flow of BGE through the capillary. However, when using fused-silica or glass capillaries, this may not be true. The wall of such capillaries in contact with a solvent is usually negatively charged due to disso-ciation (deprotonation) of surface silanol groups. This negative wall charge is compensated by an excess of cations in the vicinity of the wall, as illustrated schematically in Figure 5.4. This posi-tively charged layer is attracted towards the negative electrode and causes a so-called **electro-osmotic flow** (EOF) of the BGE in the same direction.

This is schematically illustrated in Figure 5.4. Apart from a narrow zone in the immediate vicinity of the wall (with a thickness of the order of the thickness of the electrical double layer),[3] the flow rate is the same everywhere in the capillary, as indicated by the light-blue profile in the centre of the figure. It is important to realize how different this **flow profile** is from the parabolic profile that arises from pressure-driven laminar flow in a capillary (see Section 1.6.3). The latter makes open-tubular LC virtually impossible, as it constrains the column's inner diameter to a maximum of a few

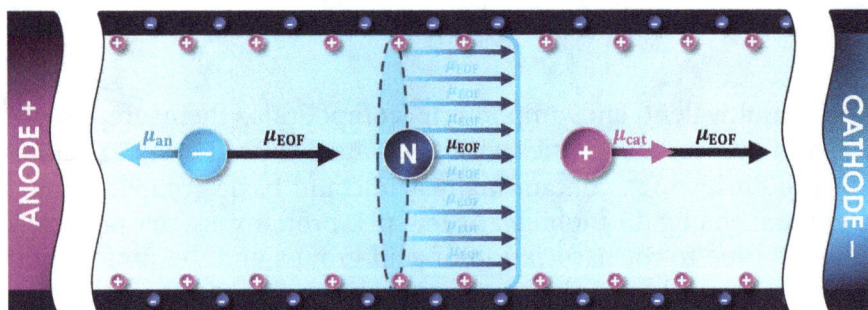

Figure 5.4 Schematic illustration of the effect of electro-osmotic flow (EOF) in capillary zone electrophoresis. Illustration of the origin of EOF due to the presence of an excess of positive ions in the solution near the capillary wall. The flow profile is illustrated in the centre of the figure, with dark arrows indicating the flow velocity. Neutral molecules (N) migrate with the EOF. Anions (light blue, left) move at an apparent mobility, which is the difference between the electrophoretic mobility (μ_{EOF}) and the ionic mobility of the anion (μ_{an}). The apparent mobility of cations (pink, right) is the sum of μ_{EOF} and the ionic mobility of the cation (μ_{cat}).

micrometres (see Sections 1.7.7 and 1.7.8). The flat EOF profile puts no constraints on the i.d. of CE capillaries. The latter is constrained by the requirement of adequate heat dissipation, and capillaries with an i.d. of 50 or 75 μm have been found perfectly suitable for CE.

Because of the EOF, the entire contents of a fused-silica capillary move towards the negative electrode with a velocity v_{EOF} that increases with the field strength E and corresponding electrophoretic mobility

$$\mu_{EOF} = \frac{v_{EOF}}{E} \tag{5.7}$$

Neutral molecules (indicated by N in the central zone of Figure 5.4) migrate through the capillary with the velocity of the EOF. For anions and cations, the migration velocity is the result of the electrophoretic mobility of the analyte ion and that of the EOF, leading to apparent electrophoretic mobilities (μ_{app}) for anions, where the mobilities of the ion and that of the EOF work in opposite directions

$$v_{an,i} = \left(\mu_{EOF} - \mu_{an,i}\right) \cdot E = \mu_{app,an,i} \cdot E \tag{5.8a}$$

and for cations, where the mobilities work in the same direction

$$v_{cat,i} = \left(\mu_{EOF} + \mu_{cat,i}\right) \cdot E = \mu_{app,cat,i} \cdot E \tag{5.8b}$$

Depending on the pH of the BGE, the electro-osmotic mobility can be greater than the electrophoretic mobility of the various analyte ions, which implies that all ions move in the direction of the EOF. Cations thus move faster than the EOF, while anions move slower, as illustrated in the electropherogram in Figure 5.5, which illustrates that capillary electrophoresis can readily be used to separate both cations and anions in a single run.

The magnitude of the EOF can be described by

$$v_{EOF} = \frac{\varepsilon_0 \varepsilon_r \zeta E}{\eta} \tag{5.9}$$

where v_{EOF} is the velocity of the EOF, ε_0 is the permittivity of the vacuum, ε_r is the dielectric constant of the BGE, η is its viscosity,

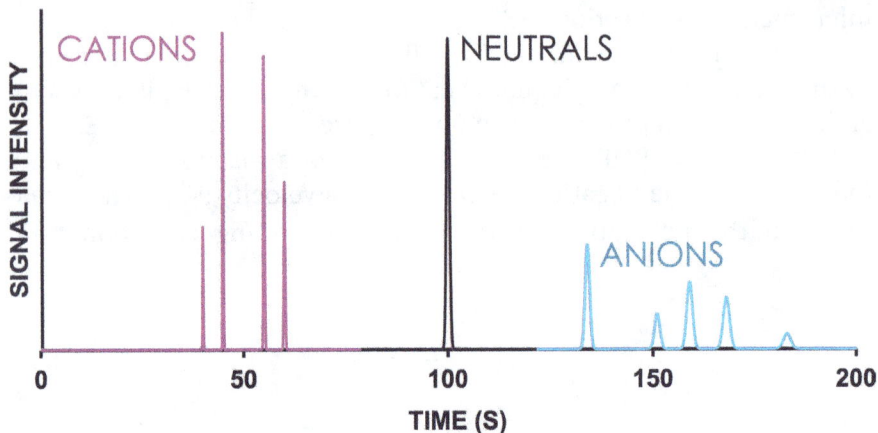

Figure 5.5 Electropherogram, showing that cations (pink) migrate faster to the detector than neutrals (dark blue), while anions (light blue) move slower.

ζ is the zeta potential of the capillary wall, and E is the electric field. The EOF depends on the zeta potential, which is affected by the net charge of the surface and, thus, in the case of, for example, a fused-silica wall, the pH of the BGE. In the case of fused silica (and aqueous BGEs), v_{EOF} is negligible below pH ≈ 2 and increases gradually with increasing pH until it reaches a maximum value around pH ≈ 7, above which it no longer increases. The zeta potential – and thus the velocity of the EOF –decreases if the ionic strength of the BGE increases.

The velocity of the EOF also changes if the condition of the inner capillary wall changes. This can be done purposefully, for example, by coating the wall with a neutral polymer to reduce the EOF or avoid analyte adsorption (see section 5.1.3.4) or with a positively charged polymer to reverse the EOF. It may also happen inadvertently, for example, by adsorption of matrix components from the sample onto the capillary wall. The latter phenomenon would affect the repeatability of CE experiments and may necessitate cleaning ("reactivation") of the capillary wall through a washing process. The EOF also changes with temperature, mainly because of changes in the viscosity of the BGE (see Section 5.1.3.2).

5.1.3 Band Broadening in Capillary Electrophoresis (B)

Figure 5.6 shows the basic layout of a CE system. Analyte ions are injected at the capillary inlet (near the anode) and migrate towards the cathode. Analyte peaks are monitored downstream, commonly using on-capillary optical detection. To describe the width of peaks in CE, the same characteristics are used as in chromatography (see Chapter 1). The width of a peak is characterized by its standard deviation (σ), either in length units (σ_z), volume units (σ_v) or time units (σ_t). The number of **theoretical plates** is determined in the same way as in chromatography, *i.e.*

$$N = \frac{L_{\text{eff}}^2}{\sigma_z^2} = \frac{V_{\text{eff}}^2}{\sigma_v^2} = \frac{t_e^2}{\sigma_t^2} \tag{5.10}$$

where L_{eff} and V_{eff} are the effective length and volume of the capillary, respectively, both measured from the point of injection at the capillary inlet to the point of detection, and t_e is the time at which the peak maximum is detected. The **resolution** between the two peaks is also measured in the same manner as in chromatography, *i.e.*

$$R_{s,j,i} = \frac{t_{e,j} - t_{e,i}}{2(\sigma_{t,i} + \sigma_{t,j})} = \frac{\Delta t_{e,j,i}}{2(\sigma_{t,i} + \sigma_{t,j})} \tag{5.11}$$

where i and j indicate the first- and last-eluting peaks of an adjacent pair.

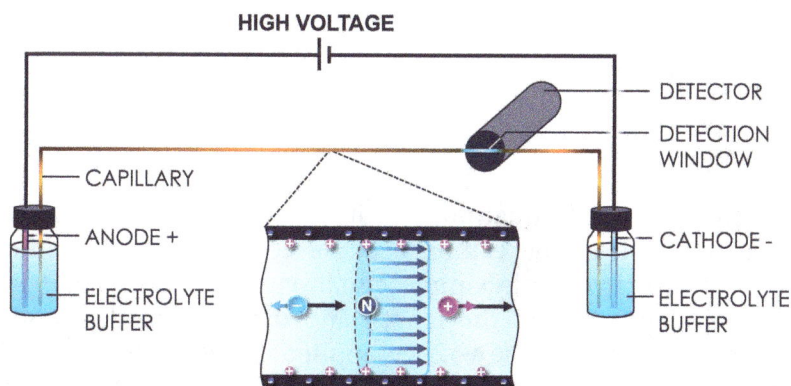

Figure 5.6 Schematic illustration of instrumentation for capillary electrophoresis. The inset illustrates the separation mechanism.

5.1.3.1 Longitudinal Diffusion

When we consider **band broadening** processes, the situation is very different from that in chromatography. If we consider *van Deemter*-type equations (Section 1.4.4) and capillary (zone) electrophoresis in an empty channel, then we can observe that there are no particles and thus no eddy diffusion (*A*-term), there is no stationary phase causing resistance to mass transfer (no C_s-term), and a very flat flow profile (no differences in the electro-osmotic velocities, no C_m-term). The only term that remains from either the *van Deemter* equation (developed for packed separation channels) or the Golay equation (developed for open capillaries; see Section 1.7.7) is the **longitudinal diffusion** term. If we use the **Einstein equation** eqn (1.88), we obtain

$$\sigma_z^2 = 2D_m \cdot t_e \tag{5.12}$$

If we have no other sources of band broadening for CE, this yields

$$N = \frac{L_{eff}^2}{\sigma_z^2} = \frac{L_{eff}^2}{2D_m \cdot t_e} \tag{5.13}$$

The time can be found from the effective distance divided by the migration velocity

$$t_e = \frac{L_{eff}}{\mu_{app,i} \cdot E} = \frac{L_{eff} \cdot L_{total}}{\mu_{app,i} \cdot V} \tag{5.14}$$

where L_{total} is the total length of the capillary (between the positive and negative electrodes). This yields

$$N = \frac{\mu_{app,i} \cdot V}{2D_m} \cdot \frac{L_{eff}^2}{L_{eff} \cdot L_{total}} \approx \frac{\mu_{app,i} \cdot V}{2D_m} \tag{5.15}$$

where the last approximation is valid if the detection window is close to the end of the capillary. Some conclusions can immediately be drawn from this simple equation. In ideal CE, the plate count is independent of the length and diameter of the capillary and is proportional to the voltage applied between the electrodes. Thus, the higher the **voltage** we can apply without sparks flying across or the capillary overheating (see Section 5.1.3.2), the higher the efficiency of a CE separation. There is also an effect from the diffusion coefficient. The lower the D_m, the higher the plate count. Mobilities of small inorganic ions in a common BGE are of the order of 50×10^{-9} m^2 V^{-1} s^{-1}, and the EOF in fused silica tubing at pH = 7 is about 100×10^{-9} m^2 V^{-1} s^{-1}, yielding apparent mobilities for cations of the order of 150×10^{-9} m^2 V^{-1} s^{-1}. If we assume D_m $=10^{-9}$ m^2 s^{-1}, eqn (5.15) suggests that for small analytes, $N \approx 75 \cdot V$, where V is the applied voltage in volts. As the voltage is easily 20 kV, plate counts over a million are anticipated for CE. For larger molecules, D_m can be an order of magnitude lower, and predicted values for N are even higher. Thus, eqn (5.15) predicts extremely narrow peaks for high-molecular-weight analytes such as DNA and proteins, provided that ideal CE conditions can be maintained. The latter caveat is not to be ignored, as we will see in the next few sections.

5.1.3.2 Joule Heating

The electrical power supplied to the electrophoretic system results in the generation of **heat**, as is evident from a combination of Joule's law and Ohm's law, *i.e.*

$$P = I^2 R = I \cdot V \qquad (5.16)$$

where P is the electrical power or heat generated, I is the current and R is the resistance. Similar to what was described for pressure-driven flow in packed columns (see Section 1.7.6), this will lead to an increase in the temperature inside the capillary when voltage is applied. The current, and thus temperature, increases when the conductivity (κ) of the BGE and the capillary diameter (d_c) increase

$$I = 0.25\pi \cdot d_c^2 \cdot \kappa \cdot E \qquad (5.17)$$

Knox and McCormack[4] measured the temperature increase in a 50 μm i.d. capillary using a borate $(Na_2B_4O_7)$ buffer of 0.083 M and a 100 μm i.d. capillary using a borate buffer of 0.02 M. The increase in the temperature in these two capillaries was comparable. With field strengths of less than 10 kV m^{-1}, the temperature only increased a few degrees above the ambient value of 20 °C, but at higher field strengths (above 25 kV m^{-1}), temperatures in excess of 50 °C were measured. Heating of the capillary will by itself lower the efficiency of CE separations because the analyte diffusion coefficients increase with increasing temperatures. Axial diffusion will increase and, therefore, N will decrease (see eqn (5.15)). In extreme cases, boiling solvents have jeopardized CE experiments under high field conditions. An increase in temperature lowers the viscosity of the BGE and thus increases electrophoretic mobility. As the conductivity is proportional to the concentration and electrophoretic mobility of the BGE ions present in the capillary, the current will increase, and consequently, more heat is produced, leading to even higher temperatures. This self-reinforcing process may lead to excessive Joule heating. When optimizing the BGE, it is good practice to construct a so-called Ohm's plot, which depicts the observed current as a function of increasing voltage. The voltage where the I–V relation deviates from linearity indicates the onset of excessive **Joule heating**.

The heat generated by the current in the capillary is dissipated through its walls. This leads to a radial temperature gradient, where the centre of the column is warmer than the regions closer to the wall. Because the product $\varepsilon_r \cdot \zeta$ in eqn (5.9) was found in practice to be approximately constant with changes in temperature,[4] the velocity of the EOF will change with the **viscosity** (η) of the BGE, which decreases with increasing temperature. Also, the electrophoretic mobility of the analyte ions will increase with decreasing viscosity. Thus, a radial temperature gradient leads to different migration velocities (apparent mobilities) at different radial positions in the capillary, and this is detrimental to the efficiency of the electrophoretic separation. The band broadening by radial temperature gradient is partially counteracted by analyte ion diffusion in the radial direction. The best way to minimize temperature increase and the effects of the radial temperature gradient is to reduce the internal diameter of the capillary, and this is why capillary electrophoresis has replaced electrophoresis in larger cells, at least for analytical purposes. In practice, the adverse effects of Joule heating can be virtually cancelled out by using capillaries with an i.d. of 75 μm or smaller and BGEs of relatively low conductivity and by efficient heat dissipation through capillary thermostating. Still, "ideal" CZE, in which axial diffusion is the only source of band broadening, is no longer on the cards.

In LC, there have been efforts to minimize the heat dissipation through the wall by insulating the column or creating a vacuum housing around the column. Such efforts are not known in CE.

5.1.3.3 Injection and Detection Band Broadening

In the first place, injection and detection contribute to dispersion in the same way as they do in other separation methods (see Section 1.7.5). We can write

$$\sigma_{z,obs}^2 = \sigma_{z,ep}^2 + \sigma_{z,inj}^2 + \sigma_{z,det}^2 = \sigma_{z,ep}^2 + \frac{l_{inj}^2}{\theta_{CE,inj}} + \frac{l_{det}^2}{\theta_{CE,det}} \tag{5.18}$$

where $\sigma_{z,obs}$ is the observed standard deviation in length units; $\sigma_{z,ep}^2$, $\sigma_{z,inj}^2$, and $\sigma_{z,det}^2$ are the contributions to the variance from the electrophoretic process, the injection and the detection, respectively; l_{inj} and l_{det} are the lengths of the **injection and detection zones**, respectively; and $\theta_{CE,inj}$ and $\theta_{CE,det}$ are constants. A common, although somewhat optimistic (see Section 3.3.3), approximation is to assume $\theta_{CE,inj} = \theta_{CE,det} = 12$. In Section 3.3.3, we learnt that as a rule of thumb, $\sigma_{z,inj} \le 0.5\,\sigma_{z,ep}$ and $\sigma_{z,det} \le 0.5\,\sigma_{z,ep}$ would be acceptable. For l_{det}, this implies that

$$l_{det} \le 0.5 \cdot \sqrt{\frac{\theta_{CE,det} \cdot L_{eff}^2}{N}} \tag{5.19}$$

For $L_{eff} = 0.5$ m, $N = 200\,000$, and the optimistic $\theta_{CE,det} = 12$, we find that the detection zone (as well as the length of an undisturbed injection zone) may be about 2 mm long. For $\theta_{CE,det} = 4$, which is a more realistic number in the case of **hydrodynamic injection** (see Section 5.2.2), this becomes about 1 mm. For $N = 1\,000\,000$, l_{det} should be shorter than 0.5 mm. Longer detection zones will significantly affect the observed bandwidths.

The effect of the injection on the observed bandwidth is more complicated as it depends not only on the length of the injection zone l_{inj} but also on the concentration of the injected analyte concentration, the composition of the injection solvent, and the concentration and composition of the used BGE. When the conductivity (ionic strength) of the injection solvent is much lower than that of the BGE, analyte band-narrowing effects may even occur. This process provides the possibility to inject relatively large volumes of a dilute sample and is known in CE as **stacking** (see Section 5.2.2). However, under certain conditions, electrophoretic band broadening and triangular analyte peaks may be observed, depending largely on the difference in electrophoretic mobility between the analyte ion (μ_X) and the corresponding BGE ion of the same charge (μ_Y). If these two are equal (*i.e.* $\mu_X = \mu_Y$), no electrophoretic band broadening is anticipated. If $\mu_X > \mu_Y$, then a fronting peak or forward-leaning triangle may be observed. In the case of $\mu_X < \mu_Y$, a tailing peak or backward-leaning triangle may occur. A quantitative explanation for these phenomena – also known as **electromigration dispersion** – was already provided by Mikkers *et al.* in 1979.[5] Their treatment was rather complex, but some quite practical guidelines were formulated. The first one is that in order to keep molecular diffusion the main source of band broadening, the concentration of the analyte ions (X_i) should be at least 100 times lower than the concentration of the corresponding ion (of equal charge; Y) in the background electrolyte (*i.e.* $X_i \ll Y_{\text{BGE}}$). The second one is that if the first condition is met, the concentration of the analyte ion in the sample must be much higher than that of the corresponding BGE ion in the sample (*i.e.* $X_i \gg Y_{\text{sample}}$). Thus, if we want to analyse benzoate and our BGE is an acetate buffer, we should ensure that the injection concentration of benzoate is much lower than the concentration of the BGE and that there is little or no acetate in the sample. In doing so, band broadening by electromigration dispersion becomes negligible, and we may actually inject larger volumes (or longer injection zones) than those derived from eqn (5.19), *i.e.* the injection zones may be longer than the detection zones.

5.1.3.4   Wall Adsorption

Analyte **adsorption** onto the inner capillary wall should be prevented in CE. Even the slightest adsorption destroys the ideal of band broadening solely due to axial diffusion.[3] According to Bruin *et al.*, the effect of adsorption on efficiency is enormous,[6] even if it barely affects the migration time. Adsorption, resulting in a retention factor of 0.1, may reduce the observed number of plates by an order of magnitude. For many low-molecular-weight analytes, interactions with the fused-silica wall are insignificant, but as we also have seen in Section 3.2.2.2, amine-containing compounds (and proteins in particular) are notorious for their tendency to adsorb on silanol groups at the wall surface. One way to reduce the effect of electrostatic interactions between basic analytes and the fused-silica wall is to work at extreme pH values. At very low pH, the silanol groups will be fully protonated, diminishing the electrostatic attraction towards positively charged analytes. At high pH, both the wall and peptides and proteins will be charged negatively, resulting in electrostatic repulsion. Fused silica is quite resistant against strong alkaline or acidic conditions;. However, extreme pH may affect the stability and conformation of analyte proteins and also cancel the possibility of pH tuning for separation optimization. More commonly, therefore, coating of the inner wall of the capillary is applied in the CZE of proteins and other adsorptive analytes. In the so-called **dynamic coating**, additives such as small (di)amines, tetraalkylammonium ions or cationic surfactants are included in the BGE to shield negative charges on the capillary wall and thus avoid band broadening by analyte adsorption. In **static coating**, a permanent chemical layer is applied to the capillary wall, which remains coated during successive rinsing and analysis cycles. Covalently attached polymers (*e.g.* polyvinyl alcohol or polyacrylamide) are used for this purpose, but physically adsorbed polymers can also be used. For example, a positively charged polymer, such as polybrene, can be strongly adsorbed onto the negatively charged fused-silica surface, leading to a reversal of the direction of the EOF and high plate numbers for proteins, which are repulsed when using an acidic BGE. Negatively charged polymers, such as poly(vinyl-sulphonic acid) or dextran sulphate, can be immobilized in a bilayer, on top of, for example, polybrene.[7] Such coatings strongly improve the migration time stability and efficiency of CE separations, particularly for proteins when analysed above their pI. In the case of CE-MS, static coatings are required to prevent analyte ionization suppression and pollution of the ion source and mass spectrometer.

5.1.3.5 Hydrodynamic Flow

If the liquid levels at the inlet and outlet of the capillary differ, a **hydrodynamic (Poiseuille) flow** with a parabolic profile will be created due to **siphoning**. As this is a pressure-driven flow, we may apply Darcy's law (eqn (1.69)) to find that a difference in BGE levels in the two vials of 1 mm (1 mm water pressure = 9.81 Pa) yields a siphon velocity of just under 0.1 mm min^{-1} for a 50 µm i.d. column. For a 75 µm i.d. column, this is about 0.2 mm min^{-1} and for a 100 µm i.d. about 0.37 mm min^{-1}. Such values do not significantly alter the measured migration times; however, as we know that radial diffusion cannot compensate for the differences in velocity between the centre of the column and the wall region, the observed band broadening may be affected, especially for high-molecular-weight analytes. Speeding up the CE analysis by deliberately applying hydrodynamic pressure on the inlet vial may be detrimental to the obtained separation efficiency.

5.1.4 Electromigration Methods

In this section, we will introduce some different electromigration techniques, limiting ourselves to separations that take place in a capillary. The domain has more than its fair share of confusion on names and classifications. One way to classify the different techniques is shown in Figure 5.7. The general term for these (and some other) separation methods is **electromigration**. The movement of analytes through a capillary – or to a specific point in the capillary – occurs due to the application of an electric field. We have grouped three techniques in the dark-blue box named "capillary electrophoresis". In these techniques, separation is based on differences in electrophoretic mobility between the different analyte ions. Some scientists may refer to all techniques – including the ones outside the dark-blue box in Figure 5.7 – as electrophoretic techniques, but the separation mechanisms are different. In **isotachophoresis**, the analytes eventually move in a train, all at the same pace. In isoelectric focussing, each analyte moves to a specific place in the capillary, where it reaches a standstill. **Capillary electro-chromatography** and **micellar electrokinetic chromatography** are chromatographic methods that may be applied to neutral molecules as well as to ions. The speed of migration of neutral molecules is based on differences

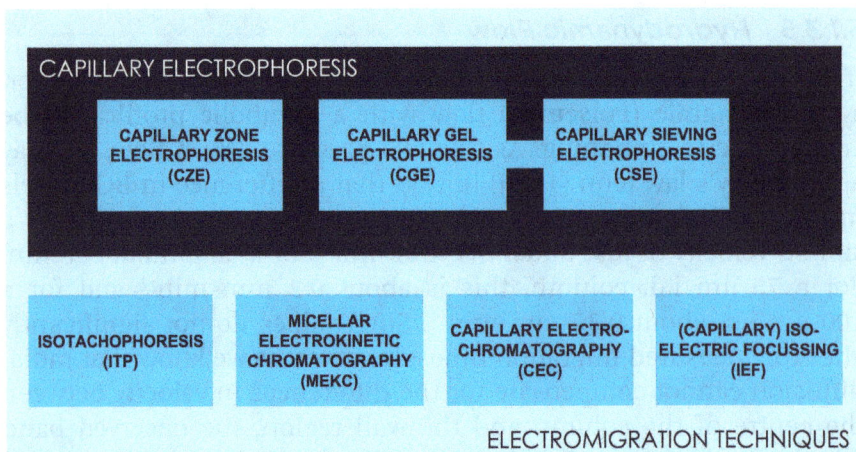

Figure 5.7 Overview of important electromigration methods.

(between different analytes) in distributions between a mobile phase and a stationary phase. Both the distribution and electrophoretic effects affect migration and separation.

Figure 5.8 shows schematic illustrations of the principles of the most important electromigration techniques, where no distinction is made between **capillary gel electrophoresis** and **capillary sieving electrophoresis** (CSE) (Figure 5.8B).

CE techniques can be used for the analysis of a wide range of charged analytes. The open tubular format puts few restrictions on the molecular size of the analysed species. The application field spans small inorganic ions, low-molecular-weight drugs, metabolites, glycans, peptides, oligonucleotides, DNA, proteins, polymers, and even supramolecular constructs such as viruses, vesicles and nanoparticles. In our discussion on the different electromigration methods, frequent reference will be made to the separation of proteins. These are, indeed, important application fields for many of the techniques shown in Figure 5.7. Interest in top-down proteomics, biotechnology and biopharmaceuticals has spurred interest in the analysis of intact proteins, especially in the study of proteins under native conditions. A protein is not a single unique molecular structure defined by a sequence of amino acids (primary structure). Various minor "post-translational" modifications occur in biological systems. During the production and processing of biopharmaceutical proteins, their structure may be modified in various ways. The bioactivity, efficacy, and toxicity of proteins may be affected by such modifications. The different modified forms of a single protein are known as *proteoforms*.

Some LC techniques can be applied to the separation of proteins under **native conditions**, where the tertiary (three-dimensional) structure of the protein remains intact. These techniques, in which the use of organic solvents can typically be avoided, include size exclusion chromatography (SEC, Module 4.2), ion exchange chromatography (IEC, Section 3.7.1), hydrophobic interaction chromatography (HIC, Section 3.15.1), and affinity chromatography (Section 3.15.2). However, LC techniques are likely to break up higher-order structures, such as non-covalent protein complexes, due to the high stress exerted on large molecules in microparticulate-packed columns (see Section 4.2.2.4), as well as the strong interactions or high salt concentrations that are often encountered. CE is an excellent technique for studying *proteins* and protein complexes under (near) native conditions. BGEs can mimic the electrolyte composition and pH of biological fluids, and under

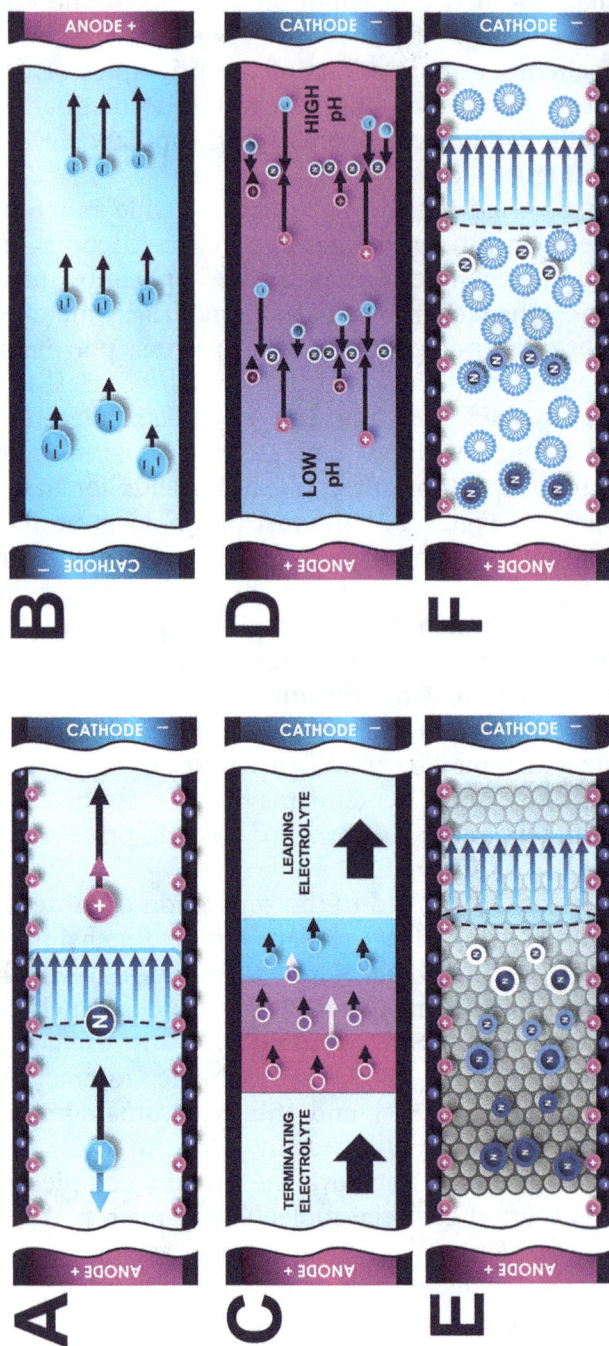

Figure 5.8 Schematic illustrations of the principles of the most important electromigration methods. (A) Capillary zone electrophoresis (CZE); (B) capillary gel electrophoresis (CGE) or capillary sieving electrophoresis (CSE). The blue shading of the contents of the capillary indicates the presence of a gel; (C) capillary isotachophoresis (cITP). In a steady state, all ions move in separate zones at identical velocities. Ions that are outside the appropriate zone will be accelerated or delayed until they fall in place, as indicated by the two white velocity arrows; (D) capillary isoelectric focussing (cIEF); (E) capillary electrochromatography (CEC); and (F) micellar electrokinetic chromatography (MEKC).

low-flow conditions in open capillaries very little stress is exerted on analyte molecules and higher-order structures. Thus, interest in electromigration techniques often concurs with interest in proteins.

Another area of great interest is *nucleic acids* (RNA and DNA). They share high molecular weight, multiple charges, and complex mixtures with proteins. However, the variations in charge are more limited because the phosphate groups that connect the (deoxy-)ribose groups in the main chain are strongly acidic. Hence, nucleic acids are negatively charged under practical conditions, and their charges increase with increasing chain length (or molecular weight). This makes them hard to separate using CZE but ideal candidates for separation using CGE (see Section 5.1.4.2).

5.1.4.1 Capillary Zone Electrophoresis

Capillary zone electrophoresis (CZE) is the most common and conceptually the most straightforward capillary electromigration method. Part of the confusion in terminology arises from the fact that, while some researchers use the term "capillary electrophoresis" to include all techniques shown in Figure 5.7, others are of the opinion that capillary electrophoresis specifically refers to capillary zone electrophoresis. In principle, CZE is also the simplest method to conduct. The capillary only contains a solution of BGE. Our discussion so far in this chapter (Sections 5.1.1 through 5.1.3) has pertained to capillary zone electrophoresis.

After introducing a sample, the different analytes move through the capillary at different velocities (*i.e.* with different apparent mobilities; see Section 5.1.2 and Figure 5.3). The BGE is important to maintain constant electrophoretic conditions throughout the capillary (fulfilling the requirements of the Kohlrausch regulation functions; see Section 5.2.3.1). Each (separated) analyte appears as a peak in the electropherogram, the area of which can be related to the concentration or the amount of the analyte (if an appropriate detector is installed). Various additives can be added to the BGE to study, for example, drug–protein interactions. High-resolution chiral separations can be achieved by adding, for example, cyclodextrins to the BGE. One interesting example is the addition of a receptor to the BGE, which enables the assessment of receptor affinity for different proteoforms on the basis of migration time shifts, in a technique known as *affinity capillary electrophoresis* (ACE).[8]

5.1.4.2 Capillary Gel Electrophoresis and Capillary Sieving Electrophoresis

Gel electrophoresis was essentially the first high-resolution electrophoresis technique. Heat-induced convectional flow could not be avoided in relatively large free-solution electrophoresis systems. Introducing the gels was a way to avoid convection. For analysis by sodium dodecyl sulphate (SDS)–poly(acrylamide) gel electrophoresis (PAGE), a mixture of proteins is first heated in an SDS-containing buffer. This is meant to produce SDS–protein complexes with equal mass-to-charge ratios and thus effective electrophoretic mobility. These are subsequently separated based on size only, employing the molecular sieving capacity of the gel through which smaller proteins migrate faster than larger proteins. Polynucleotides can be separated based on size by gel electrophoresis without pretreatment with SDS because they inherently possess virtually equal mass-to-charge ratios. Slab–gel separations that require staining for band detection have been replaced in part by capillary gel electrophoresis, particularly in environments where quantitative analysis of large numbers of samples is customary. The capillary format allows (quantitative) on-column detection, a high degree of automation, and higher voltages, leading to better resolution. The cross-linked PA gels and agarose gels used were hard to prepare *in situ* (*i.e.* in the capillary). Such capillaries with fixed gels were not robust, lasting typically for fewer than ten analyses, and the repeatability was poor. The immobilized gels have gradually been replaced by linear or slightly branched polymers (PA, PEG, dextran, and pullulan), which form entangled networks with sieving capacities. Such "liquid gels" can be easily flushed in and washed out of the capillary, providing a fresh sieving matrix for each analysis. Initially, a distinction was made between **capillary gel electrophoresis** using the traditional, cross-linked gels and **capillary sieving electrophoresis** using the readily replaceable sieving matrices. However, this is another instance where the terminology of electromigration techniques is not unambiguous. Most current contemporary methods may concern CSE, but CGE is a commonly accepted generic name covering all CE techniques that employ a sieving matrix (Figure 5.8B).

5.1.4.3 Capillary Isotachophoresis

In **capillary isotachophoresis** (cITP) (Figure 5.8C, see also Box 5.2), the different analyte bands do not move at different migration velocities. Instead, the different components form a train of zones, which all move at the same velocity. The buffer reservoir at the detection side contains a leading electrolyte that has a mobility greater than that of all analyte ions. The other reservoir contains a terminating electrolyte, with a mobility lower than that of the analytes. The sample is introduced between these two electrolytes, and under the influence of the applied voltage, it separates into a series of single-component zones, the length of which is a measure of the amount of eluent present.

Box 5.2 Hero of analytical separation science: E. Smolková-Keulemansová. Image reproduced from ref. 9 with permission from Springer Nature, Copyright 2008.

Eva Smolková-Keulemansová (1927–2024, the Czech Republic)
Eva Smolková-Keulemansová was incarnated in the Nazi concentration camps in WWII as a teenager. She managed to survive, but her parents did not. She returned to study in Prague and worked as a professor at Charles University for many years. She had a long and productive career in analytical separation science, and she was a leading advocate for the field. Eva started her research in gas chromatography but also contributed to liquid chromatography and, especially, electromigration techniques. Her use of cyclodextrin for chiral separations with capillary electrophoresis and isotachophoresis was especially eye-catching.

By working at a constant current, sharp zones emerge between the different analyte zones. For the movement of ions to remain constant, we have

$$\mu_L \cdot E_L = \mu_\alpha \cdot E_\alpha = \mu_\beta \cdot E_\beta = \cdots = \mu_\omega \cdot E_\omega = \mu_T \cdot E_T \tag{5.20}$$

where L represents the leading electrolyte (with the highest mobility), T represents the terminating electrolyte (with the lowest mobility), and α, β and ω represent the first, second and last analytes in the train. With μ_L being the highest mobility, E_L must be the lowest field strength for any of the zones. Now, if an α-ion accidentally ends up in the leading electrolyte zone, the lower field strength $(E_L \cdot E_\alpha)$ will make it slow down until it falls back in the α-zone. Likewise, if an L-ion finds itself in the α-zone, it would speed up because of the higher field strength until it catches up with the L-zone (see the white arrows in Figure 5.8C). Thus, the constant current results in self-sharpening zones. The most important aspect of developing cITP methods is a judicious choice of leading and terminating buffers. Procedures and tables exist to aid in this process.[10]

5.1.4.4 Capillary Iso-electric Focussing

Capillary iso-electric focussing (cIEF) (Figure 5.8D) is a technique by which **amphoteric molecules** are separated in a spatial pH gradient according to their isoelectric points. This implies that the technique is especially associated with the separation of proteins. The stable pH gradient is generated upon voltage application using a complex blend of amphoteric compounds ("**ampholytes**") of slightly different pI spanning a range of 2.5 up to 7 units. Prior to analysis, the protein sample is mixed with the ampholytes, and the larger part of the capillary is filled with the mixture. A low-pH solution is placed at the positive electrode and a high-pH solution at the negative one. Proteins are almost all positively charged at low pH, which implies that they move away from the positive electrode in the direction of lower pH. When a specific protein reaches the pH zone at which its net charge is zero, it will no longer experience electrophoretic mobility. If it were to move beyond this point to a point where the pH is lower, for example, by diffusion, the protein becomes negatively charged, and its electrophoretic mobility would direct it back towards the positive electrode, *i.e.* towards the point where pH = pI. Thus, as in cITP, there is a natural focussing effect in cIEF. This provides the technique with a high resolving power. In the so-called two-step cIEF equilibrium, separation is obtained in the capillary, after which the contents of the capillary are moved beyond the detection window in a second step. If this is done by applying pressure, the parabolic flow profile associated with the hydrodynamic flow may lead to significant defocusing. In one-step cIEF, proteins are focussed while the entire pH gradient is slowly moved beyond the detection point. A compound that is more basic than any of the analyte proteins, such as tetramethyl-ethylenediamine, may be added to avoid proteins focussing at a point beyond the detection window.

High concentrations of salt in the sample disturb the process, so samples are often desalted by dialysis (see Section 8.2.5) or by SEC (Module 4.2) prior to analysis. There is a risk of proteins precipitating at the focal point (pH = pI), where their solubility is the lowest and the concentration is the highest. This risk is increased if salts are present.

5.1.4.5 Capillary Electrochromatography

For about a decade (say 1995 to 2005), **capillary electro-chromatography** (CEC) (Figure 5.8E) was the most promising analytical separation method. The idea of replacing a high pressure with a high voltage to transport the mobile phase through a packed column is appealing. The movement of the eluent through the different channels in the packed bed – some a bit wider and some a bit narrower – should be much more homogeneous in the case of electro-osmotic flow than in the case of pressure-driven flow. Moreover, CEC should allow the use of smaller particles, for example with d_p = 1 µm, without the need for excessive pressures and the heat-dissipation problems associated with ultra-high-pressure liquid chromatography (see Section 1.7.6). However, significant issues have held back the successful proliferation of CEC. CEC may require very high field strength and high ionic strength buffers to create a sufficiently high EOF, and this also gives rise to heat-dissipation issues (see Section 5.1.3.2). Frits have been a frustrating issue, as these tended to create local differences in EOF, causing hydrodynamic-flow effects and sometimes bubble formation. The generation of a sufficiently high EOF on silica-based columns requires an active surface with a large number of silanol groups, which may give rise to mixed retention mechanisms and broadened peaks (see Module 1.8). Unlike the situation in LC, retention factors and flow rate cannot be controlled independently, and, more importantly, mobile-phase gradients will be (at best) very difficult to realize. Surface fouling and poor retention time repeatability complete a picture of a technique that is still immature and not routinely applicable. The advent of UHPLC since 2005 has essentially terminated interest in CEC. De Smet and Lynen[11] concluded in 2014 that contemporary UHPLC largely outperformed CEC, leaving little incentive for researchers to try and address the many practical issues that still surround the latter technique.

5.1.4.6 Micellar Electrokinetic Chromatography

Micellar electrokinetic chromatography (MEKC) (Figure 5.8F), also known as micellar electrokinetic capillary chromatography (MECC), is based on a simple modification to the operation of a CE system, resulting in dramatic changes to retention, selectivity and the applicability of the technique. Adding a surfactant to the BGE in a concentration above its **critical micelle concentration** (cmc) causes the BGE to become an emulsion of micelles. In the common situation in which the surfactant is anionic and the wall of the capillary is negatively charged (*e.g.* fused silica), the electrophoretic mobility of the micelles is opposite to the direction of the EOF, so that they migrate more slowly through the capillary. This means that the analytes that are present in the micelles are retained in comparison to the EOF, creating a (pseudo-)stationary phase for neutral analytes. Retention of neutral molecules is based on their distribution between the BGE and the micelles, so that MEKC is a chromatographic technique rather than an electrophoretic technique. In Module 5.4, we take a closer look at the equilibria that underlie MEKC separations.

5.2   Instrumentation for Capillary Electrophoresis (B)

5.2.1 General Instrumentation

We have seen in Module 5.1 that electromigration techniques hold great potential for high-resolution analytical separations. However, there are substantial obstacles to overcome, including migration-time stability, detection sensitivity, and identification capability. Some solutions to these issues will be described in this module on instrumentation (injection and detection methods) and in Module 5.3 (CE-MS).

A general scheme of a capillary electrophoresis instrument is shown in Figure 5.6. The same instrumentation is used for all electromigration techniques discussed in Section 5.1.4. The **capillary** is usually made of fused silica with a polyimide coating on the outside wall. The coating is usually removed in a small section to create a detection window (see Section 5.2.3). A typical capillary length is 0.6 m (reservoir to reservoir), with an effective length (inlet reservoir to detector) of about 0.5 m. Internal diameters of 50 or 75 μm are commonly used. A stable, low-noise, high-voltage power source is essential. **Electric fields** of 10 to 30 kV are commonly applied, which correspond (with a 0.6 m-long column) to field strengths (E) of 17 to 50 kV m^{-1} (or V mm^{-1}). Typical operation is in constant-voltage mode, but in some cases (*e.g.* cITP), constant-current operation is preferred. Typical currents are in the μA range (usually below 50 μA). By installing a low-amperage (*e.g.* 100 μA) fuse in the power supply, CE instrumentation can be quite safe. Nevertheless, the equipment should always be enclosed in a grounded box. The CE capillary is usually thermostatted to facilitate efficient and stable **heat dissipation**. This is likely prudent with respect to analytical precision (repeatability), but attempts to actively cool the capillary, the contents of which heats up due to the electrical current (see Section 5.1.3.2), may well be counterproductive because the axial temperature gradients will cause severe band broadening.

Electrical current passing between the electrodes and the buffers in the reservoirs implies that **electrochemical reactions** are taking place. Usually, this involves the decomposition of water at the positive electrode

$$2H_2O \rightarrow O_2(g) + 4H^+ + 4e^- \tag{5.21}$$

while at the negative electrode, the concurring process is

$$4H_2O + 4e^- \rightarrow 2H_2(g) + 4OH^- \tag{5.22}$$

The small amounts of gases produced during these processes should be allowed to escape to prevent a buildup of pressure, as the latter would lead to an undesirable hydrodynamic flow in the capillary. Electrolytic processes will cause the solution at the positive electrode to become more acidic and that at the negative electrode to become more basic. As can be seen from eqn (5.21) and (5.22), the number of moles of protons and **hydroxide ions** produced is equal to the number of electrons (n_e) passed, which can be calculated from

$$n_e = \frac{I \cdot t}{F} \approx 10^{-11} \cdot I \cdot t \tag{5.23}$$

where F is the Faraday constant (equal to 96 500 C mol^{-1}), I is the current in µA, and t is the time in s. For example, if $I = 25$ µA and $t = 4000$ s, 1 µmol of protons and hydroxide ions will be produced. Buffering of the electrolytes and frequent refreshing are recommended. The latter may be performed automatically on most commercial instruments.

A lack of robustness, associated with poor long-term precision and failed analyses, is the often-heard argument against the (routine) use of electromigration techniques in analytical laboratories. Good quantitative precision (repeatability down to 2% or even 1%) has been reported as a realistic expectation but with an annotation that "events on the capillary surface" (incomplete equilibration or cleaning of the capillary damaged coatings) may jeopardize the precision.[12] If the sources of such events are the samples (*e.g.* variable sample matrices), additional sample preparation procedures may be required. Good operation of CE systems arguably requires understanding and care (*e.g.* frequent regeneration of capillaries and refreshing reservoir buffers). The use of control charts (or Shewhart charts) and reference samples has been recommended to monitor system performance.[12] Because of the relatively high detection limits and the risk of overloading, the linear working range of CE is narrower than that of typical LC methods.

The **analysis time** in CE is determined by the analyte with the lowest apparent mobility, which we denote with ω and for which we can write

$$t_\omega = \frac{L_{eff}}{v_\omega} = \frac{L_{eff}}{\mu_{app,\omega} \cdot E} = \frac{L_{eff}^2}{\mu_{app,\omega} \cdot V} \qquad (5.24)$$

This implies that if we keep the voltage constant, the analysis time will be proportional to the square of the effective capillary length (inlet to detector). In theory, as there is no negative effect from shortening the capillary length on efficiency and resolution (see eqn (5.15)), there is a trend towards shorter capillaries. Some commercial instruments operate with effective lengths below 100 mm (and capillary lengths below 200 mm). However, the field strength cannot be increased freely. High E values put demands on design and materials, and due to increased currents, **Joule heating** in the capillary becomes most critical. Dunn[13] demonstrated the separation of small metal ions within 30 s in a 100 mm-long capillary of 50 µm i.d. and 80 µm o.d. at a field strength of 42 V mm^{-1}.

Effective heat dissipation through the very thin capillary wall (15 μm) prevented excessive Joule heating. Very fast CE separations (*e.g.* five amino acids within 2 s on 5 mm-long capillaries) have been demonstrated using custom-made CE systems.[14] High-throughput CE systems yielding migration times of 3 to 4 s using 40 to 75 mm-long capillaries at 3 kV cm^{-1} have also been reported.[15] For such systems, injection and detection require special measures in order to maintain high efficiency. Using commercial CE instruments (minimal effective length, 80 mm; maximal field strength, 150 V mm^{-1}), analysis times will rarely be shorter than what can be achieved with contemporary UHPLC separations (see Section 3.3.2). This suggests that conventional CE in fused-silica capillaries does not outperform LC for fast separations of relatively simple mixtures. CE is more competitive for separations of complex mixtures, which in (gradient-elution) LC often take multiple hours.

5.2.2 Injection Methods

There are two fundamentally different methods to inject a small amount of sample into a CE capillary. Hydrodynamic injection is realized by creating a (small) difference in pressure between the injection and detection ends of the capillary. Electrokinetic injection is performed by applying a voltage difference across the capillary between the sample vial and the buffer reservoir at the capillary outlet.

Pressure-driven **hydrodynamic injection** is non-discriminating, in that the composition of the sample zone will be identical to that of the sample. On the downside, the flow will have the typical parabolic profile, causing the boundary of the sample zone to be less well-defined. A very small gas pressure on the sample vial (or vacuum on the terminal reservoir) suffices, but the pressure resistance of the open capillary is so small that it may suffice to have the sample vial a few mm higher than the terminal reservoir. Reworking Darcy's law (or the equivalent Hagen–Poiseuille law; see Section 1.6.4, (eqn (1.69) and (1.70))) for an open capillary yields

$$l_{inj} = \frac{\Delta P \cdot d_c^2 \cdot t_{inj}}{32\eta_{BGE} \cdot L_{cap}} \qquad (5.25)$$

Assuming $\eta_{BGE} = 10^{-3}$ Pa s for an aqueous electrolyte solution, a capillary length (L_{cap}) of 0.5 m and an inner diameter (d_c) of 75 μm, we obtain an injection length (l_{inj}) of 0.0034 mm per mm height difference (1 mm water pressure = 9.8 Pa) between the two vials and per s of injection time (t_{inj}). This implies that elevating the sample by 3 mm and an injection time of 10 s leads to an injection plug of 1 mm. Applying a gas pressure of 0.1 bar (10^4 Pa) yields a 3.5 mm-long injection zone in 1 s. The length of the injection zone is barely affected by the viscosity of the sample. If the injection zone is 5 mm long in a 0.5 m-long capillary, then 99% of the capillary remains filled with the BGE during injection.

In **electrokinetic injection**, the injection length is different for each ionic analyte because the migration speed is determined by the EOF and the ionic mobility. The length of the injection zone is

$$l_{inj} = \frac{\mu_{app} \cdot V_{inj} \cdot t_{inj}}{L_{cap}} \tag{5.26}$$

where μ_{app} is the apparent mobility of the ion eqn (5.8a) or (5.8b) and V_{inj} is the voltage applied during the injection (which often is lower than that applied during the actual analysis). Electrokinetic injection yields a plug-flow profile rather than a parabolic profile. This implies that we may use $\theta_{CE,ekinj} = 12$ for electrokinetic injection to estimate the injection band broadening with eqn (5.18), as opposed to $\theta_{CE,hdinj} = 4$ for hydrodynamic injection.

The length of the sample zone as it travels through the column is actually different from that described by eqn (5.25) because the electric field strength in the sample zone (E_{sz}) is different if the conductivity of the sample differs from that in the rest of the column, which is filled with the BGE. For the electric current to remain constant at any position in the capillary, we have

$$E_{sz} \cdot \kappa_{sam} = E_{BGE} \cdot \kappa_{BGE} \tag{5.27}$$

where κ_{sam} and κ_{BGE} are the conductivities of the sample and the BGE, respectively, and E_{sz} and E_{BGE} are the field strengths observed in the sample- and BGE-filled zones, respectively, when a constant voltage is applied across the capillary.

As the migration velocity is proportional to the field strength, we find for analyte i that

$$v_{i,sz} = \frac{\kappa_{BGE}}{\kappa_{sam}} \cdot v_{i,BGE} \qquad (5.28)$$

where $v_{i,sz}$ and $v_{i,BGE}$ are the migration velocities in the sample zone and in the remainder of the capillary, respectively. After the sample has migrated from the sample zone to the BGE, the length of the zone will have changed to the **effective injection length**, which is

$$l_{inj,eff} = \frac{\kappa_{sam}}{\kappa_{BGE}} \cdot l_{inj} \qquad (5.29)$$

If the sample has a higher conductivity than the BGE, the sample band will get broader when it passes from the original sample zone to the BGE to start the actual separation. Conversely, when the $\kappa_{sam} < \kappa_{BGE}$, the sample band slows down and gets narrower.

According to eqn (5.29), samples with low conductivity (*i.e.* a low salt content) may be significantly compressed. In the case of hydrodynamic injection, this effect is known as **sample stacking**. In the case of electrokinetic injection, it is known as field-amplified injection. For ions with ionic mobility concurrent with the EOF (*i.e.* ions with the opposite charge to the column wall), preconcentration occurs during electrokinetic injection of the sample with a lower conductivity than the BGE. Together with the focussing effect, this may lead to a dramatic increase in sensitivity. For oppositely charged ions, the EOF would need to be reversed, or polarity switching can be used during injection.

While in hydrodynamic injection of large sample zones, the sample viscosity may start playing a role, in field-amplified electrokinetic injection, the conductivity of the sample always has a large effect. **Sample-matrix effects** may reduce the precision (repeatability) of analytical methods, and for this reason, hydrodynamic injections are usually preferred. In large sample zones with low conductivity, the EOF tends to be higher than in the BGE with a higher conductivity, but liquids are barely compressible, so that pressure effects will arise at the interface between the two zones, leading to parabolic flow profiles.

Sample introduction in CE is not as precise as in LC, but when the temperature is kept constant, the precision of hydrodynamic injections for routine applications is typically better than 1%. To compensate for injection volume variations in quantitative analysis, the use of an internal standard (IS) is recommended. The IS can be any compound that can be detected in the electropherogram and does not interfere with the separation at hand. The IS can, in principle, also be used to correct for migration time variability.

5.2.3 Detection Methods

On-capillary **UV absorption** spectrometry is the most common detection method associated with electromigration methods. By removing the external polyimide coating from a short segment of a fused-silica capillary, a detection window can be created. The optical properties of fused silica are similar to those of quartz, *i.e.* the column wall is optically transparent for wavelengths down to 190 nm. There should be no contact between the components of the UV detector and the CE system to avoid interference from the high voltage on the detection. Other aspects of the detector (grating, photomultiplier, photodiode or photodiode array) are the same as described for UV detection in LC (Section 3.9.2).

The detection path length is not in favour of on-capillary UV detection in CE. It is of the order of the column's inner diameter, *i.e.* 100 μm or less. In comparison, typical detection path lengths in conventional HPLC (10 mm) or UHPLC (5 mm) are at least 100 or 50 longer, respectively. In part, this can be compensated by sharp (*i.e.* less diluted) peaks in CE, but on the other side, the geometry of the cylindrical detector "cell" is suboptimal. Modern CE detectors include optical components such as a ball lens to focus the light beam within the inner diameter of the capillary. Other geometries, such as "bubble cells" and Z-cells, where a longer segment of the capillary is used for detection, can increase the detection sensitivity but make it difficult or expensive to replace capillaries. Detection limits for CE-UV (low μM range) are typically significantly higher than those obtained with LC-UV.

Fluorescence detection (FLD, see Section 3.9.3) is a potential way to address the sensitivity challenges posed by the capillary electromigration systems. Light from a regular (deuterium, xenon, or mercury) lamp can be restricted to a narrow wavelength range using a filter or grating and focussed on the detection region of the capillary. Higher irradiance can be achieved with laser excitation, which, due to the directionality of the beam, also makes focussing easier. Several commercial CE systems are available with a laser-induced fluorescence (LIF) detector. Fundamentally,

$$S_{fd}(\lambda_{em}) = 2.3\, k_{col}(\lambda_{em})\, I_0(\lambda_{ex})\, V_{det}\, \Phi_i(\lambda_{ex})\, a_i(\lambda_{ex})\, c_i \qquad (5.30)$$

where $S_{fd}(\lambda_{em})$ is the fluorescence signal at the emission wavelength λ_{em}, $k_{col}(\lambda_{em})$ is the collection efficiency, *i.e.* the fraction of emitted fluorescent light collected at the detector, $I_0(\lambda_{ex})$ is the intensity of the excitation light with wavelength λ_{ex} in the detection cell or region with volume V_{det}, $\Phi_i(\lambda_{ex})$ is the quantum yield of analyte i at the excitation wavelength, $a_i(\lambda_{ex})$ is its extinction (absorption) coefficient and c_i is the concentration of the analyte in the detector cell. Focussing the excitation light at the detection region increases the signal intensity, but collecting a large fraction of the emitted light is challenging. It is also challenging to limit the noise in on-capillary fluorescence detection. Noise is mainly due to the scattering of the light from the curved interfaces at the outside and inside of the capillary. Nevertheless, LIF detection provides the lowest detection limits in CE (down to femtomole levels). As native fluorescence is relatively rare, derivatization techniques (see Module 8.3) are often used in combination with CE-FLD.

5.2.3.1 Indirect Detection

In **indirect absorbance** detection, a monitoring ion of the same charge as the analyte ion(s) is added to the BGE to provide a constant high absorbance baseline. The presence of an analyte in the detection volume will typically result in a decrease in the concentration of the monitoring ion, *i.e.* in a negative peak. Indirect detection is relatively popular in combination with electromigration techniques for two reasons. Firstly, no universal detectors, such as differential refractive index detectors (Section 3.9.1), are available in the small capillary format. Secondly, the concentration of the monitoring ion is directly affected by the presence of the analyte ion, a criterion that is not fulfilled in most forms of LC (ion exchange chromatography being a notable exception). Analyte ions are not replacing monitoring ions one-to-one. The **Kohlrausch condition** and electroneutrality allow deriving an equation for the change in the concentration of the monitoring ion (Δc_M) in the presence of an analyte ion (i) and a counter ion (C). For single-charged ions, this yields[16]

$$\Delta c_M = -\frac{\mu_M(\mu_i + \mu_C)}{\mu_i(\mu_M + \mu_C)} \cdot c_i \qquad (5.31)$$

The change in monitoring ion concentration is directly proportional to the analyte concentration, but the mobilities of the analyte, monitoring, and counter ions all affect the proportionality factor. In general, (i) slow analyte ions (low μ_i) yield a higher sensitivity than fast analyte ions (high μ_i), (ii) monitoring ions with a higher mobility give rise to a higher detection sensitivity, (iii) choosing a fast counter ion (high μ_C) is advantageous for the detection of slow analyte ions, (iv) a slow counter ion (low μ_C) leads to a more uniform sensitivity, and, additionally (not based on eqn (5.31)), (v) multiple-charged analyte ions show a higher sensitivity than single-charged ions. System peaks and thermal effects may lead to baseline disturbances in indirect detection. A narrower column may yield a more stable baseline.

Indirect detection is mainly used with UV absorbance spectrometry, using monitoring anions such as benzoate, chromate or sorbate, or monitoring cations, such as imidazole or pyridine. Indirect fluorescence detection has been used with salicylate or quinine as monitoring anion and cation, respectively.

5.2.4　Microfluidics

Microfluidics and CE go together well. CE requires miniaturization to avoid excessive Joule heating and convective flow. In a chip format with typical channel dimensions of 50–100 × 10–20 μm (width × height), electro-driven flow has proven easier to realize than pressure-driven flow because the latter puts high demands on the pressure resistance of the chip itself and its connection with the outside world (*e.g.* pumps). Moreover, reducing the channel length is advantageous in CE, provided that a high voltage can be maintained. **Chip-based CE systems** have been commercialized, with shorter effective lengths (down to 24 mm) than those realized on capillary-based commercial CE instruments. On commercial microfluidic CE systems, the analysis time is down to 15 s, but much shorter times have been demonstrated in research systems. A short and narrow separation channel leads to a low sample capacity, and achieving precise injections of very small volumes is one of the challenges of microfluidic CE. To overcome channel contamination and carryover between samples, single-use chips have been suggested. However, this may make it more difficult to achieve the same constant steady-state temperature for each analysis. In addition, it goes against the trend of minimizing the use of consumables, and it wrecks the image of CE as a "green" analytical separation method. The same detectors (UV, fluorescence, and mass spectrometry) are used as in capillary systems, but there are interesting possibilities for on-chip **amperometric detection** (see Section 3.9.5.1) or **contactless conductivity detection** (C4D, Section 3.9.5.2).

5.3 Capillary Electrophoresis-Mass Spectrometry (M)

Online coupling with MS is essential if electromigration techniques are to compete as high-end analytical separation methods. High-resolution MS and MS/MS techniques may provide a great deal of information on molecular structure and are compatible with the very small amounts of analytes encountered in capillary electromigration techniques. As with LC (see Module 3.10), considerable time and effort were required for routinely applicable interfaces to emerge. In addition, CE-MS, like LC-MS, uses **electrospray ionization** as the dominant ionization method, thanks to its broad applicability, both in terms of analyte polarity and molecular weight. The main weakness of ESI, dealing with very non-polar molecules, is largely irrelevant in combination with electromigration techniques. In contrast, polar macromolecules, such as proteins, become multiply charged ions in the gas phase in an ESI interface, extending the range of applicability of ESI-MS to very large molecules and supramolecular structures. Because a single macromolecule gives rise to an envelope of different charge states, deconvolution software is essential, but the precision with which the molecular weight can be estimated is enhanced. When aiming to analyse proteins in their native (folded) state, parts of the structure are not solvent-accessible, which results in fewer sites that can be ionized and, thus, higher m/z values. This results in higher MS resolution (greater distances between peaks) but puts greater demands on the ion transfer efficiency of the mass spectrometer.

Low-concentration volatile buffers containing, for example, formate, acetate, bicarbonate, or ammonium ions are preferred for CE-MS. The two electrical circuits (CE and ESI) need to be closed. Whereas the problem in LC-MS was the high volumetric flow rate of the mobile phase, the challenge in CE-ESI-MS is the very low flow rates typically encountered in CE. For example, an electro-osmotic velocity of 1 mm s^{-1} in a 75 µm i.d. capillary corresponds to a

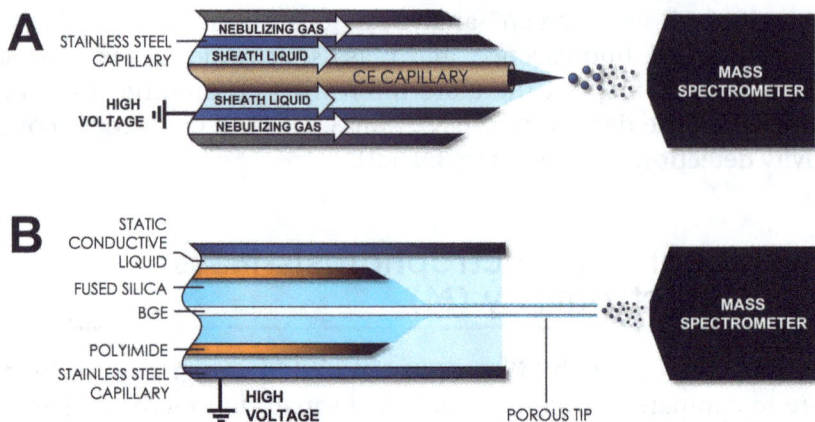

Figure 5.9 Schematic illustration of (A) sheath liquid and (B) sheathless interfaces for CE-MS. The porous tip in the bottom diagram is 30 to 40 mm long, has a wall thickness of about 5 µm, and protrudes about 5 mm from the stainless-steel capillary that is filled with a static (non-moving) electrolyte solution. Adapted from ref. 17 with permission from Springer Nature, Copyright 2019.

volumetric flow rate of 265 nL min^{-1}. In the most common and arguably the most robust interface for CE-ESI-MS, the total flow rate is increased by a **sheath liquid** (Figure 5.9A). The CE capillary is enclosed in a stainless-steel tube through which a co-axial flow of the sheath liquid (usually an aqueous-organic solvent) is added. The composition of the sheath liquid critically affects the performance of the ESI interface. The sheath liquid merges with the CE effluent at the outlet of the capillary, allowing the electrical circuit of the CE to be closed. The spray that emerges is enclosed by nebulizer gas (see Figure 5.9A). The electrical contact on the stainless-steel tube allows establishing a potential difference of 2 to 5 kV between the spray tip and the entrance of the MS. The main disadvantage of the sheath–liquid interface is that the CE effluent is diluted to reduce the detection sensitivity.

This has led to the development of **sheathless** interfaces for CE-ESI-MS (see Figure 5.9B). The key is that the CE electrical circuit is closed directly through the BGE. This may be achieved by incorporating a metal-coating electrode close to the point where the BGE leaves the capillary or – as shown in Figure 5.9B – by etching the last 30 to 40 mm of the fused-silica separation capillary with hydrofluoric acid. This can lead to a porous capillary tip with a wall thickness of about 5 µm, which is conducting when filled with electrolyte.

Through the conductive liquid, which is stationary in the case of a sheathless interface, contact is made with a surrounding metal tip. The very low electro-osmotic flow rates lead to a **nanospray**, which forms very small droplets and allows efficient ionization. The MS inlet can be closer to the spray tip than in the case of the sheath–liquid interface, and the ESI voltage can be lower (typically between 1 and 2 kV). Detection limits for proteins are about two orders of magnitude lower for sheathless interfaces than for sheath–liquid interfaces.[17]

A spectacular example of what can be achieved with CE-MS is shown in Figure 5.10. Online coupling of CE with high-resolution (time-of-flight) MS allowed the separation of recombinant human erythropoietin and the assignment of more than 250 different proteoforms,[18] including oxidation and acetylation products and 74 different glycoforms. A low pH and a separation capillary coated with polyacrylamide resulted in a near-zero electro-osmotic flow, and the sheathless interface was optimized so as to realize a stable "nanospray" at a flow rate of about 5 nL min^{-1}, with detection limits in the picomolar range.

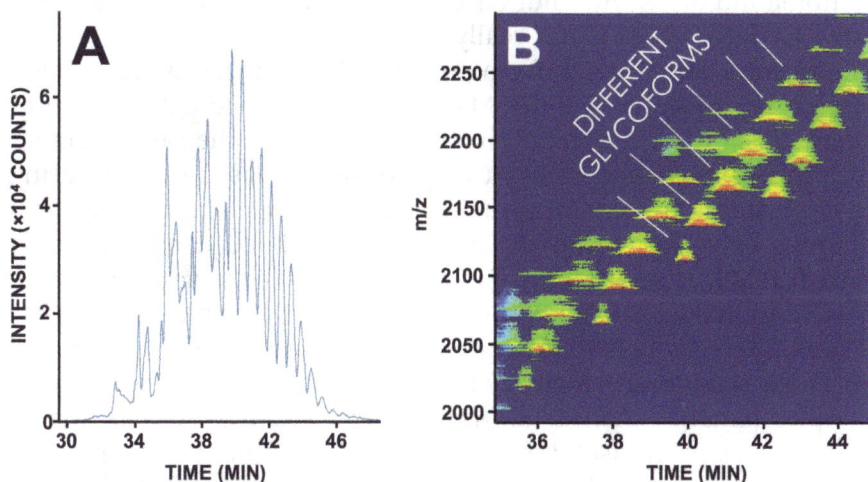

Figure 5.10 CE-ESI-MS of recombinant human erythropoietin. (A) Base-peak electropherogram (positive-ion mode). (B) Colour plot of the CE-MS data across a limited *m/z* range; each spot represents a different glycoform, including its oxidized and acetylated variants. Capillary of 1 m length × 30 µm i.d., coated with polyacrylamide to achieve a near-zero EOF. Adapted from ref. 18 with permission from American Chemical Society, Copyright 2024.

Applications of CE-MS for metabolomics have gained less traction,[19] despite the high-resolution and low-sample requirements. CE-ME allows the detection of metabolites that are missed by RPLC-MS and HILIC-MS, but it has an inherently limited metabolic coverage, with only ionic or ionogenic metabolites amenable to separation by CE. An interesting option demonstrated by Kuehnbaum *et al.* is to perform a number of **overlapping injections**, using the information from the MS to deconvolute the electropherograms.[20]

There have been a limited number of attempts to couple CE with matrix-assisted laser desorption/ionization (MALDI) MS. For example, Gasilova *et al.* realized fraction collection directly on a MALDI plate. Droplets of electrolyte solution were pre-spotted in the wells of the plate. The end of the capillary was coated with silver paint and connected to the ground, with the high voltage applied at the column inlet. Fractions were collected by inserting the end of the capillary in one of the droplets, with the aid of a fraction-collection robot. The droplet served as a temporary terminal buffer. After drying and adding a matrix, the plate was transported to the MALDI instrument.[21]

Considerable attention has been paid in the literature to the combination of CE with **inductively coupled plasma–mass spectrometry (ICP-MS)**.[22] This is not really a direct coupling between CE and MS because structural information on the analytes is completely lost in the ICP torch. CE-ICP-MS may be used for the speciation of small ions, *i.e.* for differentiating between ions of the same element with different valency states. It may not only be used for measuring "heavy" atoms present in proteins, such as metals, but also phosphorous or sulfur. Given the ease with which LC can be coupled with ICP-MS, it seems that one or more pertinent advantages are needed to make CE-ICP-MS the preferred tool. This may be the case if the CE selectivity is required for the separation query at hand, if a very high separation efficiency is needed, or if only extremely small samples (smaller than 1 μL) are available. CE-ICP-MS has shown to be particularly useful for the characterization of metal-based nanoparticles. As in CE-MS coupling, the CE electrical circuit needs to be closed, and the interface should not induce a hydrodynamic (suction) flow in the separation capillary. The latter is a potential threat because most interfaces are based on nebulization of the effluent. Most commonly, a sheath flow is used to create a stable interface, but other interface designs have also been proposed.[22]

5.4 Micellar Electrokinetic Chromatography (A)

5.4.1 Retention and Selectivity

Micellar electrokinetic chromatography (MEKC) or **micellar electrokinetic capillary chromatography** (MECC) is a brilliant invention of Shigeru Terabe, who was then at the Himeji Institute of Technology, Hyogo, Japan. We use MEKC rather than MECC, following the preferences of the inventors.[23] While CE techniques are fundamentally limited to the separation of charged analytes, MEKC made it possible to include the separation of neutral molecules. Figure 5.11 shows a reconstruction (for display purposes) of an early electrokinetic chromatogram from Terabe's team[24] that spectacularly demonstrates the possibilities of the technique. Very sharp peaks are obtained for neutral analytes across a well-defined time window. Band broadening in MEKC is largely determined by axial diffusion. Possible chromatographic contributions, such as the mass transport of analytes in and out of the micelles and the size distribution of micelles, have been found to have a minor impact on the band broadening.[23] Because the diffusion coefficients of the relatively large micelles (in MEKC) are lower than those of analyte ions (in CZE), MEKC may provide higher plate numbers than CZE. For electrokinetic chromatograms, such as the one shown in Figure 5.11, the authors estimated a plate count of 250 000 and a peak capacity of 190.[24] In comparison with (U)HPLC, the efficiency obtained by MEKC is very much higher. The micelles can be seen as stationary-phase particles that are much smaller and much more homogeneous than the packing particles used in LC. In addition, MEKC does not require the high pressures needed in (U)HPLC.[23]

In MEKC, all neutral analytes elute between the EOF marker (signal 1), for analytes that spend all their time in the BGE, and the marker for the micelles (peak 8), for analytes that are exclusively found in the (pseudo-)stationary micellar phase. The limited retention window implies that all neutral analytes can be eluted within a reasonable time. This compensates for the fact that gradient elution is not really feasible for MEKC. The retention window can be enlarged by reducing the EOF, either by lowering the pH or coating the capillary wall. However, this will significantly increase the analysis time.

In Figure 5.11, the velocity of the EOF (v_{EOF}) is about 1.7 mm s^{-1} and that of the micelles (v_{mic}) is about 0.45 mm s^{-1}. If we maintain

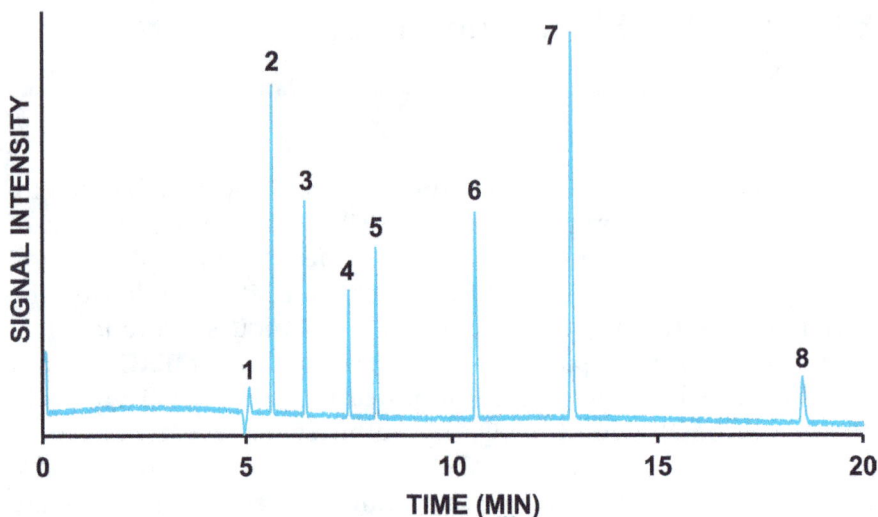

Figure 5.11 Reconstructed micellar electrokinetic chromatogram after data from ref. 24. Peaks represent (1) methanol (EOF marker), (2) resorcinol, (3) phenol, (4) *p*-nitroaniline, (5) nitrobenzene, (6) toluene, (7) 2-naphthol, and (8) Sudan III (micelle marker) micellar solution 0.05 M SDS in 0.1 M borate plus 0.5 M phosphate buffer; pH, 7; capillary, 50 µm i.d. 650 mm length, 500 mm length to detector; voltage approximately 15 kV, current, 26 µA; UV detection at 210 nm; and oven temperature, 35 °C. Adapted from ref. 24 with permission from American Chemical Society, Copyright 2024.

our definition of the retention factor as the ratio of the total amount (or the total number of molecules) of the analyte in the two phases (*i.e.* $k_{\text{MEKC},i} = q_{i,\text{mic}}/q_{i,\text{BGE}} = n_{i,\text{mic}}/n_{i,\text{BGE}}$), we have

$$v_i = \frac{k_{\text{MEKC},i}}{1 + k_{\text{MEKC},i}} \cdot v_{\text{mic}} + \frac{1}{1 + k_{\text{MEKC},i}} \cdot v_{\text{EOF}} \tag{5.32}$$

$$\frac{1}{t_i} = \frac{k_{\text{MEKC},i}}{1 + k_{\text{MEKC},i}} \cdot \frac{1}{t_{\text{mic}}} + \frac{1}{1 + k_{\text{MEKC},i}} \cdot \frac{1}{t_{\text{EOF}}} \tag{5.33}$$

which can be rearranged to yield

$$t_i = \frac{1 + k_{\text{MEKC},i}}{\left(1 + \dfrac{t_{\text{EOF}}}{t_{\text{mic}}} \cdot k_{\text{MEKC},i}\right)} \cdot t_{\text{EOF}} \tag{5.34}$$

$$k_{\text{MEKC},i} = \frac{t_i - t_{\text{EOF}}}{t_{\text{EOF}} \left(1 + \dfrac{t_i}{t_{\text{mic}}} \right)} \tag{5.35}$$

An interesting condition occurs when the migration velocity of the micelles is equal to that of the EOF but in the opposite direction. In that case, the micelles stay where they are in the capillary, and they become a true stationary phase. In that case, eqn (5.34) reduces to a familiar form, *i.e.* $t_i = (1 + k_{\text{MEKC},i}) \cdot t_{\text{EOF}}$. Analytes that reside exclusively in the aqueous buffer will elute with the EOF marker, but analytes that distribute heavily towards the micelles may stay in the capillary forever. This sounds like an awful characteristic of an analytical separation method, but it is what conventional column chromatography does to us all the time.

The relationship between $k_{\text{MEKC},i}$ and the elution time of the analyte is remarkably similar to a calibration curve in SEC (see Module 4.2), where also all analytes (from the smallest to the largest molecular weight) elute in a limited time window. The greatest selectivity is seen to occur (as in SEC) in the centre of the curve, where a relatively small difference in the retention factor between two analytes can result in a large difference in elution times. The best separation is always a trade-off between selectivity and band broadening, and this results in an optimum position for separating a pair of difficult analytes around $t = \sqrt{t_{\text{EOF}} \cdot t_{\text{mic}}}$,[25] which, in Figures 5.11 and 5.12, is around 580 s.

However, there is not a great deal of flexibility in adapting the curve of Figure 5.12 so as to obtain the best possible separation for different groups of (more or less polar) analytes. Retention factors obviously change with the volume of micelles (*i.e.* with the concentration of surfactant above the cmc). The distribution coefficient $K_{c,i} = c_{i,\text{mic}}/c_{i,\text{BGE}}$ is independent of the voltage applied (but dependent on the temperature). For the **retention factor,** we have

$$k_i = K_{c,i} \cdot \frac{V_{\text{mic}}}{V_{\text{BGE}}} = K_{c,i} \cdot \frac{v_{\text{surf}} \cdot (c_{\text{surf}} - \text{cmc})}{1 - v_{\text{surf}} \cdot (c_{\text{surf}} - \text{cmc})} \tag{5.36}$$

where V_{mic} and V_{BGE} are the volumes of the micelles and the BGE in the capillary, respectively, v_{surf} is the partial molar volume of the surfactant and c_{surf} is the total concentration of surfactant added to the BGE. For low surfactant concentrations, this becomes

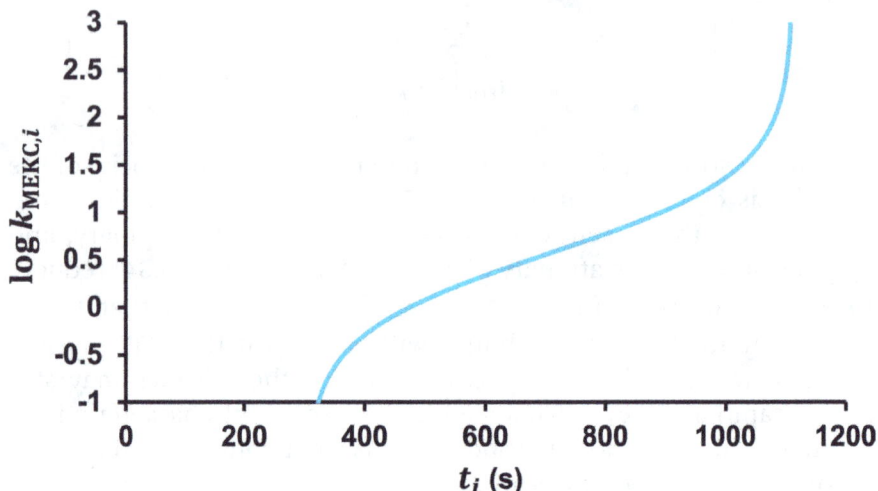

Figure 5.12   Logarithm of the retention factor $k_{MEKC,i}$ as a function of the retention time for the electrokinetic chromatogram of Figure 5.11 (t_{EOF} = 301 s and t_{mic} = 1111 s).

$$k_i \approx K_{c,i} \cdot v_{surf} \cdot (c_{surf} - cmc) \tag{5.37}$$

which shows that the concentration of the surfactant can be used to some extent to bring the retention factor of analyte i into the optimal range. The phrase "to some extent" has been added because k_i varies linearly with v_{surf}, whereas the vertical axis in Figure 5.12 is logarithmic.

Retention factors are also affected by the nature of the surfactant and field strength because a higher field strength causes a higher temperature inside the capillary.

A significant difference between SEC and MEKC is that in the former all analytes are meant to elute in the SEC window (between total exclusion and total permeation), provided that undesirable adsorption effects can be avoided. In MEKC, neutral analytes are confined to the window between the elution time of an EOF marker and that of a micelle marker, but ionic analytes can be separated by electrophoresis, simultaneously with the MEKC separation. In the usual case in which the EOF is in the direction of the negative electrode, cations can be separated before the EOF marker (if they do not interact with the negative micelles), while anions may elute anywhere within the MEKC window or later. The possibility to separate both ionic (anionic and cationic) and neutral analytes in a single run adds to the attractiveness of MEKC.

Hydrophobic analytes tend to have high $k_{MEKC,i}$ values, and they elute close to the micelle marker with very little selectivity (the vertical branch on the top right in Figure 5.12). To some extent, this may be remedied by adding an organic modifier to the BGE. However, this will increase the cmc, and at some percentage of organic solvent, it will prevent the formation of micelles altogether. Different, less hydrophobic micelle formers, such as sodium cholate or sodium deoxycholate, also yield lower $k_{MEKC,i}$ values for hydrophobic analytes.

Remarkably successful is the addition of an uncharged cyclodextrin to the BGE.[23] In such CD-MEKC (non-charged), hydrophobic analytes not only spend some time in the micelles but are also sometimes associated with cyclodextrin. During this latter time, they migrate with the velocity of the EOF, allowing the separation based partly on hydrophobicity and partly on molecular shape and structure, which determine the strength of the interaction with cyclodextrin.

Apart from anionic surfactants (most commonly SDS), cationic surfactants, such as dodecyl trimethylammonium bromide, may also be used. Bilayer adsorption of the cationic surfactant on the wall of the capillary leads to a reversal of the EOF, so that anions will elute before the EOF marker and before neutral molecules and cations. This is an important option to alter the selectivity in MEKC. The idea of electrokinetic chromatography can be extended to other pseudo-stationary phases than micelles. For example, charged derivatives of cyclodextrin allow chiral separations. Proteins may play a similar role. However, the chromatographic efficiency (plate count) is not as impressive as that observed for analysing small neutral analytes with MEKC (see Figure 5.11). Presumably, this is due to the slow association–dissociation interactions of the analyte with the pseudo-stationary phases (see Section 1.7.4.4). A partial-filling technique is required to apply proteins as pseudo-stationary phases with UV detection.

As MEKC employs non-volatile surfactants, it seems poorly – if at all – compatible with MS. ESI sensitivity is greatly impaired by the presence of surfactants. However, Somsen *et al.* succeeded in coupling MEKC online with MS, using an atmospheric pressure photo-ionization (APPI) interface and adding a dopant (acetone or toluene) to the sheath liquid to enhance sensitivity.[26]

5.4.2 Analyte Focussing

The stacking methods described in Section 5.2.2 were specifically applicable to charged analytes. Such methods cannot be used for the neutral analytes targeted by MEKC. However, the presence of a (pseudo-)stationary phase can be used to focus analyte zones. The first of these, known as **sweeping**, was published in a paper in *Science* by Quirino and Terabe in 1998.[27] The process is illustrated in Figure 5.13. A sample plug with an injection length l_{inj}^0 of the capillary is introduced hydrodynamically in a column that has been preconditioned and filled with a micellar solution as BGE. The neutral analyte molecules (marked with N) are depicted as dark circles with different outlines (Figure 5.13A). In the original sweeping procedure, the sample should have a similar conductivity as the BGE so as to maintain a constant electric field strength along the capillary. These are no longer the requirements for contemporary focussing methods in MEKC.[28] After the capillary inlet is resubmerged in the inlet buffer reservoir that contains the micellar BGE, a field is applied with – in the case of negative micelles, such as those formed by the most common surfactant, SDS – the negative electrode (cathode) at the column inlet

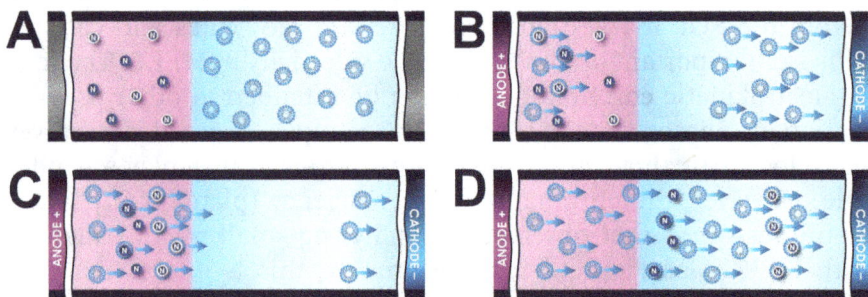

Figure 5.13 Illustration of sweeping in MEKC. Initially, the capillary is filled with a BGE containing micelles (light-blue zone). EOF is assumed to be negligible. (A) Loading of a sample solution, with roughly the same conductivity as the BGE through hydrodynamic flow (neutral analytes, dark-outlined and white-outlined circles marked with N). (B) Micelles start to enter the capillary from the anode side (micelles are moving to the right). White-outlined analytes distribute predominantly towards the micelles; dark-outlined analytes distribute more evenly. (C) Micelles have caught up with the entire sample zone. The white-outlined analyte zone is strongly focussed. The dark-outlined-analyte zone is less so. (D) A regular MEKC separation takes place.

and the positive electrode (anode) at the column outlet. Micelles are introduced into the injection zone (Figure 5.13B). The neutral analyte molecules distribute themselves between the aqueous electrolyte and the micellar (pseudo-)phase. In the example shown in Figure 5.13, the white-outlined analytes have a high retention factor (k), which implies that they are almost exclusively found in the micelles. The dark-outlined analytes are more evenly distributed $(k = 1)$.

In Figure 5.13C, micelles have penetrated throughout the injection zone. The effective injection length $(l_{inj,i}^{eff})$ becomes different for each analyte, following

$$l_{inj,i}^{eff} = \frac{l_{inj}^0}{1 + k_i} \tag{5.38}$$

In the present example, the injection lengths $l_{inj,white\text{-}outline}^{eff} \ll l_{inj}^0$ and $l_{inj,dark\text{-}outline}^{eff} \approx l_{inj}^0/2$ hold for the regular MEKC separation that started in Figure 5.13D. Analytes with high retention factors can thus be focussed strongly. Note that in this example, water is stationary, whereas the (pseudo-)stationary phase is moving. Analytes migrate when they are in the "stationary phase," and a high retention factor implies a short migration time.

Acknowledgements

Prof. Govert W. Somsen of the Vrije Universiteit Amsterdam (VU) is acknowledged for his detailed review of this chapter and for creating the course material on which this chapter was based. Janne Bolwerk is acknowledged for the useful revisions.

References

1. *Nobel Lectures*, Elsevier Publishing Company, Amsterdam, 1964.
2. J. W. Jorgenson and K. D. A. Lukacs, *Anal. Chem.*, 1981, **53**, 1298–1302.
3. M. Martin and G. Guiochon, *Anal. Chem.*, 1984, **56**, 614–620.
4. J. H. Knox and K. A. McCormack, *Chromatographia*, 1994, **38**, 207–214.
5. F. E. P. Mikkers, F. M. Everaerts and T. P. E. M. Verheggen, *J. Chromatogr. A*, 1979, **169**, 1–10.
6. G. J. M. Bruin, J. P. Chang, R. H. Kuhlman, K. Zegers, J. C. Kraak and H. Poppe, *J. Chromatogr. A*, 1989, **471**, 429–436.

7. R. Haselberg, F. M. Flesch, A. Boerke and G. W. Somsen, *Anal. Chim. Acta*, 2013, **779**, 90–95.
8. C. Gstöttner, M. Hook, T. Christopeit, A. Knaupp, T. Schlothauer, D. Reusch, M. Haberger, M. Wuhrer and E. Domínguez-Vega, *Anal. Chem.*, 2021, **93**, 15133–15141.
9. Z. Stránský, *Chromatographia*, 2008, **67**, 3–4.
10. J. Petr, V. Maier, J. Horáková, J. Ševcík and Z. Stránský, *J. Sep. Sci.*, 2006, **29**, 2705–2715.
11. S. De Smet and F. Lynen, *J. Chromatogr. A*, 2014, **1355**, 261–268.
12. S. Hartung, R. Minkner, M. Olabi and H. Wätzig, *TrAC, Trends Anal. Chem.*, 2023, **163**, 117056.
13. R. C. Dunn, *Anal. Chem.*, 2020, **92**, 7540–7546.
14. T. Zhang, Q. Fang, W.-B. Du and J.-L. Fu, *Anal. Chem.*, 2009, **81**, 3693–3698.
15. J. J. P. Mark, P. Piccinelli and F.-M. Matysik, *Anal. Bioanal. Chem.*, 2014, **406**, 6069–6073.
16. M. W. F. Nielen, *J. Chromatogr.*, 1991, **588**, 321–326.
17. E. Domínguez-Vega, R. Haselberg and G. W. Somsen, *Methods Mol. Biol.*, 2016, **1466**, 25–41.
18. R. Haselberg, G. J. de Jong and G. W. Somsen, *Anal. Chem.*, 2013, **85**, 2289–2296.
19. W. Zhang, T. Hankemeier and R. Ramautar, *Curr. Opin. Biotechnol.*, 2017, **43**, 1–7.
20. N. L. Kuehnbaum, A. Kormendi and P. Britz-Mckibbin, *Anal. Chem.*, 2013, **85**, 10664–10669.
21. N. Gasilova, A. Gassner and H. H. Girault, *Electrophoresis*, 2012, **33**, 2390–2398.
22. G. Álvarez-Llamas, M. R. Fernández de la Campa and A. Sanz-Medel, *TrAC, Trends Anal. Chem.*, 2005, **24**, 28–36.
23. S. Terabe, *Chem. Rec.*, 2008, **8**, 291–301.
24. S. Terabe, K. Otsuka and T. Ando, *Anal. Chem.*, 1985, **57**, 834–841.
25. K. Otsuka and S. Terabe, *Appl. Biochem. Biotechnol. - Part B Mol. Biotechnol.*, 1998, **9**, 253–271.
26. G. W. Somsen, R. Mol and G. J. De Jong, *Anal. Bioanal. Chem.*, 2006, **384**, 31–33.
27. J. P. Quirino and S. Terabe, *Science*, 1998, **282**, 465–468.
28. R. B. Yu and J. P. Quirino, *TrAC Trends Anal. Chem.*, 2023, **161**, 116914.

6 Supercritical Fluid Chromatography

This chapter describes chromatographic systems with mobile phases containing carbon dioxide (CO_2). When the separation is performed above the critical temperature and pressure it is defined as super-critical-fluid chromatography. When organic solvents, and possibly small amounts of water, are added to CO_2, the operation is usually performed below the critical temperature and is defined as subcritical-fluid chromatography. Both are described in Module 6.1 and covered by the term SFC. Packed-column SFC (Module 6.2) is increasingly successful for fast, high-resolution separations, especially – but not exclusively – of chiral compounds. Open-tubular SFC (Module 6.3) has proven less successful and is rarely used today. When CO_2 is added in minor amounts to liquid mobile phases, it is defined as enhanced fluidity chromatography (EFC) (Module 6.3), a technique that may potentially be faster than conventional LC due to increased diffusion coefficients and reduced viscosity (Figure 6.1).

6.1 Introduction (M)

6.1.1 Supercritical Fluids

There are three basic **states of matter**, *viz.*, gases, liquids and solids. In a simple phase diagram of a pure compound, such as that shown in Figure 6.2, we distinguish regions for each state. There are lines at which two phases coexist in equilibrium: the sublimation line between solids and gases, the melting line between liquids and solids and the evaporation line between gases and liquids. The three curves coincide at the triple point, where all three phases coexist.

Analytical Separation Science
By Bob W. J. Pirok and Peter J. Schoenmakers
© Bob W. J. Pirok and Peter J. Schoenmakers 2025
Published by the Royal Society of Chemistry, www.rsc.org

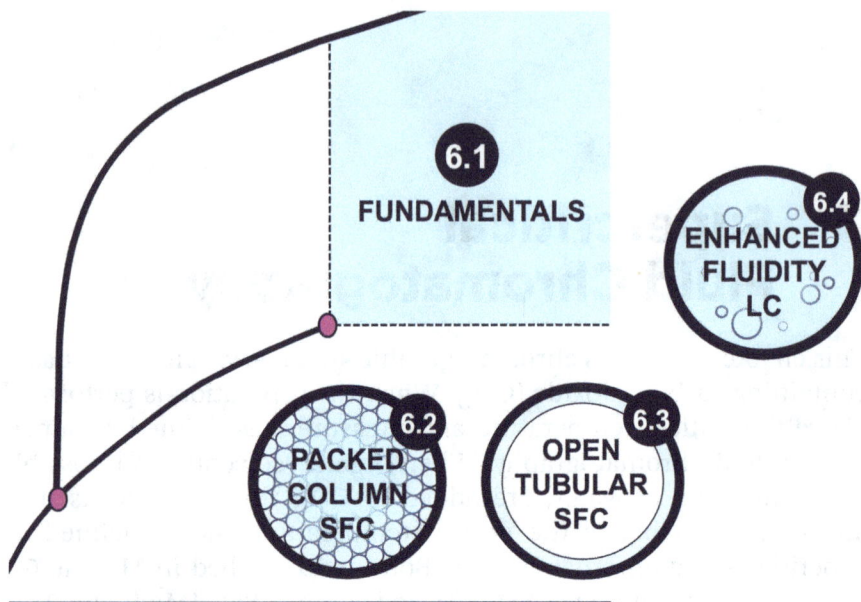

Figure 6.1 Graphical overview of the modules in this chapter.

The melting line almost always has a positive slope in the temperature–pressure diagram. This implies that a liquid may be turned into a solid by increasing the pressure. Water is one of the very few exceptions, where ice is lighter than water and floats on top. Increasing the pressure may turn ice into water.

The evaporation curve is finite. It runs from the triple point (T_{triple}, P_{triple}) to the critical point (T_{crit}, P_{crit}). Above the critical temperature, there is no distinction between a gas and a liquid. When approaching the **critical point** along the evaporation line, the differences between the gas and the liquid in terms of density, diffusion coefficient, viscosity, *etc.*, diminish. At the critical point, the meniscus disappears, and the system becomes homogeneous. From a thermodynamic perspective, liquids no longer exist. Calvin Giddings, who was among the first to explore the use of mobile phases in chromatography at conditions above their critical temperature and pressure, correctly used the term "dense gases". Unfortunately, chromatographers no longer use this term. We refer to supercritical fluids as those compounds (or mixtures) that exist under conditions above the critical temperature and pressure, *i.e.* $T > T_{crit}$ and $P > P_{crit}$. This is a matter of definition, and therefore, the term is not incorrect. However, there is a tendency to use the term supercritical phase, which is grossly incorrect. There is

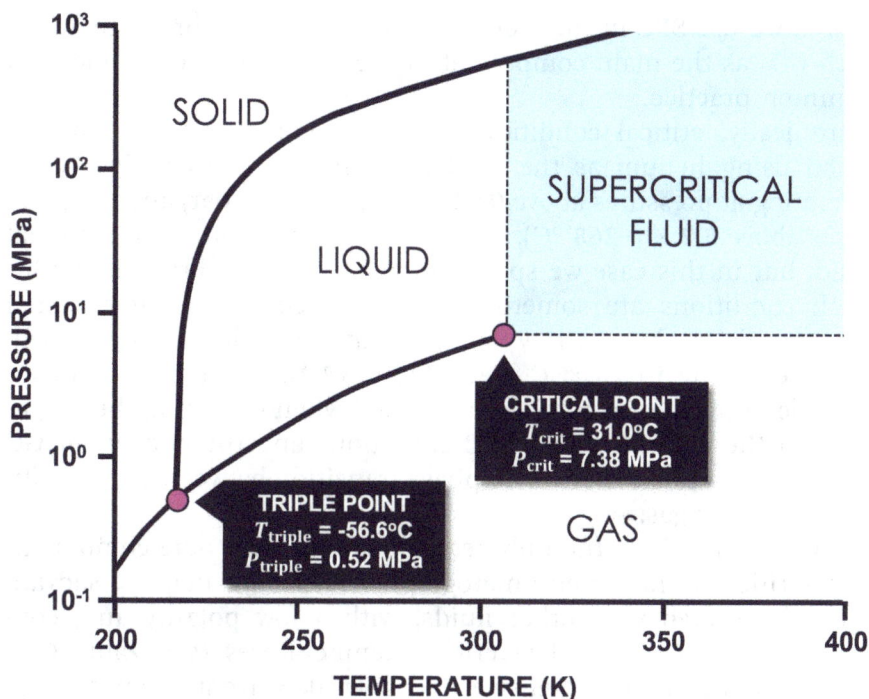

Figure 6.2 Phase diagram (temperature–pressure diagram) of pure carbon dioxide.

no phase transition between a liquid and a supercritical fluid, nor between a gas and a supercritical fluid. The situation is even more confusing if we work below (but close to) the critical temperature and/or the critical pressure. Instead of using the terms gases and liquids, chromatographers are now in the habit of using the term subcritical fluids. A sound definition of the latter does not exist. Some authors have insisted that subcritical implies $T < T_{crit}$ and $P < P_{crit}$,[1] but under such conditions, there exist only gases and liquids. When the mobile phase is a supercritical fluid, it is referred to as supercritical-fluid chromatography (SFC). When it is not (usually because $T < T_{crit}$, while $P > P_{crit}$), but the way in which we operate the system is similar, it is referred to as **subcritical-fluid chromatography**. It is convenient to abbreviate the latter also as SFC, which has come to be defined as "sub- or supercritical-fluid chromatography". In practice, SFC almost always implies a mobile phase in which carbon dioxide (CO_2) is the major component. If CO_2 is a minor component, it is termed enhanced fluidity (liquid) chromatography (EFC, see Module 6.4). With due regret, we use SFC in this book to imply a form of chromatography with CO_2 as the main component of the mobile phase, in line with common practice.

Ironically, critical conditions are most commonly encountered when using helium as the carrier gas in chromatography. When operating at pressures above 0.227 MPa (about 2.3 bar) and temperatures above 5.2 K (−268 °C), helium is, by definition, a supercritical fluid, but in this case we speak, of course, of gas chromatography. Such conditions are sometimes encountered in one-dimensional GC, especially when narrow (50 or 100 μm i.d.) capillaries are used, and frequently in GC×GC (see Module 7.2). Excellent chromatographic performance can be maintained when operating the beginning of the column under SFC conditions and the end under GC conditions because there is no phase transition between a supercritical fluid and a gas.

Carbon dioxide is the only realistic choice as a pure-component supercritical fluid for chromatography. There are major disadvantages associated with other fluids, with a low polarity and, consequently, manageably low critical temperatures (see Table 6.1). Xenon is much too expensive and a weak solvent, even at high densities. Nitrous oxide is an oxidant and, because it can be abused as a drug, is increasingly restricted. Sulfur hexafluoride is a very strong greenhouse gas, as it is a chlorofluorocarbon ("Freons"). The use of polar compounds is not practical because of their high critical temperatures, reactivity and corrosiveness. Water can be corrosive, either under supercritical or under subcritical conditions ("superheated" water). Contemporary SFC is usually subcritical-fluid chromatography, with polar solvents (**"modifiers"**) being added to CO_2, resulting in a separation system that resembles normal-phase LC, but is not limited to conventional NPLC as described in Section 3.6.1. Mobile phases consisting of CO_2, modifier and small amounts of water resemble HILIC. The possible addition of volatile buffers opens up all interactions experienced in HILIC (see Section 3.6.2). As explained in this latter section, increasing the amounts of polar modifier and water in the mobile phase may bring the phase system into the RPLC domain. Such conditions in which CO_2 is a minor mobile-phase component are referred to as enhanced-fluidity chromatography (EFC (see Module 6.4)). Caroline West has presented a more detailed discussion on SFC retention mechanisms.[3]

Table 6.1 Properties of some possible pure-component mobile phases for SFC. Carbon dioxide is commonly used in SFC and thus has been printed in bold. Based on data from ref. 2.

Compounds	T_{crit} (°C)	P_{crit} (MPa)	Practical	Inert	Non-corrosive	Polarity/solvent strength	Comments
Xenon	16.5	5.84	+	+	+	Very low	Horrendously expensive
Carbon dioxide	**31.0**	**7.38**	+	+	+	**Low**	**Pervasive choice for SFC**
Nitrous oxide	36.5	7.24	+	+	+/−	Low	Oxidizing agent; anaesthetic
Sulfur hexafluoride	45.5	3.76	+/−	+	+	Low	Very strong greenhouse effect; produces toxic gases in flame (FID)
Ammonia	132.3	11.35	−	−	−	Very high	Corrosive and potentially explosive
Methanol	**239.5**	**8.09**	−	+/−	+/−	High	Practical only as a modifier in CO_2 and is the **most common modifier**
Acetonitrile	272.3	4.83	−	+	+	High	Practical only as a modifier in CO_2
Water	374.2	22.12	−	−	−	Variable[a]	Practical only as a modifier in CO_2 (in low concentrations)

[a]Supercritical (pure) water is not thought to be corrosive for metal parts of the instrument, but it will dissolve quartz (detector cell) windows and exposed fused-silica capillaries, and probably silica-based stationary phases.

6.1.2 Why SFC?

In the vast majority of chromatographic separations, either a gas (GC) or a liquid (LC) is used as the mobile phase. Fundamentally, GC is far more attractive. This is understood most easily if we return to: (i) the reduced (dimensionless) velocity (v_0) introduced in Chapter 1, defined as $v_0 = u_0 d / D_m$, where u_0 is the linear mobile-phase velocity, d is the characteristic diameter (the column diameter for open-tubular columns or the particle diameter for packed columns), and D_m is the diffusion coefficient of the analyte in the mobile phase; and (ii) the reduced plate height (h), which is obtained from $h = H/d$, where H is the actual plate height. We can use these reduced parameters to obtain an equation for the hold-up time (t_0) of a column with length L and N theoretical plates, i.e.

$$t_0 = \frac{L}{u_0} = \frac{NH}{u_0} = \frac{Nh}{v_0} \cdot \frac{d^2}{D_m} \tag{6.1}$$

Eqn (6.1) shows that under identical conditions (N, h, and v), the hold-up time is proportional to the ratio d^2/D_m. This is also true for the retention time t_R if the retention factor (k) is equal. Thus, on a given column, retention times are shorter if D_m is higher. Alternatively, if D_m is higher, open-tubular columns with larger internal diameters or columns packed with larger particles may be used.

Table 6.2 shows an indication of the conditions used and the mobile-phase properties encountered in GC, LC and the two forms of SFC. One of these latter, "liquid-like" SFC (also known as high-density SFC), is akin to LC, typically using packed columns and CO_2 with organic modifiers and possibly other additives as the mobile phase. Another form of SFC ("gas-like" or low-density SFC) resembles GC in that open-tubular columns are used, with a preference for pure CO_2 as the mobile phase, allowing the use of flame-ionization detectors.

The diffusion coefficient D_m shown in Table 6.2 is four orders of magnitude higher in gases than in liquids. Thus, for LC to compete with GC in terms of speed in open-tubular chromatography, the internal diameter of the LC column, which is the characteristic dimension in eqn (6.1) ($d = d_c$), should be about 100 times smaller than in GC. The typical GC columns with internal diameters of 250 to 530 μm can thus be compared with open-tubular LC columns with internal diameters of 2 to 5 μm. These were shown to be highly

Table 6.2 Approximate ranges of conditions and mobile-phase properties for GC, LC and SFC. In the latter case, pure carbon dioxide is assumed to constitute the mobile phase.

	Temperature, °C	Inlet pressure, MPa	Outlet pressure, MPa	Density, kg L^{-1}	Diffusion coefficient, m^2 s^{-1}	Viscosity, µPa s
Liquid	25–60	5–150	<0.5	0.8–1.5	10^{-9}	500–2000
"Liquid-like" SFC	40–70[a]	15–40	~10	0.7–1	10^{-8}	50–100
"Gas-like" SFC	80–200	10–25	Just below inlet pressure	0.1–0.3	10^{-7}	10–40
Gas	40–350	0.12–0.25	0.1	10^{-3}–10^{-4}	10^{-5}	1–2

[a]When using modifiers, lower temperatures (25–40 °C) may also be used.

impractical in Chapter 1 (see Section 1.7.8, Table 1.3). The remedy for liquid chromatographers is to instead use very small particles, but these come at the expense of very high pressures. Hence, there exist high-pressure LC and ultra-high-pressure LC (see (eqn (1.69)), *i.e.* the Darcy equation, Sections 1.6.4 and 1.7.2). The diffusion coefficients of supercritical fluids are somewhere in between those of gases and liquids. Therefore, column diameters for open-tubular SFC may be somewhere in between those of GC and LC. The data shown in Table 6.2 suggest that for gas-like SFC columns, internal diameters of 25 to 50 µm may be appropriate (D_m 100 times lower than in GC and hence d_c 10 times smaller). For liquid-like SFC, columns of 10 or 15 µm are appropriate. When performing density programming, which is the SFC equivalent of temperature programming in GC or solvent programming (gradient elution) in LC, conditions will change during the run. As the density increases, D_m decreases, resulting in lower efficiency, unless the linear velocity is decreased proportionally, so as to keep the reduced linear velocity ($\nu = ud_c/D_m$) constant.

When using the same packed columns (where d in eqn (6.1) is the particle diameter d_p) for LC and SFC, the diffusion coefficients are

---

Box 6.1 Hero of analytical separation science: Keith Bartle. Image courtesy of The Chromatographic Society.

Keith Bartle (1939–2023, United Kingdom)
Keith Bartle was a long-time professor at Leeds University in the UK. His PhD was on NMR spectroscopy, but he got hooked on separation science when working with Milos Novotny (Indiana University Bloomington, IN) and Milton Lee (Brigham Young University, Provo, UT) in the USA. He did not have the immense focus or obdurance of his eternal hero (Sir Geoffrey Boycott – not a separation scientist). Keith was much more of an all-rounder, contributing significantly to GC, LC, electro-chromatography, and – especially – SFC. He was very inventive in finding technical solutions in the laboratory, building his own equipment for SFC and GC×GC before good commercial equipment became available.

higher in the latter case, so that proportionally faster analyses are anticipated. The gain may be up to a factor of 10 (see Table 6.2).

In summary, the promise of SFC is that it may offer a compromise between GC and LC, allowing separations of non-volatile analytes (i) faster than with LC using packed columns or (ii) using open columns with larger internal diameters than the totally impractical values ($d_c \leq 5$ μm) required for open-tubular LC. Keith Bartle (Box 6.1) was one of the scientists who contributed significantly to the development of SFC.

See the website for more applications of SFC: ass-ets.org

6.1.3   Retention in SFC

Retention in GC is determined by the pure-component vapour pressure of the analyte and by its interaction with the stationary phase, as reflected in the activity coefficient (see Module 2.2). The mobile phase does not play a role. In LC, analyte vapour pressure is irrelevant. Non-volatile analytes are eluted, thanks to their interactions with the mobile phase. For analytes that are compatible with GC, it is evident from eqn (6.1) that there is no fundamentally better alternative.

The value of open-tubular SFC may be in extending the range of analytes that can be eluted beyond the limits of (high-temperature) GC. Some analytes are not sufficiently volatile to be analysed using GC; some may degrade or decompose before the temperature reaches a point where they could have become sufficiently volatile. Significant interaction with the mobile phase – and thus a sufficiently high mobile-phase density – are required for such analytes to be eluted.

Figure 6.3 shows a schematic pressure diagram for pure carbon dioxide. At temperatures below the critical point (CP), liquids and gases coexist at a specific pressure, as indicated by the blue "isotherm" for 25 °C. The horizontal dashed line connects the gaseous and liquid phases that are in equilibrium. High liquid densities are obtained at low temperatures. This is indicated by the dark blue line on the right in the figure. This line is nearly vertical, showing the liquid to be barely compressible. The critical isotherm (corresponding to the critical temperature of 31 °C) is horizontal at the critical point. This implies that density (and other properties of the fluid) changes infinitely fast with increasing pressure at

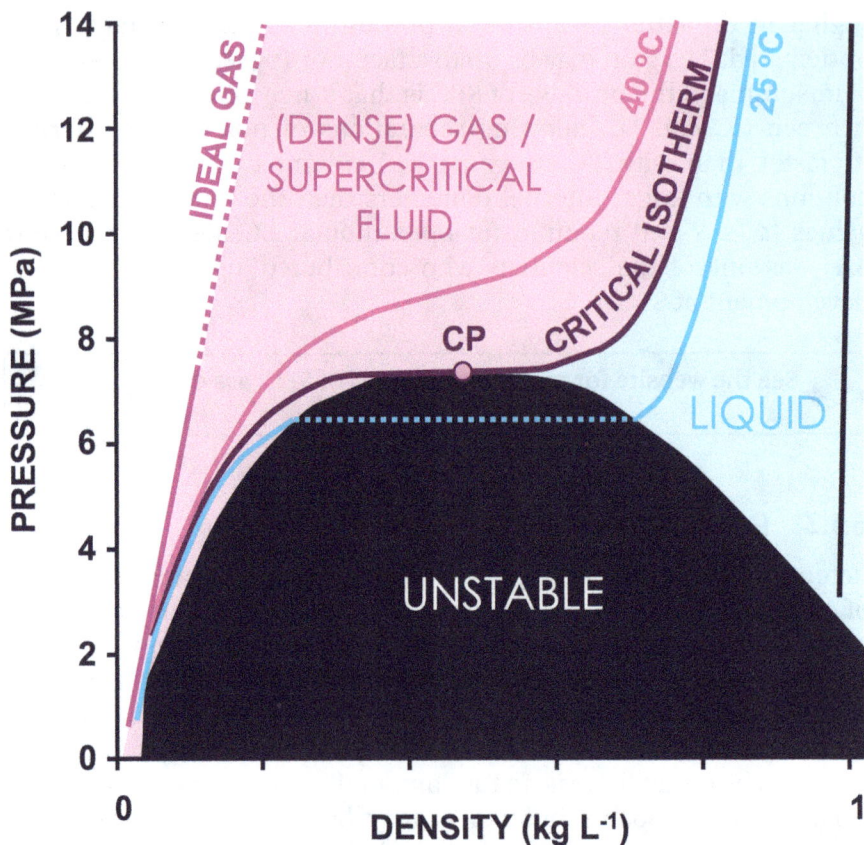

Figure 6.3 Schematic pressure–density diagram for pure carbon dioxide.

this point. Mobile-phase properties (density, diffusivity, viscosity, and the extent of interactions with analytes) change dramatically in the vicinity of the critical point. At the critical point, the diffusion coefficient of the analyte in the mobile phase equals zero, and close to the critical point diffusion is extremely slow.[4] As the pressure necessarily decreases along the length of the column (otherwise there would be no flow), operation in the region close to the critical temperature tends to result in very bad peak shapes, poor recoveries, and poor repeatability. Hence, this region close to the critical point has come to be known as the Bermuda Triangle of SFC. To avoid this area, temperatures well above the critical point (*i.e.* >40 °C for pure CO_2) may be used, resulting in dense-gas or – at pressures above the critical value – supercritical-fluid chromatography. Alternatively, temperatures below the critical value may be used, resulting in ("subcritical") liquid chromatography. At very high pressures, there is a smooth transition between gases and liquids, even around the critical temperature.

Curves displaying retention *vs.* temperature and, to a lesser extent, retention *vs.* pressure in SFC are fairly complex. Figure 6.4 shows a schematic example of the kinds of curves that may be obtained when plotting the retention factor against the temperature. At a pressure below the critical value, a pure-component mobile phase exhibits a boiling point, which is indicated by the dashed vertical line in Figure 6.4. In the region below the boiling point we have LC, and the retention almost always decreases with increasing temperature. Above the boiling point we have GC, and if the analyte is sufficiently volatile, retention decreases with temperature. A large upward jump in retention is observed, although typically not as sharp as suggested in Figure 6.4, because when operating around the boiling point, the pressure drop across the column may imply that LC is performed in the first part and GC in the second part.

At a pressure above P_{crit}, the transition from LC-like behaviour to GC-like behaviour is more gradual, as the density of the mobile phase decreases. For analytes with sufficient volatility, downward, upward and again downward regions are observed in practice. For non-volatile analytes, the behaviour is more simple. If the density of the mobile phase decreases, the retention increases. Above the

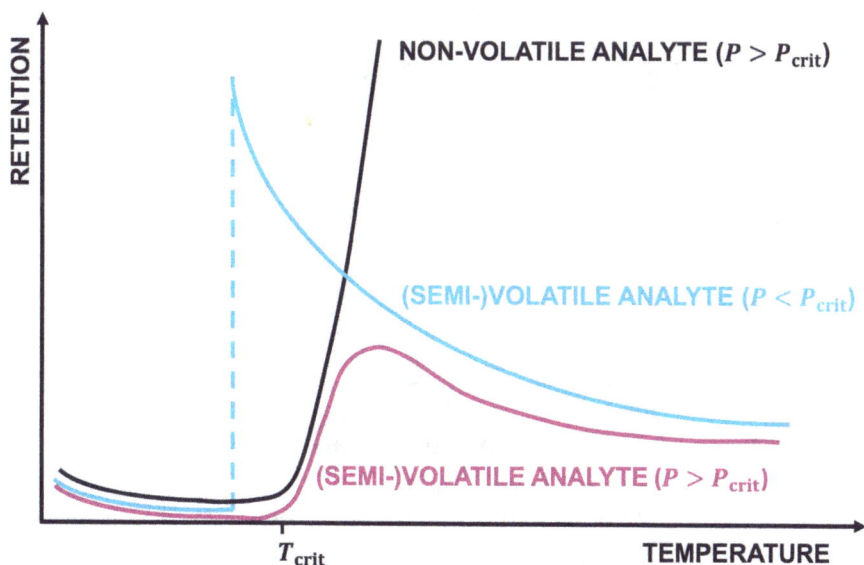

Figure 6.4   Schematic retention *vs.* temperature plot for SFC.

critical temperature, the density decreases strongly if the pressure is not far above the critical value.

Figure 6.5 illustrates the effect of pressure on the retention of naphthalene with pure CO_2 as the analyte. A comparison of the curves in this figure with those in Figure 6.3 (density vs. pressure) confirms the overriding effect of the mobile-phase density on retention.

This is clearly confirmed in Figure 6.6. While Figures 6.4 and 6.5 show retention in SFC to be a complex function of pressure and temperature, the effect of density is simple and monotonous. US President Bill Clinton famously had a sign on his desk that read "It's the Economy, Stupid!". Supercritical-fluid chromatographers should consider a sign that says "It's the Density, Stupid!".

The analogy between retention vs. density curves in SFC and retention vs. mobile-phase composition curves in RPLC (see Module 3.8) is striking. The practical consequence of this is that density programming in SFC is the equivalent of solvent programming (gradient elution) in RPLC. Implementation of density programming

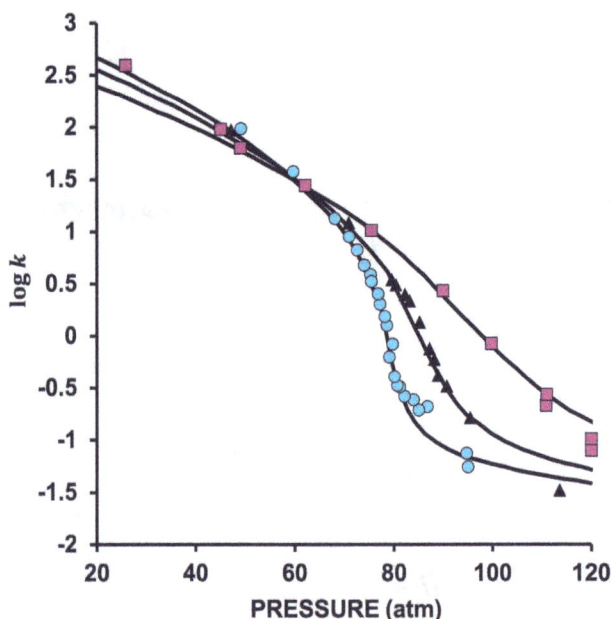

Figure 6.5 Experimental and calculated retention factors for naphthalene as a function of pressure in packed-column SFC with pure carbon dioxide as the eluent. Temperature: 35 °C (blue dots), 40 °C (dark triangles) and 50 °C (pink squares). Adapted from ref. 5 with permission from Elsevier, Copyright 1984.

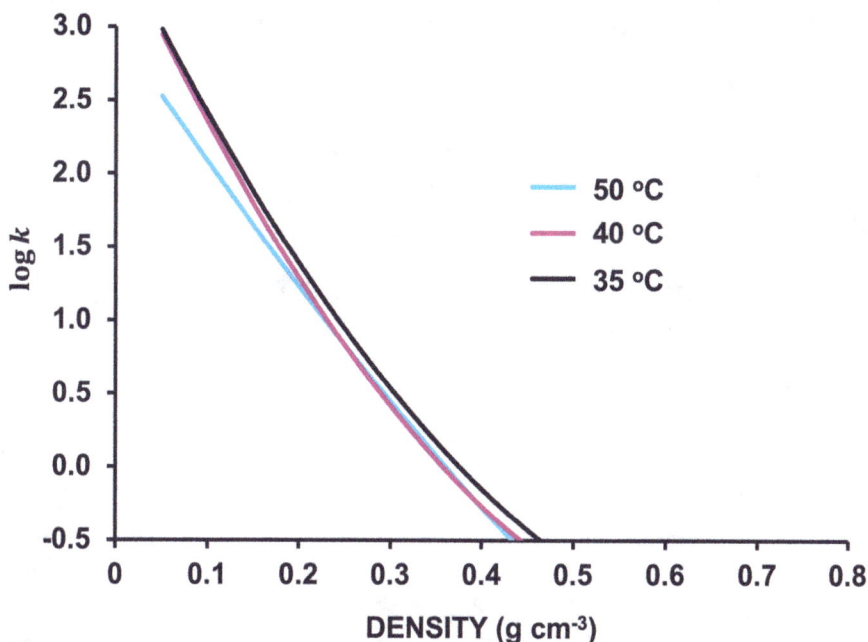

Figure 6.6 Calculated retention factors for naphthalene as a function of mobile-phase density in packed-column SFC with pure carbon dioxide as the eluent. For experimental data, see Figure 6.5 Adapted from ref. 5 with permission from Elsevier, Copyright 1984.

is not trivial. It requires an accurate equation of state that describes density as a function of pressure, temperature and mobile-phase composition (type and concentration of modifier) to be incorporated in the (software of the) instrument.

6.1.4 Viscosity and Pressure Drop

Figure 6.7 illustrates the viscosity of CO_2 as a function of temperature and pressure. Horizontal dashed lines have been added to the figure to connect points of equal density ("**isopycnic points**"). This shows that – at least in the domain of LC-like SFC – viscosity is essentially determined by the density. However, at liquid-like density, pure CO_2 has a viscosity that is 10 (at $\rho = 0.9$ g mL^{-1}) to 15 (at $\rho = 0.7$ g mL^{-1}) times lower than that of water. In GC-like SFC, the viscosity may be another two or three times lower. However, the

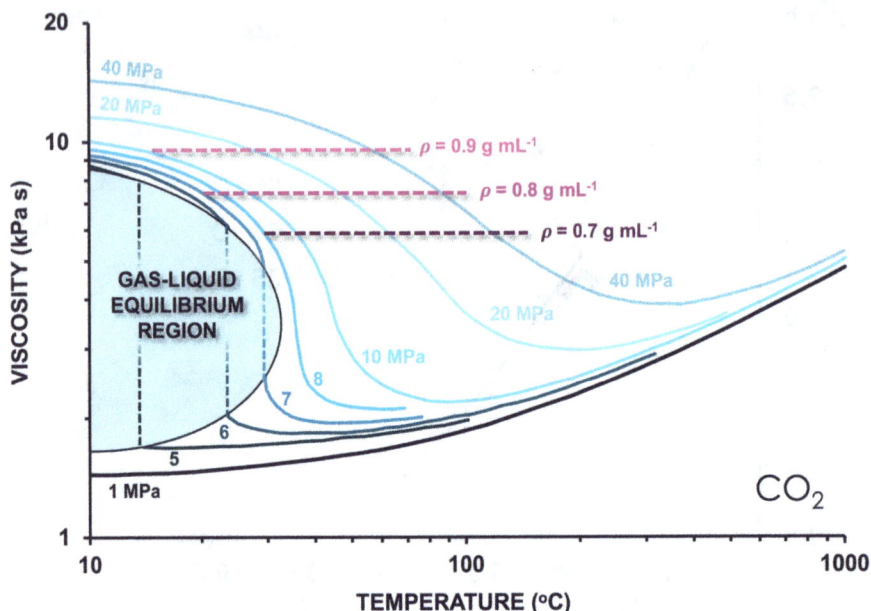

Figure 6.7 Viscosity of carbon dioxide as a function of temperature and pressure. The horizontal dashed lines connect points of equal density. Adapted from ref. 6 with permission from American Chemical Society, Copyright 1983.

viscosity of CO_2 in gas-like SFC is still about an order of magnitude higher than that of gases used in GC.

We can rewrite the **Darcy equation** (eqn (1.69)) to

$$\Delta P = \frac{\psi \cdot \eta \cdot L \cdot u_0}{d^2} = \psi \cdot N \cdot h \cdot v \cdot \frac{\eta D_m}{d^2} \tag{6.2}$$

which shows that under equal conditions (equal flow resistance factor ψ, equal plate count N, equal reduced plate height h and equal reduced velocity v), the pressure drop across the column (ΔP) is directly proportional to the viscosity of the mobile phase (η) and the diffusion coefficient (D_m) and inversely proportional to the square of the characteristic diameter d.

When performing packed-column SFC on similar columns as used in LC (same $d = d_p$), the pressure drop is seen to be proportional to the product ηD_m. Since the viscosity is lower in SFC than in LC and the diffusion coefficient $d = d_p$ is higher (see Section 6.1.4), the pressure drop may be similar in both cases. As a substantial pressure (at least 10 MPa) needs to be maintained at the end of the column, a much lower head pressure is not one of the anticipated advantages of packed-column SFC over LC. However, at equal reduced velocities, SFC separations are much faster because $u_0 \propto D_m$. In common LC practice, some of this speed gap is closed by working at higher reduced velocities, at the expense of higher pressures and reduced efficiencies. It should be noted that adding an organic solvent to the CO_2 mobile phase causes an increase in density and viscosity and a decrease in the diffusion coefficient so that the fundamental advantages of CO_2 become much smaller (see Section 6.2.1).

When performing open-tubular SFC and following the guideline that $d_c^2 \propto D_m$ (so as to obtain similar retention times, see eqn (6.1)), the pressure drop is proportional to the viscosity. This is much greater for SFC than for GC, but since pressurization is required anyway to reach a sufficient eluent strength (*i.e.* mobile-phase density) and since the flow resistance factor ψ is about 30 times lower for open-tubular columns than for packed columns, the pressure drop is not a major concern in open-tubular SFC.

6.1.5 Diffusion Coefficients

Data for diffusion coefficients in supercritical and "subcritical" mobile phases are not abundant in the literature, but for the present discussion, approximate numbers suffice. The **Wilke–Chang equation** reads[7]

$$D_{i,m} \approx \frac{7.4 \times 10^{-8}\, T \sqrt{\phi_m \cdot M_m}}{\eta_m \cdot v_i^{0.6}} \tag{6.3}$$

where $D_{i,m}$ is the diffusion coefficient of analyte i in mobile phase m, ϕ_m is the association factor of the mobile phase, M_m is its molecular weight and η_m is its viscosity, and v_i is the molar volume of the analyte. The most important aspect of this equation is that the diffusion coefficient is expected to be inversely proportional to the viscosity. Thus, a high mobile-phase density (solvent strength) comes with a high viscosity (see Section 6.1.3) and low diffusion coefficients. We cannot both have our cake (a high solvent strength) and eat it (low viscosity and high diffusivity). A supercritical fluid is not a magical mobile phase that combines the best of all worlds.

Instead, it is a compromise between the attractive transport properties of a gas and the solvent strength of a liquid.

Table 6.3 also lists the reasonable values for diffusion coefficients and viscosities shown in Table 6.2, along with reasonable values for the characteristic diameter (d_p for packed columns or d_c for open-tubular columns), the relative speed (eqn (6.1)) and the relative pressure drop across the column (eqn (6.2)). For liquid-like SFC, packed columns are the obvious choice. With pure CO_2 as the eluent, packed-column SFC can be an order of magnitude faster than LC using similar columns, but for such an extreme gain in speed, the pressure drop in SFC is higher. At an equal pressure drop, SFC is still expected to be four times faster than LC. On the other hand, open-tubular SFC cannot be performed on the same open-tubular columns as capillary GC. On a standard GC column with 250 μm i.d., SFC is seen to be 100 times slower. To achieve comparable performance, a 250 μm i.d. GC column should be compared with a 25 μm i.d. SFC column. Implementation of packed-column SFC and open-tubular SFC will be discussed in Modules 6.2 and 6.3, respectively.

Table 6.3 Indication of the relative speed and pressure drop in GC (with He as the carrier gas), LC (with an aqueous mobile phase) and SFC (using pure CO_2).

	Diffusion coefficient, $m^2\,s^{-1}$	Typical viscosity, μPa s	Characteristic dimension, μm	Relative retention time[a]	Relative pressure drop[b]
Liquid	10^{-9}	1000	$d_p = 5$	10	1000
"Liquid-like" SFC	10^{-8}	250	$d_p = 5$	1	2500[c]
"Gas-like" SFC	10^{-7}	50	$d_c = 25$	12	1.3[c]
	10^{-7}	50	$d_c = 250$	1200	0.013[c]
Gas	10^{-5}	20	$d_c = 250$	12	0.5

[a]Eqn (6.1), assuming equal N, h, and v (and k).
[b]Eqn (6.2) assuming equal N, h and v. $\psi = 32$ for open-tubular columns; $\psi = 1000$ for packed columns.
[c]Column exit pressure should exceed about 10 MPa.

6.2 Packed-column SFC (B)

6.2.1 Setup

In packed-column SFC, the mobile-phase density is close to that of a liquid. The pressure drop across the packed columns is relatively high, and a high mobile-phase density allows the eluent strength to be largely maintained along the column length. Indeed, working too close to the critical conditions at the column outlet causes considerable expansion of the mobile phase, which is instantly visible from very broad and poorly shaped peaks. Typical conditions are moderate temperatures (40 °C to 60 °C) and column-outlet pressures well above 10 MPa.

The pressure at the system outlet is controlled by a back-pressure regulator (BPR), which is the most critical part of the SFC instrument. The combination of one or two LC-type pumps delivering the mobile phase and a BPR providing a variable restriction implies that the flow rate and the density of the mobile phase can be controlled independently, which is important, because the density affects diffusivity and thus the reduced velocity v, unless the absolute velocity u_0 can be adapted simultaneously. Other important aspects of the instrument include a cooled pump, so that CO_2 can be pumped as a liquid at a temperature below ambient, and a high-pressure cell for a UV–vis detector. MS detection is definitely possible in combination with packed-column SFC, but positioning it after the BPR causes extensive band broadening and large flows of expanded CO_2. Instead, a small fraction of the effluent is typically split towards the mass spectrometer, using a capillary ending in a restrictor (see Module 6.3) at the MS inlet. Figure 6.8 shows a scheme of a packed-column SFC instrument.

Because diffusion coefficients are somewhat higher in SFC than in LC (even at high mobile-phase densities), what is good enough for LC tends to be good enough for SFC. For example, the (lengths and diameters of) connection tubing and the (cell volumes of) UV–vis detectors used in LC are well suited for SFC. Most importantly, the column dimensions and particle diameters encountered in LC are quite appropriate for SFC (if the extra-column volumes in the SFC instrument are not larger than those in the LC instrument). All of this means that packed-column SFC leans heavily on equipment and columns developed for LC, with the notable modifications described above (see the asterisks in Figure 6.8). The strong overlap with LC

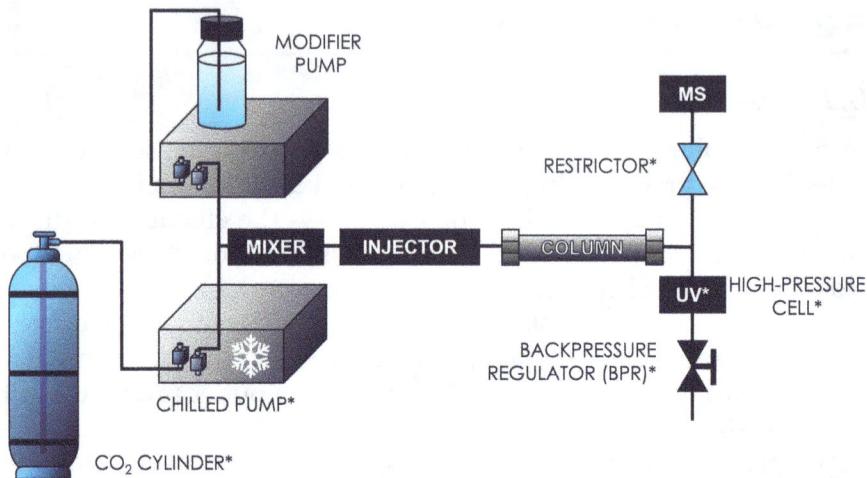

Figure 6.8 Schematic illustration of an instrument for packed-column SFC. Asterisks indicate instrument components specific to SFC. Note that due to its position in the scheme, the UV–vis flow cell must be resistant to high pressures.

technology also implies that packed-column SFC technology is quite mature and that systems tend to be robust and reliable.

> See the website for more resources on how to get started with your SFC method: ass-ets.org

---

6.2.2   The Roles of Modifiers

In traditional packed-column SFC, the mobile phase typically consists of CO_2 and small amounts (often <10%) of an organic modifier (most commonly methanol). Such a mobile-phase system is reminiscent of NPLC. Indeed, packed-column SFC was seen mainly as an attractive alternative for separations that have traditionally been performed with NPLC. The attractiveness of SFC is not only due to the fundamental advantages of higher diffusion coefficients and lower viscosities but also due to faster equilibration of the stationary phase, which allows gradient elution to be performed more easily and more rapidly. More recently, much higher percentages of modifiers have become in vogue, resulting in more-polar mobile phases. Also, HILIC-like SFC has proven successful. The addition of water as a third component (conveniently premixed with the modifier; *e.g.* 2% water in methanol) has been shown to greatly improve peak shapes for highly polar analytes.[8] The addition of water also allows higher percentages of modifiers and the addition of additives that cannot be mixed with CO_2, including MS-compatible acids and salts, such as trifluoro acetic acid, ammonium formate or ammonium acetate. This has made CO_2-based SFC feasible also for highly polar analytes, such as (basic) drugs, polar metabolites, and even proteins. In the process, SFC operating conditions are increasingly far below the critical conditions (*i.e.* $T \ll T_c$). Modern packed-column SFC is essentially subcritical-fluid chromatography – or basically a form of LC. Under such conditions, the mobile-phase compressibility is low, resulting in low baseline noise and robust experiments.

The positioning of SFC techniques relative to common LC methods is illustrated in Figure 6.9. SFC methods are seen to bridge the gap between NPLC, which is mainly used for low or moderately polar analytes, and HILIC, which is intended to separate highly polar analytes. Therefore, SFC may be attractive for samples containing a very broad range of analytes. For example, it may cover the

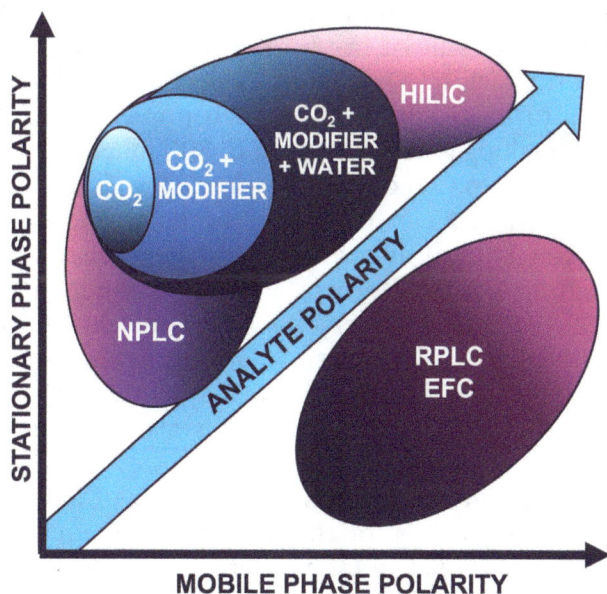

Figure 6.9   Schematic illustration of the position of CO_2-based SFC techniques in comparison with common LC methods. For EFC (enhanced fluidity chromatography), see Section 6.4.

entire range from "non-polar" metabolites (lipids) to highly polar metabolites. Selective RPLC methods rely on substantial amounts of water in the mobile phase. There are few examples of non-aqueous RPLC, and there is little potential for SFC as a genuine alternative to RPLC, but **enhanced-fluidity chromatography** (EFC; see Section 6.4) is potentially of great interest.

The first and arguably still the most important reason to add modifiers to CO_2 is to suppress strong interactions between the analytes and the surface of the particles. If small amounts of active groups cause strong adsorption, mixed retention mechanisms tend to lead to broad and tailing peaks (see the discussion on adsorption isotherms in Module 1.8). In such a case, small amounts of a modifier may successfully compete with the analytes for adsorption places and drastically improve the peak shape. Significantly increasing the overall polarity of the mobile phase may require larger concentrations of modifiers. However, adding even small amounts of organic solvent (or very small amounts of water) to CO_2 strongly increases the critical temperature. This implies that SFC systems are almost invariably operated in the subcritical (*i.e.* LC) domain. This causes densities and viscosities to be higher than in SFC with a pure CO_2 mobile phase and diffusion coefficients to be lower. As a result, the fundamental advantages of SFC in comparison with LC in terms of higher speed tend to diminish. While "subcritical"-fluid chromatography can still be faster than LC, a factor 2 or 3 is more realistic than the factor 10 shown in Table 6.3. Equilibration of the column tends to be much faster in such implementations of SFC than in NPLC, making gradient elution in SFC much more attractive.

Adding modifiers to CO_2 drastically affects the choice of detectors. Most importantly, FID can no longer be used (except, in theory, when only water is used as a modifier). In contrast, MS detection may benefit from the presence of a polar modifier in the mobile phase. In the case of UV detection, the solvent cut-off that is irrelevant with pure CO_2 (because it is more transparent than the quartz windows of the detector cells) may shift to slightly longer wavelengths (*e.g.* 200 or 210 nm). ELSD is also commonly used in combination with packed-column SFC.

The high speed and relatively high efficiency of SFC separations are used most impressively for chiral separations. Enantio-selective columns developed for use in LC have proven successful also in SFC, with cellulose- and amylose-based columns accounting for some 80% of published applications.[9] SFC has been used extensively for lipidomics, *i.e.* for studies involving metabolites of low-to-medium polarity,[10] with recent expansion to more polar metabolites.[8] Another example where a broad range of analytes may potentially be covered is the full range of vitamins, ranging from fat-soluble to water-soluble ones. The more traditional applications of SFC for hydrocarbons, such as for the group-type separations of oil fractions, are still valid but now constitute a minor fraction of all applications of SFC.

There is a major potential for SFC as a method for preparative separations because (i) CO_2 can easily be removed and (ii) it is much "greener" (*i.e.* environment friendly) than organic solvents. Many preparative SFC methods are in place for smaller molecules of pharmaceutical interest. Water-based methods, such as ion-exchange chromatography (IEC), hydrophobic-interaction chromatography (HIC), and aqueous size-exclusion chromatography (SEC), currently seem to have the edge for the purification of biopharmaceuticals. These methods are more easily applied to macromolecules, such as proteins, while ensuring the conformation and integrity (biological activity) of the analytes. Also, the scaling laws for LC separations are much simpler than those for compressible sub- or supercritical fluids, so preparative LC separations can be developed more easily and more accurately than preparative SFC methods.[4] Elsa Lundanes (Box 6.2) contributed significantly to the development of SFC and supercritical-fluid extraction (SFE).

Box 6.2 Hero of analytical separation science: Elsa Lundanes. Image courtesy of Prof. Lundanes.

Elsa Lundanes (Norway)

Elsa Lundanes (Norway) Elsa Lundanes spent almost her entire professional career at the University of Oslo (Norway), where she has been a full professor since 1999, forming a strong group with Tyge Greibrokk and later with Steven Wilson. In 2009, she was elected as a member of the Norwegian Academy of Science and Letters. She made many contributions to supercritical-fluid chromatography and extraction. She studied injection procedures for capillary SFC and developed supercritical-fluid injection. Her research focus was on hyphenated systems, on-line coupling with sample-preparation systems, and two-dimensional separations. Lately, her atttention has shifted to LC and the analysis of biomoleciles.

6.3   Open-tubular-column SFC (M)

Open-tubular SFC may be performed with pure CO_2 as the mobile phase at relatively low densities in columns with an internal diameter of 25 μm (see Table 6.3). The great attraction of such open-tubular SFC is that an FID may be used for detection. As soon as an organic solvent is added to the mobile phase, the latter tends to become impossible, but there is less need for such "modifiers" in open-tubular SFC than in packed-column SFC. The use of cross-linked polysiloxane stationary phases immobilized within fused-silica capillaries implies that few if any polar ("active") adsorption sites remain in the stationary phase. Since the main role of modifiers added in low concentrations is to block such sites, their importance diminishes. Adding a modifier will also increase the density, which is unattractive for open-tubular SFC, because it would cause 25 μm i.d. columns to become sub-optimal (*i.e.* too wide; see eqn (6.1)).

Columns with internal diameters of 25 μm are already highly challenging. They are narrower than those used in GC (or, for that matter, in capillary electrophoresis, see Chapter 5). They would need to be prepared especially for SFC. Because of the lower diffusion coefficients in supercritical fluids than in gases, the optimal stationary-phase film thickness is higher in SFC than in GC. A reduced film thickness of $\delta_f = 0.3$ (see Section 1.7.7) would result in an actual film thickness of about 0.08 μm for a 25 μm i.d. SFC column against 0.025 μm for a 25 μm i.d. GC column. Some skills would need to be developed by those manufacturing the columns, but this should be feasible. The two most attractive detectors, FID and MS, operate at low pressures. A back-pressure regulator would cause too much extra-column band broadening, and small-volume "restrictors" are required to rapidly reduce the pressure at the end of the column. Several types of restrictors have been developed and tested (see Figure 6.10), including linear restrictors (an empty – and fairly long – piece of extremely narrow fused-silica tubing), tapered restrictors (obtained by drawing out the melted end of a fused-silica capillary), frit restrictors, and integral restrictors. The latter types have proven most useful in practice, but they are not commercially available and require a good deal of skill to make in the laboratory. The end of the column is closed by melting it in a flame, in the same manner glass ampules are sealed. After cooling, a meticulous manual abrasion process starts, using the finest-grain sandpaper available. With the front end of the column being connected to a

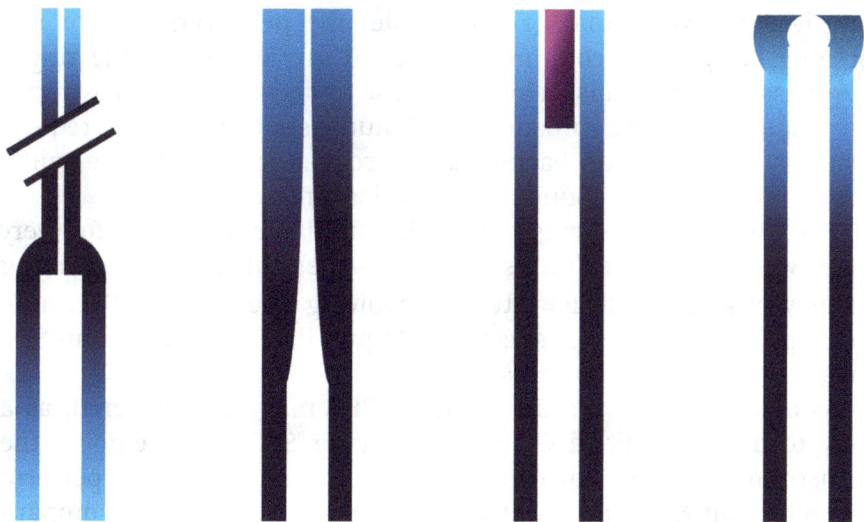

Figure 6.10 Possible restrictors for open-tubular SFC. From left to right: linear restrictor, tapered restrictor, frit restrictor, and integral restrictor.

gas (or CO_2) supply, the end is slowly polished away until a very small hole appears, causing the gas pressure to decrease or tiny gas bubbles to appear when the column end is held under water. The resulting integral restrictor can be used until its performance deteriorates and the process needs to be repeated. It is needless to say that integral restrictors cannot be made reproducibly. Because of flow-rate variations, FID and MS detectors require careful calibration and, preferably, the use of internal standards.

Density programming is not without its challenges in open-tubular SFC. When the density increases, the diffusion coefficient decreases and the optimal linear velocity decreases (the reduced velocity ν remains constant if $u_0 \propto D_m$), but the only way to increase the column pressure when using a (integral) restrictor instead of a back-pressure regulator is to increase the flow of the mobile phase. Unlike the situation in contemporary packed-column SFC, where a back-pressure regulator (variable restrictor) can be used, the flow and density of the mobile phase cannot be independently controlled when using a fixed restrictor. Hence, the chromatographic efficiency and the resolution decrease dramatically during a density program.

Injection in open-tubular SFC is also challenging. The high pressures necessitate some type of valve injection, and the volumes involved are very small. For example, the peak variance for a 5 m-long, 25 μm i.d. column is about 30 nL. Moreover, injecting a liquid sample will cause a local increase in density, which may result in additional band broadening. Volume-split injections require another restrictor or back-pressure regulator. Fast ("time-split") injections have been found more practical and more reliable.

The practical challenges and the fundamental need for very narrow ($d_c \leq 25$ μm) columns have led to open-tubular SFC becoming largely obsolete. Very few studies involving open-tubular SFC have been published in the 21st century. Liquid-like, packed-column SFC has been much more successful.

As a compromise, packed-capillary SFC may be considered, as a way to use a modified GC instrument for SFC. This removes the constraints on very narrow column diameters needed to perform open-tubular SFC under optimal conditions. However, in comparison with open-tubular columns, packed capillaries are much more likely to require modifiers in order to obtain sharp, symmetrical peaks. This precludes the use of FID detection, one of the greatest potential advantages of GC-like SFC. Moreover, if fixed restrictors need to be used, flow rate and density are again interdependent, although the negative consequences of an increase in density on the chromatographic performance are much smaller in packed-capillary SFC than in open-tubular SFC. A miniaturized BPR may allow the detection at high pressure, for example, using UV–vis spectrometry, at a risk of somewhat higher detection limits than encountered using analytical-scale UV–vis detectors. Packed-capillary SFC seems to have few advantages in comparison with analytical-scale packed-column SFC. The consumption of CO_2 or CO_2/methanol mobile phases can be reduced, but these are already among the "greenest" eluents used in chromatography in mL min^{-1} amounts.

Open-tubular SFC may potentially be used to extend the range of hydrocarbons that can be analysed beyond that of (high-temperature) GC, but the gain is marginal and the resolution is poor due to the high mobile-phase densities required. Thermo-labile compounds may be analysed by open-tubular SFC at much lower temperatures than by GC, but the advantages of the former technique are again rather marginal. Packed-column SFC is a more viable method for such analyses.

6.4 Enhanced Fluidity Liquid Chromatography (A)

In Module 6.2, we mentioned a trend towards high fractions of modifiers (plus some water) in SFC mobile phases. At some point, CO_2 may become a minor component in the mobile phase. Such conditions have been approached from the other side, *i.e.* by adding CO_2 to liquid solvents in amounts up to 30%. The pioneer of such methods, Susan Olesik from Ohio State University (Columbus, OH, USA), coined the term enhanced fluidity chromatography (EFC). The fundamental advantages are related to those discussed extensively for SFC, *i.e.* an increase in diffusion coefficients and a reduction in the viscosity of the mobile phase. For example, replacing THF in size-exclusion chromatography with a mixture of 70% THF and 30% CO_2 (by volume) resulted in an increase in the optimum linear velocity (*i.e.* an increase in the apparent mobile-phase diffusion coefficient) by about 45% and a decrease in the pressure drop at equal velocities (*i.e.* a decrease in viscosity) of about 30%.[11] Such a decrease in viscosity may imply that some applications that would otherwise require UHPLC-type equipment can be realized using conventional HPLC equipment. Similar effects on diffusivity and viscosity may be obtained by increasing the temperature, but this may lead to decreased stability of the analytes, the stationary phase, or – in the case of THF – the mobile phase. Adding CO_2 to the mobile phase may result in "greener" separations, especially if a typical HILIC mobile phase that contains 90 or 95 % acetonitrile (and 10% or 5% of water) can be replaced by a mixture of methanol, water and CO_2.

The advantages of EFC have been demonstrated for various LC techniques and for a wide range of analytes, including EFC versions of RPLC for pharmaceuticals, HILIC and HIC for proteins and peptides, HILIC for (oligo- and poly-)saccharides, nucleosides and nucleotides, IEC for amino acids, and high-temperature SEC for polystyrene (where polyolefins would be the prize ticket).

Acknowledgements

Prof. Caroline West and Dr. Paul Ferguson are acknowledged for their review of this chapter.

Recommended Reading

M. B. Hicks and P. D. Ferguson, in *Practical Application of Supercritical Fluid Chromatography for Pharmaceutical Research and Development*, Elsevier, 1st edn, 2022.

C. West, Supercritical fluid chromatography is not (only) normal-phase chromatography, *J. Chromatogr. A*, 2024, 1713, 464546.

References

1. A. M. Katti, N. E. Tarfulea, C. J. Hopper and K. R. Kmiotek, *J. Chem. Eng. Data*, 2008, **53**, 2865–2872.
2. R. C. Reid, J. M. Prausnitz and B. E. Poling, in *The properties of gases and liquids*, McGraw-Hill, New York, 4th edn, 1987.
3. C. West, *J. Chromatogr. A*, 2024, **1713**, 464546.
4. G. Guiochon and A. Tarafder, *J. Chromatogr. A*, 2011, **1218**, 1037–1114.
5. P. J. Schoenmakers, *J. Chromatogr. A*, 1984, **315**, 1–18.
6. H. H. Lauer, D. McManigill and R. D. Board, *Anal. Chem.*, 2002, **55**, 1370–1375.
7. B. E. Poling, J. M. Prausnitz and J. P. O'Connel, in *The properties of liquids and gases*, McGraw-Hill, New York, 5th edn, 2001.
8. G. L. Losacco, J.-L. Veuthey and D. Guillarme, *TrAC, Trends Anal. Chem.*, 2021, **141**, 116304.
9. K. Plachká, V. Pilařová, O. Horáček, T. Gazárková, H. K. Vlčková, R. Kučera and L. Nováková, *J. Sep. Sci.*, 2023, **46**(18), 1–28.
10. L. Laboureur, M. Ollero and D. Touboul, *Int. J. Mol. Sci.*, 2015, **16**, 13868–13884.
11. H. Yuan and S. V. Olesik, *J. Chromatogr. A*, 1997, **785**, 35–48.

7 Multi-dimensional Separations

Dramatic progress has been made in two-dimensional separations in the last few decades. Heart-cut separations with two (GC-GC, LC-LC, LC-GC, *etc.*) or more (*e.g.* LC-LC-LC) techniques are used routinely, for example, for high-resolution peak-purity testing. Comprehensive two-dimensional separations, such as GC×GC and LC×LC, offer superior peak capacities for fingerprinting and non-target analysis. Two-dimensional GC methods and two-dimensional LC methods form the bulk of this chapter, which ends with a discussion on LC-GC couplings and other two-dimensional separation techniques (Figure 7.1).

7.1 General Principles (M)

In the last few decades, the attainable performance of one-dimensional separation techniques has been enhanced in many different ways. Examples include novel column technologies (Module 3.2) and UHPLC (Section 3.3.2) for LC, long efficient columns for detailed hydrocarbon analysis with long open-tubular columns in GC, and miniaturized systems for CE (Section 5.2.4). It is fair to say that significant progress has been made over the years, but current improvements typically increase performance in small steps. A strategy aimed at achieving large improvements of at least an order of magnitude is the use of multi-dimensional separations. These separations are the topic of the present chapter.

Analytical Separation Science
By Bob W. J. Pirok and Peter J. Schoenmakers
© Bob W. J. Pirok and Peter J. Schoenmakers 2025
Published by the Royal Society of Chemistry, www.rsc.org

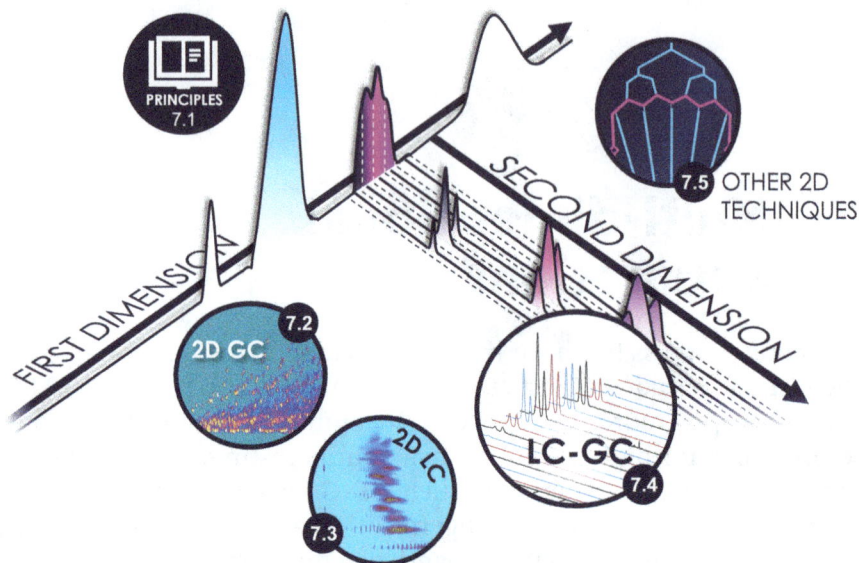

Figure 7.1 Graphical overview of the modules in this chapter.

7.1.1   Introduction to Multi-dimensional Separations

In multi-dimensional separations, fractions of the effluent from one separation are subjected to one or more additional separations. To date, the vast majority of multi-dimensional separations are two-dimensional (2D), and we will therefore focus on these in our coverage of the subject.

The principle of any multi-dimensional (or 2D) separation is that an additional dimension further refines the separation achieved in the first dimension. This is graphically expressed in Figure 7.2, where a regular one-dimensional (1D) separation is shown as the "first dimension". The peak cluster highlighted in pink comprises several co-eluting peaks. Within the confines of regular 1D separations, we would likely have adjusted the method parameters in an attempt to better separate the peak cluster. However, in a 2D separation, the first-dimension (^{1}D) effluent is fractionated. One possibility is to transfer the entire pink cluster to a second-dimension (^{2}D) separation. Another option is to transfer a number of fractions of interest – indicated in Figure 7.2 by the dashed lines – to a second-dimension (^{2}D) separation. The figure shows that the co-eluting analytes in the ^{1}D separation are well resolved in the second dimension.

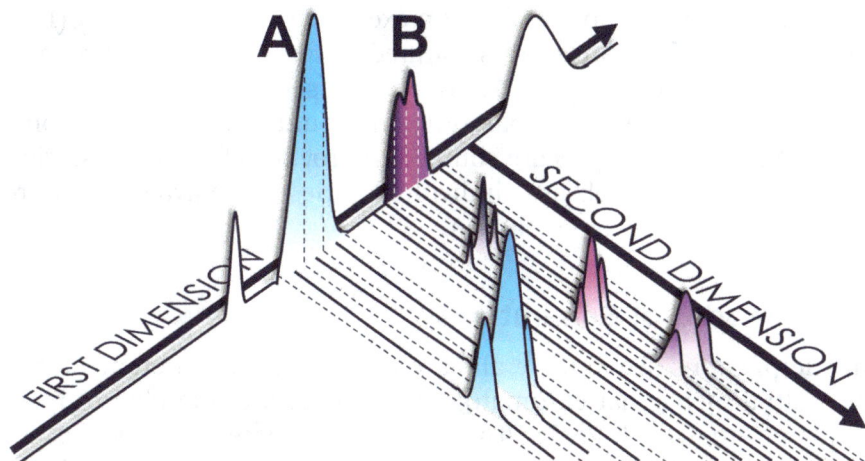

Figure 7.2 Graphical illustration of a two-dimensional separation. Fractions of the first-dimension effluent are subjected to a second-dimension separation.

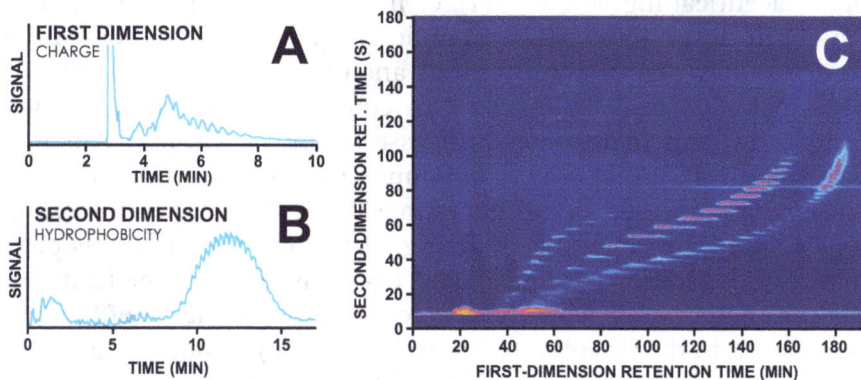

Figure 7.3 Separation of an industrial surfactant mixture using three different LC techniques. (A) Mixed-mode ion-exchange chromatography, (B) reversed-phase LC, and (C) a comprehensive two-dimensional combination of ion-exchange coupled with reversed-phase LC (IEC×RPLC). Panel C reproduced from ref. 1, https://doi.org/10.1002/jssc.201700863, under the terms of the CC BY 4.0 license, https://creativecommons.org/licenses/by/4.0/.

The final option is to divide the ^{1}D effluent into many fractions and transfer all of these to a ^{2}D separation. The added value of such a comprehensive 2D separation becomes instantly clear from Figure 7.3, which shows the separation of industrial surfactants using three different LC techniques. The one-dimensional chromatograms are obtained by ion-exchange chromatography (IEC; Figure 7.3A), where ions are separated based on their charge, and reversed-phase LC (Figure 7.3A), where separation is based on hydrophobicity. Neither separation offers much insight. It is only when the two techniques are coupled to provide a full 2D separation that we can clearly observe individual series of analytes that were previously obscured by co-elution.

7.1.2 Modes of Separations

In the previous section, we already touched upon different modes of multi-dimensional chromatography. These are intended to meet different analytical goals. To understand these, it is useful to introduce the concepts of targeted and untargeted analysis in analytical chemistry (see also Module 1.9). In **targeted analysis**, our interest is focused on a limited number of known compounds in a sample. This can be a single target analyte, such as an active pharmaceutical ingredient (API), or many target analytes, such as a list of priority pollutants. For example, the purity of a candidate drug compound is one of the essential analyses in drug development. The light blue peak in Figure 7.2A could represent such a target compound, and in that case, an assessment of the purity of this peak is crucial. While other compounds may be discerned in the ^{1}D chromatogram, there is no guarantee that the blue peak is exclusively due to the target. Re-analyzing a fraction containing this peak may reveal additional impurities. This is an example of **heart-cut** 2D chromatography, where a selected fraction of the ^{1}D effluent is transferred to the ^{2}D column, for example, using a modulation valve (Figure 7.4). In practice, we speak of "cutting the peak" or "taking one or several cuts" of the peak. Heart-cut 2D chromatography is abbreviated with a hyphen (*i.e.* LC-LC or GC-GC). Beyond **peak-purity testing**, heart-cut 2D chromatography can serve as an essential tool for compound purification in a laboratory or even on an industrial scale. In some cases, it may be of interest to re-analyze several zones from the ^{1}D chromatogram. This is referred to as **multiple-heart-cut** 2D chromatography (mLC-LC). If we analyze a sample in an **untargeted** manner, all analytes (and thus all peaks) are of equal interest. In such a case, it is more effective to divide the ^{1}D effluent into many fractions and subject them all to a ^{2}D separation. This is referred to as **comprehensive** two-dimensional chromatography and abbreviated by the multiplication sign "×" (*e.g.* LC×LC or GC×GC). Note that the symbol "×" is not the same as the letter "x". Use of the latter for

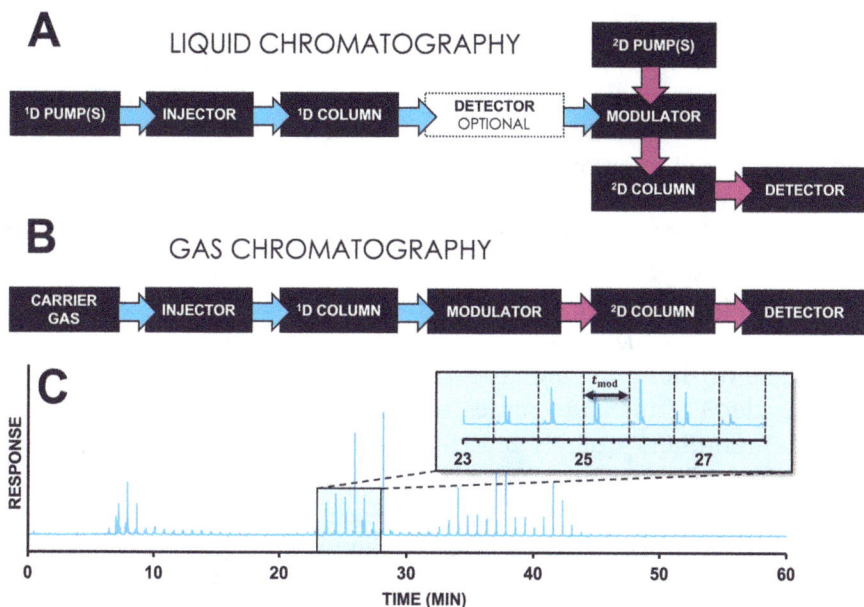

Figure 7.4 Simplified hardware setup for (A) 2D-LC separations and (B) GC×GC separations (see Module 7.5 for 2D separations other than LC or GC). Blue arrows depict the first dimension, while pink arrows represent the second dimension. The ^{1}D detector is typically only used in heart-cut 2D chromatography. (C) Example of a raw one-dimensional chromatogram obtained for a comprehensive 2D separation, with each ^{2}D chromatogram measured in series.

the present purpose must be avoided. In comprehensive two-dimensional chromatography, the term "comprehensive" refers to "two-dimensional" and not to "chromatography". It is incorrect to refer to this method as "comprehensive (liquid or gas) chromatography" instead of "comprehensive two-dimensional (liquid or gas) chromatography". Comprehensive 2D chromatography is extremely powerful, but it comes at significant costs. Aside from method complexity, the most important disadvantages of an online, real-time coupling are the constraints it imposes on the first- and second-dimension analyses. These include, for example, a maximum volume for a collected fraction of ^{1}D effluent and a maximum analysis time for the ^{2}D separation.

Comprehensive 2D chromatography can also be applied in more targeted applications, such as when the target analyte features a number of peaks. For example, when trying to establish the presence and concentration of natural vanilla extract in a food sample, it may be useful to selectively submit sections of the ^{1}D effluent for comprehensive ^{2}D analysis. We refer to this as **selective-comprehensive 2D chromatography** (sLC×LC or sGC×GC). In comparison, mLC-LC sampling of the ^{1}D effluent is conducted more frequently in sLC×LC, reducing the chance of disrupting the separation that was initially achieved on the ^{1}D column.

In heart-cut 2D chromatography, fractions of interest of the ^{1}D effluent are subjected to a ^{2}D separation, whereas in comprehensive 2D chromatography, the entire ^{1}D effluent is used for the ^{2}D analysis.

Contemporary 2D-LC systems allow users to specify which peaks should be further analyzed. The software may even feature tools that aim to detect this automatically. One mistake that experienced chromatographers quickly learn to avoid is relying fully on visual inspection of the chromatogram. Well-resolved peaks may still contain small or fully co-eluting peaks (*e.g.* blue peak, A, in Figure 7.2). The ^{1}D chromatogram in Figure 7.2 also exhibits a peak cluster (pink, B) with clearly overlapping peaks. Whether or not a peak should be fractionated for ^{2}D analysis, in practice, will partly depend on the chromatographer's expectations of the sample complexity (see Section 7.1.5).

Fraction transfer is most often performed **online** using (for example) a valve interface that is generally referred to as a transfer device or **modulator**. A single ^{2}D measurement within a 2D experiment is often called a **modulation**. The method and interface type used when transferring fractions greatly depend on the separation technique employed and are discussed in later modules for each technique. In LC, the transfer can also be accomplished **offline**. In this case, fractions of the ^{1}D effluent are collected, for example, in a sample tray using a fraction collector. After the ^{1}D analysis, this fraction tray may be transferred to the autoinjector of the same instrument (with a different mobile phase and/or a different column installed) or to a different instrument. For collecting a single fraction in heart-cut LC-LC, simpler means suffice to collect and re-inject the one fraction offline.

The nomenclature of 2D separations has been discussed several times,[2,3] leading to consensus on just about all notations. Table 7.1 provides an overview of this nomenclature.

At this point, it is useful to note that, in line with this nomenclature, the simple serial coupling of two columns is *not* a form of two-dimensional chromatography. In a coupled arrangement, the order in which the analytes elute from one column affects the order

Table 7.1 Nomenclature for two-dimensional separations (see also ref. 2 and 3).[a]

Notation	Explanation
- (hyphen)	1. Used for hyphenated separation-detection systems, such as GC-UV or LC-MS;
	2. Used for hyphenated couplings of separation systems, such as "heart-cut" GC-GC or LC-LC.
× (multiplication sign)	Used for comprehensive two-dimensional couplings, such as GC×GC or LC×LC
1D	One-dimensional (*adjective*), as in 1D-GC
2D	Two-dimensional (*adjective*), as in 2D-GC
^{1} (prefix)	Indicates a first-dimension property, such as first-dimension retention time (^1t_R) or first-dimension column diameter (^1d_c)
^{2} (prefix)	Indicates a second-dimension property, such as second-dimension retention time (^2t_R) or second-dimension column diameter (^2d_c)
^{1}D	First-dimension (*adjective*), as in ^{1}D column, ^{1}D separation
^{2}D	Second-dimension (*adjective*), as in ^{2}D column, ^{2}D separation
t_{mod} (or t_{cycle})	Modulation time (or cycle time between modulations)[a]

[a]Some authors prefer a period of modulation (P_M), but we are dealing with time, not pressure. Therefore, we base the notation on the universal symbol for time (t), as in t_{mod}.

in which they eluted from the next column. This is not the case for 2D chromatography, as discussed in this chapter.

7.1.3 Hardware for 2D Separations

Only limited hardware investments are required to establish a 2D separation. Figure 7.4A shows a simplified flowchart of the involved hardware components. The blue arrows indicate the first dimension, which is, in essence, identical to a regular 1D separation setup. For heart-cut 2D separations, a ^{1}D detector is used to determine which fractions should be subjected to a ^{2}D separation. In Modules 7.2 and 7.3, we will see how the modulator is used in such separations to store and transfer the ^{1}D fractions.

In comprehensive 2D chromatography, a ^{1}D detector is not commonly used. Instead, the ^{2}D detector measures all ^{2}D separations in series, yielding one long string of chromatograms that combines all individual ^{2}D modulations, as shown in the example in Figure 7.4C. This is further explained in Module 9.13.

> See the website for more resources on multi-dimensional separations: ass-ets.org

7.1.4   Rationale for Two-dimensional Separations

Figures 7.2 and 7.3 graphically explain the usefulness of a two-dimensional separation, but there are also fundamental reasons underlying the success of multi-dimensional techniques. These concepts are also extremely useful for method development and will be addressed in this section.

7.1.4.1 Peak Capacity

For the first perspective, we recall the concept of peak capacity from Section 1.5.2, which, in a nutshell, is a measure to quantify the number of separated peaks that can literally fit in the chromatogram. Using an additional separation dimension to further separate the peaks from the first dimension will automatically increase the peak capacity. For **heart-cut** 2D chromatography, the peak capacity is the sum of the peak capacities of the individual dimensions

$$n_{c,hc} = {}^1n_c + {}^2n_c \tag{7.1}$$

Here, ^1n_c and ^2n_c are the peak capacities of the first and second dimensions, respectively. When $^1n_c = 90$ and $^2n_c = 70$, our total peak capacity would thus be $n_{c,hc} = 160$. For **comprehensive** 2D chromatography, we, of course, have an entire plane for our peaks rather than a single dimension. As a result, the peak capacity is ideally the product of the peak capacities of the individual dimensions, *i.e.*

$$n_{c,comp} = {}^1n_c \cdot {}^2n_c \tag{7.2}$$

"Ideally" implies that none of the achieved ^{1}D separation and none of the ^{2}D efficiency is lost in the process. If such ideal conditions can be approached, the peak capacity of comprehensive 2D chromatography is vastly superior to that of 1D or heart-cut 2D chromatography. In comprehensive 2D chromatography, the ^{2}D separation is typically much faster than in heart-cut 2D chromatography, but if we have $^1n_c = 90$ (as before) and $^2n_c = 20$ for the fast ^{2}D separation, we now obtain an ideal peak capacity of $n_{c,comp} = 1800$. Note that, for various reasons, the effective peak capacity may be lower than the ideal number. This will be addressed later in this chapter (see Section 7.1.5).

For now, we will focus on the comparison of the different modes for LC. For GC, the main conclusions regarding the performance of two-dimensional *vs.* one-dimensional systems are essentially the same.

Davydova *et al.* theoretically compared the peak capacity of 1D and comprehensive 2D-LC.[4] For a whole series of variables (column dimensions, particle size, permeabilities, flow rate, gradient parameters, *van Deemter* coefficients, *etc.*), one or more reasonable values were selected, and for all the many thousands of possible combinations, the resulting peak capacities were calculated.[4] From all points, only those points were kept for which there were no points that yielded a higher peak capacity in a shorter time (so-called Pareto-optimal points). These points formed the Pareto-optimal lines shown in Figure 7.5A, displaying the peak capacity against the analysis time for 1D-LC at 100 MPa (UHPLC; solid, dark blue line), LC×LC at 40 MPa (HPLC; dotted, light blue line) and LC×LC at 100 MPa (dashed, pink line). It is observed that 1D-LC at ultra-high pressure conditions yields a theoretical peak capacity of roughly 800 in 10 h, LC×LC at 40 MPa and 100 MPa reaches 6500 and 8000 in the same time, respectively. This implies an improvement in separation power by an order of magnitude!

We can now invoke the simple result from the **statistical overlap theory** of Davis and Giddings (eqn (1.43)), which relates the expected number of singular (*i.e.* pure) peaks p for a given sample containing m different analytes to the **peak capacity** n_c,

$$\ln p \approx \ln m - \frac{m}{n_c} \tag{7.3}$$

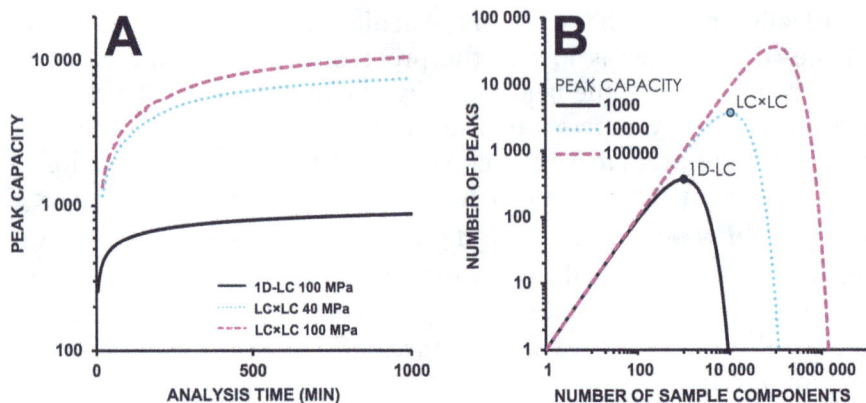

Figure 7.5 (A) Peak capacity plotted against the analysis time for 1D-LC at 100 MPa (UHPLC; solid, dark blue line), LC×LC at 40 MPa (dotted, light blue line), and LC×LC at 100 MPa (dashed, pink line). Lines are Pareto-optimal curves obtained by calculating peak capacities for a large number of situations with reasonable values for columns and conditions.[4] (B) Expected number of singular peaks as a function of the number of sample components for peak capacities of 1000 (dark blue, solid line), 10 000 (light blue, dotted line), and 1000 000 (pink, dashed line) according to the statistical overlap theory. Indicated points represent the maximum expected performance of 1D-LC and LC×LC.

In Section 1.5.2, we learned from this equation that – as a rule of thumb – the peak capacity should be a factor 20 higher than the number of analytes in our sample mixture. This is graphically captured in Figure 7.5B, where the expected number of singular peaks from the statistical overlap theory is plotted against the number of sample components for different peak capacities. For a sample of 1000 sample components, an optimized 1D-LC system with a peak capacity of 1000 can, at best, be expected to yield 400 singular peaks, whereas this is almost the full 1000 for an LC×LC separation yielding a peak capacity of 10 000. We also see from Figure 7.5B that the peak capacities offered by LC×LC allow for tackling more complex samples. In other words, the use of heart-cut and, especially, comprehensive 2D separations dramatically increases our ability to separate complex samples. More importantly, Figure 7.5 suggests that it quickly becomes more useful to solve a separation problem with a 2D separation instead of spending more time optimizing a 1D method, where we quickly approach the limit of our current possibilities.

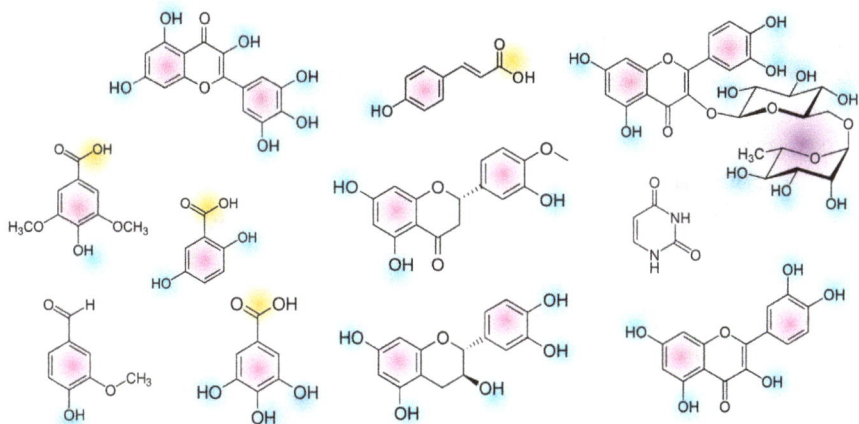

Figure 7.6 Molecular structures of several metabolites. The colours represent collections of some of the characteristic molecular moieties: alcohol (blue), acid (yellow), aromatic ring (pink), and sugar (purple).

7.1.4.2 Selectivity – Sample Dimensionality

The second perspective focuses on tuning the selectivity of our separations to the chemistry of our sample using Giddings' sample-dimensionality concept.[5] Giddings defined **sample dimensionality** as the number of parameters that would need to be defined to uniquely define the molecular structure of an analyte in the sample. To understand this concept, we may examine the molecular structures of several metabolites depicted in Figure 7.6. When trying to classify the different functional groups present in the ensemble of molecular structures, we can distinguish several groups that are predominantly present, such as alcohols (blue), aromatic rings (pink), and, to a lesser degree, acids (yellow) and sugars (purple). This classification can, of course, be further refined, as ethers, ketones, and aldehydes are also observed. Moreover, the number of rings and the position of the substitutions may also contribute to the sample dimensionality.

Giddings proposed that to fully characterize the sample, the number of **separation dimensions** (in his words, "system dimensions") should ideally be equal to the number of sample dimensions. Even for our relatively simple collection of molecules in Figure 7.6, this would mean we would need at least six separation dimensions. This seems nonsensical and not very useful. However, we may look for dominant sample dimensions that correspond to the selectivity provided by the phase system to provide some structure in the

(2D) chromatogram. The statistical overlap theory discussed in the previous section assumes that peaks randomly occupy the separation space. By establishing separation dimensions that correspond to sample dimensions, we may create the largest possible separation space for a specific sample.

In practice, this means that we should ask ourselves the question *What sample dimensions do I expect to describe the largest variation in my collection of sample molecules?* If we now regard Figure 7.6 again, we immediately notice that the alcohols and aromatic rings dominate the molecular structures. An analogy can be made to principal component analysis (Section 9.9.1) where dimension reduction is used to distil a limited number of principal components that describe the most variation. Hence, from a method-development perspective, it would be sensible to devise a separation strategy where the two separation dimensions each individually tackle one of the most important sample dimensions. For example, one could use hydrophilic interaction LC (HILIC) for the alcohols and reversed-phase LC (RPLC) for the aromatic rings.

> See the website for further example case studies of sample dimensionality: ass-ets.org

7.1.5   Key Concepts of Comprehensive 2D Chromatography

We will see in Modules 7.2 and 7.3 that heart-cut 2D chromatography is relatively easy to implement. However, for comprehensive 2D chromatography, there are a number of additional concepts that need to be addressed before we proceed.

7.1.5.1   Orthogonality and Surface Coverage

An important condition in the discussion on sample dimensionality is that the individual separation dimensions should not only tackle different sample dimensions but should also not be correlated with one another. These two requirements are not always the same. This is described by the concept of **orthogonality**. Two separations are considered orthogonal if the results (usually the retention times) are statistically uncorrelated. The retention times of the ^1D separation should not be correlated with

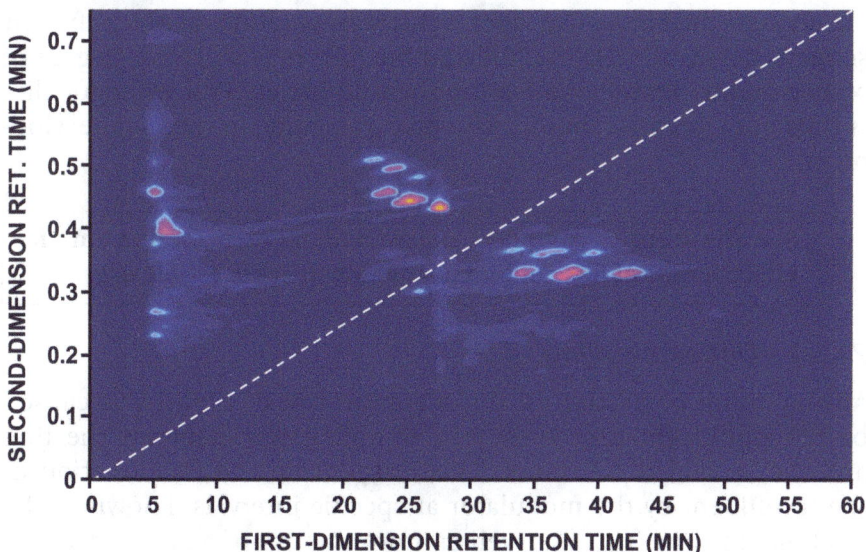

Figure 7.7 Comprehensive 2D-LC separation of a Water Blue IN dye mixture measured at 600 nm with UV–vis detection and ion-exchange chromatography as the first dimension and ion-pair RPLC in the second dimension. Adapted from ref. 6 with permission from American Chemical Society, Copyright 2019.

those of the ^{2}D separation. This is especially important in comprehensive 2D chromatography. "Zero orthogonality" corresponds to 100% correlation, causing all the peaks to elute exactly on a straight line, such as the diagonal shown in Figure 7.7. Conversely, full orthogonality corresponds to 0% correlation, and the peaks are scattered across the 2D chromatogram. Different metrics are used to compute orthogonality. These are discussed in Section 10.3.5.3. At this point, it is more useful to consider that low orthogonality is likely to reduce **surface coverage**, $f_{coverage}$. The surface coverage is the fraction of the separation surface that is occupied by peaks (see Section 10.3.5.2 for quantitative measures of $f_{coverage}$). For instance, if the orthogonality is low, then most peaks will appear in close proximity to each other. This implies that the theoretical peak capacity from eqn (7.2) will not be fully used.

If we now revisit the separation shown in Figure 7.3C, we find the orthogonality to be rather limited. Indeed, the different series of peaks elute diagonally on the separation plane, indicating that retention in the first dimension correlates with retention in the second dimension. This is in contrast to the separation shown in Figure 7.7, where this correlation is much lower. Nevertheless, the

2D analysis in Figure 7.3 does achieve separation of the different series, whereas the 1D methods do not. There is a greater degree of orthogonality between series than within series. Full orthogonality is thus not a premise, but it may serve as a guide in method development.

> See the website for an overview and more resources on the effectiveness of different orthogonality metrics: ass-ets.org

7.1.5.2 Undersampling

An important correction of the theoretical peak capacity is caused by the significant discrepancy in sampling rate between the first and the second dimensions. This arises from the fractionation of the ^{1}D effluent by the modulator at specific intervals, known as the modulation time (t_{mod}), as well as from the use of a single detector at the outlet of the second dimension. The problem is illustrated in Figure 7.8.

In Figure 7.8A, we see at the front a ^{1}D separation with dark boxes separated by white dashed lines, indicating the modulations. Both peaks are cut three to four times. In this example, the two peaks have similar retention times in the ^{2}D separation and elute close to each other. Nevertheless, the peaks are well distinguishable from each other due to the chosen sampling frequency of the ^{1}D effluent. If we now lower this frequency, say, by 50%, then we obtain the situation shown

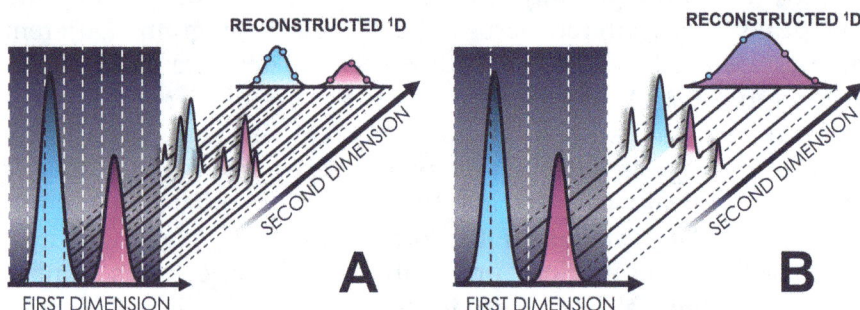

Figure 7.8 Illustration of the loss of information from the ^{1}D separation as a result of undersampling. If the ^{1}D peaks are fractionated sufficiently often, (A) the achieved separation in the first dimension is preserved and can be reconstructed. However, if the modulation time is too long, (B) the achieved separation may be lost when reconstructing the ^{1}D separation from the 2D data. Adapted from ref. 1, https://doi.org/10.1002/jssc.201700863, under the terms of the CC BY 4.0 license, https://creativecommons.org/licenses/by/4.0/.

in Figure 7.8B, where each peak is only sampled twice and the peaks can no longer be distinguished based on the ^{2}D separation. In this case, the first-dimension separation has been lost, and consequently, we lose peak capacity. This may seem like an unfortunate example, where the two peaks happen to have similar retention in the second dimension. However, both the chromatograms shown in Figure 7.3C and that in Figure 7.7 demonstrate that this is far from uncommon.

To correct for undersampling, Davis *et al.*[7] proposed the average peak-broadening factor that features the ratio of the modulation time, t_{mod}, and the ^{1}D peak width, ^1W,

$$<\beta> = \sqrt{1+3.35\left(\frac{t_{mod}}{^1W}\right)^2} \tag{7.4}$$

7.1.5.3 Peak Capacity in LC×LC

The concepts discussed above have led to the introduction of a (sample-dependent) **effective peak capacity** ($n^*_{c,2D}$) for comprehensive 2D chromatography[8]

$$n^*_{c,2D} = \frac{^1n_c \cdot {}^2n_c \cdot f_{coverage}}{<\beta>} \tag{7.5}$$

A sample-independent mathematical expression was formulated by Vivó-Truyols *et al.*[9] specifically for comprehensive two-dimensional liquid chromatography (LC×LC), correcting both for undersampling of the ^{1}D effluent and loss of ^{2}D efficiency because of an excessively large injection volume,

$$n_{c,LC\times LC} = \frac{^1t_G}{4R_s\sqrt{\left(^1\sigma_t\right)^2 + \frac{\left(^2t_{mod}\right)^2}{\delta_{det}^2}}} \cdot \frac{^2t_G}{4R_s\sqrt{\left(^2\sigma_t\right)^2 + \left(\frac{^1F}{^2F}\right)^2 \frac{\left(^2t_{mod}\right)^2}{\delta_{inj}^2}}} \tag{7.6}$$

where F represents the flow rate, t_G is the length of the gradient, R_s is the desired resolution between two peaks (typically a value of 1 is used) and σ_t is the average standard deviation of the peaks in time (*i.e.* the peak width). The term δ_{inj}^2 relates to the injection system. While a value of 4 is typically used, values can range from 4 to 12.[10] Various values can also be found for the detection factor δ_{det}^2.

Davis *et al.* used statistical overlap theory to find a value for δ_{det}^2 of 4.76.[7] Eqn (7.6) can be translated into a sample-dependent effective peak capacity by multiplication with the fractional coverage, *i.e.* $n_{c,LC\times LC}^* = n_{c,LC\times LC} \cdot f_{coverage}$.

7.1.6 Dilution Factors in LC×LC

Analytes are usually diluted upon injection into a separation environment, such as a column (see Section 1.4.5). Of course, peak focusing may counter this, but, in contrast, band broadening processes will increase dilution. In the absence of focussing, the **dilution factor** (DF) in the first dimension is calculated as

$$^1DF = \sqrt{2\pi}\frac{^1\sigma_t \cdot {}^1F}{^1V_{inj}}$$

(7.7)

where $^1\sigma_t$ is the standard deviation of the ^{1}D peak, ^1F is the ^{1}D flow rate and $^1V_{inj}$ is the injection volume. Calculation of the ^{2}D dilution factor requires not only the inclusion of the ^{2}D peak width $^2\sigma_t$ but also all band broadening in the second dimension,[9]

$$^2\sigma_{tot} = \sqrt{\left(^2\sigma_t\right)^2 + \left(\frac{^1F}{^2F}\right)^2\left(\frac{1}{FF_i}\right)^2\frac{\left(^2t_A\right)^2}{\delta_{inj}^2}}$$

(7.8)

with ^2t_A denoting ^{2}D analysis time and FF representing the **focussing factor**,[11] which is given by

$$FF_i = \frac{^2k_{1e} + 1}{^2k_{2e} + 1}$$

(7.9)

Here, $^2k_{1e}$ represents the ^{2}D retention factor in the ^{1}D mobile phase (or, in cases where the composition is altered during modulation, the ^{2}D "loading solvent"). $^2k_{2e}$ depicts the retention factor in the second dimension in the ^{2}D mobile phase. The focusing factor accounts for the contraction of the transferred volume (*i.e.* the ^{2}D injection volume) as a result of focusing at the inlet of the ^{2}D column or on a trapping unit (*e.g.* thermal modulation in GC or active modulation in LC). Such focusing will be discussed for GC and LC in later modules of this chapter.

The ^{2}D dilution factor can be calculated by substituting the result of eqn (7.8) into

$$^2DF = \sqrt{2\pi} \frac{^2F}{^1F} \frac{^2\sigma_{tot}}{^2t_A}$$

(7.10)

The total dilution is then obtained from

$$^{2D}DF = {}^1DF \cdot {}^2DF$$

(7.11)

The dilution factor is an important consideration in determining the feasibility of 1D or 2D separations for analytes expected to be present in trace concentrations. Typical values and the effect of the focussing factor are discussed in ref. 9.

> The dilution factor is one concept that has driven the development of advanced modulation interfaces. See the website for examples: ass-ets.org

7.2 Two-dimensional Gas Chromatography (M)

As described in Module 7.1, there are two main directions in which two-dimensional (2D) chromatography has evolved: comprehensive 2D chromatography and "heart-cut" 2D chromatography. We first focus our attention on heart-cut 2D gas chromatography (GC-GC).

7.2.1 Heart-cut Two-dimensional Gas Chromatography (GC-GC)

Heart-cut 2D-GC may be seen as a way to avoid (complex, laborious) sample-preparation steps, with the first dimension acting as the sample-preparation step for the ^{2}D separation. As long as the detector sensitivity suffices, GC can be regarded as a near-ideal way to isolate the analytes of interest from a complex matrix. A pre-separation with, say, 100 000 plates allows for very narrow fractions to be obtained. For example, a peak containing two **enantiomers** may be isolated, and these may then be separated on a chiral second-dimension column (*e.g.* using a cyclodextrin-modified stationary phase). In addition, GC-injection techniques may serve as a sample clean-up step. A programmed-temperature-vaporizer (PTV) injector (Section 2.4.3) can be extremely powerful, even for samples that contain non-volatile fractions. Non-volatile fractions may be dealt with in a pyrolysis step (sending the pyrolysate to waste), or liners may be replaced regularly (and possibly automatically). Headspace injectors (Section 2.4.4) only introduce volatile components or fractions into the GC.

Peak-purity testing is a potential application of GC-GC, but this is far less common than the use of LC-LC for this purpose. One reason for this is that few APIs are amenable to GC analysis. Another reason is that many questions concerning peak purity may also be addressed with GC-MS. GC-MS, when it can be applied, is a better tool for this purpose than LC-MS, thanks to the high separation power of GC (100 000 plates or more), the near universality of electron-ionization MS, and the absence of baseline disturbances or matrix effects from the mobile phase (carrier gas). GC-GC-MS may make the identification of compounds more reliable, especially if retention indices obtained on the two columns are considered in the identification or verification process. In addition, GC-GC may help obtain pure-component spectra that are easier to identify and better to store in databases. GC-GC with olfactometry (Section 2.5.8) may contribute to the identification of compounds that contribute favourably to the aroma of foods or beverages or it may be used to pinpoint bad odours to very specific compounds.

7.2.2   Hardware for GC-GC

Ideally, two independent GC ovens are used for the two columns in GC-GC. This allows for independent high-resolution operation of both dimensions. If the ^{2}D column is maintained at a much lower temperature during the transfer than the ^{1}D column, the analytes will be focussed at the top of the ^{2}D column. If insufficient focussing is achieved, a cold (or cryo-)trap can be installed.

A **Deans' switch** is an elegant way to control the flow of gases, such as a GC effluent, without (i) them getting in touch with the inside of a solenoid valve and (ii) the need to operate such a valve at the high and variable temperatures encountered in a GC oven. It is named after David Deans, who, as an industrial chemist working for what

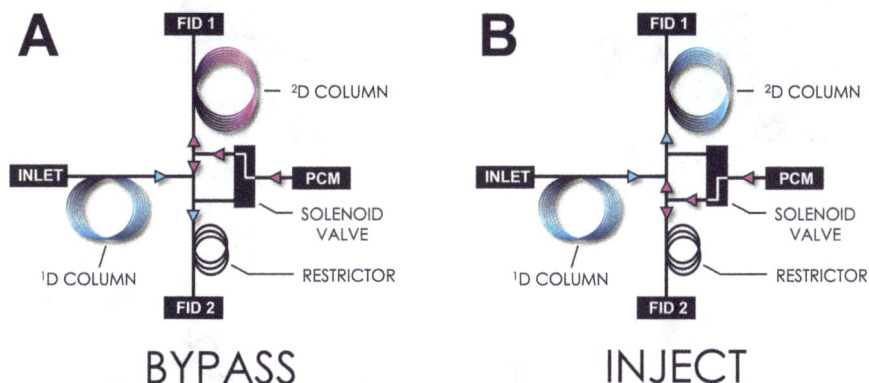

Figure 7.9 Deans' switch configuration in (A) bypass mode, and (B) inject mode, that allows the transfer of selected fractions from the ^{1}D effluent to the ^{2}D column. Adapted from ref. 13 with permission from Elsevier, Copyright 2015.

was then ICI in Billingham, UK, was the first to apply such "fluidic logic" in GC.[12] In a Deans' switch, a solenoid valve can be positioned outside the GC oven and the analytes will only come into contact with (deactivated) fused-silica tubing and, possibly, the inside of a microfluidic Deans' switch.[13] Such a setup is illustrated in Figure 7.9.

Successful operation of a Deans' switch requires an appropriate auxiliary gas pressure (supplied through the pressure-control module, PCM) and an appropriate flow restrictor. This may be similar to the ^{2}D column (in terms of length and diameter), but in that case, the secondary flame-ionization detector (FID) in the vent line provides a delayed response. Therefore, a much shorter and narrower restrictor is recommended.

To ensure that all of the ^{1}D effluent is directed to the ^{2}D column during the transfer, the flow through the ^{2}D column should be higher than that through the ^{1}D column ($^2F > {}^1F$). The functioning of the Deans' switch is illustrated in Figure 7.10. Such a switch can, for example, be created from three press-fit flow splitters. While instrument software can help establish correct settings, the necessary calculations can also be performed offline.[14]

7.2.3 Comprehensive Two-dimensional Gas Chromatography (GC×GC)

The arguments provided in Module 7.1 for the development and application of comprehensive 2D chromatography definitely apply to comprehensive two-dimensional gas chromatography (GC×GC). GC

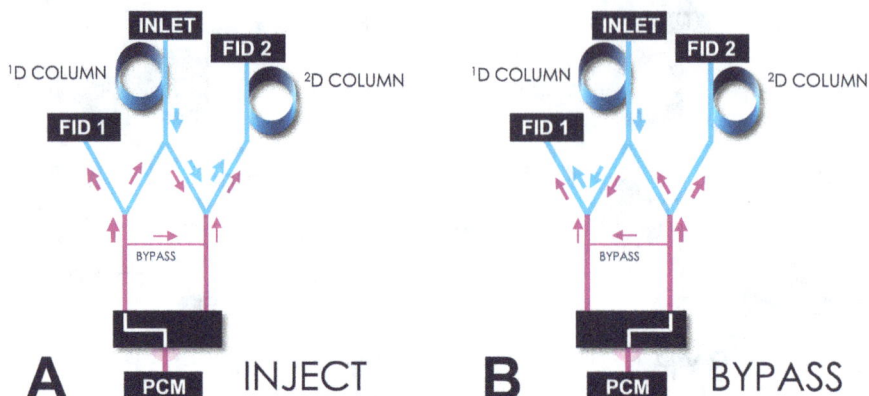

Figure 7.10   Schematic illustration of the functioning of a Deans' switch. In the left-hand position, ^{1}D effluent is transported to the ^{2}D column. In the right-hand position, it is sent to the ^{1}D detector. The narrow bypass channel serves to avoid diffusion of the ^{1}D effluent and the analytes contained in it into the blocked switching-gas line.

is a very powerful technique that can "easily" provide 100 000 plates in a (one-dimensional) run and can be equipped with near-perfect detectors for quantitation and identification, such as the FID and mass

Box 7.1 Hero of analytical separation science: J. B. Philips. Figure courtesy of John Dimandja (FDA, Stone Mountain, GA, USA)

John B. Philips (1947–1999, United States of America)

John B. Philips was a professor at the University of Southern Illinois and a true innovator. Aiming to repair the rear-window defroster in his car, he hit upon electrically conducting paint (a suspension of copper particles in ethyl acetate) and started to apply it on sections of fused silica capillaries. This allowed him to invent sample-focussing techniques first for injection in GC, but soon for a spectacular realization of comprehensive two-dimensional gas chromatography (GC×GC). His innovations heralded great developments in the technology and its application to important industrial and societal challenges.

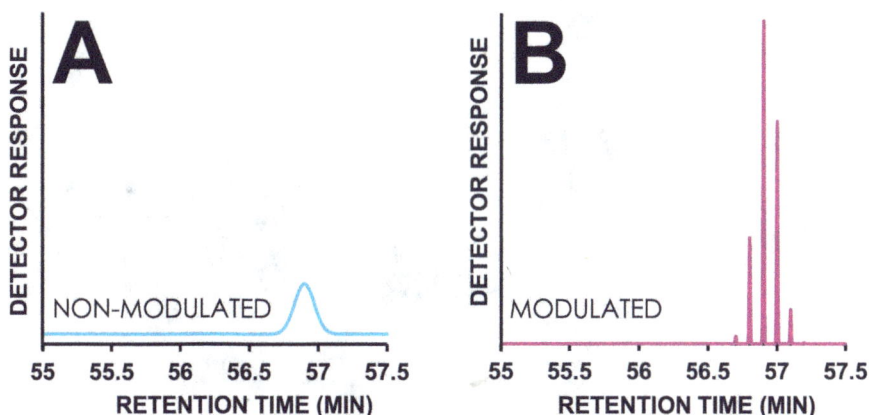

Figure 7.11 Thermal modulation of a single GC peak, with a bandwidth of about 25 s. Modulation (with t_{mod} = 6 s) results in a series of very sharp, high-concentration pulses. These serve as injection pulses when a second-dimension column is connected. Figure courtesy of Jan Blomberg.

spectrometer (**MS**). This allows a "brute-force" approach (with excessive separation power) to many applications. This is akin to the way in which we use our increasingly powerful laptops for mundane tasks, such as writing or editing a book. The tool is much too powerful, but it is readily available, so there are a few reasons to trim it down. GC×GC takes the separation power of GC a significant stride further, but it comes at the cost of an increased complexity of instrumentation and method development, and additional requirements in terms of data handling, consumables and/or energy. As much as GC is the technique to use whenever possible (see Chapter 2), GC×GC is a technique to use specifically when detailed characterization of extremely complex samples containing a large variety of volatile analytes is required.

Although the concept of comprehensive 2D chromatography in general (and that of GC×GC in particular) was well established prior to that time (see Module 7.1), it took the experimental brilliance of one man to make it practically feasible in the 1990s. In 1991, John Phillips (Box 7.1) first described GC×GC using a thermal modulator.[15] The basic idea of the modulator, which Phillips had already been using for different (sampling and injection) purposes, was to turn a peak eluting from a (first-dimension, ^{1}D) GC column into a series of sharp pulses, as illustrated in Figure 7.11. These could then be used as injection pulses in a second-dimension (^{2}D) column. If the time needed for the ^{2}D separation is shorter than the time

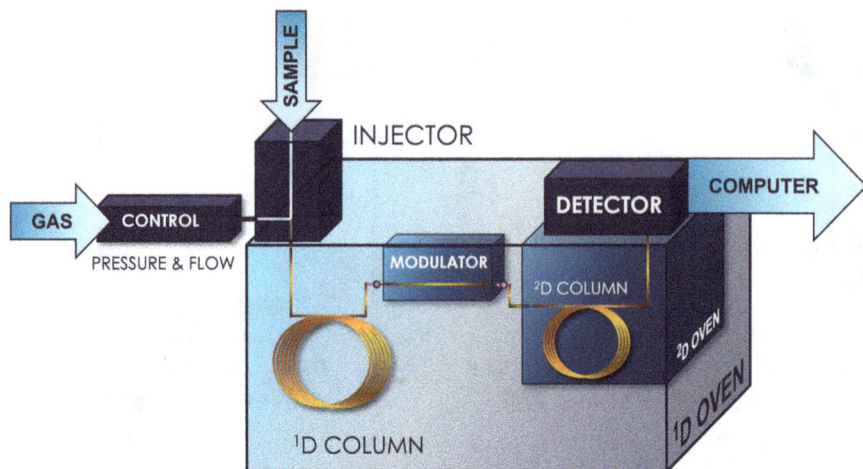

Figure 7.12 Basic setup of GC×GC. Pink dots indicate connections to the modulation capillary.

between pulses (modulation time, t_{mod}), a series of fast chromatograms is obtained, which can be converted into a colour map, as described in Module 7.1 (see also Module 9.13).

Both the ^{1}D and ^{2}D columns are connected directly to the modulator, as illustrated in Figure 7.12. This implies that all sample components injected into the ^{1}D column will eventually reach the detector, provided that they are sufficiently volatile to allow elution. Often, the ^{2}D column is in the same oven as the ^{1}D column, but in other cases, the ^{2}D column may have its own mini oven inside the larger oven, which allows a (typically small) offset between the first- and second-dimension temperature programs. The ^{2}D separation is so fast that it is essentially run isothermally.

The type of raw chromatogram obtained from the detector is shown in Figure 7.4C. A complication is that a single analyte is spread across a number of ^{2}D peaks if the modulation time is shorter than the bandwidth of the ^{1}D peak (Figure 7.8). The latter is a necessary condition for maintaining (most of) the separation obtained in the first dimension. In the chromatogram, the signals belonging to a single analyte must be merged correctly to ensure accurate quantitative analysis.[16,17]

Operating conditions for GC×GC can be estimated using the Golay equation (see Section 1.7.7). The ^{1}D column has dimensions similar to a standard column used in 1D-GC, for example, an internal diameter of 320 µm and a length of 25 m (0.32 µm film thickness).

For such a column operated at ν_{opt}, the hold-up time is about 280 s (assuming $D_m = 5 \times 10^{-6}$ m^2 s^{-1} and $D_s = 5 \times 10^{-11}$ m^2 s^{-1}, and neglecting gas compression). For a typical temperature program (50–300 °C in 45 min), the retention factor at the moment of elution (k_e) may be about 2,[18] and a fair estimate of the standard deviation of an eluting peak in time units is 2.7 s. This implies that a modulation time (t_{mod}) of 3 s will yield between three and four "cuts" (*i.e.* ^{2}D chromatograms) for each ^{1}D peak. The ^{2}D column is much smaller, for example, 1 m in length × 100 μm i.d. (0.1 μm film thickness). With the same volumetric flow rate as in the first dimension, such a column has a t_0 of about 1.1 s. This implies a window of $0 < {}^2k <$ 2.7 if "wrap around" (*i.e.* the elution of peaks in subsequent cycles) is to be avoided. Because of the continuous series of overlapping injections, no time is lost between the point of injection and t_0. This example illustrates that GC×GC is reasonably possible.

In the above example, we may also estimate the peak capacity of the GC×GC system. The ^{1}D separation offers an approximate peak capacity of 250 (calculated based on a temperature program of 2700 s, divided by 4 × 2.7 s peak width). The ^{2}D column offers a plate count of about 7300, and as the column is operated under near-isocratic conditions, we can use eqn (1.42) (with a required resolution of 1) to arrive at a peak capacity for the ^{2}D system of about 30 and an estimated peak capacity for the GC×GC system of about 7500. Especially impressive is the peak-production rate (peak capacity divided by the run time), which is almost three peaks per second.

One of the advantages of thermal modulation in GC×GC is illustrated in Figure 7.11. The pulses created after modulation can be much higher than the original ^{1}D peak. In the resulting chromatogram, the analytes will be diluted again (see Module 7.1), but because the ^{2}D column is short and narrow, so as to achieve very fast separations, the eventual signal may still be much higher than in 1D-GC. Because more separation is achieved in GC×GC than in 1D-GC, the baseline can be defined better, and the signal-to-background ratio can be drastically improved for many analytes in a complex sample.

Figure 7.13 illustrates one of the disadvantages of the basic GC×GC setup. The same mass flow of carrier gas passes through the ^{1}D and ^{2}D columns, and the gas flow is only controlled upstream of the ^{1}D column. However, the optimum gas flow is very different for the two columns. If we neglect the compressibility of the gas (which is an optimistic assumption in this case, but one that does

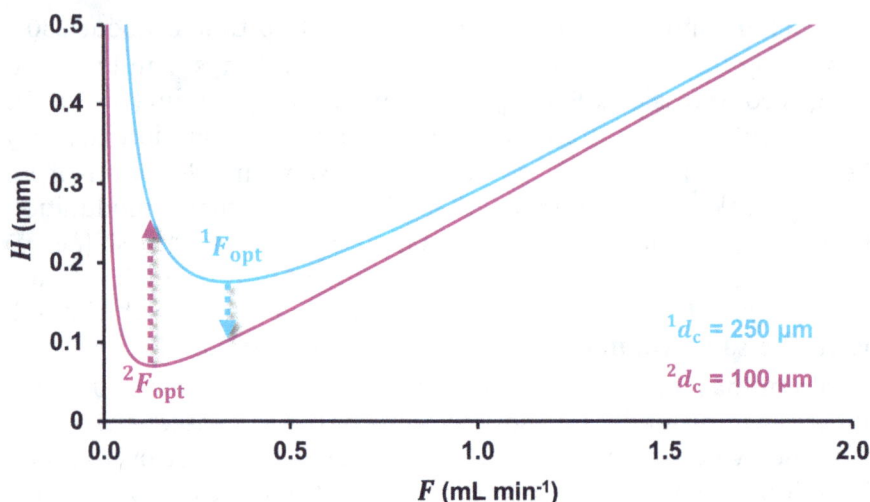

Figure 7.13 Plate-height (Golay) curves in terms of volumetric flow rate (horizontal axis). The blue arrow corresponds to the optimum flow rate in the first dimension; the pink arrow indicates the optimum flow rate in the second dimension. Mobile-phase compressibility was neglected when constructing the figure.

not invalidate the argument), the plate-height (Golay) curves for typical ^{1}D and ^{2}D columns are shown in Figure 7.13, with volumetric flow rate (F) on the horizontal axis, instead of the average linear velocity (u). It is observed that the optimum volumetric flow rate is much higher in the (slower, but wider) ^{1}D column than in the (faster, but narrower) ^{2}D column. When operating at the optimum ^{2}D flow rate ($^2F_{opt}$, pink arrow), the ^{1}D column is operated well into the "B-branch" of the plate-height curve, and efficiency is lost. When operating at the optimum ^{1}D flow rate ($^1F_{opt}$, blue arrow), the ^{2}D column is operated far above the optimum. This is favourable in terms of speed, but efficiency is lost in the second dimension and the operating pressures are relatively high (typically a few 100 kPa or a few bar excess head pressure). Thus, an uncomfortable compromise must be sought for the operating flow rate. In theory, a splitter may be installed before or after the modulator, but this would reduce analytical sensitivity and precision, and it would defeat the purpose of comprehensive two-dimensional chromatography.

Using a column with a larger internal diameter (e.g. 320 µm) offers the advantage of broader ^{1}D peaks, allowing longer modulation times. However, it becomes even more difficult to strike a

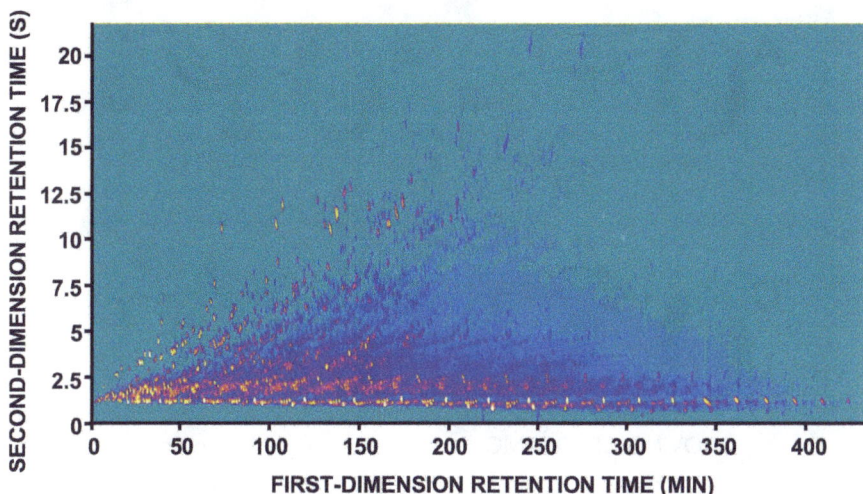

Figure 7.14 High-resolution GC×GC separation of a diesel sample. First-dimension column: 60 m × 250 µm i.d.; second-dimension column: 4 m × 100 µm i.d. More than 30 000 peaks were detected in this chromatogram. Courtesy of Jan Blomberg.

good compromise flow rate because the optimum values for the two columns are further apart. A 50 µm i.d. ^{2}D column offers barely any advantages in terms of efficiency and peak capacity when used in combination with a 250 µm i.d. ^{1}D column, while it leads to much higher operating pressures. Nevertheless, the separation power of GC×GC is impressive. Figure 7.14 shows an example of a high-resolution GC×GC separation of a diesel sample. The relatively long analysis time (7 h) and modulation time (20 s) result in thousands of resolved peaks.

7.2.4 Modulation

7.2.4.1 Thermal Modulation

The development of thermal modulators spurred the advent of GC×GC. The first working modulators were based on resistive heating of metal-clad (gold-painted or copper-painted) columns[15] (Figure 7.15A). To promote the trapping of analytes, the first part of the ^{2}D column, which served as the modulation capillary, was led out of the GC oven. Such modulators were neither very robust nor very efficient in trapping moderately volatile components, but they served to convincingly demonstrate the potential of GC×GC. A

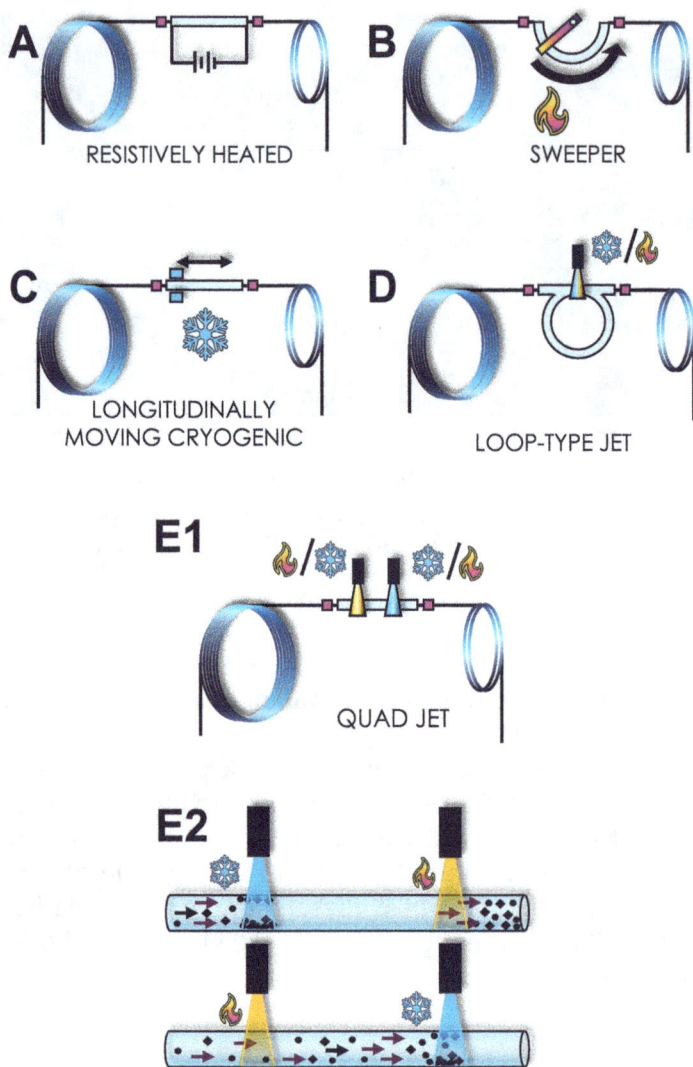

Figure 7.15 Different types of thermal modulators for GC×GC. (A) Resistively heated modulator, (B) sweeper modulator, (C) longitudinally modulated cryogenic system, (D) loop-type modulator, (E1) quad-jet modulator, (E2) close-up of two stages of the quad-jet modulator.

number of thermal modulators were subsequently developed. One of the earliest ones was the sweeper modulator, again designed by John Phillips,[19] in which a "slotted heater" (or "heated slit") was moved across a modulation capillary to "sweep up" retained analytes and move them onto the ²D column (Figure 7.15B). The modulation capillary contained a very thick (up to $d_f = 8$ μm) stationary-phase film to promote retention of analytes eluting from the ¹D column. Nevertheless, volatile analytes could not be effectively trapped. For example, the sweeper modulator was demonstrated to yield excellent separations of middle-distillate fuels (*e.g.* diesel oil) but failed for gasoline-type fuels.[20] In addition, the mechanical challenge of moving the slit neatly around the modulation capillary, without scratching or breaking it, proved too demanding for (routine) operation in many laboratories.

Marriott and Kinghorn developed the longitudinally modulated cryogenic system (LMCS, Figure 7.15C).[21] In this device, a CO_2-cooled trap is moved back and forth along the direction of the modulation capillary. If the trap is near the exit of the modulator, all analytes are collected in the cold zone. When the trap is subsequently moved to a position close to the beginning of the modulator, the trapped analytes are released to the ²D column, while a new fraction undergoes preliminary trapping at the cold spot. When the trap moves back, this fraction is trapped again. Because there is no forced heating when the cold spot moves away, the release of analyte bands may be relatively slow. In a recent variation on this theme, a modulation capillary is pulled back and forth across a cold spot, with heated zones on either side to avoid analyte retention.[22] This is a rather complex setup with sensitive carbon column guides. Moreover, different modulation capillaries are required for different ranges of analyte boiling points.

In recent years, jet modulators have become dominant. In this case, hot and cold jets can be used to alternately heat or cool segments of the column. Hot air can be used for the hot jet, while cold jets can be fuelled by expanding carbon dioxide or evaporating liquid nitrogen. The action of the quad-jet modulator (Figure 7.15E),[23] featuring two permanent (nitrogen) cold jets and two pulsed hot-air jets, is akin to that of the LMCS, but trapping and releasing may be faster and more efficient, thanks to the active heating and cooling.

A final smart twist (literally) was the invention of the loop-type interface (Figure 7.15D), which halved the number of hot and cold jets (or other means for heating and/or cooling) needed while preserving the action of the quad-jet modulator.[24] The modulation capillary has one coil (or "loop"), allowing it to be cooled or heated at two different places along its length simultaneously. A disadvantage is that columns show a tendency to break in the loop-type interface.

A significant disadvantage of jet modulators is the need for a coolant. This may incur significant costs, especially in the case of liquid nitrogen. As an alternative, nitrogen or air may be cooled in a refrigerated heat exchanger, where the coolant is contained in a closed-loop system (see *e.g.* ref. 25). In the latter case, highly volatile analytes cannot be trapped. The refrigerator approach may serve as a viable alternative for analytes with boiling points above *ca.* 150 °C (corresponding to *n*-nonane).

7.2.4.2 Flow Modulation

Modulation for achieving GC×GC operation can also be performed by manipulating carrier-gas flows. In thermal modulation, the concentration of analytes in the carrier gas is locally increased. Such a mechanism is not available for flow modulation. There are basically two ways to create narrow ^{2}D injection pulses from the ^{1}D effluent. The first, so-called **flow-diversion** approach involves sending only small fractions of the ^{1}D effluent to the ^{2}D column at regular intervals while directing most of the ^{1}D effluent (and the sample) to waste. This is schematically illustrated in Figure 7.16A. Fairly narrow pulses can be achieved, with injection bandwidths less than 50 ms being reported.[26] The concentration of the analyte in the pulses (light-blue line) approaches that of the ^{1}D effluent (dark-blue line).

In the second method, known as the **differential-flow** approach (Figure 7.16B), little or no sample is wasted. Instead, the pulses are interspersed with large amounts of added carrier gas. This causes a much higher flow rate in the ^{2}D column and, eventually, through the detector. The concentration of the pulses is not higher than that in the flow-diversion approach. However, the mass flow of analytes can be greatly increased, thanks to the higher flow rate. As a result, peaks emerging from the ^{2}D column will exhibit greater intensity if a mass-flow-sensitive detector is used, such as an FID. Detection limits are not necessarily reduced concurrently because a higher flow rate may also lead to higher noise.

Unlike in LC-MS, where the response of a mass spectrometer often balances concentration sensitivity and mass-flow sensitivity, electron ionization MS is usually considered a mass-flow-sensitive detector in GC.[27] More importantly, the very high ^{2}D flow rates encountered in differential-flow modulated GC×GC are incompatible with MS. Such

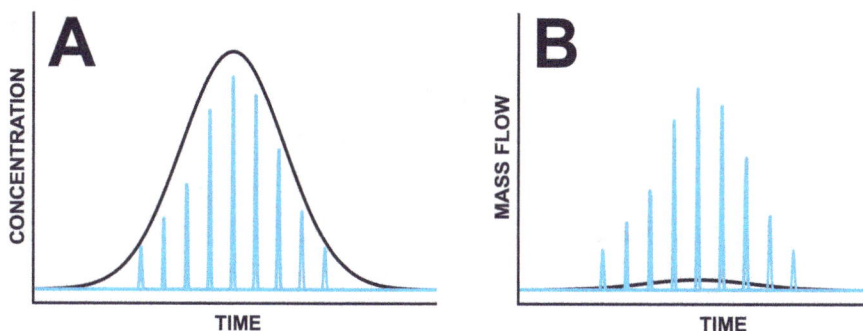

Figure 7.16 Illustration of the results of flow modulation. (A) Flow-diversion approach or differential-flow approach with concentration detection; (B) differential-flow approach with mass-flow-sensitive detection. The dark-blue line illustrates the peak eluting from the ¹D column. The light-blue line indicates the injection pulses onto the ²D column.

high flow rates cannot be accommodated in most mass spectrometers, and if they can, it will be at the expense of the MS performance.

The ²D columns used in differential-flow GC×GC are different from those used in combination with other modulators. The increased flow necessitates larger column diameters (*e.g.* 250 μm i.d.). Yet, the linear velocities will be very high, which is advantageous in terms of the ²D analysis time but detrimental to the ²D efficiency due to the *C*-term in the Golay equation (see Section 1.7.7).

Flow modulation offers some advantages. Neither the trapping of analytes nor their release is a limiting factor. Flow modulation can be used for highly volatile analytes that are difficult to trap with (some) thermal modulators, while rapid release from the trap to achieve narrow ²D-injection pulses is not difficult for high-boiling analytes. Of course, adsorption of the analytes on any solid surface should be avoided. This is especially difficult without the help of heat. Contact with valve materials can be catastrophic. Operating a solenoid valve inside a GC oven during numerous heating and cooling cycles places significant demand on the hardware. Keeping the valve outside the oven is challenging in terms of analyte adsorption. For these reasons, fluidic-logic-type switches (usually called Deans' switches in GC, see Section 7.2.2) are much preferred.

7.2.4.3   *Assessment of Modulation Strategies*

Figure 7.17 provides an overview of modulation techniques for GC×GC. Thermal modulation using nitrogen or carbon dioxide is most commonly employed in (routine) practice, but most of the research effort in recent years has been devoted to flow-modulation techniques. In addition, new coolant-free thermal modulators have emerged. In hybrid modulators, thermal effects are combined with flow manipulation.

Table 7.2 summarizes the main advantages and disadvantages of thermal modulation and the two main groups of flow modulators, *i.e.* (valve-based or pneumatic) flow-diversion techniques and differential flow modulators.

7.2.5 Selectivity in GC×GC

We have seen in Module 7.1 that successful comprehensive 2D separations require very different selectivities in the two dimensions. Ideally, the retention (or elution) times in the second dimension should be independent of those in the first dimension. In this module, we will discuss whether and how such orthogonality can be achieved in GC×GC.

In Module 2.2, we have seen that the retention factor in GC depends on various factors as follows:

$$k_i = \frac{RT\,n_s}{\gamma_{i,s}^{\infty} P_i^0 V_m} = \frac{T}{\gamma_{i,s}^{\infty} P_i^0} \cdot f_{col} \qquad (7.12)$$

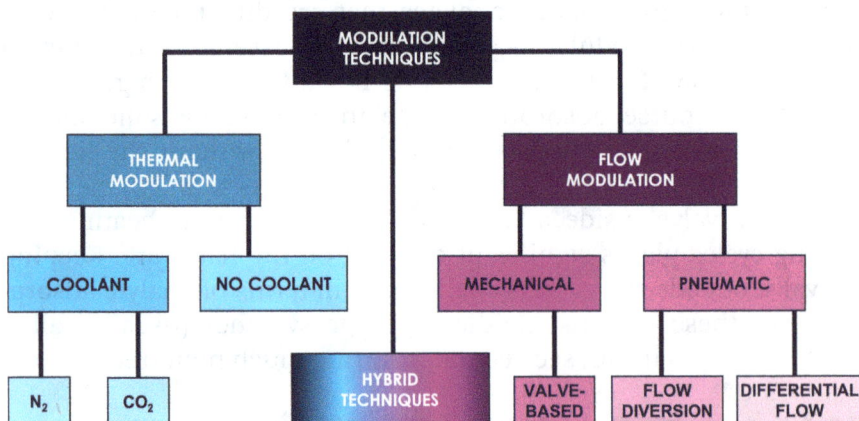

Figure 7.17   Overview of modulation techniques for GC×GC. Adapted from ref. 28, https://doi.org/10.1002/jssc.202300304, under the terms of the CC BY 4.0 license, https://creativecommons.org/licenses/by/4.0/.

Table 7.2 Main advantages and disadvantages of the various types of modulation devices.

	THERMAL MODULATION	FLOW MODULATION	
		Flow diversion	Differential flow
Advantages	• Very narrow ^{2}D injection pulses (10 ms) possible • Very high overall peak capacities possible • Established, robust technology	• No restrictions on analyte volatility • ^{2}D column can be kept narrow (*e.g.* 100 μm i.d.) to promote fast, efficient separations • ^{2}D flow rate can be different from (smaller than) ^{1}D flow rate	• No restrictions on analyte volatility • High signals in terms of mass flow • Very stable and robust once optimized
Disadvantages	• Deep cooling needed to trap highly volatile analytes • Limited flexibility in choosing column diameters and operating conditions • Cost (and footprint) of coolants	• Low sensitivity (most of the sample goes to waste) • Injection pulse width (≥40 ms) limits ^{2}D performance	• High ^{2}D flow rate • Poor compatibility with MS • Relatively wide ^{2}D column offering limited peak capacity • Requires optimization of conditions and hardware (capillaries) • High carrier-gas consumption

where $\gamma_{i,s}^{\infty}$ is the activity coefficient of the analyte at infinite dilution, P_i^0 is the pure-component vapour pressure of the analyte and f_{col} is a column-dependent constant. In a typical temperature-programmed run, the retention factor is high upon injection. With increasing temperature, the exponentially increasing vapour pressure starts to

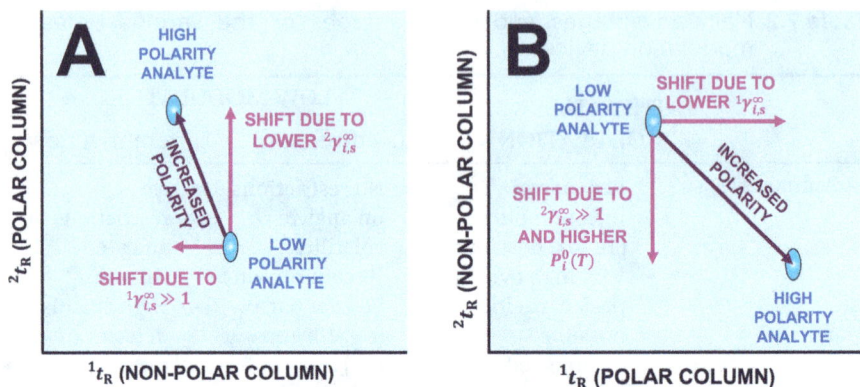

Figure 7.18 Schematic illustration of the behaviour of different (polar and apolar) compounds of equal volatility on GC×GC systems with non-polar (^{1}D)×polar (^{2}D) columns (A) and polar (^{1}D)×non-polar (^{2}D) columns (B). Pink arrows indicate the shifts induced by increased analyte polarity.

catch up with the linearly increasing temperature, and it is a matter of time – or, better, temperature – until k_i becomes small and the analyte elutes from the column as a narrow band due to the low retention factor $k_{i,e}$ at the moment of elution. Clearly, volatility (pure-component vapour pressure) plays a crucial role in determining retention in GC.

As explained in Module 2.2, the activity coefficient takes on different types of values for different combinations of analytes and stationary phases. On a non-polar column, non-polar analytes tend to exhibit smaller values for $\gamma_{i,s}^{\infty}$ than polar compounds. However, due to the weakness of dispersion forces between non-polar molecules, the effects of the activity coefficients on retention times (or elution temperatures) are relatively minor. For example, benzene with a boiling point of 80.1 °C elutes between *n*-hexane (boiling point 68.7 °C) and *n*-heptane (boiling point 98.4 °C) (see Table 2.7). Indeed, it has long been common to refer to a separation on a non-polar column as a boiling-point separation. Thus, in GC×GC chromatograms obtained with a non-polar ^{1}D column and a polar ^{2}D column, the main differences between the locations of polar and non-polar compounds are observed in the vertical (^{2}D) direction. On the polar ^{2}D column, polar analytes exhibit lower activity coefficients than apolar analytes and, therefore, they are eluted later (eqn (7.12)). This is illustrated in Figure 7.18A.

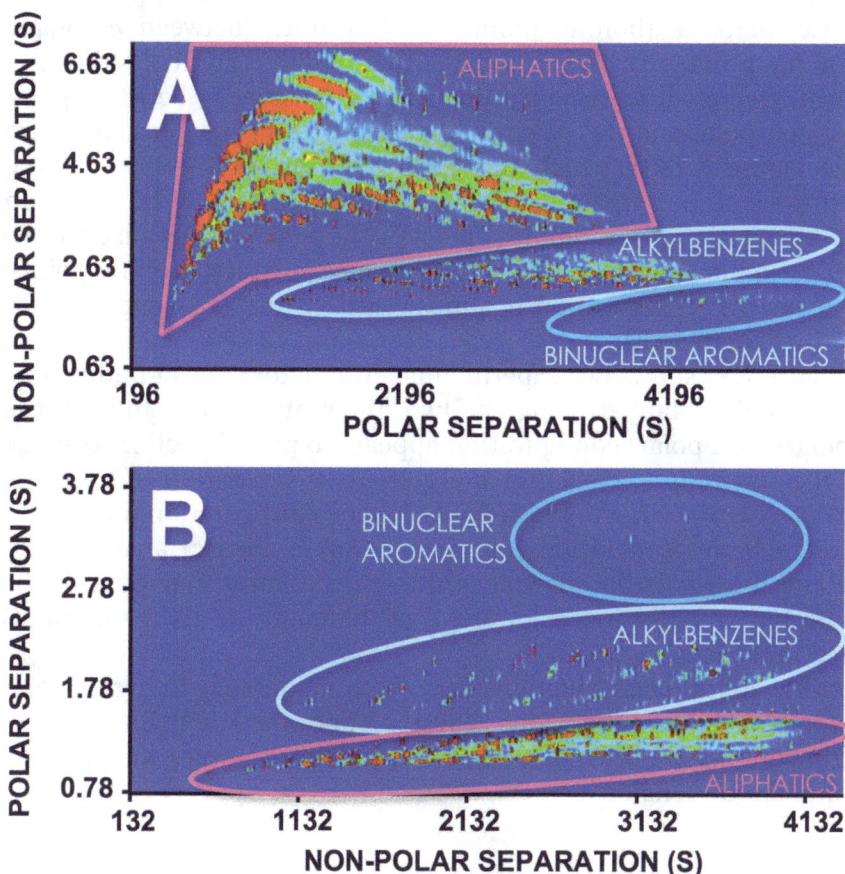

Figure 7.19 GC×GC of a synthetic jet fuel using a polar×non-polar column combination (top) and a conventional non-polar×polar column combination (bottom). Reproduced from ref. 31 with permission from Elsevier, Copyright 2011.

A number of scientists have studied the combination of a polar column in the first dimension and a non-polar column in the second dimension. This has confusingly been called "reversed-phase comprehensive two-dimensional gas chromatography", with the conventional non-polar – polar combination being referred to as "normal-phase GC×GC".[29] We will refrain from using such terminology. One possible way to refer to the ^{1}polar×^{2}non-polar combination is to call it a "transposed" phase system.

On a polar column, the effects of the activity coefficient on the elution temperature are larger than on non-polar columns. For example, on a Carbowax 20M (poly(ethylene glycol)) stationary phase, benzene (boiling point 80.1 °C) elutes between n-nonane (boiling point 151 °C) and n-decane (boiling point 174.1 °C).[30] These greater differences result in larger distances between polar and non-polar analytes of similar volatility in the 2D chromatogram (see Figure 7.18B), which may lead to improved coverage of the entire separation space. However, orthogonality may be larger in the conventional (^{1}non-polar×^{2}polar) configuration, because the ^{1}D separation is almost entirely due to analyte volatility (boiling point), whereas the ^{2}D separation is almost entirely due to analyte polarity (activity coefficient).

Figure 7.19 shows two experimental chromatograms of middle-distillate fuels obtained using the two different setups.[31] Indeed, the (^{1}polar×^{2}non-polar) configuration appears to provide better coverage of the separation space and more effective separation of aliphatic analytes. However, the (^{1}non-polar×^{2}polar) configuration appears to achieve better separation for more polar (aromatic) analytes and for the most volatile analytes, although the latter may possibly be improved by selecting a lower initial temperature than the 40 °C used in ref. 31.

Other column combinations exist, with moderately to highly polar columns in both dimensions. These do not offer the highest overall orthogonality, but they may be appropriate for samples with specific dominant dimensions.

7.2.6 Structured Chromatograms

Both chromatograms in Figure 7.19 reveal a great deal of structure. There are many more than two sample dimensions, and the treatment of Giddings (see Section 7.1.4.2) does not instantly suggest a structured chromatogram.[5] However, there are a few dominant dimensions, which have the greatest effect on retention, *i.e.* the number of aromatic rings and the number of carbon atoms in mineral oils. These dominant dimensions give rise to the evident structure. Within the main groups, Figure 7.19 shows additional fine structure in the form of a series of aligned peaks due to other sample dimensions, such as aliphatic rings and branches.

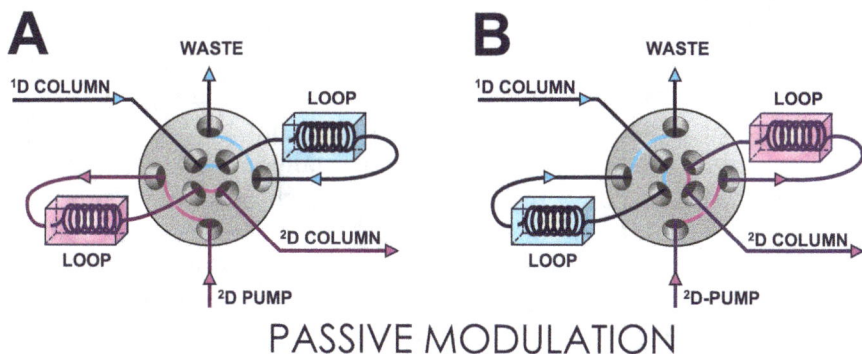

PASSIVE MODULATION

Figure 7.20 Schematic of a typical modulation interface used for 2D-LC. See text for explanation. Light blue depicts the first dimension, whereas pink represents the second dimension. Note that (especially for heart-cut applications) the ^{1}D effluent may pass a detector prior to arriving at the valve. Adapted from ref. 32 with permission from American Chemical Society, Copyright 2019.

7.3 Two-dimensional Liquid Chromatography (M)

7.3.1 Introduction

In two-dimensional liquid chromatography (2D-LC), fractions of the first- dimension (^{1}D) effluent are collected using a modulation valve (the modulator). An example is shown in Figure 7.20. At any moment, the ^{1}D effluent is collected by a storage loop, or simply "loop" (Figure 7.20A). Switching the rotor connects the contents of this loop to the second dimension (Figure 7.20B). The ^{2}D pump(s) displace the loop volume into the ^{2}D column, while the other loop is gradually filled with the next fraction of ^{1}D effluent. Thus, by switching the valve, any or every fraction of ^{1}D effluent can be transferred to the ^{2}D column.

For heart-cut 2D-LC, typically just one fraction is transferred. For this purpose, a 6-port valve suffices. For comprehensive 2D-LC (LC×LC), the setup shown in Figure 7.20 allows for continuous operation, with the valve alternatingly switching positions to allow the entire ^{1}D effluent to be subjected to ^{2}D separations.

While 2D-GC (GC-GC, as well as GC×GC) is established as a robust technique that is suitable for routine use, this is not yet the case for 2D-LC. LC×LC, in particular, is at this point an emerging technique that is still developing significantly.

Heart-cut LC-LC was first demonstrated by Huber *et al.* in 1973.[33] Comprehensive LC×LC was described already by Erni and Frei in 1978,[34] but it was not until 1990 that Bushey and Jorgenson[35] described an automated system for LC×LC that paved the way for modern applications of the technique. Initially, 2D-LC was met with little enthusiasm. For complex mixtures, such as natural products (plant extracts, herbal medicines), many more compounds could be separated or, in chromatographic terms, many more peaks could be produced. However, scientists were not attracted by the prospect of identifying large numbers of unidentified compounds with laborious offline techniques and without the help of computers and large databases. Polymer scientists contributed to the proliferation of LC×LC, because they already had applications ready. Copolymers (*i.e.* polymers built up from two or more different monomers) were known to feature distributions of both composition and molecular weight. Such issues were addressed with time-consuming cross-fractionation methods,[36] in which polymers were first fractionated according to one property (*e.g.* polarity, as reflected in solubility in a progressive series of solvent mixtures), after which the fractions were separated according to another property (*e.g.* molecular weight, by size-exclusion chromatography (SEC), see Module 4.2). LC×LC made such characterization methods much easier and faster,[37-39] and the usefulness of LC×LC has never really been questioned in polymer science and industry.

Other applications of LC×LC started to emerge at a gradually increasing rate, but at the same time, the LC-MS field showed dramatic improvements and revolutionary changes, justifiably capturing most of the attention. Good commercial equipment for heart-cut LC-LC has existed for quite some time, but for LC×LC, it only started to emerge in the last decade. Dedicated software for both types of methods has only started to appear recently.

The focus of the present module is on the essentials of 2D-LC. Readers interested in a more in-depth treatment and the many applications of these techniques are referred to the website that accompanies this book or dedicated treatments.[40]

See the website for searchable databases on published 2D-LC methods and technical literature: ass-ets.org

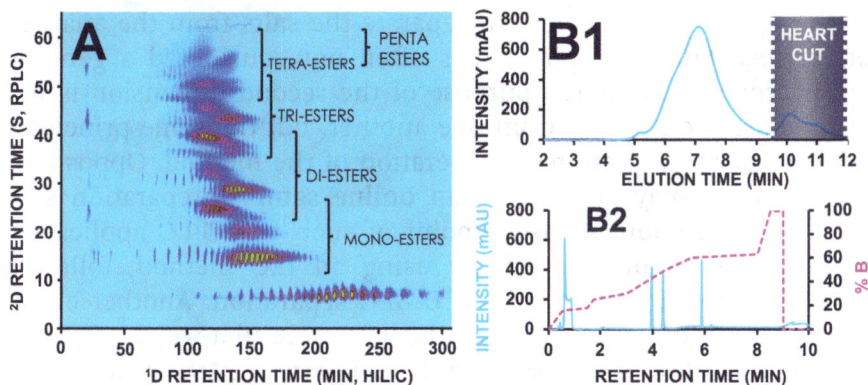

Figure 7.21 (A) LC×LC MS separation of castor oil ethoxylates ("polyols"). Adapted from ref. 41 with permission from Elsevier, Copyright 2019. (B1) Separation of smokeless powder constituents of a forensic sample by size-exclusion chromatography. The highlighted fraction contains the co-eluting small-molecule additives, which are injected into a ^{2}D RPLC separation (B2). Figures B1 and B2 are adapted from ref. 42 with permission from Elsevier, Copyright 2022.

7.3.2 Target Applications of LC-LC and LC×LC

The most important application area for 2D-LC has traditionally been the characterization of **complex mixtures,** benefiting from the high peak capacity offered by 2D-LC (see Module 7.1). To resolve truly complex mixtures, 1D methods would require impractically long analysis times to yield a peak capacity in the same order of magnitude as that of LC×LC. In such cases, the 1D methods are not very informative, as illustrated in Figure 7.3. Another example is the separation of a castor oil ethyloxylate sample, shown in Figure 7.21A. Such a material is a so-called polyol, which is a pre-polymer used, for example, in the preparation of polyurethanes. In the highly orthogonal separation shown in Figure 7.21A, the two separation dimensions (HILIC and RPLC) specifically target dominant sample dimensions (hydrophilic and hydrophobic groups, respectively). As a result of this synergy, the chromatogram offers a structured characterization of a very complex mixture.

A second well-established application of 2D-LC is the use of the second dimension as an **online clean-up method** between a ^{1}D separation and a detection system. A popular example is the removal of non-volatile salts after a ^{1}D ion-exchange separation, so as to allow online coupling with a mass spectrometer. The sole purpose of the second dimension is to separate the salts from the analytes of interest. Such an approach is often not considered a genuine LC×LC separation, as the purpose of the second dimension is not to separate the analytes from one another, but the same principles apply to instrumentation and operation of the method. Oppositely, a ^{1}D column may also act as an **online sample-preparation** step for a ^{2}D separation. A case-in-point of such an LC-LC application is the isolation of an antibody using an IEC method, followed by its selective injection into a ^{2}D SEC separation. Another example concerns the separation of the low-molecular-weight fraction following a SEC pre-separation (see Figure 7.21B1 and B2). A SWOT (strengths, weaknesses, opportunities and threats) analysis of heart-cut LC-LC is presented in Table 7.3.

A SWOT analysis of LC×LC is presented in Table 7.4. One goal of LC×LC that is reputedly gaining popularity in industry, but not extensively documented in the scientific literature, is the use of 2D-LC for **profiling** a sample by two mechanisms (*i.e.* two LC methods) simultaneously, thus replacing two LC methods with a single analysis. The scope of such a method can be to simultaneously determine the chemical properties of the sample constituents and optionally explore the correlation between these. We also observe that LC×LC is used for **method selection,** *i.e.* to establish which of two 1D-LC methods is better suited for the eventual analysis. In this approach, LC×LC is used as a scouting method in the analytical laboratory.

An obvious fourth aim is to add **selectivity** or **separation power** to an existing separation. This can be either because (i) a selection of analytes is very similar or, occasionally, (ii) the sample contains clusters of very dissimilar analytes. An example of (ii) is shown in Figure 7.21B, which shows the SEC separation of a forensic sample of smokeless powder in panel B1. These smokeless powders largely comprised heavy nitrocellulose polymers but also contain small-molecule additives that co-elute. Using LC-LC, this peak cluster can be transferred to a ^{2}D method, where separation of the low-molecular-weight constituents can be achieved (Figure 7.21B2). Category (i) is much more common and a popular motivation for LC-LC, in cases where different peak clusters in the ^{1}D chromatogram require further separation (*e.g.* Figure 7.2). Peak-purity testing (see Module 7.1.2) is perhaps the most popular reason to perform (heart-cut) 2D-LC separations.

Tables 7.3 and 7.4 list the strengths and weaknesses of heart-cut and comprehensive 2D-LC, respectively. In summary, LC-LC

Table 7.3 SWOT analysis of heart-cut 2D liquid chromatography (LC-LC). Adapted from ref. 32 with permission from American Chemical Society, Copyright 2019.

Strengths	Weaknesses
• Very high resolving power • Added selectivity from the second ("orthogonal") dimension • Choice from many retention mechanisms • Enhanced purification or purity assessment of target analytes • Preparative separations possible • Greatly reduced uncertainty of peak assignments (in comparison with 1D-LC) • Readily combined with MS and MS/MS techniques • Method development is relatively straightforward	• Somewhat increased conceptual and instrumental complexity • Analysis time is increased (especially when multiple fractions are selected for analysis in the second dimension) • Possibly reduced detection sensitivity due to analyte dilution[a] • Phase system incompatibility issues[a]

Opportunities	Threats
• Rigorous assessment of peak purity is crucial in (bio-)pharmaceutical industries • "Spatial" comprehensive 2D (and 3D) LC[43]	• For qualitative analysis high-resolution hyphenated techniques (LC-MS or LC-MS/MS) are usually preferred

[a] These issues may (largely) be addressed by incorporating active modulation techniques (see Section 7.3.4).

methods are powerful tools for targeted analysis, assessing the purity of a compound, or obtaining singular ("pure") peaks. LC×LC, on the other hand, is the method of choice for both the comprehensive characterization of very complex mixtures and exploratory analysis. The separation power offered by (online, real-time) LC×LC comes at the cost of stringent conditions imposed on the second-dimension separation because the ^{1}D bandwidth and the ^{2}D analysis time are coupled. Multiple heart-cut (mLC-LC) and, especially, selective-comprehensive (sLC×LC) approaches are used only occasionally for specific applications.

Table 7.4 SWOT analysis of comprehensive 2D liquid chromatography (LC×LC). Adapted from ref. 32 with permission from American Chemical Society, Copyright 2019.

Strengths	Weaknesses
• High peak capacities (1000–10 000) are routinely possible • High peak-production rates (typically 1 peak per second) • Choice from many different retention mechanisms • Added selectivity from second ("orthogonal") dimension • Structured, readily interpretable chromatograms[a] • "Group-type" separations of classes of analytes • Readily combined with MS and MS/MS techniques • Greatly reduced uncertainty of peak assignments (in comparison with 1D-LC)	• Added conceptual and instrumental complexity • Rather long analysis times (typically 30 min to 2 h) • Possibly reduced detection sensitivity due to analyte dilution[b] • Phase-system incompatibility issues[b] • Data-analysis software needed • Difficult and time-consuming method development[c]
Opportunities	**Threats**
• Increased need for detailed characterization of complex samples from many fields[d] • "Spatial" comprehensive 2D (and 3D) LC[43]	• High-resolution hyphenated techniques (LC-MS, LC-IMS-MS, IMS-MS) may compete for certain applications[e]

[a]In the case of low sample dimensionality or dominant dimensions (see Module 7.1).
[b]These issues may (largely) be addressed by incorporating active modulation techniques (see Section 7.3.4).
[c]This may be overcome by using advanced method-development software (see Module 10.4).
[d]See 2dlcdatabase.com for an overview of these.
[e]IMS = ion-mobility spectrometry.

7.3.3 Instrumentation

The versatility of LC is also evident from the different possible hardware setups. The instrumentation used for 2D-LC depends highly on the application, the mode of operation, and the modulation interface. Table 7.5 lists typical hardware components used for different modes of 2D-LC. Note that the table is neither exhaustive nor definitive. For instance, two column ovens can be employed to control the temperature of the columns individually, when desired. In addition, not all heart-cut applications use two detectors.

Table 7.5 Overview of hardware typically used for different modes of 2D-LC.

Mode	Hardware components
Offline 2D-LC	Standard 1D-LC setup (pump, injector, column, detector), fraction collector.
Online LC-LC	Two pumps (^{1}D + ^{2}D), injector, column oven, 6-port valve, two detectors (^{1}D + ^{2}D)
Online mLC-LC	Two pumps (^{1}D + ^{2}D), injector, column oven, one or two multi-loop heart-cut valve assemblies, modulation valve (8- or 10-port, or 2 × 6-port valve), two detectors (^{1}D + ^{2}D)
Online LC×LC (including sLC×LC)	Two pumps (^{1}D + ^{2}D), injector, column oven (optionally two), 8- or 10-port valve (or 2 × 6-port valve), detector

The simplest way to conduct 2D-LC involves injecting a collected fraction of the effluent from a 1D-LC system into another 1D-LC system (**offline 2D-LC**). While this method is laborious, it requires no additional hardware. Manipulation of the fraction (*e.g.* changing the solvent) is relatively easy.

Typically, however, chromatographers prefer to invest in dedicated instrumentation and automation. An example of a setup dedicated to 2D-LC is shown in Figure 7.22. Readers interested in embarking

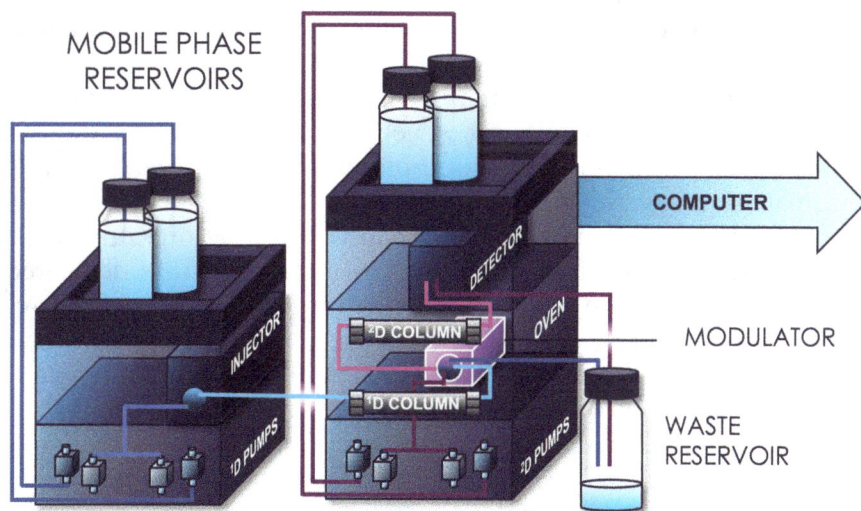

Figure 7.22 Schematic example of a hardware setup used for 2D-LC.

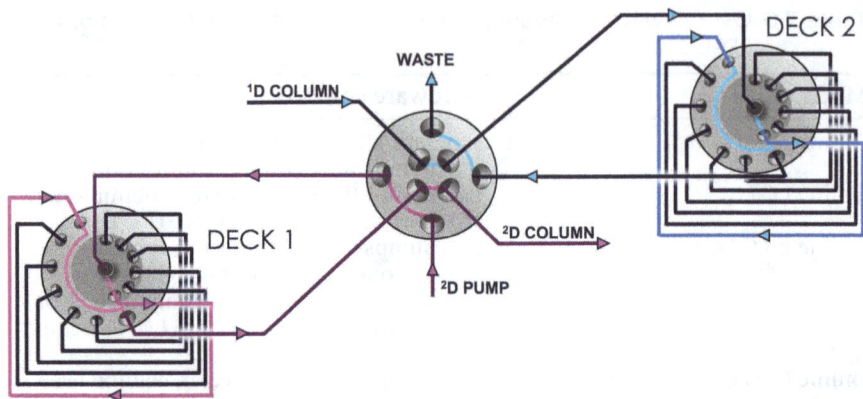

Figure 7.23   Example of a 2D-LC interface capable of multiple heart-cut
2D-LC. Only one valve position is shown. Blue indicates the
first dimension, and pink depicts the second dimension. Due
to the implementation of two decks, the two dimensions can
be operated independently. The fractions contained in one
deck can be analyzed by the second dimension, while the ¹D
effluent can be sampled by the second deck.

on an exploration of 2D-LC in their laboratory may be content with a
second pump and a valve that can be used for modulation.

For heart-cut 2D-LC, a second detector dedicated to monitoring
the ¹D effluent is useful, and the valve can be a two-position 6-
port valve. For comprehensive 2D-LC, two 6-port valves, one 8-port
valve, or one 10-port valve may be employed. Dedicated software
to automate the modulation-valve switches is desirable. Significant
ease-of-use functions have been implemented since the introduction
of instrumentation dedicated to 2D-LC. Contemporary systems allow
fully automated LC×LC as well as the easy implementation of heart-
cut 2D-LC methods. The latter is achieved by dedicated software,
where the user can simply click on one or multiple peak(s) on a ¹D
chromatogram on the screen to have the corresponding fractions
stored in loops automatically and injected one by one onto a ²D
column. To facilitate this process, such systems feature dedicated
mLC-LC valve assemblies that contain multiple storage loops. An
example is shown in Figure 7.23, where the central modulation valve
is surrounded by two mLC-LC valve assemblies, which are typically
referred to as "decks". Such an interface allows simultaneous ¹D
sampling by one deck and ²D separation of the contents of the loops
from the other deck. Modulation interfaces and connections are
further discussed in Section 7.3.4.

For comprehensive 2D-LC, no such mLC-LC valves are required. However, due to the stringent time constraints of the second dimension, high-speed ^{2}D separations are generally required, while significant efficiency (and peak capacity) should be maintained. To achieve this, it is beneficial to use UHPLC instrumentation for the second dimension.

7.3.4 Modulation

7.3.4.1 Passive Modulation and Solvent Incompatibility

The modulator can truly be considered the heart of a 2D-LC instrument, as it connects the two separation dimensions. From another perspective, the modulator is the automatic fraction collector of the ^{1}D separation and the autosampler of the ^{2}D separation. The ^{1}D effluent becomes the ^{2}D sample and, thus, the ^{1}D mobile phase must be compatible with the ^{2}D separation. This is where most problems in 2D-LC method development arise. The valve setups shown in Figures 7.20 and 7.23 directly transfer the ^{1}D effluent to the ^{2}D column, without altering the composition. This is commonly referred to as **passive modulation**.

However, in Section 3.5.5, we addressed that incompatible injection solvents can have unwelcome effects in 1D-LC. We are confronted with the same issues in the second dimension of a 2D-LC separation. The most pressing issue is the risk of analyte **breakthrough**. We have discussed this in Section 4.3.2 for "large" molecules, but breakthrough is also a potential pitfall in small-molecule separations.

In the early days of modern 2D-LC, much attention was focussed on attempts to combine normal-phase LC (NPLC) with RPLC. However, the popularity of "classical" NPLC (Section 3.6.1) has dwindled with the advent of HILIC (Section 3.6.2) and the resurgence of supercritical-fluid chromatography (SFC; Chapter 6). A more contemporary example of incompatibility issues is the use of organic THF in a ^{1}D size-exclusion chromatography method, with the effluent being transferred to an aqueous-organic ^{2}D RPLC separation. Almost all analytes may distribute into a high-concentration THF plug in favour of interacting with a C18 stationary phase. All peaks corresponding to such analytes will show poorly retained peaks until the THF plug has been sufficiently diluted, with analyte peaks likely tailing due to molecules slowed down in the tail of the THF peak.

Breakthrough can also occur in more common combinations of LC modes, such as HILIC with RPLC and even RPLC with RPLC. In the latter case, the later part of a ^{1}D gradient, containing high levels of a modifier (*e.g.* acetonitrile), may be problematic as the injection solvent at the start of a ^{2}D gradient that commences with a weak solvent (high concentration of water).

To mitigate such injection effects, different groups have developed modulation interfaces that actively alter the composition of the ^{1}D effluent prior to injection onto the ^{2}D column. We refer to these as active modulation, and various **active modulation** techniques will be the subject of the remainder of Section 7.3.4.

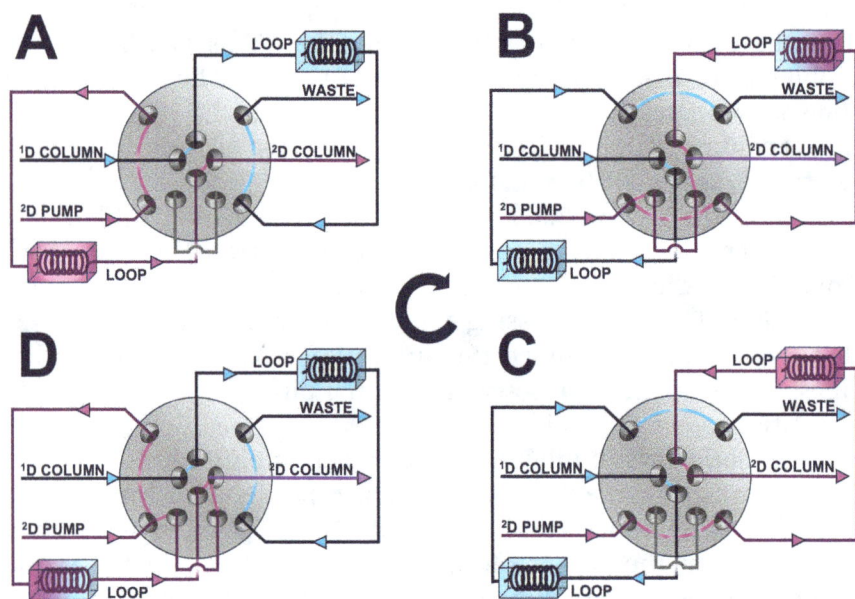

Figure 7.24   Schematic of the valve setup used for ASM. Blue depicts the first dimension, and pink depicts the second dimension. Each modulation cycle follows steps A, B, C, and D, as indicated by the circular arrow in the middle of the figure. See text for further explanation. In FSM, the ^{2}D pump flow is typically split prior to entering the valve, with the bypass being recombined just prior to the ^{2}D column. This action is comparable to alternating between panels B and D (ignoring positions A and C). Schematic: Adapted from ref. 32 with permission from American Chemical Society, Copyright 2017. Concept: adapted from ref. 44 with permission from American Chemical Society, Copyright 2019.

7.3.4.2 Active Solvent Modulation

The first active-modulation techniques to be discussed are **active-solvent modulation** (ASM) and fixed-solvent modulation (FSM). In these modulation strategies, a fraction of the ^{2}D eluent is used to actively dilute the ^{1}D effluent as or before it is injected onto the ^{2}D column. The concept is based on the reasonable assumption that actively diluting the ^{1}D effluent with a solvent that is a weaker ^{2}D eluent will mitigate breakthrough and facilitate focusing at the entrance of the ^{2}D column.

Figure 7.24 shows an 8-port, 4-position valve that can be used for ASM. The positions shown in panels A and C are identical to those of passive modulation (Figure 7.20). However, in positions B and D, a fraction of the flow from the ^{2}D pump bypasses the storage loop (through the little bridge shown at the bottom of the valve) and joins the loop effluent stream to actively dilute it prior to entering the ^{2}D column.

In **fixed-solvent modulation**, the bypass setup is permanent. Typically, this is accomplished by simply splitting the flow from the ^{2}D pump before it enters the valve and creating a bypass for one of the connections to recombine with the valve effluent prior to entering the ^{2}D column. The resulting action is akin to alternating between positions B and D in Figure 7.24 (and skipping positions A and C).

FSM is simpler to implement because it requires only two T-pieces and a bypass capillary. However, the bypass cannot be turned off. As a consequence, the ^{2}D analysis time will increase, as it will take longer to flush out the modulation loop. Stoll *et al.* also identified a risk of additional disturbances in the baseline when using FSM.[44]

Both FSM and ASM rely on the assumption that the ^{2}D mobile phase is a weak eluent – at least during the flush-out phase. This is often the case when gradient-elution RPLC is used for ^{2}D separation, with an initial mobile phase rich in water. The power of ASM to mitigate breakthrough becomes apparent from Figure 7.25, which shows a separation without (Figure 7.25A) and with (Figure 7.25B) ASM. The inset in panel C, which displays the ^{2}D chromatograms along the dashed line, convincingly demonstrates that breakthrough is avoided.

Both FSM and ASM can usually be made to work when RPLC is used in the second dimension. For other ^{2}D modes, the techniques are more difficult to envisage. For example, in ^{2}D HILIC, a high concentration of weak solvent (such as acetonitrile) does not usually lead to very strong retention (as does water in RPLC). ^{2}D SEC will

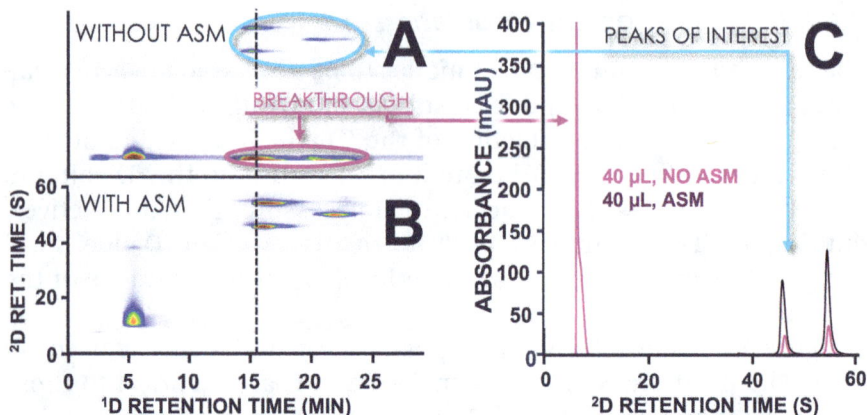

Figure 7.25 HILIC×RPLC of monoclonal-antibody fragments without (A) ASM and (B) with ASM. The pink arrows and circle indicate undesired peaks as a result of breakthrough. (C) 1D chromatograms corresponding to the LC×LC modulation indicated by the dashed line in panels A and B. Adapted from ref. 45 with permission from American Chemical Society, Copyright 2017.

suffer from injection effects due to the presence of a weak solvent in the trap and the potential slow redissolution of polymer analytes.

Recently, the group of Oliver Schmitz (University of Duisburg-Essen, Germany) introduced a setup called **at-column dilution (ACD)**, based on an original idea of Uwe Neue (Waters, Milford, MA, USA) for injection in 1D-LC.[46] Instead of the ²D pump, a separate pump ("transfer pump") is used to flush out the modulation loop towards the ²D column. Prior to the ²D column, this flow is combined with the actual ²D pump flow using a T-piece.[47] In the case that a "binary pump" (high-pressure gradient mixing; see Section 3.3.1.2) is used in the second dimension, one of the two pumps may be used for the purpose. While ACD requires an additional accurate high-pressure pump for the transfer, the main advantages are (i) accurate control of the dilution ratio and (ii) shorter gradient delay by avoiding the modulation loop.

7.3.4.3 Stationary-phase-assisted Modulation

A completely different method to achieve active modulation relies on the use of small packed columns containing a stationary phase instead of loops. This concept, named **stationary-phase-assisted modulation** (SPAM), is illustrated in Figure 7.26. The columns are typically small guard columns, often referred to as **traps**, that are

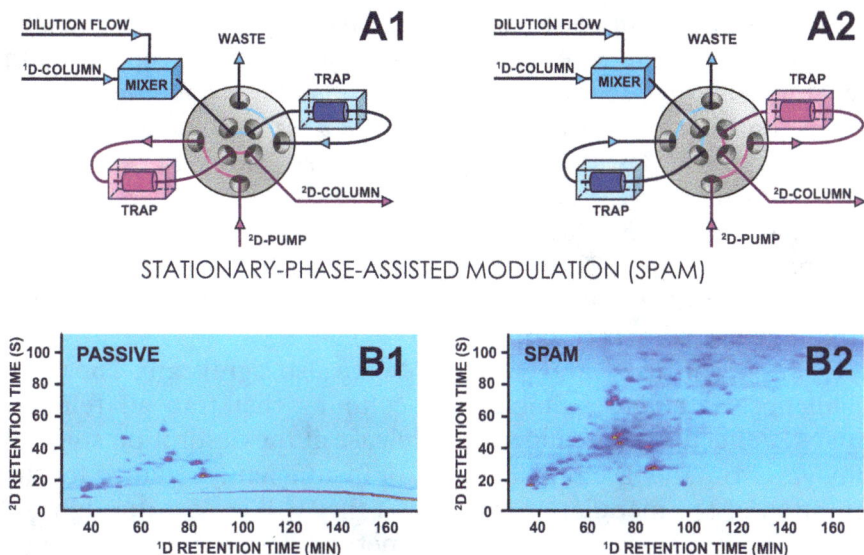

Figure 7.26 (A) Schematic representation of the valve setup used for stationary-phase-assisted modulation in the two positions. Blue depicts the first dimension, and pink depicts the second dimension. Adapted from ref. 32 with permission from American Chemical Society, Copyright 2019. (B) LC×LC MS separation of steroids from bovine urine without (B1) and with (B2) stationary-phase assisted modulation. A significant gain in sensitivity is obtained. Adapted from ref. 48 with permission from American Chemical Society, Copyright 2018.

envisaged to retain all analytes from the ^{1}D effluent, thereby focusing the analyte bands while removing the (incompatible) solvents. To facilitate retention, the ^{1}D effluent is sometimes actively diluted using a mixer prior to trapping.

Upon switching the valve, analytes that are retained on one of the traps are injected into the ^{2}D column as a sharp band. Because the trap and associated capillary connections typically comprise a much smaller volume (<10 µL) than regular sampling loops (≥20 µL), the injection volume is also much lower. The latter significantly reduces the likelihood of solvent incompatibility effects, such as breakthrough, and reduces ^{2}D injection band broadening and dilution factors to improve chromatographic efficiency and detection sensitivity. An example is shown in Figure 7.26B, where an LC×LC-MS separation of steroids in bovine urine is shown.[48] The chromatogram in panel B1 was obtained without SPAM, whereas panel B2 features a chromatogram obtained with SPAM. A significant improvement in sensitivity can be observed.

The advantages of SPAM are evident. As long as the trap chemistry, conditions, and volume are tuned properly, significant gains in sensitivity and reduction of detection limits may be obtained, while mitigating solvent effects. SPAM requires a strong eluent to rapidly and completely desorb the trapped analytes. This can be realized with ^{2}D techniques such as RPLC, HILIC and SEC. In the case of RPLC, a gradient may focus the analytes even more when they enter the main ^{2}D column. The trap can be kept in line with the ^{2}D column during the gradient, and it can then be considered a genuine guard column before the ^{2}D column.

However, the disadvantages of SPAM are also significant. The main challenge is to find suitable trap columns that trap all relevant analytes effectively and allow rapid release at the desorption stage. If analytes from the ^{1}D effluent are not or insufficiently retained for the duration of the modulation time, then they will be lost through the ^{1}D waste line. The implications for method development are large, especially for complex mixtures of unknown composition, for which premature elution must be thoroughly investigated. Other disadvantages include the lack of robustness of guard columns, which are not typically designed to last very long under high-pressure conditions.

FSM, ASM and ACD do not suffer from these disadvantages and are more robust techniques. Not surprisingly, SPAM finds its application mainly for complex mixtures, for which retention on the trap is not envisaged to be challenging, such as polymers.

7.3.4.4 Thermal Modulation

In Section 7.2.4.1, we saw how thermal modulation using hot and cold jets allows for elegant modulation in 2D-GC. However, the effects of temperature on retention are typically much smaller in LC than in GC. In addition, rapid changes in temperature are more difficult to realize for liquids because of the much higher heat capacities for liquids than for gases, along with low thermal conductivities and heat transfer coefficients. In packed columns, the numerous heat transfers between the mobile phase and packing material, especially, thwart rapid temperature control. Thus, the successful implementation of thermal modulation in LC is very challenging. Nevertheless, several groups have attempted to realize this seemingly attractive process, with varying degrees of success.

The most obvious approach is to do the same as in GC, *i.e.* to cool the ^{1}D effluent to promote retention, as shown in Figure 7.27A. Such an approach is unlikely to work for the reasons described above. For

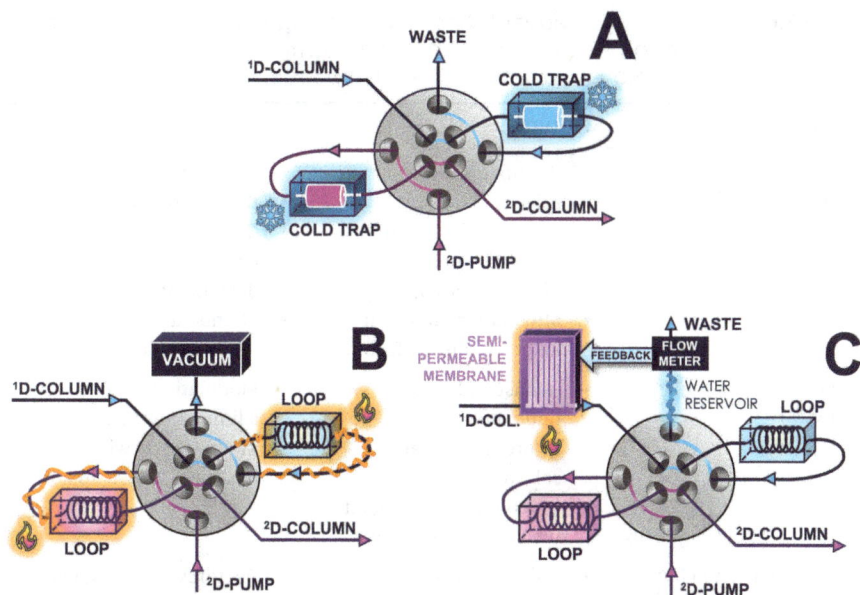

Figure 7.27 Schematics of several thermal modulation approaches. (A) Cold trapping,[49] (B) vacuum evaporation modulation,[50] and (C) evaporative membrane modulation.[51] Panel A adapted from ref. 49 with permission from Elsevier, Copyright 2021. Panel B adapted from ref. 50 with permission from John Wiley & Sons, Copyright 2008 WILEY-VCH Verlag GmbH & Co. KGaA, Weinheim. Panel C adapted from ref. 51 with permission from Elsevier, Copyright 2017.

small molecules, the effect of temperature on retention is indeed (too) small, but a straightforward thermal modulation interface was designed and demonstrated by Niezen *et al.* for polymers, where the effect of temperature on retention is much greater.[49] To achieve sufficient – and sufficiently rapid – cooling, chilled isopropyl alcohol and mineral oil were flushed through an aluminium block that featured holes to provide enough surface area for the trap column. For high-molecular-weight synthetic polymers, this approach even seems to work with commercial column ovens. Proteins may exhibit equally large temperature effects on retention, but their thermal stability may be a limiting factor.

A more sophisticated example of thermal modulation is **vacuum-evaporation modulation** (VEM), which is illustrated in Figure 7.27B. The sample loops are wrapped in heating tape that is used to facilitate evaporation of volatile solvents.[50] A vacuum is used to aid the evaporation and to remove the vapours. An ingenious approach is **evaporative membrane modulation** (EMM, Figure 7.27C), where the

Table 7.6 Overview of advantages and disadvantages of different modulation techniques. Adapted from ref. 52 with permission from the Royal Society of Chemistry.

Technique	Strengths	Weaknesses
Stationary-phase-assisted modulation (SPAM)	• Eliminates incompatible solvent • Improves detection sensitivity (reduces dilution factors) • Modulation volume no longer a limiting factor.	• Trap robustness • Discrimination • Operation and optimization is sample dependent • Method development may be challenging
Active-solvent modulation (ASM)	• Robust and easy to implement • Dilutes incompatible solvent • On-column focusing on the second dimension	• Modulation volume is still a limiting factor. • Mainly useful with RPLC in second dimension
At-column dilution (ACD)	• Robust and easy to implement • Dilution ratio can be controlled • ^{2}D dwell time may be reduced (if mixing is adequate)	• Requires an additional pump • T-piece may not suffice for mixing solvent flows (post modulation)
Thermal modulation (without evaporation)	• Conceptually straightforward • May be used to augment other approaches (e.g. SPAM)	• Only possible for (thermally stable) macromolecules
Thermal modulation (with evaporation)	• Removal of incompatible solvents through evaporation • Fast operation appears possible (under vacuum conditions, VEM)	• Possible discrimination (loss of volatile analytes during evaporation, VEM) • Some analytes (e.g. polymers) may redissolve slowly • Possible additional band broadening (EMM)

^{1}D effluent is flushed through a microchannel chip, which is heated using an infrared LED.[51] The effluent in the channel passes along a semipermeable membrane. A gas flow passes on the other side of the membrane. This allows for the evaporation of low-molecular-weight, volatile solvents. A calorimetric flow meter at the ^{1}D waste line is used to tune the heating rate to account for changes in solvent composition. The setup was used to remove acetonitrile from an RPLC effluent, thereby reducing the peak volume. Unfortunately, for the described prototype, this gain was negated by band broadening due to the chip volume.[51]

7.3.4.5 Summary

An overview of the strengths and weaknesses of different modulation techniques is provided in Table 7.6. While many strategies are ingenious, only ASM and ACD offer reliability and robustness, be it mainly with RPLC as the ^{2}D-separation mode. The other strategies are still undergoing significant development.

7.3.5 Mobile Phase Composition Programs for LC×LC

In Section 1.5.1, we learned that the resolution between peaks and, on a broader scale, the entire separation can be optimized by altering retention, selectivity, or efficiency. We have seen in Module 7.1 how selectivity-related concepts, such as sample dimensionality and orthogonality, as well as efficiency-related concepts, such as peak capacity, are indeed key principles that strongly affect the ability to separate mixtures using two-dimensional techniques. In this section, we will see how special mobile-phase-composition programs may also play a key role in method development, especially for LC×LC.

7.3.5.1 Second-dimension Gradient Programs

We will see in Section 7.3.6 that most 2D-LC methods feature RPLC in the second dimension. This is not surprising due to its high peak capacity, good compatibility with detectors, and short equilibration times. Because we explicitly aim to combine orthogonal separation modes (see Section 7.1.5.1), the analytes that elute together at a given time from the ^{1}D column are expected to yield very different retention times in the ^{2}D system. Therefore, it is also not surprising that ^{2}D RPLC methods predominantly use gradient elution in the second dimension. In principle, each ^{2}D modulation may require a unique optimal gradient, but this poses unrealistic demands on method development and does not result in robust separation methods. In the most common approach, each ^{2}D separation is conducted using the same ^{2}D gradient program, as shown in Figure 7.28A. Such a method is referred to as **full-in-fraction gradients**. If the ^{1}D and ^{2}D systems are fully orthogonal, then this is statistically the best approach. However, when full orthogonality cannot be

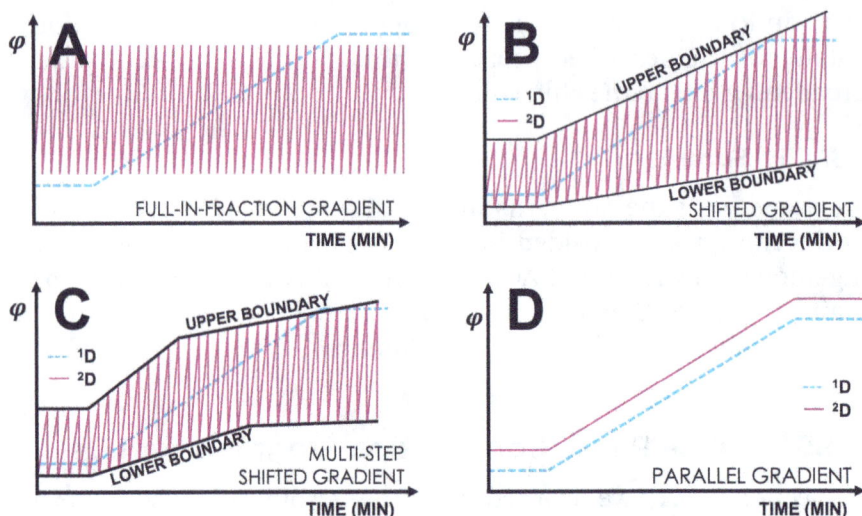

Figure 7.28 Examples of different LC×LC gradient programs. (A) Full-in-fraction, (B) shifted gradient, (C) multi-step shifted gradient, and (D) parallel gradient. The dashed light-blue line shows the ^{1}D gradient. The pink lines represent the composition of the ^{2}D mobile phase as a function of time. Adapted from ref. 1, https://doi.org/10.1002/jssc.201700863, under the terms of the CC BY 4.0 license, https://creativecommons.org/licenses/by/4.0/.

realized, which is often the case in practice, some degree of correlation can be expected between the ^{1}D and ^{2}D systems. In such situations, modern 2D-LC systems allow for the programming of **shifted ^{2}D gradients**, as shown in Figure 7.28B. Typically, the user defines the boundaries between which the ^{2}D mobile-phase gradients are run. Such systems often follow the ^{1}D mobile-phase program, but, in principle, this is not necessary. In practice, shifted gradient programs can be finely tailored as desired to reach the best possible results. An example is the multi-step gradients shown in Figure 7.28C. Such a method may require significant method-development time (unless automated method-development software is available; see Module 10.4).

The various ^{2}D mobile-phase-composition programs do much to make the inevitably somewhat correlated RPLC×RPLC separations more attractive (see Section 7.3.6). The lack of orthogonality can be compensated by a shifted-gradient program, allowing the advantages of RPLC to be exploited in both dimensions. Shifted gradients have been criticized for being less robust due to the challenges imposed on the solvent-delivery system to accurately reproduce such complex gradient assemblies on a run-to-run basis.[53] However, contemporary commercial LC×LC systems seem up to this task. One very different strategy is the use of **parallel ²D gradients**, in which the ²D mobile-phase program closely follows that of the first dimension. Its potential has been demonstrated for heavily correlated RPLC×RPLC separations. Van den Hurk *et al.* recently compared the use of shifted gradients with parallel gradients for the separation of a complex peptide mixture.[54]

7.3.5.2 Discontinuous First-dimension Elution

One main challenge in LC×LC method development is the discrepancy between the timescales of the first and second dimensions. In conventional LC×LC applications, the ²D method is designed to be completed (and the column re-equilibrated) before the next modulation fraction is transferred. Some researchers have, however, explored the use of discontinuous elution from the first dimension, and these approaches are briefly discussed here. The first is the **stop-flow** approach, where, as the name suggests, the ¹D flow is stopped after a fraction has been injected onto the ²D column.[55] This strategy decouples the ¹D and ²D analysis times, and there is no longer any limit on the duration of the ²D method. However, the longitudinal diffusion process (see Section 1.7.4.1) inevitably continues while the flow is stopped, and this may give rise to significant additional band broadening. Xu *et al.* determined that for a total stop-flow time of about 1.4 h, roughly 50% of the theoretical plates are lost for low-molecular-weight analytes. High-molecular-weight analytes exhibit much lower diffusion coefficients (and thus much less longitudinal diffusion), but they still show a loss of 20% in efficiency.[56]

A different method to decouple the ¹D and ²D analysis times through the use of ¹D mobile-phase-composition programs is the concept of **pulsed elution**, which was introduced recently by Jakobsen *et al.*[57]. This approach is based on modulation of the ¹D eluent strength, rather than modulation of the ¹D flow rate. An example of a mobile-phase composition program is shown in Figure 7.29, where the ¹D and ²D gradient programs are shown in blue and pink, respectively. The ¹D eluent strength is periodically elevated for a relatively short time to accelerate the migration of weakly adsorbed analytes through the column. As the ¹D pulse travels through the column, it will slowly deform and sweep

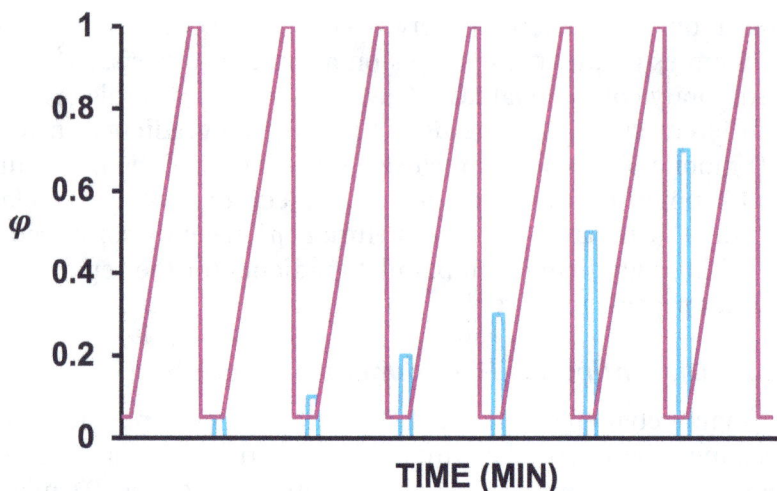

Figure 7.29 Example of a ¹D pulsed-elution mobile-phase-composition program (blue) in combination with a conventional ²D gradient program (pink).

up weakly retained analytes along a distance depending on their retention behaviour as a function of mobile-phase composition. Over time, the solvent strength of these pulses is increased to facilitate elution of analytes that are more strongly retained. The periods of low eluent strength between pulses provide sufficient time for a ²D run to be completed in an optimal fashion. Unlike in the stop-flow approach, the flow is kept constant in pulsed-elution LC×LC. While analytes are retained on the stationary phase, they exhibit much slower longitudinal diffusion. Analytes eluting from the ¹D column are retained on a trap in a manner similar to that in SPAM (see Section 7.3.4.3) prior to their transfer to the ²D column. The use of pulsed-elution LC×LC requires analytes to be retained on the ¹D column during periods of low eluent strength.

A remarkable (pseudo) 2D-LC method that uses just one column (or an ion-exchange column and an RPLC column connected directly in series) deserves mentioning here. It is the LC component of the ***Multidimensional-Protein-Identification-Technology*** (or MudPIT) approach that stood at the basis of bottom-up proteomics.[58] In this field, the proteins in very complex biological samples are digested to yield a very large number of different peptides. This entire sample is injected into the LC system, and successive fractions are eluted by a stepwise increase in the ionic strength of the mobile phase. The series of lengthy RPLC gradients does not quite represent an

LC×LC separation (for that, the number of steps is much too small), but very large numbers of peptides (and from these, proteins) can be identified using HRMS detection, suitable software, and large databases.

7.3.6 Achieving Selectivity

We can now apply all the concepts of two-dimensional LC to establish an effective 2D-LC method by selecting appropriate retention mechanisms or "selectivities".

7.3.6.1 Tuning Selectivity to Sample Dimensions

The different retention mechanisms and selectivities, and the different molecular forces involved, have been extensively discussed in Chapter 3. We summarize the most important separation modes and associated selectivities in Table 7.7. The challenge is now to decide between the different modes of LC based on the dimensions of the sample. In doing so, we must also consider the compatibility of the different modes with each other, detection methods, and the modulation interface. The enormous toolbox of different LC separation systems may seem daunting, but it offers immense flexibility in finding suitable tools for each separation problem. In 2D-LC, the number of options increases dramatically compared to conventional 1D-LC.

Aside from aspects such as solvent compatibility, it is also important to prevent mixed-mode and/or unintended interactions from thwarting the intended orthogonality of the two separation dimensions. Let us consider a mixture of small molecules with different hydrophobicities and anionic groups. In terms of sample dimensionality and orthogonality, it may be prudent to combine an anion-exchange separation with a reversed-phase LC separation.

However, charged groups can strongly affect the ability of the hydrophobic parts of the molecules to interact with a C18 stationary phase. To avoid the charge from dominating the RPLC separation, it may be necessary to add an ion-pairing reagent to the mobile phase (see Section 3.7.3). The type and concentration of the ion-pairing reagent – not to mention the pH – may also significantly alter retention and selectivity. This was illustrated in Section 3.7.3 (Figure 3.30), where tetrabutylammonium hydroxide (TBA) and tetramethylammonium hydroxide (TMA) were compared. TBA features C_4

Table 7.7 Overview of separation modes in liquid chromatography. Adapted from ref. 1, https://doi.org/10.1002/jssc.201700863, under the terms of the CC BY 4.0 license, https://creativecommons.org/licenses/by/4.0/.

Mechanism	Acronym	Selectivity	Common stationary-phase (SP) selectors
Reversed phase	RP	Hydrophobicity, chain length, carbon skeleton	Alkyl (hydrocarbon: C1 to C30; most commonly C18), cyano (π–π)*, phenyl (π–π)*, carbon-clad zirconia (or graphitized carbon), PGC
Ion pairing	IP	Hydrophobicity, suppression of analyte ionization (acid/bases)	Alkyl (hydrocarbon)
Hydrophobic interaction	HIC	Hydrophobicity	Short-chain alkyl hydrocarbons (C4 to C8)
Normal phase	NP	Polarity, functional groups	Bare silica, amino-propyl, diol, cyano
Argentation	AgLC	Degree of saturation, *cis–trans* isomers	IEX columns (*e.g.* sulfonic acid) or bare silica loaded with silver ions
Hydrophilic interaction	HILIC	Hydrophilicity, polar character	Zwitterionic: sulfobetain, phosphocoline; basic: amino propyl; neutral: diol, amide
Ion exchange	IEX	Charge, ionic interactions	Strong cation exchangers (SAX): sulfonic acid; weak cation exchangers (WCX): carboxylic acid; weak anion exchangers (WAX): triethyl amine; strong anion exchangers (SAX): quaternary amine
Size exclusion	SEC	Molecular size, molecular weight	Crosslinked poly(styrene-divinyl-benzene) or methacrylate porous beads (SEC organic solvents); Polar-functionalized porous silica (SEC aqueous)
Mixed mode	MM	Combination of retention mechanisms	Anion-exchange/reversed-phase (AEX/RP), cation exchange/reversed-phase (CEX/RP), anion-exchange/cation-exchange/reversed-phase (AEX/CEX/RP); AEX/HILIC, CEX/HILIC, AEX/CEX/HILIC
Chiral	Chiral	Selector-specific chirality	Variety of selectors depending on the application, with the most common based on polysaccharide derivatives (chiral carbamate/ benzoate polymers of cellulose and amylose)
Affinity	Affinity	Selector-specific affinity	Stationary phases with chemically bonded antigens or proteins

chains that can act as hydrophobic anchors, whereas such chains are not present in TMA. When using TBA, additional hydrophobic interactions are obtained for anions that show little retention when using TMA. As a result, compared to IEC×RPLC, the use of TMA results in a more orthogonal IPC×RPLC separation, whereas using TBA results in lower orthogonality.

Other chromatographic aspects may simplify or complicate 2D-LC. An overview of these is provided in Table 7.8.

> See the website for further information on compatibility issues, along with additional material and examples: ass-ets.org

7.3.6.2 Notes on Coupling Selectivities

Figure 7.30 organizes and summarizes all the above concepts and considerations into a matrix that lists the strengths, weaknesses, and challenges associated with each possible combination of separation modes. Table 7.8 provides clarifications of the symbols used in Figure 7.30. The figure is intended to be used as a guide, and it is not exhaustive. Figure 7.30 may be used to provide a starting point for method development in 2D-LC or to narrow down the number of potential options to a manageable level. It is good to emphasize that a combination that we deem challenging may still be fruitful for specific applications. Moreover, some combinations listed as challenging may be (much) less difficult in heart-cut LC-LC than in LC×LC or easier with specific detectors or for specific samples. Pursuing the application of challenging combinations has spurred innovation in modulation technology. Therefore, we encourage researchers not to shy away from challenging LC×LC systems.

> See the website for a searchable database of published 2D-LC methods since 1978: ass-ets.org

Most separation modes are discussed in detail in earlier modules. In the remainder of this section, we will only highlight some key points related to the different separation modes that are particularly relevant for 2D-LC. In this discussion, we will also refer toFigure 7.31low-molecular-weight, which is based on the information provided in more than 240 applications of 2D-LC that were published between 2019 and 2023.

Table 7.8 Overview of symbols used in Figure 7.30. Adapted from ref. 1, https://doi.org/10.1002/jssc.201700863, under the terms of the CC BY 4.0 license, https://creativecommons.org/licenses/by/4.0/.

Symbol	Meaning	(Context) Definition
S	Selectivity/specificity	Capability of the separation method to differentiate analytes based on unique chemical characteristics (Chapter 3)
B	Breakthrough/peak distortion	Effects of elution strength of ^{1}D effluent relative to ^{2}D eluent, especially when the injection solvent is a strong/immiscible solvent in the ^{2}D system (Section 4.3.2)
A	Adsorption effects	Specific for SEC: effect of injection solvent and eluent on size discrimination (Section 4.3.2)
F	Fast separation	Possibility to develop method with short analysis times
H	High resolving power	Method capable of providing high peak capacities (Section 1.5.2 and Module 7.1)
M	MS compatible	Possibility of using volatile mobile-phase additives and obtaining good MS sensitivity using common ionization strategies (Module 3.10)
Q	Post-method equilibration	Time required after running a gradient to obtain repeatable elution times over many consecutive runs
O	Orthogonality	Degree of independence of two separation mechanisms (note: the score is attributed under the assumption that the sample dimensions correspond with the chromatographic methods) (Section 7.1.5.1)
X	Mobile-phase compatibility	Parameter reflecting the degree of miscibility of two mobile phases
P	Practicality	Usefulness of a particular combination
E	Easy (active) modulation	Easiness of developing methods using active modulation strategies, such as SPAM, ASM, and ACD
I	Isocratic mode	Possibility of (easily) running isocratic methods
▣	Published combination	—
🞄	Polymer compatible	—
🞄	Protein compatible	—
⮎	Reversing order recommended	—

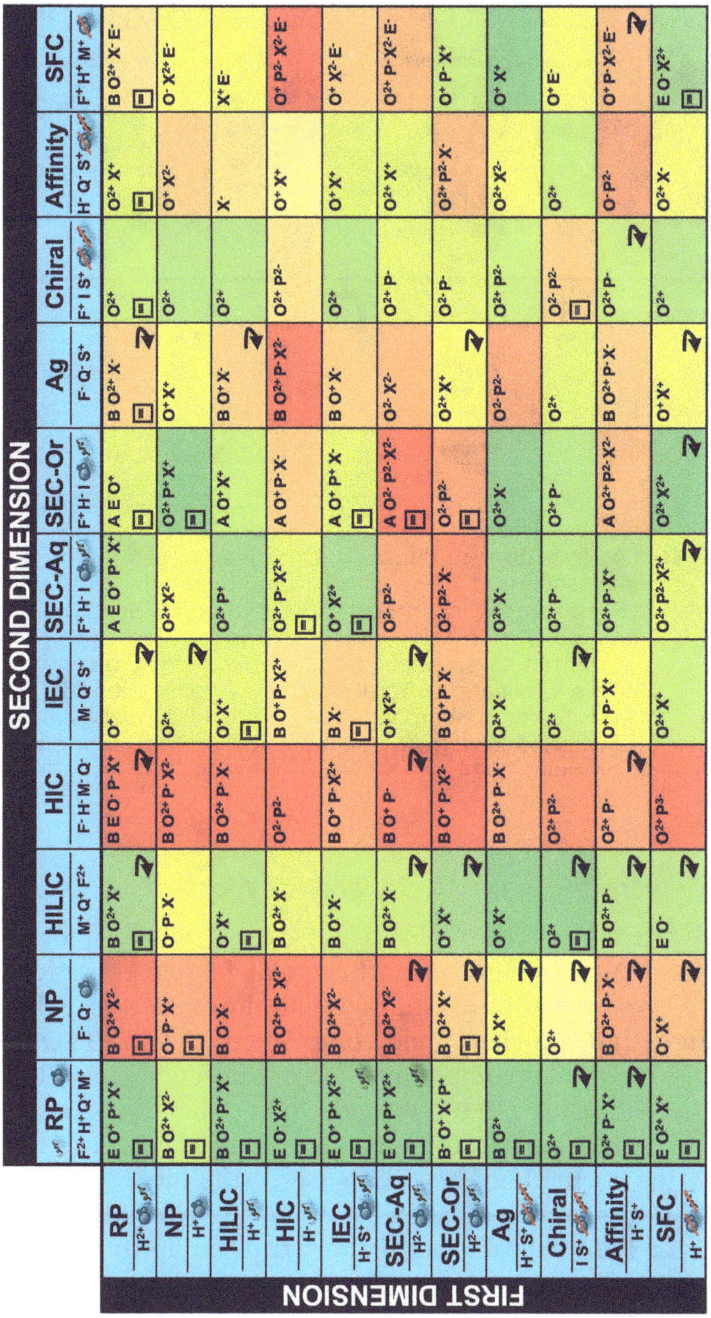

Figure 7.30 Overview of the possible online LC×LC combinations. Adapted from ref. 1, https://doi.org/10.1002/jssc.201700863, under the terms of the CC BY 4.0 license, https://creativecommons.org/licenses/by/4.0/.

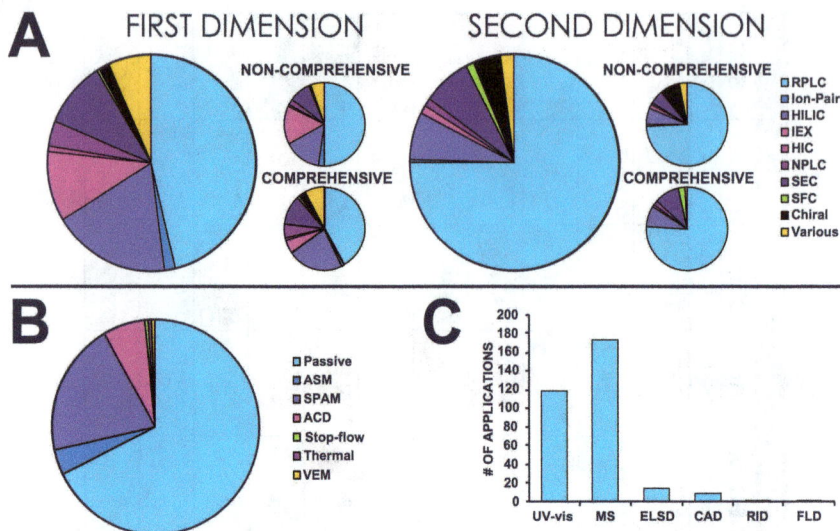

Figure 7.31 (A) Inventory of separation modes applied in the first (left) and second (right) dimensions between 2019 and 2023. (B) Inventory of modulation strategies used in all online applications. (C) Inventory of applied detection techniques (note that several detection techniques may be employed in a single application). Total number of applications: 240. See the website for a searchable database of all published applications. Adapted from ref. 59 with permission from Elsevier, Copyright 2024.

7.3.6.3 Normal-phase LC Separation Modes

In the category of normal-phase LC separation modes, **HILIC** (Section 3.6.2) is by far the most popular for incorporation in 2D-LC. It allows separation based on small differences in polarity and structure. HILIC provides good complementarity to RPLC methods because highly polar compounds that are unretained in RPLC show strong retention in HILIC, whereas non-polar analytes are unretained in HILIC but strongly retained in RPLC. HILIC is typically used as the ^{1}D method, both because of the attractive features of RPLC as a ^{2}D system and concerns about the re-equilibration times needed for HILIC separations (reported to be up to 20–30 column volumes, as opposed to one column volume for RPLC). However, such concerns have recently been rebutted.[60] A more significant concern has always been the compatibility of the HILIC effluent containing high fractions of acetonitrile, with the water-rich RPLC eluent. However, active-modulation techniques, such as SPAM and

ASM, have been shown to be useful for minimizing incompatibility effects. Figure 7.31 confirms the impression from Figure 7.30 that HILIC is a very attractive ^{1}D separation mode. An inventory of applications published between 1978 and 2019[1] revealed that 15% of LC×LC methods concerned HILIC×RPLC, while 7% of non-comprehensive methods involved HILIC-RPLC. Due to the choice of different stationary-phase chemistries available, there have even been applications that combine HILIC in both dimensions.

A much smaller number of recent applications (see Figure 7.31) employ **NPLC** (Section 3.6.1). When LC×LC just started to emerge, several research groups specifically targeted combinations of RPLC and NPLC due to the perceived high orthogonality of the two methods. The solvent incompatibility rendered this a very challenging combination, and this remained the case until the introduction of active modulation techniques. Since then, HILIC has gained popularity, and interest in "classical" NPLC has gradually dwindled. Today, NPLC is typically combined with SEC or with (non-aqueous) RPLC. Particularly well-known applications include lipids and other types of analytes with significant polar and non-polar moieties. NPLC×SEC is used for determining the mutually dependent functionality-type distributions and molecular weight distributions of synthetic polymers. While NPLC has been applied in both the first and second dimensions, it remains largely a ^{1}D method due to long re-equilibration times.

It is these same re-equilibration concerns that also cause **argentation LC** (AgLC, see Table 7.7), a special form of NPLC, to only be practical as a ^{1}D mode. The silver ions are non-covalently associated with the stationary phase, and this dynamic equilibrium can be disturbed by small quantities of various solvents, rendering the separation mode too unstable to be used as a ^{2}D method.

7.3.6.4 Two-dimensional Reversed-phase LC

Another clear conclusion we can draw from Figure 7.30 is that RPLC (Module 3.5) is an excellent ^{2}D-separation mode. This is attributed to its high resolution, good compatibility with mass spectrometry, and relatively short equilibration times. RPLC is the most popular separation mode in 2D-LC – as it is in 1D-LC. This is clearly apparent from Figure 7.31, where roughly 50% of all 2D-LC methods employ RPLC as the ^{1}D method, while as much as 75% use RPLC as the ^{2}D system. RPLC is particularly suited as a ^{2}D method due to the short equilibration times (as short as a single column volume), fast and high-efficiency separations, and compatibility with MS.

Roughly 35% of all published LC-LC and LC×LC methods up to 2019 were found to employ RPLC in both dimensions.[1] The rationale for this is that it is relatively easy to tailor RPLC selectivities to specific sample dimensions. In practice, orthogonality can be enhanced through the use of shifted ^{2}D gradients (see Section 7.3.5), different stationary phases (see Table 7.7), or different organic modifiers (*e.g.* acetonitrile in one dimension and methanol in the other). For ionogenic solutes, different values of the pH in the two dimensions may dramatically affect retention times and overall orthogonality. The charge state of the analytes and that of the stationary phase may be affected by changes in the pH.

7.3.6.5 Ion-exchange Chromatography

Ion-exchange chromatography (IEC, Section 3.7.1) shares many benefits and drawbacks with NPLC. When analytes in a sample feature a charge distribution, IEC is a very powerful separation method that is complementary to RPLC. However, the re-equilibration problems that hamper NPLC are also a notorious handicap of IEC. Often, a salt gradient is required to fully re-equilibrate (or "regenerate") the stationary phase. Another reason for IEC to be mainly a ^{1}D method is its incompatibility with several popular detectors (*e.g.* MS; evaporative light-scattering detector, ELSD). In this context, the second dimension, which is often RPLC, can be highly useful in enabling the hyphenation of IEC with MS. In the ^{2}D RPLC separation, all salts are typically unretained, and they can be flushed away and separated from the (retained) target analytes.

Combinations of IEC with separation modes with mobile phases that are high in organic solvents must be considered with great care due to incompatibility issues between (many) salts and (many) organic solvents.

7.3.6.6 Size-exclusion Chromatography

Size-exclusion chromatography (SEC, see Module 4.2) is attractive for 2D-LC applications involving the separation of large molecules. Since **SEC** is an isocratic method and since, therefore, no column equilibration is required between runs, it can be used as a ^{2}D system. This approach requires the use of short, efficient columns, which until recently were quite uncommon in SEC. Conventional large SEC columns are better suited for use in a ^{1}D method. Problems often arise when ^{1}D SEC is combined with an "interactive" ^{2}D system (see Module 4.3) due to the inherent mismatch between the strong solvent that is a prerequisite for SEC and the weak solvents encountered at the start of a (gradient-elution) ^{2}D run. Breakthrough effects (see Section 4.3.2) may be hard to avoid. Conversely, the presence of water in the ^{1}D effluent may cause adsorption effects when SEC with an organic solvent is used as a ^{2}D method.[61] When SEC is used in the second dimension, it is fruitful to deliberately shorten the modulation time to create overlapping injections because of the limited window in which peaks can appear in genuine SEC.[62]

7.3.6.7 Other Separation Modes

Chiral chromatography (Module 3.13) is a highly attractive mode for 2D-LC applications. Due to its ability to separate enantiomers and – in general – structurally similar molecules, chiral LC is highly complementary to other LC modes. Given the limited peak capacity obtained (and needed) from a chiral separation, it is a logical choice as a ^{2}D method. Such an application may be challenged by slow desorption kinetics, but recent reports suggest progress in this regard. The speed of the ^{2}D separation is much less of an issue in heart-cut 2D-LC, where chiral columns can be elegantly used for enantiomer separations. **Affinity chromatography** separations (Section 3.15.2) offer both very high selectivity and compatibility with other LC modes, but, like chiral separations, they suffer from slow desorption kinetics.

SFC (see Chapter 6) has also been combined several times with LC in a 2D-LC format. SFC can offer NPLC-type and HILIC-type selectivities without the typical hazardous solvents and with fewer compatibility issues. SFC has proven extremely useful for chiral separations. Moreover, thanks to the low viscosity of the mobile phase and the high mobile-phase diffusion coefficients, SFC allows fast and efficient separations, but as a ^{2}D method, SFC is susceptible to solvent mismatch effects. If the challenges associated with the interfacing of SFC to other separations can be overcome, it may gain popularity as one of the modes in 2D separations.[63]

Hydrophobic interaction chromatography (HIC, Section 3.15.1) is useful for the separation of proteins, but the high and variable salt concentrations render it only feasible as a ^{1}D method if the ^{2}D separation is to be combined with MS. HIC×RPLC is actually an elegant way to make HIC compatible with MS, with the salts employed in HIC gradients flushed away from the RPLC effluent before entering the MS. An attractive feature – aside from its unique selectivity – is that HIC is compatible with other LC modes that employ strictly aqueous eluents, such as IEC and aqueous SEC.

7.4   LC Hyphenated with GC (LC-GC) (A)

The advent of large-volume on-column injection (LVI) in GC, as described in Section 2.4.2.1, made it possible to hyphenate LC with GC (LC-GC). Ingenious coupling injection methods were developed with a well-controlled speed of injection (typically the flow rate from a narrow, *e.g.* 1 mm i.d., LC column), well-controlled conditions (gas flow rate, temperature) in a large retention gap, and concurrent evaporation of most or all of the injection volume (*i.e.* the LC mobile phase). Such an approach is quite elegant but hard to use on a routine basis. This is due to the intricate control needed, the contamination of the GC system with (non-volatile) components from the LC effluent, and, mostly, the desire of liquid chromatographers to use gradient elution RPLC for most of their separations. Online LC-GC with an on-column interface can only be optimized for a particular mobile-phase composition, and dealing with water is very difficult. Liquid water should absolutely be kept from the column, but even if it can be contained in the retention gap and good peak shapes are obtained, the long-term stability of such a system is questionable.

A PTV injector (see Section 2.4.3) offers a much easier and more robust way to realize an online coupling between LC and GC. The speed of injection is not critical and, importantly, PTV injectors can deal with aqueous samples, provided that the volatility of the analytes is sufficiently low to keep these in the PTV injector while water vapour is vented. Contamination of the system can be counteracted by heating the injector to a high temperature after a number of injections. In the worst case, which is encountered with non-volatile, highly stable contaminants, such as salts, it suffices to occasionally replace the liner. Changes in the mobile-phase composition (GC injection solvent) can be counteracted by programming different PTV conditions for different fractions – if necessary.

In all cases, irrespective of the interface between the GC and the LC, contaminants in the LC effluent with a similar volatility as the target analyte will show up in the chromatogram. When the target analytes are present at trace levels, contaminant peaks may dominate the chromatograms. Highly pure solvents are required for LC-GC. One example is the analysis of the low-molecular-weight fraction (residual monomers, additives) in online size-exclusion chromatography-GC (SEC-GC). THF is a preferred solvent, but it is hard to purify and sensitive to oxidation so that impurities are hard to avoid. This is less of an issue for the high-molecular-weight fraction (the polymer) analyzed with pyrolysis (see below), because in that case volatile contaminants can be vented off together with the eluent.

Since online LC-GC-MS is a demonstrated possibility and since (when using a PTV injector) it is not very difficult to perform, the main question that remains is when the technology is an attractive solution to analytical problems. With the second dimension being GC, all target analytes are necessarily volatile so that 2D-GC (GC-GC or GC×GC) is a realistic option. Therefore, we may rephrase the question to when online LC-GC (or comprehensive two-dimensional LC×GC) is attractive in comparison with GC-GC or GC×GC. We identify two situations in which this may be the case: (i) when LC provides unique (or at least special) selectivity, and (ii) when the analytes are transformed between the LC and GC stages so that non-volatile analytes are separated in the first dimension and volatile analytes in the second dimension.

The first category includes the use of LC as an online sample-preparation technique, for example, for determining low-molecular-weight compounds in polymers with SEC-GC-MS or hydrocarbon contaminants in vegetable oils with NPLC-GC-FID. More generally, LC-GC-FID is recommended for determining mineral-oil contamination in food samples. Another case in point for the special selectivity offered by LC is group-type separations (GTS; see Module 1.9). LC offers unique possibilities to highlight certain aspects of the molecule (or sample dimensions; see Section 7.1.4.2) that dominate the selectivity, while the effects of other sample dimensions are suppressed. Analyte volatility is always a dominant factor in determining retention in one-dimensional GC so that a complex sample always yields many peaks. A typical example is the separation of hydrocarbon samples by LC in one fraction (often in the form of a single peak) for alkanes ("saturates"), one for aromatics, and one for polyaromatics. Similar group-type separations may be realized by SFC. The three fractions can then be separated by GC (or, in a three-dimensional setup, by GC×GC), facilitating a clear interpretation of the data and accurate quantitation with flame-ionization detection (Section 2.5.1). There are compelling arguments for LC-GC (or SFC-GC) with a group-type separation in the first dimension, but there are no compelling reasons for performing such separations online. Unless many samples need to be processed and full automation is desirable, it is easy to collect fractions using a fraction collector and transport these to the autosampler of the GC. Ideally, the sample tray of the fraction collector should be interchangeable with that of the autosampler.

A key example of analyte transformation between a first-dimension LC separation and a second-dimension GC seperation includes pyrolysis (Py) as a form of reaction modulation. Pyrolysis will be discussed in Section 8.3.5 as a technique to break down high-molecular-weight samples, such as polymers, into characteristic fragments that allow qualitative analysis (usually with mass-spectro-metric detection) or quantitative analysis (often with FID). When separating a polymer by size-exclusion chromatography (see Module 4.2), the many peaks for molecules of a specific size are not fully separated and a broad envelope (*i.e.* distribution of peaks) is obtained. To obtain a quantitative picture of the variation in polymer composition (and even the sequence of monomers in the chain), many fractions from this fraction can be analyzed by Py-GC[64] (Figure 7.32). The wish to analyze many fractions for each sample and the concomitant desire to perform fast SEC-Py-GC are arguments in favour of an online coupling.

A few groups experimented with the hyphenation of LC and GC×GC, which is one of the few realistic three-dimensional chromatographic separations. For complex hydrocarbon samples, such as oil fractions[65] or coal tar,[66] LC can perform a group-type separation, with the resulting fractions analyzed in detail by GC×GC. This approach creates a set of structured chromatograms and generates a wealth of information, but the number of relevant fractions is such that offline coupling between the LC and GC parts of the system will usually be the most sensible approach.

7.5   Other 2D Techniques (A)

7.5.1   Two-dimensional Electrophoresis

Spatial two-dimensional gel electrophoresis 2D-GE (sometimes called **2D-PAGE**) is a prime example of a high-resolution spatial separation. In the first dimension, charged analytes (typically proteins) are separated according to their isoelectric point using

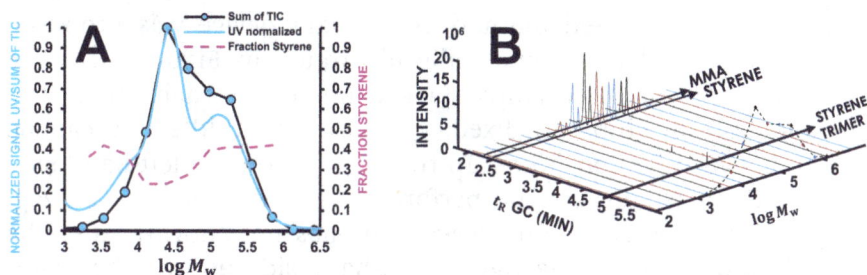

Figure 7.32 Example of the use of online comprehensive SEC×Py-GC-MS for determining the composition of a styrene–methyl methacrylate block copolymer (nominal composition: 24% styrene) as a function of molecular weight. (A) SEC chromatogram (light blue line), reconstructed chromatogram from (total-ion-current) MS data (dark interpolation line with points for each fraction analyzed), and the fraction of styrene as a function of molecular weight, calculated from the MS data. (B) Pyrolysis chromatograms for each fraction (direction from top left to bottom right) and superimposed molecular weight distribution determined from SEC (direction: from bottom left to top right). Adapted from ref. 64 with permission from Elsevier, Copyright 2022.

isoelectric focussing (IEF). After equilibrating the gel, the second-dimension separation takes place by polyacrylamide-gel electrophoresis (PAGE), often under denaturing conditions adding sodium dodecyl sulfate to the buffer (SDS-PAGE) or without a denaturing agent (native PAGE). The separated proteins are then visualized with silver or Coomassie (Brilliant) Blue. 2D-GE is an impressive technique that offers peak capacities of several thousand and allows hundreds of proteins to be detected. It also has limitations. Not all proteins (hydrophilic and hydrophobic, low molecular weight and high molecular weight, *etc.*) may be detected in a single run. The dynamic range of detection is limited to about three orders of magnitude, which makes it difficult to detect low-abundance proteins. Significant time is needed for sample preparation and analysis, typically amounting to 12 h per sample. Finally, coupling with MS is difficult and laborious. Separated proteins need to be collected from the spots, and the staining may complicate the MS analysis. This implies that protein separation methods that are easily compatible with MS are of interest, even if their overall separation power (peak capacity) does not match that of 2D-CE.

One important goal of the development of novel 2D-CE methods is to replace the conventional 2D-GE method with a more convenient and, ideally, automated method. An elegant device was proposed by Lu *et al.*[67] and is schematically illustrated in Figure 7.33. The device consists of three chips stacked on top of each other. The top and bottom chips are fixed, whereas the middle one can be moved between two positions. In the position on the left-hand side of the figure, ^{1}D IEF can be performed in the light blue zig-zagging channel (reservoirs and power supplies not shown). After the middle chip is slid to the right (right-hand-side figure), the major segments of the IEF channels are connected by the pink lines that allow eight ^{2}D capillary gel electrophoresis (CGE) separations to be performed simultaneously. While elegant, such a device cannot match the separation power and peak capacity of planar 2D-GE. In addition, moving the centre chip to exactly the right position proved an intricate process, for which a "micropositioner" was used.

Many capillary 2D-CE methods have been described, with different combinations of CE modes. The most common "modulator" in such systems is a gated interface, where the temporary passage of ^{1}D effluent to the ^{2}D capillary is controlled by switching voltages. Norman Dovichi's group at Notre Dame University (IN, USA) has successfully   demonstrated   a   variety   of   different   two-

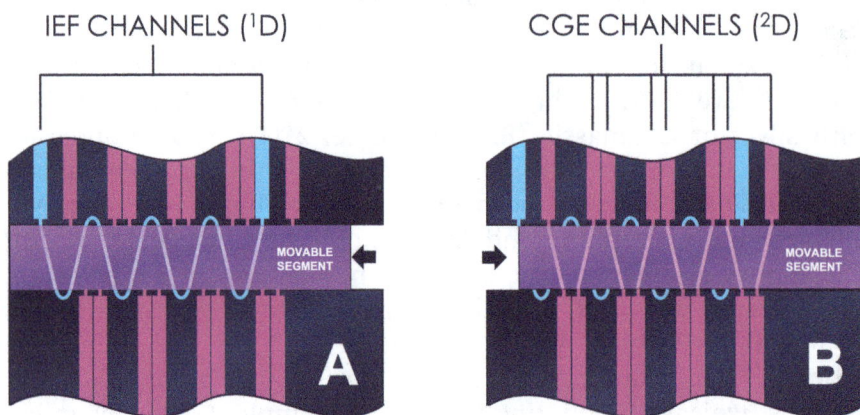

IEF CHANNELS (^{1}D)   CGE CHANNELS (^{2}D)

MOVABLE SEGMENT

A   B

Figure 7.33 Schematic representation of the microfluidic device by Lu *et al.* It consists of three chips, the top and bottom ones of which (black boxes) are fixed, while the middle one (purple) can be moved left and right between two positions (indicated by the arrows). In the position on the left, isoelectric focussing can be performed in the blue zig-zagged channel. In the position on the right, capillary gel electrophoresis (CGE) can be performed in eight parallel channels (dark pink). Adapted from ref. 67 with permission from American Chemical Society, Copyright 1995.

dimensional systems, including IEF in combination with capillary zone electrophoresis (IEF×CZE), capillary sieving electrophoresis (CSE) in combination with micellar electrokinetic chromatography (MEKC), *i.e.* CSE×MEKC, CSE×CZE, and more. Some of their applications involved single-cell analysis, exploiting one of the strengths of electromigration techniques in the analysis of very small samples. Two-dimensional protein profiles of single cells were obtained by CSE×MEKC. The typical detection method in such systems is fluorescence, after tagging of the proteins, but online 2D-CE-MS has also been demonstrated.

Two-dimensional separation methods may also be used to render CE compatible with ESI-MS. In this case, the ^{2}D separation mainly serves to separate analytes from components present in the ^{1}D separation buffer. Schlecht *et al.* described a conceptually very simple system, akin to valve-based heart-cut 2D-LC.[68] However, they needed to develop a special valve to transfer volumes as small as 4–20 nL. Metal components had to be avoided near the electrolyte solutions. Polyether ether ketone (PEEK) and polytetrafluoroethylene (PTFE) were used as materials to achieve electrical isolation, allowing the authors to apply potentials up to 15 kV. Successful direct coupling with ESI was demonstrated for a first-dimension CE method that contained tricine or ε-aminocaproic acid, both of which are incompatible with MS without a second-dimension cleanup stage.

Another strength of electromigration techniques has been exploited, in particular, by Hervé Cottet's group from the University of Montpellier (France). CE allows analyte zones to move back and forth to well-defined positions in the capillary, and Cottet's group devised several ways in which this could be used to realize single-column heart-cutting 2D-CE. A general approach,[69] which, in principle, works for all kinds of buffers and any kind of sample, consists of four steps: (i) A CZE separation is performed until the target zone almost reaches the end of the column. In this way, all analytes that migrate faster than the target analyte are washed away. (ii) The ^{1}D buffer in the outlet reservoir is replaced by the ^{2}D buffer, and the contents of the capillary are flushed away to the inlet capillary until the target zone almost leaves the column. All analytes that migrated slower than the target in the ^{1}D separation are washed away and heart-cutting has been achieved. (iii) The buffer in the inlet reservoir is replaced by the ^{2}D buffer. (iv) The ^{2}D separation is performed exclusively on the target zone.

7.5.2 Coupling Liquid Chromatography with Electromigration

In this section, we will focus on combinations of LC and CE. LC is the obvious choice for a first-dimension separation prior to CE (other than the 2D-CE methods discussed above) because GC and SFC options seem quite far-fetched. Using CE in the first dimension is extremely challenging. The low amounts of sample involved and the dilution of CE fractions in a subsequent LC separation would constitute a great challenge for LC detection systems (including MS; see ref. 68). There are some obvious advantages to using CE as a second separation stage in a heart-cut or comprehensive two-dimensional separation. CE and common LC techniques, such as RPLC, are quite orthogonal. In addition, CE can be extremely fast. One electromigration technique that may benefit strongly from a subsequent ^{2}D separation is (capillary) isoelectric focussing (see Section 5.1.4.4), which is a technique that allows filling the entire capillary with the sample (plus ampholyte solution) prior to starting the separation. IEF itself is barely compatible with MS detection due to the presence of ampholytes, acids, and bases. If it is followed by a second-dimension technique such as RPLC MS detection becomes instantly feasible.

7.5.2.1 Offline Chromatography-electromigration Systems

Offline combinations of various electromigration techniques are abundant in the scientific literature.[70] First-dimension separation by RPLC, size-exclusion chromatography (see Module 4.2), or ion-exchange or ion-pair chromatography (see Module 3.7) is followed by ^{2}D separations with CE, micellar electrokinetic chromatography (see Sections 5.1.4.6 and 5.4), capillary isoelectric focussing (CIEF, see Section 5.1.4.4), or capillary gel electrophoresis (CGE, see Section 5.1.4.2). Offline ^{1}D separations with IEF are usually followed by ^{2}D separations with RPLC-MS. Like all other offline approaches, the collected fractions can be treated prior to the second-dimension run. The most obvious action is the evaporation of the first-dimension effluent and reconstitution of the sample in the CE buffer. This will eliminate organic modifiers from RPLC effluents, which are generally unwelcome in electromigration methods. It should also help alleviate sensitivity issues. If enough samples are available, there is no need to miniaturize the (column diameter of) the LC system, and larger amounts of proteins may be obtained in each fraction. In sample-limited situations, large-volume stacking (see Section 5.2.2) may be another option to achieve sufficient sensitivity in the second dimension. The presence of (buffer) salts or ion-pair reagents in the ^{1}D effluent may affect the injection and the separation by electro-migration in the second dimension. In this context, IPC may be more attractive than IEC because the ionic strength of the eluent is generally lower. Like in all other offline couplings, automation or robotization is highly desirable. The time scales of the ^{1}D and ^{2}D separations are disconnected, so lengthy, high-resolution ^{2}D separa-tions are possible. These will, of course, lead to longer total analy-sis times. An interesting option to reduce the analysis time is to perform ^{2}D CE separations in parallel, in what has been dubbed capillary array electrophoresis (CAE).[71] Because CE is a high-voltage but low-amperage technique, a single power supply may serve a number of capillaries. However, the issue of the (lack of) robustness of the CE separations is aggravated. If there is a 90% chance that there are no capillary-related issues (and if we assume such issues to be independent), there is less than a 30% chance that an array of twelve capillaries will all work properly at the same time.

7.5.2.2 On-line Chromatography-electromigration Systems

Online comprehensive two-dimensional RPLC×CE was already described by Bushey and Jorgenson in 1990.[72] They used two 6-port valves for the interfacing. The effluent from the RPLC system filled a loop installed on the first valve. A second (low-pressure) pump then flushed the contents of this loop along the inlet (grounded side) of the CE capillary. Injection into the CE relied on the elec-tromigration of analytes from this flowing stream. Comprehensive two-dimensional separation of fluorescently labelled peptides from a tryptic digest of ovalbumin was demonstrated. A 250 mm × 1.0 mm RPLC column was operated at 10 µL min^{-1}, and a solvent gradient was run in 5 h. For analyzing a tryptic digest, a ^{2}D capillary with an internal diameter of 41 µm and an effective length (inlet to detector window) of 65 mm was used. An injection into the ^{2}D CE system was made every minute between 95 and 275 min, and 180 electrophero-grams were recorded.

To speed up the ^{2}D CE runs without losing separation power and peak capacity required a faster and more efficient way of injecting (a fraction of) the LC effluent into the CE system. To this end, valve-based interfaces have largely been replaced by gating systems that only allow the LC effluent to pass into the CE capillary for small periods of time when the "gate" opens. One example makes use

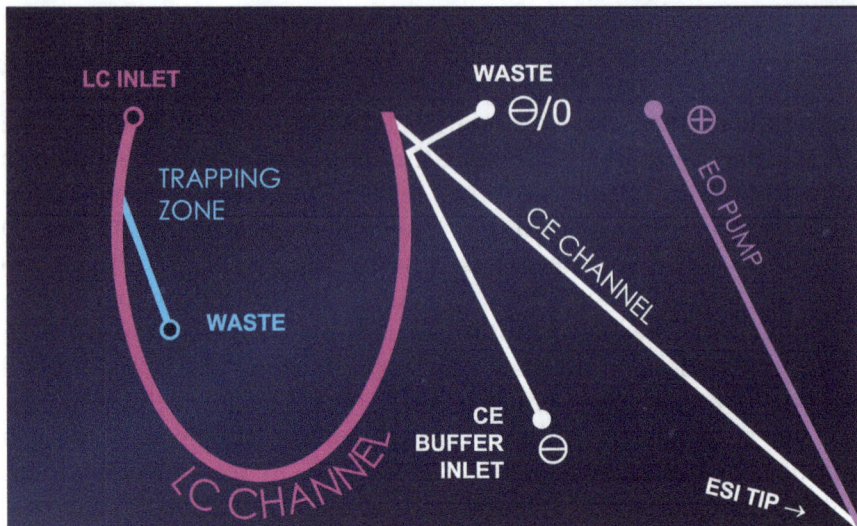

Figure 7.34   Schematic presentation of a glass microfluidic device suitable for online LC-CE–ESI-MS, as well as for LC-MS. The channels on the LC side of the chip are 25 µm deep × 250 µm wide; the channels on the CE side are 8 µm × 50 µm. The LC column is 10 mm (trapping zone) + 100 mm long; the CE channel is 50 mm long. The LC column is packed with 3.5 µm particles; the CE channels are coated with polyamine to obtain a stable (reversed) osmotic flow and to avoid the adsorption of positively charged peptides. The bottom-right corner of the tip is directly connected to a mass spectrometer. Adapted from ref. 73 with permission from American Chemical Society, Copyright 2011.

of a transverse flow. The effluent of the ^{1}D LC and the entrance of the ^{2}D CE are positioned directly opposite to each other in close proximity in a four-way junction. Most of the time, a relatively large transverse flow passes through the other two ports of the four-way junction, creating a strong flow between the LC and the CE. When the transverse flow is stopped for a short time, an injection is made into the CE. A disadvantage of gated systems is that most of the ^{1}D effluent, carrying most of the sample, is wasted. This causes fairly high mass detection limits.

As is the case with one-dimensional CE (see Section 5.1) and two-dimensional CE (see Section 7.5.1), online LC-CE systems can be elegantly implemented on a chip. A good example originates from the group of Mike Ramsey (University of North Carolina, Chapel Hill, NC, USA).[73] When the sample is loaded on the packed LC column (dark pink in Figure 7.34), the sample solvent and loading mobile phase exit through the light blue channel to waste (valve on the waste line open). This allows the concentration of the analytes in the trapping zone. Ramsey's group performed gradient-elution LC with the waste line closed at a flow rate of 65 nL min^{-1} and a pressure below 5 MPa after splitting the flow before the chip. The exit of the LC column was connected to the CE capillary. Rapid gated injections can take place by switching power at the waste line (top centre in Figure 7.34). When the voltage is zero, electro-osmotic flow takes the LC effluent together with the CE buffer to waste. When a negative voltage is briefly applied to the waste line, injection takes place. The bottom-right corner of the chip was cut to provide an electrospray ionization tip at the end of the CE channel. This allowed the chip to be directly connected to a mass spectrometer. The potential applied to the 50 mm long CE channel (from the junction where the injection took place to the ESI tip) was 5.6 kV. The chip is simple and elegant, but the LC pump and injector, the CE power supply, and the mass spectrometer are not integrated on the chip.

Electromigration methods can also be elegantly used for spatial separations in microfluidic devices or on TLC plates. IEF, free-flow electrophoresis (FFE), thin-layer electrophoresis, and thin-layer electrochromatography have all been studied.[70] An elegant example is the cyclic-olefin-copolymer (COC) IEF-LC chip described by Liu *et al.*[74] (Figure 7.35). The separation in the (dark pink) IEF channel results in five fractions that can be transported online to five LC columns operated in parallel. Matrix-assisted laser-desorption/ionization (MALDI) characterization of resulting peaks can be performed offline. The IEF channel used was 70 mm long, 180 µm deep, and 254 µm wide and was wall-coated with hydroxypropyl-methylcellulose prior to each analysis. The LC columns were 42 mm long, 130 µm deep, and 254 µm wide and were packed with 5 µm particles. The IEF separation was developed in 30 min at 1 kV. LC was performed with 30 min long gradients at a total flow rate of 2 µL min^{-1} (for all five columns combined). Detection was performed with a fluorescence microscope.

7.5.2.3 Comprehensive Three-dimensional SEC×RPLC×CE

The one three-dimensional (3D) liquid-phase separation that deserves attention in this textbook is an amazing experimental piece of work by Moore and Jorgenson.[75] The authors describe the separation of a very complex mixture of peptides (tryptic digest of hen ovalbumin). They performed a very slow SEC separation in about 8 h

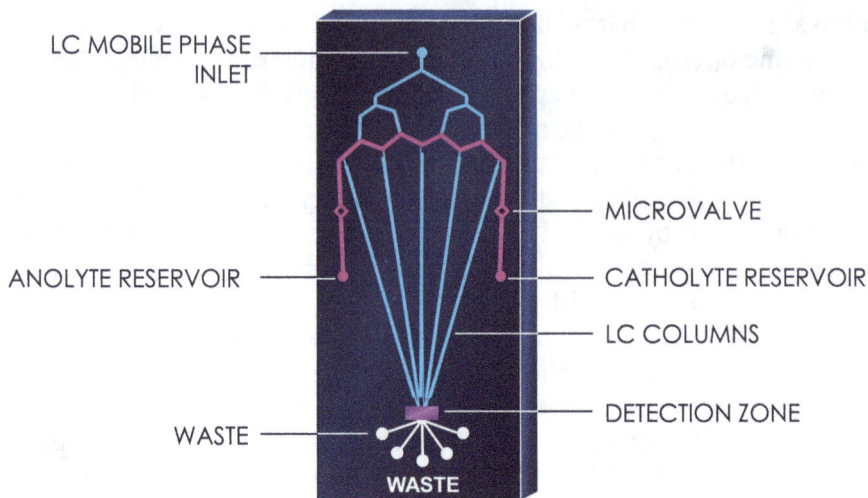

Figure 7.35 Schematic presentation of a microfluidic device suitable for online IEF-LC. IEF is first performed in the dark pink channel. After the 1D separation has been developed, five fractions are sent to parallel LC columns (light blue). Reproduced from ref. 74 with permission from the Royal Society of Chemistry.

(250 mm × 4.6 mm i.d. column, packed with 5 μm particles, operated with a mobile phase of 85% methanol in water at 11 μL min^{-1}, resulting in a backpressure of a mere 300 kPa) in the first dimension. They used gradient-elution RPLC in the second dimension with a total analysis time of 6.35 min (50 mm × 2.1 mm i.d. column, 5 μm particles), and a 10 : 1 flow splitter after the SEC column sent about 1 μL min^{-1} to the RPLC column (and 10 μL min^{-1} to waste). To avoid the breakthrough in the RPLC dimension, the ^{1}D effluent was diluted with 4 μL min^{-1} of water, resulting in an injection composition of about 17% methanol. The sample entered the CE continuously, but optical gating was realized by using a laser to destroy the fluorescent labels on the peptide almost all the time, except for a short period. The CE analysis time was about 2 s, but peaks eluted almost exclusively between 1.4 and 2 s (zone shown in Figure 7.36). The splitter between the first and second dimensions and the gating between the second and third dimensions imply that only a fraction of the injected peptides was ultimately detected. However, on-capillary fluorescence did provide sufficient sensitivity to obtain Figure 7.36, which shows a small fraction of the results. At each time point from the SEC effluent, a two-dimensional RPLC×CE picture is obtained, with every dark spot representing a peak. Peaks from successive time

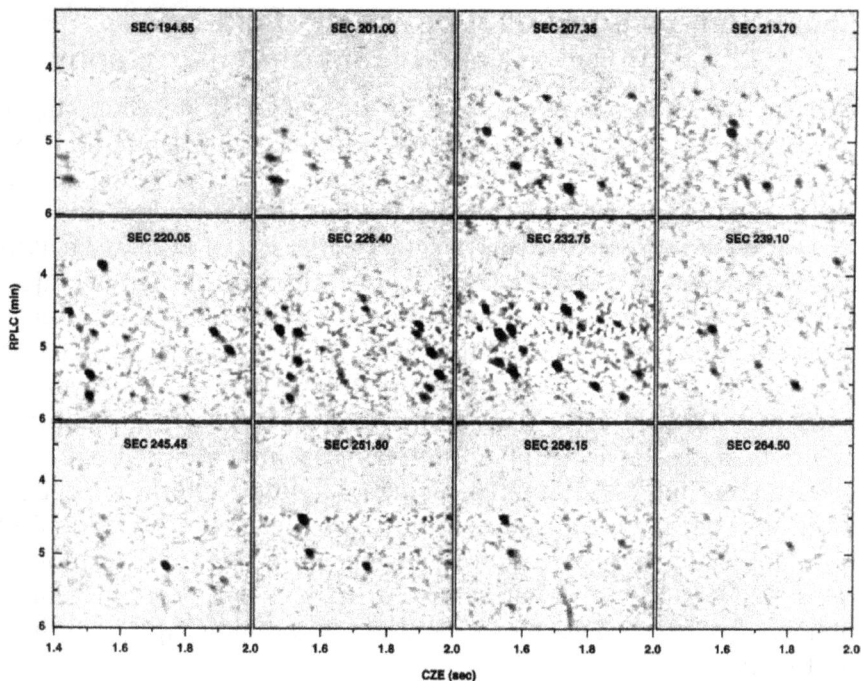

Figure 7.36 Selection of two-dimensional RPLC×CE plots obtained at twelve different time points (^{1}D SEC times listed at the top of each image). Adapted from ref. 75 with permission from American Chemical Society, Copyright 1995.

points need to be merged to obtain three-dimensional sample zones, with data analysis posing serious challenges.

Despite a relatively modest peak capacity of about 2800, the comprehensive 3D SEC×RPLC×CE separation achieved by Moore and Jorgenson is a formidable achievement. They managed to find ingenious solutions for many instrumental challenges, not least the coupling between the ^{2}D RPLC and ^{3}D CE separations, in a way that allowed very fast CE runs. In addition, completing the analysis of a single sample required some 8000 CE runs, and completing these is also a remarkable feat. It is hard to emulate, let alone improve such a separation.

7.5.3 Combinations of Supercritical-fluid Chromatography and Liquid Chromatography

SFC may be an interesting option for one of the separation stages in a heart-cut or comprehensive two-dimensional separation. As explained in Section 7.1.5.1, very different ("orthogonal") selectivities are required in two-dimensional separations. As explained in Chapter 6 (see Figure 7.9), the selectivity offered by SFC may be akin to that of NPLC or HILIC. Water may be tolerated in the SFC injection solvent (or in the first-dimension effluent in LC×SFC), and water may be a co-solvent in HILIC-like SFC. The interfacing of ^{1}D RPLC to ^{2}D SFC may be easier than it is for RPLC×HILIC and is certainly easier than RPLC×NPLC. In addition, SFC can be very fast, adding to its attractiveness as a ^{2}D separation. A disadvantage of the straightforward coupling of LC and SFC using a loop-type interface (Section 7.3.4.1) is that large transfer volumes easily lead to the broadening or deformation of the second-dimension peaks. Sun *et al.* exploited the use of a stationary-phase-assisted modulation (Section 7.3.4.3) interface for RPLC×SFC, but they found the trapping capacity to be a limiting factor.[76]

Given the success of SFC in chiral separations (see Section 6.2.2), combining RPLC and SFC for non-chiral and chiral analysis in one run appears an attractive option. Venkatramani *et al.* signalled that such a combination could be quite advantageous in the screening of active pharmaceutical ingredients (APIs) for both non-chiral and chiral impurities, and they developed a multiple-heart-cut RPLC-SFC system.[77] These authors also used a trapping-column (SPAM) modulation interface, transferring the analytes to the second-dimension column in backflush mode. This allowed them to reduce the transfer volume to 40 µL and to obtain good second-dimension peak shapes when using a first-dimension column of analytical dimensions (3 mm i.d.).

SFC as a first-dimension separation offers the potential advantage of easy evaporation of most of the first-dimension solvents in the modulator. In addition, the special group-type selectivity of SFC may be exploited. Isabelle François, then from Ghent University (Belgium), performed pioneering work on comprehensive two-dimensional separations involving SFC. SFC×RPLC MS was used for the detailed characterization of triacylglycerols (triglycerides) in fish oil.[78] GTS in the first dimension was according to the number of double bonds. Online comprehensive coupling was successfully realized, but in the end offline coupling was preferred, as it allowed longer analysis times for the ^{2}D non-aqueous reversed-phase (NARP) separations, resulting in a higher overall peak capacity.

Acknowledgements

Prof. Dr. Jan Christensen, Prof. Dr. Peter Tranchida and Prof. Dr. Dwight Stoll are acknowledged for their review of the chapter, feedback, and fruitful discussions.

Recommended Reading

D. R. Stoll and P. W. Carr, in *Multi-Dimensional Liquid Chromatography Principles, Practice, and Applications,* CRC Press, Boca Raton, 2023.

References

1. B. W. J. Pirok, A. F. G. Gargano and P. J. Schoenmakers, *J. Sep. Sci.*, 2018, 41.
2. P. Schoenmakers, P. Marriott and J. Beens, *LC-GC Eur.*, 2003, **16**, 1–4.
3. P. J. Marriott, P. Schoenmakers and Z. Y. Wu, *LC-GC Eur.*, 2012, **25**(5), 266–275.
4. E. Davydova, P. J. Schoenmakers and G. Vivó-Truyols, *J. Chromatogr. A*, 2013, **1271**, 137–43.
5. J. C. Giddings, *J. Chromatogr. A*, 1995, **703**, 3–15.
6. B. W. J. Pirok, M. J. den Uijl, G. Moro, S. V. J. Berbers, C. J. M. Croes, M. R. van Bommel and P. J. Schoenmakers, *Anal. Chem.*, 2019, **91**, 3062–3069.
7. J. M. Davis, D. R. Stoll and P. W. Carr, *Anal. Chem.*, 2008, **80**(2), 461–473.
8. D. R. Stoll, X. Wang and P. W. Carr, *Anal. Chem.*, 2008, **80**, 268–278.
9. G. Vivó-Truyols and P. J. Schoenmakers, *Anal. Chem.*, 2010, **82**, 8525–8536.
10. P. J. Schoenmakers, G. Vivó-Truyols and W. M. C. Decrop, *J. Chromatogr. A*, 2006, **1120**, 282–90.
11. J. Lankelma and H. Poppe, *J. Chromatogr. A*, 1978, **149**, 587–598.
12. D. R. Deans, *Chromatographia*, 1968, **1**, 18–22.
13. M. R. Jacobs, R. Gras, P. N. Nesterenko, J. Luong and R. A. Shellie, *J. Chromatogr. A*, 2015, **1421**, 123–128.
14. P. Boeker, J. Leppert, B. Mysliwietz and P. S. Lammers, *Anal. Chem.*, 2013, **85**, 9021–30.
15. Z. Liu and J. B. Phillips, *J. Chromatogr. Sci.*, 1991, **29**, 227–231.
16. G. Vivó-Truyols and H.-G. Janssen, *J. Chromatogr. A*, 2010, **1217**, 1375–1385.
17. S. Peters, G. Vivó-Truyols, P. J. Marriott and P. J. Schoenmakers, *J. Chromatogr. A*, 2007, **1156**, 14–24.
18. L. M. Blumberg and M. S. Klee, *Anal. Chem.*, 1998, **70**, 3828–3839.
19. J. B. Phillips and E. B. Ledford, *Field Anal. Chem. Technol.*, 1996, **1**, 23–29.
20. J. Beens, J. Blomberg and P. J. Schoenmakers, *J. Sep. Sci.*, 2000, **23**, 182–188.
21. P. J. Marriott and R. M. Kinghorn, *Anal. Chem.*, 1997, **69**, 2582–2588.

22. B. Giocastro, M. Zoccali, P. Q. Tranchida and L. Mondello, *J. Sep. Sci.*, 2021, **44**, 1923–1930.
23. J.-F. Focant, A. Sjödin and D. G. Patterson, *J. Chromatogr. A*, 2003, **1019**, 143–156.
24. R. B. Gaines and G. S. Frysinger, *J. Sep. Sci.*, 2004, **27**, 380–388.
25. M. J. Wilde and S. J. Rowland, *Anal. Chem.*, 2015, **87**, 8457–8465.
26. A. Ghosh, C. T. Bates, S. K. Seeley and J. V. Seeley, *J. Chromatogr. A*, 2013, **1291**, 146–154.
27. G. Hopfgartner, K. Bean, J. Henion and R. Henry, *J. Chromatogr. A*, 1993, **647**, 51–61.
28. N. B. L. Milani, B. W. J. Pirok and P. J. Schoenmakers, *J. Sep. Sci.*, 2023, 46.
29. C. Vendeuvre, R. Ruiz-Guerrero, F. Bertoncini, L. Duval, D. Thiébaut and M. C. Hennion, *J. Chromatogr. A*, 2005, **1086**, 21–28.
30. L. Rohrschneider, *J. Chromatogr. A*, 1966, **22**, 6–22.
31. M. Ajam, J. Beens and P. Sandra, *J. Chromatogr. A*, 2011, **1218**, 4478–4486.
32. B. W. J. Pirok, D. R. Stoll and P. J. Schoenmakers, *Anal. Chem.*, 2019, **91**, 240–263.
33. J. F. K. Huber, E. Ecker and M. Oreans, *J. Chromatogr. A*, 1973, **83**, 267–277.
34. F. Erni and R. W. Frei, *J. Chromatogr. A*, 1978, **149**, 561–569.
35. M. M. Bushey and J. W. Jorgenson, *Anal. Chem.*, 1990, **62**, 161–167.
36. S. T. Balke, *Sep. Purif. Methods*, 1982, **11**, 1–28.
37. P. Kilz, *Chromatographia*, 2004, **59**, 3–14.
38. J. Adrian, D. Braun and H. Pasch, *Die Angew. Makromol. Chem.*, 1999, **267**, 82–88.
39. P. J. Schoenmakers, *J. Chromatogr. A*, 2003, **1000**, 693–709.
40. D. R. Stoll and P. W. Carr, in *Multi-Dimensional Liquid Chromatography Principles, Practice, and Applications*, CRC Press, Boca Raton, 2023.
41. G. Groeneveld, M. N. Dunkle, M. Rinken, A. F. G. Gargano, M. Pursch, E. P. C. Mes and P. J. Schoenmakers, *J. Chromatogr. A*, 2018, **1569**, 128–138.
42. R. S. Van Den Hurk, N. Abdulhussain, S. A. Van Beurden, M. E. Dekker, A. Hulsbergen, R. A. H. Peters, B. W. J. Pirok and A. C. Van Asten, *J. Chromatogr. A*, 2022, **1672**, 463072.
43. B. Wouters, E. Davydova, S. Wouters, G. Vivo-Truyols, P. J. Schoenmakers and S. Eeltink, *Lab Chip*, 2015, **15**, 4415–4422.
44. D. R. Stoll, K. Shoykhet, P. Petersson and S. Buckenmaier, *Anal. Chem.*, 2017, **89**, 9260–9267.
45. D. R. Stoll, D. C. Harmes, G. O. Staples, O. G. Potter, C. T. Dammann, D. Guillarme and A. Beck, *Anal. Chem.*, 2018, **90**, 5923–5929.
46. C. R. Mallet, Z. Lu, J. Mazzeo and U. Neue, *Rapid Commun. Mass Spectrom.*, 2002, **16**, 805–813.
47. Y. Chen, L. Montero, J. Luo, J. Li and O. J. Schmitz, *Anal. Bioanal. Chem.*, 2020, **412**, 1483–1495.
48. A. Baglai, M. H. Blokland, H. G. J. Mol, A. F. G. Gargano and P. J. Schoenmakers, *Anal. Chim. Acta*, 2018, **1013**, 87–97.
49. L. E. Niezen, B. B. P. Staal, C. Lang, B. W. J. Pirok and P. J. Schoenmakers, *J. Chromatogr. A*, 2021, **1653**, 462429.
50. H. Tian, J. Xu and Y. Guan, *J. Sep. Sci.*, 2008, **31**, 1677–1685.
51. E. Fornells, B. Barnett, M. Bailey, E. F. Hilder, R. A. Shellie and M. C. Breadmore, *Anal. Chim. Acta*, 2018, **1000**, 303–309.
52. G. Groeneveld, B. W. J. Pirok and P. J. Schoenmakers, *Faraday. Discuss.*, 2019, **218**, 72–100.
53. S. Chapel, F. Rouvière, S. Heinisch and J. Chromatogr, *J. Chromatogr. B*, 2022, **1212**, 123512.

54. R. S. van den Hurk, B. Lagerwaard, N. J. Terlouw, M. Sun, J. J. Tieleman, A. X. Verstegen, S. Samanipour, B. W. J. Pirok and A. F. G. Gargano, *Anal. Chem.*, 2024, **96**, 9294–9301.
55. F. Bedani, W. T. Kok and J. G. M. Janssen, *J. Chromatogr. A*, 2006, **1133**, 126–134.
56. J. Xu, D. Sun-Waterhouse, C. Qiu, M. Zhao, B. Sun, L. Lin and G. Su, *J. Chromatogr. A*, 2017, **1521**, 80–89.
57. S. S. Jakobsen, J. H. Christensen, S. Verdier, C. R. Mallet and N. J. Nielsen, *Anal. Chem.*, 2017, **89**, 8723–8730.
58. B. Zhan, J. R. Yates, M.-C. Baek, Y. Zhang and B. R. Fonslow, *Chem. Rev.*, 2013, **113**, 2343–2394.
59. R. S. van den Hurk, M. Pursch, D. R. Stoll and B. W. J. Pirok, *TrAC, Trends Anal. Chem.*, 2023, **166**, 117166.
60. C. Seidl, D. S. Bell and D. R. Stoll, *J. Chromatogr. A*, 2019, **1604**, 460484.
61. E. Reingruber, J. J. Jansen, W. Buchberger and P. Schoenmakers, *J. Chromatogr. A*, 2011, **1218**, 1147–52.
62. B. W. J. Pirok, N. Abdulhussain, T. Aalbers, B. Wouters, R. A. H. Peters and P. J. Schoenmakers, *Anal. Chem.*, 2017, **89**, 9167–9174.
63. M. Burlet-Parendel and K. Faure, *TrAC, Trends Anal. Chem.*, 2021, **144**, 116422.
64. W. C. Knol, J. P. H. Smeets, T. Gruendling, B. W. J. Pirok and R. A. H. Peters, *J. Chromatogr. A*, 2023, **1690**, 463800.
65. R. Edam, J. Blomberg, H.-G. Janssen and P. J. Schoenmakers, *J. Chromatogr. A*, 2005, **1086**, 12–20.
66. M. Zoccali, P. Q. Tranchida and L. Mondello, *Anal. Chem.*, 2015, **87**, 1911–1918.
67. J. J. Lu, S. Wang, G. Li, W. Wang, Q. Pu and S. Liu, *Anal. Chem.*, 2012, **84**, 7001–7007.
68. J. Schlecht, K. Jooß and C. Neusüß, *Anal. Bioanal. Chem.*, 2018, **410**, 6353–6359.
69. S. Anouti, O. Vandenabeele-Trambouze, D. Koval and H. Cottet, *Anal. Chem.*, 2008, **80**, 1730–1736.
70. L. Ranjbar, J. P. Foley and M. C. Breadmore, *Anal. Chim. Acta*, 2017, **950**, 7–31.
71. C. A. Emrich, H. Tian, I. L. Medintz and R. A. Mathies, *Anal. Chem.*, 2002, **74**(19), 5076–5083.
72. M. M. Bushey and J. W. Jorgenson, *Anal. Chem.*, 1990, **62**, 978–984.
73. A. G. Chambers, J. S. Mellors, W. H. Henley and J. M. Ramsey, *Anal. Chem.*, 2011, **83**, 842–849.
74. J. Liu, C.-F. Chen, S. Yang, C.-C. Chang and D. L. DeVoe, *Lab Chip*, 2010, **10**, 2122.
75. A. W. Moore and J. W. Jorgenson, *Anal. Chem.*, 1995, **67**, 3456–3463.
76. M. Sun, M. Sandahl and C. Turner, *J. Chromatogr. A*, 2018, **1541**, 21–30.
77. C. J. Venkatramani, M. Al-Sayah, G. Li, M. Goel, J. Girotti, L. Zang, L. Wigman, P. Yehl and N. Chetwyn, *Talanta*, 2016, **148**, 548–555.
78. I. François, A. Dos Santos Pereira and P. Sandra, *J. Sep. Sci.*, 2010, **33**, 1504–1512.

8 Sample Preparation

In this chapter we present an overview of important sample preparation techniques. Note that this is necessarily incomplete and that, even more than for analytical separations, the nature of the sample dictates the required actions and procedures. After a general introduction, the main part of the chapter is devoted to extractions using various types of extractants (gas, liquid, and supercritical fluid) and in various implementations. Solid-phase extraction (SPE) and solid-phase micro-extraction (SPME) are major examples of conventional and miniaturized extractions, respectively. The "greenness" of extractions is discussed in the overview. Derivatization (*i.e.* chemical modification) and fragmentation (or "breakdown") of analytes are discussed in the final module (Figure 8.1).

8.1 Introduction to Sample Preparation (B)

8.1.1 Aim of Sample Preparation

This book deals with analytical separations. At various points, it also describes the analytical measurements that take place after the separation, because without good detectors (see, for example, Modules 2.5 and 3.9) or hyphenation with other analytical instruments (see Modules 2.6, 3.10 and 4.6), analytical separations are essentially meaningless. After the separation, a method is needed to identify and/or quantify the analytes. Such measurements are much easier – and often better – if the analytes are first separated. Figure 8.2 schematically illustrates the roles of analytical separations and sample preparation. The objective of analytical chemistry is to provide qualitative and quantitative information on a sample

Analytical Separation Science
By Bob W. J. Pirok and Peter J. Schoenmakers
© Bob W. J. Pirok and Peter J. Schoenmakers 2025
Published by the Royal Society of Chemistry, www.rsc.org

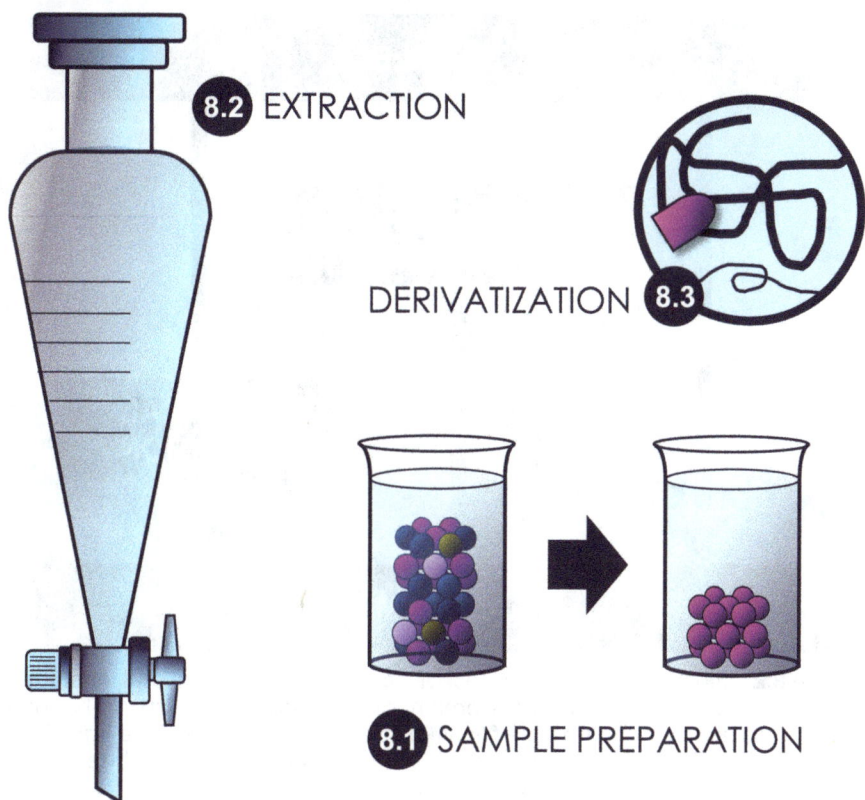

Figure 8.1 Graphical overview of the modules in this chapter.

and its constituents. This can be achieved by performing a direct measurement of the sample, as indicated by the left downward arrow in Figure 8.2. Such direct measurements may be performed on relatively simple samples, such as active pharmaceutical ingredients (APIs) or industrial chemicals. In such cases, many physical (*e.g.* spectroscopic) methods can be used to characterize the sample. Direct measurements may also be successful if they are highly selective or even specific (*i.e.* 100% selective) and sufficiently sensitive for certain analytes. For example, high-energy elemental analysis methods, such as atomic-absorption spectroscopy (AAS), inductively coupled plasma-optical emission spectroscopy (ICP-OES) or ICP-mass spectrometry (ICP-MS), neutron activation analysis (NAA), and, with good software in place to correct for matrix effects, X-ray fluorescence spectroscopy (XRF), are sufficiently selective to allow direct measurement of the concentrations of many specific elements.

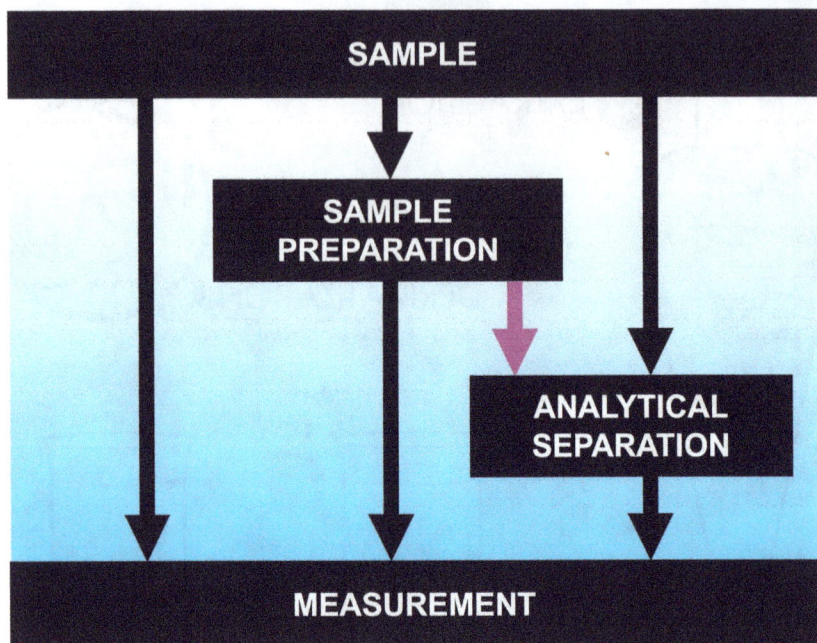

Figure 8.2   Illustration of the roles of sample preparation and analytical separations in the domain of analytical chemistry. The pink arrow indicates the focus of the present chapter. The light-blue shading indicates a rough scale from the sample (white) to the final results of the analysis (blue).

There has also been a trend to introduce complex samples directly into mass spectrometers. This, in principle, allows very fast analyses of complex samples. One reason why such direct introduction MS methods may be successful is that mass spectrometry can also play the role of an analytical separation method. If soft ionization methods are employed, so that the analyte molecules are ionized but otherwise remain intact, it may be argued that the MS instrument separates the analytes based on the mass-to-charge (m/z) ratio of the corresponding ions. Their molar mass may then be measured accurately, and additional characterization may be performed by subsequent fragmentation and characterization of the ions in tandem MS (MS/MS) or multi-stage (MS^n) methods. One issue with this is that calibration is required, usually with the aid of reference standards. This is not unlike the needs of other measurement methods, except that MS methods may require more frequent calibration. This is because MS sensitivity depends on a chemical process – the ionization stage – with a significant inherent variation.

In contrast, spectroscopic measurements rely on highly reproducible physical phenomena. MS methods can be highly selective, thanks to high-resolution mass spectrometry (HRMS) instruments, MS/MS and MS^n methods, and the use of ion-mobility spectrometry (IMS) as a selective filter for ions of equal mass but with different geometric conformations (collision cross-sections). However, the ionization stage is a weak point in direct MS analysis. All ionization methods suffer from significant bias if many different analytes or analytes in very different concentrations are simultaneously introduced. Such effects, known generally as **matrix effects**, or more specifically as ion suppression (or the opposite, ion enhancement) effects, are difficult to correct for and may be hard to diagnose without resorting to sample preparation strategies.

The vertical arrow in the centre of Figure 8.2 indicates the second option, which includes a sample preparation stage before the analytical measurement (detection). The most important sample preparation options will be discussed in this chapter. On the right-hand side are analytical separations, which can be seen as the most advanced and most selective collection of sample preparation methods. In this chapter, we will use the word "sample" to describe the untreated sample and the word "extract" to refer to the result of the sample preparation process. The chromatographic and other analytical separation methods described in this book have the potential to narrow down the **extract** submitted for actual measurement to one or a few analytes simultaneously, thereby drastically reducing the demands on the measurement methods and the chance of bias in the results. For very complex samples, such as whole blood, soil, or combined food products, it may be necessary to perform initial sample preparation steps prior to chromatographic separation. This is indicated by the pink arrow in Figure 8.2. Processes that follow this pink arrow are the main subject of this chapter. The chapter is necessarily brief. For more detailed descriptions, the reader is referred to pertinent literature.[1-3] The main objectives of sample preparation and analytical separation, following the path indicated by the pink arrow in Figure 8.2, are summarized in Table 8.1.

The separation goal is much more ambitious for the analytical separation stage because of the much greater separation power (efficiency). The separation time can be reduced if highly retained compounds are removed at the sample preparation stage. Removing unwanted compounds at the sample preparation stage can also increase the robustness of the analytical separation. Likewise,

Table 8.1 Objectives of sample preparation and analytical separation strategies, following the path in Figure 8.2 indicated by the pink arrow. Commonly, the analytical measurement after the separation takes place in a detector.

Sample preparation	Analytical separation
Create an "extract" in a suitable state for introduction into the analytical separation system	—
Simplify the sample prior to analytical separation	Separate all relevant analytes, ideally into single-component peaks
Remove matrix components that are detrimental to the subsequent analytical separation (*e.g.* strong acids and bases), build up on the analytical column (strong adsorbers), or increase the analysis time	Remove matrix components that are detrimental to the subsequent analytical measurement (*e.g.* salts)
Maximize the robustness of the analytical separation	Maximize the accuracy and robustness of the analytical measurement
Increase the concentration of relevant analytes in the "extract" to submit for analytical separation	Maximize the concentration of analytes in the effluent (peak height)

removing unwanted compounds at the analytical separation stage can avoid complications and downtime at the measurement stage, for example, by keeping salts out of a mass spectrometer. The accuracy of the analytical measurement may be improved by removing interferences. In the case of MS measurements, matrix effects (*i.e.* ion suppression or ion enhancement) may be reduced. The precision of the measurement may be increased, but the overall precision of the method is determined by variations in the sample preparation, analytical separation, and the measurement. Enriching the sample to increase the concentration ("**preconcentration**") of the relevant analyte(s) is one of the main objectives of sample preparation. During an analytical separation, analyte concentrations typically decrease, but it is possible to increase analyte concentrations, especially during the injection, for example, by injecting large samples in a weak eluent prior to the start of a gradient in liquid chromatography (LC).

In assessing and optimizing an analytical method, it is fair to include the necessary sample preparation. The time needed per analysis is the sum of the time required for sample preparation and for analytical separation, including the analytical measurement (online detection or offline characterization). In addition, the time needed to process and interpret the data must be considered. Direct MS measurements may seem fast, but – apart from a possible bias – they may produce complex results that require time and expertise to interpret. However, this latter time is often hard to quantify. It may also depend greatly on available software or computational tools and skills.

8.1.2 Sampling and Sample Pre-treatment

Ideally, samples are homogeneous. This may be the case for gaseous samples from a confined space because of the high diffusion coefficients in gases. However, the composition may vary with time, and in large systems (the ambient atmosphere), the composition clearly varies with location. Air samples may be taken in two fundamentally different ways. In **passive gas sampling**, an adsorbent contained in a tube or disc (or an SPME fibre, see Section 8.2.3.1) is exposed to air for a typically long period of time. Such devices may, for example, be used to monitor the total exposure of factory workers to toxic gases during the sampling period. Sampling is very easy and non-invasive, but a disadvantage is that incidental, very high concentrations may go unnoticed in a long-term average. In **active gas sampling,** a pump is used to suck a measured amount of gas through a filter (for characterizing particulate matter) or through an adsorbent (to capture analyte molecules). The collected samples can be extracted from the sorbent with a liquid or brought into a headspace (see Section 2.4.4) or thermal desorption unit (see Section 2.4.5). Polyurethane foams (PUFs) are a specific type of adsorbent with a high permeability and a large surface area, which make them attractive for use in air sampling devices. However, PUFs occupy a relatively large volume, and they need to be extracted to recover the analytes (*e.g.* using Soxhlet extraction; see Section 8.2.1.5). An SPME fibre (housed in an appropriate syringe) allows direct injection of the collected sample into a gas chromatography (GC) or MS instrument.

When analysing bottled water for mineral content, **homogeneity** of the sample also seems like a fair assumption, although even for such a seemingly clear sample, adsorption on or ion exchange with the wall of the bottle may affect the analysis. If the water sample originates from a river (*i.e.* the river is the subject for analysis), there is likely spatial variation (across the length of the river, the distance from the wall, and the depth of sampling) and temporal variation (*e.g.* between seasons and between dry and rainy periods).

More often than not, the subject of our sampling should be treated as heterogeneous. It is important to choose and document the time and place of sampling carefully. To get an impression of the entire subject, such as the pollution of a river, we may need to take many "grab samples" in a judicious manner. In a more controlled environment, such as a wastewater treatment plant, we may accumulate a sample over a longer time. A flow-proportional sample, accumulating a constant amount of sample per unit volume, is most representative of a given (large) volume of the wastewater stream, but it is difficult to design and implement methods to collect such samples. A time-proportional sample, accumulating a constant amount of sample per unit time, is easier to take. Together with the total flow across the sampling period, it provides a reasonable indication of the total amounts of analytes that passed the measurement point during the period. It may be necessary to **stabilize the sample** to avoid the development of (increased) **heterogeneity** (*e.g.* through precipitation or coagulation) or the occurrence of chemical reactions. Examples include the addition of buffers for pH control in water samples or the addition of anticoagulants to blood samples (see below).

The samples we take from the subject under investigation may, in turn, also be heterogeneous. Unlike the initial subject, these samples are likely small enough to be homogenized before analysis. For example, food samples may be homogenized in a blender. Even then, the resulting sample is likely heterogeneous (*e.g.* an emulsion or a suspension), and the next sample preparation steps need to be conducted rapidly, before large-scale phase separation (*e.g.* flocculation, creaming, or sedimentation) occurs. If solid particles are not relevant, they may be removed by **filtration**. This is one of the most frequent sample preparation steps in conjunction with analytical separations because introducing particles may perturb or clog up the separation system. However, the assumption that the solid particles are irrelevant for the analysis may be invalidated by adsorption or inclusion of analytes on or in the solid particles.

When the solid particles differ in density from the liquid, they may be left to rise to the top ("creaming") or sink to the bottom ("sedimentation") of the sample. Both processes can be drastically accelerated by centrifugation. The extent to which such acceleration occurs is – apart from the density difference – determined by the speed (in rotations per minute, rpm). At extremely high speeds, even molecules can be separated by centrifugation, a process known as ultracentrifugation. Ultracentrifugation can be used to separate polymers based on molecular weight, but the process is very slow (*e.g.* 24 h), and the characterization of the sample after completion of the ultracentrifugation process is cumbersome. Ultracentrifugation is used to enrich uranium by separating $^{235}UF_6$ from $^{238}UF_6$ molecules, but this process is not intended for analytical purposes and is far outside the scope of the present textbook.

Biofluids are extremely important samples for medical research, pharmaceutical testing, and clinical diagnosis. Blood is often the best sample to address the relevant questions, but it is extremely complicated. Whole blood is very rarely subjected to analytical separations. Most realistically, field-flow fractionation (FFF) techniques have been applied for separating and studying blood cells, with the addition of an anticoagulant being the only required sample pre-treatment step. Serum or plasma samples may be subjected to chromatographic separation. **Serum** is obtained after clotting is allowed to occur by exposing a fresh blood sample to air. The red and white blood cells and platelets form a solid mass together with the clotting factors. Usually, subsequent centrifugation is performed to obtain serum as a clear supernatant. Clotting does not occur when an anticoagulant is added to the blood sample. Centrifugation also results in a clear, yellowish supernatant, known as blood **plasma**. In contrast with serum, plasma still contains clotting factors, such as fibrinogen and proteins involved in the clotting process. Plasma may also be obtained by filtration of the sample. This is a more realistic option for rapid analysis, such as point-of-care diagnosis.

Both serum and plasma are still very complicated samples that may require further preparation steps, as they contain salts, lipids and other metabolites, peptides, proteins, antibodies, *etc.* Depending on the objective of the analysis and the limitations of the method, some of these components may need to be removed. However, each sample preparation step entails a risk of altering the sample (*e.g.* through oxidation) and affecting the presence or concentration of relevant analytes. For example, proteins can be removed from the sample through **precipitation** by adding a water-miscible organic solvent, such as methanol or acetonitrile, or a water-soluble polymer, such as poly(ethylene glycol) (PEG). However, small molecules, such as drugs, may be attached to proteins and may thus be removed from the sample. When the aim is to study proteins, the concentration of albumin tends to be so high that it obstructs the study of other proteins. Therefore, an albumin-depletion step may be included, for example, using free-flow electrophoresis (see Chapter 5).

In recent years, the **dried-blood-spot** (DBS) sampling method has received considerable renewed interest, thanks in large part to the advent of highly sensitive and miniaturized methods for separation and characterization. The latter are needed because very small amounts of blood are collected – most famously from the heels of babies. Drops of blood are collected and left to dry on filter paper, after which the collected samples can be easily transported (*e.g.* sent by mail to a laboratory). The spots may then be extracted with a suitable liquid for injection into a chromatographic instrument. The dried-blood-spot method shows lower quantitative precision and accuracy than liquid sampling methods but often suffices as an indicative method for diagnosis. For clinical studies, more-quantitative data are usually required.

To subject **solid samples** to analytical separations, they either need to be dissolved or the analytes must be extracted. If strong acids or bases are required for complete dissolution, the structure of many organic analytes may be altered. For such analytes, extraction methods (Module 8.2) are preferred. Polymeric materials may often be dissolved and subjected to separation methods (Chapter 4), but some polymers (*e.g.* polyethene and polypropene) require high dissolution temperatures and high-boiling solvents, such as 1,3,5-trichlorobenzene, while cross-linked polymers cannot be dissolved without breaking chemical bonds. To facilitate the extraction of analytes from polymeric materials, it is essential to create a large surface area by forming small particles. Grinding of polymeric materials is typically possible below their glass transition point to avoid dealing with elastic, rubbery particles. Grinding at low (*e.g.* liquid-nitrogen) temperatures, known as "cryo-grinding", is an efficient sample preparation step.

Distillation is a "unit operation" that is extensively used in industry. However, it is much less in vogue in analytical chemistry. As explained in Section 1.4.3, chromatography is a much more powerful (analytical) separation method than distillation based on a comparison between the numbers of (practical) plates in distillation columns and (theoretical) plates in gas chromatography columns. Distillation-like processes, such as venting off the most volatile components in a sample or, more commonly, leaving behind the "heavy" non-volatile residue, are often performed in GC injectors (see Module 2.4). Programmed-temperature vaporizer (PTV) injectors (Section 2.4.3) are suitable for this purpose.

8.2 Extraction Methods (B)

Extractions are arguably the most important suite of sample preparation methods. Many different extraction techniques exist, the most important of which are listed in Table 8.2. Headspace methods can be seen as extractions into a gas phase, but they are typically integrated with GC separations. Therefore, they have been discussed with GC injectors in Module 2.4. Dynamic **headspace** (purge-and-trap) methods may be used to extract volatile analytes from liquid samples. In the case of solid samples, we may speak of headspace or thermal desorption.

8.2.1 Extractions with Liquids or Supercritical Fluids

Extraction with a liquid is arguably the most classical operation. Both the extraction of analytes from a liquid with another, non-miscible liquid (**liquid–liquid extraction;** LLE) and extraction from a solid with an extractant that does not dissolve the sample matrix (**liquid–solid extraction;** LSE) have found ample use during many decades. Both processes are governed by the same type of distribution coefficient as LC, but while chromatography works optimally when analytes are present in both (mobile and stationary) phases in significant concentrations ($0.1 < K_{LC} < 10$), extraction is aimed at concentrating 100% of all the analytes of interest in the extractant while leaving 100% of all other components in the sample residue. The former goal may be approached, but in doing so, the latter goal usually becomes unrealistic. Because many (biological and environmental) samples are aqueous, many extractions take place with relatively non-polar solvents, such as low-molecular-weight *n*-alkanes (pentane, hexane, and heptane), ethyl acetate or dichloromethane. Such solvents are very compatible with analysis by GC, rendering LLE a suitable method for volatile and semi-volatile analytes. For non-volatile analytes, the non-water-miscible solvents are not very attractive, as they are incompatible with the most common LC method (RPLC). Hydrophilic interaction liquid chromatography (HILIC) is not compatible with non-polar solvents and relatively non-polar analytes. Normal-phase liquid chromatography (NPLC) is feasible, with alkanes being ideal injection solvents, but solvents like ethyl acetate and dichloromethane may need to be diluted with (or replaced by, after evaporation) a less polar solvent. A more diverse range of solvents is encountered in the extraction

Table 8.2 Overview of important extraction methods.

Extractant	Liquid sample	Solid sample
Gas	• Headspace extraction • Headspace solid-phase micro-extraction (HS-SPME)[a] • Headspace liquid-phase micro-extraction (HSLPME)[a]	• Headspace extraction • Thermal desorption • Headspace solid-phase micro-extraction (HS-SPME)[a] • Headspace liquid-phase micro-extraction (HS-LPME)[a]
Supercritical fluid[b]	N/A	• Supercritical-fluid extraction (SFE)
Liquid	• Liquid–liquid extraction (LLE) • Liquid-phase micro-extraction (LPME)[a]	• Liquid–solid extraction (LSE) • Pressurized solvent extraction (PSE) • Microwave-assisted extraction (MAE)
Solid	• Solid-phase extraction (SPE) • Dispersive solid-phase extraction (dSPE) • Magnetic solid-phase extraction • Microwave-assisted (solid-phase) extraction (MASE) • Stir-bar sorptive extraction • Solid-phase micro-extraction (SPME)[a]	• Matrix solid-phase dispersion (MSPD)

[a]Micro-extraction method.
[b]"Supercritical" or "subcritical" fluid; usually carbon dioxide with organic modifier(s) below the critical temperature and/or pressure of the mixture (see Module 6.1).

of solid samples, but extractions with highly aqueous solvents that are directly compatible with reverse-phase liquid chromatography (RPLC) are rare.

Relatively non-polar analytes are easiest to extract from an aqueous sample, and extraction becomes more difficult as the polarity of the analytes increases. However, when compounds contain functional groups that can be ionized, they can be extracted from an aqueous to an organic phase as ion pairs. Basic analytes can typically be extracted at low pH (for example, with a tetra-alkyl ammonium hydroxide).

8.2.1.1 Enrichment Factors

The **enrichment factor** (E_f) of any type of exhaustive extraction that starts and ends with a liquid solution can be described as

$$E_f = \frac{V_{sample}}{V_{extract}} \cdot R_{ex} \tag{8.1}$$

where V_{sample} is the original volume of the sample, $V_{extract}$ is the final volume of the extract, and R_{ex} is the recovery of the extraction. For example, if 100 mL of the sample is extracted into a volume of 1 mL extract with a 90% yield, the enrichment factor is 90.

A frequent way of calibrating is to spike a sample with known amounts of one or more internal standards. However, this assumes that the analogues behave the same as the analytes present in the sample. This assumption may not be valid if analyte molecules interact with other molecules or with particles present in the sample. The closer the internal standards resemble the analytes, the greater the chance that they behave in a similar manner. For example, deuterated analogues of the analytes are especially suitable in combination with MS detection.

8.2.1.2 Extraction with Ionic Liquids

Efforts to reduce hazards in the laboratory have led to several categories of solvents with low vapour pressures and low flammability. One such category of solvents that has been intensively investigated is ionic liquids.[4] These liquid (or "molten") salts exhibit low vapour pressures and are essentially non-flammable, but many of them are highly toxic. The structure of anions and cations that together constitute ionic liquids can be tuned to achieve desirable properties. Hydrophilic ionic liquids can be used for LLE [or liquid-phase micro-extraction (LPME); see Section 8.2.3.2] of polar analytes from low-polarity organic solvents; hydrophobic ionic liquids can be used for LLE (or LPME) of low- or medium-polarity analytes from aqueous samples. The structure of ionic liquids can be tuned to maximize interactions with specific (groups of) target analytes. Magnetic ionic liquids can be created by incorporating a paramagnetic component in the structure. Ionic liquids can also be immobilized on solid surfaces for application in SPE (or SPME; see Section 8.2.3.1).

8.2.1.3 Extraction with Deep-eutectic Solvents

The high toxicity of ionic liquids has spurred interest in **deep-eutectic solvents** (DESs), which are formed by two or three non-volatile, typically solid compounds that form a eutectic mixture with a much lower melting point.[5] Strong hydrogen-bonding interactions are required to make this happen. A DES is easily formed by warming a mixture of a hydrogen bond acceptor, such as choline chloride (a quaternary ammonium salt), and a hydrogen-bond donor, such as an amine or an alcohol, in a specific ratio. By choosing the components and (in the case of ternary mixtures) varying the composition, the properties of the DES can be tailored to the application. These ingredients are cheap, readily available and biodegradable.

8.2.1.4 Extraction with Sub- or Supercritical Fluids

Thermal methods for extracting analytes from solids, *i.e.* headspace analysis and thermal desorption, have been discussed as injection techniques for GC in Module 2.4. This is because gas-phase samples cannot easily be stored or transported, so their integration with the measurement device (*e.g.* GC or MS) is highly desirable.

Extraction with a supercritical (or "subcritical") fluid (**supercritical fluid extraction**, SFE) is similar in operation to supercritical fluid chromatography (SFC; see Chapter 6), except that (i) an extraction vessel is used instead of a chromatographic column and (ii) there is a facility for collecting fractions of the extract after the back-pressure regulator (BPR).[6] The equipment for SFE is schematically illustrated in Figure 8.3. Extraction can be performed in static mode, with the BPR temporarily closed, or in dynamic mode under a flow of the extractant. As in SFC, carbon dioxide is by far the most common base solvent. The extracts are typically collected in an icy spray of CO_2. When the extract warms to room temperature, any remaining CO_2 evaporates quickly, and no evaporation step is needed. If a small

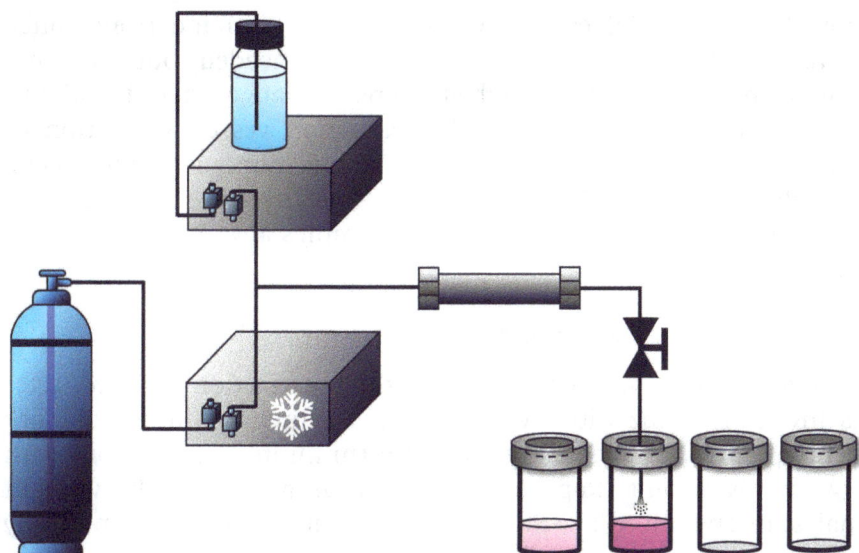

Figure 8.3 Schematic illustration of supercritical fluid extraction (SFE). Carbon dioxide (bottom left) and an (optional) organic co-solvent (top left) are pumped into a temperature-controlled extraction vessel. After the back-pressure regulator, several fractions of the extract can be collected, with increasing extraction pressure and/or percentage of co-solvent.

percentage of an organic solvent, such as methanol, is added to the extractant, this may be left in the sample submitted to analytical separation or measurement, or only a very small amount of solvent needs to be evaporated. Other advantages of CO_2 include its low toxicity, low corrosivity (in the absence of water), and high chemical inertness. Moreover, CO_2 is non-flammable and available in high purity at a cost far below that of high-purity organic solvents. Like SFC, SFE can benefit from the higher diffusion coefficients in supercritical fluids as compared to liquid extractants. This may lead to fast extractions if the mass transport in the extractant is a limiting factor. In some cases, the mass transport inside a solid sample can be enhanced by pressurized CO_2. For example, certain polymers can be made to swell in sub- or supercritical fluids. In addition, SFE offers great flexibility in extraction conditions. Pressure, temperature, and the type and concentration of co-solvent can all be chosen. As a result, high extraction yields can often be achieved. SFE can also be coupled online to LC or SFC.

One important application of SFE is the extraction of fragrances or bioactive compounds from plants and other food-related matrices.

Most famously, SFE can be used to extract caffeine from coffee beans. Drying food samples is not typically needed, but non-solid samples may need to be mixed with a particulate carrier (adsorbent) before being introduced into the extraction vessel. Extraction of low-molecular-weight components (residual monomers, oligomers, and additives) is usually fast above the glass transition temperature of the polymer. The grinding of solid samples always speeds up the extraction.

8.2.1.5 Soxhlet Extraction

Among liquid–solid extractions, **Soxhlet extraction**, named after its inventor, Franz Ritter von Soxhlet, is arguably the benchmark technique. It is an elegant way to perform multiple extractions with a warm extractant. Repeated cycles occur automatically, without analyst intervention. The Soxhlet instrument, pictured schematically in Figure 8.4, is a straightforward laboratory reflux setup (flask and condenser) with the Soxhlet unit inserted in the middle. The extractant is brought to a boil in a round-bottomed flask (extractant reservoir in Figure 8.4), and its vapour passes through the vapour tube to the condenser, from which it drips down into the sample cup, which is permeable to the extractant and the analytes. The solid sample (*e.g.* food, soil, or polymer) is placed in the sample cup. Once the liquid rises to the drain level in Figure 8.4, the siphon tube causes the liquid, now containing extracted analytes, to drain from the Soxhlet unit into the flask. As long as the heating of the flask continues, the process is repeated, and the Soxhlet unit, containing the sample cup, is filled again until it reaches the drain level. As long as the analytes are much less volatile than the extractant, the pure solvent is evaporated from the flask, and each new cycle features extraction of the sample with pure solvent, irrespective of the presence or concentration of analytes in the extract. The boiling point of the extract may increase slightly, but the condensation temperature of the pure solvent remains the same. The speed of extraction can be tuned by supplying more or less power to the heating unit. Typically, the power is tuned to achieve a desired cycle time (*e.g.* 10 or 15 min). Because many repeat extractions with clean solvent can be performed automatically, Soxhlet is perceived as an exhaustive extraction. However, this is not necessarily always the case. Extraction yields can still be improved by grinding samples to reduce the particle size.

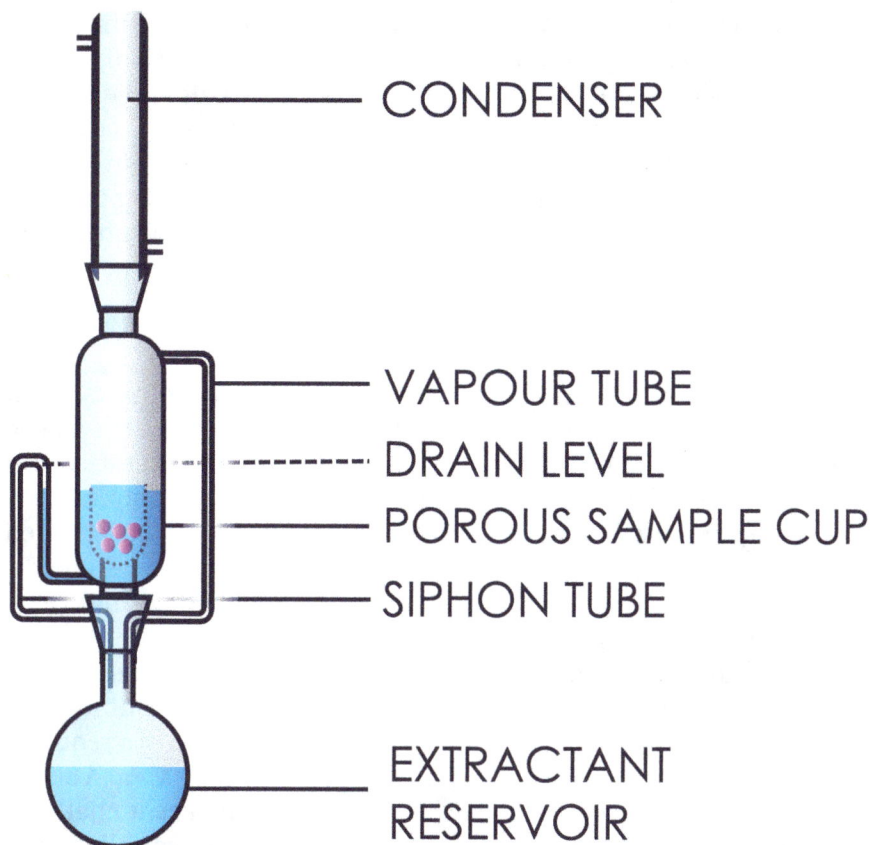

Figure 8.4 Schematic representation of a Soxhlet apparatus. Clean solvent evaporates from the extractant reservoir at its boiling point and rises through the vapour tube into the condenser, from which it drips down into the sample cup. When the liquid containing extracted analytes rises to the "drain level", it drains through the siphon tube back into the reservoir.

Soxhlet extraction is most commonly performed with a low- or moderate-polarity solvent, such as toluene (boiling point 110.6 °C), to extract relatively hydrophobic analytes, such as lipids (oils and fats) from dried food samples. When mixtures of solvents are used in Soxhlet extraction, the gas-phase composition, which, after condensation, becomes the composition of the liquid extractant in the sample cup, is likely different from the solvent composition in the flask. Only in the case of eutectic mixtures will the gas-phase and liquid-phase compositions be the same. Soxhlet extractions are usually carried out for a long time (*e.g.* 8 h). They can be left to run unattended overnight, provided this is permissible within the safety rules of the laboratory. Apart from the long extraction times, the primary disadvantage of Soxhlet extraction is the amount of solvent required (typically at least 100 mL). Usually, most or all of the solvent is evaporated after extraction (of course, the solvent can be collected and disposed of as a liquid, for example, using a rotary film evaporator). The sample and analytes are exposed to high temperatures for a long time, causing a risk of thermal degradation or unwanted chemical reactions.

8.2.1.6 *Pressurized Liquid Extraction*

In **pressurized liquid extraction,** common extraction solvents, such as *n*-hexane, dichloromethane, or ethyl acetate, are heated above their atmospheric boiling points in a pressure-resistant vessel (extraction cell).[7] Mixtures of water and an organic solvent (*e.g.* methanol) are used to extract hydrophilic analytes. The elevated extraction temperatures result in fast extractions with high yields, especially for difficult samples, such as soil, solid food and polymers. Dedicated commercial systems exist for the purpose, but the hardware can also be assembled from standard components. Typical cell volumes (10 to 100 mL) are much larger than the volume of an LC column. Therefore, thick-walled stainless steel cells are required for safe operation at pressures in the range of 10 to 20 MPa. A series (*e.g.* five) of fast extraction cycles (a few minutes) makes it challenging to reach a homogeneous extraction temperature (typically in the range of 70 °C to 150 °C) throughout the entire cell and the typically 5 to 50 g of sample. Both static and dynamic extractions are possible. The elevated temperatures increase the risk of analyte degradation and co-extraction of matrix components. Method development for PLE has received significant attention because of the multiple parameters involved.

8.2.2 Solid-phase Extraction

Solid-phase extraction (**SPE**) has become arguably the most popular method for extracting analytes from liquid samples due to the broad range of adsorbents available and the wide choice of desorbing solvents. Moreover, SPE can often combine analyte extraction with sample clean-up. Some of the main formats of SPE are illustrated in Figure 8.5. SPE is most commonly performed in polypropene (PP) cartridges or in stainless steel pre-columns. The latter more readily allow complete automation and are designed to be used multiple

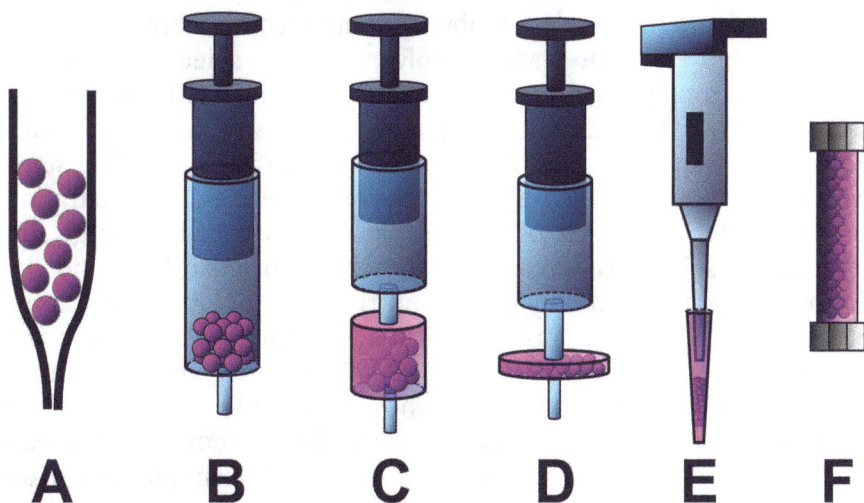

Figure 8.5 Schematic illustration of different solid-phase extraction (SPE) formats: (A) open cartridge, (B) syringe-type cartridge, (C) mount-on syringe cartridge, (D) disc-type SPE, (E) pipette-tip SPE and (F) high-pressure pre-column for SPE.

times. Although larger cartridges are available, the main focus for PP cartridges has shifted to tubes of 1 mL or less in volume, containing 100 mg or less of packing material. The packed bed is typically held in place by frits at the top and at the bottom. The particles are larger than those used in high-performance liquid chromatography (HPLC) (*e.g.* 40 μm) to allow flushing of the cartridge with very little pressure, either through gravity or applied from a syringe (Figure 8.5B–C). Sometimes a (rough) vacuum is applied at the outlet of the cartridge to speed up the process. An alternative format is an extraction disc (Figure 8.5D), which is wider than it is high. This allows the use of smaller particles (*e.g.* 8 μm), while still maintaining a high volumetric flow rate when using a syringe to load the sample. The relatively high efficiency and relatively large surface area allow for a high sample load, despite the short bed height.

The SPE process is schematically illustrated in Figure 8.6. It starts with a fresh cartridge (left) packed with a suitable sorbent to extract the analytes from the loading solvent. In the conditioning step (C), the adsorbent is flushed and wetted with the loading solvent. Next, a sample mixture is loaded onto the column under conditions such that components X and Y adsorb on the SPE particles. At this stage, some undesired sample constituents (*e.g.* salts) may already be washed away to clean up the sample. The sample can then be cleaned up further by removing matrix components (M) in a washing step (W). The washing solvent can be selected such that the least strongly adsorbed target analyte is just left on the cartridge. Thereafter, different (groups of) analytes (X, Y, and possibly additional fractions) may be desorbed in different elution steps with successively stronger elution solvents. However, the separation power of SPE is limited because separation is based on large differences in selectivity (created by the choice of adsorbent and the various solvents), with only minor contributions of chromatographic efficiency. SPE is a separation with a low number of theoretical plates. In comparison with LLE, SPE offers much more flexibility. Many different adsorbents can be chosen, as well as many different elution solvents. Unlike in LLE, the elution solvent is not necessarily immiscible with the loading solvent. For example, the loading solvent can be water, and the elution solvents can be various

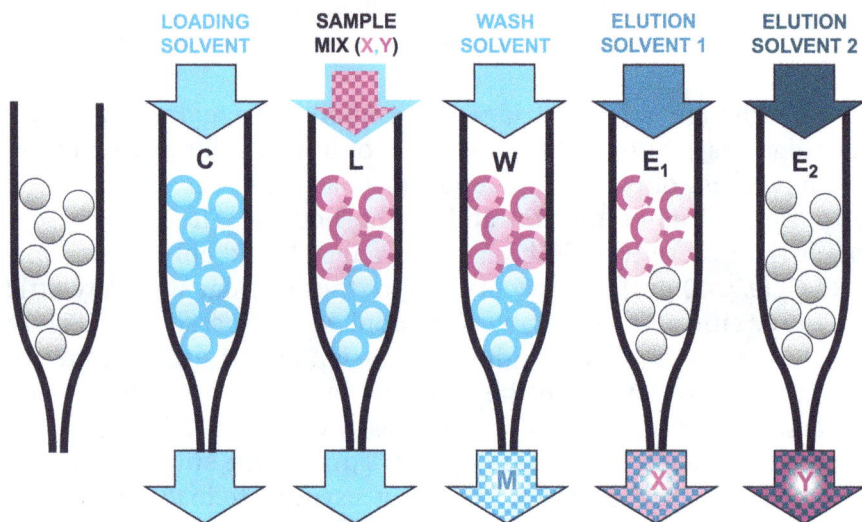

Figure 8.6   Schematic illustration of the process of solid-phase extraction (SPE). On the left is a fresh (dry) cartridge. The first step (C) involves conditioning of the adsorbent, usually with the sample solvent (loading solvent, L). Next, a sample mixture is loaded onto the cartridge; the solvent is chosen such that components X and Y adsorb onto the SPE particles. The sample is then cleaned up further in a washing step (W) by flushing the cartridge with a washing solvent to remove matrix components (M). Different desorption solvents can finally be used in different stages. Elution solvent 1 elutes analyte(s) X in elution step E_1; elution solvent 2 elutes analyte(s) Y in elution step E_2.

mixtures of methanol or acetonitrile with water. The washing step adds flexibility and a possibility to remove unwanted matrix components, an option which does not exist in LLE. The enrichment factor in SFE can be high if a large volume of the sample can be loaded onto the cartridge and if a small volume of elution solvent suffices. Miniaturization of SPE can result in low amounts of (organic) solvents used, but to make SPE a "**green**" technique, it is essential that cartridges (or pre-columns) can be reused multiple times.

8.2.2.1 Adsorbents for SPE

The adsorbents used in SPE appear to cover most of the range of materials used as stationary phases in LC. Octadecyl silica is arguably the most used material, but since separation efficiency is not a critical aspect, strongly adsorbing materials, such as polystyrene-divinylbenzene (PSDVB), are also popular. PSDVB has the advantage of much greater pH stability than silica-based materials ($1 < \text{pH} < 13$ as compared to, typically, $2 < \text{pH} < 8$). "Bare" silica is the traditional material for extracting polar analytes. Apart from the above general-purpose materials, many more specific materials are commercially available, and even many more have been created and studied by researchers (such as Marie-Claire Hennion, see Box 8.1).

Carbon-based materials, including carbon nanotubes (CNTs) or nanofibres, fullerenes, and graphene oxide, usually adsorbed on

Box 8.1 Hero of analytical separation science: Marie-Clair Hennion.

Marie-Claire Hennion (France)
Marie-Claire Hennion has long been a professor of analytical chemistry at the École Supérieure de Physique et de Chimie Industrielles de la Ville de Paris (or, more briefly, ESPCI Paris-Tech) in France, where she directed the Laboratory for Environmental and Analytical Chemistry. She contributed to many aspects of analytical separation science and sample preparation. Especially eye-catching were the new materials she developed for solid-phase extraction and subsequent ultra-trace analysis. These included highly selective adsorbents based on antibodies to concentrate environmental pollutants from water samples or toxins from human serum.

carrier materials, have received significant interest in the literature. Metal-organic frameworks (MOFs) have been studied more recently. Among the more specific SPEs are **molecularly imprinted polymers (MIPs)**, immobilized monoclonal or polyclonal antibodies, and aptamer-based adsorbents.[8] MIPs are created by performing polymerization reactions in the presence of the (non-reactive) target analytes.[9] The idea is that after washing away the analyte, perfectly fitting cavities remain. Practice seems less perfect than theory. Possible reasons for this include diffusional constraints and shrinkage of the material during polymerization and the cross-linking necessary to obtain materials that are stable during the loading and elution stages.

Restricted-access (RAM) materials, championed by Karl-Siegfried Boos,[10] may still be the most promising material for extracting low-molecular-weight analytes from untreated body fluids, such as **whole blood**. However, published examples focused on plasma and urine as samples. A hydrophilic layer surrounding the particles avoids the adsorption of particles and macromolecules from the biofluids. Small, moderately polar compounds do not adsorb in the hydrophilic layer either, but unlike larger analytes, they can penetrate into the pores, the inside of which has a hydrophobic (usually octadecyl) coating. Boos succeeded in using such particles for analysing whole blood with online SPE, but a robust, reliable system did not emerge. More recently, the RAM principle has been combined with that of MIPs.

Ion exchange can be seen as a special form of SPE, with an ion exchanger taking the place of the solid adsorbent. The laws of ion exchange are different from those of absorption into a liquid and are more similar to those of adsorption onto a solid phase. Ion exchange for sample preparation is fully analogous to ion exchange chromatography (IEC), as described in Section 3.7.1. Ion exchangers can be used not only to concentrate analyte ions but also to clean up samples.

In some configurations, particularly when using HPLC-type pre-columns or guard columns for SPE, it is possible to connect two different SPE materials in series. This allows extracting a broader range of analytes from the sample. In more advanced online setups,[11] akin to those described in Chapter 7, the analytes can be desorbed from the different pre-columns with different elution solvents and analysed with different GC or LC methods.

An important application of SPE is the detection of pollutants in **surface water,** and a spectacular example is the detection of (illicit) drugs at the low ng L^{-1} level.[12] Mixed-mode polymeric SPE material can be used for sample clean-up and analyte pre-concentration. Almost all drugs are basic and non-volatile, making LC a more suitable analytical separation method than GC. Tandem MS (typically in SRM) mode provides the required selectivity and suitably low detection limits. Adding greatly to the reliability of the results is the simultaneous detection of metabolites, for example, benzoylecognine as the primary metabolite of cocaine.

Wastewater is a more logical sample to study consumption in communities in which (almost) all of it is collected through the sewage system.[13] In wastewater, concentrations are significantly higher, and consumption may be more clearly related to the population served by the wastewater treatment plant. Wastewater(-based) epidemiology has become a well-established method for studying the consumption of medicines. It offers a more honest picture of drug or alcohol consumption than population surveys do. In addition, market data cannot replace wastewater monitoring. For example, the consumption of Viagra has proven to be much higher than indicated by the official market data (medical prescriptions and pharmacist sales) due to a separate market powered by the internet.

Wastewater(-based) **epidemiology** has also become the generally accepted method to monitor the level of infections in the population with COVID-19 and other viruses. However, this typically relies on the measurement of RNA using quantitative polymerase chain reaction (PRC) methodology for the analytical measurement. In this case, ultracentrifugation appears to be the most common method for sample preparation and analyte enrichment.

8.2.2.2 *Dispersive Solid-phase Extraction and QuEChERS*

In **dispersive solid-phase extraction** (dSPE), a particulate adsorbent is added to a liquid sample. By stirring or shaking the solution, intensive contact between the sample and the surface of the adsorbent can be promoted. The mechanical actions must be more vigorous if the densities of the adsorbent and the sample solution are very different. The extraction process can also be aided by sonication or microwave radiation, resulting in examples of ultrasound-assisted extraction (UAE). Alternatively it can be facilitated by ultrasound-assisted solid-phase extraction (UASE) and microwave-assisted extraction (MAE), or microwave-assisted solid-phase extraction (MASE). After establishing equilibrium, the particles, which now contain the target analytes, are separated from the sample by filtration or centrifugation. Filtration offers the advantage of a possible washing step to remove matrix components. An elegant option is magnetic solid-phase extraction (MSPE), a dSPE technique that uses magnetic particles, which can be removed from the suspension – after extracting the analytes from the liquid sample – using an electromagnet. The particles are usually back-extracted with another solvent. MSPE with biotinylated magnetic beads is used, for example, for the extraction of specific proteins from blood (serum) samples.

All these various dSPE techniques are potentially fast because the intensive stirring enhances the interaction of the analytes with the surface. The entire surface of the adsorbent can be used homogeneously, unlike the situation in SPE, where an analyte concentration gradient is formed during the loading of the sample.

The Quick, Easy, Cheap, Effective, Rugged and Safe (**QuEChERS**) procedure deserves special mention as one of the most successful extraction schemes. It is based on an initial liquid-phase extraction, followed by a clean-up of the extract in a dSPE step. The QuEChERS strategy was initially devised for determining pesticides in fruits and vegetables,[14] but it has since found many other applications.[15] The initial extraction of the largely aqueous processed sample is performed with acetonitrile. Recovery of a diverse range of analytes is enhanced by adding salt (*e.g.* a mixture of magnesium sulfate and sodium chloride or sodium citrate). PSA (primary–secondary amine) sorbent, a modified silica with $\equiv Si-(CH_2)_3NH(CH_2)_2NH_2$ ligands, is used for dSPE. The expansion of the QuEChERS approach to different domains has led to a number of modifications, such as more careful control of the pH during the extraction, but the essence of the process is still largely the same.

Extraction of a solid with a solid seems unattractive, given the slow diffusion of analytes in or on solid matrices, but **matrix solid-phase dispersion** (MSPD) comes close.[16] In this technique, a "solid" sample is blended with a particulate adsorbent. The blend of solids is subsequently extracted with a liquid. The solid samples studied are biological or food samples that must be disrupted to mobilize the analytes, creating a heterogeneous rather than a solid sample.

8.2.3 Micro-extraction Methods

One option to miniaturize SPE is to immobilize a packed bed or create a monolith within a **pipette tip** (PT-SPE; see Figure 8.5E), where the volume of the packed bed (or monolith) can be as low as 1 μL. The sample solution can be aspired and dispensed multiple times. If adsorption is very strong (approaching 100%) and analyte concentrations are low, the sum of the aspiration volumes equals the sample volume. At high analyte concentrations, equilibrium between the surface and the solution may be reached, or the surface may become saturated. PT-SPE can be very fast and largely automated.

Miniaturization of SPE and LLE processes has also led to new concepts, such as SPME and LPME.

8.2.3.1 Solid-phase Micro-extraction

The SPME process is illustrated in Figure 8.7. At the heart of the SPME process is a fibre that can be withdrawn into a syringe needle. When the fibre is exposed (Figure 8.7A), it can adsorb or absorb analytes from a sample (Figure 8.7D) or allow the analytes to be desorbed (Figure 8.7G), most commonly into a GC injector, where desorption can be realized by increasing the temperature. When the fibre is withdrawn into the needle (Figure 8.7B, C, E and F), it is protected. This allows the needle to penetrate through the septum of a sample vial (Figure 8.7C) or, for example, a GC injection port (Figure 8.7F), without damaging the sorptive fibre. With the needle withdrawn (Figure 8.7E), the adsorbed or absorbed analytes can also be safely moved from the point of sampling to the point of analysis, even if the distance is large. In the latter case, a protective holder around the needle is advisable.

Since its inception by Janusz Pawliszyn in 1989,[17] SPME has achieved significant success.[18] Figure 8.7 shows a vial containing a liquid sample, but SPME is also frequently applied to sampling from the gas phase, such as the headspace in a vial, or for *in situ* sampling of air in various indoor or outdoor settings. Likewise, SPME has been applied to sampling from a variety of liquids, including *in vivo* sampling of body fluids. When applying SPME *in situ*, it combines the sampling and sample preparation stages, which is an attractive feature. The wide range of fibres used for SPME has made the technique extremely flexible. SPME is a practical method. It requires no equipment, apart from the dedicated SPME syringe and needle. Extractions can be attempted with little method development and low risk – although fully developing, optimizing, and validating an extraction method is a sizeable task.

Typically, only a small fraction of the total amount of each analyte is extracted because the amount of extractant (volume in the case of

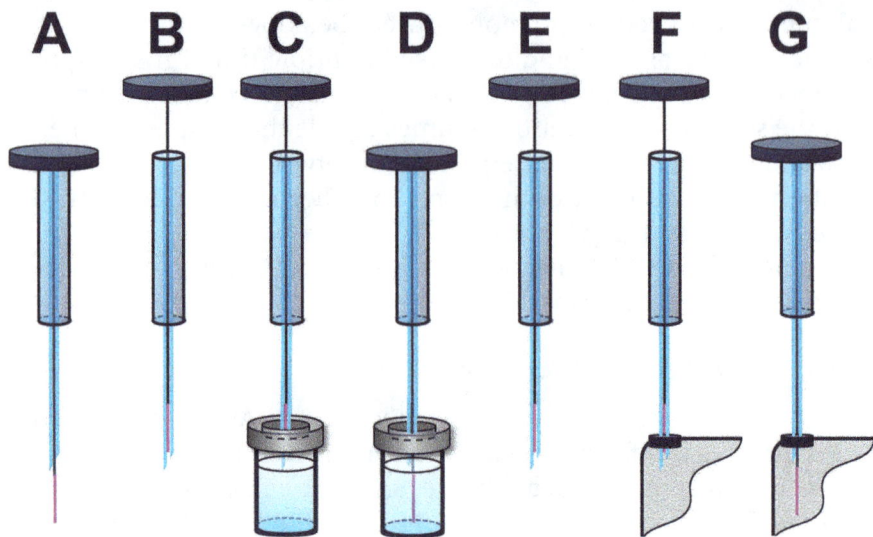

Figure 8.7 Schematic illustration of the process of solid-phase micro-extraction (SPME): (A) needle with the absorptive fibre (pink) exposed; (B) SPME needle with the absorptive fibre withdrawn; (C) needle inserted into a sample vial; (D) absorptive fibre exposed to the sample; (E) fibre withdrawn into the needle and needle withdrawn from the sample vial, allowing transport of the extracted analytes; (F) insertion of the needle into an analytical instrument (*e.g.* the injection port of a GC); and (G) desorption of the analytes from the fibre into the analytical instrument.

absorption; surface in the case of adsorption) is very small. SPME is a non-exhaustive equilibrium method, rather than an exhaustive method, which, for example, Soxhlet extraction and SPE aim to be. As a result, SPME is typically not the most efficient method to reach the lowest possible analyte detection limits. The SPME process can be quantitatively described as follows.

Assuming the extraction medium is a thin, homogeneous film with volume V_{extr} (*i.e.* we assume absorption rather than adsorption), defining an equilibrium coefficient as

$$K_{i,extr} = \frac{c_{i,extr}}{c_{i,sample}} \tag{8.2}$$

where $c_{i,extr}$ and $c_{i,sample}$ are the concentrations (in mol L^{-1}) of analyte i in the extraction phase and the sample, respectively, and assuming equilibrium is achieved (and that the sample volume, V_{sample}, is not affected by the transfer of analytes to the extractant), we can write the mass balance as

$$c_{i,sample}^0 V_{sample} = c_{i,sample} V_{sample} + c_{i,extr} V_{extr} \tag{8.3}$$

Combination of eqn (8.2) and (8.3) yields after some reshuffling

$$n_{i,extr} = c_{i,extr} V_{extr} = \frac{K_{i,extr} V_{sample} c_{i,sample}^0}{K_{i,extr} + \dfrac{V_{sample}}{V_{extr}}} = \frac{K_{i,extr} n_{i,sample}^0}{K_{i,extr} + \dfrac{V_{sample}}{V_{extr}}} \tag{8.4}$$

where $n_{i,extr}$ is the number of moles of analyte i extracted and $n_{i,sample}^0$ is the original amount of analyte in the sample. The maximum (equilibrium) recovery (R_{ex}) follows from eqn (8.4) as

$$R_{ex} = \frac{n_{i,extr}}{n_{i,sample}^0} = \frac{c_{i,extr} V_{extr}}{c_{i,sample}^0 V_{sample}} = \frac{K_{i,extr}}{K_{i,extr} + \dfrac{V_{sample}}{V_{extr}}} \tag{8.5}$$

It follows that the recovery is always less than 100%. Typically, in SPME, $V_{sample} \gg V_{extr}$ so that a high value of $K_{i,extr}$ is vital to achieve any kind of recovery. The **enrichment factor** (E_f) can be high if it is defined as

$$E_f = \frac{c_{i,extr}}{c_{i,sample}^0} = \frac{K_{i,extr} \cdot \dfrac{V_{sample}}{V_{extr}}}{K_{i,extr} + \dfrac{V_{sample}}{V_{extr}}} = \frac{1}{\dfrac{V_{extr}}{V_{sample}} + \dfrac{1}{K_{i,extr}}} \tag{8.6}$$

which may be high if both the ratio V_{sample}/V_{extr} and the distribution coefficient $K_{i,extr}$ are high. However, this is a bit deceptive because the final analytical measurement does not take place in the small extractant volume. In case GC follows SPME, $n_{i,extr}$ is the parameter that determines the final analytical sensitivity. If LC – or another method in which liquid samples are introduced – is used for the analytical measurement, the volume of liquid used to desorb the analytes essentially determines the enrichment factor (see eqn (8.1)).

An interesting feature of eqn (8.6) is that the enrichment factor does not depend on the volume of the sample. When analysing a very large sample, say, a lake, $V_{sample} \gg V_{extr}K_{i,extr}$, which reduces eqn (8.6) to $E_f = K_{i,extr}$ and eqn (8.4) to

$$c_{i,extr} \approx K_{i,extr} \cdot c_{i,sample}^0 \tag{8.7}$$

i.e. the concentration (and the amount) extracted are directly proportional to the concentration of the analyte in the sample. Note that this result is equivalent to the assumption that the concentration of the analyte in a large sample is unaffected by the SPME sampling, which is obviously correct for (very) large samples.

To perform quantitative analysis, $K_{i,extr}$ must be determined in a reference experiment. It must then be assumed that any variations in the matrix relative to this reference experiment do not affect the value of $K_{i,extr}$, which is not necessarily a valid assumption. For example, the ionic strength of the sample is likely to affect the distribution coefficient. It is also assumed that the large sample is homogeneous so that the extract obtained is representative of the entire sample.

Non-equilibrium (diffusion-limited) calibration in SPME is theoretically feasible but is surrounded by uncertainties. Diffusion coefficients of analytes are not typically known, and estimating these, for example, using the Wilke–Chang equation (see Section 6.1.4) introduces a source of error. Diffusion also depends on the composition and viscosity of the matrix, the interaction of the analyte with matrix components, and the temperature. Diffusion in liquids is very slow, and any convective flow may speed up the extraction considerably. In that case, the diffusion coefficient no longer determines the rate of extraction. In **non-equilibrium calibration,** any stirring or agitation needs to be tightly controlled, and the effects must, ideally, be understood.[18] However, even then, calibration factors cannot be predicted *a priori* and must be determined experimentally. **Equilibrium SPME** is recommended for quantitative purposes; however, SPME is often slow, with extraction times of 1 h not uncommon, even when stirring or agitation is used to greatly improve the kinetics. The kinetics of the extraction may be described by

$$n_{i,\text{extr}}(t) = \left\{1 - e^{\left(\frac{-t}{\tau_{\text{extr}}}\right)}\right\} \frac{K_{i,\text{extr}}\, n_{i,\text{sample}}^0}{K_{i,\text{extr}} + \dfrac{V_{\text{sample}}}{V_{\text{extr}}}} = \left\{1 - e^{\left(\frac{-t}{\tau_{\text{extr}}}\right)}\right\} n_{i,\text{extr}}^{\text{eq}} \qquad (8.8)$$

where τ_{extr} is the time constant of the extraction and $n_{i,\text{extr}}^{\text{eq}}$ is the number of moles of analyte i that can be extracted at equilibrium, corresponding to eqn (8.4). The latter value can only be reached after a lengthy extraction. A yield $(n_{i,\text{extr}}(t)/n_{i,\text{extr}}^{\text{eq}})$ of 95% is achieved when $t \approx 3\tau$. Ai measured τ_{extr} values for small analytes and found these to be of the order of 20 to 50 min, with either ultrasonic or stir-bar agitation.[19] This implies that a time of 1 to 2.5 h is required to approach equilibrium. For headspace SPME, the time constants were approximately 10 times lower, suggesting that equilibrium may be reached in 5 to 15 min. It should be noted that these are indicative values based on a small number of data points.

The most common **fibre coatings** for SPME are the same as those used in open-tubular columns for GC, ranging from low-polarity PDMS to polar PEG or Carbowax. Polystyrene and cross-linked polystyrene-*co*-divinylbenzene coatings are also used, but at low temperatures, such phases may behave more as solid adsorbents than as liquid-like absorbents because the glass transition temperature of polystyrene is about 100 °C. Apart from such common materials, many other materials have been investigated and applied for specific applications. As was the case for SPE, CNTs, MPIs, and MOFs have been studied as sorbing materials for SPME. New materials include carbon-nitride and boron-nitride nanotubes. The thermal stability of such materials may be advantageous if an SPME needle is repeatedly used for thermal desorption injections in GC.

There is a lower threshold for using new materials as extractants than as stationary phases in chromatography. In the former case, it is essential that analytes can be adsorbed or absorbed from the sample under sorption-promoting conditions and quantitatively desorbed under desorption-promoting conditions. The two conditions can be very different. Efficient chromatography requires fast transfer of analytes from the mobile phase to the stationary phase and *vice versa* under compromise conditions, ensuring that the retention factor (k) takes on a reasonable value (*e.g.* $k = 1$). A stationary phase that can only be operated at very high or near-zero retention factors is not very useful in chromatography.

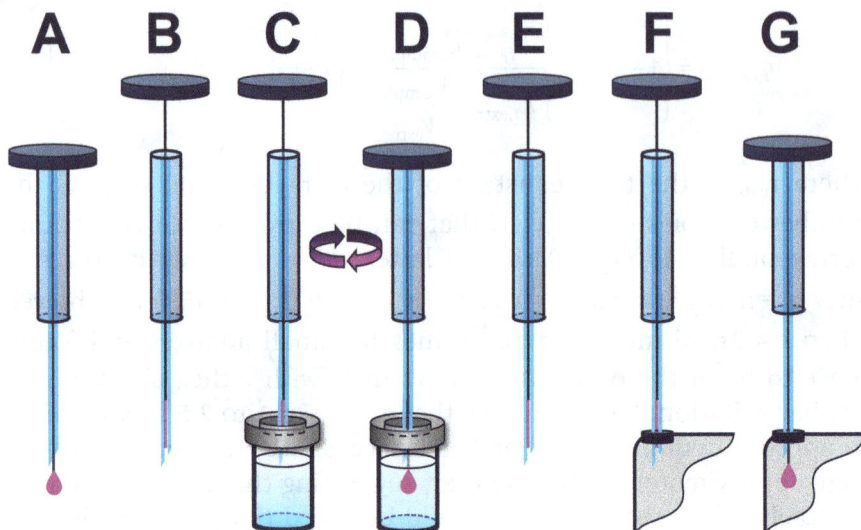

Figure 8.8 Schematic illustration of the process of single-drop micro-extraction (SDME): (A) needle with a pending droplet of extractant (pink) exposed; (B) LPME needle with the extractant aspired into the syringe barrel; (C) needle inserted into a sample vial; (D) droplet of the extractant exposed to the sample. The syringe plunger can be moved up and down (alternating between stages C and D) to accelerate the extraction; (E) extractant aspired into the needle and needle withdrawn from the sample vial, allowing transport of the extractant with extracted analytes; (F) insertion of the needle into an analytical instrument (*e.g.* the injection port of a GC); (G) injection of the extract into the analytical instrument.

8.2.3.2 Liquid-phase Micro-extraction

Alongside SPME, many liquid-phase micro-extraction (LPME) techniques have emerged. One of these is **single-drop micro-extraction** (SDME),[20,21] which is schematically illustrated in Figure 8.8. In other implementations, hollow fibre membranes or (supported) flat membranes are used to confine and protect the liquid extractant. LPME bears a close resemblance to SPME. However, instead of a solid or polymeric coating on a needle, LPME employs a liquid extractant. In the case of a liquid sample, the extractant must be immiscible with the sample solution, creating a micro-LLE kind of system.

In SDME, a small droplet of extractant liquid (typically 1–3 μL) can form at the end of a syringe needle without being dispensed (Figure 8.8 A, D and G), or it can be withdrawn into the syringe (Figure 8.8 B, C, E and F). Extraction occurs when the droplet is exposed to the sample (Figure 8.8 D). The liquid can be repeatedly withdrawn, and a new droplet can be formed to mix the analyte into the extraction solvent and to repeatedly expose a fresh surface (*i.e.* a surface with low analyte concentrations) to the sample. This speeds up the extraction, as does the gentle stirring of the sample solution. eqn (8.2) through eqn (8.7) are valid for both SPME and LPME.

A fundamental advantage of LPME in comparison with SPME is that mass transfer within the extractant can be much faster because (i) diffusion coefficients are much higher in a liquid than in a polymeric coating and (ii) agitation is possible in both phases. Another advantage of LPME is that it is extremely simple and extremely cheap. However, there is much less choice in immiscible solvents for LPME than there is in sorption materials for SPME. Several solvents can be used to extract low- or medium-polarity analytes from aqueous samples. Chloroform has frequently been used, but viscous solvents with a low volatility, such as *n*-octanol, are generally preferred as extractants. However, the extraction of polar analytes is still challenging. When injecting the extract in, for example, a GC instrument, LPME produces a solvent peak, whereas SPME does not. In principle, LPME is quite compatible with LC and with capillary electrophoresis (CE). Direct injection would imply that eqn (8.6) describes the actual enrichment factor in LPME. However, if the sample is aqueous and the extractant is immiscible with water, the latter cannot directly be injected in a water-based LC mobile phase. Combining LPME of aqueous samples with RPLC is akin to the comprehensive two-dimensional coupling of NPLC and RPLC (NPLC×RPLC). This coupling was discussed in Module 7.3, and the solutions presented there can also be used to combine LPME with RPLC. Given the small volume encountered in CE, evaporation of the solvent and reconstitution in the CE buffer is a more practical approach. Headspace-SDME is well documented, but it is typically confined to highly volatile analytes. If an increase in temperature is desired to increase the concentrations of analytes in the headspace – as is customary in HS-SPME – the extractant should have very low volatility. High-boiling, water-immiscible solvents used in LPME include chlorobenzene (boiling point 132 °C), *n*-undecane (196 °C), and di-*n*-hexyl ether (226 °C). Analytes should generally be more volatile than the solvent if GC is the intended analytical separation method.

8.2.4   Overview of Extraction Techniques

Table 8.3 provides an overview of some of the strengths and weaknesses of many of the most common extraction methods. A more detailed discussion is presented in Section 8.2.2 for conventional (analytical-scale) extractions and in Section 8.2.3 for micro-scale extractions.

The "greenness" of various extraction methods is a very hot topic. Just about every advancement is being advocated in the literature as (much) more **"green"** than previous methods. To some extent, this may be objective. For example, the amounts of solvents used may be quantified. To some extent, it is subjective. In addition, there is usually a range within which individual methods are used. For example, in LLE, a litre of water may be extracted with 50 mL of *n*-hexane, but it is also quite feasible with conventional means to perform the extraction with 10 mL of this solvent. In Figure 8.9, we attempt to give an indication of the "greenness" of some of the most important extraction methods. As a (very) approximate indication, the range along each axis covers an improvement (from the outside to the centre) by a factor of 100. Soxhlet requires sizeable amounts of solvents (top of the diagram) and energy (above right). In contrast, it requires no disposable consumables (bottom right) other than the solvents and sample cups that are usually made from cellulose. Some hazardous vapours (bottom left) may escape, especially due to volatile components in the sample. Highly toxic solvents (above left) can usually be avoided. LLE requires no disposables, but it is otherwise not very green. Some gains are offered by PLE, which takes place in a closed vessel, but the method consumes more energy. LLE with an ionic-liquid extractant offers greener methods in terms of disposables and hazardous vapours, but the toxicity of the ionic liquids is a drawback. In most cases, SPE is performed in disposable cartridges, but the use of disposables may be reduced by using stainless steel (pre-)columns instead of cartridges. This is just one indication of the tentative character of Figure 8.9. Not tentative is the conclusion that miniaturization offers by far the best path towards greener extraction methods. For example, in SPME, the main improvement that can still be made is more robust fibres and syringes that can be used indefinitely.

Table 8.3 Summary of the strengths, weaknesses and specific characteristics of major extraction techniques.

Technique	Extractant	Samples	Solvent usage	Instrument costs	Consumables	Speed	Automation	Online coupling	Comments
Headspace (HS)/ thermal desorption (TD)	Gas	Headspace of solid/ liquid	+	+/-	+	+/-	+/-	+	• Used in combination with GC • Dynamic HS is faster and more precise than static HS
Super-critical-fluid extraction	Super-critical fluid	Solid (liquid)	+	-	+/-	+	+/-	+	• Carbon dioxide can easily be removed from the extract
Liquid–liquid extraction (LLE)	Liquid	Liquid	-	+	+	+/-	+/-	-	• Limited choice of immiscible solvents • Extraction of polar compounds from water is not feasible
Dispersive liquid–liquid extraction (DLLE)	Liquid	Liquid	+/-	+	+	+	-	-	• Small droplets formed by the rapid injection of extractant with a disperser • High surface area leads to rapid and efficient extraction

(continued)

Table 8.3 (*continued*)

Technique	Extractant	Samples	Solvent usage	Instrument costs	Consumables	Speed	Automation	Online coupling	Comments
Liquid-phase microextraction (LPME)/single-drop microextraction (SDME)	Liquid	Liquid	+	+	+	−	+/−	−	• Limited choice of immiscible solvents (as in LLE) • Preference for solvents with high surface tension (or high interfacial tension when surrounded by the sample) • Facile direct injection of extract into an analytical system (GC, LC, MS)
Soxhlet extraction	Liquid	Solid	−	+	+/−	−	+/−	−	• Automatic multistage extraction • Reference method in many official procedures • No automatic sample change • Reputedly exhaustive • Requires thermally stable analytes

(*continued*)

Table 8.3 (*continued*)

Technique	Extractant	Samples	Solvent usage	Instrument costs	Consumables	Speed	Automation	Online coupling	Comments
Microwave-assisted extraction (MAE)/ultrasonic-assisted extraction (UAE)	Liquid	Solid	+/−	+/−	+/−	+	+/−	+/−	• Limited choice of solvent (polar, requires dipole moment) • Needs an open extraction vessel to avoid pressure build-up (unless combined with PLE)
Pressurized liquid extraction (PLE)	Liquid	Solid	+/−	−	+/−	+	+	+/−	• Fast and efficient • Requires thermally stable analytes • Reactions with the matrix may occur
Solid-phase extraction (SPE)	Solid (sorbent)	Liquid	+/−	+	+/−	+	+	+	• Vast choice of formats and sorbents • Possibly fully inline in a high-pressure format
Headspace solid-phase microextraction (HS-SPME)	Solid/polymer film	Headspace of solid/liquid	+	+	−	−	+	+	• Choice of different sorbents • High precision requires long equilibration

(*continued*)

Table 8.3 (continued)

Technique	Extractant	Samples	Solvent usage	Instrument costs	Consumables	Speed	Automation	Online coupling	Comments
Solid-phase micro-extraction (SPME)	Solid/polymer film	Liquid	+	+	−	−	+	+	• Needles are vulnerable • Choice of different sorbents • High precision requires long equilibration • Needles are vulnerable
Stir-bar sorptive extraction (SBSE)	Solid/polymer film	Liquid	+	+	−	−	+/−	+/−	• More robust than SPME • (Mini) stir bar can be transferred into the GC injector
Dispersive solid-phase extraction (dSPE)	Solid (particles)	Liquid	+	+	+	+	−	−	• High speed, thanks to large surface area and convection • Requires separation of particulate extract from the sample • Requires back extraction from the ad-/absorbent
Magnetic solid-phase extraction (MSPE)	Solid (magnetic particles)	Liquid	+	+	+/−	+	−	−	• High speed, thanks to large surface area and convection

(continued)

Table 8.3 (*continued*)

Technique	Extractant	Samples	Solvent usage	Instrument costs	Con-sum-ables	Speed	Automa-tion	Online coupling	Comments
Microwave-assisted (solid-phase) extraction (MASE)	Solid (particles)	Liquid	+	+/–	+	+	–	–	• Limited choice of solvent (polar, requires dipole moment) • Needs an open extraction vessel to avoid pressure build-up • Requires separation of particulate extract from the sample • Requires back extraction from the adsorbent/absorbent

• Easy separation of particulate extract from the sample
• Requires back extraction from the ad-/absorbent

SOLID-PHASE
MICRO-EXTRACTION

LIQUID-LIQUID
EXTRACTION

PRESSURIZED-LIQUID
EXTRACTION

IONIC-LIQUID-LIQUID
EXTRACTION

SOLID-PHASE
EXTRACTION

SOXHLET

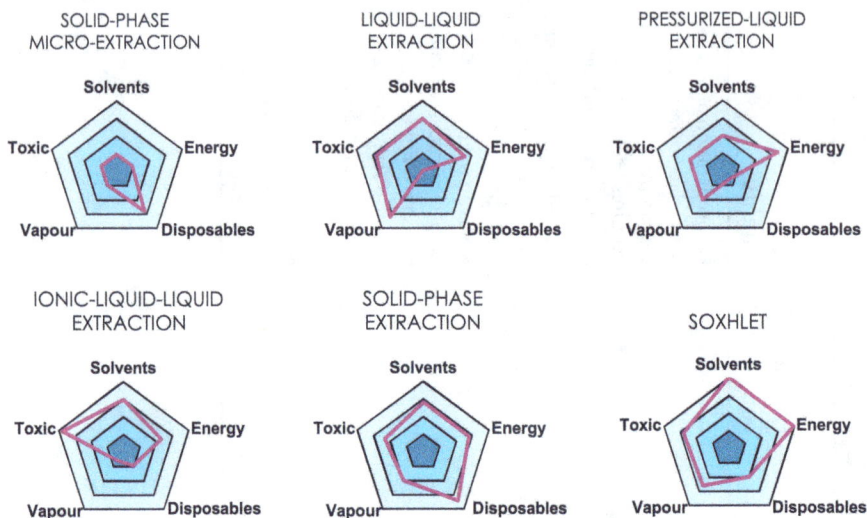

Figure 8.9 Indication of the "greenness" of some major extraction methods along the axes of five parameters: solvent consumption,
energy consumption, use of disposables, hazardous vapours,
and toxic materials (clockwise from the top). Closer to the
centre implies more green.

8.2.5 Alternatives to Extraction

Apart from the collection of extraction methods that dominate the
field of sample preparation, many other sampling, sample pretreatment, or sample preparation methods exist. Two membrane-based
techniques are briefly described in this section, and these are
schematically illustrated in Figure 8.10. **Microdialysis** (Figure 8.10A)
is a process in which a dialysis probe (catheter) is immersed in
or inserted into a sample and flushed with a solvent (usually an
aqueous solution) that is akin to the sample matrix. The outside
of the probe – which often takes the shape of a needle, without a
hole at the tip – is a semipermeable wall, through which compounds
up to a certain molecular weight ("cutoff") can pass. Fresh solvent,
the so-called **perfusate**, can be pumped into the probe from the
centre and then flow back alongside the membrane. Microdialysis
is a continuous process, with the pump operating at a very low
flow rate (*e.g.* 1 μL min^{-1}). When microdialysis is used for analytical
purposes, the concentration of analyte is higher outside the probe
than inside it, and analytes cross the membrane and leave the
probe in what is then called the *dialysate*. The perfusate may also
contain a high concentration of a low-molecular-weight compound.

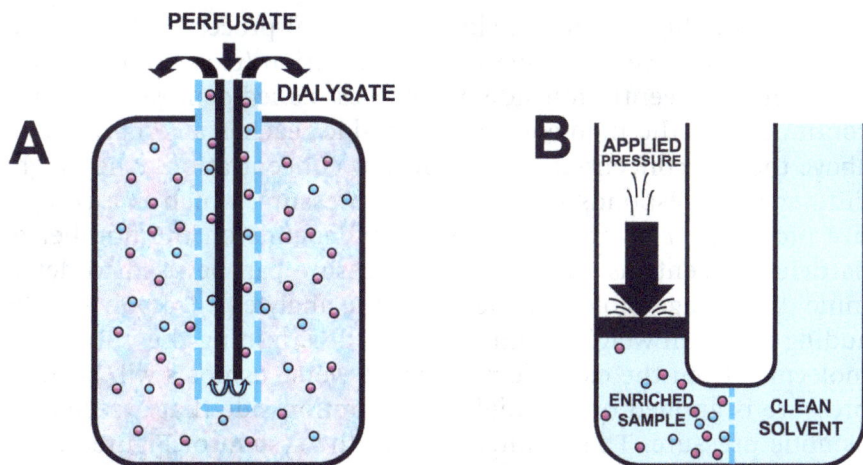

Figure 8.10 Schematic illustration of microdialysis (left) for sampling (and sample preparation) and reverse osmosis (right) for sample enrichment.

In such a "retrodialysis" mode, it may be used as a delivery tool, for example, for administering insulin to diabetes patients in a real-time controlled system. High-concentration perfusate may also be used to calibrate the microdialysis process of exogenous substances, assuming that the mass transport is the same for a negative concentration difference as for a positive concentration difference.

Microdialysis is often used to investigate the organs of test animals with minimal invasion. In such *in vivo* studies, the perfusate mimics the extracellular fluid. For example, when monitoring the brain, the perfusate is an (artificial) cerebrospinal fluid (CSF). It serves as both a sampling and a sample preparation tool. The dialysate does not contain cells or high-molecular-weight compounds, such as proteins. Thus, it can be injected directly into an LC-MS or CE-MS system. Microdialysis probes can remain in place and in operation for at least 1 or 2 days, allowing time series to be recorded. For example, drug concentrations in the extracellular fluid can be monitored and connected with drug dosages. When the probe is operated for longer times, the tissue, and thus the dialysis efficiency, may change.

Reverse osmosis (Figure 8.10B) is a process used on a large scale to purify water. In that case, the remainder is water that is concentrated with impurities. This implies that reverse osmosis can, in principle, be used for sample enrichment. When two reservoirs (left and right) are separated by a semipermeable membrane and the concentration of analytes is higher on one side, the natural process is for solvent molecules to move from the low-concentration (clean solvent) side to the high-concentration side in a process called osmosis. This will continue until the high-concentration side reaches a certain height above the low-concentration side. At low concentrations, the height difference corresponds to the osmotic pressure, which is a colligative property, *i.e.* a property that solely depends on the number of particles present. As such, osmotic pressure can be used to determine the average molecular weight of the analyte (*e.g.* a polymer) by adding a certain weight of analytes and determining the number of molecules from the osmotic pressure. Reverse osmosis will occur if pressure is applied on the high-concentration side that exceeds the osmotic pressure. This is illustrated in the system of Figure 8.10B. The excess pressure will cause the solvent to move from the high-concentration side to the low-concentration side, opposite to the natural direction (hence the name reverse osmosis). Reverse osmosis can be used to enrich a sample in analytes that cannot pass through the separating membrane. While a sample can be concentrated, it is not typically cleaned up, apart from low-molecular-weight analytes that may pass the membrane along with the solvent. In the case of water samples, the concentration of all high-molecular-weight analytes and lipophilic compounds tends to increase, together with the concentration of particles such as bacteria.

8.3   Derivatization and Fragmentation (B)

8.3.1   Derivatization

Derivatization entails a change in chemical structure (forming *derivatives*) to enhance steps in the analytical chain. In this module, we consider changes that break or create chemical bonds. Ion exchange and ion pairing interactions are discussed in Module 3.7, and various other complex-forming interactions have been discussed in other parts of this book. As schematically illustrated in Figure 8.11 (left-hand side), chemical derivatization may be used during sampling, prior to or during the extraction (or other sample preparation) stage, or prior to injection in an analytical separation system. At the sampling and sample preparation stages, derivatization is typically aimed at enhancing the extraction yield and/or the selectivity. Before injection into the analytical separation system, the goals may include enhancing the chromatographic selectivity, efficiency (peak shape and width), or detectability. Moving to the right in

Figure 8.11, chemical reactions are seen to be integrated with the analytical separation, especially at the injection stage. Reactions are mostly performed prior to the injection in the analytical separation system or after the separation ("post-column"), but prior to detection. In LC, reactions can be performed after the separation to improve the detectability of analytes and to enhance detection sensitivity. Such reactions can also be performed between the two separation stages of (comprehensive or heart-cut) two-dimensional LC (LC×LC or LC-LC), as described in Module 7.3. Derivatization is often needed to successfully apply CE, either by creating sufficient (differences in) electrophoretic mobility or enhancing the detectability of the analytes. However, capillary electrophoresis is performed in very efficient, miniaturized systems. There is no tolerance for any processes that may result in additional band broadening. Therefore, the only online derivatization reactions reported for CE involve integrated microfluidic systems.[22] In GC, certain detection methods, such as the **methanizer** and sulfur-specific or nitrogen-specific **chemiluminescence** (see Section 2.5.4.4), rely on chemical reactions of the analytes. GC requires significant analyte volatility, while highly polar analytes often yield poor peak shapes. Derivatization reactions can be performed prior to or during the injection. Reactions are facilitated by the high (or programmed) temperatures at which injections can be performed. Some GC detectors (methanizer and chemiluminescence) rely on chemical reactions taking

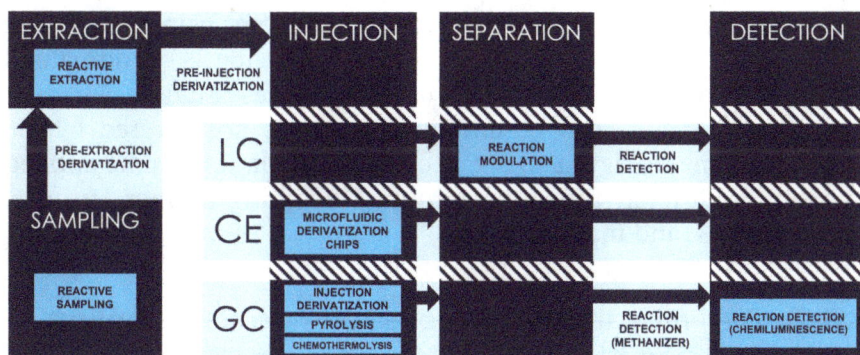

Figure 8.11 Overview of the use of chemical reaction (derivatization) approaches in the context of analytical separation science. Derivatizations are used in different ways and to different extents in LC, CE, and GC. Therefore, these techniques are represented by separate horizontal bars in this scheme (after the universal extraction column). Reactions occurring in mass spectrometers are not considered in this figure.

place before or during the actual detection. Pyrolysis is a destructive process in which high-molecular-weight, non-volatile analytes are broken down into smaller molecules, which are amenable to GC separation. In chemothermolysis, pyrolysis is combined with a derivatization reaction to reduce the polarity of the product molecules. Pyrolysis, chemothermolysis, and other methods that involve breaking down large molecules are discussed in Section 8.3.5.

Reactions that take place in an ionization chamber or a collision cell of a mass spectrometer are not considered in this chapter. They are discussed in other places in this book, especially in Modules 2.6 (GC-MS), 3.10 (LC-MS), 5.3 (CE-MS), and 6.2 (SFC-MS). Table 8.4 summarizes the general advantages and disadvantages of derivatization reactions.

A myriad of chemical reactions has been used for derivatization in conjunction with analytical separations. Indeed, a number of books exist that are solely devoted to the subject (see ref. 23 for a recent example). In this book, we present a concise treatment of the main points. A limited number of important derivatization reactions are indicated in Figure 8.12. Alcohols, phenols, and carboxylic acids can all be derivatized to esters using acid anhydrides. For phenols and carboxylic acids, electrophilic alkylating reagents, such as dimethyl sulfate, can also be used to produce esters. For carboxylic acids, amidation reagents are a third option. Alcohols, phenols, and carboxylic acids can also be derivatized to silyl ethers using, for example, chlorosilanes. Amines react with acid anhydrides to form amides (a path indicated in pink in Figure 8.12). Aldehydes and, less easily, ketones can be transformed into oximes using hydrazines as reagents. When the treatment with a hydrazine is accompanied by a borohydride reduction, it produces amines. Many other derivatization reactions can be employed that are not covered by the simple scheme in Figure 8.12. For example, silylation can be applied to many additional functional groups, including amines, amides, imines, thiols, and many more.

8.3.2 Derivatization Reactions to Improve Sampling and Sample Preparation

In **solid-phase analytical derivatization** (SPAD), extraction is combined with derivatization on the surface of an adsorbent. A classical example is the derivatization of carboxylic acids on a cross-linked polystyrene-*co*-divinylbenzene resin impregnated with pentafluorobenzyl bromide. Many other examples of SPAD have been published,

Table 8.4 General advantages and disadvantages of derivatization reactions.

Advantages	Disadvantages
• May enhance the selectivity and recovery of the sample preparation (extraction) process	• Extra step in the workflow • Possibly not (fully) automated
• May improve the compatibility of analytes with the analytical separation system • May improve peak shape, reduce peak width, and enhance precision	• Incomplete reactions affect precision and accuracy • Possible formation of different isomeric/tautomeric forms
• May enhance detection selectivity and reduce detection limits	• Impurities and side reactions may give rise to interferences
• Reaction modulation allows measuring different analyte properties in a single analysis	• Use of toxic chemicals and solvents

for example, for the simultaneous adsorption and derivatization of carbonyl compounds from water or air samples with 2,2,2-trifluoro-ethyl hydrazine (TFEH) or 2,4-dinitro-phenyl hydrazine (DNPH). SPE platforms (see Section 8.2.2) lend themselves excellently to the implementation of SPAD methods, as do SPME fibres (Section 8.2.3.1). When using SPME fibres impregnated with a reagent, extraction and derivatization can be carried out simultaneously, after which the fibre can be used directly for injecting the derivatives into a GC.

Derivatization of highly polar compounds in water, sometimes called *in situ* **derivatization**, can drastically simplify their extraction. For example, phenols can be derivatized with acetic anhydride (at pH > 10). The resulting esters are much less polar and can be extracted more easily with LLE (or LLME) or with SPE (or SPME). Similar *in situ* derivatizations at high pH have been demonstrated for soil samples. Carboxylic acids may be amidated in water with, for example, 2,2,2-trifluoroethylamine. The greatest limitations for such *in situ* derivatizations are that (i) both the reactant and the product must be stable in water, (ii) reaction yields may be low without a large excess of reagent, (iii) the excess may need to be removed prior to analysis, and (iv) a high concentration of reagent increases the risk of side reactions with other components of the matrix rather than those targeted.

Not all reactants and reaction products are soluble in water (up to the concentrations encountered). **Dispersive liquid–liquid extraction**

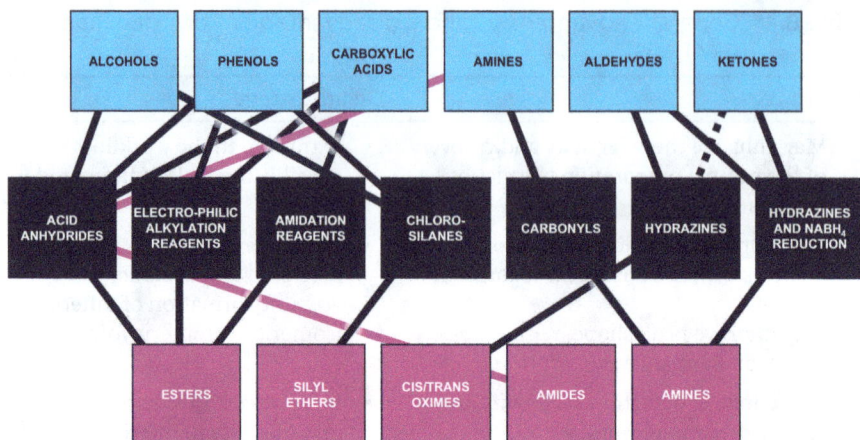

Figure 8.12 Summary of some of the most important derivatization reactions. All paths are downward from analytes (blue, top), through reagents (black, middle) to products (pink, bottom). Acid anhydrides yield esters, except in the case of amine analytes, when they yield amides. This exceptional path is indicated in pink.

(DLLE) may be elegantly combined with derivatization. The reactant and the reaction product are confined to the organic (droplet) phase, into which the analyte molecules diffuse. Because the latter react to form derivatives, the concentration gradient remains intact. By creating many tiny droplets, a large contact area can be obtained, resulting in fast extractions. Phase separation can be achieved by centrifuging based on a density difference between the extractant (*e.g.* chlorinated solvents) and the sample solvent.

8.3.3 Derivatization Reactions to Improve Liquid-phase Separations

LC imposes no requirement on the thermal instability, volatility, molecular weight, or polarity of the analytes, thanks to the wide selection of different phase systems and retention mechanisms (RPLC, NPLC, HILIC, IEC, SEC, *etc.*; see Chapter 3) available. As a result, separations do not usually require the derivatization of analytes to achieve sufficient retention, selectivity, or efficiency.

One significant exception is the formation of **diastereomers** (also called diastereoisomers), which allows the separation of chiral analytes, such as amino acids, on non-chiral (reversed-phase) columns. In contrast with the dynamic, reversible interactions between chiral analytes and chiral selectors on the stationary or mobile phase, which are responsible for selectivity in chiral LC (Module 3.13), derivatization reactions aim to create stable chemical bonds between chiral analytes and chiral reagents. The derivatization reaction requires a **chiral derivatization agent** (CDA) with high enantiomeric purity that allows easy, fast, and quantitative reactions with all analytes of interest. It is crucial that no racemization occurs during the process. The formation of diastereomers makes the LC separation easy to develop. The elution order can be predicted and inverted if both enantiomers of the CDA are available. This allows choosing conditions, such that an enantiomeric form that is present in a low concentration is eluted before the high-concentration form so as to obtain more reliable quantification. Due to the increased availability of chiral columns, the direct method (without derivatization) has gained popularity over the indirect method (with derivatization) for separating enantiomers.

The lack of near-perfect detectors, such as those available for GC (Module 2.5), has led to a proliferation of post-column derivatization reactions in various modes of LC. Post-separation reactions to enable detection are rooted in the heyday of thin-layer chromatography (TLC). Spraying the plate after the separation (or "development") with a derivatization reagent, dipping it in a solution, or exposing it to a vapour-phase visualization reagent, such as iodine, are still often used in conjunction with TLC. The derivatization of primary amines with fluorescamine and that of amino acids with ninhydrin are just two of many examples. Interestingly, fluorescamine derives its name from the derivatization reaction. It is not an amine itself, nor is it fluorescent, but it makes amines exhibit fluorescence. While a greater fraction of all analytes may be detected in HPLC than in TLC, using UV, differential refractive index (DRI), evaporative light scattering (ELSD), or MS detectors, the detection limits and selectivity or specificity of detection leave much to be desired in LC. As a result, numerous LC methods have been developed that incorporate post-column derivatization reactions[23] (see also Box 8.2). The immense success of LC-MS in the 21st century has somewhat reduced the need for derivatization reactions, but many of these methods are still commonly employed.

Many aliphatic molecules are non-UV active, and post-column derivatization reactions of alcohols with phenyl isocyanate, carboxylic acids with phenacyl bromide, or amines with 1-fluoro-2,4-dinitrobenzene, to name just a few examples, alleviate this issue. Fluorescence detection is attractive because of its high selectivity

Box 8.2 Hero of analytical separation science: Roland Frei. Photo reproduced from ref. 24 with permission from Taylor & Francis, Copyright 2010.

Roland W. Frei (1936–1989, Switzerland)
Roland Frei studied and worked in many beautiful places, from Genève, *via* Hawaii, Samoa, Japan, Halifax (Nova Scotia) and Basel, to finally Amsterdam, where he was professor of analytical chemistry at the Vrije Universiteit from 1977 to 1989. As a scientist, he was as inspiring as these places. He excelled in integrating chemical processes for both sample preparation and detection with HPLC separations in complete (automated) systems. Frei demonstrated many practical applications of such systems for biomedical and environmental analysis that were spectacular at the time – and arguably still are today.

and low attainable detection limits (see Section 3.9.3). However, performing quantitative derivatization reactions on target analytes present in very low concentrations in very complex matrices can be quite challenging. Fewer examples exist where chemiluminescence is used. The use of derivatization reactions to aid in MS detection is dwindling due to advances in ionization methods (see Module 3.10). Electrochemical detection can be sensitive and selective, but in terms of robustness, the combination of post-column derivatization reactions followed by electrochemical detection seems like a case of double jeopardy. In contrast, electrochemistry is increasingly used as a derivatization method prior to MS, for example as a way to mimic the formation of drug metabolites.

8.3.4 Derivatization Reactions to Improve GC

GC is more compatible with low- to medium-polarity compounds than with polar compounds, such as carboxylic acids, amines, alcohols, and phenols, for a number of reasons. Polar compounds tend to be less volatile than non-polar compounds of similar molecular weight. For example, methanol (nominal molecular weight 32 Da, atmospheric boiling point 65 °C) and water (18 Da, 100 °C) are liquids, whereas propane (44 Da, −42 °C) and *n*-butane (58 Da, −1 °C) are gases. Strong dipole moments and, especially, hydrogen-bonding functionalities add to the cohesive energy of polar compounds. The same functional groups may give rise to adsorption on instrument components (*e.g.* injection liners) or active sites in the column. This results in tailing peaks and reduced quantitative precision. Very polar compounds are also predominantly present in aqueous samples and cannot easily be extracted from there. Finally, polar compounds have a greater reactivity and a greater tendency to decompose at elevated temperatures. For such polar compounds, the choice is often between direct analysis by LC or analysis by GC after derivatization. Theoretically, GC can be much faster than LC because of the higher diffusion coefficients, but more commonly, the fundamental advantages of GC (higher diffusion coefficients and lower viscosities) are used to create highly efficient separations (often $N \geq 100000$). Arguably, the most important advantage of GC in practice is that it can be equipped with a fantastic choice of detectors, offering very low detection limits and high selectivity or specificity for many classes of analytes (see Module 2.5).

Although definite progress has been made in dealing with aqueous samples (*e.g.* using PTV injection; see Section 2.4.3) and high-polarity stationary phases (*e.g.* free fatty acid columns), reducing the polarity of polar compounds through derivatization techniques is still a commonly employed strategy. Historically, reducing detection limits has also been a strong driver for derivatization reactions prior to GC analysis. For example, the introduction of halogen-containing groups enables highly specific and sensitive detection using an electron-capture detector (ECD); see Section 2.5.4.3. However, such detectors have become less common in large part because they introduce a radioactive source in the laboratory that puts demands on safety procedures and training of personnel. Derivatization reactions can also be beneficial in combination with GC-MS. For example, fluorinated groups have been used to enhance the sensitivity of EI-MS.

In injection derivatization (also known as **injection-port derivatization**, IPD, or *direct in-port derivatization*, DIPD or DID), the sample and reactants are mixed and introduced into the injector, or they are injected sequentially (under stop-flow conditions). The reaction takes place in the injector (sometimes in specifically designed liners) under an inert (carrier gas) atmosphere. Somewhat confusingly, this process is sometimes referred to as on-column derivatization. Technically, this term is correct only in the case of on-column injection, although, even in that case, reactions should preferably take place in a retention gap rather than in the column. On-column injectors are not the preferred instrumentation to accommodate derivatization reactions. The column should be protected from any reactive chemicals as much as possible, but any excess reagent will enter the column. The derivatization reaction should not interfere with the analysis or alter the properties of the stationary phase. A PTV injector (see Section 2.4.3) offers the best option to control (and program) the temperature, split flow, and column flow. This allows venting solvents and side products to waste, provided that these are much more volatile than the derivatized analytes. A PTV injector also offers the easiest option for injecting large volumes of samples, including aqueous ones. A disadvantage is that the number of parameters to optimize is increased further in comparison with the already challenging method development for large-volume injections using a PTV.

Alkylation (for example, with diazomethane, alcohols, or alkyl halides), acylation (for example, with acid anhydrides), and especially silylation reactions are popular. For the latter type of reaction, hexamethyldisilazane (HMDS), *N,O*-bis-trimethylsilyl-trifluoroacetamide (BSTFA) and *N*-methyl-*N*-trimethylsilyl-trifluoroacetamide (MSTFA) are examples of commonly used reagents. A combination of ion-pair extraction with quaternary amines, followed by an alkylation reaction, all in the GC injector, has also been described. Silylation of alcohol, phenol, or carboxylic acid groups (for example, with HMDS) tends to yield stable, volatile products, which can be efficiently separated by GC and can be detected by MS with great sensitivity. Disadvantages of silylation reactions are the sensitivity of the reagent to moisture and the high risk of side reactions. Performing the derivatization reaction in the injector lowers these obstacles. It also requires much smaller amounts of reagents, minimizing the production of toxic waste and improving safety in the laboratory.

Derivatization reactions may also be performed after the analytes are adsorbed on a solid sorbent, either from the gas phase (passive or active air sampling; see Section 8.1.2) or from the liquid phase (SPE; see Section 8.2.2). In that case, a thermal desorption system can be used for injection in a GC instrument.

8.3.5 Breaking Down Molecules

Lysis is a word used for "breaking down", especially of cells. Breaking down the cell membrane to obtain the fluid from within (the *lysate*) is often a necessary first step when analysing biological samples. This can be done through osmotic forces (*cytolysis*) or, more commonly, enzymatically, with *lysozyme* being the best-known catalyst. Adding "-lysis" to one of the various prefixes is also used to form other words that break down samples or molecules. Unlike the many derivatization reactions discussed before (*e.g.* those summarized in Figure 8.12), enzymatic degradation, hydrolysis, and pyrolysis can be used for purposes other than slightly altering the analytes to improve their properties for extraction, analytical separation, or detection. Such reactions can instead be used to break down large molecules or nanoparticles into small fragments, which can then be analysed with GC or LC (or often GC-MS or LC-MS).

A good example of **enzymatic degradation** is the **digestion of proteins** into peptides. This reaction, usually catalysed by trypsin, is at the heart of *bottom-up proteomics*, a mode of characterization inspired by the relative ease of (LC) separation and (MS) characterization of peptides, as opposed to proteins. **Immobilized-enzyme reactors** (IMERs) allow (near-)quantitative inline digestion of a mixture of proteins in (much) less than 1 min, whereas off-line liquid-phase reaction takes many hours. This is due to the large surface area of the solid carrier material and the high local concentrations of (immobilized) enzymes, which can be realized without self-digestion of the catalyst. IMERs seem extremely attractive, but researchers in the life sciences appear reluctant to embrace them due to questions surrounding IMER-to-IMER repeatability and (interlaboratory) reproducibility. Recent studies suggest that this issue is being resolved.[25]

There is also increased interest in using enzymes to degrade other macromolecules, not only to improve the (bio)degradability of plastics, but also to aid in their analytical characterization. Enzymes that specifically break certain bonds offer opportunities for elucidating macromolecular structures.

Hydrolysis, which involves breaking chemical bonds through a reaction with water, can also be used to break down macromolecules. The process can occur with various catalysts (acids, bases, and enzymes). Any kind of polymer formed by a (poly-)condensation reaction can, in principle, be broken down by hydrolysis. These include polyesters (including polyethylene terephthalate, PET), polyamides (including nylon, but also proteins and peptides), polyethers (including polysaccharides, such as starch and cellulose), and polyurethanes. By tuning the conditions, some selectivity can be introduced. For example, in the case of cellulose ethers, such as methyl cellulose or hydroxypropylmethyl cellulose (HPMC), a careful process using hydrobromic or hydroiodic acid (HBr or HI) allows breaking down glucose–glucose bonds in the main chain without destroying the substitution pattern on the ring.

The **pyrolysis** (Py) process occurs at high temperatures (450 °C or higher) in the absence of oxygen. Occasionally, pyrolysis is coupled directly with MS, and sporadically it has been used in combination with LC. However, in the vast majority of cases, pyrolysis is coupled with GC or GC-MS. Usually, the medium of the reaction is the GC carrier gas (H_2 or He). Pyrolysis may take place with the sample deposited on a filament or a ferromagnetic wire or with the sample being introduced in some kind of micro-furnace or micro-oven. When a ferromagnetic metal or alloy is heated by high-frequency induction, it reaches an exact temperature, known as the Curie point. It has long been thought that this was the ideal way to heat samples for Py-GC. However, the exact temperature refers to the alloy and not to a sample (*e.g.* a layer of polymer) deposited on it. Rapidly heating the alloy may actually cause a very large temperature gradient in the sample. Similar temperature gradients may arise in a resistively heated metal (*e.g.* platinum) filament, although the latter is a bit more flexible because the current through the filament may be related to its temperature. Small ("micro")furnaces maintained at a constant high temperature are also frequently used for Py-GC. In such microfurnaces, a cold sample is introduced into a small metal (*e.g.* platinum) cup. The better heat conduction and heat-transfer coefficients cause the metal to warm up more quickly than the sample, again giving rise to heterogeneous heating of the sample. The liner of a PTV injector can also be seen as a "micro-oven". The introduction of samples in the liner as dilute solutions and the good control of the temperature make PTV injectors quite suitable for Py-GC experiments. PTV injectors have proven to be much more precise than other pyrolysis systems, allowing quantitative analysis of polymer content and composition.[26] A PTV injector also allows treating the sample in different stages, with a thermal desorption step to analyse volatile components in the sample (residual monomers or solvents, oligomers, additives, and contaminants) before progressing to a pyrolysis stage. A disadvantage of PTV injectors is their relatively low maximum temperature (often around 600 °C).

Pyrolytic decomposition of polymers may occur through depolymerization (resulting in monomers), chain scission (resulting in larger fragments), loss of side groups (resulting in polyunsaturated fragments), or any combination thereof. Because the process involves the formation of radicals, all kinds of other reactions may also occur, and complex pyrograms may be obtained. Complete degradation to stable monomers offers the best option for accurate quantitation. The formation of substantial amounts of dimeric and trimeric fractions allows for studying the chain sequence of copolymers.[27] The method is much faster than established NMR methods; it allows for distinction between closely related monomers and measuring concentrations of monomers, branching agents, or end groups present at low levels. Complex polymers that contain multiple monomers may provide clearly interpretable pyrograms, as opposed to very complex NMR spectra.

The combination of pyrolysis with GC and GC-MS is quite straightforward. The pyrolysate may be focussed to some extent at the top of the column, but for highly volatile fragments, the focussing effect is limited. Performing the pyrolysis process rapidly reduces injection band broadening, but installing a cryo-trap between the pyrolyzer and the column offers a more rigorous solution to the problem. Polar polymers yield fragment molecules that are typically not amenable to GC, so Py-GC is combined with derivatization reactions. Pre-injection formation of trimethyl-silyl derivatives or methyl esters may yield fragments that remain stable during the pyrolysis process. *In situ* derivatization, for example, with tetramethyl-ammonium bromide or hexamethyl-disilazane, is a convenient and efficient process known as **chemothermolysis**. It greatly expands the number of macromolecules that can be studied with Py-GC or Py-GC-MS.

Acknowledgements

Dr. Lourdes Ramos is acknowledged for her extensive review of this chapter. Annika van der Zon is acknowledged for her review of this chapter.

References

1. A. B. Kanu, *J. Chromatogr. A*, 2021, **1654**, 462444.
2. L. Ramos, *J. Chromatogr. A*, 2012, **1221**, 84–98.
3. D. A. V. Medina, A. T. Cardoso, E. V. S. Maciel and F. M. Lanças, *TrAC, Trends Anal. Chem.*, 2023, **165**, 117120.
4. K. D. Clark, M. N. Emaus, M. Varona, A. N. Bowers and J. L. Anderson, *J. Sep. Sci.*, 2018, **41**, 209–235.
5. B. Hosseininezhad, M. Nemati, M. A. Farajzadeh, E. M. Khosrowshahi and M. R. A. Mogaddam, *TrAC, Trends Anal. Chem.*, 2023, **169**, 117346.

6. M. Herrero, J. A. Mendiola, A. Cifuentes and E. Ibáñez, *J. Chromatogr. A*, 2010, **1217**, 2495–2511.
7. V. Andreu and Y. Picó, *TrAC, Trends Anal. Chem.*, 2019, **118**, 709–721.
8. V. Pichon, F. Brothier and A. Combès, *Anal. Bioanal. Chem.*, 2015, **407**, 681–698.
9. V. Pichon, N. Delaunay and A. Combès, *Anal. Chem.*, 2020, **92**, 16–33.
10. S. Caglar, R. Morello and K. S. Boos, *J. Chromatogr. B: Anal. Technol. Biomed. Life Sci.*, 2015, **988**, 25–32.
11. A. del P. Sánchez ‑ Camargo, F. Parada ‑ Alonso, E. Ibáñez and A. Cifuentes, *J. Sep. Sci.*, 2019, **42**, 243–257.
12. C. J. E. Davey, M. H. S. Kraak, A. Praetorius, T. L. ter Laak and A. P. van Wezel, *Water Res.*, 2022, **222**, 118878.
13. R. Steenbeek, E. Emke, D. Vughs, J. Matias, T. Boogaerts, S. Castiglioni, M. Campos-Mañas, A. Covaci, P. de Voogt, T. ter Laak, F. Hernández, N. Salgueiro-González, W. G. Meijer, M. J. Dias, S. Simões, A. L. N. van Nuijs, L. Bijlsma and F. Béen, *Sci. Total Environ.*, 2022, **847**, 157222.
14. S. J. Lehotay, M. O'Neil, J. Tully, A. V. García, M. Contreras, H. Mol, V. Heinke, T. Anspach, G. Lach, R. Fussell, K. Mastovska, M. E. Poulsen, A. Brown, W. Hammack, J. M. Cook, L. Alder, K. Lindtner, M. G. Vila, M. Hopper, A. de Kok, M. Hiemstra, F. Schenck, A. Williams and A. Parker, *J. AOAC Int.*, 2007, **90**, 485–520.
15. R. Perestrelo, P. Silva, P. Porto-Figueira, J. A. Pereira, C. Silva, S. Medina and J. S. Câmara, *Anal. Chim. Acta*, 2019, **1070**, 1–28.
16. L. Ramos, *TrAC, Trends Anal. Chem.*, 2024, **172**, 117601.
17. R. P. Belardi and J. B. Pawliszyn, *Water Pollut. Res. J. Can.*, 1989, **24**, 179–191.
18. N. Reyes-Garcés, E. Gionfriddo, G. A. Gómez-Ríos, M. N. Alam, E. Boyacl, B. Bojko, V. Singh, J. Grandy and J. Pawliszyn, *Anal. Chem.*, 2018, **90**, 302–360.
19. J. Ai, *Anal. Chem.*, 1997, **69**, 1230–1236.
20. M. A. Jeannot and F. F. Cantwell, *Anal. Chem.*, 1996, **68**, 2236–2240.
21. Y. He and H. K. Lee, *Anal. Chem.*, 1997, **69**, 4634–4640.
22. A. I. Shallan, R. M. Guijt and M. C. Breadmore, *Bioanalysis*, 2014, **6**, 1961–1974.
23. I. S. Krull, in *Reaction detection in liquid chromatography*, CRC Press, 2017.
24. D. Klockow and J. Albaigés, *Int. J. Environ. Anal. Chem.*, 2009, **89**, 557–558.
25. A. A. M. van der Zon, J. Verduin, R. S. van den Hurk, A. F. G. Gargano and B. W. J. Pirok, *Chem. Commun.*, 2024, **60**, 36–50.
26. E. Kaal and H.-G. Janssen, *J. Chromatogr. A*, 2008, **1184**, 43–60.
27. W. C. Knol, S. Vos, T. Gruendling, B. W. J. Pirok and R. A. H. Peters, *Anal. Chim. Acta*, 2023, **1238**, 340635.

9 Data Analysis: Chemometrics and Statistics

Studying analytical separation science is not complete without developing an understanding of the data obtained, their interpretation, and the ways to draw correct conclusions and obtain relevant information. These are the goals of the present chapter as depicted in Figure 9.1. It starts with a discussion on quantitative analysis, followed by a focus on statistics and hypothesis testing. Regression analysis is essential for calibration and quantitative analysis, retention modelling, and curve resolution of chromatographic peaks. Multivariate statistics can be applied, for example, to analyse samples or chromatographic phase systems. The last two modules of this chapter are devoted to signal (pre-)processing, describing essential tools for obtaining the best possible results from analytical separations.

9.1 Introduction to Quantitative Analysis (B)

9.1.1 The Need for Chemometrics and Statistics in Separation Science

In the previous chapters, we have learned about the methods used to achieve the separation of a sample mixture. What remains is converting the raw data into useful information and assessing the value thereof. An advanced, heavily optimized, high-resolution separation system is useless without a complementary data analysis strategy to convert complex raw signals into useful information.

Analytical chemistry traditionally focuses on two existential questions: "What?" and "How much?". Although modern analytical

Analytical Separation Science
By Bob W. J. Pirok and Peter J. Schoenmakers
© Bob W. J. Pirok and Peter J. Schoenmakers 2025
Published by the Royal Society of Chemistry, www.rsc.org

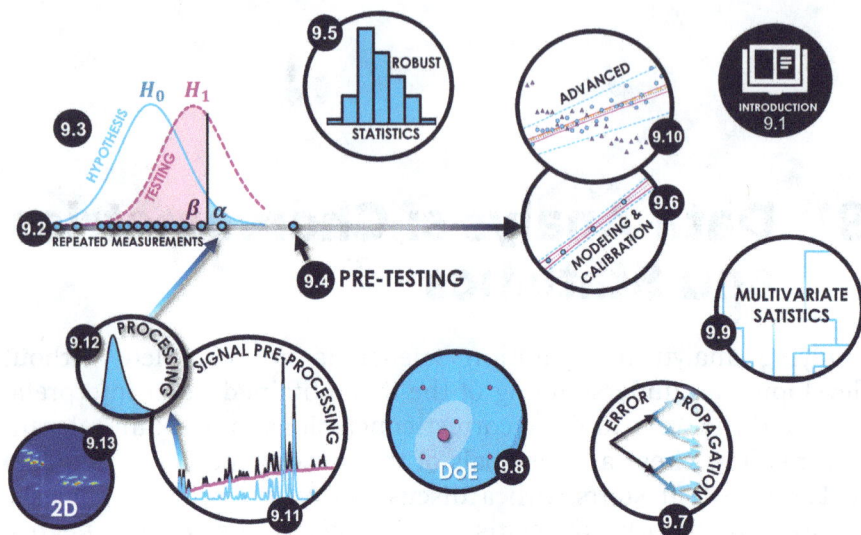

Figure 9.1 Graphical overview of modules in Chapter 9.

chemists are increasingly aware of, and involved in, the context of these questions ("What for?"), the basic questions remain extremely important. To answer the "What?" question (more formally called **qualitative analysis**), separation scientists usually collaborate with other scientists, such as mass spectrometrists (see, for example, Modules 2.6 and 3.10). Arguably, the most important field of applications of analytical separation science involves answering the "How much?" question. A peak represents the number of molecules of a given analyte. Using a calibration curve, we can convert the area under this peak into a concentration. This is known as **quantitative analysis** (see Figure 9.2).

Unfortunately, no analytical method is perfect, and errors, however small, are unavoidable throughout the entire analytical workflow. For example, the LC pump may not displace precisely the programmed volume; the injector may not inject exactly the set injection volume; the detector may – for various reasons – give a slightly different relative response; and the operator may perform the sample preparation procedure slightly differently for each sample.

Clearly, from the hardware used to record the chromatogram to the operator weighing the sample, errors can and do arise everywhere. It is thus paramount that quantitative analysis includes an estimation of the error to place the experimental values in context.

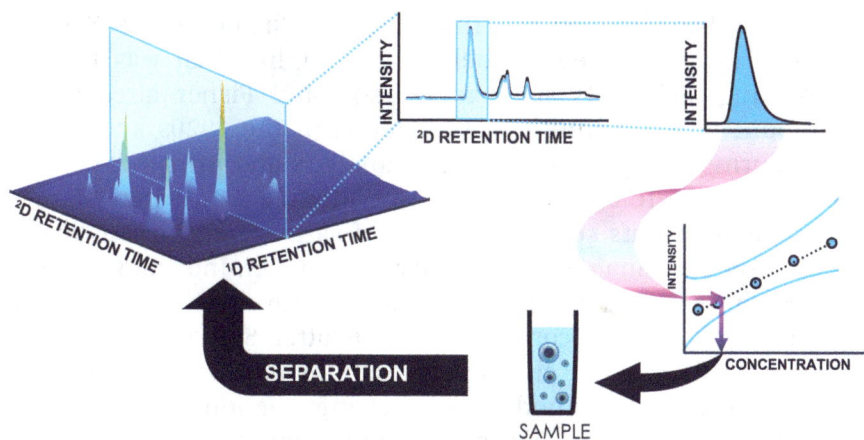

Figure 9.2 When data are generated by a separation method, a number of steps are required to obtain the peak area of each individual component. Using chemometrics and statistics, analytical methods can be calibrated to convert raw data into useful information. This information yields new insights into our sample that may spark new (separation) questions. In this book, we have thus far only focused on the experimental separation part of our workflow (represented by the arrow at the bottom left). We will now focus on the computational part that is illustrated in the remainder of the figure.

The result of a separation experiment should never be "52 mg L^{-1}", but, for example, "52 ± 4 mg L^{-1}". Here, the latter value (±4 mg L^{-1}) represents the confidence interval, indicating the (maximum) error we may expect around the value of 52 mg L^{-1}. Such an estimate of the error can be obtained through the use of **statistics**, which is an integral part of any separation method and will thus be treated in detail in this chapter. Statistics allow us to describe a sample (a set of repeated measurements) and to infer conclusions about the population from which the measurements originate. They can also be used to draw objective conclusions about the sample and help us to answer analytical questions, such as "Are these two samples identical?", "Is my instrument defective?", or "Does the concentration exceed the maximum value?". Today, the use of statistics in analytical chemistry is also often referred to as chemometrics.

The word **chemometrics** can reasonably be defined as the use of mathematical and statistical methods to describe the state of a chemical system (*e.g.* a sample) and design or select optimal measurement procedures and experiments. While the word chemometrics was coined in 1971 by Svante Wold, who, in 1974, together with

Bruce Kowalski founded the International Chemometrics Society, it should be considered a label for a discipline that was long in the making. Scientists such as Pearson and Fisher already laid the groundwork for multivariate analysis in the 1920s and 1930s. It was during this period that factor analysis and principal component analysis were conceptualized (see Module 9.9). Later, Mandel introduced concepts such as analysis of variance and least-squares regression to the analytical chemistry community. Indeed, statistical methods have been a cornerstone of quantitative analysis in analytical chemistry since the conception of the latter. Some contemporary chemometricians define chemometrics more narrowly as the intense use of multivariate methods to extract information from chemical systems by data-driven approaches (although this is sometimes referred to as "chemoinformatics"). For our purposes, the definition of chemometrics provided at the start of this section is more appropriate. Luc Massart (Box 9.1) did much to bridge the gap between chemometrics and analytical separation science.

Besides the statistical analysis of experimental results, we will also devote significant attention in this chapter to **signal processing** (Modules 9.11 and 9.12). A chromatogram or electropherogram of a specific sample may feature a number of (partially) overlapping

---

Box 9.1 Hero of chemometrics in the context of analytical separation science: D. L. Massart. Photo courtesy of Prof. Yvan Vvander Heyden, Department of Analytical Chemistry, Applied Chemometrics and Molecular Modelling (FABI), Vrije Universiteit Brussel (Belgium).

D. (Luc) Massart (1941–2005, Belgium)

Luc Massart was professor of Analytical Chemistry at the Pharmaceutical Institute of the Vrije Universiteit Brussels (Belgium) since 1974. At that point he had already become one of the instigators of, and driving forces behind, the emerging field of chemometrics. He must be credited for bringing chromatography and chemometrics together, applying chemometric techniques to data analysis and optimization.

Apart from his numerous scientific achievements, Luc Massart was a brilliant teacher and lecturer. He taught the fundamentals and applications of chemometrics to numerous analytical chemists and separation scientists.

peaks. The full signal may be affected by various background and system disturbances. To allow statistical analysis, a signal-processing strategy that robustly extracts the required information from the signal is essential.

9.1.2 Variables and Data Order

Before we can start understanding how to use chemometrics and statistics to convert raw data into useful information, we must first establish several essential analytical definitions. When confronted with an analytical problem, the task of a scientist typically is to assess a property. For example, we might be asked to measure the concentration of a pesticide in a bottle of spring water using RPLC. In this example, all the water in that specific bottle represents the **population**. Of course, it is rather difficult to inject the entire litre of water into an LC system, and we thus take a **sample** from this bottle. The sample represents the part of the population that we study. At this stage, we must note the slightly conflicting definitions of the word "sample" in analytical chemistry compared to statistics. From the point of statistics, had we studied a batch of bottled spring water, then our statistical sample may have comprised several bottles, and the population would be all of the bottled spring water in this batch. In such cases, each individual bottle is considered an **object**. The pesticide concentration to be determined is an example of a **variable**. A **univariate method** measures one variable per object (*e.g.* one pesticide), whereas a **multivariate method** measures multiple variables per object (*e.g.* different pesticides).

The pesticide concentration is an example of an **interval variable**, as it can be described by a number and the zero point is clearly defined. Analytical methods may also use **ordinal variables**, which can be described only in words, but can be ranked (*e.g.* "good", "average", "poor"), and **nominal variables**, which can also be described with words, but not be ranked (*e.g.* "blue", "red").

Instruments produce data, which, through the use of models and other statistical techniques, can be converted into variables of interest (*e.g.* the concentration of a specific pesticide). Hence, it is also useful to understand the type of data, or **tensor order**, that an instrument produces. This is strongly related to the **order of the instrument**, and examples are shown in Table 9.1. A zero-order instrument, such as a pH meter or a weighing balance, produces a single number. A first-order instrument, such as a UV–vis spectrophotometer, yields a **vector**, a series of logically ordered numbers

Table 9.1   Examples of the type of data tensors produced by instruments of different orders.

Instrument order	Tensor order	Example
Zero	Zero (number)	pH meter
First	First (vector)	UV–vis spectrometer
Second	Second (matrix)	GC-MS
Third	Third (block)	LC×LC-MS

that represent a spectrum (*i.e.* the absorbance at a series of wavelengths). The next level is a second-order instrument, which produces data in the form of a **matrix**. An example is GC-MS, where, for each retention-time point, a full mass spectrum is recorded. Third-order instruments, such as LC×LC-MS, yield a third-order tensor, a block (or "cube") of data. For each first-dimension fraction, a second-dimension chromatogram is recorded, and for each discrete time point, a mass spectrum is measured.

9.1.3   Calibration Curve

The relationship between a variable (*e.g.* the amount or concentration of an analyte) and the observed signal is at the core of quantitative analysis. We do not measure concentration directly; instead, we measure a related property, which, in separation science, is usually peak height or area. Its graphical representation is preferably called a **calibration curve**, which is a more general term than "calibration line". An example is shown in Figure 9.3. Ideally, the calibration curve passes through zero, although there are many reasons why it may not. For example, the presence of a background signal may cause a positive intercept. The slope of the curve is, by definition, the **sensitivity** of the analysis. This should not to be confused with the **detection limit**, which depends not only on the magnitude of the signal but also on that of the noise. Commonly, the correlation coefficient is used to describe the quality of the fit, but **standard errors** (explained in later sections) are more useful to determine the analytical **precision**. The use of statistics allows you to inspect residuals and confidence intervals around the curve (see Module 9.6).

Important guidelines for the use of calibration curves are the following:

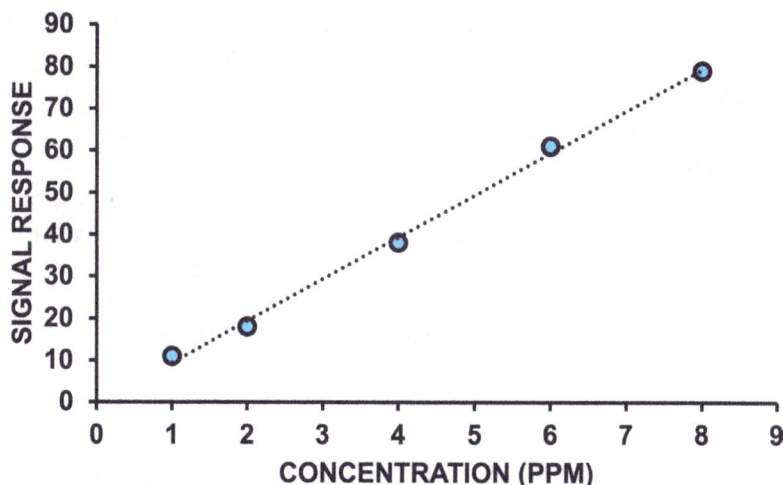

Figure 9.3   Schematic example of a calibration curve.

- Perform a blank measurement (zero analyte concentration).
- Spread the calibration points evenly across the calibration range.
- Make sure that the unknown concentration is within the range of calibration points and perform additional calibration measurements if this proves not to be the case.
- Do not assume a non-zero intercept (*i.e.* test the statistical significance of a non-zero intercept).

9.1.4   Peak Height *Versus* Peak Area

When performing chromatographic measurements, the quantitative information is contained in the **peak height** and **peak area**. Generally, peak area measurements tend to be more accurate, provided that the peaks are sufficiently resolved, with or without the use of selective detection. Peak area is less affected by variations in peak shape. For example, if the peak shape is affected by the amount of sample injected or the concentration of the analyte, this may affect the peak height but not the area. Peak height may be a better measure if (i) peaks are not fully resolved or (ii) the volumetric mobile phase flow rate is not well controlled and a concentration detector is used. In (U)HPLC analysis, it is prudent to monitor the pressure as an indirect indication of the flow (especially in the case of non-programmed – isocratic and isothermal – analyses). Shifts in retention time are also indicative of changes in flow rate. However, when performing gradient-elution experiments, the relative changes in the peak area can be more significant than the relative changes in the retention time.

Chromatographers almost always rely on electronic means (integrators, data stations) to compute peak heights and areas. Correct settings are important for obtaining correct results, and the inspection of the integration markers and the drawn baselines is important to verify your results. In the case of incompletely resolved peaks, drawing a correct baseline is always critical, whether peak height or peak area is used. Some systems may allow peak fitting or curve resolution, but it is prudent to always inspect the results of the process.

It may be argued that the zeroth statistical moment of the peak (see Section 9.12.1) is the best way to compute its area. However, this is true only for well-resolved peaks, a sufficiently large number of data points and sufficiently low noise.

9.1.5 Internal Standard

Chromatographers usually perform only one or two repeat measurements on a sample. Compared to other methods of analysis, we tend to have long measurements and high amounts of information produced each time we make an injection. With the possible exception of (fully filled) fixed-loop injection of non-compressible fluids, the precision of chromatographic injections is a limiting factor in the precision of our measurements. Automatic sample handling and injection tend to be more precise than manual operation, but they are still imperfect. Therefore, we often use internal standards to compensate for variations in the injected volume. This is based on the assumption that our sample preparation procedures (*e.g.* the addition of controlled amounts of solvents) are more reliable than the injection process in the chromatograph. This is reasonable, if only because the volumes involved in sample preparation are larger and, therefore, easier to control. Gas chromatography with a conventional split/splitless injector is not only imprecise but also suffers from discrimination (see Section 2.4.1.1). In this case, several different internal standards may be added to correct for peak areas in different regions of the chromatogram.

If the concentration of the internal standard is identical in each calibration standard and sample, the simplest way to account for variations in the injection volume is to use the peak height or peak area relative to that of the internal standard peak as the observed signal when constructing the calibration curve. If the sample preparation procedure is such that the concentration of the internal standard is different in the reference (calibrants) from that in the sample, for example, because different amounts of solvent containing a known concentration of internal standard are added to the sample and the references, then this may be accounted for. In the simplest case, by performing a one-point calibration and validating that there is no significant (non-zero) intercept of the calibration curves for the analyte and the internal standard, we obtain

$$x_i = \frac{A_i \, A_{st}^0 \, x_{st}}{A_i^0 \, A_{st} \, x_{st}^0} x_i^0 \tag{9.1}$$

where x denotes the concentration, A denotes the peak area, the superscript 0 indicates the reference, and the subscripts i and st refer to the analyte and the internal standard, respectively (so that, for example, x_i is the concentration of the analyte in the sample and x_i^0 is that in the reference).

There are a number of considerations in selecting an appropriate internal standard. An ideal internal standard

- elutes within the same range as the analytes of interest, but
- is baseline separated from the analyte peaks and possible matrix signals,
- is chemically similar to the analytes,
- is highly stable, and
- is readily available in high purity.

9.1.6 Detector Linearity

So far, we have assumed a linear calibration curve (see Figure 9.3). Unfortunately, non-linear calibration curves are frequently encountered in chromatography. If peak height is used as a measure of analyte concentration, these may be the result of variations in peak shape (width, symmetry) with the amount of analyte injected, for example, due to a non-linear distribution isotherm (see Module 1.8). Such effects can usually be accounted for by using peak area rather than peak height as the signal intensity parameter. When the relationship between the signal and the concentration (or mass flow) of the analyte is fundamentally non-linear, but the relationship is known (as, for example, encountered for flame photometric detectors in gas chromatography, see Section 2.5.4.2), this can be accounted for by correcting each data point in the chromatogram before proceeding with the data analysis.

A much more difficult situation occurs when the relation between signal and concentration is empirically found to be non-linear and, potentially, different at each point in the chromatogram. This situation is encountered when using an evaporative light-scattering detector in liquid chromatography (see Section 3.9.4). In such a case, it is virtually impossible to perform rigorously correct calibration.

9.2 Repeated Measurements (B)

In the previous module, we saw that quantitative analysis requires an estimation of the error. We will now learn about the types of errors that may occur and see that repetition of measurements is the only way to assess the random error. We will do this by drawing from descriptive statistics and learning how it can be used to present resulting values as well as the numerical context.

9.2.1 Errors

We return to the example of the determination of a pesticide in a bottle of spring water by LC from Section 9.1.2. For method development, we inject a standard mixture in which our pesticide of interest has a concentration of exactly 9 ppm. We conduct several repeat injections and obtain the chromatograms shown in Figure 9.4A, where the peak area of the peak of interest (indicated by an arrow) is highlighted. Using a calibration curve, the peak area is translated into concentration, and the different concentrations determined are depicted by blue dots in Figure 9.4B.

9.2.1.1 *Random and Systematic Errors*

At this stage, we can make two observations. The first is that all the points lie close to each other, albeit not on top of each other. Indeed, there is a variation in the repeated measurements, a **random error**, that results in a spread of values around the sample mean, $\bar{x}$, of the repeated measurements. The random error affects the **precision (repeatability, reproducibility)** of our method. The only way

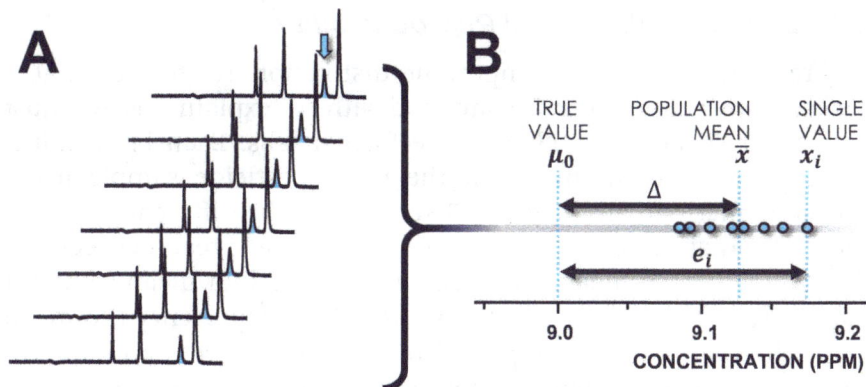

Figure 9.4 (A) Eight repetitions of the separation of several pesticides by LC. The pesticide of interest is marked by an arrow and blue colouring of its area. The peak areas are converted into concentrations using a calibration method. (B) Different concentration values (blue dots) obtained for the pesticide of interest from the eight repeat measurements. See the text for explanation of the symbols.

to quantify the random error is through repeated measurements. Our second observation is that the entire set of experimental values systematically deviates from the **true value** μ_0, *i.e.* the 9 ppm of our standard. This is referred to as the **systematic error**, or bias, (Δ) which causes the entire population of repeated measurements to be shifted in either direction from the true value.

We can also capture the present discussion mathematically. The total error e_i for a given repeated measurement x_i is the difference between the true value μ_0 and x_i.

$$e_i = x_i - \mu_0 \tag{9.2}$$

A distinction between the two types of errors can be made by introducing the mean of the repeated measurements, $\bar{x}$. For measurement x_i, the difference with $\bar{x}$ represents the random error $(x_i - \bar{x})$, and the difference between $\bar{x}$ and the true value represents the systematic error $(\bar{x} - \mu_0)$. The **total error**, e_i, is the difference between a measurement and the true value, also referred to as the **accuracy**, which – from eqn (9.2) – can be written as

$$e_i = (x_i - \bar{x}) + (\bar{x} - \mu_0) \tag{9.3}$$

9.2.1.2 Repeatability and Reproducibility

For random errors, one important distinction relates to what a repeated measurement x_i is compared with. To explain this, we must widen our scope and include three further labs. Each lab conducted repeated measurements of the same pesticide sample using the same method, and Figure 9.5 shows how the different measurements are distributed. For Lab A, we again see a spread between the different repeated measurements, culminating in a mean value for Lab A, $\bar{x}_A$. This $\bar{x}_A$ is distinct from μ, which represents the mean of the full population of repeated measurements from *all* labs. We refer to the **repeatability** as when all measurements are conducted under the exact same operating conditions over a short interval of time (*i.e.* same day, same lab, *etc.*). It is also known as intra-assay precision. We can also define **intermediate precision**, which is the precision within one laboratory but under different conditions (*e.g.* different analyst, different days, different instrument). In contrast, we refer to **reproducibility** when measurements obtained from different laboratories are compared. Figure 9.5 refers to the case where measurement x_i is compared to the population mean μ, which comprises data obtained from different labs.

It should be noted that in this text we have followed the European Medicines Agency (EMA) regulatory definitions, but other definitions may exist outside the pharmaceutical field.

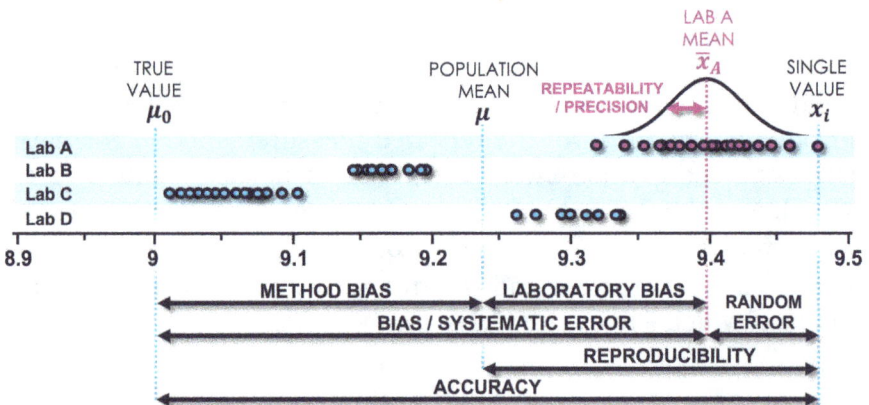

Figure 9.5 Graphical representation of various key definitions within quantitative analysis. See the text for a full discussion.

9.2.1.3 Sources of Bias

Systematic errors can arise from various sources. For example, suppose five laboratories measure the retention factor for a reference compound using an LC column from the same batch, yet one lab obtains consistently higher retention factors. In such cases, when the systematic error arises due to an effect intrinsic to the laboratory, we speak of **lab bias**. Similarly, a systematic error that can be attributed to the method is referred to as the **method bias,** for example, if volatile compounds are lost during Soxhlet extraction (Section 8.2.1.5).

The determination of the bias requires comparison of the results with the true value. But how do we know the true value? If a reference sample or a – even better – certified standard is available, then the bias can be measured. If this is not the case, then an interlaboratory trial of a sample can be conducted in which different labs will run the same sample. This is also shown in Figure 9.5, where the bias for each lab can be determined.

It is important to note that such interlaboratory trials can be of wider scope. For example, five different labs could measure a sample using a CE method, another five different labs would use LC, and another 5 labs a spectroscopic method. Regardless, **interlaboratory trials** are the ultimate way to obtain information about the bias of a method. However, it should be mentioned that such reproducibility studies tend to require much time and effort and are therefore used sparingly.

In the remainder of this chapter, we will assume that systematic errors are absent. We will now focus on using statistics to deal with random errors.

9.2.2 Descriptive Statistics

We have thus far learned the importance of repeated measurements to assess the random error in quantitative analysis. We will now proceed with the application of statistics to chromatographic data.

To this end, we will invoke 100 repeated measurements from our pesticide analysis. The values are shown in Table 9.2. The values appear rather similar, but it is difficult to quantify this feeling by just looking at the table. The total set of values is a sample of 100 determinations from a population of an infinite number of determinations that could have been performed. When describing the population, we use the **Greek alphabet** (*e.g.* μ, σ), whereas we use the

Table 9.2 Representation of a sample of 100 repeated measurements of the concentration of a pesticide by RPLC. In a different situation, the values can also be pictured as 10 samples (see leftmost column) of 10 repeated measurements each. The rightmost columns present the sample mean and sample standard deviation for each set (*i.e.* row). The row-based values are used from Section 9.2.3.1 onward.

SET											$\bar{x}$	s
#1	9.08	9.13	9.11	9.13	9.10	9.13	9.15	9.12	9.14	9.10	9.119	0.0213
#2	9.12	9.15	9.13	9.14	9.11	9.13	9.11	9.13	9.13	9.12	9.127	0.0125
#3	9.11	9.09	9.11	9.14	9.11	9.12	9.15	9.14	9.16	9.14	9.127	0.0221
#4	9.13	9.14	9.16	9.08	9.14	9.10	9.14	9.09	9.12	9.13	9.123	0.0254
#5	9.16	9.12	9.12	9.11	9.15	9.13	9.17	9.12	9.15	9.11	9.134	0.0217
#6	9.12	9.10	9.11	9.13	9.12	9.17	9.11	9.14	9.12	9.12	9.123	0.0200
#7	9.10	9.13	9.14	9.12	9.11	9.13	9.16	9.12	9.13	9.15	9.129	0.0179
#8	9.12	9.14	9.13	9.12	9.14	9.13	9.12	9.13	9.13	9.12	9.128	0.0079
#9	9.13	9.12	9.13	9.13	9.09	9.18	9.13	9.11	9.14	9.11	9.127	0.0236
#10	9.12	9.10	9.15	9.11	9.14	9.12	9.10	9.14	9.13	9.14	9.125	0.0178

Latin alphabet (*e.g.* $\bar{x}$, s) for descriptions of the sample. The average $\bar{x}$ is an estimate of μ. Here, each datapoint x_i can be considered an object (see Section 9.1.2).

Repeated measurements come in one of the two forms, *viz.* central tendency measurements and dispersion measurements.

> See the website for detailed tutorials to conduct most of the descriptive statistics for your data: ass-ets.org

9.2.2.1 Central Tendency Measurements

The perhaps most common measurement is the central ten dency measurement, which can be carried out in different ways. The **arithmetic mean**, $\bar{x}$, or simply the **mean**, is the sum of all measurements, $\Sigma_{i=1}x_i$, divided by the number of measurements, n.

$$\bar{x} = \frac{\Sigma_{i=1}^{n}x_i}{n} \tag{9.4}$$

Note that eqn (9.4) specifies that the sum Σ is taken of all values x_i, where i is the index. The $i = 1$ means that the index starts with the first value and ends at the final value n. This sum is then divided by n. Alternatively, we can use the **median**, $\tilde{x}$, which is the value that separates the higher half of the data from the lower half. In the event of an odd number of values, the median is literally the middle value of the sorted vector. In the event of an even number, it is the mean of the middle two values of the sorted vector.

$$\tilde{x} = \begin{cases} x_{(n+1)/2} & \text{if odd number} \\ \dfrac{x_{(n/2)} + x_{(n/2)+1}}{2} & \text{if even number} \end{cases} \tag{9.5}$$

9.2.2.2 Robust and Non-robust Methods

At this stage, it is useful to note that statistics is a very powerful tool to distil useful information from large datasets and obtain objective answers to analytical questions. However, the decision on the type of statistics employed depends heavily on the type of data to which the statistical methods are applied. We refer to this as robust and non-robust statistics.

Parametric statistical methods operate on the assumption that the data follow a known (mathematical) distribution. In many cases, a

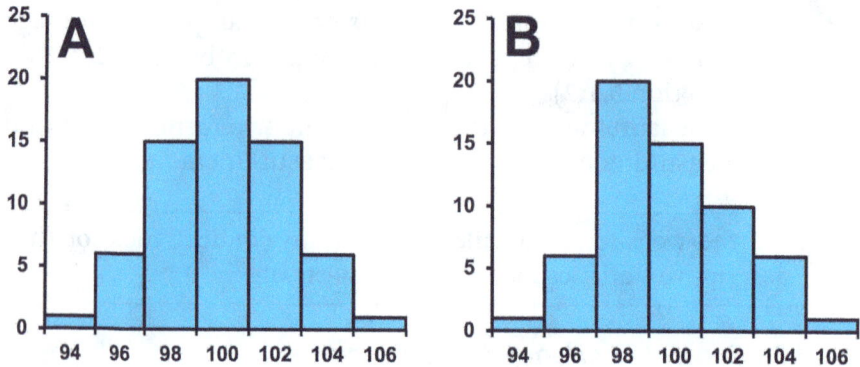

Figure 9.6 Examples of two datasets which are (A) normally distributed and (B) not-normally distributed (*i.e.* skewed).

normal distribution is assumed. This is in contrast to **non-parametric statistical methods,** in which there is no assumption that the data follow a distribution. Consequently, non-parametric statistical methods can be applied to data with a skewed distribution that may even contain outliers (*e.g.* Figure 9.6). Parametric statistics is considered non-robust because it relies on the assumption that data must follow a specific distribution. For non-parametric statistics, no such assumption is made and such methods are considered robust.

The mean and median are examples of non-robust and robust methods, respectively. If we, for example, take the vector [1; 2; 3; 4; 5; 6; 42], then we observe that the value 42 clearly deviates from the other values. The mean for these seven values is 9, which clearly is strongly affected by the value 42. In contrast, the median value is 4, which represents the values 1 through 6 much better. Here, the median is robust because it was unaffected by the presence of an outlier and by how the different values were distributed. The use of the median comes at the cost of completely ignoring the remaining information (the other six values). In contrast, the non-robust mean was applied to data that were not normally distributed and thus yielded a value that was not descriptive of most of the values. While parametric methods are sensitive to the distribution of the data, they tend to be statistically much more powerful. We will use power analysis in Module 9.3 to illustrate this.

One of the early proponents of statistics was Florence Nightingale (Box 9.2).

Box 9.2 Hero of chemometrics: Florence Nightingale. Image by Henry Hering, copied by Elliott & Fry as half-plate glass copy negative, late 1856–1857. Image in Public Domain.

Florence Nightingale (1820–1910, United Kingdom)

Florence Nightingale is well known as a nurse who cared for soldiers during the Crimean War (1853–1856). Less well known is her role as a pioneer of statistics and data visualization. She used graphics to bring her findings across, which reputedly was revolutionary at the time among mathematicians and statisticians. She compiled data on seasonal variations in patient mortality in pie charts with segments of equal angles, but variable length. She called such plots "cockscombs" (after the crest of a domestic rooster). Among all her other impressive activities, Florence Nightingale was an early proponent of info-graphics.

9.2.2.3 Dispersion Measurements

A non-robust dispersion measurement is the well-known **standard deviation**, s, of n measurements, which quantifies the amount of variation or dispersion. The standard deviation describes the average deviation of x_i from the average.

$$s = \sqrt{\sum_{i=1}^{n} \frac{(x_i - \overline{x})^2}{n - 1}} \tag{9.6}$$

Note that eqn (9.6) describes the computation of the standard deviation for the sample. It should not be confused with the standard deviation for the population, which is given by

$$\sigma = \sqrt{\sum_{i=1}^{n_{\text{pop}}} \frac{(x_i - \mu)^2}{n_{\text{pop}}}} \tag{9.7}$$

where n_{pop} is the number of objects in the entire population. The variance s^2 or σ^2 is the square of the standard deviation. In some cases, it can be useful to calculate the relative error, for example, to compare the precision of results with different units or magnitudes. This can be done using the relative standard deviation (RSD), which is equal to

$$\text{RSD} = \frac{s}{\bar{x}} \cdot 100\% \qquad (9.8)$$

or the coefficient of variation (CV), which is simply s divided by the mean (CV = $s/\bar{x}$).

For data that are not normally distributed, the standard deviation may be of less use to analytical chemists. Instead, a robust method is the inter-quartile range (IQR). It is defined as the value of x so that 25% of observations are larger (Q3) minus the value of x so that 25% of observations are smaller (Q1). The IQR is used in the **box-and-whisker plots**, an example of which is plotted in Figure 9.7. The box always has a length equal to the IQR, with the median

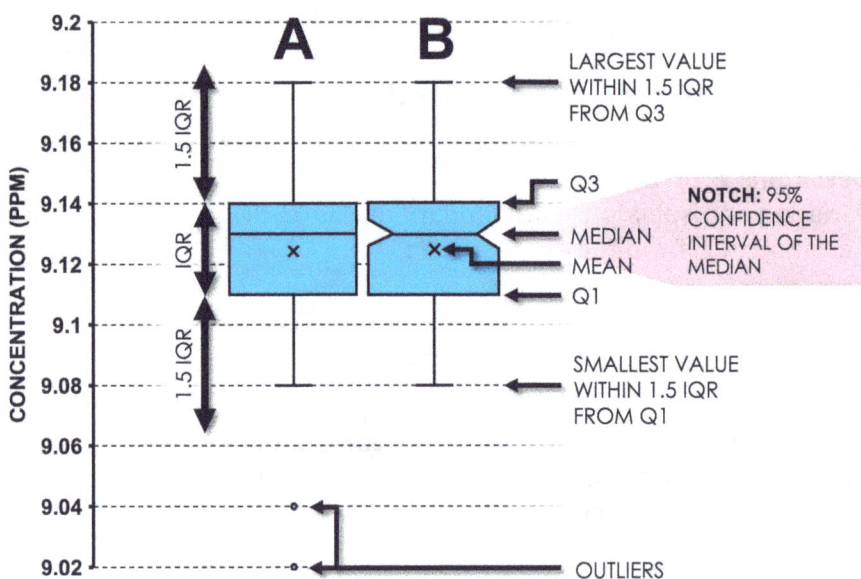

Figure 9.7 The box-and-whisker plot of the 100 pesticide measurements from Table 9.2. For series A, two additional outliers were added. Series B represents the unmodified data. This type of plot is an example of a robust descriptive statistical method in that it does not rely on the normality of the data. Despite the presence of two outliers in series A, the IQR and median are still the same, which is not the case for the mean. For series B also the confidence interval of the median is indicated by the notches.

indicated within the box, whereas the whiskers are never larger than 1.5 times the IQR. Any value beyond the whiskers is considered an outlier. A box-and-whisker plot is a quick and easy tool to obtain an overview of the data. You may also find box-and-whisker plots in literature with "notches" (*i.e.* indentations) that represent the 95% confidence interval of the median defined as *x*. Plot B in Figure 9.7 is an example featuring notches.

9.2.2.4 Probability Distributions: The Probability Density Function (PDF)

The box-and-whisker plot is one method to visualize the data of Table 9.2. Another method to visualize it is by plotting a **histogram,** as is done in Figure 9.8A. The histogram displays the frequency of encountering a value using **bins** (see the arrow in Figure 9.8A). Each bin captures the number of measurements with values falling within the domain of the bin in the case of continuous data or the number of discrete observations in the case of non-continuous data. The number of bins is typically equal to $\sqrt{n}$. In contrast to the box-and-whisker plot, we now see a detailed graphical visualization for the entire population of datapoints.

The frequency can also be expressed relative to the total number of observations. This is depicted by the secondary right-hand *y*-axis shown in Figure 9.8A. The **relative frequency** is the number of times that the number appears between the limits of the bin divided by the total number of measurements. Given that we conveniently have 100 measurements, this translates into the frequency *y*-axis simply being divided by 100 to obtain the relative frequency. The bins marked in pink denote the frequency at which a result of 9.14 or 9.15 was found (*i.e.* 23 times). In terms of relative frequency we could say that – based on this sample of 100 measurements, the probability of finding a value of 9.14 or 9.15 is 23%.

We have thus far used discrete values, but, in theory, a concentration could take any value. A continuous curve is needed to describe the population (of an infinite number of measurements) from which the sample is taken. This is shown in Figure 9.8B where the curve represents the full population of an infinite number of measurements. This curve is what is known as a **probability density function (PDF),** which describes the probability of a number *x* occurring in an infinitesimally small bin *x* + d*x*. The integral of the PDF describes the probability of *x* being above or below a certain value (or between two specified values).

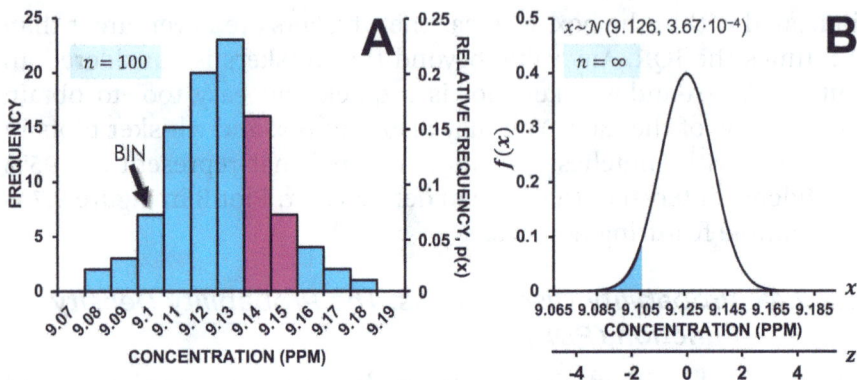

Figure 9.8 (A) The histogram plot of the 100 repeated measurements from Table 9.2. The histogram shows the frequency of finding a result or a range of results that are captured by the different "bins". The secondary y-axis depicts the relative frequency of finding a particular value or range of values. As an example, the pink coloured bins show that the probability of finding a value of 9.14 or 9.15 is 23%. (B) Normal distribution $x \sim N(\mu, \sigma^2)$ of the population of data if an infinite number of measurements were to be obtained. The blue coloured area depicts the integrated area under the curve. This is an example of a probability density function.

In this specific case, our PDF is the well-known normal distribution, which is described by

$$f(x) = \frac{1}{\sigma\sqrt{2\pi}} e^{-\frac{(x-\mu)^2}{2\sigma^2}} \tag{9.9}$$

Eqn (9.9) is also known as the Gaussian distribution and sometimes referred to as a bell curve although many other distributions are also of bell shape, so the latter is not a preferred term. Processes and calculations tend to accommodate well-known distributions, such as the normal distribution in this case, but also in the case of the assumptions made when computing the plate number (see Section 1.4.3). Other well-known distributions in chemometrics and statistics include the Student's t distribution, the F distribution, the χ^2 distribution, and the Poisson distribution. We will cover most of these as we move forward in this chapter.

An important property of a PDF, and therefore also of the distribution in Figure 9.8B, is that the area is equal to 1. In other words, there is a 100% probability to obtain some value. When a certain variable follows a distribution, it is possible to calculate the probability of that variable between certain limits. This will be important when we start to consider hypothesis tests and calculate confidence limits.

Any distribution needs a number of parameters to be defined, and we see from eqn (9.9) that for the normal distribution these are μ and σ, although it is more custom to use the variance σ^2. We will follow this convention and denote the distribution as $x \sim N(\mu, \sigma^2)$, which means that the property x follows the normal distribution, denoted by N which depends on parameters μ and σ^2. For a normal distribution, approximately 68% of the population values lie within 1σ of the mean, approximately 95% within 2σ of the mean (actually 1.96σ, see next sections), and 99.7% within 3σ of the mean. This would mean that for our normal distribution with a μ of 9.126 ppm and a σ^2 of 3.67×10^{-4} ppm^2, 95% of the values lie within the range of 9.088 and 9.164 ppm.

9.2.2.5 Standard Normal Distribution

Calculation of a characteristic or probability of the normal distribution, for example, for our data in Figure 9.8B, requires integrating the PDF for a given interval. This can be tedious. It is therefore useful to convert the data to the **standardized normal distribution**, with a mean of 0 and a standard deviation (and variance) of 1. This is accomplished by expressing our data x into a **standard normal variable** z through

$$z = \frac{x - \mu}{\sigma} \tag{9.10}$$

Figure 9.8B shows the expression of our original distribution $x \sim N$ (9.126 ppm, 3.67×10^{-4} ppm^2) as $z \sim N(0,1)$ with the secondary x-axis. We will see below how the standard normal distribution allows probabilities and characteristics to be calculated more easily. For example, the 95% confidence interval now always lies between $z = -1.96$ and $z = 1.96$. In fact, all the probabilities can now be easily tabulated.

9.2.2.6 Cumulative and Inverse Cumulative Distribution Function

As we build towards the capability to calculate confidence intervals and run significance tests, we must first learn to interact with the

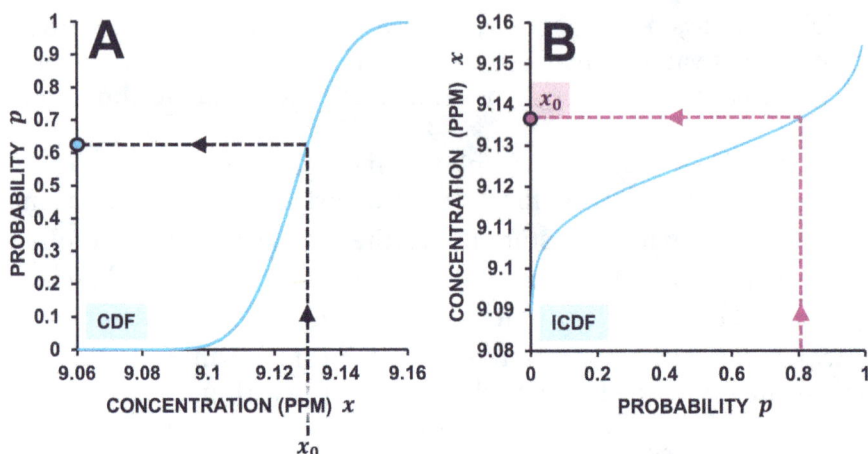

Figure 9.9 The cumulative distribution function (A) and inverse cumula-
tive distribution function (B) plotted for the pesticide data.
Note that the shown examples of x_0 are two different cases.
See the text for a discussion.

distributions. Two important representations of the PDF are the
cumulative distribution function (CDF) and the inverse cumulative
distribution function (ICDF). We will use these functions to tackle
various problems throughout this chapter.

The **CDF** shown in Figure 9.9A (blue dashed line) expresses the
probability of finding a value x_0 or lower, following eqn (9.11).

$$p(x < x_0) = \int_{-\infty}^{x_0} f(x) \, dx \tag{9.11}$$

For example, the figure shows that the probability of finding
a value of $x_0 = 9.13$ or lower is 62.4% (dark blue dashed lines).
The second tool is the **ICDF**, which is also known as the **quantile
function**. The ICDF yields the value of the variable x so that its
probability is less than or equal to an input probability value. For
example, if we want to know for which value of x_0, the probability
of finding $x < x_0$ is 80%, we would obtain $x_0 = 9.136$ ppm. This is
indicated by the pink arrows in Figure 9.9, with the ICDF plotted
separately in panel B.

The CDF and ICDF will become essential when we start to leverage
statistics for significance tests (Module 9.3).

> 🔘 **The website contains several exercises to practice these calculations: ass-ets.org**

9.2.3 Confidence Intervals

We have learned so far that repeated measurements of a sample allow us to calculate a mean, $\bar{x}$, as an approximation of μ_0 in the presence of random errors. However, given that the individual measurements vary, it is also unlikely that $\bar{x}$ will be equal to μ_0. This is a problem for quantitative analysis because industry, society, and policymakers rely on analytical scientists to provide reliable and objective information on the samples submitted for analysis. If 100 measurements cannot provide us with a true answer, then what can? In this section, we will learn that it is more useful to calculate a range of values, which is likely to include the true value μ_0. Such a range is called a **confidence interval**. Intuitively, it concurs with laboratory practice. The more repeated measurements we perform, and the more precise they are, the more confident we become about the result (in the absence of systematic errors).

9.2.3.1 Sampling Distribution of the Mean

To explore this, we return to our example of 100 pesticide measurements of the sample in Table 9.2. In conventional practice, we would conduct a much smaller number of repetitions. For example, we could execute 5 or 10 repeats from the first row and obtain a mean. Unfortunately, given that the individual measurements vary due to random errors, it is also likely that the resulting mean, $\bar{x}$, differs. Indeed, if we consider all the rows in Table 9.2 as different samples from the same population, then we see that the values of each row yield a slightly different $\bar{x}$. This suggests that $\bar{x}$ also follows a distribution.

Figure 9.10A again shows the distribution of all 100 values, x, from Table 9.2 as if we would run 100 repetitions of one sample from the population of origin. In contrast, Figure 9.10B only shows the distribution of the means, $\bar{x}$, of the rows shown in the same table. In chromatographic terms, this corresponds to repeating the 10-repeat RPLC method sequence a total of 10 times. We see that the means from Figure 9.10B are clustered much closer to one another.

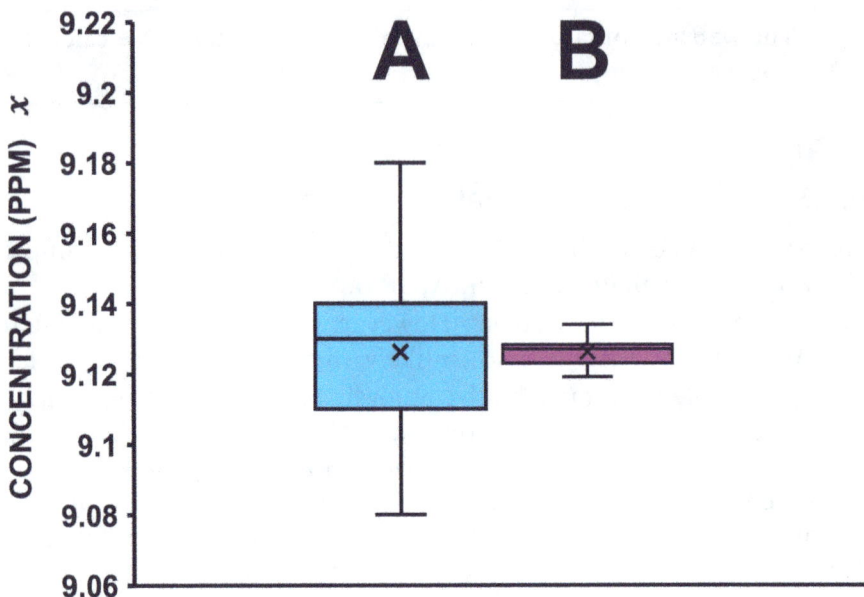

Figure 9.10 Left: Distribution of all 100 values from the pesticide determinations listed in Table 9.2. Right: Distribution of means from each set of 10 measurements (*i.e.* each row). This is the sampling distribution of means and its mean is the same as the mean of the original distribution.

If we would add an infinite number of sample means, then we would obtain the distribution of all possible sample means, $\bar{x}$, which is called the **sampling distribution of the mean**. Its mean $\mu_{\bar{x}}$ is equal to the mean of the original population μ, *i.e.*

$$\mu_{\bar{x}} = \mu \tag{9.12}$$

and its standard deviation, known as the **standard error of the mean (SEM)**, $\sigma_{\bar{x}}$, is directly related to the standard deviation of the original population, σ, through

$$\sigma_{\bar{x}} = \frac{\sigma}{\sqrt{n}} \tag{9.13}$$

where n is the number of measurements underlying each mean value (in our case $n = 10$).

Eqn (9.13) can be perceived as $\sigma_{\bar{x}}$ representing the degree of variability of $\bar{x}$. Important is that the sampling distribution of means gives a good indication of the true mean in the absence of systematic errors.

9.2.3.2 Central Limit Theorem

Another property of the sampling distribution of means is that it tends to follow the normal distribution, regardless of the distribution shape of the original population. This is known as the **central limit theorem**, which states that if we take random samples of size n from a distribution following *any* shape, the distribution of means of these samples *will* follow the normal distribution.

A classic example concerns dice rolls. A six-sided die can fall on six different faces. If we take such a die and roll it 1000 times, then we may expect a block-shaped distribution (Figure 9.11A, emphasized by the pink dashed line). Indeed, there is an equal chance for the die to fall on each of the sides. However, if we now take 3 dice and roll them together and take the average result (*i.e.* the mean), then repeating this 1000 times results in a bell-shaped distribution of the means (Figure 9.11B). This distribution of means increasingly progresses towards a narrower distribution as the number of dice rolled simultaneously increases (Figure 9.11C–F). In Figure 9.11F, a normal distribution is approached. The scale does not run from $-\infty$ to $+\infty$, but the limiting values (1 and 6) seem infinitely far from the distribution.

What this shows us is that, if we take random samples of size n from a distribution following *any* shape, the distribution of means of these samples *will* approach the normal distribution. In practice, a lot of the data we use are a result of averaged measurements. It is therefore **reasonable to assume that the data obtained follow the normal distribution.**

9.2.3.3 Confidence Intervals for Large Sample Sizes

We can now use all the above concepts to establish a confidence interval. Let us first summarize.

1. We have established that the sampling distribution of means represents all possible sample means that we could obtain when we sample our original population (Section 9.2.3.1). We can picture this distribution (Figure 9.12A), which represents

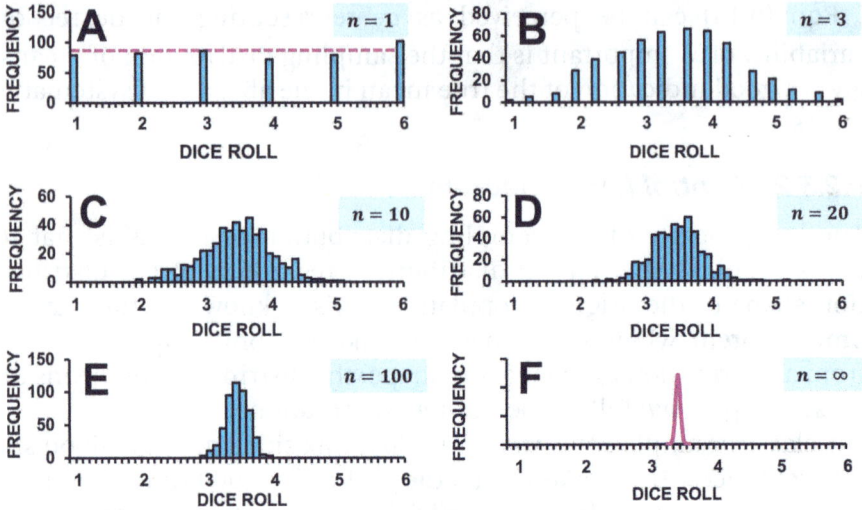

Figure 9.11 Distribution of mean results after randomly rolling n dice simultaneously 1000 times, with (A) $n = 1$, (B) $n = 3$, (C) $n = 10$, (D) $n = 20$, (E) $n = 100$, and (F) $n = \infty$. For example, for panel B, three dice are rolled and the mean value is noted. This is then repeated 999 additional times, yielding the distribution shown.

all values we can possibly obtain if we assume that systematic errors are absent.

2. As explained by the central limit theorem, this sampling distribution of means follows the normal distribution (Section 9.2.3.2).

3. We have learned earlier that a normal distribution can be described by μ and σ, which for this distribution are equal to $\mu_{\bar{x}}$ (eqn (9.12)) and $\sigma_{\bar{x}} = \sigma/\sqrt{n}$ (eqn (9.13)), respectively.

4. Given a probability p, the ICDF can be used to find the value of x for which the corresponding probability of finding that value or lower is p (Section 9.2.2.6).

The confidence interval is about determining a range that is likely to include the true mean μ. This range depends on the **significance level** α, which is 0.01 for a 99% confidence interval, 0.05 for 95%, 0.1 for 90%, *etc.* In other words, for a **95% confidence interval**, 95% of the sample means $\bar{x}$ will lie between $\mu - 1.96\sigma/\sqrt{n}$ and $\mu + 1.96\sigma/\sqrt{n}$ (Figure 9.12A).

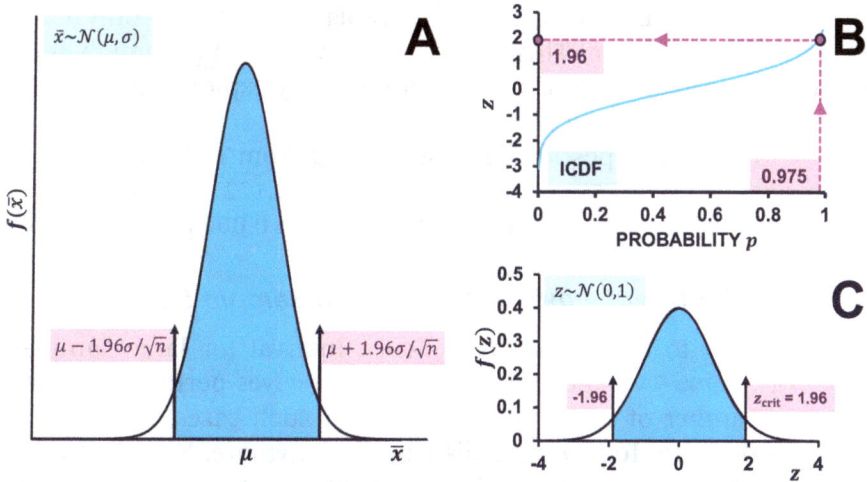

Figure 9.12 (A) Sampling distribution of means with the 95% confidence interval indicated. (B) Finding the z-value (1.96) for which the probability of finding it or lower is 0.975. (C) Standard normal distribution with the critical values corresponding to a confidence interval of 95%.

$$\bar{x} = \mu \pm z\frac{\alpha}{2}\frac{\sigma}{\sqrt{n}} \qquad (9.14)$$

Here, $z\frac{\alpha}{2}$ is the **critical value**, or critical z-value, also written as z_{crit}. The value 1.96 arises from the fact that 95% of the population is covered from the centre of the distribution, and thus the significance α is divided across the two sides ($\alpha/2$ on either side). We can use the ICDF (Section 9.2.2.6) function to determine that a probability of 0.975 corresponds to a z-value of 1.96 (Figure 9.12B). For the left side, one can either again use ICDF to determine the z-value for a probability of 0.025 or simply take its negative given the symmetry of the normal distribution.

Given that we usually only have one sample with repeated measurements with a single mean $\bar{x}$, we can rearrange eqn (9.14) to arrive at the **confidence interval for the mean** μ

$$\mu = \bar{x} \pm z\frac{\alpha}{2}\frac{\sigma}{\sqrt{n}} \qquad (9.15)$$

As σ is often unknown, it can be replaced by s, but only if the sample size n is large. This typically means $n > 30$ (although some scientists suggest $n > 50$). We will numerically explore the rationale for this in the next section.

For our 100 RPLC pesticide measurements from Table 9.2, we find, from eqn (9.15), 95% confidence limits of 9.122 and 9.130 for a mean of 9.126, which we can present as 9.126 ± 0.004 ppm.

9.2.3.4 Confidence Intervals for Small Sample Sizes

We saw how to establish a confidence interval for large samples, but – of course – we more typically find ourselves performing a very limited number of repeat experiments. In such cases, the value of $(\overline{x} - \mu)/s$ does not follow the z-distribution anymore. Note that these s values are based on (much) less than 30 repeats (see the previous section). Typical laboratory practice yields data based on five or even just three repeats. For such small sets, the standard deviations will differ even more. To establish a confidence interval for smaller samples, we cannot employ the normal distribution, as it does not sufficiently express our increasing uncertainty with decreasing sample size. Instead, we can employ **the Student's t-distribution**. It owes its name to William Sealy Gosset, who introduced the t-distribution under the pseudonym "Student". The PDF of this distribution is described by

$$f(x) = \frac{\Gamma\left(\dfrac{\nu+1}{2}\right)}{\sqrt{\nu \cdot \pi} \cdot \Gamma\left(\dfrac{\nu}{2}\right)} \left(1 + \frac{x^2}{\nu}\right)^{-\frac{(\nu+1)}{2}} \tag{9.16}$$

with $x = t$, of which t is given by

$$t = \frac{\overline{x} - \mu}{s/\sqrt{n}} \tag{9.17}$$

Here, the result is the t, the t-statistic is a mere abbreviation from "test statistic". The gamma function (Γ) is a mathematical function that is commonly available in software tools. For positive integers, it can also be computed from $\Gamma(x) = (x - 1)!$. In eqn (9.16), ν is the number of **degrees of freedom,** which – in this case – is equal to the number of repetitions minus one, *i.e.*

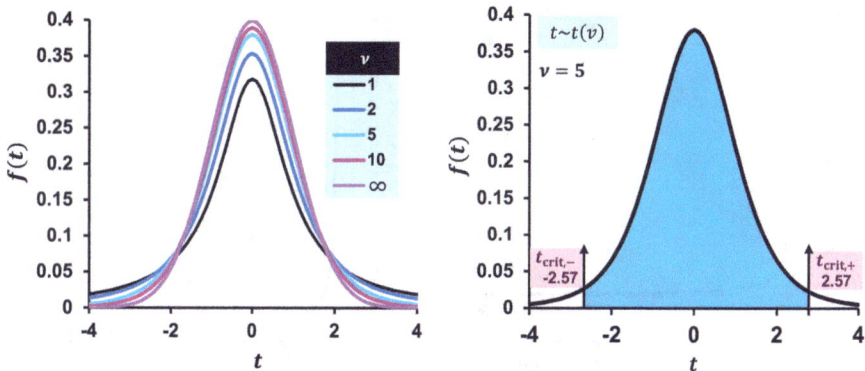

Figure 9.13 Plot of the t-statistic probability density function for different degrees of freedom (left). The distribution becomes wider as the number of degrees of freedom decreases. With $n = \infty$, the distribution becomes the normal distribution. On the right-hand side, the 95% confidence interval of the distribution is illustrated for $v = 5$ ($n = 6$).

$$v = n - 1 \tag{9.18}$$

The degrees of freedom express how certain we are in computing s. A careful inspection of eqn (9.16) reveals that v is the only parameter that must be defined for the t-distribution and we see its influence in Figure 9.13.

The t-distribution becomes much wider as v decreases, thus reflecting the increasing uncertainty. This is also reflected by the critical values for the 95% confidence interval, which are 12.71 ($v = 1$), 4.3 ($v = 2$), 3.18 ($v = 3$), and 1.96 for ($v = \infty$). Indeed, for an infinite number of datapoints ($v = \infty$) the t-distribution is identical to the standard normal distribution.

The **confidence interval of the mean** for small samples can now be defined as

$$\mu = \bar{x} \pm t_{((\alpha/2),v)} \frac{s}{\sqrt{n}} \tag{9.19}$$

where $t_{(\alpha/2),v}$, or t_{crit}, is the critical value at significance α with degrees of freedom v.

If we apply eqn (9.19) to the first row of RPLC measurements from Table 9.2, then we find, for $\bar{x} = 9.119$, $s = 0.0213$, and $t_{(\alpha/2),v} = 2.262$ at $v = 9$ and $\alpha = 0.05$, that the concentration is 9.119 $\pm$

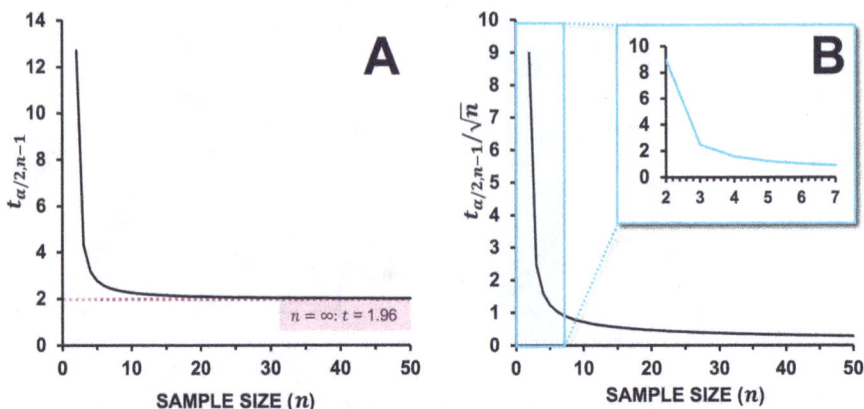

Figure 9.14   A plot of the number of measurements against (A) the critical
t-value plotted for a 95% confidence interval as a function of
n and (B) the same critical t-value but now divided by the
square root of the number of measurements. A dramatic
improvement is obtained for the first five measurements.
Beyond $n > 30$, the critical t-value is almost identical to the
critical z-value, and the normal distribution may be used.

0.0152 ppm. This is a much wider range than that provided for
the 100 measurements using the normal distribution in Section
9.2.3.1.

   This prompts the question of what number of repeated measurements, *i.e.* the sample size, n, is sensible. We see that eqn (9.19)
comprises several components: (i) the precision, (ii) the critical
t-value $(t_{(\alpha/2),v} = t_{crit})$, and (iii) $\sqrt{n}$. Figure 9.14A shows the component t_{crit} plotted for a 95% confidence interval as a function of n.
We can observe a dramatic narrowing of the confidence interval as
n increases with the improvement diminishing around $n = 10$. It
is up to each individual person to decide whether an additional
experiment is worthwhile. However, Figure 9.14B, where both the
components t_{crit} and $\sqrt{n}$ are included, shows that having five
instead of two measurements leads to a seven-fold narrowing of
the confidence interval. In other words, the first five experiments
definitely pay off.

   We furthermore note that beyond $n = 30$, and certainly for $n = 50$,
the critical t-value is very close to 1.96, the value we use for normal
distribution. Hence, it is reasonable to employ the normal distribution and use z-values once n becomes very large ("large samples").

As the number of degrees of freedom is crucial in determining the confidence interval, it is of interest to maximize it. The first method to do so is to, of course, increase the number of experiments – at the expense of time and resources. The second, more attractive, method is to pool the variance from different sets of experiments. This can only be done if the variances are homogeneous (*i.e.* if the standard deviations of the different groups are similar). We will learn how to check for the homogeneity of variances, also referred to as homoscedasticity, in Section 9.4.3. If the variances are homogeneous, they may be pooled using

$$s_{pooled}^2 = \frac{v_1 s_1^2 + v_2 s_2^2 + \dots}{v_1 + v_2 + \dots} \tag{9.20}$$

Here, v_x and s_x^2 are the degrees of freedom and variances for the different sets of experiments. If the variances are not homogeneous, the **Welch–Satterthwaite equation** may be used, which is given by

$$s_r^2 = \sum_{i=1}^{n} k_i s_i^2 \tag{9.21}$$

where $k_i = (\partial r / \partial x_i)^2$ and $r = f(x_1, x_2, \dots, x_n)$. The associated degrees of freedom can be calculated through

$$v_R = \frac{\left(\sum_{i=1}^{n} k_i s_i^2 \right)^2}{\sum_{i=1}^{n} \frac{\left(k_i s_i^2 \right)^2}{v_i}} \tag{9.22}$$

where v_R may be a non-integer value.

9.3 Hypothesis Testing (M)

9.3.1 Introduction to Hypothesis Testing

We have learned several statistics to describe our sample, such as the mean, median, and variance, and we learned how these can be used to infer information about the population of origin. For example, the mean pesticide concentration from the

100 RPLC measurements in Table 9.2 was $\bar{x} = 9.126$ ppm with a standard deviation of $s = 0.0192$ ppm. Suppose that the vial contained a standard pesticide solution certified to be $\mu_0 = 9.124$ ppm, do we find that our data support that the method works correctly? Ultimately, the descriptive statistics do not allow us to draw evidence-based conclusions. To ensure objectivity, a "feeling" of the analytical chemist does not suffice. We need numerical information to support the hypothesis that we are testing for. Hypothesis testing is an **inferential statistical method** that can be used to determine whether there is sufficient evidence in the measured data to support conclusions about the population.

Examples of analytical questions include

- Does the concentration of a contaminant exceed the maximum limit?
- Is the new method with a new column providing the same result?
- Are the product specifications within acceptable limits?
- Does the type of buffer affect the retention in RPLC?
- Is the instrument/column/pump defective?
- Is the ion-exchange method as precise as the CE method?
- Is this chromatogram an outlier in my set of repetitions?
- Are the data normally distributed?

Hypothesis tests will never answer these questions directly, but indicate whether there is evidence in the data, represented as a probability distribution, to support conclusions about the population. Hypothesis testing is a cornerstone of analytical chemistry and many other disciplines, with applications including the calibration of instruments, adherence to specifications or regulations, and development of guidelines or policies.

9.3.2 Comparison of a Sample Mean with a Reference: One-Sample *t*-test

We will first focus on the simple case of comparing our data with a reference value, which is commonly done in separation sciences to obtain method and instrument diagnostics or to answer analytical questions.

For example, suppose that we are identifying a compound in a gas chromatogram using retention indices after a comparison of mass spectra was inconclusive. According to the library, the compound should have a retention index of 1290. Six repeated GC measurements yield values of 1293, 1291, 1285, 1287, 1291 and 1283. A hypothesis test can now be used to check whether the retention index matches.

> See the website for tutorials on how to conduct these hypothesis tests with different software packages: ass-ets.org

9.3.2.1 Step 1: Checking the Requirements of the Statistical Test

Statistical tests typically rely on several requirements that must be met. For example, parametric tests (the tests discussed in this module) all require the data to be normally distributed. If a test is carried out, these conditions are assumed to be fulfilled, because – if they are not – the results of the conducted statistical test are unreliable and invalid, up to the point where they may very well be nonsense. In such cases, a non-parametric test (Module 9.5) must be used.

It is thus imperative that these requirements be considered and – if possible – tested before conducting the actual statistical test. For the comparison of a mean with a reference value, the assumptions include that (i) the means of the data are normally distributed and (ii) the data are randomly sampled. The latter assumption is difficult to verify retrospectively and must be taken into account when the data are obtained. The first assumption can be tested. Such pre-testing will be the topic of Module 9.4. For now, we proceed assuming that these requirements are met.

9.3.2.2 Step 2: Formulating Hypotheses

As the name implies, hypotheses are formulated in a hypothesis test. The **null hypothesis**, H_0, expresses the assumption that the **effect** under investigation does not exist. In analytical chemistry, this effect is typically the deviation encompassed in the analytical question. In our case, this means that H_0 expresses that there is no difference between our observed data and the reference value that cannot be explained by random variation (*i.e.* no effect; the retention index matches). The **alternative hypothesis**, H_1, expresses that the effect is significant. This is mathematically expressed in eqn (9.23a) and eqn (9.23b).

$$H_0: \mu = \mu_0 \tag{9.23a}$$

$$H_1: \mu \neq \mu_0 \tag{9.23b}$$

9.3.2.3 Step 3: Decide on the Significance Level

The goal of the hypothesis test is to assess the data (*i.e.* the measurements) and determine whether there is enough evidence to reject H_0. The word enough here signifies the need of a threshold, which is the significance level, α, which was discussed in Section 9.2.3.3. The significance level is typically set to 0.05 (*i.e.* 5%), but different values may be used in specific fields. The value of 0.05 may feel arbitrary, but it is not without consequence. If we choose a value of 0.05, then we accept that in 5% of the cases we will wrongly reject H_0, whereas in fact it was true. The significance level is also known as the probability of making a **type-I error**. If we increase α, then we increase the risk of making type-I errors, also known as false positives. Hence, it may appear intuitive to minimize α, but we will see in Section 9.3.5 that this comes at a great cost.

9.3.2.4 Step 4: Calculate the Test Statistic

With the hypotheses formulated and the significance level set, the next step is the calculation of the test statistic. This is the value used to summarize the sampled ("observed") data from the population. For our specific case of comparing the mean of our data to a reference value, the *t*-statistic is employed, similar to what we saw for the confidence intervals. It is calculated as

$$t_{\text{obs}} = \frac{|\bar{x} - \mu_0|}{s/\sqrt{n}} = \frac{r_A}{r_B} \tag{9.24}$$

r_A and r_B are graphically illustrated in Figure 9.15B. Panel A shows the boxplot representation of our six repeated retention index measurements and the reference value of 1290 depicted by the light dotted line. In panel B, the same dataset is shown, but now the values are significantly offset by subtracting 10 from each measurement. We can see that in Figure 9.15A, the value 1290 falls within the range, while it clearly does not in panel B. The *t*-statistic in eqn (9.24) mathematically captures this. The numerator expresses the distance, caused by the effect under investigation (depicted in panel B as r_A), between the mean of our measurements and the

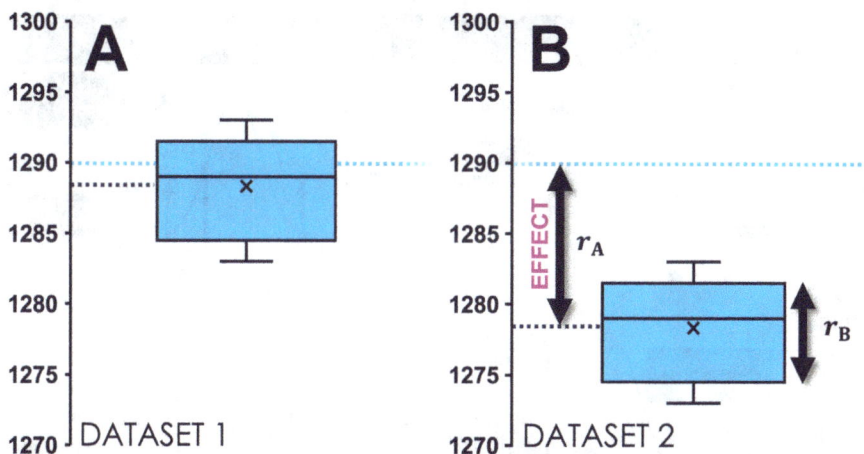

Figure 9.15 Box-and-whisker plot of six repeated retention index measurements in panel A, and the same values negatively offset in panel B. The light dotted line represents the reference value to which the dataset is compared. r_A and r_B graphically express the numerator and denominator of eqn (9.24), respectively. See the text for further explanation.

reference value that we compare to, in this case 1290. The denominator estimates the uncertainty in the mean, the random error, depicted as r_B.

In essence, the hypothesis test of comparing a mean with an absolute value boils down to determining whether any deviation between the experimental mean and the reference value (r_A, the "effect") can be explained by the random variation in the data (r_B). In statistics, this is rephrased as the probability of finding the reference value or more extreme in our data (*i.e.* measurements) assuming that H_0 is true. If the probability, often referred to as the *p*-value, is sufficiently large (*i.e.* larger than the significance level), then the deviation is not significant and H_0 is accepted; if this is not the case, then H_1 is accepted.

9.3.2.5 Step 5: Conducting the Statistical Test

This final step can be done in two ways that will yield the same outcome (Figure 9.16, top).

1. We employ the CDF for our *t*-distribution and calculate the probability (*i.e.* *p*-value) of finding our *t*-value ($t_{obs} = 1.038$; eqn (9.24) with $\bar{x} = 1288.333$ and $s = 3.933$; see introduction of Section 9.3.2 for values) or higher ("more extreme"). Note that

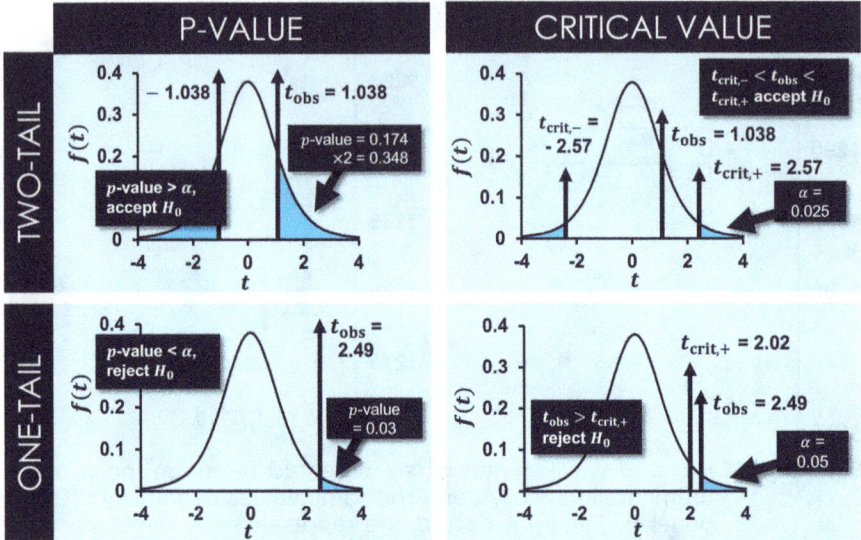

Figure 9.16   Two strategies to conduct the two-tailed t-test (top) and one-tailed t-test (bottom) using the p-values (left) and critical t-values (right). A hypothesis test will yield the same outcome regardless of whether p-values or critical t-values are used. See the text for discussion. The two-tailed test pictures an example in which H_0 is accepted, whereas the one-tailed test displays a different example for which H_0 is rejected.

the CDF will return the probability to find t_{obs} or *lower* (p = 0.8266). We must take the total areaof the distribution (equal to 1) and subtract 0.8266 to find the probability of finding t_{obs} or *higher,* which is p = 0.1734. Note that, due to the fact that the t-distribution is symmetrical, we could also have used CDF with t_{obs} = -1.038 and obtained the same answer.

However, the way our $H_1 : \mu \neq \mu_0$ is formulated, it expresses that the retention index does not exactly match; it does not express whether any deviation is positive or negative. In other words, we are conducting a two-tailed test where we check for any deviation, that is for both a positive *and* negative deviation. We therefore must multiply our p-value by two and end up at p = 0.3468. This means there is a 34.68% chance to observe t_{obs} = 1.038 or a more extreme value under the supposition that the null hypothesis is true. We compare this probability to the significance level, for example 5% and conclude that the chance of finding our value is larger than the significance level, thus accepting H_0.

2. We construct ICDF for our t-distribution and calculate the **critical t-value**, t_{crit}, that corresponds to our significance threshold. For a significance level of 5%, this would be finding the t-value for 0.025 and 0.975 (remember, we check for both a positive *and* negative deviation), which corresponds to the t_{crit} values of –2.57 and 2.57, respectively. Our t_{obs} of 1.038 lies between these two limits and thus we accept H_0.

Both strategies yield the same answer, as graphically depicted in Figure 9.16 (top). We see that, in either case, we arrive at the conclusion that we have no evidence to reject H_0. Had we run the hypothesis test on the dataset shown in Figure 9.15B (where t_{obs} = 7.267, far greater than t_{crit} = 2.57), we would have seen that H_0 would have been rejected. It is important to note that if H_0 is accepted, it does not mean that it is true. It just means that the present data provide insufficient evidence to reject it.

9.3.3 Tail Testing

In the discussion of the previous section, we were interested in *any* deviation from the mean of our data with respect to the reference value. This yielded what is known as a **two-tailed test**, or a two-sided test, where the probability is computed for both the left and right sides of the distribution. However, in some cases, we may want to check only at one side. For example, we may have had reasons to assume that the deviation was only in one direction or we may want to check whether our sample concentration exceeds a specific threshold. In such cases, we can do a **one-tailed test** or one-sided test.

Typically, the hypotheses will already express our desire to do a one-tailed test if applicable. For example, imagine that we have six concentration determinations of a contaminant in a pharmaceutical product. The regulatory document states that the concentration in the drug cannot exceed μ_0. We can then test whether the mean concentration for our pharmaceutical product (μ) exceeds μ_0. Our hypotheses now become:

$$H_0: \mu \le \mu_0 \tag{9.25a}$$

$$H_1: \mu > \mu_0 \tag{9.25b}$$

Here, the alternative hypothesis is one-directional; we want to check whether there is a positive deviation that cannot be explained by the random variation in our data. Consequently, we conduct a **right-tailed *t*-test**. An easy way to remember this is to consider the sign of H_1, which – in eqn (9.25b) – points to the right; hence, this results in a right-tailed test. Had the sign been in the other direction in both equations (*i.e.* $H_0: \mu \ge \mu_0$ and $H_1: \mu < \mu_0$), then we would conduct a left-tailed test. This also underlines why we multiply the *p*-value by 2 in two-tailed tests, as $H_1: \mu \ne \mu_0$ captures both $H_1: \mu > \mu_0$ and $H_1: \mu < \mu_0$.

A one-tailed *t*-test is identical to the two-tailed *t*-test, except that (i) the full significance is now considered only at the relevant side of the distribution (*i.e.* there is only one critical value) and (ii) the numerator of t_{obs} now does not use the absolute difference between $\bar{x}$ and μ_0. Instead, it is calculated as

$$t_{\text{obs}} = \frac{\bar{x} - \mu_0}{s/\sqrt{n}} \tag{9.26}$$

The graphical example of a one-tail test in Figure 9.16 (bottom) demonstrates once more that the use of a *p*-value and critical *t*-value will result in the same conclusion.

9.3.4 Comparison of Two Sample Means

A *t*-test can also be used for the comparison of two means that belong to two different datasets. The execution of the test depends on the way the data were obtained. When the two datasets belong to the same sample (*i.e.* the same sample is measured before and after the studied effect), we refer to a **paired test**, whereas an **unpaired** test concerns the comparison of two datasets that were independently obtained.

9.3.4.1 Unpaired Comparison

For instance, suppose that an analytical department is tasked with checking whether a packaging plant is causing an increase in the concentration of hazardous compounds in a soup product. Two datasets are acquired. One dataset of six repeated measurements of a batch of soup prior to packaging, yielding a mean value, $\bar{x}_{before}$ (which is an estimation of the true mean, μ_2). For the other dataset, six packages of soup are measured, yielding $\bar{x}_{after}$ (which is an estimation of the true mean, μ_1). This is an example of an unpaired comparison, because $\bar{x}_{before}$ and $\bar{x}_{after}$ were not obtained using the same sample.

For an unpaired t-test, in addition to the normality criterion and the random sampling process, an additional requirement is that the variances of these datasets must be homogeneous. In other words, the precision of both datasets must be similar. We will learn how to test for this in Module 9.4.

Our hypotheses become H_0: $\mu_1 \leq \mu_2$ (the concentration did not increase) and H_1: $\mu_1 > \mu_2$ (the concentration did increase). For a one-tailed **unpaired equal-variance comparison** of two datasets, t_{obs} is defined as

$$t_{obs} = \frac{\bar{x}_1 - \bar{x}_2}{s_{pool}\sqrt{\dfrac{1}{n_1} + \dfrac{1}{n_2}}} \tag{9.27}$$

where $\bar{x}_i$ and n_i refer to the mean and number of datapoints of each dataset, respectively. For two-tailed tests eqn (9.27) is still valid, but sometimes the numerator is written as $|\bar{x}_1 - \bar{x}_2|$.

Regardless, in any case, the standard deviations of the two datasets must be pooled using eqn (9.20). The number of degrees of freedom is now calculated as $\nu = n_1 + n_2 - 2$, *i.e.* the total number of datapoints minus two. The remainder of the t-test is similar to that of a comparison with a reference value.

In the event that the variances are not homogeneous, the **Welch's test** or **unequal variance** test is used and calculations are slightly different, with

$$t_{obs} = \frac{\bar{x}_1 - \bar{x}_2}{\sqrt{\dfrac{s_1^2}{n_1} + \dfrac{s_2^2}{n_2}}} \qquad (9.28)$$

using the Welch–Satterthwaite equation (eqn (9.22)), and realizing that $\nu = n_i - 1$, to calculate the number of degrees of freedom

$$\nu = \frac{\left(\dfrac{s_1^2}{n_1} + \dfrac{s_2^2}{n_2}\right)^2}{\left(\dfrac{s_1^4}{n_1^2(n_1-1)} + \dfrac{s_2^4}{n_2^2(n_2-1)}\right)} \qquad (9.29)$$

For the unpaired test, the calculated variance is the sum of two effects: (i) the variance intrinsic to the measurements and (ii) the variance due to any differences in sampling of the two datasets (*e.g.* n_1 vs. n_2). The latter can be eliminated by aiming for a paired test.

9.3.4.2 Paired Comparison

To understand the paired comparison, we consider the dataset provided in Table 9.3. To compare two CE systems, a scientist first analyses a sample of eight different objects on System 1. Later, the same sample of objects are measured on System 2. In contrast to the unpaired setup, we can pair the two datasets because identical objects were used to generate both. The differences of each object are now considered, resulting in a new series of values (*e.g.* "Difference" column in Table 9.3) that reflect the differences between the two datasets. If the two datasets are similar, the mean of the difference should be close to 0. In other words: $H_0: \mu_{\text{difference}} = 0$ and

Table 9.3 Peak areas of a compound of interest measured in a sample of eight objects on two systems. The final column represents the row-based difference of the two datasets.

Object	System 1	System 2	Difference
1	27 185	25 904	1281
2	20 068	21 207	−1139
3	32 593	35 582	−2989
4	176 438	172 343	4095
5	29 319	28 466	853
6	21 634	18 502	3132
7	3387	3273	114
8	28 181	29 889	−1708

Table 9.4 Summary of measured concentrations of a compound of interest in water using RPLC before and after heating.

	n	$\bar{x}$	s
Before heating	12	21.95	2.699
After heating	12	22.65	2.283
Difference	12	−0.81	0.604

$H_1: \mu_{\text{difference}} \neq 0$. From this point, the test becomes a regular t-test comparison of a mean with an absolute value (*i.e.* 0).

Paired designs are favourable in comparison with unpaired designs because the use of identical objects causes the two samples to be less likely different from each other intrinsically. Consequently, any difference between the two datasets is no longer due to the differences within the sample objects (because they are the same) and more likely to be caused by the effect that is being studied. This allows the test to become more sensitive to the effect under investigation. As a direct consequence, paired designs require less resources (*i.e.* fewer experiments) to offer the same statistical power (see Section 9.3.5). Note that factors such as the time at which an object was measured can still significantly affect the outcome. For example, suppose that a mixture is subjected to an intense source of light for 30 min and analysed prior to this exposure, and once more 24 h after exposure, then the statistical test cannot discern between any effect due to the exposure of light and the 24 h of storage time.

We can explore the differences in statistical power with the example outlined in Table 9.4, where summarized data are shown on measured concentrations of an analyte of interest in water using RPLC before and after heating the samples for one hour. If we subject these data to an unpaired t-test comparison, we can use eqn (9.28) and arrive at $t_{\text{obs}} = 0.69$ and a p value of 0.5 (two-sided test). Treating the data as a paired design yields $t_{\text{obs}} = 4.65$ and p-value = 0.0007. In other words, with $\alpha = 0.05$, $H_0: \mu_{\text{difference}} = 0$ is accepted in the unpaired t-test, yet rejected in the paired design! Only if we replace n with 100 for each dataset (*i.e.* if we greatly increase our sample size, but assume we keep the same averages and standard deviations), we arrive at the same conclusion with an unpaired design, with $t_{\text{obs}} = 1.98$ and p-value = 0.049.

> See the website for further examples of paired and unpaired designs: ass-ets.org

Clearly, paired designs should be used in favour of unpaired approaches, if allowed by the nature of the problem. Paired methods are far more efficient and require fewer resources (*i.e.* fewer measurements) to detect a significant difference. In addition, when large differences between samples are expected, a paired design is generally more prudent to ensure that such differences do not obscure the ability of the method to detect any difference due to the effect under investigation.

9.3.5 Power Analysis

An unsettling observation of the previous section is that two different, yet valid, t-tests, under different assumptions, yielded opposing conclusions about the data in Table 9.4. This raises the concern about how reliable these hypothesis tests actually are. We will learn in this section how statistical power determines the reliability of such statistical methods.

We have already seen how the set significance level, α, equals the probability of a type-I error. In this case, also known as a **false positive**, we reject H_0 whereas in fact it is true (*i.e.* there is no effect). In contrast, wrongly accepting H_0, whereas in fact it is false (*i.e.* there is an effect) is referred to as a **type-II error,** and also known as a **false negative,** for which the probability is β. The statistical power of a method, also referred to as the **statistical sensitivity** or the probability of a **true positive,** equals $1 - \beta$. These different definitions are often summarized in the so-called confusion matrix as is shown in Figure 9.17.

In addition to the t-distribution PDF that we have been using thus far to represent H_0 being true, we can also construct the distribution representing H_1 being true. For t-tests, H_1 follows the non-central t-distribution. The equation for this is too complicated for the scope of this book, but it is useful to understand that it depends not only on v but also on the non-centrality parameter θ. For the comparison of a mean with a reference value (Section 9.3.2), θ is given by

		TRUTH	
		H_0	H_1
TEST RESULT	H_0	**TRUE NEGATIVE** Specificity Probability: $1 - \alpha$	**FALSE NEGATIVE** Type-II Error Probability: β
	H_1	**FALSE POSITIVE** Type-I Error Probability: α	**TRUE POSITIVE** Power, Sensitivity Probability: $1 - \beta$

Figure 9.17 The "confusion matrix" clarifying the definitions of key concepts in hypothesis testing.

$$\theta = \frac{\mu - \mu_0}{\sigma/\sqrt{n}} \tag{9.30}$$

For the (unpaired) comparison of two means, it is given by

$$\theta = \frac{\mu_1 - \mu_2}{\sqrt{\dfrac{\sigma_1^2}{n_1} + \dfrac{\sigma_2^2}{n_2}}} \tag{9.31}$$

Such a non-centrality parameter expresses how much the true mean deviates from the expected value, or how much the two means deviate, in comparison with their expected dispersions (standard deviation). This is in line with intuition. The probability of falsely rejecting the alternative hypothesis will logically depend on the expected deviations. The importance of the type-II error becomes dramatically apparent in Figure 9.18, where the PDF for both H_0 and for H_1 are plotted for the unpaired treatment of the data in Table 9.4 in the previous section. We can now see why the p-value of 0.5 was of little use; the probability of making a type-II error is a dramatic 0.89, which is also reflected by the pink-shaded area in the plot. In other words, there is an 89% probability of wrongly accepting H_0 whereas in fact it is false! Its statistical power is thus 100 – 89 = 11%. For the paired treatment of Table 9.4, β equals 0.012 and the statistical power is almost 99%.

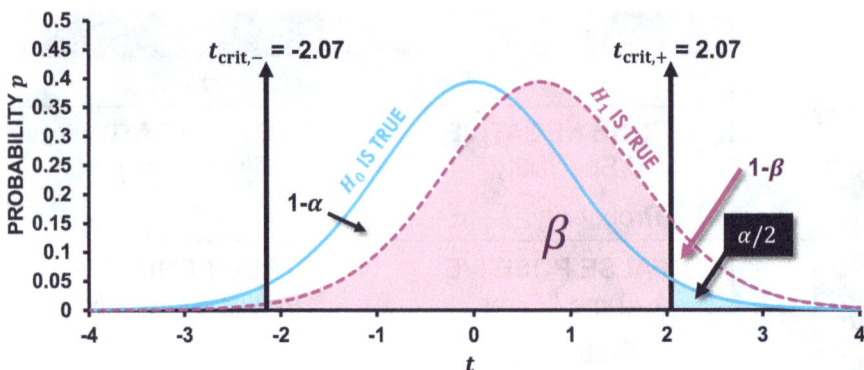

Figure 9.18 Power analysis of the non-matched-pairs treatment of the data in Table 9.4. Plots of the probability density function of the *t*-distribution supporting H_0 (blue), and the PDF of the non-central *t*-distribution supporting H_1.

> 🖱 **See the website for tutorials to calculate the statistical power for different cases: ass-ets.org**

9.3.5.1 *Factors Affecting β and the Statistical Power*

We have seen on several occasions that α equals the area under the curve of the PDF of H_0 beyond the critical limits (*i.e.* t_{crit}). Similarly, β is the area under the curve supporting H_1 that is *less* extreme than t_{crit}. There is thus a direct relation between α and β. As forewarned at the end of Section 9.3.2.3, reducing α (*e.g.* from 0.05 to 0.01) will directly increase β, and *vice versa*.

Other factors that affect β are the sample size and the effect size. A larger sample size will reduce the standard deviation of both distributions and result in less overlap (Figure 9.19A). This is also in agreement with the intuitive thought that an improved confidence in the data reduces the likelihood of falsely accepting H_0.

The effect size expresses the magnitude of the effect causing a difference between the datasets or towards the reference value, irrespective of the sample size. To separation scientists, it is best compared with the concept of chromatographic resolution. Jacob Cohen proposed an equation for the effect size, d, given by

$$d = \frac{|\bar{x}_1 - \bar{x}_2|}{s_{pool}} \tag{9.32}$$

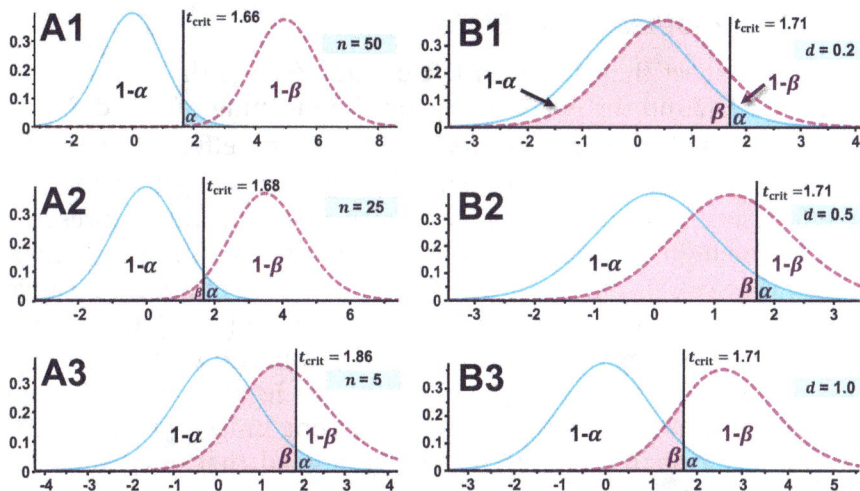

Figure 9.19 (A) Effect of sample size on β when the effect size and α are kept constant, with n = 50 (A1), n = 25 (A2), and n = 5 (A3). (B) The effect of effect size on β when the sample size and α are kept constant, with d = 0.2 (B1, small effect), d = 0.5 (B2, medium effect), and d = 1.0 (B3, large effect). Note the different domains of the x-axis, and consequently the fact that the width of the distribution changes.

With the pooled standard deviation computed through eqn (9.20). The effect size can also be calculated for the population through $d = (\mu - \mu_0)/\sigma$. The effect size intuitively captures the relative difference between the two datasets. An effect size of 0.2 is considered small, 0.5 is considered medium, and 1.0 is considered large. Figure 9.19B shows that larger effect sizes yield a higher statistical power. When the effect size becomes even larger, this can arguably be interpreted as the differences being so large as not to require a statistical test.

While **power analysis** allows the statistical power to be determined *post hoc*, it can also be used prospectively. By deciding on the desired statistical power, the significance level, and with some knowledge of the effect size, the required sample size to obtain that statistical power can be computed. Such calculations require the use of the more complicated non-central distributions and are beyond the scope of this book. Readers are referred to helpful tools that exist online.

9.3.5.2 Hypotheses and Ethics

One final aspect that has been left unaddressed is the formulation of H_0 and H_1, and the effect of this on the meaning of α and β. The convention is that H_0 represents there being no effect, and H_1 the opposite. In this case, α represents falsely concluding that there is an effect, whereas in fact there isn't (false positive), and β represents falsely concluding that there is no effect, whereas in fact there is (false negative). By choosing $\alpha = 0.05$, the chance of a false positive is 5%, and by ensuring sufficient statistical power $(1 - \beta)$, the chance of a false negative may be minimized. This convention is sometimes also referred to as reject-support testing. In legal terms, this convention corresponds to the adage "innocent unless proven guilty beyond a reasonable doubt". It can be argued that good (ethical) science should rely on the similar approach of **falsifiability**, *i.e.* the presumed effect is absent, unless there is statistically significant evidence of the opposite.

9.3.6 Comparison of a Standard Deviation with a Reference Value

While we have thus far focused the hypothesis tests on comparisons involving the mean, it is often useful to focus on the standard deviation. An example is the comparison of the precision of an instrument, s, to the precision specified by the manufacturer, σ_0. The χ^2 ("chi squared") **test for the variance** can be used for this purpose. As the name suggests, the probability density function for this test employs the χ^2 distribution (Figure 9.20), which is given by

$$f(x) = \begin{cases} \dfrac{x^{\left(\frac{\nu}{2}-1\right)} \cdot e^{-\left(\frac{x}{2}\right)}}{2^{\left(\frac{\nu}{2}\right)} \cdot \Gamma\left(\frac{\nu}{2}\right)} & x > 0 \\ 0 & \text{otherwise} \end{cases} \tag{9.33}$$

where $x = \chi^2$, ν is the number of degrees of freedom, which is calculated as $\nu = n - 1$, and $\Gamma\left(\frac{\nu}{2}\right)$ is the gamma function, as was the case with the t-distribution.

The χ^2 statistic can be calculated by

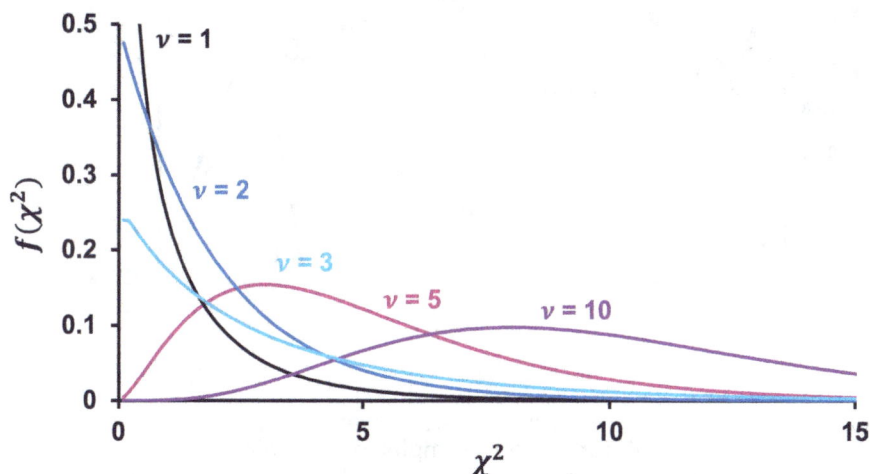

Figure 9.20 The χ^2 distribution plotted for different numbers of degrees of freedom.

$$\chi_{obs}^2 = \frac{(n-1)s^2}{\sigma_0^2} \tag{9.34}$$

Similar to the *t*-test, H_0 is rejected if $\chi_{obs}^2 > \chi_{crit}^2$ or if the *p*-value is smaller than the significance level. For example, suppose that the manufacturer of an LC autosampler injection system capable of injecting typical volumes of 10 μL guarantees that the precision (characterized by the standard deviation) is 0.07 μL or less. By measuring a calibration standard 10 times, we find for our system $s = 0.1$. Is the precision of our instruments worse than 0.07? To answer this, we first formulate the hypotheses tests H_0: $\sigma^2 \leq \sigma_0^2$ and H_1: $\sigma^2 > \sigma_0^2$ and obtain a right-tailed test. (Remember: a higher precision is a smaller standard deviation and variance). We can now use the CDF or ICDF to calculate the *p*-value (Figure 9.21A) or critical value (Figure 9.21B), respectively, and upon comparing with our $\chi_{obs}^2 = 18.37$ (eqn (9.34)) find that H_0 is rejected ($p < \alpha$).

9.3.6.1 Confidence Interval of the Standard Deviation

The χ^2 statistic can also be used to calculate the confidence interval for the standard deviation using eqn (9.35). Here, $\chi_{1-(\alpha/2),n-1}^2$ and $\chi_{\alpha/2,n-1}^2$ refers to the lower and upper critical values, respectively, where $n-1$ represents the degrees of freedom, and $\alpha/2$ the

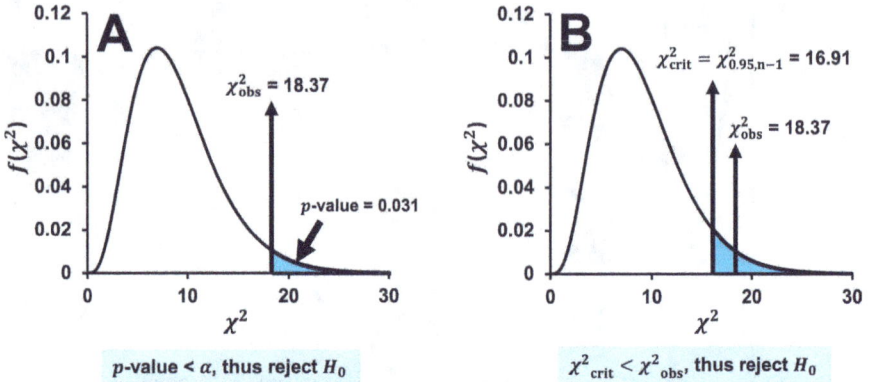

Figure 9.21 Comparison of a standard deviation with a reference value at $\alpha = 0.05$ for the autosampler precision case discussed in the text. (A) Using the p-value and (B) using the critical value. Both cases yield the same answer and H_0 is rejected.

significance level (*e.g.* $\alpha = 0.05$ for 95% confidence interval) divided over both sides.

$$\frac{(n-1)s^2}{\chi^2_{1-(\alpha/2),n-1}} < \sigma^2 < \frac{(n-1)s^2}{\chi^2_{(\alpha/2),n-1}} \tag{9.35}$$

9.3.7 Comparison of Two Standard Deviations

If estimated standard deviations from two samples are compared, the **F distribution** – named after British scientist Ronald Aylmer Fisher – can be used. Because each distribution has its own number of degrees of freedom, the F distribution requires both to be specified, as can also be seen from its PDF.

$$f(x) = \frac{\sqrt{\dfrac{(v_A \cdot x)^{v_A} \cdot v_B^{v_B}}{(v_A x + v_B)^{v_A+v_B}}}}{x \cdot B\left(\dfrac{v_A}{2}, \dfrac{v_B}{2}\right)} \tag{9.36}$$

Where $x = F$, B is the beta function with Γ again denoting the gamma function, as is seen by

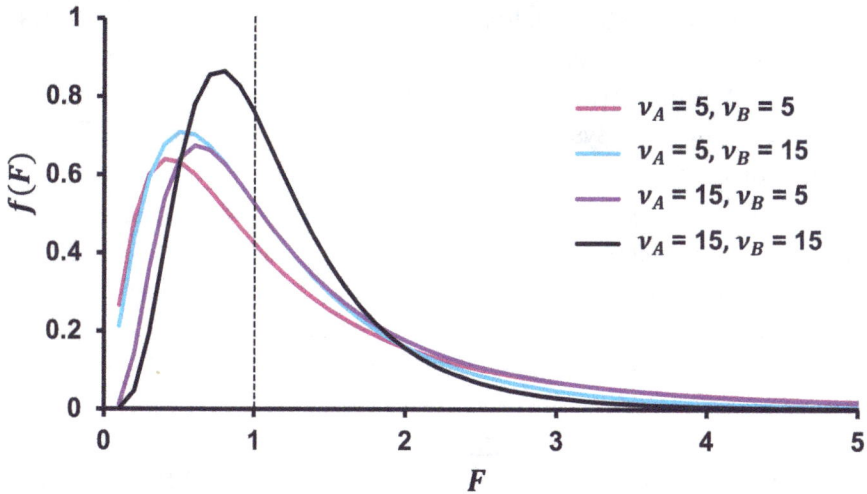

Figure 9.22 Plot of the *F* distribution for different combinations of the number of degrees of freedom of the two datasets. For statistical tests, the *F* statistic is calculated so that *F* is always equal to or larger than 1 (dashed line).

$$B(x, y) = \frac{\Gamma(x) \cdot \Gamma(y)}{\Gamma(x + y)} \tag{9.37}$$

Note that in eqn (9.36) there is a distinction between B (the beta function) and sample B. The dependence of the *F* distribution on the degrees of freedom of both sets is shown in Figure 9.22.

The *F* statistic can be calculated through

$$F_{obs} = \frac{s_A^2}{s_B^2}, F_{obs} \geq 1 \tag{9.38}$$

Note that the ratio is always formulated so that F_{obs} is equal to or larger than 1 to increase the statistical power (indicated by the dashed line in Figure 9.22). An F_{obs} statistic closer to 1 indicates that the variances are similar, which is in support of H_0 ($\sigma_A^2 = \sigma_B^2$ for two sided, or $\sigma_A^2 \leq \sigma_B^2$ for 1 sided). In contrast, larger F_{obs} values cast doubt about H_0. Again, if $F_{obs} > F_{crit}$, or if the *p*-value associated with F_{obs} is smaller than the significance level, we reject H_0.

For example, suppose we want to compare the stability of the retention time for an analyte on an old system relative to a newer system. The old system yields a mean retention time of $\bar{x}_{\text{old}} = 1.37$ min, with $s_{\text{old}} = 0.033$ based on 10 repetitions. For the other system, $\bar{x}_{\text{new}} = 1.42$ min with $s_{\text{new}} = 0.028$ is obtained using 7 repetitions. Now, does the new system yield better retention time stability? In other words, $H_0: \sigma_{\text{old}}^2 \leq \sigma_{\text{new}}^2$ (the newer system has a similar or worse precision; *i.e.* no effect) and $H_1: \sigma_{\text{old}}^2 > \sigma_{\text{new}}^2$ (the newer system has a better precision; *i.e.* significant effect). We can use the CDF of the *F*-distribution (with $\nu_{\text{old}} = 9$ and $\nu_{\text{new}} = 6$ degrees of freedom) to obtain a *p*-value of 0.356. This is larger than $\alpha = 0.05$, and we thus accept H_0 and conclude that with the present data there is no evidence to reject H_0.

> **See the website for further practical examples and their calculations: ass-ets.org**

9.3.8 Multiplicity Problem

Our tests thus far have always been limited to one or two datasets on which statistical inferences were made. Comparison of a larger number of datasets is also possible, but before we explore the statistical methods for such comparisons, we must first understand the multiplicity problem.

Suppose that we have three methods that we would like to compare, *i.e.* Method 1, Method 2, and Method 3. We could now employ the comparison of two means several times and pairwise compare Method 1 with 2, 1 with 3, and 2 with 3. For each of these tests, there is a significance level, α, which is the probability of wrongly concluding that this particular pair of means is different. As the number of comparisons is increased, the likelihood of at least once making a type-I error also increases. We can also picture this when imagining the comparison of a chromatographic method applied in two different laboratories. Even though both labs will follow the method to the letter, the likelihood increases that we will find at least one difference by pure chance as we invoke more attributes in the comparison (*e.g.* buffer concentration, retention time stability, detector sensitivity, laboratory-specific effects). As the number of simultaneous comparisons increases, it becomes more likely that the compared datasets will differ at least in one attribute. As a consequence, confidence will be reduced.

The error rate of the family of tests can be calculated through

$$\alpha' = 1 - (1 - \alpha)^k \tag{9.39}$$

Here, k is the number of simultaneous comparisons made (in the above example $k = 3$). If we do 10 tests simultaneously, each at $\alpha = 0.05$, then using eqn (9.39) we can calculate that we have 40% chance of making a type-I error! The **Bonferroni correction** adjusts for multiple comparisons by setting the individual test significance level to α/k. Re-using eqn (9.39) but now with the corrected significance level yields for $k = 10$ an acceptable α' of 0.049.

However, due to the much lower corrected α' we must now accept that for each test β will be higher and that consequently our statistical power decreases.

9.3.9 Analysis of Variance (ANOVA)

9.3.9.1 One-way ANOVA

The multiplicity problem discussed in the previous section renders it undesirable to conduct multiple comparisons at the same time. Then what if our analytical question does require the comparison of several groups?

For example, the concentration of a peptide is quantified using three different LC-MS systems. The data obtained are shown in

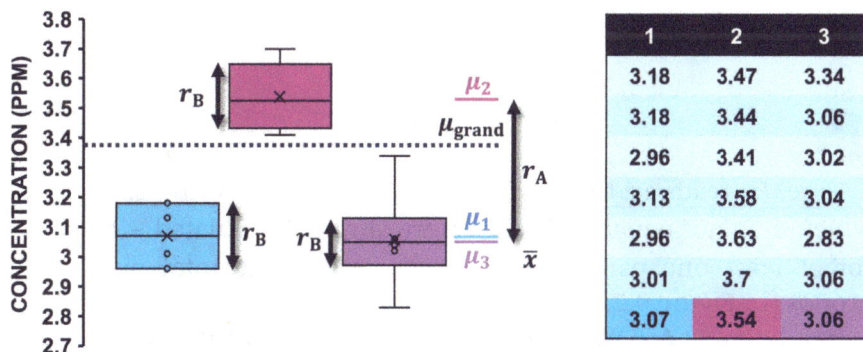

Figure 9.23 Box-and-whisker plot representations of the data shown in the table. An ANOVA study compares the within-group variation (r_B) with the between-group variation (r_A).

Figure 9.23 with a box-and-whisker plot representation of each of the systems.

To statistically answer the question "Which means are different?" we could now do a pairwise comparison of each method and see for which pair a difference is found. We would have to run these against a corrected significance level (Section 9.3.8) and lose statistical power for each. Instead, we can also change our question to "Is the factor instrument significant?" or, in other words, "Do any of the different instruments yield different results?".

To understand why this affects our computations, we remember from eqn (9.2) that each value $x_{i,j}$ of each group j (*i.e.* the number of the instrument) and each repetition i in our table is the sum of the true value μ and an error $e_{i,j}$ as $x_{i,j} = \mu + e_{i,j}$. However, if there is an effect of the factor under investigation (in this case, the different instruments used to obtain the data), then it is reasonable to assume that it would impact the $x_{i,j}$ as

$$x_{i,j} = \mu + a_j + e_{i,j} \tag{9.40}$$

Here, a_j is the effect of instrument j relative to the overall mean μ. Eqn (9.40) implies that the variation in $x_{i,j}$ around μ is not exclusively determined by the random error $e_{i,j}$, but inflated by a_j if there is an effect. Analysis of variance, often abbreviated as ANOVA, is a useful way of finding out whether the factor under investigation has an effect. If the factor has no effect, then the **within-variability** in each group (r_B; Figure 9.23) is similar to the **between-variability** between the means of the groups (r_A; Figure 9.23). ANOVA thus considers the hypotheses

$$H_0: a_1 = a_2 = a_k = 0 \tag{9.41a}$$

$$H_1: \exists 1, k: a_j \neq a_j' \tag{9.41b}$$

Here, H_1 reads as: there exists ($\exists$) at least one group $j = 1, 2, ..., k$ for which a_j is not equal to 0, or – more simply formulated – $H_1: a_j \neq 0$ for at least one pair $j = 1, 2, ..., k$; k depicts the total number of groups (in Figure 9.23, $k = 3$).

In a **one-way ANOVA**, one factor is studied and the data structure is as shown in Figure 9.24. Here, n_j is the number of replicates within group j. Note that the numbers of replicates do not need to

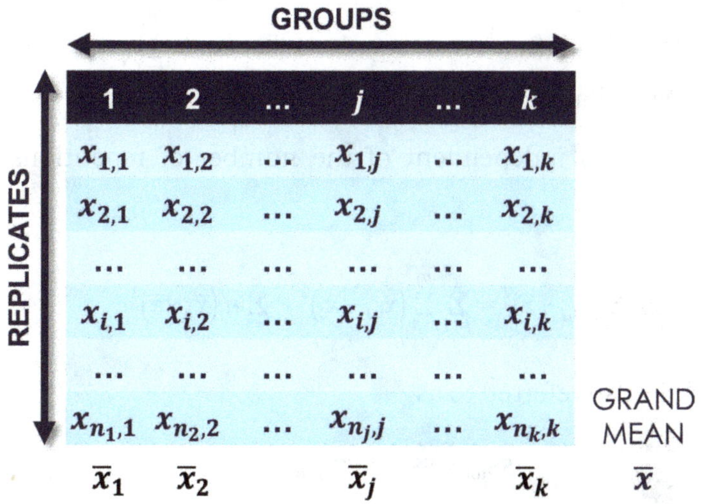

GROUPS

Figure 9.24 Data structure of a one-way ANOVA, where k is the number of groups, and n_j the number of repetitions for group j.

be equal for the different groups (such a case is also referred to as "unbalanced").

For the comparison of the within-group variance with the between-group variance, we must first break down the total variance into the respective components. We start by noting that the total error in a datapoint $x_{i,j}$ is the difference between $x_{i,j}$ and the mean of the data in that group plus the difference between this group mean $(\bar{x}_j)$ and the grand mean $(\bar{x})$, or

$$e_{i,j} = x_{i,j} - \bar{x} = \left(x_{i,j} - \bar{x}_j\right) + \left(\bar{x}_j - \bar{x}\right) \tag{9.42}$$

Errors can be both positive and negative. We square them to avoid them cancelling each other out. If we now sum the squared errors for all repetitions i in all groups j, we obtain a measure for the variability in the total error referred to as the **sum of squares (SS)**, as $\sum_{j=1}^{k} \sum_{i=1}^{n_j} \left(x_{i,j} - \bar{x}\right)^2$, which can be further broken down into

$$\sum_{j=1}^{k} \sum_{i=1}^{n_j} \left(x_{i,j} - \bar{x}\right)^2 =$$

$$\sum_{j=1}^{k} \sum_{i=1}^{n_j} \left(x_{i,j} - \bar{x}_j\right)^2 + \sum_{j=1}^{k} \sum_{i=1}^{n_j} \left(\bar{x}_j - \bar{x}\right)^2 + 2\sum_{j=1}^{k} \sum_{i=1}^{n_j} \left(x_{i,j} - \bar{x}_j\right)\left(\bar{x}_j - \bar{x}\right) \tag{9.43}$$

However, the latter term equals 0 because $\sum_{i=1}^{n_j}\left(x_{i,j}-\overline{x}_j\right)=0$ due to the cancellation of the deviations of the mean. Moreover, $\sum_{j=1}^{k}\sum_{i=1}^{n_j}\left(\overline{x}_j-\overline{x}\right)^2$ is independent of the number of repetitions in the group, thus

$$\sum_{j=1}^{k}\sum_{i=1}^{n_j}\left(x_{i,j}-\overline{x}\right)^2 = \sum_{j=1}^{k}\sum_{i=1}^{n_j}\left(x_{i,j}-\overline{x}_j\right)^2 + \sum_{j=1}^{k}n_j\left(\overline{x}_j-\overline{x}\right)^2 \qquad (9.44a)$$

Which is also referred to as

$$\text{SS}_{tot} = \text{SS}_{res} + \text{SS}_{group} \qquad (9.44b)$$

Here, SS_{tot} is again the total sum of squares, which we now have broken down into two components: SS_{res} is the within-group sum of squares, and SS_{group} is the between-group sum of squares. The latter is the sum of squares due to the effect of the factor under investigation. We can now distill a measure of variation for the different sum of squares by dividing by the number of degrees of freedom resulting in the **mean squares** ($\text{MS} = \text{SS}/\nu$), so that

$$\text{MS}_{tot} = \frac{\sum_{j=1}^{k}\sum_{i=1}^{n_j}\left(x_{i,j}-\overline{x}\right)^2}{n-1} \qquad (9.45a)$$

$$\text{MS}_{res} = \frac{\sum_{j=1}^{k}\sum_{i=1}^{n_j}\left(x_{i,j}-\overline{x}_j\right)^2}{n-k} \qquad (9.45b)$$

$$\text{MS}_{group} = \frac{\sum_{j=1}^{k}n_j\left(\overline{x}_j-\overline{x}\right)^2}{k-1} \qquad (9.45c)$$

Here, MS_{tot} is the total variance, MS_{res} is the within-group variance, and MS_{group} is the between-group variance (due to the factor a_j under investigation). It is useful to appreciate that the within-group variance is equal to the pooled standard deviation (eqn (9.20)), which also induces the requirement for ANOVA that the variances of the groups must be homogeneous (*i.e.* comparable). We can now calculate the F statistic for ANOVA

Table 9.5 Example of a one-way ANOVA table.

Source	ν	SS	MS	F	p-value
Between groups	$k - 1$	SS_{group}	$SS_{group}/(k - 1)$	MS_{group}/MS_{res}	p
Residual	$n - k$	SS_{res}	$SS_{res}/(n - k)$	—	—
Total	$n - 1$	SS_{total}	—	—	—

$$F_{obs} = \frac{MS_{group}}{MS_{res}} \qquad (9.46)$$

The hypothesis test is completed as is custom for F-tests by comparing either the associated p-value of F_{obs} with the significance level with $k - 1$ and $n - k$ degrees of freedom in the numerator and denominator, respectively, or by comparing F_{obs} with F_{crit}. Note that the ANOVA test is always one sided as it otherwise would imply that the between-group variance was smaller than the within-group variance. Tools to conduct ANOVA typically return a table such as Table 9.5. If the p-value is larger than the significance (α), H_0 is accepted, suggesting that there is no effect of the studied factor.

By avoiding the Bonferroni correction, ANOVA yields greater statistical power than multiple pairwise comparisons of the means. It is important to note that ANOVA assumes (i) random sampling of data, (ii) variances of the groups being homogeneous, (iii) the data being normally distributed, and (iv) no outliers being present. While statistical tests should be used to check this, it is useful to first visualize the data using a box-and-whisker plot to gauge the danger of violating these requirements.

It should be noted that depending on the actual question of interest, different models of ANOVA can be used. When the interest is specifically to check whether the effects of one or more groups a_j are deviating from 0 (*e.g.* "Are any of the columns from different batches influencing the results?"), we speak of a **fixed-effect model**, also known as Model I. In the event that H_0 is rejected, it can then be useful to do a *post hoc* analysis (*e.g.* "Which of these groups caused H_0 to be rejected?") and see which means actually are deviating using pairwise t-tests. While these strictly speaking should be conducted using the Bonferroni correction, it can be argued that the ANOVA can be considered as a pre-test protecting against the family-wise error rate (eqn (9.39)).

When the interest is in checking whether there is an underlying effect that renders a group deviate from 0 (*i.e.* "Is the sample homogeneous?"; with sampling at different locations), then we speak of a **random-effect model**, also known as Model II. Given that the effect – if present – is random, no *post hoc* analysis is needed when H_0 is rejected.

> See the website for guides on how to conduct an ANOVA with different software tools. The guide will also address conducting ANOVA when the number of datapoints is different for each group: ass-ets.org

9.3.9.2 Two-way ANOVA

It is also possible to test multiple factors at the same time in what is known as *n*-way ANOVA. Given the complexity of higher-order ANOVA tests, we will stick to a two-way ANOVA.

An example of such data is shown in Table 9.6, where the retention time is measured using different buffer types at various concentrations. ANOVA can now be used to answer different questions: (i) "Is the buffer type influencing the results?", (ii) "Is the buffer concentration influencing the results?", and, interestingly, (iii) "Is there interaction between the buffer concentration and buffer type?". The term **interaction** refers to the dependency of the two factors under investigation. In this case, interaction would mean that the effect of buffer concentration depends on the effect of buffer type.

A single value $x_{i,j,h}$ belonging to repetition i, level j in a, and level h in b now comprises several contributions

$$x_{i,j,h} = \mu + a_j + b_h + (ab)_{j,h} + e_{i,j,h} \tag{9.47}$$

where $(ab)_{j,h}$ is the interaction, a_j is the effect of group j of a, and b_h is the effect of group h of b. In terms of hypotheses, we now have up to three sets of hypotheses: $H_0: a_1 = a_2 = a_k = 0$ ("factor a is not significant") and $H_1: a_j \neq 0$ for at least one j; $H_0: b_1 = b_2 = b_1 = 0$ ("effect b is not significant") and $H_1: b_h \neq 0$ for at least one h; and $H_0: ab_1 = ab_2 = ab_{h*j} = 0$ ("the change in a does not depend on the change in b") and $H_1: ab_j, \neq 0$ for at least one j.

Computationally, it is beyond the scope of this book to detail the exact calculations, but 9.44b is now

Table 9.6 Example of two-way ANOVA data with three repetitions for each combination of the two factors under investigation (buffer concentration and buffer type). Values are retention times obtained in a HILIC method.

	Buffer 1	Buffer 2	Buffer 3
Concentration 1	3.11	3.10	3.21
	3.09	2.99	3.18
	3.24	3.33	2.90
Concentration 2	2.99	3.17	3.10
	2.90	3.17	3.15
	3.17	3.01	3.11
Concentration 3	3.01	2.90	3.17
	2.90	3.00	3.01
	3.33	2.98	3.02

$$SS_{tot} = SS_a + SS_b + SS_{ab} + SS_{res} \tag{9.48}$$

In a two-way ANOVA, these SS are compared using several F tests. It is important to note that there are several methods to compute the SS. In this book, we employ the common method (known as type III).

Execution of the two-way ANOVA test will consume more degrees of freedom at the cost of statistical power due to the fact that more comparisons are made. The assessment of the interaction also consumes degrees of freedom. Most software tools allow omitting the interaction to preserve statistical power.

9.4 Pre-testing (M)

The majority of the statistical methods that were covered in Modules 9.2 and 9.3 required specific requirements to be met. For example, the unpaired t-test comparison of two means as well as ANOVA requires the variances to be homogeneous, whereas almost all tests – and, incidentally, the calculation of the plate number according to (eqn (1.27)) – require the data to be normally distributed. Connected to both aspects is that the presence of outliers can cause data that are in fact normally distributed to appear not normal. This is similarly true for regression (Module 9.6) and analytical methods in general. In this module, different pre-testing methods will be covered that can be used to verify whether the requirements for parametric statistical methods are met.

9.4.1 Testing for Normality of Data: Goodness-of-fit Tests

The tools discussed here to test for the normality of data have in common that they compare the observed distribution of a dataset with what would be expected when the data were to follow a specific distribution. While this is the normal distribution in this case, it is important to note that the tests discussed here can also be used to test the validity of other distributions.

9.4.1.1 Graphical Methods

We start with a simple, illustrative method that will also help us understand the other tools. Suppose that 18 repeated measurements were made of the concentration of an additive in water. The resulting values are 2.603, 2.530, 2.732, 2.690, 2.600, 2.790, 2.722, 2.655, 2.595, 2.724, 2.455, 2.465, 2.585, 2.852, 2.710, 2.657, 2.670 and 2.720 ppm. When plotted as a box-and-whisker plot such as in Figure 9.25A, the distribution looks symmetrical and possibly normal at first glance. This is in contrast to the histogram in Figure 9.25B,

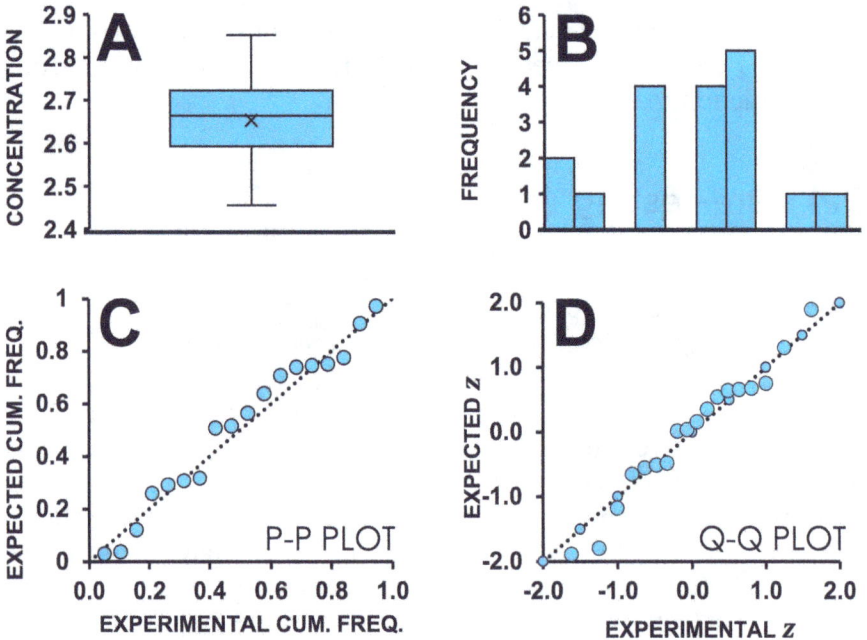

Figure 9.25 The series of repeated measurements of the additive concentration (see the text) expressed as a (A) box-and-whisker plot, (B) histogram, (C) P-P plot, and (D) Q-Q plot.

which suggests otherwise. While insightful, these methods are not designed to check whether the data fit the distribution of interest.

A more useful graphical approach to assess how well the data fit to a distribution is the **probability–probability (P–P) plot**. Here the empirical CDF of the data is plotted against the CDF of the distribution to which the data are assumed to adhere to.

For our data, the steps to create a P–P plot to check for the normality of the data are as follows.

1. All values are sorted.
2. The sorted values are ranked. This creates a second vector of rank numbers.
3. To create the empirical CDF, the ranks (*i.e.* the rank numbers, not the actual values!) are converted into the cumulative frequency using eqn (9.49), where n is the total number of datapoints.

$$\text{cumulative frequency} = \frac{\text{rank}}{n+1} \tag{9.49}$$

4. To create the theoretical CDF of the normal distribution, the sorted values (x) are first standardized through $z = (x - \mu)/\sigma$ (eqn (9.10)). This is our PDF where we assume that $\mu = \bar{x}$ and $\sigma = s$.
5. The PDF is converted to the CDF. This is our theoretical CDF.

The result of creating the P–P plot for the additive data is shown in Figure 9.25C, where the empirical CDF and theoretical CDF of the normal distribution are plotted against each other. If the data are perfectly normal, all datapoints should appear on the diagonal.

Another variation of the P–P plot is the **quantile–quantile (Q–Q) plot,** where the empirical CDF values are also converted into z-values using eqn (9.10) and plotted against the theoretical expected values when assuming that the data follow the normal distribution. This is shown in Figure 9.25D.

9.4.1.2 Kolmogorov–Smirnov Test

The graphical tools presented in the previous sections are useful, but do not express the result as a number. For continuous distributions, the Kolmogorov–Smirnov test can be used. In this test, the differences between the expected and observed CDFs are compared. The

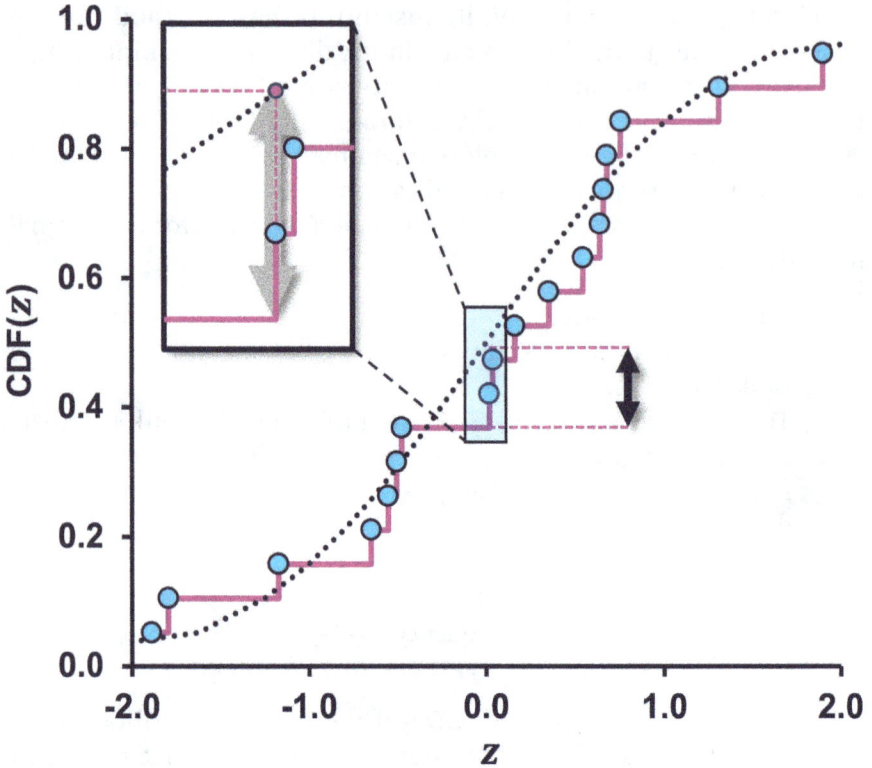

Figure 9.26 In the Kolmogorov–Smirnov test, the largest difference between the expected and experimental values is used as the test statistic.

largest difference (indicated in Figure 9.26 using an arrow) is used as a test statistic in a hypothesis test with H_0: the data are normal, and H_1: the data are not normal. This is then compared to a tabulated critical value (Appendix A.1).

The Kolmogorov–Smirnov test is in principle only applicable to large samples ($n > 50$). As the sample size decreases, it becomes less likely that s sufficiently captures σ (and $\bar{x}$ captures μ). In such cases, it is necessary to use the **Lilliefors correction**, which employs the same computations to acquire the statistic, but more stringent tabulated critical values that render it more sensitive to deviations from normality when the sample size is low (Appendix A.2).

For small sample sizes, the Shapiro–Wilk test may be preferred as it has been shown to yield better statistical power than the Kolmogorov–Smirnov test with a Lilliefors correction.[1]

It is important to note that goodness-of-fit tests such as the normality tests mentioned above are vulnerable to outliers.

9.4.1.3 χ^2 Test

For non-continuous large sets of data, the χ^2 goodness-of-fit test compares the frequency distribution with a known distribution, such as the normal distribution. Note the word frequency, which here refers to binned – and hence non-continuous – data. An example may be the categorical distinction of the degree of degradation in a sample over time through the detection of a marker compound (*e.g.* 0–10 ppm loss of a marker analyte, 10–20 ppm, 20–30 ppm, *etc.*). The statistic is computed as

$$\chi_{obs}^2 = \sum_i \frac{(O_i - E_i)^2}{E_i} \tag{9.50}$$

Here, O_i is the observed frequency for each bin (*i.e.* the original data in each category i) and E_i is the expected frequency. E_i is obtained in a similar manner as the expected CDF values used for the P–P plot (Section 9.4.1.1) by establishing the z-values for the data using $\bar{x}$ and s. The relative frequency for each bin is then computed, which can be converted to the expected frequency E_i by multiplying each relative frequency with the total number of measurements, n. χ_{obs}^2 can then be obtained from eqn (9.50) and compared with χ_{crit}^2 at $k - 3$ degrees of freedom, where k is the number of bins. Alternatively, the p-value can be computed. The χ^2 test is analogous to that discussed in Figure 9.21.

> See the website for practical examples of normality tests: assets.org

9.4.2 Outlier Testing

A question often asked by an analytical chemist is: "Is this datapoint an outlier?" The presence of outliers has a significant impact on the outcomes of statistical tests, such as the goodness-of-fit tests discussed above.

Several outlier tests have been developed, but a serious problem is that these rarely give the same result. One outlier test may conclude that there are two outliers, whereas another test does not detect any. Inappropriate removal of a suspected outlier can lead to significant underestimation of the variance. Several outliers can indicate problems with the analysis method.

   The decision on how to handle outliers is thus a sensitive topic. It is arguably threatened by subjectivity and thus requires careful consideration. This is especially true in regulated environments, where the removal of an outlier can be regarded as omission of evidence. The removal of outliers should thus never be considered a routine operation. The main purpose of outlier tests should be to detect anomalies in the results, rather than to discard unwelcome datapoints. We believe that any discarded datapoints must always be documented.

   Where relevant, the hypotheses for the tests in this section are always H_0: no outliers present, and H_1: outliers present.

9.4.2.1 Dixon's Test

One of the classical outlier tests is the Dixon's Q test. This test considers the outermost points and compares these to their neighbours. It can be used when σ is not known. To apply the test, the values (x) in the dataset are first sorted. Depending on the number of datapoints, the Q statistic is calculated differently. For $3 \leq n \leq 7$ it is calculated for the smallest (x_1) or largest value (x_n), respectively, using

$$Q_{10,x_1} = \frac{x_2 - x_1}{x_n - x_1} \tag{9.51a}$$

$$Q_{10,x_n} = \frac{x_n - x_{n-1}}{x_n - x_1} \tag{9.51b}$$

For $8 \leq n \leq 12$ it is calculated through

$$Q_{11,x_1} = \frac{x_2 - x_1}{x_{n-1} - x_1} \tag{9.52a}$$

$$Q_{11,x_n} = \frac{x_n - x_{n-1}}{x_n - x_2} \tag{9.52b}$$

And when $n > 12$ through

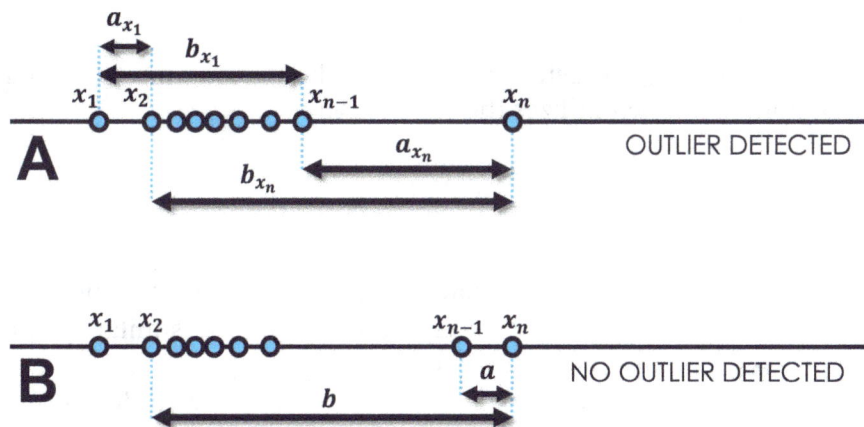

Figure 9.27 Two depictions of the Dixon's Q test to check whether x_n is an outlier using eqn (9.52b). The arrows with designations a and b are graphical representations of the numerator and denominator of this equation. The test is vulnerable to the presence of two outliers, which for the data shown in panel (B) will not be detected, in contrast to the case of a single outlier in panel (A). See the text for further clarification.

$$Q_{22,x_1} = \frac{x_3 - x_1}{x_{n-2} - x_1} \tag{9.53a}$$

$$Q_{22,x_n} = \frac{x_n - x_{n-2}}{x_n - x_3} \tag{9.53b}$$

Figure 9.27A expresses the Q test graphically for a case of 8 datapoints. Eqn 9.52a and b are graphically expressed by the arrows with designation a and b representing the numerator and denominator, respectively. While the statistic can be calculated for both the left (x_i) and the right side (x_n) simultaneously, the largest of the two values will indicate the most likely candidate to be an outlier as it depicts a larger difference of x_i with respect to the rest of the datapoints. This observed Q_{obs} value is then compared to a tabulated Q_{crit} value (Appendix A.3). If $Q_{obs} < Q_{crit}$, then H_0 is accepted.

The vulnerability of Dixon's method is depicted in Figure 9.27B. If there are two outliers present on the same side, the Dixon's test is unlikely to detect any of them.

9.4.2.2 Grubbs Test

A different strategy to detect an outlier is the maximum normalized deviation test, which utilizes the calculation of

$$G_{\text{obs}} = \frac{|x_i - \bar{x}|}{s} \tag{9.54}$$

Here, x_i is the suspected datapoint, either x_1 or x_n. The G_{obs} statistic is then compared to a tabulated G_{crit} value similar to the Dixon's test (Appendix A.4, Table A.4).

However, more interesting is the Grubbs' statistic to detect two outliers, which compares the sum of squares of the full dataset $\left(\text{SS}_0 = \sum(x_i - \bar{x})^2\right)$ to that of the data without the first two $\left(\text{SS}_{x_1,x_2}\right)$ orthe final two $\left(\text{SS}_{x_{n-1},x_n}\right)$ datapoints, such that

$$G_{\text{obs},x_1,x_2} = \frac{\text{SS}_{x_1,x_2}}{\text{SS}_0} \tag{9.55a}$$

$$G_{\text{obs},x_{n-1},x_n} = \frac{\text{SS}_{x_{n-1},x_n}}{\text{SS}_0} \tag{9.55b}$$

Also here, G_{obs} is compared with G_{crit} (Appendix A.4, Table A.5), but in this specific case, H_0 is rejected if $G_{\text{obs}} < G_{\text{crit}}$, because the test statistic now is a ratio of variances. The variance shrinks dramatically if influential outliers are removed.

9.4.2.3 Critical-range Method

The Dixon's and Grubbs' tests allow for the detection of outliers when σ is not known. When σ is known, it is possible to use the critical range method. This test rejects H_0 when the following criteria are met:

$$(x_{\text{max}} - x_{\text{min}}) > \text{CR}_{\alpha,n} \cdot \sigma \tag{9.56}$$

Here, $\text{CR}_{\alpha,n}$ is a tabulated critical-range value (Appendix A.5), depending on the probability (α) and the number of data points (n), and x_{max} and x_{min} are the maximum and minimum values of the dataset, respectively.

9.4.2.4 Other Outlier Tests

Examples of other known outlier-detection methods include those that measure the deviation of a datapoint towards a certain point. One instance utilizes the **median absolute deviation** (MAD), which is a robust measure, sometimes used as a robust alternative to the standard deviation, to quantify the variability of univariate data and is the median of the absolute differences of each value and the median value, given by

$$\text{MAD} = \text{median}(|x_i - \tilde{x}|) \tag{9.57}$$

Mathematical outlier detection tools can check for each datapoint in a series whether the value is more than a certain number (*e.g.* 3) of MAD away from the median of the data. For example, for a series of repeated measurements, the MAD can be calculated. For each value, it could then be tested whether it falls beyond 3 times the MAD and, if so, classify it as an outlier. If it is reasonable to assume that the data are normally distributed, the MAD can be scaled so that it is conveniently related to the standard deviation as is shown in eqn (9.58).

$$\text{scaled MAD} = \frac{1}{\text{erfinv}(0.5)\sqrt{2}} \cdot \text{MAD} \tag{9.58}$$

where **erfinv** is the inverse error function. Consequently, the MAD is equal to 0.6745σ.

> 👆 See the website for practical examples of outlier tests: ass-ets.org

9.4.3 Homoscedasticity

Testing for the homogeneity of variances, or homoscedasticity, is of course a comparison of variances. The F-test can be used for the comparison of two variances (Section 9.3.7). In this section, we will treat Bartlett's test for the comparison of more than two variances. However, given the vulnerability of these parametric tests to deviations from normality, as well as outliers, it is often safer to conduct a more robust (*i.e.* non-parametric) alternative test (see Section 9.5.4).

9.4.3.1 Bartlett's Test

For the comparison of several variances, the Bartlett's test can be used. In this test, the χ^2 statistic is computed.

$$\chi_{obs}^{2} = \frac{\left[\nu_{pool} \cdot \log(s_{pool}^2) - \Sigma_{i=1}^{k} \nu_i \cdot \log(s_i^2)\right]}{C} \tag{9.59}$$

Here, k is the total number of groups, ν_{pool} is the total number of degrees of freedom $\nu_{pool} = \sum_{i=1}^{k} \nu_i, s_{pool}^2$ is the pooled variance $s_{pool}^2 = \sum_{i=1}^{k} \nu_i s_i^2 / \nu_{pool}$, and C is given by

$$C = 1 + \frac{\left(\Sigma_{i=1}^{k} \frac{1}{\nu_i}\right) - \frac{1}{\nu_{pool}}}{3(k-1)} \tag{9.60}$$

The observed χ_{obs}^2 value is compared to χ_{crit}^2 at $k - 1$ degrees of freedom. Bartlett's test is very sensitive to deviations from normality. Therefore, in practice, non-parametric alternatives, such as Levene's test, are often used (see Section 9.5.4).

9.5 Non-parametric Statistics (M)

9.5.1 Introduction to Robust Methods

The majority of the tests treated in the modules thus far require a number of criteria to be met. Examples include the normality of data, the absence of outliers, and the homogeneity of the variances of the different sets to be compared. In the previous module, we treated a number of tools that can be used to test such assumptions. Now, what if these tests reveal that the requirements for the parametric test are not met?

This module will focus on methods from non-parametric statistics, which are also known as **robust statistics**. Whereas in parametric statistics the data or mathematical representations thereof are assumed to follow a specific distribution (and in some cases the variances are assumed to be homogeneous), this is not the case for non-parametric statistics. The latter field is therefore also known as

Table 9.7 Overview of the parametric and non-parametric tests for the different objectives in statistical analysis. The relevant section is indicated in parentheses.

Objective	Parametric	Non-parametric
Describing a sample	Mean (9.2.2) Standard deviation (9.2.2)	Median (9.2.2) IQR (9.2.2)
Comparison of a sample mean with a reference Comparison of two sample means (paired)	*t*-test (9.3.2)	Sign test (9.5.2) Wilcoxon's test (9.5.2)
Comparison of two sample means (unpaired)	*t*-test (9.3.4)	Mann–Whitney *U* test (9.5.3)
Comparison of standard deviations	*F*-test ($k = 2$, 9.3.7) Bartlett's test ($k \geq 2$, 9.4.3)	Levene's test ($k \geq 2$, 9.5.4)
Comparison of several means	ANOVA (9.3.9)	Kruskal–Wallis (9.5.5)
Measuring the relationship between variables	Pearson correlation (9.6.2)	Spearman correlation (9.6.2)

distribution-free statistics, as we do not need to assume that the data follow a specific distribution.

The non-parametric counterparts to the various methods discussed thus far are tabulated in Table 9.7 for the different objectives. This module will address each of these tests and explain how they differ fundamentally from the parametric methods. As we cover the different methods, we will quickly develop a fundamental understanding of how the improved robustness comes at the cost of a loss of statistical power.

Non-parametric methods often abstract the data to some degree to distil an answer. A good example is the median, which was treated in Section 9.2.2. Where the mean is strongly affected by an outlier, the median does not change. However, in contrast to the mean, the median ignores the actual values of the rest of the data.

This will be the common thread throughout this module. With rarely any criteria to fulfil for their use, non-parametric methods have the advantage that they can almost always be applied and are easy to compute.

Had this been the only difference, scientists would resort exclusively to non-parametric methods, but of course there is a disadvantage. Because actual information is often discarded or significantly abstracted, the statistical power of non-parametric statistics is often lower than that of their parametric counterparts. It is thus more likely to make a type-II error, *i.e.* to accept H_0 when it in fact is not true.

See the website for practical examples of non-parametric tests for different software packages, as well as further reading: ass-ets.org

---

9.5.2 Comparison of a Sample Mean with a Reference

The first test that will be covered is the comparison of a sample mean with a reference. The parametric *t*-test uses the *t*-distribution to approximate the population; the non-parametric options amplify the direction of deviation.

9.5.2.1 Sign Test

The sign test offers an alternative to the comparison of a sample mean with a reference as well as the unpaired comparison of two sample means.

Suppose we have nine repeated measurements of the retention time of an analyte. A protocol specifies that the median retention time should be $\tilde{x}_0 = 3.5$ min. The obtained measurements are 3.3, 3.2, 3.4, 4.7, 3.7, 3.3, 3.4, 3.2, and 3.5 min. The question is: "Is the median of the population of data $\tilde{x}_0 = 3.5$?". Our hypotheses will be $H_0: \tilde{x} = \tilde{x}_0$ and $H_1: \tilde{x} \neq \tilde{x}_0$. This is a two-tailed test. The sign test follows the following steps.

1. **Subtract the supposed median (*i.e.* $\tilde{x}_0 = 3.5$) from each value in the dataset.** We obtain now $-0.2, -0.3, -0.1, 1.2, 0.2, -0.2, -0.1, -0.3$, and 0.
2. **Count the number of positive and negative values (*i.e.* consider the signs).** There are 6 negative values, 2 positive, and one neutral (we ignore the latter).
3. **Take the minimum of the two values.** This is the test statistic (r_{obs}): 2 is lower than 6, so our statistic is $r_{obs} = 2$.

4. **Compare the r_{obs} with a tabulated r_{crit} value (Appendix A.6).** If $r_{obs} > r_{crit}$ then H_0 is accepted: assuming that $\alpha = 0.05$, we see that for our case the tabulated value for $n = 9$ datapoints is $r_{crit, \alpha=0.05, n=9} = 1$. This is smaller than r_{obs}, and thus we accept H_0.

In our case, the sign test essentially tested whether too few positive differences occurred with respect to the median that was supposed to be true.

In contrast to its parametric counterpart, the sign test exclusively considers the direction of deviation with the median. It is thus not surprising that the statistical power of this test is not high. The sign test can also be used for a paired comparison of two sample means; in this case, the two datasets are first subtracted instead of step 1, and the test continues from step 2 with the differences between the two sets.

9.5.2.2 Wilcoxon's Signed-Rank Test

In addition to the direction of the deviation with the median that is used by the sign test, the Wilcoxon Signed-Rank test also considers the magnitude through the ranks of the numbers. An example is shown for a paired comparison of two datasets. Like with the equivalent parametric test (see Section 9.3.4), the two datasets are subtracted from each other, and the test continues with the differences (d). Of course, the test can also be used as a comparison of a sample median with a reference; in this case, the median is first subtracted from all values similar to the sign test (see previous section). From this point on, a vector of data $(d$, Table 9.8) is obtained and the test is identical.

The steps are as follows:

1. **Note the sign for each value in** (d), *i.e.* whether the number is positive or negative.
2. **Rank the data, ignoring the signs.** Coincident values rank as half (see Table 9.8).
3. **Sum the ranks separately for the positive and negative values.** In the case of Table 9.8, this results in $T^+ = 25.5$ and $T^- = 19.5$.
4. **Take the minimum of T^+ and T^-, which becomes T_{obs}.** In this case: $T_{obs} = 19.5$.
5. **Compare T_{obs} with the tabulated T_{crit} value (Appendix A.7).**
6. **If $T_{obs} > T_{crit}$, then H_0 is accepted.**

The Wilcoxon sign test offers more statistical power than the sign test, but requires the distribution to be symmetrical. This could be conducted by comparing the data to a symmetrical distribution function as well as through other methods that will not be treated in this book.

9.5.3 Comparison of Two Sample Means (Unpaired): Mann–Whitney U Test

For the equivalent to the parametric t-test unpaired comparison of two sample means (Section 9.3.4), the Mann–Whitney U test is an option, albeit with some remarks discussed below. This test was developed by Henry Mann and Donald Ransom Whitney, and is also known as the **Wilcoxon's Rank-Sum test**. While the test employs a summation of the ranks, it is not the same as Wilcoxon's Signed-Rank test. Instead, it is slightly more elaborate than the non-parametric test of the previous section.

As an example, two samples are compared. Repeated measurements were conducted for each yielding for sample A $(n_A= 9)$: 28.1, 25.9, 27.8, 27.2, 27.7, 27.3, 26.8, 27.4, and 28.5;. for sample B $(n_B= 8)$: 26.6, 26.7, 29.4, 30.3, 27.0, 27.9, 27.8, and 29.3. The question is whether the two samples belong to the same population, *i.e.* H_0: the two samples belong to the same population and H_1: the samples do not belong to the same population.

To conduct the test, the **Mann–Whitney U test** follows the following steps:

Table 9.8 Example results for a Wilcoxon Signed-Rank test for the paired comparison of two sample means.

Object	x_1	x_2	d	Sign	Rank
1	27 185	25 904	1281	+	4
2	20 068	21 207	−1139	−	5
3	32 593	35 582	−2989	−	7
4	176 438	172 343	4095	+	9
5	29 319	28 466	853	+	3
6	21 634	18 502	3132	+	8
7	3387	3273	114	+	1.5
8	28 181	29 889	−1708	−	6
9	3466	3352	114	+	1.5

1. **Combine, sort, and rank the data.** Coincident values rank half. See Table 9.9.
2. **Sum the ranks for both groups.** In the example, this yields $R_A = 72.5$ and $R_B = 80.5$.
3. **Compute and use eqn (9.61a and b).** Resulting in $U_A = 44.5$ and $U_B = 27.5$.

$$U_A = n_A n_B + \frac{n_A(n_A + 1)}{2} - R_A \tag{9.61a}$$

$$U_B = n_A n_B + \frac{n_B(n_B + 1)}{2} - R_B \tag{9.61b}$$

4. **Select the lowest of U_A and U_B as U_{obs}.** Thus, $U_{obs} = 27.5$.

5. **Compare U_{obs} with a tabulated U_{crit} value (Appendix A.8).** In our example, $U_{crit} = 15$.

6. **Accept H_0 if $U_{obs} > U_{crit}$. Reject H_0 if $U_{obs} \leq U_{crit}$.** Here, H_0 is accepted.

As can be seen from the table with critical values (Appendix A.8), the Mann–Whitney U test cannot be executed with a low sample size. Strictly speaking, the Mann–Whitney U test does not test whether

Table 9.9 Example of ranked data used for Mann-Whitney U unpaired comparison of two samples.

Sample	Value	Rank
A	25.9	1
B	26.6	2
B	26.7	3
A	26.8	4
B	27	5
A	27.2	6
A	27.3	7
A	27.4	8
A	27.7	9
A	27.8	10.5
B	27.8	10.5
B	27.9	12
A	28.1	13
A	28.5	14
B	29.3	15
B	29.4	16
B	30.3	17

the medians are equal, but whether the probability distribution of one randomly drawn set of measurements is identical to the probability distribution of another set of measurements.[2,3] It has been shown that the spread of the data matters a great deal, and that, ideally, the distributions should be of similar shape.[4] The test can, however, be employed for the hypotheses as outlined above. It is also possible to use the test for larger sample sizes.[5]

9.5.4 Comparison of Variances: Levene's Test and Brown–Forsythe Test

For the comparison of two or more variances, we have thus far addressed the F-test to compare two variances (Section 9.3.7) and Bartlett's test to compare multiple variances (Section 9.4.3). When we assume that the data are normally distributed, it is easy to imagine outliers constituting a significant problem. It is therefore often safer to opt for a non-parametric test.

Levene's test was designed by Howard Levene as a robust test that is insensitive to deviations from normality and to outliers.[6] This test computes the W statistic as

$$W_{\text{obs}} = \frac{n_{\text{tot}} - k}{k - 1} \cdot \frac{\Sigma_{j=1}^{k} n_j \cdot \left(\bar{z}_j - \bar{z}\right)^2}{\Sigma_{j=1}^{k} \Sigma_{i=1}^{n_j} \left(z_{i,j} - \bar{z}_j\right)^2} \tag{9.62}$$

where $z_{i,j} = |x_{i,j} - \bar{x}_j|$, $\bar{z}_j$ is the mean z for group j, and $\bar{z}$ is the overall mean z, n_{tot} is the total number of datapoints, and k is the number of groups that are compared. Note that with this definition, Levene's test is computationally equivalent to a one-way ANOVA and W approximately follows the F-distribution. To conduct the test, W_{obs} is compared to F_{crit} with $k - 1$ and $n_{\text{tot}} - k$ degrees of freedom.

A quadratic variant exists which employs the squared deviations of the data using $z_{i,j}^2 = \left(x_{i,j} - \bar{x}_j\right)^2$ instead of the absolute deviations $z_{i,j} = |x_{i,j} - \bar{x}_j|$. This addition may increase the statistical power for certain types of data. Interestingly, eqn (9.62) shows that Levene's test employs the mean, which was concluded not to be a robust statistic in Module 9.2. To improve the robustness, the Brown–Forsythe test, named after Morton Brown and Alan Forsythe, considers the median instead of the mean,[7] using $z_{i,j} = |x_{i,j} - \tilde{x}_j|$.

Further variants and alternatives have been developed as well. The choice of which type of test to use generally depends on whether the data can reasonably be expected to be normally distributed (at the cost of reduced robustness) and different statistical powers may be attained depending on the type of data and their properties.

9.5.5 Analysis of Variance

9.5.5.1 One-way ANOVA: Kruskal–Wallis Test

The alternative to one-way ANOVA has strong parallels with Levene's test. Named after William Kruskal and Wilson Wallis, the Kruskal–Wallis test is an extension of Wilcoxon's rank-sum test.[8] Although the test is typically used to assess whether H_0: all group medians are equal, against H_1: at least one group median is not equal, the test actually compares whether the probability distributions of the different groups are similar[4] (as does the Mann–Whitney U test).

The test proceeds as follows:

1. **Rank the data from all combined groups.** An example is shown in Table 9.10.
2. **Compute the test statistic H_{obs}.**

$$H_{obs} = \frac{12}{n_{tot}(n_{tot} + 1)}\left(\sum_{j=1}^{k} \frac{R_j^2}{n_j}\right) - 3(n_{tot} + 1) \tag{9.63}$$

where k is the number of groups, n_{tot} is the total number of datapoints, n_j is the number of datapoints in group j, R_j is the sum of all ranks in group j, and equal to

$$R_j = \sum_{i=1}^{n_j} r_{i,j} \tag{9.64}$$

Here, $r_{i,j}$ is the rank in this group for repetition i.

In contrast with one-way ANOVA, H_{obs} does not follow the F statistic but the χ^2 statistic if the number of repetitions per group n_j is larger than five.

Table 9.10 Example of the conversion of data from Figure 9.23, shown here in the left panel, to ranks in the right panel for a Kruskal–Wallis test.

Parametric (ANOVA)			Non-parametric (Kruskal–Wallis)		
1	2	3	1	2	3
3.18	3.47	3.34	10.5	15	12
3.18	3.44	3.06	10.5	14	7.5
2.96	3.41	3.02	2.5	13	5
3.13	3.58	3.04	9	16	6
2.96	3.63	2.83	2.5	17	1
3.01	3.7	3.06	4	18	7.5

3. Compare H_{obs} with H_{crit} at $k - 1$ degrees of freedom in a one-tailed χ^2 test.

4. Reject H_0 if $\chi_{obs}^2 > \chi_{crit}^2$. The output is a table similar to that of the parametric ANOVA test (see Table 9.5), but now with a χ^2 statistic.

See the website for an example Kruskal–Wallis result table and its interpretation: ass-ets.org

---

9.6 Regression: Modelling and Calibration (M)

9.6.1 An Introduction to Regression Analysis

In this chapter, we have thus far exclusively focused on describing the sample data and we used these descriptions to infer conclusions about the population from which the sample originated. Another important application of statistics is studying the relationship between two or more variables of our sample. Regression analysis can be used to mathematically express this relationship in a **model**. The model can then be used to predict one variable based on the information on other variables.

Establishing a relationship between a variable of interest, such as the amount or concentration of an analyte, and an observed variable (or collection of variables), such as the peak area, is referred to as **calibration**. Calibration is of paramount importance in quantitative analysis. An example is shown in Figure 9.28. The line in this figure is typically referred to as a **calibration curve** (see also Section 9.1.3).

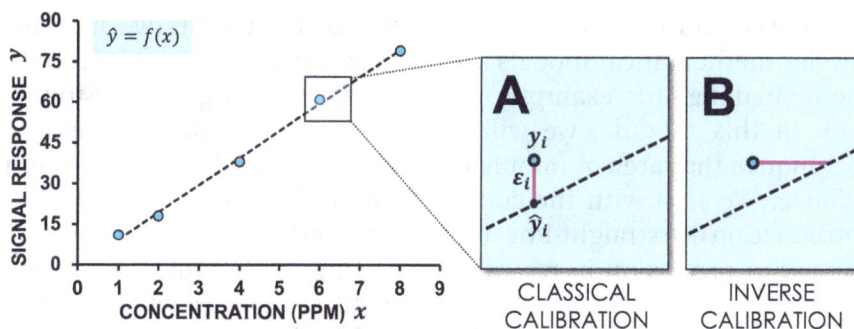

Figure 9.28 Example of a calibration curve that relates the property of interest (concentration, x) to the measured property (signal response, y). The mathematical model ($\hat{y}$, dashed line) is obtained through regression. In classical calibration (A), it is assumed that the error is in the measured property (*e.g.* the signal response), whereas in inverse calibration (B), the error is assumed to be in the property of interest (*e.g.* the concentration). See the text of Section 9.6.2 for the data used.

The calibration curve can be established by measuring reference solutions with known concentrations. As these reference solutions are typically either certified or known, it is reasonable to assume that the error will not be in the concentration, but in the instrumental signal response. This strategy is highlighted in Figure 9.28A as **classical calibration** and is also known as **model-I regression**. Here, we assume that all the errors are in the measured or dependent variable (*i.e.* the signal response). For classical calibration, it is needed to first fit the model ($\hat{y} = f(x)$) and then invert it to predict the variable of interest as $x = f^{-1}(\hat{y})$. In **inverse calibration** (Figure 9.28B), this inversion of the model is not needed as we predict x from measures of y. With **model-II regression**, the errors are assumed to be in both variables. Classical calibration is typically (but not exclusively) used for **univariate calibration**, *i.e.* calibration in which there is only one independent variable (x). Multivariate calibration is referred to as 1st order, 2nd order, or higher-order calibration and more commonly (but again not exclusively) performed using inverse calibration.

An important task for the analyst is to ensure that the measured variable does not respond to other variables that do not relate to the property of interest. For example, this could be the case when different analytes contribute to a change in absorbance, resulting in an increased peak area.

While calibration curves are a cornerstone of quantitative analysis, similar mathematical models are also essential for method development, making, for example, use of retention modelling (Module 3.8). In this Module, we will only cover the essential regression techniques that are commonly encountered in analytical separation science. We start with the common calibration curve, as a suitable introduction to straight-line regression, and then expand towards the use of regression in retention modelling. In Module 9.9, we will introduce the reader to multivariate methods, but we will not cover these in great detail. Readers interested in further deepening their understanding are recommended to study other texts.[5,9]

9.6.2 Covariance and Correlation

When conducting regression analysis, it is useful to assess the covariance and correlation of the data. We start with the **covariance** between x and y, which is the degree of linear association between x and y. It is defined as

$$\text{cov}(x,y) = \frac{\sum_{i=1}^{n}(x_i - \bar{x})(y_i - \bar{y})}{n-1} \tag{9.65}$$

where n is the total number of objects (*i.e.* datapoints). We can apply this to the different datasets shown in Figure 9.29. When comparing the covariance of the data shown in Figure 9.29A–C, we see an increasing value. This is in stark contrast with the variance of these same datasets, which in each case is always roughly the same (*i.e.* $s^2 \approx 0.0756$ for each x and y in each dataset). The variance is also approximately 0.0756 for both x and y in Figure 9.29D, yet the covariance now is negative. Figure 9.29B and C are examples of **positive covariance**, whereas Figure 9.29D is an example of **negative covariance**.

One important aspect is that the covariance is not dimensionless; instead, it is the unit of x multiplied by the unit of y. Figure 9.29E shows this with a very high covariance value due to the different scaling relative to the other datasets. To properly study the covariance, it is thus useful to divide the data by the standard deviations of x and y. This yields what is known as the **correlation coefficient**, which is defined as

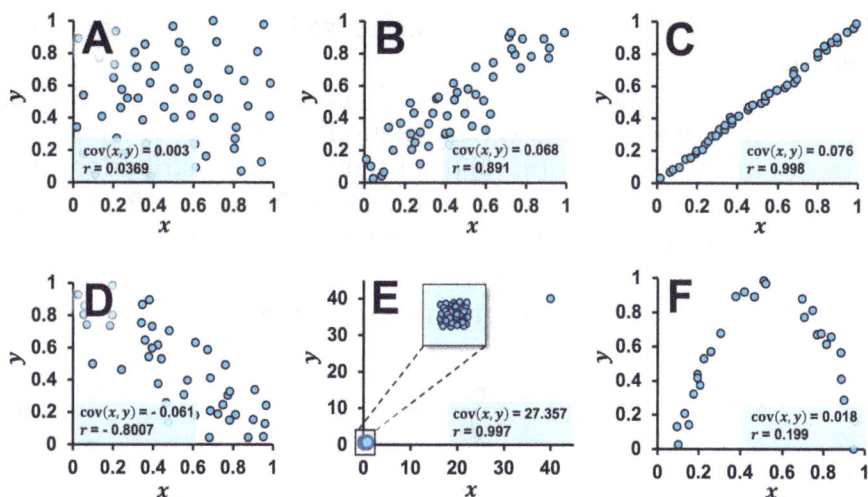

Figure 9.29 The covariance and correlation coefficient for different distributions of data. For panels A–D the standard deviation in both X and Y is roughly 0.275.

$$r = \frac{\text{cov}(x,y)}{s_x \cdot s_y} \tag{9.66}$$

Eqn (9.66) is also known as the **Pearson correlation coefficient**. It is a measure of the degree of linear association between x and y, relative to their variation. Application of this concept to the datasets in Figure 9.29A–C indeed shows a strongly increasing correlation coefficient. The positive and negative covariance in Figure 9.29B and D yields a similar correlation coefficient. Note that eqn (9.66) yields a unitless result, as both $\text{cov}(x, y)$ and $s_x \cdot s_y$ have the same units and thus cancel each other out.

As both the correlation coefficient and covariance express the linear association, they are not useful for non-linear relationships such as that shown in Figure 9.29F. In addition, the correlation coefficient is highly vulnerable to extreme so-called leverage points, as illustrated in Figure 9.29E. Here, one datapoint was added to the dataset of Figure 9.29A. While the correlation coefficient was 0.037 in Figure 9.29A, it is almost 1 in Figure 9.29E, following the addition of a single datapoint.

The Pearson correlation coefficient, r, is based on the sampled data points. Similar to the relation between μ and $\bar{x}$, and σ and s, it could be that, were we to have had an infinite number of datapoints (*i.e.* the full population), the population-based correlation coefficient

(ρ) would provide a different characterization of the correlation between x and y.

To determine whether the sampled correlation coefficient (r) is valid to discern any true linear association from noise, it is possible to use a t-test, with $t \sim t(n-2)$ (*i.e.* t follows the t-distribution with $\nu = n - 2$ degrees of freedom, see also Section 9.2.3). In this case, the hypotheses are H_0: there is no correlation (*i.e.* $\rho = 0$) and H_1: there is correlation (*i.e.* $\rho \neq 0$). To execute this test, t_{obs} is compared with t_{crit} at $\nu = n - 2$ degrees of freedom (similar to Section 9.3.2) using

$$t_{obs} = \frac{r\sqrt{n-2}}{\sqrt{1-r^2}} \tag{9.67}$$

Similar to earlier treatments of t-tests, it is also possible to conduct one-sided t-tests to specifically investigate positive or negative correlations.

See the website for an example of the statistical test.

An alternative to the Pearson correlation coefficient is the Spearman rank correlation coefficient, which is a non-parametric alternative that first reduces the values of x and y to ranks $R(x)$ and $R(y)$, respectively.

$$r_s = \frac{\mathrm{cov}(R(x), R(y))}{S_{R(x)} \cdot S_{R(y)}} \tag{9.68}$$

The Spearman correlation is essentially the Pearson correlation of the ranks of x and y and evaluates a monotonic relationship, even if they are non-linear. By considering the rank, the Spearman correlation will increase when two variables for a given datapoint have a similar rank. Or, in other words, when the position of the rank in vector **x** is similar to its position in vector **y**.

Similar to the Pearson correlation coefficient, eqn (9.68) can be used to compute a significance test for the Spearman (Box 9.3) correlation.

> **Box 9.3** Clarification on vectors.
>
> The concept of a vector that was described in Section 9.1.2 will now become important in this Module. A vector is a one-dimensional series of data. In equations they are typically depicted through a **bold**, non-*italic* symbol (*e.g.* x as value is $\mathbf{x}$ as vector). Vectors can be a **row-vector** (*e.g.* $\mathbf{x} = [1, 2, 3]$), or a **column vector** (*e.g.* $\mathbf{x} = \begin{bmatrix} 1 \\ 2 \\ 3 \end{bmatrix}$, or $\mathbf{x} = [1; 2; 3]$).

9.6.3 Least-squares Regression

We start by adding some mathematical definitions to the concepts illustrated in Figure 9.28. As discussed above, the figure shows a scatterplot of several measured signal response values (y_i) for different concentrations (x_i): $\mathbf{x} = [1, 2, 4, 6, 8]$ and $\mathbf{y} = [11, 18, 38, 61, 79]$. Given that the errors made while preparing the standards are usually negligible in comparison with the measurement errors, it is reasonable to assume that x is known and that the entire error is in y. At this point, x is an **independent variable**, and y is a **dependent variable** as it depends on x. The calibration curve is a model $\hat{y} = f(x)$ that is fitted through the data so that

$$y_i = \hat{y}_i + \varepsilon_i = f(x_i) + \varepsilon_i = f(x_i, \beta) + \varepsilon_i \tag{9.69}$$

We will get to the meaning of β in a moment. Eqn (9.69) states that any value of y can be obtained from the model $\hat{y}$ (*i.e.* $f(x)$) plus an error ε. In other words, for each datapoint $y_i - \hat{y}_i = \varepsilon_i$ (see also Figure 9.28A). When we believe the relation to be a straight line, then we can fit this as our model through our data, such as is the case in Figure 9.28. In accordance with

$$y = \beta_0 + \beta_1 \cdot x + \varepsilon_i \tag{9.70}$$

Here, β_0 and β_1 are the model parameters. Note that these parameters are in no way related to the likelihood of a type-II error, which happens to also be denoted by β. The model parameters are depicted using Greek characters β_0 and β_1 as they are based on the availability of the full population (*i.e.* an infinite number of datapoints). Together the parameters are depicted in 9.69 as β, which is the column vector $\beta = \begin{bmatrix} \beta_0 \\ \beta_1 \end{bmatrix}$. Of course, in practice, we only have a sample, *i.e.* a limited number of datapoints (*e.g.* five in Figure 9.28). Thus, we do not know β_0 and β_1 and, instead, estimate them through b_0 and b_1 (similar to estimating σ with s, see Section 9.2.2). We will proceed later with estimating the standard error and subsequent confidence intervals for **b** in its approximation of β. Thus, we describe our model of eqn (9.70) henceforth as

$$\hat{y} = b_0 + b_1 \cdot x \tag{9.71}$$

Here, b_1 and b_0 are the parameters for which the following equations can be derived for the **slope**

$$b_1 = \frac{\sum_{i=1}^{n}(x_i - \bar{x})(y_i - \bar{y})}{\sum_{i=1}^{n}(x_i - \bar{x})^2} \tag{9.72}$$

and for the **intercept**

$$b_0 = \bar{y} - b_1 \bar{x} \tag{9.73}$$

of the model, respectively.

9.6.3.1 Fitting of a Linear Model Using Least-squares Regression

The calibration curve shown in Figure 9.28 is obtained by fitting the straight-line model from eqn (9.71) through the data. This is done by calculating its parameters b_0 and b_1 such that the total squared error for the sum of all y_i (*i.e.* $\sum \varepsilon_i^2$) is minimized (hence "**least squares**"). This total squared error is

$$\sum \varepsilon_i^2 = \sum (y_i - \hat{y}_i)^2 = \sum (y_i - b_0 - b_1 \cdot x_i)^2 \quad \text{for all } i = 1...n \tag{9.74}$$

where the sum is taken over all measurements. Here, we consider only models that are linear, for which the solution can be directly computed. Linearity does not refer here to the fact that the model is a straight line, but to the model being linear in terms of the coefficients b_j. The model must meet the condition

$$\frac{\partial \hat{y}}{\partial b_j} \neq f(b_0, b_1, ..., b_m) \qquad \text{for all } j=0...m \qquad (9.75)$$

This means that both eqn (9.71) and, for example, the model $\hat{y} = b_0 + b_1 \frac{1}{x} + b_2 \cdot \log(x)$ are linear, but that $\hat{y} = b_0 \cdot x + \exp(b_1 \cdot x)$ is not. If the model is linear, it can be expressed as

$$y = \begin{bmatrix} \dfrac{\partial \hat{y}}{\partial b_0} & \dfrac{\partial \hat{y}}{\partial b_1} & ... & \dfrac{\partial \hat{y}}{\partial b_m} \end{bmatrix} \begin{bmatrix} b_0 \\ b_1 \\ \vdots \\ b_m \end{bmatrix} + \epsilon \qquad (9.76)$$

In this regard, the straight-line case, *i.e.* eqn (9.71), can be expressed as $y_i = \begin{bmatrix} 1 & x_i \end{bmatrix} \begin{bmatrix} b_0 \\ b_1 \end{bmatrix} + \varepsilon_i$. We can also write eqn (9.76) for all x_i by rewriting it as a matrix $\mathbf{X}$, so that $\hat{y}$ is defined as $f(\mathbf{X}, \mathbf{b})$

$$\begin{bmatrix} y_1 \\ y_2 \\ \vdots \\ y_n \end{bmatrix} = \begin{bmatrix} \dfrac{\partial f}{\partial b_0}\Big|_{x_1} & \dfrac{\partial f}{\partial b_1}\Big|_{x_1} & \cdots & \dfrac{\partial f}{\partial b_m}\Big|_{x_1} \\ \dfrac{\partial f}{\partial b_0}\Big|_{x_2} & \dfrac{\partial f}{\partial b_1}\Big|_{x_2} & \cdots & \dfrac{\partial f}{\partial b_m}\Big|_{x_2} \\ \vdots & \vdots & & \vdots \\ \dfrac{\partial f}{\partial b_0}\Big|_{x_n} & \dfrac{\partial f}{\partial b_1}\Big|_{x_n} & \cdots & \dfrac{\partial f}{\partial b_m}\Big|_{x_n} \end{bmatrix} \cdot \begin{bmatrix} b_0 \\ b_1 \\ \vdots \\ b_m \end{bmatrix} + \begin{bmatrix} \varepsilon_1 \\ \varepsilon_2 \\ \vdots \\ \varepsilon_n \end{bmatrix} \qquad (9.77)$$

Where the partial derivatives in the matrix are referred to as the $\mathbf{X}$ matrix, which is denoted with a non-italic bold capital character $\mathbf{X}$ to indicate that it is a matrix. For our straight line this would result in

$$\begin{bmatrix} y_1 \\ y_2 \\ \vdots \\ y_n \end{bmatrix} = \begin{bmatrix} 1 & x_1 \\ 1 & x_2 \\ \vdots & \vdots \\ 1 & x_n \end{bmatrix} \begin{bmatrix} b_0 \\ b_1 \end{bmatrix} + \begin{bmatrix} \varepsilon_1 \\ \varepsilon_2 \\ \vdots \\ \varepsilon_n \end{bmatrix} \qquad (9.78)$$

> See the website for useful resources on the linear algebra that is used throughout this module: ass-ets.org

Least-squares regression is all about finding **b**, so that the sum of squares error $\Sigma\varepsilon^2$ is minimized. This is accomplished through

$$\widehat{\mathbf{b}} = (\mathbf{X}^T \cdot \mathbf{X})^{-1} \cdot \mathbf{X}^T \cdot \mathbf{y} \tag{9.79}$$

where the superscript T depicts the transposed matrix and $\widehat{\mathbf{b}}$ is the optimal solution if the errors (ε) are homoscedastic (meaning they have a constant variance across all measurements) and normally distributed, *i.e.* $\varepsilon \sim N(0, \sigma^2)$. Doing so for our calibration data yields $b_0 = -0.5488$ and $b_1 = 9.9878$.

9.6.3.2 Confidence Interval of the Model Parameters

The standard error of each of the model parameters can be computed by using the **variance–covariance matrix** of the regression coefficients, which presents the covariance of the different pairs of parameters

$$\Sigma = \begin{bmatrix} s_{b_0}^2 & \text{cov}(b_0, b_1) & \cdots & \text{cov}(b_0, b_m) \\ \text{cov}(b_1, b_0) & s_{b_1}^2 & \cdots & \text{cov}(b_1, b_m) \\ \vdots & \vdots & \ddots & \vdots \\ \text{cov}(b_m, b_0) & \text{cov}(b_m, b_1) & \cdots & s_{b_m}^2 \end{bmatrix} \tag{9.80}$$

The variance–covariance matrix has the convenient property that the variance in each of the model parameters is encompassed by its diagonal. It can be estimated as

$$\widehat{\Sigma} = s_e^2 (\mathbf{X}^T \cdot \mathbf{X})^{-1} \tag{9.81}$$

where

$$s_e^2 = \frac{\Sigma_{i=1}^n (y_i - \widehat{y}_i)^2}{n - (m + 1)} \tag{9.82}$$

Here, $n - (m + 1)$ are the degrees of freedom, with n referring to the number of rows of $\mathbf{X}$, *i.e.* the number of datapoints, and $m + 1$ referring to the number of columns of $\mathbf{X}$, *i.e.* the number of parameters. Note that while eqn (9.81) yields a variance–covariance matrix similar to eqn (9.80), the variances are estimates of the population and thus written as $s_{b_j}^2$ for all $j = 0...m$. For our straight line, we then obtain

$$s_{b_0}^2 = \frac{s_e^2 \cdot \Sigma_{i=1}^n (x_i)^2}{n \cdot \Sigma_{i=1}^n (x_i - \overline{x})^2} \tag{9.83}$$

$$s_{b_1}^2 = \frac{s_e^2}{\Sigma_{i=1}^n (x_i - \overline{x})^2} \tag{9.84}$$

We can now calculate the confidence intervals using the t-distribution similar to Section 9.2.3.4 with $t_{\text{obs}} = (b_j - \beta_j)/s_{b_j}$ so that

$$\beta_j = b_j \pm t_{(\alpha/2),\ (n - (m + 1))} \cdot s_{b_j} \tag{9.85}$$

where $t_{(\alpha/2),\ (n-(m+1))}$ is the critical t-value for a t-distribution $t \sim t(n-(m+1))$. Doing so for our calibration data in Figure 9.28 yields, with $t_{\text{crit}} = 3.1824$ as 95% confidence interval, $\beta_0 = -0.5488 \pm 4.7857$ and $\beta_1 = 9.9878 \pm 0.9728$.

9.6.3.3 Confidence Interval of the Predicted Response

Figure 9.30 again shows the calibration curve (pink, dotted line) fitted through the data (blue dots). From the curve, we can now predict the response at different concentrations. For example, we may predict the response at a concentration of interest, say, $x_0 = 5$ ppm, through $\hat{y}_0 = f(x_0, \mathbf{b}) = \mathbf{x}_0 \cdot \mathbf{b} = \begin{bmatrix} 1 & x_0 \end{bmatrix} \begin{bmatrix} b_0 \\ b_1 \end{bmatrix}$, which results in 49.39. However, we have just assessed the variance in the model parameters. It is thus likely that this variance propagates into a standard error of the predicted value.

The variance of the predicted value $\sigma_{\hat{y}_0}^2$ can be calculated through

$$\sigma_{\hat{y}_0}^2 = \mathbf{J}_0 \cdot \mathbf{\Sigma} \cdot \mathbf{J}_0^{\mathrm{T}} \tag{9.86}$$

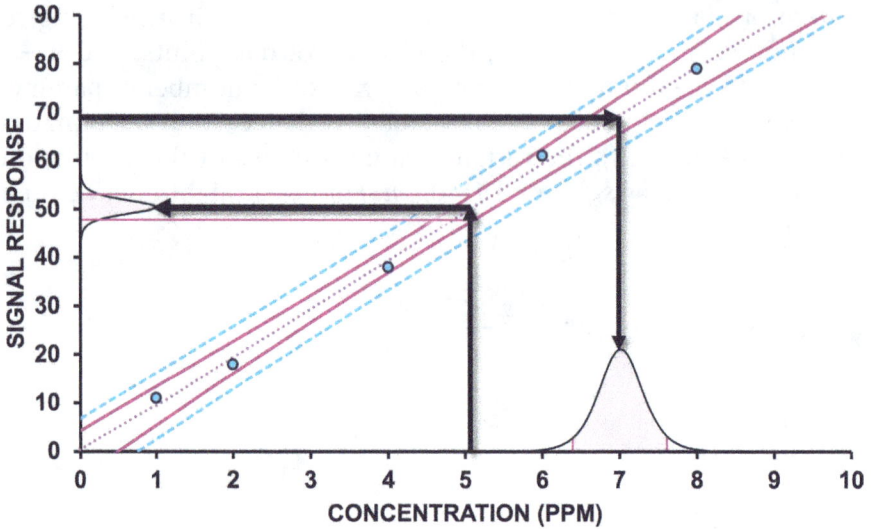

Figure 9.30 Calibration curve (purple, dotted line) fitted with least-squares regression through the datapoints (blue dots), along with the confidence intervals for the predicted response (pink, solid lines), and experimental response (blue, dashed lines). The confidence interval for the predicted response at a concentration of 5 ppm is indicated by the arrows.

where $\mathbf{J}_0$ is the Jacobian, a vector of partial derivatives of the regression function with respect to its parameters, evaluated at x_0, and $\mathbf{\Sigma}$ is the variance–covariance matrix (eqn (9.80))

$$\mathbf{J}_0 = \left[\left. \frac{\partial f}{\partial b_0} \right|_{x_0} \quad \left. \frac{\partial f}{\partial b_1} \right|_{x_0} \quad \cdots \quad \left. \frac{\partial f}{\partial b_m} \right|_{x_0} \right] \tag{9.87}$$

Here, $\mathbf{J}_0$ is equivalent to $\mathbf{x}_0$ (*i.e.* $\mathbf{J} \equiv \mathbf{x}_0$). For instance, if we execute the partial derivatives for our straight-line model we obtain $\mathbf{J}_0 = [1 \quad x_0] = \mathbf{x}_0$. Using $\mathbf{J} \equiv \mathbf{x}_0$ and eqn (9.81), and realizing that we have a limited number of data points (*i.e.* we actually calculate $s_{\hat{y}_0}^2$ not $\sigma_{\hat{y}_0}^2$), we can rewrite eqn (9.86) as

$$s_{\hat{y}_0}^2 = \mathbf{x}_0 \cdot s_e^2 \left(\mathbf{X}^{\mathrm{T}} \cdot \mathbf{X} \right)^{-1} \cdot \mathbf{x}_0^{\mathrm{T}} \tag{9.88}$$

which for our straight-line case results in

$$s_{\hat{y}_0}^2 = s_e^2 \cdot [1 \quad x_0](\mathbf{X}^\mathsf{T} \cdot \mathbf{X})^{-1}\begin{bmatrix} 1 \\ x_0 \end{bmatrix} \qquad (9.89)$$

Similar to eqn (9.85), we can now calculate the confidence interval in $\hat{y}$ through

$$\mathrm{CI}_{\hat{y}_i} = \hat{y}_i \pm t_{(\alpha/2),(n-(m+1))} \cdot s_{\hat{y}_i} \qquad (9.90)$$

which, for $x_0 = 5$ ppm, results in $y_0 = 49.39 \pm 2.61$. We can apply the same principle for the full calibration curve (*i.e.* for all x) and obtain the solid pink lines as plotted in Figure 9.30. Note that with the increasing sparsity of data, the confidence intervals become wider towards the outskirts of the curve.

9.6.3.4 Confidence Interval of the Experimental Response

Next to the standard error of the predicted response, it is also useful to consider the standard error in the experimental response. The variance in the experimental response can be calculated as

$$s_{y_0}^2 = s_e^2 \cdot \mathbf{x}_0(\mathbf{X}^\mathsf{T} \cdot \mathbf{X})^{-1} \cdot \mathbf{x}_0^\mathsf{T} + \frac{s_e^2}{g} \qquad (9.91)$$

where g is the number of repetitions of x_0 for which the experimental response is calculated, resulting in

$$\mathrm{CI}_{y_i} = \hat{y}_i \pm t_{(\alpha/2),(n-(m+1))} \cdot s_{y_i} \qquad (9.92)$$

For the straight-line case, the resulting experimental response is plotted along the entire model in Figure 9.30 (blue, dashed lines) with $g = 1$. In contrast to the confidence interval of the predicted response, the experimental response is purely related to the precision of the measurements and thus is constant regardless of the value of x.

9.6.3.5 Confidence Interval in x Through Inverting the Model

We have thus far examined the standard error in the model and predicted response. However, ultimately, calibration models are used to convert new sample measurements (*i.e.* peak areas, *y*) into estimates of the concentration *x*.

In these cases, the selected model must be **inverted**. For the straight-line model, this means that eqn (9.71) is written as

$$\hat{x}_{sample} = \frac{\bar{y}_{sample} - b_0}{b_1} \tag{9.93}$$

where $\bar{y}_{sample}$ is the mean of the *g* repetitions (or objects) of the measured sample. We can then assess the standard error in $\hat{x}_{sample}$ through

$$s_{\hat{x}_{sample}} = \frac{s_e}{b_1}\sqrt{\frac{1}{g} + \frac{1}{n} + \frac{(\bar{y}_{sample} - \bar{y})^2}{b_1^2 \Sigma_{i=1}^{n}(x_i - \bar{x})^2}} \tag{9.94}$$

This allows the calculation of the confidence interval of $\hat{x}_{sample}$

$$CI_{\hat{x}_{sample}} = \hat{x}_{sample} \pm t_{(\alpha/2),(n-2)} \cdot s_{\hat{x}_{sample}} \tag{9.95}$$

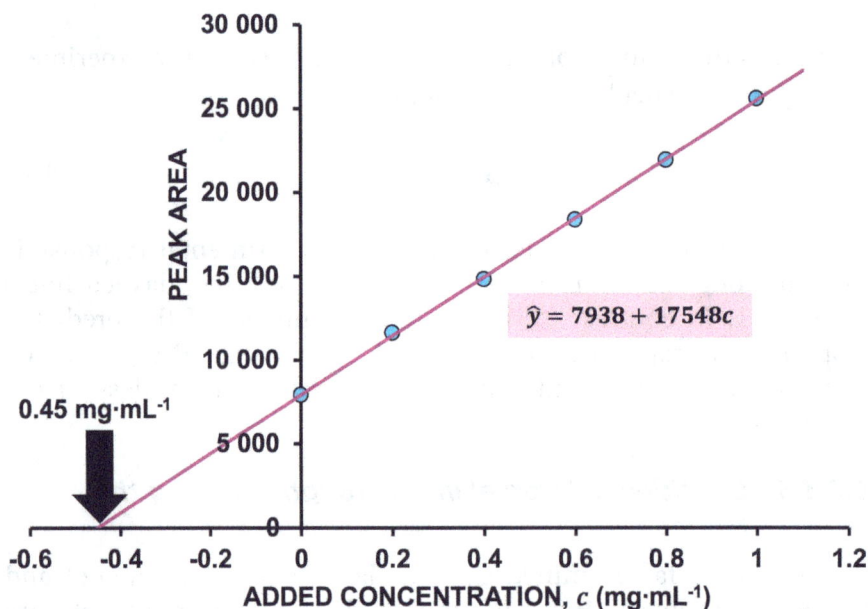

Figure 9.31 Illustration of the standard addition procedure; for explanation, see the text.

$s_{\hat{x}_{\text{sample}}}$ is one of the most useful results of a statistical treatment of the calibration data, as it yields the estimate of the **precision** (or a **confidence interval**) for the determined concentration. Note that b_1 is the slope of the calibration curve, *i.e.* the sensitivity of the analytical method, as it represents the change in instrument response (y) per unit change in concentration (x).

9.6.3.6 Standard Addition

When significant matrix effects disturb the response, the standard addition method may be used. The sample is then spiked with known concentrations of the analyte. The intercept represents the response of the original sample without added standard. Its nonzero value reflects the presence of analyte in the sample before spiking. Standard addition assumes that matrix effects influence both the sample and spiked solutions in the same way. The idea is then that through extrapolation the intercept of the calibration line with the x-axis (which depicts the concentration of spiked analyte) may be located, which will represent the amount of analyte that was originally present in the sample. The process is illustrated in Figure 9.31. A stock solution of 2 mg mL^{-1} of the analyte is added to five aliquots of 5 mL sample solution in volumes of 1, 2, 3, 4, and 5 mL, respectively. To the respective test samples 4, 3, 2, 1, and 0 mL volumes are added, so that the volume of each test sample is 10 mL. A final test sample is made up of 5 mL sample and 5 mL pure solvent. The concentrations added to the six test samples are 0, 0.2, 0.4, 0.6, 0.8, and 1.0 mg mL^{-1}. The concentration in the original sample can be found by dividing the intercept (signal obtained from the unspiked sample) by the slope (response factor) of the regression line, shown in Figure 9.31. The concentration in the diluted sample is found to be 0.45 mg mL^{-1} and in the original sample 0.90 mg mL^{-1}. In practical samples, it is recommended to combine the standard-addition procedure with specific detection, such as selected ion monitoring (SIM), selected reaction monitoring (SRM), or multiple reaction monitoring (MRM) mass spectrometry. The confidence interval may be calculated using eqn(9.94), with $g = \infty$.

9.6.4 Validation

A successfully fitted model is not necessarily a good model and it is important to validate the model accordingly. Model performance generally hinges on two factors, *viz.* the data and the mathematical

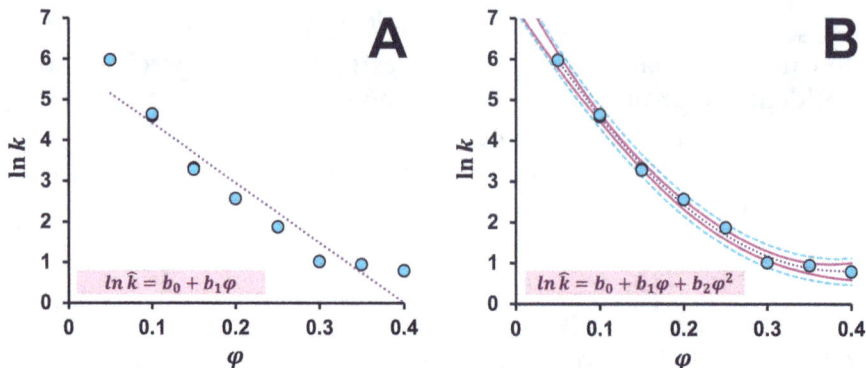

Figure 9.32 Scatter plots of retention data for an analyte in LC. Using least-squares regression, a (A) first-order polynomial and (B) second-order polynomial were fitted to the data (purple, dotted lines). The equations present the fitted model. For panel B, the confidence intervals of the predicted (pink, solid line) and experimental (blue, dashed line) response are shown. Retention data: ϕ = [0.05, 0.1, 0.1, 0.1, 0.15, 0.15, 0.15, 0.2, 0.25, 0.3, 0.35, 0.4] and ln k = [5.97, 4.59, 4.62, 4.63, 3.31, 3.3, 3.28, 2.56, 1.87, 1.01, 0.94, 0.79]. Note that in contrast to the discussion in Section 9.6.2, here, $x = \phi$ and $\hat{y}$ = ln $\hat{k}$. Note that these models are linearized versions of a non-linear model (see Section 9.10.2).

equation used for regression. In this section, we learn about different methods to evaluate these factors.

To facilitate the discussion, we will leave the calibration curve and switch to another application of fitting that is highly relevant to separation science, *i.e.* retention modelling. Figure 9.32 displays two different retention models: a straight-line model and a second-order polynomial, fitted to the same retention data. A visual inspection of these plots suggests that the second-order polynomial describes the retention data better. In this section, we explore whether it is possible to establish this conclusion objectively, using scientific methods.

Note that the retention models of $\ln \hat{k}$ are in fact linearized versions of actual non-linear models (*e.g.* $\hat{k} = \exp(b_0 + b_1\phi + b_2\phi^2)$). The linearization allows the least-squares regression to be applied as discussed above, but results in $\mathrm{SS_{res}} = \sum_{i=1}^{n}\left(\ln k_i - \ln \hat{k}_i\right)^2$ being minimized rather than $\mathrm{SS_{res}} = \sum_{i=1}^{n}\left(k_i - \hat{k}_i\right)^2$. The implications of this and possible alternatives are discussed in Section 9.10.2.

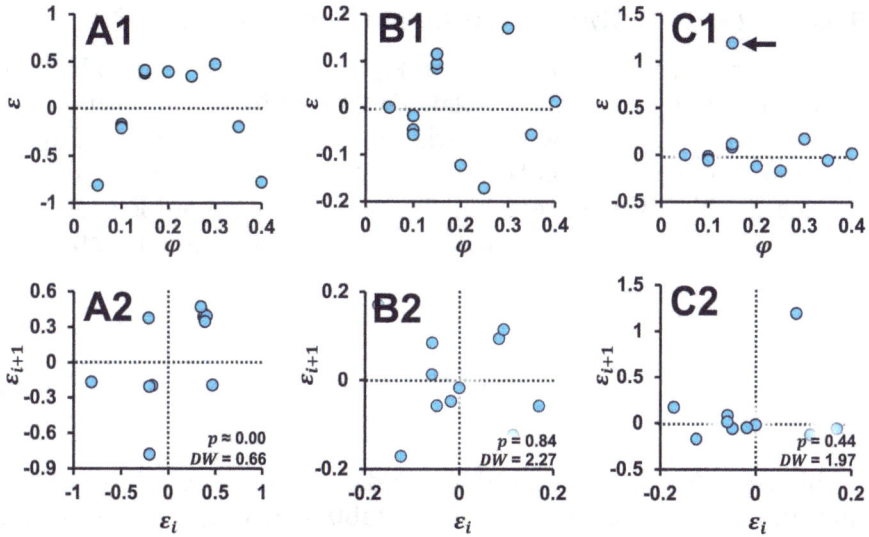

Figure 9.33   Residual plot ($\varepsilon = y - \hat{y}$) of a (A1) straight-line and (B1) second-order polynomial fitted to the retention data of Figure 9.32. Panel C1 reflects the same as panel B1, but one datapoint was adjusted to be an outlier (datapoint ln $k = 3.3$ was changed to 2.2). The dotted line represents a perfect description of the plotted data by the model. Panels A2–C2 reflect the Durbin–Watson test for serial correlation or autocorrelation; See the text.

9.6.4.1 Inspection of Residuals

One of the easiest methods to inspect a model is by evaluating its residuals with respect to the training data (*i.e.* $\varepsilon = y - \hat{y}$). Figure 9.33A1 and B1 illustrate this for the data and models shown in Figure 9.32A and B, respectively. Our visual observation that the straight-line model does not describe the data well is confirmed by the clear pattern that is apparent in Figure 9.33A1. In contrast, the data in Figure 9.33B1 do not feature a distinct pattern and the points look randomly distributed, which is indicative of a good fit. Residuals should not only be randomly distributed but should also exhibit constant variance across all values of x (homoscedasticity). For Figure 9.33C1, one original datapoint from panel B was, for educational purposes, turned into an outlier (datapoint ln $k = 3.3$ was changed to 2.2). The presence of this outlier becomes clear from the plot. When outliers are suspected, it is possible to follow-up the graphical inspection of the data with an outlier test (see Section 9.4.2).

9.6.4.2 Durbin–Watson Test for Autocorrelation

The earlier noted distinct pattern in Figure 9.33A1 appears to be that the error in one datapoint correlates with that in neighbouring data points. This is also known as **serial correlation** or **autocorrelation**. To check whether there is any such autocorrelation, it is possible to use the **Durbin–Watson test**, which tests for H_0: no autocorrelation in the residuals, and H_1: autocorrelation in the residuals. The test statistics is computed by

$$\text{DW}_{\text{obs}} = \frac{\Sigma_{i=2}^{n}\left(\varepsilon_i - \varepsilon_{i-1}\right)^2}{\Sigma_{i=1}^{n}\varepsilon_i^2} \tag{9.96}$$

This test does not assume that the data follow a certain distribution and DW_{obs} is compared with a tabulated DW_{crit} value, which depends on the number of datapoints, n, and the number of model parameters $(m + 1)$ minus one $(i.e.\ m)$.

The DW_{obs}-statistic is designed to result in a value between 0 and 4. $\text{DW}_{\text{obs}} = 2$ indicates no autocorrelation, $\text{DW}_{\text{obs}} < 2$ indicates positive autocorrelation, and $\text{DW}_{\text{obs}} > 2$ negative autocorrelation. Generally, a value between 1.5 and 2.5 is considered to indicate a good model, and only values lower than 1 or larger than 3 give rise to concern. Most software packages supporting this test also produce a p-value. Model validity should also be assessed based on residual patterns, leverage, and Cook's distance.

Figure 9.33A2 confirms the observation from panel A1 that there is autocorrelation, with DW = 0.66 and a p-value well below $\alpha = 0.05$. In contrast, no serial correlation is observed for the data in panels B, which appears to capture all the variation in the data. The test assumes independent residuals, and is highly unreliable for data with outliers, and the values shown in panel C2 should be ignored. The test may not be reliable in non-linear settings or for heteroscedastic data.

9.6.4.3 Cook's Distance: Influence and Leverage of Individual Datapoints

One method to detect outliers such as the one present in Figure 9.33 C1 and C2 involves the use of **Cook's distance,** which is a measure of how **influential** a datapoint is $(i.e.$ how significantly the regression coefficients are affected by it). Here, the model constructed using all data is compared to a model with – each time – one of the datapoints absent. Cook's distance CD_r^2 for point r is calculated as

$$CD_r^2 = \frac{\sum_{i=1}^{n}\left(\hat{y}_i - \hat{y}_{i,\neg r}\right)^2}{(m+1)\,s_e^2} \tag{9.97}$$

where $\hat{y}_i$ is the predicted value for point x_i when all datapoints are used, $\hat{y}_{i,\neg r}$ is the value for point x_i when the model obtained without using point r ("¬r") during regression is used, n is the number of datapoints, and $(m+1)$ is the number of parameters in the model. When $n \leq 10$ a value of $CD_r^2 > 1$ for datapoint r indicates that this datapoint is influential. Other threshold definitions, such as $CD_r^2 > \frac{n}{4}$, are also in use.

In this context, it is also worthwhile to revisit the concept of leverage observed in Figure 9.29E. **Leverage** indicates whether datapoint r is located relatively far from the other datapoints (and thus has a large impact on the fit). It is defined for datapoint r as the r,r diagonal element of the hat matrix (or projection matrix), **H**, which is defined as

$$\mathbf{H} = \mathbf{X}\left(\mathbf{X}^{\mathrm{T}}\mathbf{X}\right)^{-1}\mathbf{X}^{\mathrm{T}} \tag{9.98}$$

Point r is thought to have high leverage if value $h_{r,r}$ from **H** meets the condition

$$h_{r,r} > 2\frac{m+1}{n} \tag{9.99}$$

For the data shown in Figure 9.29E, for which $n = 57$ and $m = 1$ (*i.e.* $\hat{y} = b_1 x + b_0$), the point in the upper right corner has a leverage of almost 1. This is well above the threshold value of $2(m+1)/n = 0.07$, thus indicating high leverage of this point, while the points in the lower left corner have a leverage well below 0.03.

Both the influence and leverage of a datapoint play important roles. These properties are interrelated. A datapoint with high leverage is usually also highly influential. When the leverage of a datapoint is low, it is more difficult to detect an outlier using Cook's distance. Similarly, a true high-leverage point is sometimes mistakenly identified

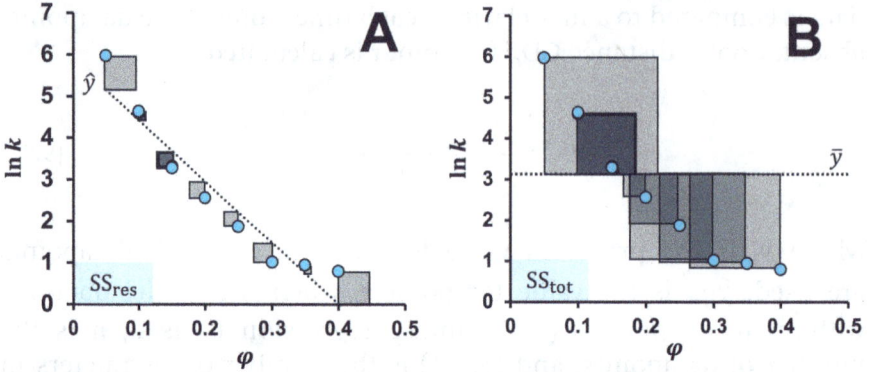

Figure 9.34   Graphical expression of the SS_{res} (A) and the SS_{tot} (B) components of the coefficient of determination (R^2).

as an outlier due to its high influence. Another disadvantage of Cook's distance is that, similar to Dixon's test, it is vulnerable to the presence of more than one outlier close to each other.

9.6.4.4   Coefficient of Determination (R^2)

In the earlier parts of this section, we have addressed means to validate the data used to construct the model. However, it is also useful to evaluate the mathematical equation used for the regression. A very common metric for this purpose is the **coefficient of determination**, R^2

$$R^2 = \frac{SS_{reg}}{SS_{tot}} = \frac{\sum_{i=1}^{n}(\hat{y}_i - \bar{y})^2}{\sum_{i=1}^{n}(y_i - \bar{y})^2} = 1 - \frac{SS_{res}}{SS_{tot}} \qquad (9.100)$$

The R^2 quantifies the fraction of the total variation explained by the model by comparing the squared residuals of the model (Figure 9.34A) *vs.* the total sum of squares of the datapoints compared to the mean (Figure 9.34B).

One major flaw of R^2 is that it will always increase when the used mathematical model features more parameters, even if they do not improve the model's predictive power. For example, a cubic model $\hat{y} = b_0 + b_1x + b_2x^2 + b_3x^3$ will always be able to capture more of the variation in the data than a straight line $\hat{y} = b_0 + b_1x$. To somewhat correct for this, the **adjusted coefficient of determination** R_a^2 can be used, which accounts for the number of terms, preventing artificial increases due to unnecessary terms. This is achieved by incorporating the degrees of freedom. The result is that the addition of further parameters is penalized.

$$R_a^2 = 1 - \frac{SS_{res}/(n - (m+1))}{SS_{tot}/(n-1)} \tag{9.101}$$

$$R_a^2 = 1 - \frac{SS_{res}/(n - (m+1))}{SS_{tot}/(n-1)}$$

9.6.4.5 F-test of Significance: Lack-of-Fit Test

A more sensitive tool to evaluate the mathematical model used for regression is the **F-test of significance**. We already concluded in this section that the quadratic model $(\hat{y} = b_0 + b_1 x + b_2 x^2)$ better describes the retention data in Figure 9.32 than a straight line $(\hat{y} = b_0 + b_1 x)$. The difference between these models is the quadratic term $b_2 x^2$. The F-test of significance can be used to assess the significance of this additional term, with H_0: addition of the extra term is not significant, and H_1: addition of the extra term is significant.

To compute the test, the F-statistic compares the reduction in residual variance due to the additional term with the residual variance of the more complex model and is calculated as

$$F_{obs} = \frac{MS_{res,adding|b_{new}}}{MS_{res|b_{new}}} \tag{9.102}$$

where $MS_{res,adding|b_{new}}$ is the mean square of the residuals obtained by the addition of the extra term b_{new}, where the model with b_{new} has a number of $q_{with|b_{new}}$ parameters, and the model without b_{new} features $q_{without|b_{new}}$ parameters

$$MS_{res,adding|b_{new}} = \frac{\Sigma_{i=1}^{n}\left(\hat{y}_{i,without|b_{new}} - \hat{y}_{i,with|b_{new}}\right)^2}{q_{with|b_{new}} - q_{without|b_{new}}} \tag{9.103}$$

and $MS_{res|b_{new}}$ is the residual mean square of the model including the extra term.

$$\text{MS}_{\text{res}\,|\,b_{\text{new}}} = \frac{\Sigma_{i=1}^{n}\left(\hat{y}_{i,\text{with}\,|\,b_{\text{new}}} - y_i\right)^2}{n - q_{\text{with}\,|\,b_{\text{new}}}} \tag{9.104}$$

Under the usual assumption that H_0 is true, F follows the F distribution $F \sim F\left(\left(q_{\text{with}\,|\,b_{\text{new}}} - q_{\text{without}\,|\,b_{\text{new}}}\right), \left(n - q_{\text{with}\,|\,b_{\text{new}}}\right)\right)$. As with the earlier F-tests (see Section 9.3.7), F_{obs} is compared with F_{crit} at the desired significance level. Doing so for our comparison of the quadratic model with $q_{\text{with}\,|\,b_{\text{new}}} = 3$ parameters, and the straight-line model with $q_{\text{without}\,|\,b_{\text{new}}} = 2$ parameters, yields for $F \sim F(1,9)$ that $F \approx$ 180.82 and consequently a p-value of 2.9×10^{-7}, which confirms our visual analysis that the addition of the $b_2 x^2$ term is significant.

Ultimately, establishing the appropriate mathematical equation should be approached with care. As the complexity of the model increases, so does the risk of **overfitting**. In that case, the model is customized too much to the training data and it will work less well with different datasets. Conversely, simpler models may not sufficiently describe the data, resulting in bias (underfitting).

9.6.5 Figures of Merit: Characterizing Analytical Methods

Establishing a calibration model allows for determining several performance characteristics of a method (*i.e.* "analytical figures of merit") that are of high importance in regulated environments. A well-known characteristic is the **sensitivity** of the method, which is equal to the slope of the straight-line calibration curve, *i.e.* b_1. It is unrelated to the sensitivity of the statistical test (*i.e.* $1 - \beta$).

However, arguably a more important characteristic is the **limit of detection** (LOD, x_{D}), which is depicted in Figure 9.35. Here, panel A displays a section of the calibration curve from Figure 9.30 with reversed axes (for educational purposes).

To understand the LOD, we start by considering the **blank** measurement (or "blank"). A blank is identical to the sample measurement with respect to the matrix, but without the analyte(s) present. It features the baseline (μ_{bl}, Figure 9.35D), along with a random variation (*i.e.* the noise, σ_{bl}, Figure 9.35D). Like any measurement, the blank is subject to errors, which – after sufficient repeats – can be captured by a normal distribution $y \sim N(\mu_{\text{bl}}, \sigma_{\text{bl}})$ as seen on the leftmost distribution in Figure 9.35B and C.

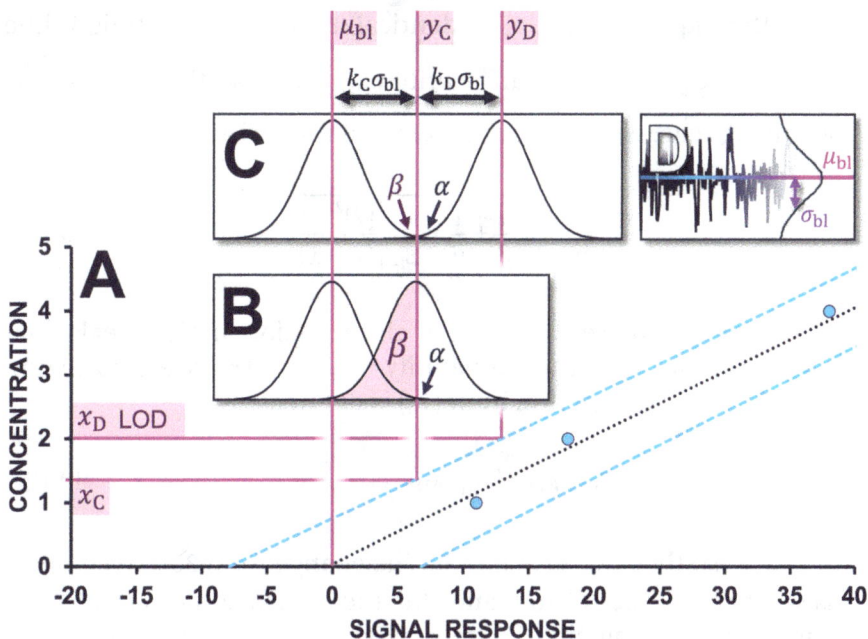

Figure 9.35 Graphical clarification of the relation between the decision limit and detection limit. (A) Inverted calibration curve with the *x*- and *y*-axes swapped. (B) Decision limit, (C) detection limit, and (D) noise captured in terms of μ_{bl} and σ_{bl}.

The **decision limit** (y_C, also known as L_{crit} or $CC\alpha$) is defined as the lowest concentration level of the analyte that can be detected in a sample with a chance α of a false-positive decision. It is hence analogous to a critical value such as t_{crit} and F_{crit}, as we have encountered in the tests in the previous modules. The decision limit is defined as

$$y_C = \mu_{bl} + k_C \cdot \sigma_{bl} \tag{9.105}$$

where k_C is a constant (unrelated to the retention factor) that is typically set to 3, so that, with y_C defined as eqn (9.105), the probability of wrongly concluding that there is an analyte, when in fact there is not, is $\alpha = 0.0013$. This significance level is currently employed in the IUPAC guidelines, but it remains policy driven and can differ across fields. For example, decision 2002/657/EC of the European Commission sets $\alpha = 0.01$ for certain compounds.

Note that μ_{bl} is essentially identical to b_0 of the straight-line model. Consequently, we can rewrite eqn (9.71) so that $x_C = \dfrac{k_C \cdot \sigma_{bl}}{b_1}$. When using regression, we can write in analogy with eqn (9.94)

$$x_C \approx \frac{t_{\alpha,n-2} \cdot S_e}{b_1} \sqrt{\frac{1}{g} + \frac{1}{n} + \frac{(\bar{x})^2}{\Sigma_{i=1}^{n}(x_i - \bar{x})^2}} \tag{9.106}$$

Unfortunately, power analysis of the decision limit reveals that while α is rather low, the probability of β is 50% (see also Figure 9.35B). Hence, the **detection limit** (y_D) is defined as

$$y_D = \mu_{bl} + (k_C + k_D) \cdot \sigma_{bl} \tag{9.107}$$

with $k_D = 3$, so that $k_D' = k_C + k_D = 6$. The change from $3\sigma_{bl}$ to $6\sigma_{bl}$ now causes β to be at 0.0013 too, and this the statistical power is 0.9987 (Figure 9.35C). Using similar steps as above we can also define the detection limit in terms of concentration as

$$x_D \approx \frac{t_{\beta,n-2} \cdot S_e}{b_1} \sqrt{\frac{1}{g} + \frac{1}{n} + \frac{(2x_C - \bar{x})^2}{\Sigma_{i=1}^{n}(x_i - \bar{x})^2}} \tag{9.108}$$

With these limits defined, we arrive at the **quantification limit** (y_Q), which is defined as

$$y_Q = \mu_{bl} + 10\sigma_{bl} \tag{9.109}$$

so that in concentration units this equals

$$x_Q = \frac{10\sigma_{bl}}{b_1} \tag{9.110}$$

The rationale is that if one were to calculate the relative precision at the detection limit, a value of 16% would be found, which is considered too high. The quantification limit (x_Q) is the minimum concentration that can be determined with a relative precision of 10%. Eqn (1.110) can also be calculated similar to eqn (1.6) if regression models are used.

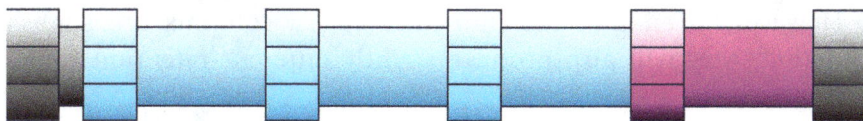

Figure 9.36 Schematic example of a POPLC column. The three cyano columns are depicted as blue, and the C18 column is depicted as pink.

9.7 Error Propagation (M)

The confidence intervals discussed in Module 9.2 concerned variables that relied on only a single independent variable. For example, the measurement of the concentration (the dependent variable) discussed in that module relied only on the measured signal response (the independent variable). However, what if the dependent variable of interest relies on more than one independent variable? How would we calculate the confidence interval in this case? This module treats the propagation of errors from different sources into the dependent variable.

As an example, we will use the concept of combined-column selectivities. Rather than optimizing method parameters for a given column selectivity, this approach combines a series of smaller columns with different selectivities that are chosen such as to achieve complete separation of the sample. To accomplish this, the user can measure the retention times of the different compounds on columns with different selectivities. One of the most notable examples is the phase-optimized LC (POPLC) approach.[10] Figure 9.36 shows a schematic example of such a system, where three 50 mm long cyano columns (depicted as blue) are combined with a single 50 mm C18 column (depicted as pink).

To predict the retention time for one of the analytes in a mixture, its retention time would be measured on one of the cyano columns and on the C18 column. This means that the total retention time $(t_{R,tot})$ is the sum of three times the retention time on a cyano column $(3t_{R,cyano})$ plus one time that on the C18 column $(t_{R,C18})$, *i.e.*

$$t_{R,tot} \approx 3t_{R,cyano} + t_{R,C18} \tag{9.111a}$$

However, given that the individual $t_{R,cyano}$ columns are slightly different, they will also produce slightly different retention times, so that

$$t_{R,tot} = t_{R,cyano,1} + t_{R,cyano,2} + t_{R,cyano,3} + t_{R,C18} \qquad (9.111b)$$

When using the same mobile phase, the retention on the cyano column is lower. This explains why it is desirable to use three cyano columns. From multiple identical C18 or cyano columns, a standard deviation for the retention time can be determined. In absolute terms, the variation of the retention time on cyano columns ($\sigma_{t_{R,cyano}}^2$) is lower than the variation of the retention time on the C18 column ($\sigma_{t_{R,C18}}^2$), even though the relative (percentage-wise) precision of the retention times is similar on the two columns. This difference in variation is further enhanced by other effects, such as differences in the stability of the stationary phase.

We can summarize the above as that the random errors (variances) of both retention times (*i.e.* $\sigma_{t_{R,C18}}^2$ and $\sigma_{t_{R,cyano}}^2$) are different, and that, therefore, their respective errors **propagate** differently into the total retention time ($t_{R,tot}$). This affects the calculation of the confidence interval for the dependent variable $t_{R,tot}$ as it depends on two independent variables with different precisions. In the context of error propagation, we can write eqn (9.111b) in generic terms as

$$y(x_1, x_2, ..., x_m) = ax_1 + bx_2 + ... + cx_m \qquad (9.112)$$

where x_1, x_2, and x_m are independent variables, and a, b, and c are their associated coefficients. If we recall eqn (9.86) ($\sigma_{\hat{y}_0}^2 = \mathbf{J}_0 \cdot \Sigma \cdot \mathbf{J}_0^T$) and we assume that only the diagonal elements of Σ are non-zero (*i.e.* the independence between different variables results in the absence of covariance terms), we can write for the variance

$$\sigma_y^2 = \left(\frac{\partial y}{\partial x_1}\right)^2 \sigma_{x_1}^2 + \left(\frac{\partial y}{\partial x_2}\right)^2 \sigma_{x_2}^2 + ... + \left(\frac{\partial y}{\partial x_m}\right)^2 \sigma_{x_m}^2 \qquad (9.113)$$

which for our POPLC example becomes

$$\sigma_{t_{R,tot}}{}^2 = \left(\frac{\partial t_{R,tot}}{\partial t_{R,cyano,1}}\right)^2 \sigma_{t_{R,cyano}}{}^2 + \left(\frac{\partial t_{R,tot}}{\partial t_{R,cyano,2}}\right)^2 \sigma_{t_{R,cyano}}{}^2$$

$$+ \left(\frac{\partial t_{R,tot}}{\partial t_{R,cyano,3}}\right)^2 \sigma_{t_{R,cyano}}{}^2 + \left(\frac{\partial t_{R,tot}}{\partial t_{R,C18}}\right)^2 \sigma_{t_{R,C18}}{}^2 \qquad (9.114)$$

$$= (1)^2 \sigma_{t_{R,cyano}}{}^2 + (1)^2 \sigma_{t_{R,cyano}}{}^2 + (1)^2 \sigma_{t_{R,cyano}}{}^2 + (1^2) \sigma_{t_{R,C18}}{}^2$$

$$= 3\sigma_{t_{R,cyano}}{}^2 + \sigma_{t_{R,C18}}{}^2$$

Note that because three different cyano columns (which are not exactly the same) are used, each of them appears in the error propagation eqn (9.114). When all variances ($\sigma_{t_{R,cyano}}{}^2$ and $\sigma_{t_{R,C18}}{}^2$) are known, $\sigma_{t_{R,tot}}$ can be calculated. The confidence interval is then, in general terms

$$CI_y = y \pm z\frac{\alpha}{2} \cdot \sigma_y \qquad (9.115)$$

where $z\frac{\alpha}{2}$ refers to z_{crit} for a significance level of α (see Section 9.2.3.3). Applying this to the POPLC example, if known that for cyano column our variance is $\sigma_{t_{R,cyano}} = 0.012$ min and for the C18 column $\sigma_{t_{R,C18}} = 0.018$ min, then for a determined retention time of $t_{R,tot} = 5.611$ min, we could calculate that $\sigma_{t_{R,tot}}{}^2 = 1^2 \times 0.012^2 + 1^2 \times 0.012^2 + 1^2 \times 0.012^2 + 0.018^2 = 0.000756$, culminating in $\sigma_{t_{R,tot}} = 0.0275$. The 95% confidence interval is then $t_{R,tot} = 5.611 \pm 0.054$ min.

Another example is the computation of the retention factor according to (eqn (1.13)).

$$k = \frac{t_R}{t_0} - 1 \qquad (9.116)$$

$$\sigma_k^2 = \left(\frac{\partial k}{\partial t_R}\right)^2 \sigma_{t_R}^2 + \left(\frac{\partial k}{\partial t_0}\right)^2 \sigma_{t_0}^2 = \left(\frac{1}{t_0}\right)^2 \sigma_{t_R}^2 + \left(\frac{-t_R}{t_0^2}\right)^2 \sigma_{t_0}^2 = \frac{\sigma_{t_R}^2}{t_0^2} + \frac{\sigma_{t_0}^2 t_R^2}{t_0^4} \qquad (9.117)$$

so that

$$\sigma_k = \sqrt{\frac{\sigma_{t_R}^2}{t_0^2} + \frac{\sigma_{t_0}^2 t_R^2}{t_0^4}} = \frac{1}{t_0}\sqrt{\sigma_{t_R}^2 + \frac{t_R^2}{t_0^2}\sigma_{t_0}^2} \qquad (9.118)$$

From eqn (9.118), it is clear that the error in k becomes large if t_0 is small, and that errors in t_0 weigh heavier than errors in t_R. The associated confidence interval for k then becomes.

$$CI_k = k \pm z\frac{\alpha}{2} \cdot \sigma_k \qquad (9.119)$$

Due to the errors and their propagation, it is important to round values to remove figures that are not significant. The guiding concept to decide the number of significant figures is that the first figure of absolute uncertainty is the last significant figure of the answer. For example, we found above a confidence interval for the t_R in 5.611 ± 0.079, which then becomes 5.61 ± 0.08.

9.8 Design of Experiments (M)

9.8.1 Introduction

In Module 9.6, we discussed fitting a model to the results of a number of measurements. We did not spend much time discussing the positioning of these datapoints. When creating a calibration curve, there is not much wrong in spreading the datapoints evenly along the concentration (or amount injected) axis, provided that the calibration points span the expected range of concentrations, as is confirmed by Figure 9.30. The situation changes drastically if we aim to describe a dependent variable (or "response") y that is a function of two independent variables, x_1 and x_2. In that case, it is outright dangerous – and sometimes grossly incorrect – to create a graph of y as a function of x_1 for some specific values of x_2; select the most-desirable value of x_1^* from this plot, and then prepare a graph of y as a function of x_2 at the selected value of x_1^*.

This is illustrated in Figure 9.37, for the example of developing a derivatization reaction to obtain better chromatographic selectivity and detection sensitivity. The yield of the reaction depends on the temperature and on the pH. The old-fashioned **one-variable-at-a-time (OVAT)** approach would have you find the optimum pH at some arbitrary temperature (or *vice versa*). Performing measurements at six different pH values yields a good impression of the effect of the pH at 60 °C, as shown in the top-right figure. This suggests that pH = 4.8 (point 1) is an optimum value *at this temperature*. The last three words are often

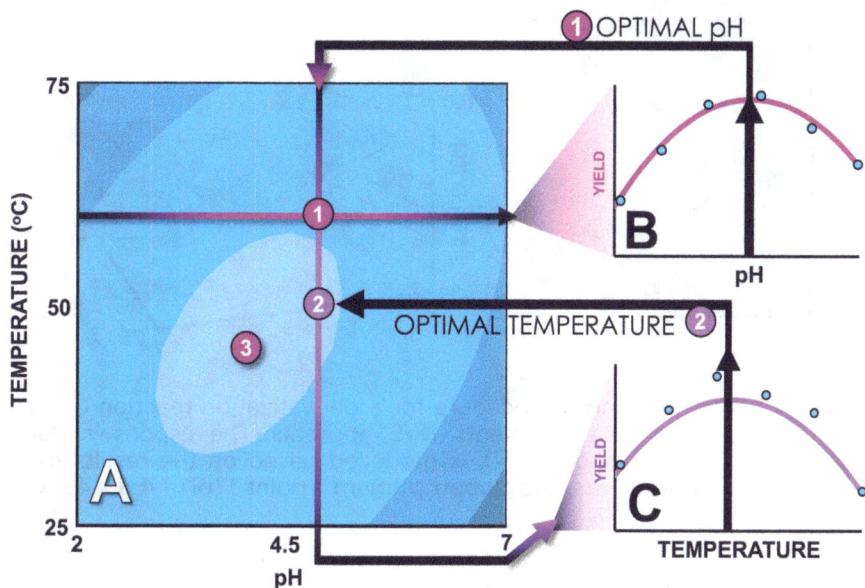

Figure 9.37 Developing a procedure for a derivatization reaction, using the one-variable-at-a-time (OVAT) approach. Curve B depicts the yield as a function of pH at a temperature of 60 °C. The optimum found from this graph is point 1 (pH = 4.8). Curve C depicts the yield as a function of temperature at the pH of point 1. This yields an optimum at point 2 (pH = 4.8, T = 49 °C), which differs from the global optimum at point 3 (pH = 4, T = 45 °C).

omitted, but they do matter if the effects of pH and temperature on the yield are not independent. The figure on the bottom-right shows an additional line that depicts the yield at pH = 4.8, and this suggests an optimum temperature of 49 °C (point 2). Throughout the process, the blue response surface (lighter colours signify a higher yield) is not known, but it is shown here to illustrate the process. The twelve measurements performed do not provide a complete impression of the surface, and point 2 does not represent the global optimum.

Figure 9.38 illustrates what can be achieved with a systematic experimental design. In the example, it required one more measurement (13 instead of twelve), but one less buffer to be prepared (5 instead of 6), so that the effort required is similar. However, if we fit a quadratic response surface of the form Yield $= b_0 + b_1pH + b_2T + b_{11}pH^2 + b_{22}T^2 + b_{12}pH \cdot T$ through the data (and we verify the validity of the model), we (i) obtain a good impression of the entire response surface, (ii) can locate the global optimum

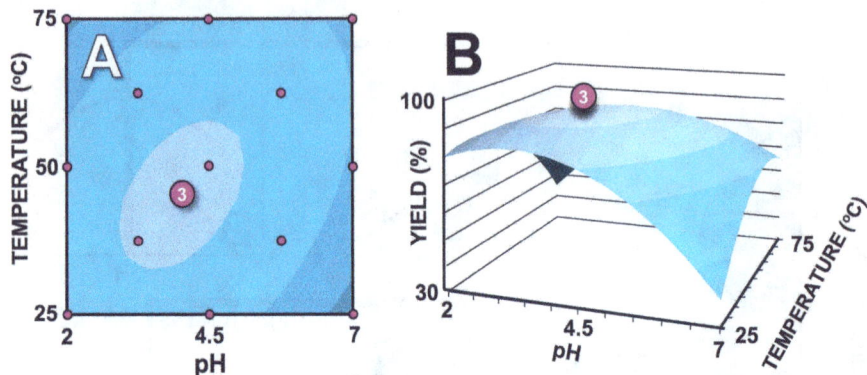

Figure 9.38 Developing a procedure for a derivatization reaction using a design-of-experiments (DoE) approach. The response surface, (A) 2D, and (B) 3D, is modelled based on the results of 13 experiments. The global optimum is point 3 (pH = 4, T = 45 °C).

at pH = 4 and T = 45 °C, and (iii) can establish robust conditions, under which the yield is hardly affected by small errors in the pH or temperature. Thus, **design of experiments (DoE)** is greatly preferred to an OVAT approach.

An important aspect that we did not mention explicitly is the goal of the optimization process. In the above example, maximizing the yield of the derivatization reaction was the obvious goal. There are other clear-cut goals, such as optimizing the various parameters for PTV injection in GC (see Module 2.4) so as to achieve the highest possible plate count. However, as we will see in Chapter 10, it may be challenging to define the goal of a chromatographic method-development process in simple mathematical terms. Next, we need to define our parameter space, *i.e.* which variables we will consider and within which limits. In the above example, chemical knowledge and practical considerations may have been used to arrive at 2 < pH < 7 and 25 °C < T < 75 °C. Without such knowledge, we may also use **screening designs** to establish the most-relevant parameters.

9.8.2 Factorial Designs

Factorial designs are the best-known systematic approach to experiments. In a full-factorial design, m parameters (or factors) are studied at n levels and all n^m combinations of the different levels for the factors are studied. Several examples are shown in Figure 9.39. The levels for each factor are scaled between −1

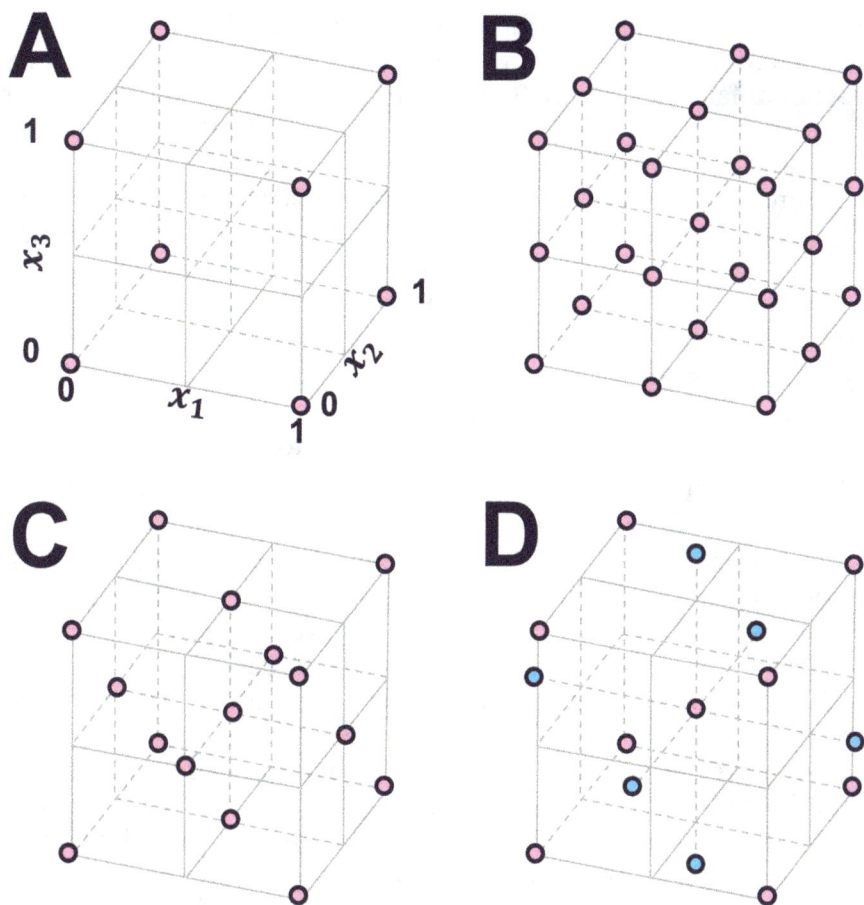

Figure 9.39 Illustration of full-factorial 2^2 (A) and 3^3 (B) designs, and a fractional-factorial (C) and a central-composite (D) design for 3 parameters at 3 levels in which the blue points are positioned outside the cube.

and +1, so that, for example, three levels take on only the values −1 (low), 0 (medium), or 1 (high). This is good practice to avoid complications in fitting a model with vastly different ranges for the different coefficients. We could (and perhaps should) have scaled the pH in Figures 9.37 and 9.38 through $x_1 = (\text{pH} - 4.5)/2.5$ and $x_2 = \{T(\text{Co}) - 50\}/25$.

On the top left (Figure 9.39A) is a 2^3 **full-factorial design** for three factors at two levels (8 experiments) and on the top right (Figure 9.39B) a 3^3 full-factorial design for three factors at three levels (27 experiments). Clearly, more levels and more factors imply more experiments in the design and a full-factorial design soon becomes unattractive. The figure on the bottom left (Figure 9.39C) shows a **fractional-factorial design**, for which only a fraction of the number of experiments need to be performed. The design in Figure 9.39C is a fractional (half) 3^3 design with a centre point added.

Figure 9.39D shows a **central-composite design**, which differs from the design in Figure 9.39C in that the face-centred points are moved outside the cube, with normalized parameter values (x_1 and x_2) higher than 1 or lower than −1. Mathematically, such a design is expected to yield a better description of a quadratic model. Fewer experiments limit the maximum complexity of the model and the statistical information, but the 15 points in the designs of Figure 9.39C and D include three levels for each factor, which allows fitting all linear terms (b_1x_1, b_2x_2, b_3x_3), quadratic terms ($b_1x_1^2$, $b_2x_2^2$, $b_3x_3^2$), and second-order interaction terms ($b_{12}x_1x_2$, $b_{13}x_1x_3$, $b_{23}x_2x_3$), plus an intercept (b_0). The significance of all terms may be investigated. Often one or more of the quadratic and of the interaction terms prove insignificant, which simplifies the eventual model. Factorial designs with two levels are of limited use for modelling response surfaces, because these cannot describe curvature, and they are mainly used as screening designs (see Section 9.8.3).

Figure 9.40 shows an example of a mixture design that can be used if the sum of the levels of the factors is constrained, such as for the optimization of the composition of a ternary mobile-phase mixture. In that case, the sum of the volume fractions of the three constituents is always unity (*i.e.* $\varphi_1 + \varphi_2 + \varphi_3 = 1$). The composition of a ternary mixture can be depicted in a triangle, and the design shown (10 datapoints) can be used to fit a quadratic or cubic "retention model".

9.8.3 Screening Designs

Screening designs can be used to explore many variables for their main effects only. Interaction effects are ignored when such designs are used. There are two instances in which they may be of use during the development of analytical separation methods. The first is when many possible factors may affect, for example, the yield of a derivatization reaction, the efficiency of an injection in GC, or the preparation of a monolithic column. In such situations, a screening design may be used to determine which parameters have a significant effect on the outcome. The other instance is

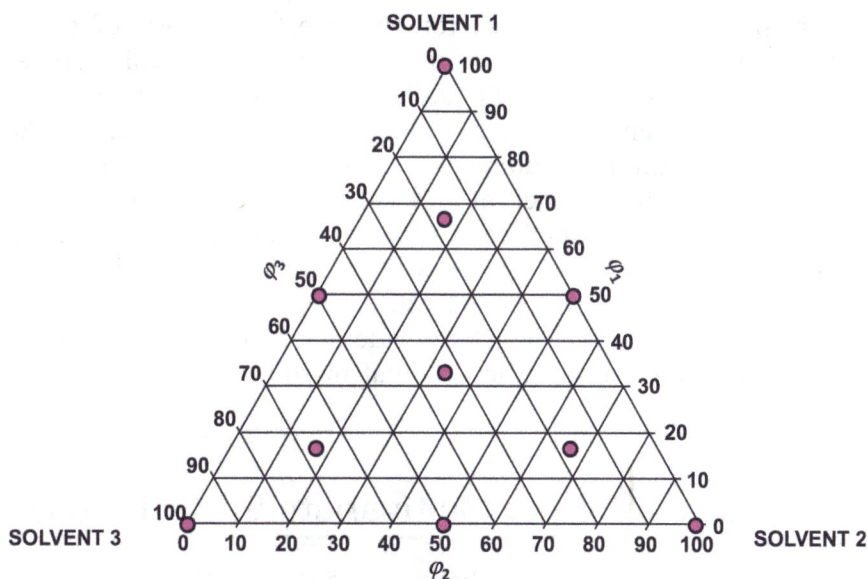

Figure 9.40 Mixture design that can be used to optimize the composition of a ternary mobile-phase mixture in LC.

at the method-validation stage, to explore the robustness of a method. The **robustness** characterizes the resistance of the system to change without adaptation of its configuration. **Ruggedness** quantifies the reproducibility of the results when conditions other than the laboratory are changed.

A robustness test can be very meaningful prior to the exact method description and the roll-out of a method to a number of different users (*e.g.* production sites of a company), or prior to a (round-robin) reproducibility test, which is likely to be long and costly. A robustness test may indicate that a method is robust, or it may signal that certain parameters require special attention. For example, the outcome may be that pH must be tightly controlled. How that can be achieved must then be described carefully in the method. An example of a ruggedness test is provided in ref. 11.

There are two types of screening designs, *viz.* fractional-factorial designs, such as 2^{8-4} designs that explore the main effects of 8 parameters through 16 experiments. The other type of design used for the purpose are **Plackett–Burman** (PB) designs, which were pioneered for the purpose of method-robustness testing by Mary Mulholland and Janette Waterhouse.[12] They demonstrated that robust ranges for operating an LC method could be established

for six parameters (including, for example, percentage of acetonitrile, temperature, and flow rate) based on a small number
of experiments following a PB design. PB designs exist for numbers of experiments that are a multiple of 4 (*e.g.* 12, 16, 20
experiments), and the number of factors that can be tested is
one less than the number of experiments. For example, robust
ranges can be established for up to 11 variables from a design
with 12 experiments. In case we want to explore the main effects
of only 9 factors, we can apply the 12-experiment design with
two "dummy variables", which provide some information on the
confidence interval under the nominal method conditions (centre
of the design).

9.9   Introduction to Multivariate Statistics (A)

Most of the modules in this chapter have thus far been limited to
univariate methods, where one variable is evaluated at a time. In
many cases, it is of interest to study multiple variables simultaneously. Multivariate methods are often known for the classification
tools that leverage the simultaneous assessment of several variables,
but multivariate methods are also of interest for other purposes,

Box 9.4 Hero of chemometrics: Lutgarde Buydens. Image courtesy of
Prof. Buydens.

Lutgarde Buydens (Belgium)
Lutgarde Buydens obtained her PhD from Luc
Massart at the Vrije Universiteit Brussels
(Belgium), from where she moved to become
Professor of Analytical Chemistry and later
Dean of the Faculty of Science at the Radboud
University in Nijmegen (The Netherlands).
Throughout her scientific career, she was one
of the leaders in the field of chemometrics,
with special focus on multivariate analysis of
analytical-chemical data to address a range of
industrial and societal challenges. Lutgarde was one of the pioneers
of the application of artificial intelligence in analytical spectroscopy
and separation science.

including data pre-treatment, outlier detection, calibration, and explorative analysis. Multivariate statistics play an important role in distilling useful information from multi-dimensional high-resolution separations. Lutgarde Buydens (Box 9.4) has been among those who paved the way for the application of multivariate statistics in analytical chemistry and analytical separation science.

Multivariate statistics is an entire discipline of its own, and a comprehensive treatment is beyond the scope of this book. Instead, this module focuses on introducing the reader to the essential concepts of multivariate methods by treating well-known techniques that are relatively easy to access and have proven their usefulness for separation scientists. Other multivariate methods are summarized as much as possible so as to inform the reader of their existence and purpose.

9.9.1 Principal-component Analysis

Principal component analysis (PCA) is an effective method that can be used to compress correlated data to fewer dimensions while minimizing the loss of information. The concept of PCA is based on the assessment of the covariance between different variables and the subsequent calculation of eigenvectors and eigenvalues to identify the principal components, *i.e.* new variables that are constructed as linear combinations of the initial variables. The technique can be used for various purposes, ranging from exploratory analysis to data pre-processing and estimation of missing data.

To understand the concept, we regard the information shown in Table 9.11, which contains selected multivariate data from the Periodic Table of Elements. The table lists five different variables (melting point, boiling point, density, oxidation number, and electronegativity) for a selection of elements.

We have seen in Section 9.6.2 that covariance is not dimensionless and it is thus important to first adjust the data, especially when the different variables have vastly different scales. This is true for our data in Table 9.11, where the temperatures are expressed with a significantly different magnitude compared to the oxidation number or electronegativity.

It is thus common practice to perform **mean centring** and **autoscaling** or **standardizing** the data for each variable by subtracting the mean of all values for each variable and dividing by the standard deviation of that variable. This is analogous to z-scaling $(z = (x - \mu)/\sigma)$. Mean centring corresponds to repositioning the

Table 9.11 Selected chemical data from the Periodic Table of Elements. Data and example from ref. 13, reproduced from ref. 13 with permission from John Wiley & Sons. Copyright 2003 John Wiley & Sons, Ltd.

Element	Melting point (K)	Boiling point (K)	Density	Oxidation #	Electronegativity
Li	453.69	1615	534	1	0.98
Na	371	1156	970	1	0.93
K	336.5	1032	860	1	0.82
Rb	312.5	961	1530	1	0.82
Cs	301.6	944	1870	1	0.79
Be	1550	3243	1800	2	1.57
Mg	924	1380	1741	2	1.31
Ca	1120	1760	1540	2	1
Sr	1042	1657	2600	2	0.95
F	53.5	85	1.7	−1	3.98
Cl	172.1	238.5	3.2	−1	3.16
Br	265.9	331.9	3100	−1	2.96
I	386.6	457.4	4940	−1	2.66
He	0.9	4.2	0.2	0	0
Ne	24.5	27.2	0.8	0	0
Ar	83.7	87.4	1.7	0	0
Kr	116.5	120.8	3.5	0	0
Xe	161.2	166	5.5	0	0
Zn	692.6	1180	7140	2	1.6
Co	1765	3170	8900	3	1.8
Cu	1356	2868	8930	2	1.9
Fe	1808	3300	7870	2	1.8
Mn	1517	2370	7440	2	1.5
Ni	1726	3005	8900	2	1.8
Bi	544.4	1837	9780	3	2.02
Pb	600.61	2022	11 340	2	1.8
Tl	577	1746	11 850	3	1.62

coordinate system so that the origin is moved to the centre of the data.

> See the website for a detailed description of the process of PCA as it is applied to the data shown in Table 9.11: ass-ets.org

Note that, as explained in Section 9.6.2, covariances that are positive indicate correlation in which the variables both increase or decrease, whereas negative covariances imply inverse correlation, where one variable decreases while the other variable increases.

One of the successes behind PCA lies in dimensionality reduction. However, there are at most as many principal components (PCs) as there are variables or observations, and it is thus important to ensure that the first PC captures the largest possible variance in the dataset. This is accomplished by drawing a line through the origin in the direction in which the collection of points is the most spread out. The next PC is then calculated similarly, but perpendicular to the first PC. This process continues, for each PC until the number of PCs is equal to m.

Mathematically speaking, PCA operates through **eigen-decomposition** of the variance–covariance matrix or single-value decomposition (SVD) to the data matrix. Figure 9.41A-D1 shows the boiling points of the different elements in Table 9.11 plotted against their melting points with different orientations of the candidate principal component (blue scatter points, y_i). If these would be our only two variables, then the first principal component (PC1) is a new latent variable (presented in the figure by an axis, $\hat{y}$) that crosses the origin so that the projected points on this axis (pink dots, $\hat{y}_i$ capture the maximum variance of the ensemble of original datapoints.

To aid understanding, this may be imagined as maximizing the degree of variance, depicted in the figure by ζ. Inversely, when fitting PC1 through the data, it is done so to minimize the distances between the original datapoints and their projected points on the candidate PC (*i.e.* minimize $\sum_{i=1}^{n} \varepsilon^2 = \sum_{i=1}^{n} (\hat{y}_i - y_i)^2$, the pink lines in the panels). The variance along PC1 is the magnitude component, its **eigenvalue**, and the orientation of PC1 with respect to the original variable axes is its **eigenvector**, its direction component. Different orientations are depicted in the various panels, with Figure 9.41D1 showing the best solution. The eigenvector with the highest eigenvalue will be the first PC; the second-highest eigenvalue will represent the eigenvector that represents the second PC, and so on.

Interestingly, we can see from Figure 9.41D1 that where Direction 1 (*i.e.* where y represents the first PC, or PC1) captures a large amount of variance in the data; this is not the case for Direction 2, which captures only very little variance. One could thus state that it is possible to omit the second direction and consequently that the data have been reduced by one dimension.

The distance of each projected data point on a PC ($\hat{y}_i$) to the origin is called its **score**, and the covariances between the original variables and the unit-scaled (*i.e.* not mean-centred) PCs are the **loadings**. The

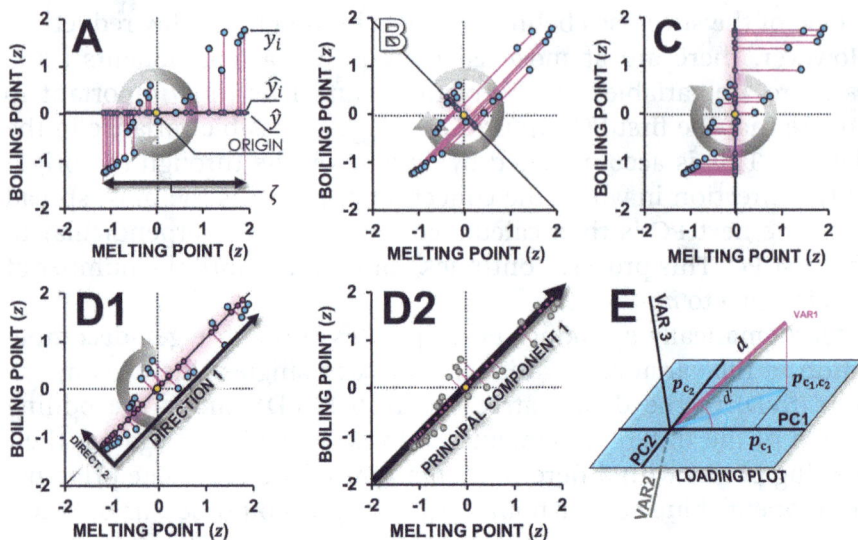

Figure 9.41 (A–D2) Different orientations of a line drawn through the origin through mean-centred and autoscaled data of the elements from Table 9.11 for the variables "Melting Point" and "Boiling Point". The first principal component is the line that describes the largest variance, *i.e.* the spread of the pink dots is the largest and ζ is maximized. (E) Schematic depiction of a loading plot and its relation to the principal components and three original variables.

loadings express how much each variable contributes to a particular PC. A schematic depiction is shown in Figure 9.41E, where three variables and two constructed PCs are shown. The coordinates p_{c_1}, p_{c_2} are the loadings of variable 1 with PC1 and PC2, respectively; d is the variance of variable 1, and d' is the variance explained by PC1 and PC2 together.

In matrix terms, the original data $\mathbf{X}$ (n objects $\times$ m variables) have been decomposed into a score matrix ($\mathbf{T}$) and a loading matrix ($\mathbf{P}$) for the PCs and the remaining unexplained residuals (ε), related through

$$\mathbf{X} = \mathbf{T}\,\mathbf{P}^{\mathrm{T}} + \varepsilon \qquad\qquad (9.120)$$

where $\mathbf{T}$ contains the scores for k principal components and n datapoints

$$\mathbf{T} = \begin{bmatrix} t_{1,1} & t_{1,2} & \cdots & t_{1,k} \\ t_{2,1} & t_{2,2} & \cdots & t_{2,k} \\ \vdots & \vdots & \ddots & \vdots \\ t_{n,1} & t_{n,2} & \cdots & t_{n,k} \end{bmatrix} \tag{9.121}$$

and $\mathbf{P}$ contains the loadings for k principal components and m variables.

$$\mathbf{P} = \begin{bmatrix} p_{1,1} & p_{1,2} & \cdots & p_{1,k} \\ p_{2,1} & p_{2,2} & \cdots & p_{2,k} \\ \vdots & \vdots & \ddots & \vdots \\ p_{m,1} & p_{m,2} & \cdots & p_{m,k} \end{bmatrix} \tag{9.122}$$

The remaining distance between each datapoint and its location in the new space with a given number of PCs is represented by ε.

Once PCA has been applied, various types of information can be retrieved. Several examples are shown in Figure 9.42. Panel A shows the **scores plot,** for which the scores from $\mathbf{T}$ for the first two PCs (*i.e.* the first two columns of $\mathbf{T}$) are plotted against each other. We observe that the samples are largely grouping according to the groups of the periodic table, with noble gases, alkali metals, halogens, and alkaline earth metals clearly grouped. The transition metals and post-transition metals ("weak metals") occupy a larger space, but can still be distinguished from the rest. Note that distances between clusters here should not be taken as absolute for classification due to the mean centring and autoscaling, as well as the fact that the different PCs capture different degrees of variance.

Figure 9.42B shows the **loadings plot** for the different variables. The further a variable is distant from the origin in a loading plot, the more it contributes to the variance in the original data. Variables that project in the same direction, such as the melting point and the boiling point, are positively correlated. Conversely, variables in opposite directions are negatively correlated. We see in particular that the electronegativity is distanced. It is also possible to directly plot the loadings (*i.e.* a column of $\mathbf{P}$) as a function of the variables. Loadings with a high value for a given variable indicate a strong influence of a variable on the PCA model (Figure 9.42B).

Another important plot is the **scree plot** (Figure 9.42C), which displays the amount of unexplained variance after including each subsequent PC. For our case, we see that the first PC captures 65% of

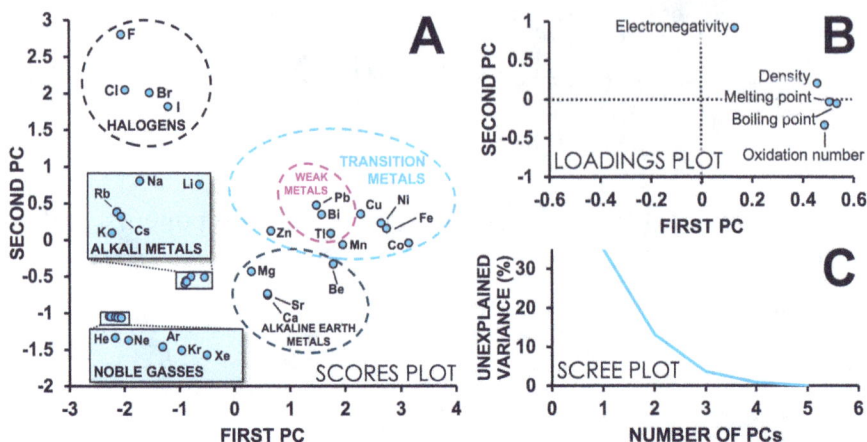

Figure 9.42 PCA results of the mean-centred and autoscaled data shown in Table 9.11. (A) Scores plot, (B) loadings plot, and (C) scree plot. Data and example from ref. 13, reproduced from ref. 13 with permission from John Wiley & Sons, copyright 2003 John Wiley & Sons, Ltd.

the variance in the data, the second PC another 23%, and the third PC 9%. If the number of PCs is equal to the number of variables, 100% of the variance is explained. If one would accept losing 12% of the information, then one could reduce the dimensionality of this dataset by 3! To determine how many PCs to retain it is possible to find a natural break in the curve (elbow method), or regard the cumulated variance threshold.

Sometimes the scores and loadings are plotted in the same plot called a **biplot**. To accomplish this, the scores are scaled (depicted with w) through $\left(t_{i,j|w} = t_{i,j}/\left(\sum_{i=1}^{n} t_{i,j}^2/n\right)\right)$ with $i=1...n$ and $j=1...k$. When a variable then is projected into the same direction as a group, it may indicate that the variable contributes relatively strongly to the data in that group.

These tools allow PCA to be used to explore the data and the variables that strongly contribute to the model, detect outliers (*i.e.* scores far away from the remaining scores), reduce noise in data pre-processing (*i.e.* by removing the final, least-descriptive PCs that may reflect noise), classification (*i.e.* by regarding the clusters using k-means clustering or logistic regression), and process control, where new datapoints (*e.g.* new samples) are compared to an earlier population of datapoints to check whether they meet the profile (and thus cluster as expected).

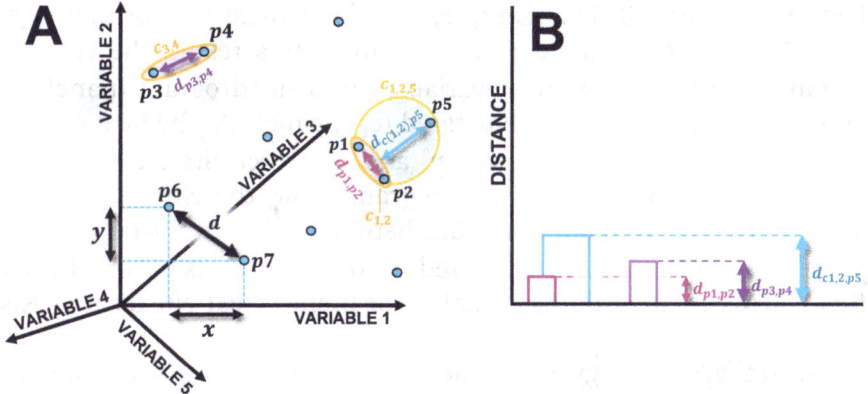

Figure 9.43 (A) Datapoints in a multi-dimensional space comprising several variables. Different datapoints (denoted by *p*) are depicted and discussed in the text. The *d* values represent distances. The circles represent clusters. (B) Schematic example of how the combination of evaluated distances and linking strategies results in a hierarchical dendrogram.

9.9.2 Hierarchical Cluster Analysis

A different method to quickly explore multivariate data and find hidden clusters relies on the use of (dis)similarity descriptors for pairs of objects. An example is shown in Figure 9.43A where data have been plotted in a multi-dimensional space. The thick arrow denotes the distance *d* between the two points *p*6 and *p*7. This similarity can be calculated using several methods, including the **Euclidean distance**

$$d_{i,j} = \sqrt{x^2 + y^2 + \ldots} \tag{9.123}$$

and the **Manhattan distance**

$$d_{i,j} = |x| + |y| + \ldots \tag{9.124}$$

where $x = x_i - x_j$, $x = y_i - y_j$, etc.

Other distances such as the Mahalanobis and Canberra distances are also sometimes used. Again, it is important to autoscale the data to avoid variables with large magnitude dominating the clusters.

After the computation of similarities between all possible pairs of objects, the next step involves a linkage strategy. This is depicted in

Figure 9.43A and B. For example, first the closest pair of points p^1 and p^2 is linked (denoted as $d_{p1,p2}$; pink). As a result, the covered distance is plotted for these variables in a dendrogram (panel B), and the two datapoints are clustered ($c_{1,2}$, panel A). The next closest pair is that between datapoints p^3 and p^4, which are subsequently clustered and noted in the dendrogram. Now, the closest distance is not between two datapoints, but between the first cluster $c_{1,2}$ and p^5, which are linked and clustered accordingly, as is shown in the figure. This is the guiding principle of **hierarchical cluster analysis** (HCA).

The **linkage strategy** to connect a cluster can involve different computations. For example, the nearest neighbour involves computing the distance between two clusters by forming a pair between the two clusters that are closest to each other. Conversely, the furthest neighbour considers the largest distance possible between any datapoint in both clusters. It is also possible to calculate the average distance between all of the possible combinations of datapoints of the two clusters. Ward's linkage considers the increase in the total within-cluster sum of squares when two clusters are joined and is generally a safe option.

Figure 9.44 shows the results of HCA on the data from Table 9.11. A similar pattern can be observed with the different groups being separated from each other. Interestingly, the alkali metals and alkaline earth metal clusters are clustered together at a higher level. The transition metals and weak metal clusters are also clustered together, but are clearly separated from all other groups. The dendrogram shows how the data are structured, with objects that all share short distances demonstrating high consistency between the objects, whereas larger cluster separation shows that the clusters are not similar to each other. Deciding on the number of clusters considered is an important issue and can be done, for example, through the dendogram cutoff method, comparison of cluster fusion distances, or the silhouette score.

9.9.3 Multivariate Calibration

Multivariate regression methods allow establishing a relationship between a collection of measured variables and the property of interest. When the latter is a nominal or ordinal variable, we speak of multivariate classification, and when it is an interval variable, we refer to it as multivariate calibration. An example of multivariate

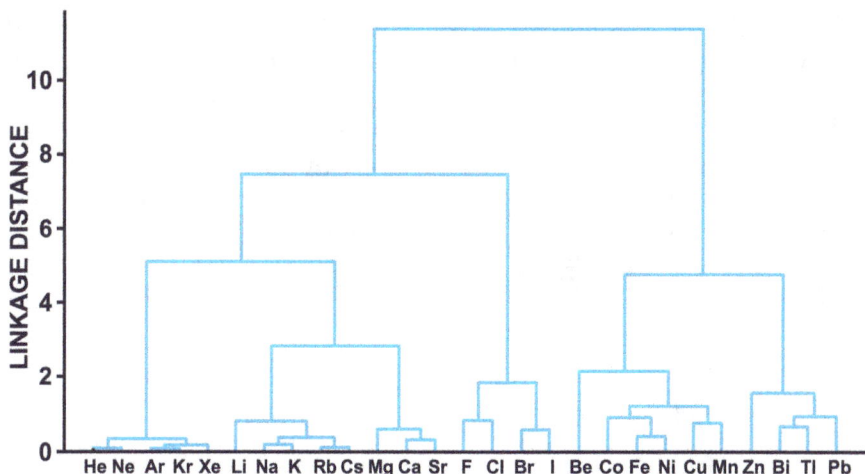

Figure 9.44 Hierarchical cluster analysis of the data from Table 9.11 of the periodic table. Data and example from ref. 13, reproduced from ref. 13 with permission from John Wiley & Sons. copyright 2003 John Wiley & Sons, Ltd.

classification is supervized pattern recognition, which is beyond the scope of this book.

Multivariate calibration is typically performed using inverse calibration, where the response variable is regressed directly against multiple predictors, without inverting the model. We now assume that the measurement errors primarily affect the response variable (i.e. property of interest). In our example of a univariate calibration curve where absorbance was related to concentration (Figure 9.30), we assumed that all the error would be in the measured absorbance.

For multivariate calibration, we can write eqn (9.70) as

$$\mathbf{y} = \widehat{\mathbf{y}} + \epsilon = \mathbf{f}(\mathbf{X}, \mathbf{b}) = f(x_1, x_2, ..., x_m, \mathbf{b}) + \epsilon \tag{9.125}$$

while the concentration of interest was x in eqn (9.69), it is now represented by y. Instead, x_1 through x_m represent the absorbance at all m measured wavelengths, *i.e.* the spectra. As a result, it is not needed to invert the model and we can directly use it to assess the concentration.

Our models now also can become more complex, such as, for example, the model used to fit the surface in Module 9.8.

$$\widehat{y} = b_0 + b_1 x_1 + b_2 x_2 + b_3 x_1^2 + b_4 x_2^2 + b_5 x_1 x_2 \tag{9.126}$$

As an application, we may envisage a model constructed based on n UV–vis spectra of a complex dye mixture to quantify the concentration of a degradation product. In matrix notation, we now have the extended matrix $\mathbf{X}$ with, next to a column of ones (for b_0), m columns, each representing one variable. In our example, each row is then a full UV–vis spectrum – with the exception of the constant in the first column – with each value in each column presenting the absorbance at a measured wavelength; $\mathbf{b}$ contains an $m + 1$ number of coefficients.

$$
\begin{bmatrix} y_1 \\ y_2 \\ \vdots \\ y_n \end{bmatrix} = \begin{bmatrix} 1 & x_{1,1} & x_{1,2} & \cdots & x_{1,m} \\ 1 & x_{2,1} & x_{2,2} & \cdots & x_{2,m} \\ \vdots & \vdots & \vdots & \ddots & \vdots \\ 1 & x_{n,1} & x_{n,2} & \cdots & x_{n,m} \end{bmatrix} \begin{bmatrix} b_0 \\ b_1 \\ \vdots \\ b_m \end{bmatrix} + \begin{bmatrix} \varepsilon_1 \\ \varepsilon_2 \\ \vdots \\ \varepsilon_n \end{bmatrix}
\tag{9.127}
$$

Given that b_0 presents an offset of the entire model, it may be moved out of the $\mathbf{X}$ matrix so that $\mathbf{y} = b_0 + \mathbf{X} \cdot \mathbf{b} + \varepsilon$. At this point, we are ready to proceed with regression as described in Module 9.6.

It will, however, become more difficult to find $\mathbf{b}$. Using least squares will require two conditions to be met. First, the number of objects should be greater than the number of variables. This is increasingly difficult to satisfy as the number of variables becomes large, such as the case with many wavelengths measured in a full UV–vis spectrum (e.g. from 190 to 640 nm with steps of 2 nm results in 226 variables!). Second, there should be little to no co-linearity in $\mathbf{X}$, which for spectra from similar samples is difficult to satisfy, as these are often also similar. Moreover, there is a strong correlation between the absorptions at adjacent wavelengths.

To resolve this, one could opt to pick only a limited number of sufficiently selective variables in $\mathbf{X}$ to reduce m below n. However, this variable selection, which is complicated in itself, is likely to be based on the selected objects and thus there is a risk of overfitting the model to the specific samples.

Another option is to perform dimensionality reduction by identifying a useful number of principal components that capture most of the information. Similar to PCA, the $\mathbf{X}$ matrix is mean centred and split into an $n \times k$ matrix ($\mathbf{T}$) with the scores of n samples and k principal components, and an $m \times k$ matrix ($\mathbf{P}$) with the loadings for m variables (see eqn (9.120)). The resulting scores and loadings depend on the techniques used to calculate these. **Principal**

component regression (PCR) computes these in a manner similar to PCA by solely maximizing the explained variability. As a consequence, the new latent variables (*i.e.* the PCs) will be those with the maximum explained variance. In contrast, **partial least squares** (PLS) balances the selection of latent variables by also attempting to maximize the correlation with the dependent variable. The resulting information thus describes both the original data and the dependent variable. These techniques are not further discussed here, and the reader is referred elsewhere.[5,9]

9.10 Advanced Regression Techniques (A)

9.10.1 Weighted Regression

The regression methods discussed thus far in this chapter assume that the error is constant or, as it is referred to in chemometrics, it is homoscedastic. We visually saw this in Figure 9.30 with the confidence interval of the experimental response (blue dashed lines) being independent of x (*i.e.* the concentration). Sometimes, this is not the case. For instance, some LC detectors can respond heavily to the composition of the mobile phase, in terms of organic modifier fractions (*e.g.* ELSD or CAD, see Section 3.9.4) or the presence of any salts or additives (*e.g.* MS, see Section 3.10.1). Similarly, the repeatability of retention times may depend on the type and stability of the stationary phase.

In cases where the variance is not constant (*i.e.* heteroscedastic), it can be useful to apply weights to the datapoints during least-squares regression. This is referred to as **weighted least-squares regression (WLS)**. The coefficients are then calculated as

$$\mathbf{b} = \left(\mathbf{X}^{\mathrm{T}} \cdot \mathbf{W} \cdot \mathbf{X}\right)^{-1} \mathbf{X}^{\mathrm{T}} \cdot \mathbf{W} \cdot \mathbf{y}\} \qquad (9.128)$$

where $\mathbf{W}$ is a diagonal matrix

$$\mathbf{W} = \begin{bmatrix} w_1 & 0 & \cdots & 0 \\ 0 & w_2 & \cdots & 0 \\ \vdots & \vdots & \ddots & \vdots \\ 0 & 0 & \cdots & w_n \end{bmatrix} \qquad (9.129)$$

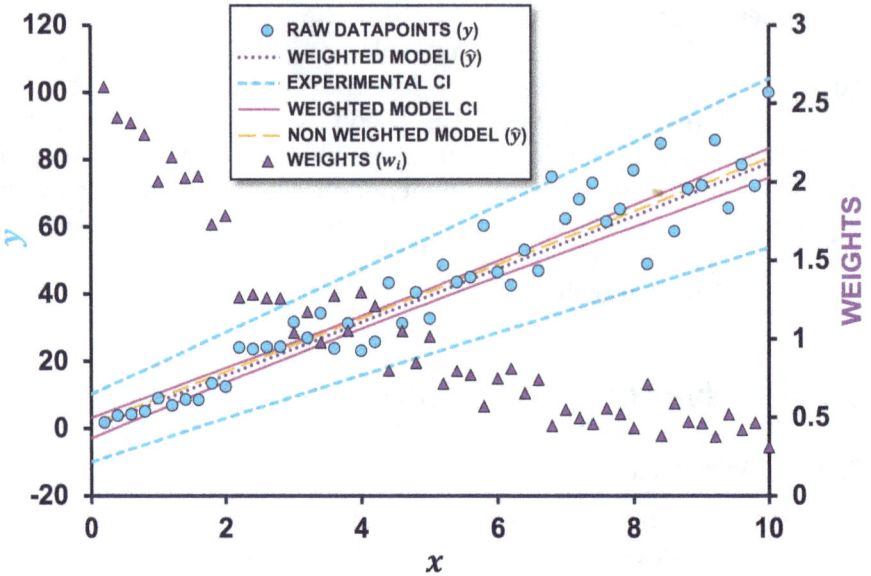

Figure 9.45 Example of heteroscedastic data (the variance in y is not constant over x). The dotted purple line is the weighted model fitted through the data with the triangles indicating the weights per datapoint. The orange long-dashed line in the centre is the model without weights (*i.e.* ordinary least squares). The blue dashed lines are the weighted confidence intervals of y, whereas the solid pink lines are the confidence interval (CI) of the weighed $\hat{y}$.

that contains the weights computed such that higher errors reduce the weight

$$w_i = \frac{1}{\sigma_i^2} \cdot \frac{n}{\sum_{i=1}^{n} \frac{1}{\sigma_i^2}}$$

(9.130)

The last factor is constant and it does not affect the outcome of eqn (9.128). Therefore, $w_i = 1/\sigma_i^2$ may be used instead. An example is shown in Figure 9.45 where the relative weights are shown by the triangles. The resulting confidence intervals in $\hat{y}$ and in y are consequently adjusted and now take into account the change in variance with x. Note that the use of WLS changes the standard errors of the regression coefficient.

> 🖐 See the website for further details on how the least-squares
> calculations are affected: ass-ets.org

9.10.2 Non-linear Regression

In many cases, there is an interest in using regression on non-linear models. We already saw the retention model in Section 9.6.4. Another example is the fitting of a distribution function to a chromatographic peak (eqn (1.24), analogous to eqn (9.9)), which can also be written as

$$\hat{y} = \frac{b_0}{b_2\sqrt{2\pi}} \cdot \exp\left(\frac{-1}{2}\left(\frac{t - b_1}{b_2}\right)^2\right) = \frac{A_{peak}}{\sigma\sqrt{2\pi}} \cdot \exp\left(\frac{-1}{2}\left(\frac{t - t_R}{\sigma}\right)^2\right) \tag{9.131}$$

where $A_{peak}/\sigma\sqrt{2\pi} = h_{peak}$. The latter equality is useful for signal pre-processing and processing purposes. In this section, we will briefly discuss methods to perform non-linear regression.

One method to deal with non-linear models is linearization, such as was done for the retention models in Section 9.6.4, where $\hat{k} = \exp(b_0 + b_1\varphi + b_2\varphi^2)$ is linearized to $\ln \hat{k} = b_0 + b_1\varphi + b_2\varphi^2$. Unfortunately, this causes the weight of the error to reside in $\ln \hat{k}$, rather than in $\hat{k}$, and thus – during least-squares regression – we minimize $S_{res} = \sum_{i=1}^{n}\left(\ln k_i - \ln \hat{k}_i\right)^2$ instead of $SS_{res} = \sum_{i=1}^{n}\left(k_i - \hat{k}_i\right)^2$. One way to compensate for this is by using weighted regression.

As an alternative to linearization, it is also possible to approximate **b** using iterative methods. These approaches employ a solver strategy that iteratively attempts to minimize $\sum_{i=1}^{n}(\varepsilon)^2$. An initial guess of **b** is required. The **Gauss–Newton** approach employs

$$\Delta \mathbf{b} = \left(\mathbf{X}^T\mathbf{X}\right)^{-1}\mathbf{X}^T\Delta \mathbf{y} \tag{9.132}$$

where Δb quantifies the gain with respect to the residuals Δy. The **Levenberg–Marquardt** approach also employs eqn (9.132), but using gradient descent drives the convergence using the sum of squares, making it more robust for highly non-linear models.

> See the website for further details on non-linear regression and optimization functions: ass-ets.org

Regardless of the employed strategy, this form of non-linear regression suffers from a number of disadvantages. The iterative minimization of $\sum_{i=1}^{n}(\varepsilon)^2$ requires significantly more computational resources, especially because the solution for **b** is not unique. Multiple solutions may exist, and to reduce the intensity of the computations, it is important to use a tolerance parameter. Because there is no unique solution, it is also possible that the employed approaches never achieve convergence.

Despite these disadvantages, these strategies do allow the regression of non-linear models to a parametric model. These models can, similar to ordinary least-squares regression, be validated using the diagnostic tools reviewed in Section 9.6.4, as well as placed into context with confidence intervals for the prediction (provided Taylor expansion is a good approximation around the optimum).

9.10.3   Robust Regression

The presence of outliers during the fitting process was explored in Section 9.6.4 with an eye on their influence on the model, as well as on methods to detect them. In cases where an outlier cannot be eliminated, or where the potential presence of undetected outliers cannot be allowed to exceedingly influence the fit, we can apply robust regression where we no longer assume that the noise of the data is normally distributed. This is analogous to the use of non-parametric statistics in cases where the data do not meet the normality criterion.

The concept of robust regression relies on **loss functions** (also known as cost functions). We will explore this concept in this section. An example is shown in Figure 9.46A, where data with an outlier (denoted by the arrow) are plotted. With ordinary least-squares regression, the model (dashed line) is significantly affected by the presence of the outlier. However, using the loss function, the model (dotted line) is barely affected. Here, robust regression is achieved using the bi-square loss function, which is illustrated in Figure 9.46B. The **bi-square loss function**, also known as the Tukey loss function, is given by

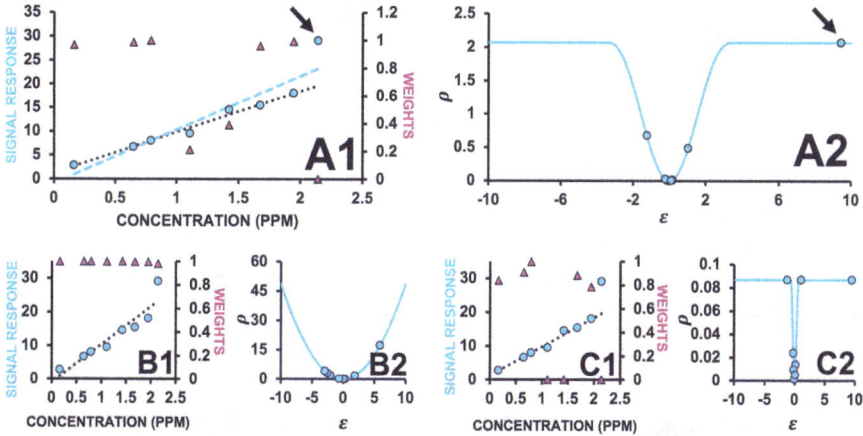

Figure 9.46 Robust regression of data using the bi-square loss function with tuning parameter (A) tune equal to (A) 4.895, (B) 2, and (C) 20. Panels A1, B1, and C1 reflect the fit (dotted line) using the shown weights (pink triangles, secondary *y*-axis). Panels A2, B2 and C2 show the resulting bi-square loss functions, with the residuals for each point plotted against ρ. The light-blue dashed line in panel A1 shows the fit obtained using ordinary least-squares regression, where the outlier data-point (depicted with an arrow) has a strong influence on the regression.

$$\rho(\varepsilon) = \begin{cases} \dfrac{k^2}{6}\left(1 - \left(1 - \left(\dfrac{\varepsilon}{k}\right)^2\right)^3\right) & \text{if } |\varepsilon| \le k \\[2em] \dfrac{k^2}{6} & \text{if } |\varepsilon| > k \end{cases} \quad (9.133)$$

where k can be approximated as $k = \text{tune} \cdot s_{e,\text{rob}}$. Here, $s_{e,\text{rob}}$ refers to a robust estimation of the standard error (see Section 9.4.2.4) and **tune** is a tuning constant. To improve the fit, k is, however, typically computed as $k = \sqrt{1 - h} \cdot \text{tune} \cdot s_{e,\text{rob}}$, where h represents the leverage (see Section 9.6.4.3). High-leverage points will yield an effect similar to a lower **tune** constant.

In robust regression, $\sum_{i=1}^{n} \rho(\varepsilon_i)$ is minimized rather than $\sum_{i=1}^{n} \varepsilon_i^2$. Figure 9.46A2 shows the residuals for the fit from panel A1 plotted on top of the loss function. We see that all low residuals contribute very little to ρ. However, depending on k, the bi-square function features a cut-off threshold after which any increasing residual does not further contribute to ρ. This is the case for our data in Figure 9.46A2, with the point that meets this threshold indicated by an arrow. As a

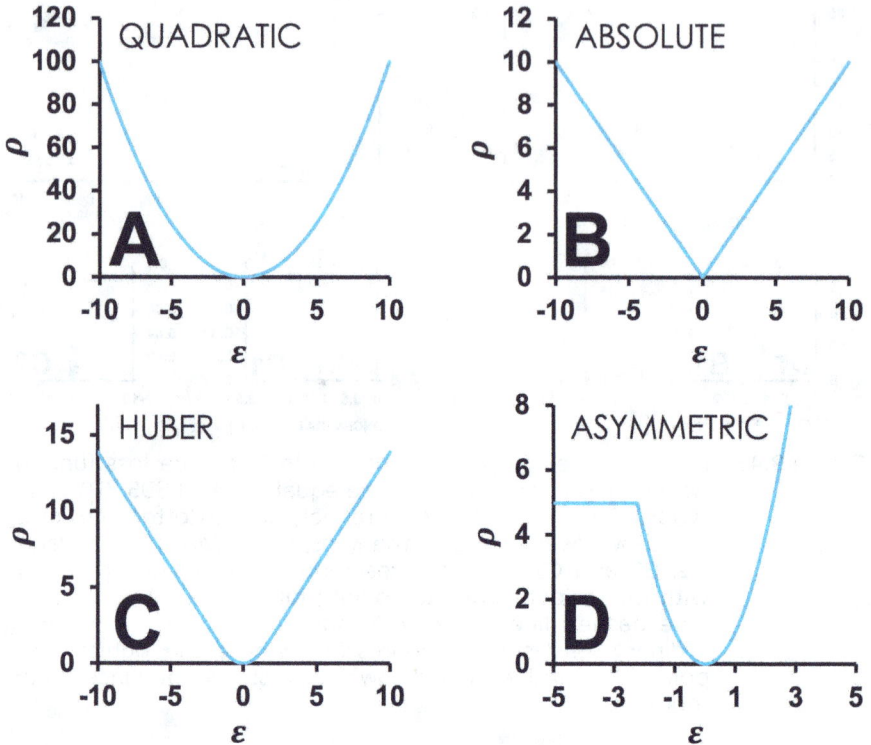

Figure 9.47 Plots of the (A) quadratic, (B) absolute, (C) Huber, and (D) asymmetric truncated quadratic loss functions.

result, the regression weights are adjusted, as is also shown in Figure 9.46A1 (secondary y-axis).

The **tune** constant can be adjusted to increase or decrease the sensitivity of the loss function to outliers. For the bi-square function, it is typically 4.895. Figure 9.46B1–C2 shows the effect of decreasing (B1–B2 with **tune** = 2) or increasing (C1–C2 with **tune** = 20) this constant.

Other loss functions exist. Both the **quadratic** (Figure 9.47A)

$$\rho(\varepsilon) = \varepsilon^2 \tag{9.134}$$

and **absolute loss function** (Figure 9.47B)

$$\rho(\varepsilon) = |\varepsilon| \tag{9.135}$$

are well-known. However, the quadratic function cannot easily cope with data that contain multiple outliers and does not cut off ρ for higher residuals. The absolute error essentially conducts regression around the median (also known as quantile regression), but it gives rise to convergence problems, as numerical derivatives of the loss function with respect to ε become unstable close to zero.

The **Huber loss function** (Figure 9.47C) is given by

$$\rho(\varepsilon) = \begin{cases} \dfrac{1}{2}\varepsilon^2 & \text{if } |\varepsilon| \le k \\[2mm] k|\varepsilon| - \dfrac{1}{2}k^2 & \text{if } |\varepsilon| > k \end{cases} \tag{9.136}$$

with $k = \text{tune} \cdot s_{e,\text{rob}} \cdot \sqrt{1-h}$ where h is the leverage (see Section 9.6.4.3), and **tune** is the tuning constant, which typically is 1.345 for the Huber loss function. In comparison with the bi-square loss function, which can handle severe presence of outliers, the Huber loss function is suitable when outliers are less severely present.

Robust regression significantly reduces the effects of outliers and can be applied to any model. Similar to non-linear regression, there is no unique solution for **b**, and regression is conducted iteratively (often by iterative-least-squares regression). For the loss function

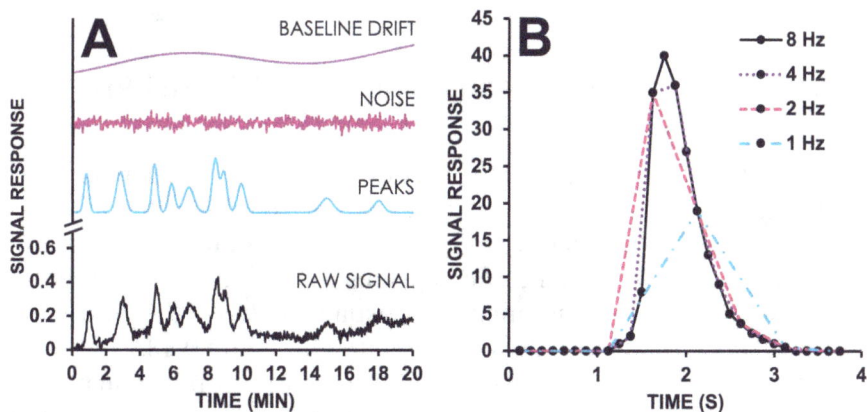

Figure 9.48 A raw chromatographic signal (A) decomposed into different frequency components: the high-frequency noise, the low-frequency baseline drift, and the medium-frequency chromatographic peaks. (B) The effect of detector sampling frequency on the captured information of a peak. Image from panel B adapted from ref. 14, https://doi.org/10.1002/jssc.201700863, under the terms of the CC BY 4.0 license, https://creativecommons.org/licenses/by/4.0/.

to work effectively, a sufficient number of datapoints are required. Another point to note is that the Durbin–Watson (see Section 9.6.4.2) test cannot be used for robust regression.

In some cases, it is of interest to consider negative deviations stronger than positive deviations or vice versa. An example is baseline correction, where positive chromatographic peaks must be ignored as they do not contribute to the baseline. Here, asymmetric functions, such as the **asymmetric truncated quadratic loss function** (Figure 9.47D) may be of use. This function can – depending on the direction of focus – be given any form, but an example is

$$\rho(\varepsilon) = \begin{cases} \varepsilon^2 & \text{if } \varepsilon < -k \\ k^2 & \text{otherwise} \end{cases} \tag{9.137}$$

9.11 Signal Pre-processing (M)

Until now, we have mainly focused on the handling of processed data in this chapter. However, with the advent of ever-more powerful separation systems, often combined with increasingly sophisticated mass spectrometers, extracting all relevant information from the generated signals is among the greatest challenges we currently face in analytical separation science. Therefore, it is worthwhile to examine the pre-processing of raw data from a separation system.

9.11.1 Properties of a Signal and Aim of Signal Pre-processing

Any raw signal comprises several components or frequencies. For analytical separation science, a signal can be roughly categorized as follows: (i) a high-frequency component that contains the noise, (ii) a low-frequency component that captures the baseline drift, and (iii) a medium-frequency component that usually includes the chromatographic peaks of interest. This is illustrated in Figure 9.48A.

Of course, whether the chromatographic or electrophoretic peak is captured by the medium-frequency category hinges on the detector sampling rate. A sufficiently large number of data points (acquisition frequency) are required to distinguish between high-frequency and medium-frequency signals. In addition, a sufficient number of datapoints are required to properly describe the peak. This is indicated in Figure 9.48B.

Signal pre-processing involves all post-analysis corrections to the signal to resolve undesired background signals, baseline drift, noise, retention-time shifts and insufficiently resolved peaks.

9.11.2 Noise Filtering

Noise is the random fluctuation of a signal and can have various sources. For example, in LC, noise generally originates from small fluctuations in the mobile-phase composition, temperature, and flow rate, but it can also originate from the detector and/or its electronics. In LC, **red noise** (which is positively autocorrelated) can be observed, due to composition, temperature and pump fluctuations, as well as **pink noise** (due to drift, gradient elution issues and detector instability), but we will limit the discussion to random noise, also known as white noise, where every point is independent of the previous one. **White noise** is typically normally distributed (although shot noise can be Poisson-distributed) and its average intensity is by definition zero. It follows that the noise intensity can be characterized through the standard deviation (σ_{noise}) of a section of the chromatogram containing only noise. This is graphically represented in Figure 9.49. The total amplitude of the noise is then roughly $4\sigma_{noise}$ and sometimes referred to as the peak-to-peak noise (N_{pp})). In this context, chromatographers often compare the height of a peak (h_{peak}) and N_{pp} to calculate the signal-to-noise ratio (S/N), which is given by

$$S/N = \frac{h_{peak}}{N_{pp}} \tag{9.138}$$

Unlike the limit of decision (Section 9.6.5), S/N plays no formal role in analytical method development. However, it is a convenient tool for the chromatographer to gauge whether a peak is sufficiently prominent from the baseline. The rule of thumb is generally that the S/N should be at least between 3 and 7 for a signal fluctuation to be designated as peak.

Various approaches can be used to remove the noise from a signal. The simplest is the **moving-average filter (MA)**, which is given by

$$S_{i,out} = \frac{S_{i-(k-1)/2,in} + \dots + S_{i-1,in} + S_{i,in} + S_{i+1,in} + \dots + S_{i+(k-1)/2,in}}{k} \tag{9.139}$$

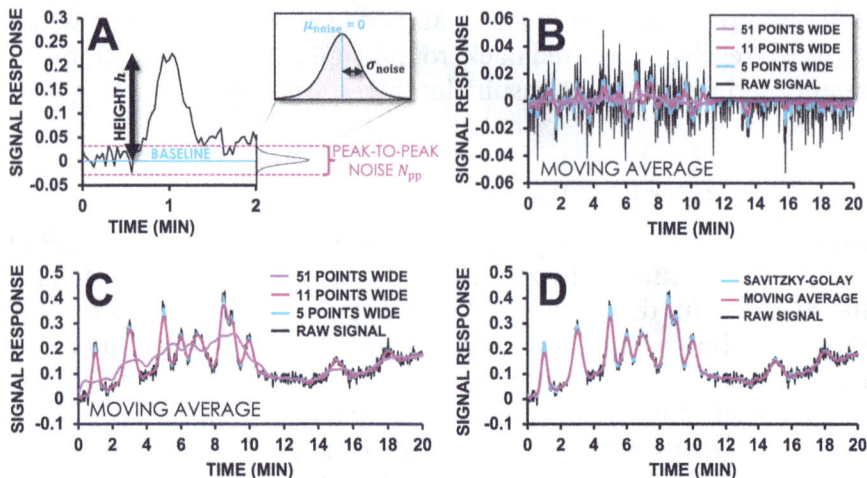

Figure 9.49 (A) Characteristics of noise to compute the signal-to-noise ratio, (B) application of the moving-average filter to the noise component of Figure 9.48 with various filter widths. (C) Application of the moving-average filter to the total signal of Figure 9.48 with various filter widths. (D) Comparison of the 11-point moving-average and Savitzky–Golay filters, where the Savitzky–Golay filter employs a second-order polynomial.

Here, k is the width of the filter, which is an odd number, given that the moving window is centralized around a datapoint. In essence, the k-width window "moves" past each datapoint $S_{i,\text{in}}$ and calculates the equal-weighted average of datapoints in its domain as $S_{i,\text{out}}$. As can be seen in Figure 9.49B, where different widths are applied, the moving average filter is quite effective in removing noise. The signal-to-noise ratio is expected to increase by a factor $\sqrt{k}$. However, especially for narrow peaks (*e.g.* sparsely sampled by the detector and/or with relatively high plate numbers), there is a serious risk of losing information when applying a moving-average filter. This effect is aggravated when the filter is taken too wide (*i.e.* k is too large). The appropriate width depends on the detector sampling frequency and the width of the peaks and must be selected carefully.

A far more popular filter is the **Savitzky–Golay (SG) filter**, named after Abraham Savitzky and Marcel Golay (the same from the Golay equation in Module 1.7). It requires a constant time interval between data points. Unlike the MA filter, the SG filter employs relative weights to the datapoints. These weights are obtained by fitting an mth-order polynomial to a window (*i.e.* filter width) of data points using least-squares regression (see Module 9.6). The resulting weighting factors have been tabulated for a large number of polynomials. Modern software tools generally contain Savitzky–Golay functions by which the weighting coefficients are automatically computed. For example, applying a second-order polynomial over a 7-point window results in

$$S_{i,out} = \frac{-2}{21}S_{i-3,in} + \frac{3}{21}S_{i-2,in} + \frac{6}{21}S_{i-1,in} + \frac{7}{21}S_{i,in}$$
$$+ \frac{6}{21}S_{i+1,in} + \frac{3}{21}S_{i+2,in} + \frac{-2}{21}S_{i+3,in}$$

(9.140)

As a consequence of the relative weights, the SG filter is better suited to handle narrower peaks that are described by a relatively sparse number of datapoints. This is illustrated in Figure 9.49D, where the 11-point MA filter is compared to the 11-point quadratic SG filter (also using a second-order polynomial). For the relatively broad peaks around 15 and 18 min, both filters perform similarly, but for the peaks around 9 min, we see a dramatic loss in resolution for the MA filter. Moreover, a significant loss of peak height and area can be observed for the earlier peaks when using the MA filter. An important exception to this is polymer separations, which typically feature a broad distribution described by very many datapoints. Consequently, the MA filter can be a better option there as it is better at removing the actual noise.

A more recent, but increasingly common, strategy to reduce noise employs penalized least squares, based on the **Whittaker smoothing** function. Here, a vector y is fitted (see Module 9.6) through the original data y (*i.e.* the signal). With this strategy, two goals are balanced: (i) the ability of $\hat{y}$ to fit to the real data with (ii) the roughness of $\hat{y}$. As $\hat{y}$ becomes a smoother curve, it will deviate more from the original data y (*i.e.* $\sum_{i=1}^{n} (y_i - \hat{y}_i)^2$ becomes larger), and if it becomes rougher it will be able to capture the original data better, causing $\sum_{i=1}^{n} (y_i - \hat{y}_i)^2$ to decrease. The roughness of y can be described locally by the deviation of data points from their neighbours, *i.e.* $\Delta\hat{y}_i = \hat{y}_i - \hat{y}_{i-1}$, and thus for the entire vector as $\sum_{i=1}^{n} (\Delta\hat{y}_i)^2$. The goal (Q) is to create a smooth baseline (*i.e.* to minimize the roughness) that still captures the original data sufficiently (*i.e.* to minimize the sum of squares residuals while avoiding a poor description of the real baseline). In matrix terms, this means minimizing the cost function Q.

$$Q = |\mathbf{y} - \widehat{y}|^2 + \lambda|\mathbf{D}\widehat{y}|^2 \tag{9.141}$$

where λ is the smoothing parameter and $\mathbf{D}$ is a matrix such that $\mathbf{D}\widehat{y} = \Delta\widehat{y}$. To find the minimum, the partial derivative of eqn (9.141) yields

$$\frac{\partial Q}{\partial \widehat{y}} = -2(\mathbf{y} - \widehat{y}) + 2\lambda\mathbf{D}^{\mathrm{T}}\mathbf{D}\widehat{y} \tag{9.142}$$

so that, when $\partial Q/\partial\widehat{y} = 0$ we obtain

$$\left(\mathbf{I} + \lambda\mathbf{D}^{\mathrm{T}}\mathbf{D}\right)\widehat{y} = \mathbf{y} \tag{9.143}$$

with $\mathbf{I}$ the identity matrix.

Other well-known filters exist, such as the Kalman filter, each with their pro and cons, but these are not discussed in the book.

> See the website for further materials on smoothing strategies and how to apply these.

9.11.3   Baseline Correction

Aside from the noise, a chromatographic signal can feature a further array of different signal distortions, such as baseline drift, artefacts, and other deformations. For example, drift can occur in LC due to a gradient programme of the mobile phase and in GC as a result of temperature-induced bleeding of the stationary phase. The presence of such background distortions can significantly impact the characterization of a peak, and the aim of the baseline correction is thus to remove all such unwanted deformations to facilitate subsequent signal processing.

In the past few decades, a vast number of strategies have been developed to address the challenge of baseline correction. Unfortunately, no single strategy serves all purposes, as recent studies suggest that the appropriate baseline-correction strategy depends highly on the properties of the signal.[15] In this book, we will thus focus on the core concepts that guide these different strategies, illustrated with several examples. For a more elaborate description of the different background-correction tools and the methods to use these, we refer to additional literature and our website.

> 🔵 See the website for further reading material and software functions for many of the discussed background-correction strategies: ass-ets.org

With the large number of signal properties adding to the challenge of baseline correction, it is difficult to formulate a complete list of tasks that must be achieved. Any approach to background correction must be able to (i) distinguish a peak from baseline, (ii) sufficiently capture the shape of the baseline, even when the number of datapoints is sparse, (iii) handle noise, (iv) account for sudden baseline distortions, (v) interpolate the baseline at peak locations, and – perhaps the most challenging – (vi) robustly maintain the information of interest.

Baseline-correction strategies can – similar to statistical methods – be roughly classified as either parametric, relying on specific assumptions regarding the background signal, or non-parametric, without such assumptions.

One method to both determine the shape of the baseline and distinguish it from chromatographic peaks relies on the use of **local minimum values** (LMVs), which are datapoints (y_i) in the signal for which $y_{i-1} > y_i$ and $y_i < y_{i+1}$. Figure 9.50A shows all LMVs as a scatter plot over the signal. By comparing each LMV with the relative noise level (*e.g.* through S/N), a distinction can be made as to whether an LMV belongs to a peak (pink) or to baseline (blue). Any peak LMVs are then replaced by the median of the baseline LMVs near p_i (see Figure 9.50A1 and A2) to allow interpolation of the baseline. The result is shown in Figure 9.50B. LMV is very suitable for highly structured baselines.

A very common strategy to remove the baseline is **asymmetric least squares** (AsLS), which employs the penalized least squares approach of the Whittaker smoother discussed in the previous section. To leverage it for baseline correction, weighted regression (Section 9.10.1) is implemented, using

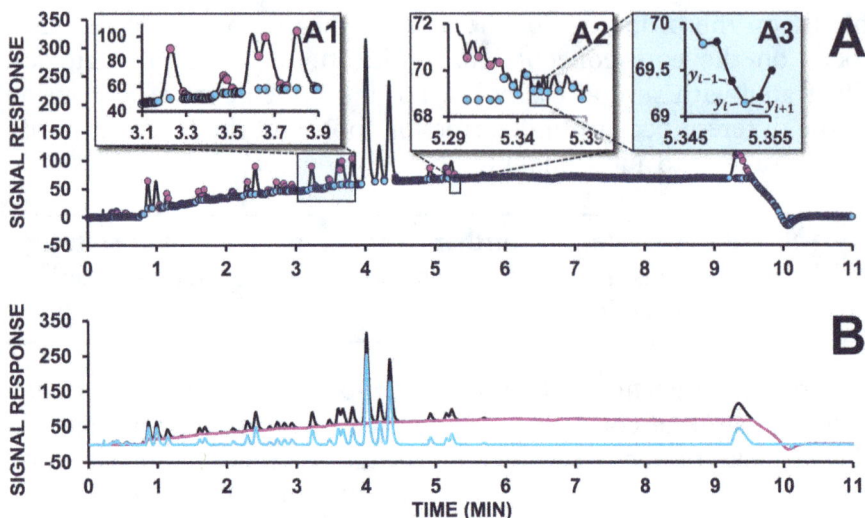

Figure 9.50 Application of the local minimum value (LMV) baseline-correction strategy to a chromatographic signal. (A) Local minimum values recognized as those belonging to the baseline (blue) or as outliers (pink). (B) Original signal (dark blue), baseline (pink) and corrected chromatogram (light blue). Adapted from ref. 16 with permission from the Authors.

$$\left(\mathbf{W} + \lambda\mathbf{D}^{\mathrm{T}}\mathbf{D}\right)\hat{y} = \mathbf{W}y \tag{9.144}$$

Here, $\mathbf{W}$ contains the weights w_i that describe whether a point y_i deviates from the baseline based on an asymmetry parameter As that can vary between 0 and 0.5. A value of 0.5 indicates a normal fit.

$$w_i = \begin{cases} \text{As} & \text{if } y_i > \hat{y}_i \\ 1 - \text{As} & \text{if } y_i > \hat{y}_i \end{cases} \tag{9.145}$$

AsLS is relatively easy to use, provided that suitable As and λ (eqn (9.143)) values are found. Further adaptations have been developed such as adaptive iteratively reweighted PLS (airPLS), where the regions of the baseline can be penalized differently, as well as asymmetrically reweighted PLS (arPLS), modified adaptive iteratively reweighted PLS (mairPLS), and morphologically weighted penalized least squares (MPLS), which mainly solve the impact of noise on the performance. Other approaches exist, including strategies based on multivariate statistics, Bayesian statistics and artificial neural networks. PLS methods are very suitable for signals with sharp artifacts.

> See the website for further reading on baseline correction methods.

9.11.4 Retention-time Alignment

Once noise and baseline deformations have been removed, it may sometimes still be needed to correct for small shifts in elution times. This is especially important when a large number of separations are compared simultaneously, such as the comparison of different studied objects in a sample (*i.e.* repeated measurements) or the comparison of different samples. This concept is illustrated in Figure 9.51. Procedures to deal with this issue are generally referred to as retention-time alignment. In this section we will briefly discuss strategies to correct for errors in the time domain.

An example of a method to conduct the alignment is **correlation-optimized warping** (COW). This strategy divides the signal into several segments. For each region, the Pearson correlation coefficient (see Section 9.6.2) is then maximized by stretching or compressing the relative regions to match with one another.

In some cases, an approach may use several stages to accomplish the retention alignment. An example is the **automatic time-shift alignment** (ATSA) method, where – after dividing the chromatograms into segments similar to COW – a total-peak-correlation coefficient is computed as

$$\text{TPC} = \left(\frac{\sum_{i=1}^{n_{\text{pks}}} w_i \cdot r_i}{\sum_{i=1}^{n_{\text{pks}}} w_i} \right) \frac{n_{\text{pks}}}{n_{\text{ref,pks}}} \tag{9.146}$$

where n_{pks} is the number of matched peaks in the current chromatogram, $n_{\text{ref,pks}}$ is the total number of peaks in the reference chromatogram, r_i is the correlation coefficient of peak i, and w_i is the weights, which refers to

$$w_i = \frac{A_i}{n_{\text{pts},i}} \tag{9.147}$$

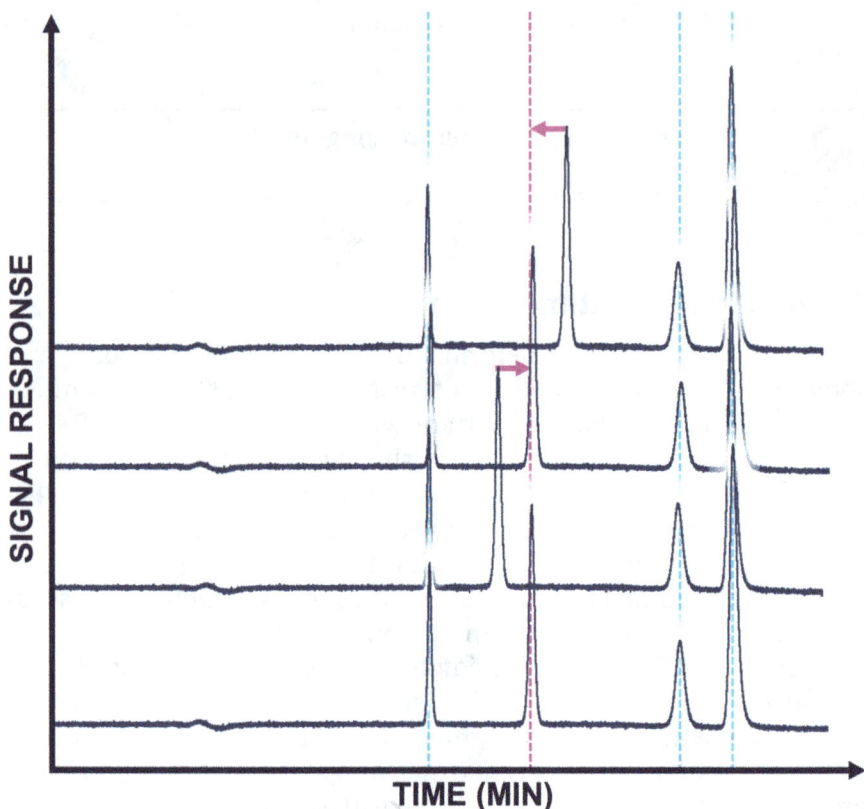

Figure 9.51 Schematic illustration of the concept of retention-time alignment. In this example, the elution times of the second peak (denoted by pink dashed lines) are corrected for two of the chromatograms.

with A_i as the area of peak i and $n_{\text{pts},j}$ the number of datapoints of peak i.

The second stage involves further refinement with new segmentation. Any gaps as a result of the shifting are reconnected using warping of the signal. Metrics such as ATSA require the initial time shift and segment size to be specified *a priori*.

It is important to be aware that, depending on the employed strategy, the retention-time alignment may influence peak areas and thus quantitation.

Methods also exist that leverage available multi-channel-detector data (such as mass spectra), but these are not discussed here.

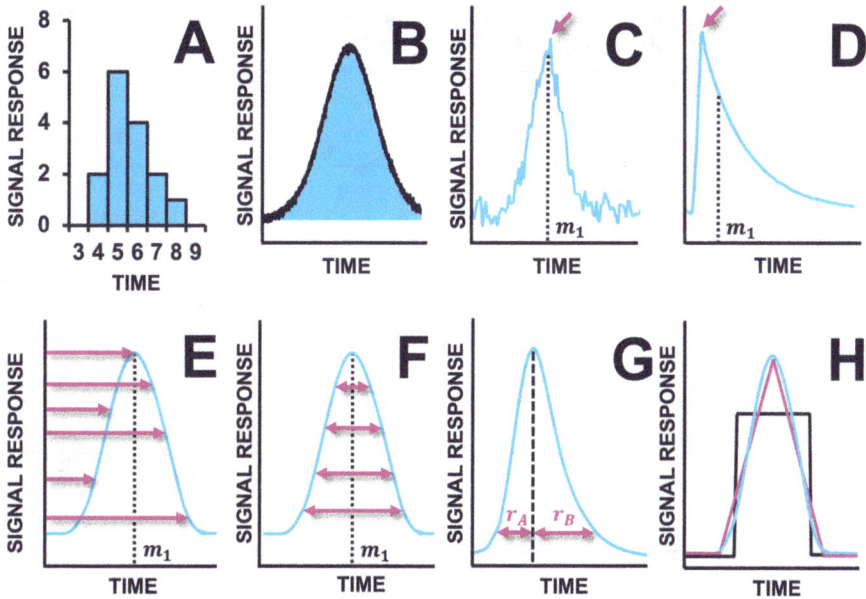

Figure 9.52 Schematic representations of peaks to illustrate the function of statistical moments. (A) The zeroth moment depicted as the area in a binned-histogram representation of a peak. (B) The zeroth moment depicted as the area under the peak. (C) and (D) depict cases where the peak apex differs from the normalized first moment. (E) Normalized second moment, (F) Centralized normalized second moment. (G) Peak assymmetry, (H) Kurtosis depicted for three peaks with values of K = 1.8 (dark blue; block shape), K = 2.4 (triangle; pink), K = 3 (light blue; Gaussian). See the text for further explanation.

9.12 Signal Processing (M)

In this module, we discuss approaches to obtain useful information from pre-processed signals. We will focus on peak detection and quantitation, as well as – to a limited degree – classification.

9.12.1 Characterization of a Peak Using Statistical Moments

Before we can address the detection of peaks and the handling of the information, we must first understand how peaks can be characterized and what information can be distilled from a signal. This will help us to gauge the benefits and risks of the various methods.

Table 9.12 Data and calculations to determine the statistical moments for the distribution shown in Figure 9.52A.

DATA

t (s)	I (mV)	$t \cdot I$	t_{rel}	$t_{rel}^2 \cdot I$	$t_{rel}^3 \cdot I$	$t_{rel}^4 \cdot I$
3	0	0	−2.6	0	0.00	0.00
4	2	8	−1.6	5.12	−8.19	13.11
5	6	30	−0.6	2.16	−1.30	0.78
6	4	24	0.4	0.64	0.26	0.10
7	2	14	1.4	3.92	5.49	7.68
8	1	8	2.4	5.76	13.82	33.18
9	0	0	3.4	0	0.00	0.00

MOMENTS

Moment (x)	0	1		2	3	4
Gross (M_x)	15	84	—	17.6	10.08	54.85
Normalized (m_x)	—	5.6	—	—	—	—
Centralized (μ_x)	—	—	—	1.17	0.67	0.65

One useful method to characterize a peak is by considering its **statistical moments**. This starts with the zeroth gross statistical moment, M_0, or **0th statistical moment**, which is the **area** of a peak. It is given by

$$M_0 = \int_{-\infty}^{+\infty} f(t)\, dt = \sum_{i=1}^{n} f(t_i) \tag{9.148}$$

where $f(t)$ is the distribution function that describes the peak. The area can be calculated by integrating the peak, *i.e.* taking the sum of all its datapoints. This is schematically depicted in Figure 9.52A and B. An example for calculation is shown in Table 9.12, where we see that the resulting area is 15.

The **statistical first moment**, m_1, is the centre of gravity of the peak, or in the context of separation science, the **elution time** (or retention time). It is given by

$$m_1 = \frac{M_1}{M_0} = \frac{\int_{-\infty}^{+\infty} t\, f(t)\, dt}{\int_{-\infty}^{+\infty} f(t)\, dt} = \frac{\sum_{i=1}^{n} t_i \cdot f(t_i)}{M_0} \tag{9.149}$$

In order to compute m_1, the gross first moment is divided by M_0, the area. If we regard the example calculation in Table 9.12, we see that the determined M_1 of 84 mV·s in itself would make little sense. Dividing this by M_0 yields a retention time, m_1, of 5.6 s, which is more logical. This process is called **normalization**, which is applied for all moments beyond M_0, it can generally be written as

$$m_n = \frac{M_n}{M_0} \tag{9.150}$$

At this point, it is useful to make the comparison of m_1 with the peak apex (*i.e.* the highest point of a peak). Two examples are shown in Figure 9.52C and D. In panel C, we see that noise causes the apex (indicated by the arrow) to be different from m_1, the latter of which is less influenced by the noise. Panel D shows an extremely tailing peak that may be encountered, for example, due to sample overloading. In such cases, m_1 is a better representation of the mean elution time of the ensemble of analytes than the peak apex.

The **second centralized moment**, μ_2, is a measure of the **width of the peak** and is given by

$$\mu_2 = \frac{\int\limits_{-\infty}^{+\infty} (t - m_1)^2 \, f(t) \, dt}{\int\limits_{-\infty}^{+\infty} f(t) \, dt} = \frac{\sum_{i=1}^{n} (t_i - m_1)^2 \cdot f(t_i)}{M_0} \tag{9.151}$$

It represents the variance. Thus, in analytical separation science, it is a measure of the lateral spreading of the peak. For a Gaussian (*i.e.* normal distribution), $\mu_2 = \sigma^2$, while for an exponentially modified Gaussian (EMG) distribution (see eqn (9.158) below) $\mu_2 = \sigma^2 + \tau^2$.

The nth moment is **centralized** by the generic formula

$$\mu_n = \frac{\int t_{rel}^n \cdot f(t) dt}{M_0} \tag{9.152}$$

where t_{rel}^n is the relative time given by $t_{rel} = t - m_1$. This is indicated in the example calculation in Table 9.12. For the distribution shown in Figure 9.52A, the centralized normalized second moment (μ_2) is 1.17 s². The rationale behind centralization is depicted in Figure

9.52E and F, where panel E depicts that in the non-centralized case, m_2 is technically assessing the average t^2 for the different molecules in the analyte distribution. In separation science, the concept of zone spreading can be imagined as originating from the centre of the peak. Consequently, the assessment according to Figure 9.52F, $(\Delta t)^2 = \sigma^2$, is more appropriate.

The **third statistical moment** is the **vertical asymmetry** of the peak.

$$\mu_3 = \frac{M_3}{M_0} = \frac{\int_{-\infty}^{+\infty}(t - m_1)^3 f(t)\mathrm{d}t}{\int_{-\infty}^{+\infty} f(t)\mathrm{d}t} \tag{9.153}$$

It is a measure of the departure of the peak shape from the Gaussian standard. For a symmetrical peak, μ_3 is 0. For a fronting peak, the distribution has more weight before the centre of the peak (*i.e.* $\mu_3 < 0$), whereas for a tailing peak this is the opposite ($\mu_3 > 0$).

As an alternative, the **asymmetry factor** is sometimes used, which is defined as the relative width with respect to the peak centre at 0.1 h (see Figure 9.52G).

$$A_s = \frac{r_B}{r_A} \tag{9.154}$$

For distributions for which the second centralized moment is equal to σ^2 (*e.g.* the EMG or Gaussian), the **skew** is also commonly computed, which is defined as

$$S = \frac{\mu_3}{(\mu_2)^{3/2}} = \frac{\mu_3}{\sigma^3} \tag{9.155}$$

As shown in eqn (9.155), the skew increases when the third moment increases.

Finally, the fourth statistical moment is the **kurtosis** of the distribution, defined as

$$\mu_4 = \frac{\int(t - m_1)^4 \cdot f(t)\mathrm{d}t}{M_0} \tag{9.156}$$

which is then typically used to compute the kurtosis factor

$$K = \frac{\mu_4}{(\mu_2)^2} = \frac{\mu_4}{\sigma^4} \qquad (9.157)$$

Figure 9.52H shows different distributions. The block shape distribution (dark blue) will have a $K = 1.8$, for the triangle (pink) this is $K = 2.4$, and the Gaussian (light blue) $K = 3$. As can be seen, the kurtosis is a measure of the compression or stretching of the peak along a vertical axis. It can be visualized by moving in or pulling apart the sides of Gaussian peaks while maintaining a constant area.

Higher-order statistical moments exist but contain no additional useful information to characterize the peak in the context of analytical separation science.

9.12.2 Peak Detection

The first process in obtaining information from a chromatogram or electropherogram is detecting all peaks in the signal. This process is called peak detection. Similar to baseline correction, we once again deal with a field that is continuously developing and different strategies exist. In this book, we will only focus on the key concepts that are at the core of many strategies.

The easiest method to detect peaks involves finding the **local maxima;** this is analogous to the concept of local minima to identify the baseline in the LMV baseline-correction strategy (Section 9.11.3). Here, all points p_i for which $p_i > p_{i-1}$ and $p_i > p_{i+1}$ is true are considered peaks. Minimum thresholds in peak heights or widths can be set to avoid small distortions to influence the detection of peaks. Figure 9.53C shows an example of characterization of peaks using this method.

Another traditional method to detect peaks and their start and end points is based on amplifying the variation in the signal using its **derivatives**. This is shown in Figure 9.53A, where the peak around 5 min in Figure 9.50 is shown after strong smoothing using a moving average filter. If we now take the second derivative (scattered points in the figure, secondary axis), we can observe that it reaches its minimum at the location of the peak apex. One issue is that maxima from the noise are also identified as peaks, with the second derivative finding minima throughout the signal (Figure 9.53B). This strategy thus relies on the removal of the background, at the risk of removing information. The latter would result in false negatives (*i.e.* undetected components

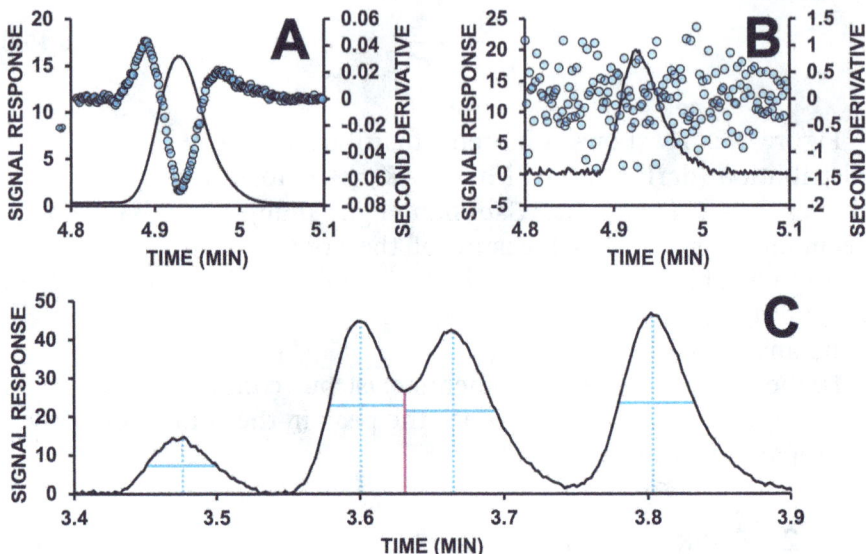

Figure 9.53 Enlarged signal of the peak around 5 min in Figure 9.50 min (A) with significant smoothing and (B) without any signal pre-processing. The scatter points are the second derivatives of both signals. In the case of a smoothened peak, the second derivative shows a clear minimum at the location of the peak, whereas it is undetectable for panel B due to the noise. (C) Peak detection and characterization by the local-maxima strategy. The dotted lines indicate the centre of the peak, the horizontal blue lines their half height, and the pink vertical line the border between the co-eluting peaks. Adapted from ref. 16 with permission from the authors.

at trace concentrations). Another issue is that the second derivative becomes challenging for co-eluting peak clusters.

The final key strategy discussed in this chapter relies on the use of least-squares regression (Module 9.6), often referred to as curve fitting, or **matched-filter responses**. The concept is based on fitting a peak distribution function (*e.g.* a Gaussian, eqn (9.131)) to a section of the chromatogram or electropherogram and considering the fitted peak height parameter in the distribution function and possibly the residuals. If the distribution function is fitted to a chromatographic peak, then the residuals will be unstructured and very low relative to when the distribution function is fitted to a section with baseline or other distortions.

An example of such regression (see also Module 9.6 and Section 9.10.2) is shown in Figure 9.54. In panel A, a distribution function

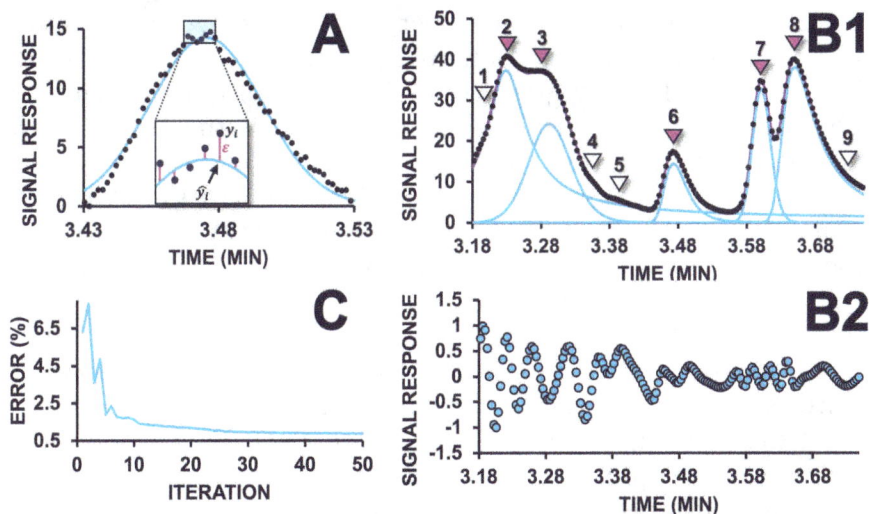

Figure 9.54 A) Gaussian distribution function (blue line) fitted on a signal (dots) using iterative non-linear regression. (B) Section of a chromatogram (dots) with the peaks indicated above by the triangles. Unfilled triangles are undetected peaks. The light blue lines depict the distribution functions that were fitted (eqn (9.159)) simultaneously through detected peaks. (C) Plot of the total residual errors for the fit in panel B as a function of iteration number. (D) Residuals for the fit shown in panel B. Adapted from ref. 16 with permission from the authors.

$(\hat{y})$ is fitted through the datapoints (y). The pink lines depict the residuals for each datapoint. This technique is also frequently used to deconvolute co-eluting peaks. This is also known as resolution enhancement, and it will be discussed further in the next section. Techniques such as multivariate curve resolution (MCR) and parallel factor analysis (PARAFAC) are also methods of choice for curve resolution of higher-order data, such as those obtained from a separation combined with mass spectrometry.

> See the website for further reading and methods for peak detection: ass-ets.org

9.12.3 Curve Resolution

Unfortunately, it is very common for separation scientists to encounter co-elution. While the practitioner will typically resort to practical methods to improve resolution, it is also possible to improve the resolution using computational methods. This practice is referred to as **resolution enhancement**. In this section, we will illustrate this using non-linear regression. Curve-fitting procedures, such as those discussed in the previous section, can aid in the case of severe co-elution, although this often relies on an *a priori* estimate of the number of peaks. To understand this, Figure 9.54B1 shows a larger segment of a chromatogram in which nine peaks are located. Using non-linear regression, we can fit a distribution function for each peak through this signal simultaneously. However, in practice, the algorithm must first determine how many peaks should be fitted. Typically, this strategy first involves a derivative or local-maxima-based assessment of the number of peaks (see Section 9.12.2). Doing so for the chromatogram shown in Figure 9.54B1 yields five peaks, indicated by the pink triangles, which are then fitted through the signal. If we regard the residuals (Figure 9.54B2) of the fit, we find an error of just 0.8% of undescribed signal. This may suggest that the fit went well. However, with four peaks having been ignored (white triangles), we also see from panel B1 that, in particular, the fits of peaks #2, 3 and 8 are significantly affected, as they overcompensate to include the areas of the undetected peaks. Consequently, we have not just missed four peaks, but our area assessment of the detected peaks is also wrong.

When multi-channel data are available (*e.g.* mass spectra), it will become easier to correctly estimate the number of peaks, but ultimately computational methods cannot robustly compensate for a complete lack of separation resolution. It is thus important that there is at least some degree of **prominence** of the peak to allow peak-detection techniques to locate it. The prominence can be considered a measure of how much a peak stands out due to its intrinsic height and location relative to neighbouring peaks.

Another important aspect is the distribution function (*i.e.* the model) that is fitted through the signal. To understand this, we can regard Figure 9.55, where a section of a different chromatogram is shown. We have four peaks, of which two are co-eluting. Figure 9.55A1 shows the results of fitting four Gaussian distributions through the four peaks, with panel A2 showing the residuals. The total error and, especially, the autocorrelation (see Section 9.6.4.2) of the datapoints in the residual plot suggest that patterns in the peak shape are not sufficiently captured by the Gaussian function.

It is useful to note that iterative fitting procedures, such as those employed here, often require several iterations to sufficiently minimize the residuals. The required number highly depends not

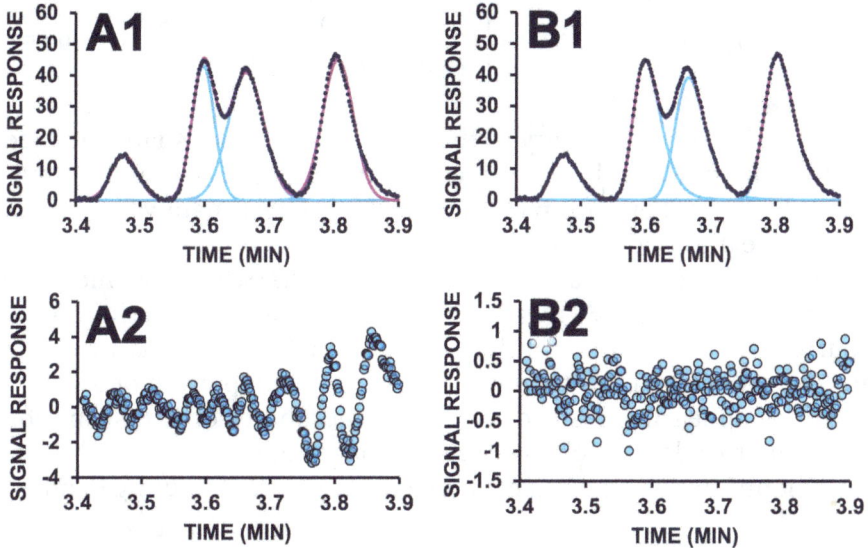

Figure 9.55 Curve-fit results for the simultaneous fitting of 4 distribution functions on the peaks using (A) the Gaussian distribution, and (B) the modified Pearson VII distribution. In both cases, the top panels depict the original data (dots), with the four individual distribution functions (light blue) and their sum (pink). The bottom panels show the residuals of the total fit. Adapted from ref. 16 with permission from the authors.

only on the complexity of the fitting problem, but also on the starting values supplied by the user for each parameter. This is shown in Figure 9.54C for the signal deconvolution shown in panel Figure 9.54B1.

As peaks in separation science rarely are truly symmetrical, chromatographers typically use different distribution functions that can capture the lack of symmetry and other shape features. A very common distribution function is the **EMG** which is given by

$$f(t; h_{\text{peak}}, \mu, \sigma, \tau) = \frac{h_{\text{peak}} \cdot \sigma}{\tau} \sqrt{\frac{\pi}{2}} \exp\left(\frac{1}{2}\left(\frac{\sigma}{\tau}\right)^2 - \frac{t-\mu}{\sigma}\right) \text{erfc}\left(\frac{1}{\sqrt{2}}\left(\frac{\sigma}{\tau} - \frac{t-\mu}{\sigma}\right)\right) \quad (9.158)$$

with a scaled error function of $\text{erfcx}\, t = \exp t^2 \cdot \text{erfc}\, t$. This function has an additional parameter τ that complicates the iterative fitting process.

Another alternative is the **modified Pearson VII** function, which is given by

$$f\left(t;\ h_{\text{peak}},\mu,\sigma,E,M\right) = \left(1 + \frac{(x - \mu)^2}{M(\sigma + E(x - \mu))^2}\right)^{-M} \qquad (9.159)$$

Here, E is the asymmetry of the peak and M is the shape, defined so that the function is a modified Lorentzian at $M = 1$, and a Gaussian at $M = \infty$. For chromatography, suitable estimate values for E and M are 0.15 and 5.

Figure 9.55B1 shows the regression results when the modified Pearson VII (eqn (9.159)) is used as the distribution function for all four peaks. More importantly, we find that the residuals in Figure 9.55B2 now yield a low total error and feature no pattern or autocorrelation, appearing instead as noise. The ability of a distribution function to adapt itself to the chromatographic peak is thus of paramount importance to accurately deconvolute co-eluting bands.

9.12.4 Determination of Peak Areas for Quantitation

While the statistical moments allow us to characterize the peaks accurately, their use is limited to cases where the start and end points of each peak can be accurately determined. We also have seen that resolution enhancement is not always straightforward.

Once all the data processing has been conducted, peak areas may be obtained by integrating the area of each peak. In the case of co-elution, the approach depends on the chosen data-treatment method. If the peak is not deconvoluted, such as in the case of Figure 9.53C, then the area is typically divided by drawing a vertical line from the saddle point between the two peaks to the baseline. The resulting values are shown in the second column of Table 9.13.

If the peak is deconvoluted using a curve-fit approach, then the individual distribution functions may be integrated. The values for this from Figure 9.55A1 and B1 are shown in the last two columns of Table 9.13.

The difference in peak areas for peaks #2 and #3 is dramatic. If one would assume that the well-fitted modified Pearson VII gives an accurate representation of the areas, then we notice that the areas for peak #2 deviate 15–20% for the other two approaches! It is beyond the scope of this book to indicate which method would be best suited generally, because many factors would need to be considered. For example, the modified Pearson VII distribution describes the data very accurately, but this strategy is at the moment difficult to implement robustly for datasets with severe co-elution,

Table 9.13 Determined peak areas for the peaks in Figures 9.53C and 9.55 using different peak detection and deconvolution strategies. Adapted from ref. 16 with permission from the authors.

Peak #	Local maxima	Curve-fit (eqn (9.131))	Curve-fit (eqn (9.159))
1	0.670	0.747	0.752
2	2.044	1.849	2.368
3	2.435	2.831	2.361
4	2.481	2.703	2.755

due to the dependency on a good estimate of the number of peaks in convoluted signals. This is the reason why many software tools rely on the method shown in Figure 9.53C, which is perhaps not perfect, but may yield repeatable results.

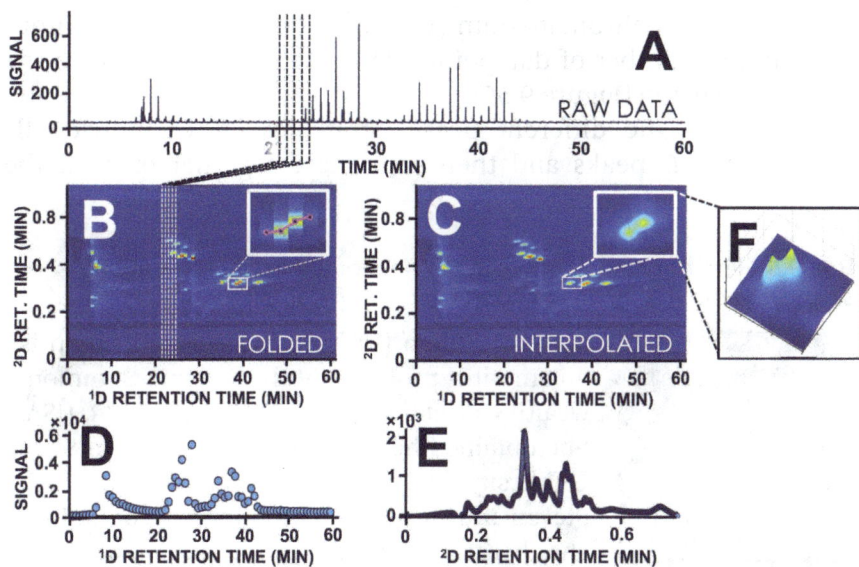

Figure 9.56 (A) Example of a raw LC×LC chromatogram. Dashed lines depict ^{2}D modulations. (B) Folded 2D plot of raw data shown in panel A. (C) Interpolated version of data shown in panel B. (D) ^{1}D chromatogram by summing all ^{2}D datapoints, and (E) ^{2}D chromatogram by summing all ^{1}D datapoints. (F) Small shifts in retention time can result in the detection of two peaks. Adapted from ref. 16 with permission from the Authors.

9.13   Data Analysis for Comprehensive 2D Separations (A)

For two-dimensional separations, the available means to deal with data are more complicated and still under development. Readers interested in this are referred to other studies.[17] Instead, we will briefly discuss the general procedure and challenges associated with this type of data.

As explained in Chapter 7, comprehensive 2D separations typically rely on the second-dimension detector to record the full 2D chromatogram. This is depicted in Figure 9.56A where a raw comprehensive 2D-LC chromatogram is shown, essentially comprising a series of second-dimension (^{2}D) chromatograms. Using the modulation time, this linear vector can be divided into the different modulations, which can then be stacked together in a surface plot or colour plot, such as is shown in Figure 9.56B.

With the first dimension (^{1}D) generally being undersampled, scientists often interpolate the first dimension to create the plot shown in Figure 9.56C. This also becomes apparent if we regard the reconstructed ^{1}D chromatogram (Figure 9.56D) which only comprises a limited number of datapoints. This is in stark contrast to the second dimension (Figure 9.56E).

To integrate the different peaks, strategies either individually integrate the ^{1}D peaks and then sum these together or treat the

---

Box 9.5 Hero of chemometrics in the field of analytical separation science: Sarah Rutan. Image courtesy of Prof. Rutan.

Sarah C. Rutan (United States) Sarah Rutan is an emeritus professor at the Virginia Commonwealth University (VCU) in Richmond, VA, USA. She obtained her PhD at the Washington State University (Pullman, WA, USA) working with Steven Brown on Kalman filtering. Soon after, she became a professor at VCU. She used chemometric techniques and solvatochromism to contribute to the fundamental understanding of reversed-phase liquid chromatography. She has also contributed to the development of multivariate methods for data handling and optimization in multidimensional chromatography.

interpolated landscape so that a 3D peak is integrated as a whole. Both strategies come with their disadvantages. In the case of summing the different ¹D peaks, it is identifying which actually belong together to the same peak (also known as clustering, see the inset of Figure 9.56B). Whereas considering the peaks as being three-dimensional in the interpolated signal can easily yield incorrect results or even split peaks (see Figure 9.56F). Sarah Rutan (Box 9.5)has contributed to data analysis techniques for comprehensive two-dimensional chromatography.

> See the website for a more elaborate discussion, and some examples, of data processing for 2D separations: ass-ets.org

Acknowledgements

Prof. Dr. Attila Felinger is acknowledged for his review on the chapter and the fruitful discussion. Dr. Johan Westerhuis is acknowledged for his review of the chapter and contribution to several modules. Dr. Gabriel Vivo-Truyóls is acknowledged for his review of the chapter and for laying the foundations for the Chemometrics & Statistics course of the MSc program Analytical Chemistry of the University of Amsterdam. This has formed the basis for the current course and the current chapter. Dr. Tijmen Bos is acknowledged for his review of the chapter. Ebru Kara Chasan Memet is acknowledged for her extensive contributions. Janne Bolwerk, Lucas Bervoets, Carmen Pale and Boaz Geurtsen are acknowledged for their useful contributions.

Recommended Reading

N. R. Draper and H. Smith, in *Applied Regression Analysis*, 3rd edn, 1998.
D. L. Massart, *et al.*, in *Handbook of Chemometrics and Qualimetrics, Part A*, Elsevier Science & Technology, 1997.

References

1. N. M. Razali and Y. B. Wah, *J. Stat. Model. Anal.*, 2011, **2**, 21–33.

2. R. M. Conroy, *Stata J. Promot. Commun. Stat. Stata*, 2012, **12**, 182–190.
3. G. W. Divine, H. J. Norton, A. E. Barón and E. Juarez-Colunga, *Am. Stat.*, 2018, **72**, 278–286.
4. A. Hart, *BMJ*, 2001, **323**, 391–393.
5. D. L. Massart, B. G. M. Vandeginste, L. M. C. Buydens, P. J. Lewi and J. Smeyers-Verbeke, in *Handbook of Chemometrics and Qualimetrics: Part A*, Elsevier, Amsterdam, the Netherlands, 1997.
6. In *Levene, in Contributions to probability and statistics; essays in honor of Harold Hotelling*, Stanford University Press, Stanford, 1960, pp. 278–292.
7. M. B. Brown, *J. Am. Stat. Assoc.*, 1974, **69**, 364–367.
8. W. H. Kruskal and W. A. Wallis, *J. Am. Stat. Assoc.*, 1952, **47**, 583–621.
9. N. R. Draper and H. Smith, in *Applied Regression Analysis*, Wiley & Sons, 3rd edn, 1998.
10. S. Lamotte, R. Brindle and K. D. Bischoff, *CLB Chem. Lab. Biotech.*, 2006, **57**, 349–351.
11. L. M. C. Buydens, B. G. M. Vandeginste, G. Kateman, P. J. Schoenmakers and M. Mulholland, *Chemom. Intell. Lab. Syst.*, 1991, **11**, 337–347.
12. M. Mulholland and J. Waterhouse, *J. Chromatogr. A*, 1987, **395**, 539–551.
13. R. G. Brereton, in *Chemometrics: Data Analysis for the Laboratory and Chemical Plant*, John Wiley & Sons, Ltd, 2003.
14. B. W. J. Pirok, A. F. G. Gargano and P. J. Schoenmakers, *J. Sep. Sci.*, 2018, 41.
15. L. E. Niezen, P. J. Schoenmakers and B. W. J. Pirok, *Anal. Chim. Acta*, 2022, **1201**, 339605.
16. B. W. J. Pirok and J. A. Westerhuis, *LCGC N. Am.*, 2020, **6**, 8–14.
17. B. W. J. Pirok, S. C. Rutan and D. R. Stoll, in *Multi-Dimensional Liquid Chromatography*, CRC Press, Boca Raton, 2022, pp. 233–272.

10 Method Development and Optimization

Method development is one of the most important and most challenging tasks of chromatographers. It is where all knowledge and experience come together. There are strategies and tools to facilitate the process, and these are the subjects of this chapter. First, we describe the principles and formal aspects of method development, followed by a discussion of column screening and column selection. Significant attention is paid to measures for the quality of separation, as expressed through chromatographic response functions. Contemporary strategies and algorithms for method optimization are discussed, followed by a brief description of the method validation process. We pay special attention to peak tracking, which is an essential step in computational optimization approaches. The chapter ends with an outlook on the future. An overview is presented in Figure 10.1.

10.1 Introduction to Method Development (B)

The goal of analytical method development can be formulated as "to obtain an analytical procedure fit for the intended purpose".[1] The intended purpose can be described in terms of target analytes or groups of analytes (*e.g.* total fats, total saturated fats in foodstuffs) and the target matrix. Target specifications of the method can be phrased in terms of specificity (absence of interferences), accuracy (correctness of the results), precision (random variations in the results), applicable range (minimum and maximum concentrations expected), *etc.* The collection of such target specifications

Analytical Separation Science
By Bob W. J. Pirok and Peter J. Schoenmakers
© Bob W. J. Pirok and Peter J. Schoenmakers 2025
Published by the Royal Society of Chemistry, www.rsc.org

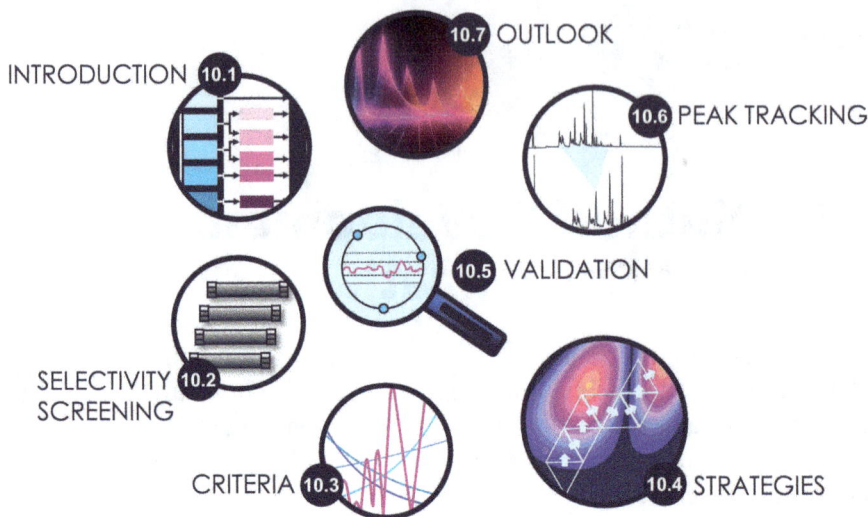

Figure 10.1 Graphical overview of the modules in this chapter.

may be referred to as the **analytical target profile** (ATP).[1] Methods that apply to a range of different samples (in terms of analytes, matrices, products, *etc.*) are attractive from an operational perspective. However, such **generic methods** or **platform methods** tend to offer a compromise solution instead of an optimal one for most of the samples to which they are applied. Additional method validation (see Module 10.5) may be required when expanding the scope of a method.

Much of the information provided in earlier chapters of this book is relevant in the context of method development, but a start can be made without first studying all of it. We will frequently refer to earlier modules and sections in this book to allow the reader to acquire or refresh relevant background knowledge.

We first (re)consider the role of analytical separations in the world of chemical measurements and analysis. Figure 10.2 provides an overall picture of the field.

Analytical measurements (bright blue ellipse) may be performed on a sample. Examples include spectroscopic and electrochemical measurements, as well as any other detection principle discussed in this book. This will work well for "pure" samples. For example, a student who has synthesized a compound in an organic chemistry lab may record an NMR (nuclear magnetic resonance) spectrum to verify the success of the experiment. In the vast majority of cases, a sample is (much) too complex to allow direct measurement, and

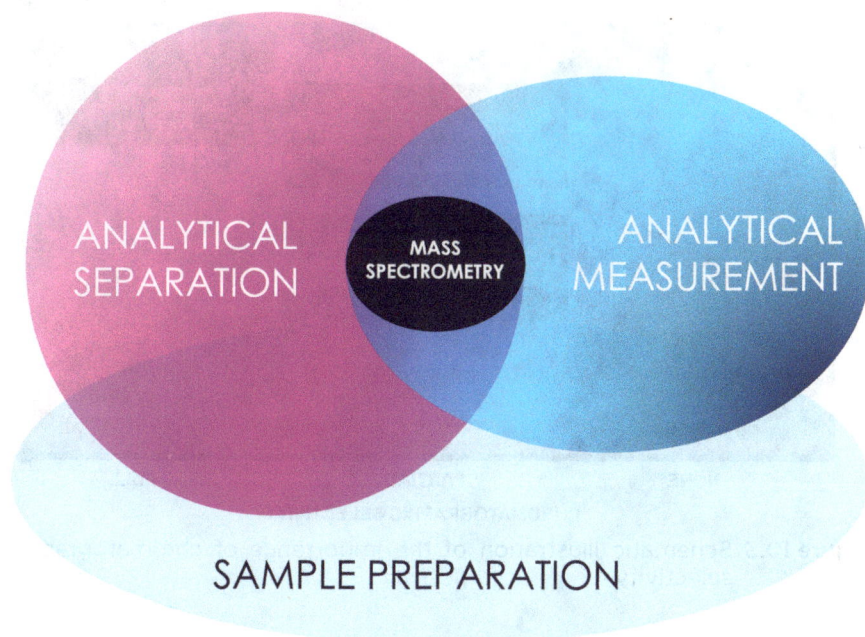

Figure 10.2 Overall summary of the field of analytical chemistry. Many analytical methods involve sample preparation, followed by separation and measurement (often referred to as "detection" by separation scientists), but analytical measurements may also be performed without a prior separation step. Mass spectrometry combines separation and measurement.

sample preparation and/or **analytical separations** are called for. The distinction between sample preparation and analytical separations is diffuse. From a spectroscopy perspective, separations may be viewed as sample preparation. From an analytical separation perspective, analytical measurements are considered detection. An analytical method can be described as

$$\begin{bmatrix}\textbf{Analytical} \\ \textbf{Method}\end{bmatrix} = \begin{bmatrix}\text{Sample} \\ \text{Preparation}\end{bmatrix} + \begin{bmatrix}\text{Analytical} \\ \text{Separation}\end{bmatrix} + \begin{bmatrix}\textbf{Analytical} \\ \textbf{Measurement}\end{bmatrix} \qquad (10.1)$$

The steps that are in bold are indispensable, while the other steps are optional. Analytical separations without a measurement step are feasible but are referred to as preparative separations.

Mass spectrometry (MS) can be seen as combining separation and measurement. The separator and detector are two different parts of the instrument, but – unlike analytical separation systems – they are

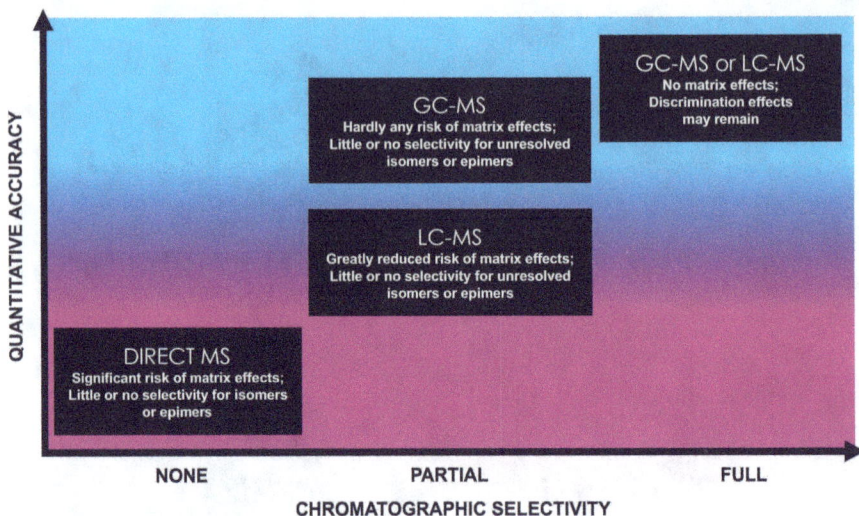

Figure 10.3 Schematic illustration of the importance of chromatographic selectivity.

not typically seen as interchangeable modules. However, the inlet system is the weak point in MS of mixtures (ion suppression, matrix effect, *etc.*). Off-line sample preparation is sometimes necessary, while analytical separations are used pervasively in combination with MS. Hence, the enormously successful combinations of gas chromatography (GC) and MS (GC-MS, Module 2.6) and liquid chromatography (LC) and MS (LC-MS, Module 3.10) receive ample attention in this book, as does the combination of capillary electrophoresis (CE) with MS (Module 5.3). In this book, and the remainder of this chapter, we are focussing on analytical separations, with (non-exclusive) emphasis on the most important techniques, GC and LC.

The importance of chromatographic selectivity is illustrated in Figure 10.3. When directly introducing a complex mixture into a mass spectrometer, there is a substantial risk of bias in the observed signal intensities due to matrix effects. In addition, there is little selectivity for isomers and hardly any for epimers and stereoisomers, even when applying multiple reaction monitoring (MRM) tandem MS (MS/MS) techniques.[2] The latter problem may, to some extent, be alleviated by inserting an ion mobility stage (IMS) into the mass spectrometer, but this does nothing to repair biases introduced at the ionization stage. When introducing chromatographic selectivity, the risk of ionization basis is greatly reduced. Separation of the analytes from the matrix already has a significant positive effect (partial selectivity in Figure 10.3; analytes separated from the matrix, but not all from each other). In LC-MS, co-eluting ions may still give rise to ionization bias, especially when the concentrations are vastly different. This is much less so in the case of GC-MS. When all peaks are separated by GC, LC, or another analytical separation method (full selectivity in Figure 10.3), the different ions (analytes or matrix components) will not affect each other in the MS source, but there may still be discrimination effects because the ionization and transmission efficiencies may be different for different analytes.

10.1.1   Overview of GC Methods

GC is one of the most important and, in many ways, most attractive chromatographic methods. Fundamentally, GC benefits from high (analyte) diffusion coefficients and low (mobile-phase) viscosity. This combination allows fast separations or high-resolution separations using long, open-tubular columns. This leads to a recommendation that GC should be the first-choice separation method – if possible. The possibilities of using GC are illustrated in Figure 10.4.

Gaseous samples can be introduced directly into a GC system using a gas sampling valve. Volatile samples can be injected as liquids using a split-splitless injector (Section 2.4.1) at a high temperature, but preferably at a low temperature using a programmed-temperature vaporizer (PTV) injector (Section 2.4.3) or – less conveniently – using cold-on-column injection (Section 2.4.2). When the matrix contains non-volatile components, on-column injection cannot be used, but PTV injection can be used. Headspace injection (Section 2.4.4) is another option, which is especially useful for determining volatile analytes in solid, non-volatile samples. For these latter types of samples, thermal desorption techniques (Section 2.4.5) may also be considered. Non-volatile, low-molecular-weight analytes tend to be polar, since polar (hydrogen-bonding) groups increase the cohesive energy and boiling points of compounds. The polar groups (hydroxyl, amino, carboxylic acids, *etc.*) can be derivatized to increase analyte volatility and stability at elevated temperatures (Section 8.3.4). Even macromolecules can be analyzed by GC after hydrolysis or pyrolysis (Section 8.3.5). In that case, GC provides information on the structure (building blocks, such as monomers) of the molecules, but not on the intact molecules (chain length or molecular weight). The fact that derivatization

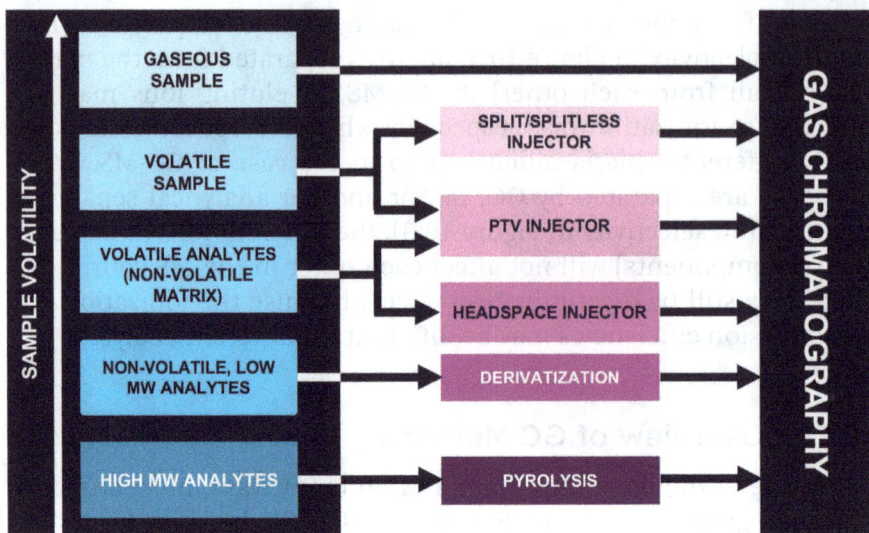

Figure 10.4 Ways to use gas chromatography (GC) for a variety of samples.

and pyrolysis methods are commonly used underlines the attractiveness of GC as an analytical separation method.

The method optimization strategies described in the present chapter can definitely be applied to GC methods. However, GC is so powerful (100 000 theoretical plates or more on a routine basis), and the detection techniques (such as GC-MS) are so well developed that optimization is often considered unnecessary. Nevertheless, when developing routine methods, optimization may result in faster, more economical methods. For comprehensive characterization or fingerprinting of volatile (fractions of) complex samples, comprehensive two-dimensional gas chromatography (GC×GC, Module 7.2) is a mature method.

10.1.1.1 Starting Conditions for GC

Some starting conditions for GC experiments are outlined in Table 10.1. The reader should be aware that this is a rough guide that ignores much of the detail provided in Chapter 2 and elsewhere in this book. In the rare event that nothing is known about the sample, a 5% diphenyl column may be the default choice, but if there is an indication that the analytes of interest or the sample matrix are highly polar, a wax column may be used from the outset. If initial experiments on a non-polar column yield distorted peaks, a polar column should be tried.

Table 10.1 Starting conditions for GC.

Parameter	Settings
Stationary phase	For low-polarity analytes, use a low-polarity column • 100% dimethylpolysiloxane (*e.g.* DB-1, HP-1, ULTRA-1, CP-Sil 5 CB, RTX-1, SPB-1, MDN-1, BP-1, InertCap-1, ZB-1) • 5% diphenyl–95% dimethylpolysiloxane (*e.g.* DB-5, HP-5, ULTRA-2, CP Sil 8 CB, RTX-5, SPB-5, MDN-5, BP-5, InertCap-5, ZB-5) For polar analytes, use a polar column, for example, • polyethylene glycol (PEG) (*e.g.* DB-WAX, CPWax 52-CB, RTX-Wax, SupelcoWax-10, BP-10, InertCap-Wax, ZB-wax)
Column internal diameter	250 or 320 µm
Column length	20–30 m
Film thickness	0.25 or 0.32 µm (internal diameter/1000)
Gas flow rate	1–2 mL min⁻¹
Initial temperature program	On low-polarity column: 40–300 °C On high-polarity column: 40–250 °C
Program duration	20–30 min
Injector	Split injection (PTV injection if available)

Columns with internal diameters of 250 or 320 µm are standard in GC. Narrower or wider columns should only be used if there is a strong indication to do so (*e.g.* short 100 µm i.d. columns for very fast analysis). Columns of 20–30 m in length are common, but this advice is less strict than that regarding column diameters. A 10 m long column may, for example, be used for screening experiments (with, for example, a 10 min temperature program). Selecting a film thickness equal to one-thousandth of the column's internal diameter (reduced film thickness of 0.3; see Section 1.7.7) is a strong recommendation. Thicker films may be used to increase the retention of low-boiling compounds (allowing the GC oven to operate above ambient temperatures) or to increase the sample loadability, but this will be at the expense of a loss in efficiency. Conversely, however, little efficiency is gained by using thinner stationary-phase films. Thin film columns may be used to extend the working range of GC to stable, low-volatile compounds. One example is special high-temperature-GC columns (*e.g.* metal instead of fused silica) for "simulated distillation" of hydrocarbons (see Section 2.1.5).

A split injector may be used during initial experiments, but once the method development progresses to quantitative analysis, a PTV injector is recommended (see Section 2.4.3). Aqueous samples usually require extraction (see Module 8.2), but if the analyte volatility is sufficiently low, they can be analyzed directly using a PTV injector.

10.1.2   Overview of LC Methods

Figure 10.5 shows an overview of LC methods. While in GC, volatility is the prevailing parameter, the most important parameters in LC are analyte polarity and molecular weight. The overwhelmingly favourite choice is reversed-phase liquid chromatography (RPLC). For many reasons described in Module 3.5, RPLC is the preferred option whenever it is applicable. This is very often, as shown by the (approximate!) outline in Figure 10.5. For very non-polar analytes, non-aqueous RPLC (NARP) may be used, but more selectivity may be provided by normal-phase liquid chromatography (NPLC, Section 3.6.1). Supercritical-fluid chromatography (SFC, Chapter 6) is hardly ever – if any time at all – the only method that can be applied, but it can be considerably faster and more attractive than NPLC or – less often – RPLC. For extremely polar or ionic analytes, which do not show sufficient retention in RPLC, hydrophilic-interaction liquid chromatography (HILIC, Section 3.6.2), ion-exchange

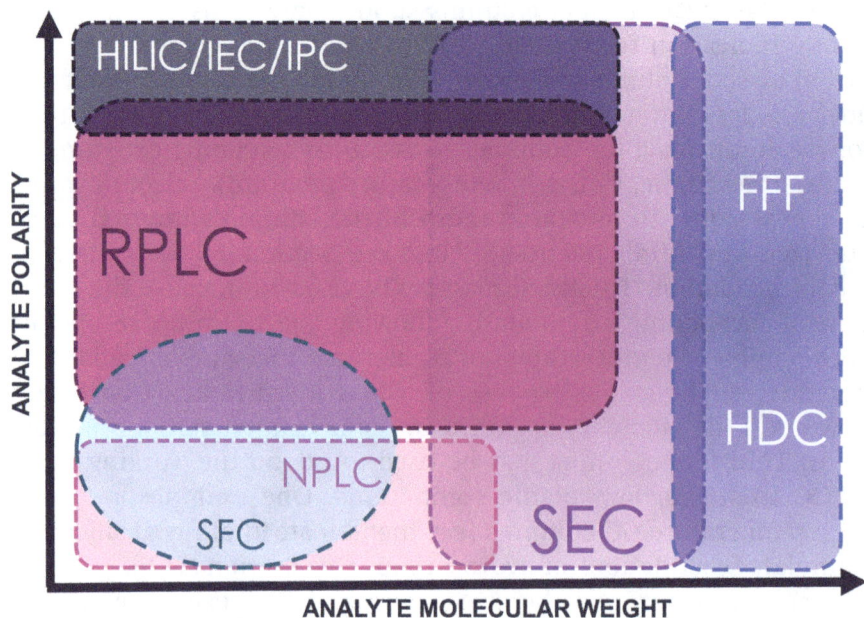

Figure 10.5 Overview of approximate application regions for liquid chromatography (LC) techniques.

chromatography (IEC, Section 3.7.1), or ion-pair chromatography (IPC, Section 3.7.3) may be used. Size-exclusion chromatography (SEC, Module 4.2) is a broadly applicable technique, especially for molecules of high molecular weight. It offers less resolution than the previously mentioned methods. For extremely large molecules (molecular weight $\gg$ 1 MDa) or nanoparticles, or sensitive (large) molecules or molecular aggregates, field-flow fractionation (FFF, Module 4.5) and hydrodynamic chromatography (HDC, Module 4.4) become more suitable than SEC.

The variety of electromigration techniques (Chapter 5) cannot easily be collected in an overview diagram, but they deserve to be mentioned. Capillary electrophoresis (Chapter 5) and capillary isoelectric focussing (cIEF, Section 5.1.4.4) are especially suitable for high-resolution separations of proteins, while capillary gel electrophoresis (CGE, Section 5.1.4.2) is arguably the benchmark method for separating nucleic acids (DNA, RNA). All of these methods rely on electric charges for the migration of analyte ions. Neutral molecules can be separated using micellar electrokinetic chromatography (MEKC, Section 5.1.4.6, Module 5.4). Despite the abundance of excellent high-resolution separations demonstrated in the literature, electromigration techniques are not used nearly as often as LC methods. The most likely reason for this is the greater robustness and reliability of LC.

10.1.2.1 Starting Conditions for LC

Possible starting conditions for LC experiments are outlined in Table 10.2. Again, this is a rough guide only. Much background is provided in Chapter 3 and elsewhere in this book. The default mode is RPLC – if no information is available to the contrary. The default stationary phase is octadecyl silica. The mobile phase may be acidified or buffered at a pH between 2 and 7. Reasonable flow rates are indicated in the table. If the flow rate is increased (if pressure permits) or decreased, the gradient duration should be adapted accordingly (inversely proportional to the flow rate). If the analytes elute around the t_0 marker, or if it is known beforehand that the analytes are highly polar, then HILIC may be attempted. If the analytes are ions, IEC or IPC are possible choices.

UV detection is the easiest choice, but non-UV-absorbing analytes ("non-chromophores") may be overlooked. Some analytes may also be overlooked if electrospray ionization (ESI) MS is used in total-ion-current or base-peak mode.

Table 10.2 Starting conditions for LC.

Parameter	Settings
Stationary phase	Octadecyl silica
Column dimensions	HPLC: 100 mm × 4.6 mm i.d.; 3 µm particles
	UHPLC: 50 mm × 2.1 mm i.d.; 1.7 µm particles
Mobile phase	Water to acetonitrile (ACN)
Gradient	5–100% ACN
Gradient duration	HPLC: 5 min
	UHPLC: 1.5 min
Mobile phase flow rate	HPLC: *ca.* 2 mL min^{-1}
	UHPLC: *ca.* 0.7 mL min^{-1}
Detection	UV at 190 or 200 nm

10.1.3 Systematic Approaches to Method Development

A distinction can be made between a classical or "minimal" approach to method development and an "enhanced approach".[1] In the minimal approach, the target specifications of the method should be reached, but the path to reach this point is not specified. This reflects a trial-and-error approach based on the knowledge and experience of the analyst. Good analysts translate their experience into an effective, intuitive approach, and they are masters at finding solutions to every problem and developing methods that meet the target specifications. However, such methods tend to be variations on a theme, and other solutions may be superior in terms of analysis time, working range, *etc.*

Enhanced approaches (i) make use of prior knowledge or risk assessment to identify the most relevant parameters, (ii) apply a systematic **design-of-experiments** (DoE) approach (see Module 9.8) and model the results to identify optimal or permissible ranges of the parameters and the interactions (correlation) between the different parameters, and (iii) define a control strategy of the analytical method based on the results of the systematic experiments.

10.1.3.1 Interpretive Methods

Interpretive methods for method development are strategies in which the chromatogram is interpreted as the sum of the signals

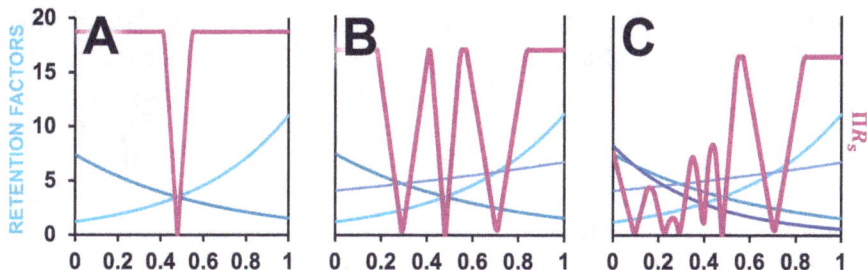

Figure 10.6 Examples of retention models (blue lines) and response surfaces (pink lines) for mixtures of (A) two, (B) three, and (C) four analytes. The response surfaces correspond to the product of all R_S values, capped at a maximum of $R_S = 1.5$. A low plate count ($N = 1000$) is assumed for illustration purposes. The pink axis has an arbitrary scale.

(peaks) of the individual analytes. These methods rely on modelling of retention surfaces using a retention model. As such, interpretive methods are a typical example of an enhanced approach to analytical method development. The underlying philosophy is that retention surfaces tend to be smooth and monotonous, whereas **response surfaces** that are indicative of the overall quality of the chromatogram are highly complex. This is illustrated in Figure 10.6. In Figure 10.6A, two retention surfaces are shown (light and dark blue lines). These are smooth curves (drawn on the basis of a log–linear relationship between the retention factor and the composition parameter on the horizontal axis; the lines are curved in the figure because the vertical axis is not logarithmic). The composition parameter can, for example, be the volume fraction of one component of a binary mixture of two eluotropic LC mobile phases (see Section 3.5.3) or that of one stationary phase in a binary mixture of two GC stationary phases. The pink response surface represents the product of all resolution values (limited to a maximum resolution of 1.5).

In Figure 10.6A, there is only one resolution value for the two analytes. At the intersection point, the response function is zero, and around it, the values are low. On the left and right are plateaus ($R_S \geq 1.5$) that are befitting of the working ranges obtained from enhanced method development approaches. In Figure 10.6B, one analyte has been added to the mixture (intermediate blue colour). All three retention curves remain very simple, but the response surface already becomes much more complicated. The two stable plateaus are at the edges (close to 100% A or 100% B). There are also two locations close to the 50/50 point that yield adequate resolution at shorter analysis times (lower retention factors for the last-eluting analyte). However, these are narrow ranges, so the method conditions are less robust. All this information can be deduced from the simple retention models. Figure 10.6C shows example retention models and a response surface for a mixture of four analytes. It is obvious that retention models are simple and can be established from a few data points (*i.e.* two for a linear or log–linear model), while the response surface is very complex. It can be calculated indirectly from the retention models, but it would require many experiments to characterize the surface directly without interpreting the chromatogram in terms of retention data for individual analytes. Many of the samples that chromatographers are confronted with are much more complex than just four analytes (plus matrix compounds that need to be separated from the target analytes). For such samples, interpretive ("enhanced") method development is the way to go. Trial-and-error ("minimum") approaches are highly unlikely to yield the best operating conditions (*i.e.* the global optimum). Lloyd Snyder (Box 10.1) was one of the driving forces behind enhanced method development in LC for many years.

The price to pay for using interpretive methods is that either analytes (standards) need to be injected individually to build retention models, or the peaks for the different analytes need to

---

Box 10.1 Hero of analytical separation science: Lloyd Snyder. Photo reproduced, with permission, from Ref. 3.

Lloyd R. Snyder (1931–2018, United States of America)
Lloyd Snyder obtained his PhD from UC Berkeley (CA, USA) and started his life-long career at Standard Oil (CA), before moving to Technicon (NY) on the East Coast, to finally return to California, where he joined forces with John Dolan in LC Resources. Lloyd Snyder worked on GC, flow-injection analysis and, especially, LC. Grasping complex processes, such as gradient-elution LC or preparative LC, in simple models to gain understanding and to achieve optimal results were Lloyd's unique strengths. He was also a master at explaining LC in very clear terms. The nine (!) books he wrote on the subject, with Jack Kirkland and others are still worth reading.

be correctly labelled in each chromatogram. This so-called peak tracking is addressed in Module 10.6.

10.1.4 Optimization

10.1.4.1 Definition of Optimization

Once initial method development has yielded a separation method, we face the decision of whether to accept the resulting method or to continue method development. This process is typically referred to as **optimization**. The definition of the word optimization is as vague and situation-dependent as the process it represents. In analytical separation science, it is widely used to describe very different procedures. In computer science, the concept of optimization typically refers to rewriting an algorithm so that its efficiency and speed are maximized. In analytical separation science, we also often seek the best possible combination of parameters, but this does not necessarily entail maximum speed. The actual motive for the development of a separation method is, after all, the separation of a sample. In mathematics, optimization entails establishing conditions that represent the maximum of minimum value of a specific function. For separations, such a function should express the critical information that the developed method aims to yield.

10.1.4.2 Cost–Benefit Character of Optimization

To decide whether optimization should be pursued, the value of the information gained should be weighed against the additional method development time and effort. This cost–benefit balance depends strongly on the purpose of the method and the employed optimization strategy. If the required critical information can be obtained from a rudimentary, sub-optimal method, it greatly depends on the intended usage of the method whether further development and optimization are warranted. For example, if the method is part of an effort to solve a rare problem, it may only need to be used again occasionally–if at all. In such a case, a long analysis time would not be a major obstacle. However, if large numbers of samples are anticipated, further optimization will be worthwhile. The aim of the optimization may then be to maintain sufficient separation while minimizing method-performance characteristics such as analysis time, solvent consumption, and possibly detection limits. For conventional, one-dimensional (1D) separations, the

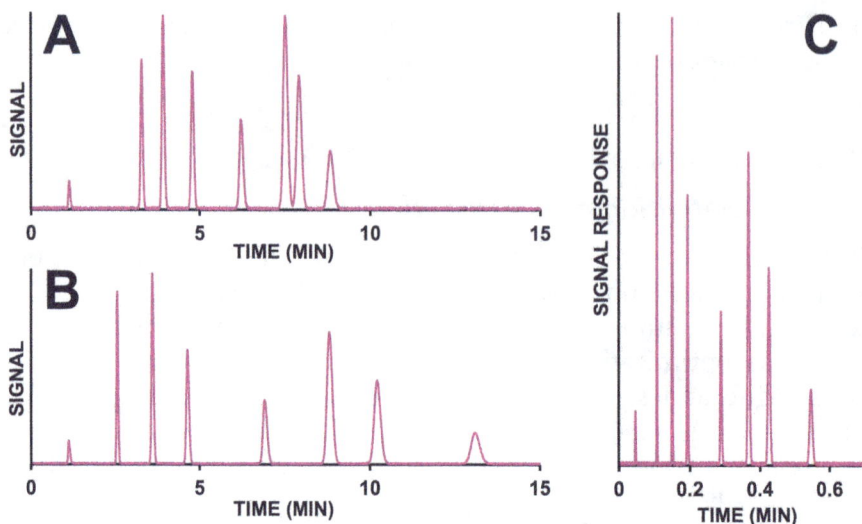

Figure 10.7 Three stages of developing an LC method. (A) Result of minimum method development on a 100 mm × 4.6 mm i.d. RPLC column packed with 3 μm particles at 1 mL min⁻¹. (B) Result of enhanced optimization on the same column at the same flow rate. (C) Translating the separation with the same stationary and mobile phases as chromatogram B to a UHPLC column (20 mm × 2.1 mm i.d.; 1.8 μm particles). All three chromatograms run at a flow rate of 1 mL min⁻¹.

costs of method optimization, in terms of time and effort, are often quite limited. An example is shown in Figure 10.7.

A seven-component sample is separated by RPLC on a standard column, and after tweaking the ACN–water ratio ("minimum" method development), the chromatogram in Figure 10.7A is obtained. If only one or a few samples need to be analyzed, without a prospect of more samples to follow, this method may well be considered adequate. Enhanced method development in this case may entail the optimization of the composition of a ternary mobile-phase mixture of ACN, water, and methanol, following a procedure akin to that illustrated in Figure 10.6. The result is shown in Figure 10.7B. Resolution is seen to be improved, but the analysis time is increased. The real gain of the optimization becomes apparent if we transfer this high-resolution to a very short (commercially available) ultra-high pressure LC (UHPLC) column. The analysis can then be at least 15 times faster and the solvent consumption 15 times lower, which are significant benefits if the method is to be applied numerous times. The analysis time may be further shortened by increasing the flow rate, but this may not be deemed necessary.

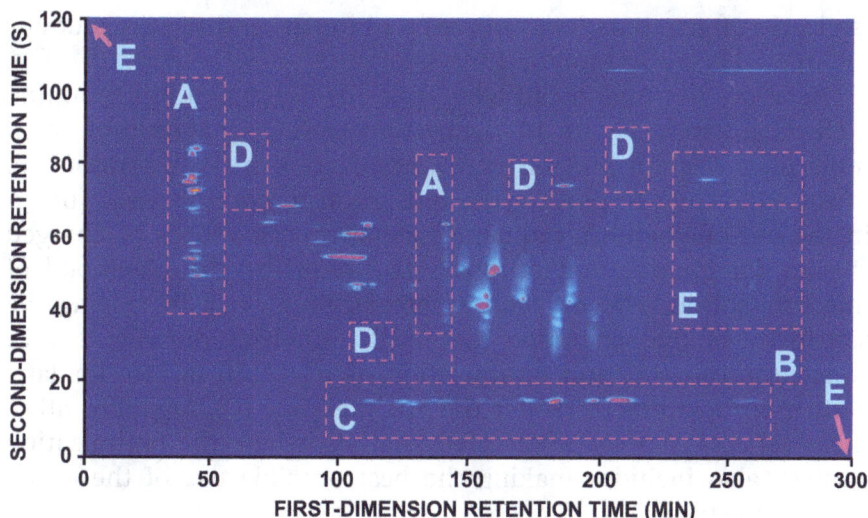

Figure 10.8 Separation of aged synthetic dyes by comprehensively coupled strong-anion-exchange chromatography and reversed-phase LC with diode-array UV detection (SAX×RPLC-DAD). Adapted from ref. 4 with permission from Elsevier, Copyright 2016.

Developing and optimizing two-dimensional (2D) methods tend to require more time and effort. The cost–benefit discussion can be illustrated using the example of the comprehensive two-dimensional LC (LC×LC) method encountered in Module 7.1 (Figure 7.3). In this example, different homologous series of surfactants were well separated, allowing accurate quantitative analysis of the sample. The critical information, group-type analysis of the different series of surfactants, could readily be obtained from the LC×LC chromatogram. However, the orthogonality of the two underlying LC methods was limited, as was the surface coverage. The peak capacity was lower than expected, and peak shapes were imperfect. In the industrial setting of this application, further optimization was not warranted.

We can distinguish two types of optimization strategies. **Targeted** (or **sample-dependent**) optimization enhances the capability of a separation method to provide specific information on known target (groups of) analytes. **Untargeted** (or **sample independent**) optimization strategies are used to maximize the likelihood of separation for as many peaks as possible. Targeted optimization may seem a more cost-efficient option, but the choice depends largely on the goal of the analysis. Quantifying concentrations of priority pollutants is clearly targeted analysis, whereas a biomarker-discovery study is untargeted (see also Module 1.9).

The above discussion can be graphically illustrated by examining a chromatogram. Figure 10.8 shows an LC×LC separation of small molecules. Examples of targeted optimization include some or all of the following: Improving the resolution in specific areas (box A in the chromatogram); reducing excessive peak widths of charged species for quantitation purposes (B); reducing the likelihood of injection-solvent effects, such as breakthrough (C); increasing the sensitivity of the separation method for analytes present at trace concentrations (D); and optimization of retention factors for late-eluting compounds to reduce the analysis time (E). Arguably, all of these aspects improve untargeted analysis, where the optimization strategy also includes making the best possible use of the entire separation space.

10.2 Selectivity Screening (M)

In Module 10.1 (Figure 10.3), we explained that chromatographic selectivity remains essential, despite the significant progress made in mass spectrometry in the last few decades. We have also explained in Section 1.6.1 that extremely small differences in interactions between analytes and the phase system (stationary and mobile phases) suffice to achieve chromatographic separation (see the discussion of eqn (1.57)). This makes it very difficult to use any kind of algorithm to predict selectivity based on the molecular structures of the analytes. Therefore, collecting some experimental data is typically needed to start using the approach of "enhanced" or "interpretive" method development outlined above (see Section 10.1.3). Column-characterization methods, such as those described in Module 2.8 for GC and in Section 3.8.2.2 for LC, sometimes in combination with chemometric classification techniques, aid in the selection of columns that promise different (or similar) selectivities. These and more elaborate models (see Section 3.8.2) do not allow accurate prediction of retention data for analytes that were not in the training set.

10.2.1 Selectivity Screening in GC

For a given set of analytes (*i.e.* the sample we need to separate), almost all selectivity in GC arises from the stationary phase. The temperature may affect not only the selectivity but also the retention. Variations in the near-inevitable temperature program used in open-tubular (capillary) GC do not strongly affect the selectivity. Different mobile phases (hydrogen, helium) have a marginal effect on the selectivity. Thus, if selectivity must be improved, it is necessary to test different columns. The choice is limited by the need to achieve good peak shapes and high efficiencies, which usually implies using polar columns for polar analytes and non-polar columns for non-polar analytes. A range of analytes with diverse polarities in the sample may narrow the choice considerably. As a result, selectivity tends not to be the most studied parameter in GC. Analysts are often content with good peak shapes and then resort to longer columns if resolution needs to be improved.

Chiral separations (see Section 2.1.4) form a major exception to this rule. Different columns with different chiral selectors may provide different selectivities. In addition, a pair of isomers can usually be separated under isothermal conditions, so that striving for an optimal temperature becomes sensible. Possibly, an isothermal separation of a critical pair of enantiomers may follow a separation from the remainder of the sample by a heart-cut 2D-GC approach (Section 7.2.1). When screening the selectivity of different chiral columns, it is recommended to explore the effect of temperature by performing at least two experiments at different temperatures. In an interpretive temperature-optimization approach, a plot of $\ln (k/T)$ against $1/T$ is expected to yield straight-line retention models (see eqn (1.55)). In optimizing the resolution, one should be aware that the efficiency of open-tubular (capillary) columns decreases rapidly with increasing retention factors (see Section 1.7.7).

10.2.2 Selectivity Screening in LC and SFC

Selectivity screening in liquid chromatography (LC) and supercritical-fluid chromatography (SFC) is more involved because selectivity can be influenced in multiple ways, *viz.*, by varying (i) the stationary phase, (ii) the mobile-phase composition (*e.g.* the nature of the modifier), and (iii) the temperature. For ionogenic analytes (weak ions) and ionic analytes (strong ions), we may add (iv) the pH, (v) the ionic strength, and (vi) the type and concentration of an eventual ion-pair reagent. We have discussed the effects of these parameters extensively in Chapter 3 for LC and in Chapter 6 for SFC. It is remarkable that the first parameter in the list is not

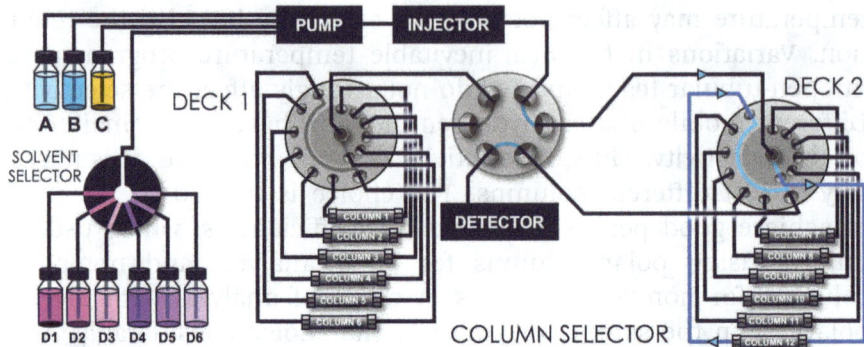

Figure 10.9 Schematic example of a setup for the automatic screening of selectivity provided by multiple combinations of stationary phases (columns), mobile phases (modifiers), and possibly temperatures.

usually optimized. Often, method development in LC starts with the RPLC column with the dimensions and from the manufacturer with which the laboratory is comfortable. As much as there is to say for such an approach, it may not yield the best results, and if a method is meant to be used many times, for example, when drug development reaches the clinical trial stages, the cumulative time and consumables used may add up to significant costs of a sub-optimal method. Therefore, especially in (pharmaceutical) industry, instrumentation and protocols have been developed to automatically screen a number of different columns in search of the best possible selectivity. A schematic illustration of a setup for (column) selectivity screening is provided in Figure 10.9.

One example of a systematic selectivity-screening strategy has been described by Hetzel *et al.*, who screened 20 different columns (mostly alkyl-modified RP columns, but also some more polar or mixed-mode columns) with two different modifiers (methanol and acetonitrile) and at two different temperatures (30 °C and 50 °C).[2] Coverage of the chromatogram was used as one of the criteria, but the comparison is inevitably incomplete because, in practice, coverage is easily increased by adapting the range (initial and final composition of the gradient).

There has been considerable attention in the scientific literature for selectivity screening in SFC. Several reasons for this can be identified. The column selectivity plays a very important role in SFC, as the stationary phase tends to be more polar than the CO_2-based mobile phase. This offers possibilities for relatively strong, selective

interactions, more than in, for example, RPLC. In addition, arguably, the most successful application domain for SFC is in chiral
separations, relying on chiral selectors that are immobilized on
the stationary surface. Finally, SFC separations, and re-equilibration
of SFC columns in the case of gradient elution, can be very fast,
allowing very rapid column-screening protocols.[5]

Several screening protocols that have been developed and
employed in the pharmaceutical industry have been described in
the scientific literature. Losacco *et al.* from Merck (Rahway, NJ, USA)
describe a protocol in which they studied ten different columns and
five different modifiers (each added to carbon dioxide) for a total of
50 different combinations.[6] Five columns were run in parallel, with a
five-way splitter installed after the injection valve. Five UV detectors
were used simultaneously, one for each column. The modifier pump
was equipped with a selection valve, allowing a choice between five
different modifiers. Testing five columns (all 250 × 4.6 μm i.d.; 5 μm
fully porous particles) with one modifier could be performed in 10
min. Testing five columns with five different modifiers (25 combinations) took 50 min. Only one gradient-elution experiment was
performed with each column-modifier combination, which would
not allow for determining the optimum isocratic composition. The
latter would be relevant since the goal of the screening effort was a
subsequent upscaling to preparative separations.

In an LC study from the same location, a diverse set of 14 chiral-
selective columns was installed in two column ovens and screened
successively with nine mobile-phase combinations and just one set of
detectors (DAD and MS).[7] The entire screening process took 6 h for a
pair of enantiomers. In this case, subsequent interpretive optimization
(using quadratic two-parameter retention models) was performed on
the most promising candidate columns to identify the best isocratic
composition and temperature for the separation.

Lin *et al.*[8] have described an extensive study into selectivity-screening strategies for drug compounds with single or multiple chiral
centres. They compared different small-particle columns (3 μm
vs. sub-2 μm particles), separation modes (isocratic *vs.* gradient),
and separation techniques (SFC *vs.* LC). On the latter subject, LC
was performed in reversed-phase mode. Since SFC is essentially
a normal-phase technique (see Module 6.2 and Figure 10.9), the
comparison of the two techniques was not on equal terms. Eight
commercial columns were tested. All of these were polysaccharide
derivatives, which are the most common and most successful types
of chiral stationary phases (CSPs). Four of the columns tested

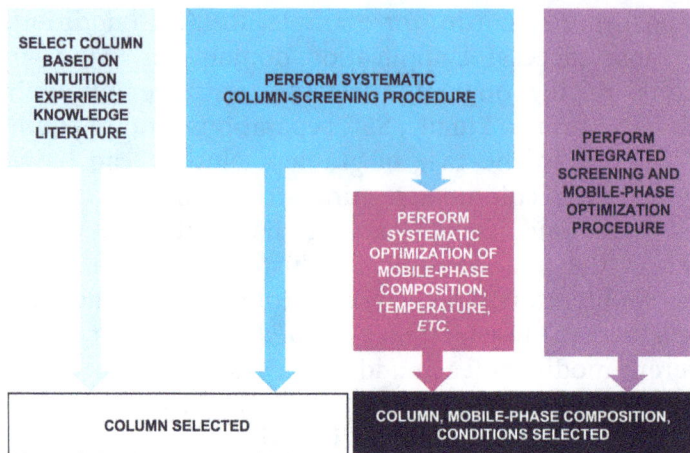

Figure 10.10 Summary of approaches that lead to the selection of an LC or SFC column.

featured a polysaccharide derivative that was physically coated on silica particles ("coated" CSPs), whereas in the other four columns, the polysaccharide was covalently bonded to the packing material ("immobilized" CSPs). The authors identified small sets of complementary columns that together yielded a very high chance of success for a large selection of chiral analytes, with success being identified as yielding "optimizable resolution" ($R_s \geq 0.5$). The separate optimization stage entailed establishing the best possible isocratic mobile-phase composition and temperature, and eventually the column length. RPLC was found to outperform SFC for the separation of analytes with multiple chiral centres.

Figure 10.10 provides a schematic overview of approaches that lead to the selection of a column for an LC or SFC method. The first option (light blue box on the left) is that columns are selected based on intuition, or, more scientifically, based on knowledge and experience of the analyst with related samples. In addition, literature reports may provide indications of suitable columns. The second option is to make use of a systematic column-screening procedure, such as those discussed in this module (large blue box at the top centre). This will also lead to the selection of a column. Several published approaches are two-stage processes in which subsequent (selectivity) optimization is performed on the most promising columns. Ideally, these two steps are integrated (purple box), for example, by performing two different gradients (of different durations; see Section 3.17.2), possibly at two different temperatures and making a rapid prediction of the attainable resolution on each column before making a selection.

10.3 Chromatographic Response Functions (M)

We can distil from the discussions in Sections 10.1.3 and 10.1.4 that there is a clear need for optimization to be guided by a clear objective in the form of a mathematical expression that quantifies the desirable outcome. This is especially important when computational methods are used for optimization. This field has received attention from various disciplines (including computer science, engineering, and analytical chemistry), and different terms are used in the literature for such mathematical expressions, including **chromatographic response functions, optimization criteria, objective functions,** and chromatographic optimization functions.

10.3.1 Elemental Criteria

Chromatographic response functions quantify the quality of a separation. The building blocks of such response functions are **elemental criteria** such as the resolution. In this section, we will revisit the resolution function and discuss other elementary criteria.

10.3.1.1 Resolution

The resolution was introduced in Section 1.5.1. It quantifies the degree of separation between two peaks and thus embodies the aim of separation science. The resolution is one of the most important elementary criteria. Eqn (10.2) (previously (eqn (1.35))) expresses the resolution ($R_{S,j,i}$) between two peaks (i and j in order of elution) in terms of retention times (t_R) and band widths (in time units, σ_t)

$$R_{S,j,i} = \frac{t_{R,j} - t_{R,i}}{2(\sigma_{t,i} + \sigma_{t,j})} \qquad (10.2)$$

We also learned in Section 1.5.1 that, by assuming the chromatographic peaks to be of Gaussian shape with equal plate counts N, eqn (10.2) can be expressed in terms of selectivity ($\alpha_{j,i} = k_j/k_i$), average retention factor ($\bar{k} = (k_i + k_j)/2$), and efficiency (N).

$$R_{S,j,i} = \frac{\alpha_{j,i} - 1}{\alpha_{j,i} + 1} \cdot \frac{\bar{k}}{1 + \bar{k}} \cdot \frac{\sqrt{N}}{2} \tag{10.3}$$

Eqn (10.3) relates the extent of separation to fundamental chromatographic parameters. While such a criterion appears of great use from a method-development perspective, there are several characteristics that may render the resolution less useful.

One general complication with the resolution is its reliance on accurate determination of peak widths. This is challenging and less robust in cases of co-elution, when deconvolution strategies (Section 9.12.3) may be required for this purpose. Eqn (10.3) allows the peak width to be approximated through the plate number (e.g. from independent measurements on well-separated analytes). However, this approach may lead to inaccuracies when peaks are not symmetrical.

Another challenging issue is that the resolution, as defined above, is independent of the relative heights of the two peaks, while this is not true for the visually observed resolution. To understand this, it is important to realize that, in principle, the peak width at half height is independent of concentration, as long as non-ideal effects (e.g. overloading) are absent. However, when, in a partly resolved pair, one peak is significantly smaller than the other one, the width of the larger peak is significantly larger at the half height of the smaller peak. Consequently, the extent of co-elution (peak overlap) will be more severe for the smaller peak than suggested by the resolution value. This is illustrated in Figure 10.11A where two slightly tailing peaks are shown with $R_{S,j,i} = 1.0$ and equal heights (i.e. $h_i = h_j$). In Figure 10.11B, the resolution according to eqn (10.2) is the same, but the height of peak i is 10 times larger than that of peak j (i.e. $h_i = 10h_j$). The apparent overlap is much more severe as the tail of peak i inflates the area of peak j.

To compensate for this, eqn (10.2) can be adjusted to be applied at $h = 0.135h_i$ of each peak, providing useful resolution values for varying peak heights or areas. Within a peak pair, each peak then has its own **peak-height-corrected resolution** value $^iR'_S$. For peak i in pair i,j, this yields[9]

$$^iR'_{S,j,i} = \frac{t_{R,j} - t_{R,i}}{2\sigma_i + \sigma_j\sqrt{4 + 2\ln\left(\dfrac{h_j}{h_j}\right)}} \tag{10.4}$$

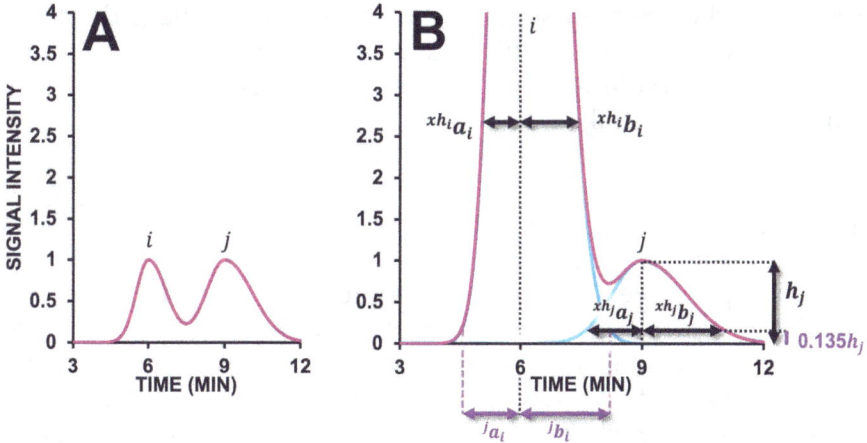

Figure 10.11 Two peaks at a resolution of $R_{S,i,j} = 1.0$ with a relative height ratio of (A) $h_i = h_j$ and (B) $h_i = 10h_j$. Additional details for the peak asymmetry are depicted in panel B; see the text for clarification. The indicated peak widths a and b are taken at 13.5% of the height of the respective peak ($x = 0.135$).

Note that when the two peaks are equal in height (*i.e.* $h_i = h_j$), eqn (10.4) reduces to eqn (10.2). Eqn (10.4) can also be extended by introducing the plate number, using either separate values N_i and N_j for each peak, or by assuming $N_i = N_j = N$. For peak i, this yields[9]

$$^iR'_{S,j,i} = \frac{(t_{R,j} - t_{R,i})\sqrt{N_i \cdot N_j}}{2t_{R,i}\sqrt{N_j} + t_{R,j}\sqrt{N_i}\sqrt{4 + 2\ln\frac{h_j}{h_j}}} \approx \frac{(t_{R,j} - t_{R,i})\sqrt{N}}{2t_{R,i} + t_{R,j}\sqrt{4 + 2\ln\frac{h_j}{h_j}}} \quad (10.5a)$$

and for peak j, this yields

$$^iR'_{S,j,i} = \frac{(t_{R,j} - t_{R,i})\sqrt{N_i \cdot N_j}}{2t_{R,j}\sqrt{N_i} + t_{R,i}\sqrt{N_j}\sqrt{4 + 2\ln\frac{h_j}{h_j}}} \approx \frac{(t_{R,j} - t_{R,i})\sqrt{N}}{2t_{R,j} + t_{R,i}\sqrt{4 + 2\ln\frac{h_j}{h_j}}} \quad (10.5b)$$

Eqn (10.5a) and eqn (10.5b) show that the resolution is always the largest for the largest peak, which is in agreement with the observation that it is more prominently separated from its smaller neighbour. The extent to which the area of the larger peak experiences peak overlap is lower or, in other words, the peak purity is higher. Note that each peak in a chromatogram now has two resolution values, one describing its separation from the previous peak and one describing the extent of separation from the next peak. If the small peak is lower than 0.135 times the height of the large peak it does not affect the resolution of the latter. In that case the square root function should be taken to be zero.

A further complicating factor in using resolution values is that eqn (10.3) is valid only for Gaussian peaks. While peaks ideally meet this requirement, in practice, they rarely do. Thus, eqn (10.3) yields only approximate values for resolution. However, it reveals trends on what to expect when changing retention, selectivity, and efficiency. eqn (10.2) and (10.4) are valid for symmetrical peaks that are of equal height or that differ in height, respectively. Eqn (10.5a) and eqn (10.5b) are valid for Gaussian peaks that differ in height.

For asymmetric peaks, the width at half-height does not sufficiently capture the peak width as it pertains to resolution. The **asymmetry factor** from Chapter 9 (eqn 9.154) can be used to describe the asymmetry of a peak in terms of the ratio of the widths of the front (^{x}a) and tail (^{x}b) of the peak at the fractional height $x \cdot h$,

$$^{xh}A_s = \frac{^{xh}b}{^{xh}a} \tag{10.6}$$

Compared to (eqn 9.154), eqn (10.6) contains an additional clarification of the fractional height $(x \cdot h)$ at which the widths b and a are determined. Often, a value of $x = 0.1$ is used (the asymmetry of 10% of the peak height). In the context of the present discussion, $x = 0.135$ (the height at which a Gaussian peak has a width of 4σ) is more appropriate because eqn (10.2) is based on this "4σ height". The denominator of eqn (10.2) $(2\sigma_i + 2\sigma_j)$ is the average width of peaks i and j at the 4σ height (of each peak) $(2\sigma_j)$, and unit resolution corresponds to a difference in retention times that equals this average 4σ width. To more accurately calculate the resolution for asymmetrical peaks, the present discussion suggests that this should involve the tail of peak i (b_i) and the front of peak j (a_j),

$$R_{s,j,i} = \frac{t_{R,j} - t_{R,i}}{^{xh}b_i + {}^{xh}a_j} \tag{10.7}$$

However, it is not immediately clear at which fractional height (xh) b_i and a_j should be determined. This is illustrated in Figure 10.11B, where the indicated width of peak i at 13.5% of its peak height

$^{0.135h_i}b_i$ is seen to underestimate the width of peak i as experienced by the much smaller peak j. In contrast, peak i is not affected by peak j at 13.5% of its peak height $(0.135h_i)$ because there simply is no peak j at this height. Thus, for the purpose of this discussion, we set our reference peak as j, and our observation height is thus $0.135h_j$. At this height, we can calculate the apparent plate number using

$$^{0.135h_j}N_j = \left(\frac{4 \cdot t_{R,j}}{^{0.135h_j}a_j + {^{0.135h_j}b_j}}\right)^2 \tag{10.8}$$

Eqn (10.8) is analogous to the normal computation of the plate number. However, this approach changes when we calculate the apparent plate number of one peak from the perspective of another peak. From the perspective of the lower second peak j (*i.e.* at a height of $0.135h_j$ from the baseline), the apparent width of the higher first peak is

$$^{0.135h_j}N_i = \left(\frac{4 \cdot t_{R,i}}{^{0.135h_j}a_i + {^{0.135h_j}b_i}}\right)^2 \tag{10.9}$$

If we combine eqn (10.6–10.8), we obtain

$$^jR_{s,j,i} = \frac{(t_{R,j} - t_{R,i})(1 + {^jA_{s,j}})(1 + {^jA_{s,i}})\sqrt{{^jN_i} \cdot {^jN_j}}}{4\,^jA_{s,i} \cdot t_{R,i}(1 + {^jA_{s,j}})\sqrt{{^jN_j}} + 4t_{R,j}(1 + {^jA_{s,i}})\sqrt{{^jN_i}}} \tag{10.10}$$

In practice, it is more customary for (automated) peak analysis to be conducted for individual peaks. This means that only $^iN_i = N_i$, $^jN_j = N_j$, $^iA_{s,i} = A_{s,i}$ and $^jA_{s,j} = A_{s,j}$ are available in contemporary workflows. Under the assumption that peak asymmetry is independent of the relative peak height, we can then approximate eqn (10.10) using these more accessible values. This yields

$$^jR_{s,j,i} = \frac{(t_{R,j} - t_{R,i})(1 + A_{s,i})(1 + A_{s,j})\sqrt{N_i \cdot N_j}}{4t_{R,i}A_{s,i}(1 + A_{s,j})\sqrt{N_j}\sqrt{1 + 0.5\ln\left(\dfrac{h_j}{h_j}\right)} + 4t_{R,j}(1 + A_{s,i})\sqrt{N_i}} \tag{10.11}$$

or if we assume that $N_j = N_i = N$ and $A_{s,j} = A_{s,i} = A_s$,

$$^{j}R_{s,j,i} = \frac{\left(t_{R,j} - t_{R,i}\right)\left(1 + A_s\right)\sqrt{N}}{4A_s t_{R,i}\sqrt{1 + 0.5\ln\left(\dfrac{h_j}{h_j}\right)} + 4t_{R,j}} \tag{10.12}$$

The above resolution equations show a selection of many possible corrections for various cases.[9] In general, eqn (10.10) can be expected to be the most accurate, provided that accurate values of the parameters can be obtained. This requires a robust data-analysis strategy, such as a peak-deconvolution approach, the performance of which is likely to hinge on the severity of co-elution in the chromatogram. For automated interpretive optimization, eqn (10.12) offers the most convenient route, under the assumption that plate numbers and peak asymmetries are constant throughout the chromatogram. If this is not a reasonable assumption, eqn (10.11) is recommended. In the event that peaks are symmetrical but have different areas, eqn (10.5) is best suited.

10.3.1.2 Peak-to-valley Ratio

A very different method to characterize the separation between two peaks is by considering the so-called **peak-to-valley ratio**. Here, the depth of the valley between the two (co-eluting) peaks is considered and related to a measure of peak height. Several approaches have been proposed, some of which are depicted in Figure 10.12.

We see in Figure 10.12 that some of these methods consider the interpolated height between the two peaks, as depicted by the pink solid line. One method compares the depth of the valley by considering the distance between the saddle point of the peaks (at t_v) and the interpolated height (at t_v) as f, with the height of the interpolated peak relative to the baseline, g.

$$P = \frac{f}{g} \tag{10.13}$$

A very similar approach aligns the depth of the valley precisely in the middle of the interpolated peak-height line (*i.e.* at t_{mid}), yielding

$$P_m = \frac{f_m}{g_m} \tag{10.14}$$

The two methods (P and P_m) will yield similar results for Gaussian peaks, even for peaks that differ substantially in height. Both P and P_m range from 0 (no valley), to 1 (baseline separation). In contrast, the resolution criteria discussed in Section 10.3.1.1 are not necessarily equal to zero if no valley is observed. Resolution (R_s-type) criteria are better at describing marginal separations.

A third method computes a peak-to-valley ratio for each peak by relating the distance between the saddle point and the baseline to the peak height.

$$P_{v,i} = 1 - \frac{v}{h_i} \qquad (10.15)$$

In contrast to P and P_m, the $P_{v,i}$ criterion has different values for each peak in a pair, and each peak (except the first and last one in the chromatogram) has two $P_{v,i}$ values, one describing its separation from the preceding peak and another describing its separation from the following peak. Similar to the computation of the resolution criteria for individual peaks (eqn (10.4), eqn (10.5a), eqn (10.5b), and eqn (10.10–10.12)), the peak-to-valley ratio according to eqn (10.15) will be largest for the larger peak in a pair of peaks of unequal heights.

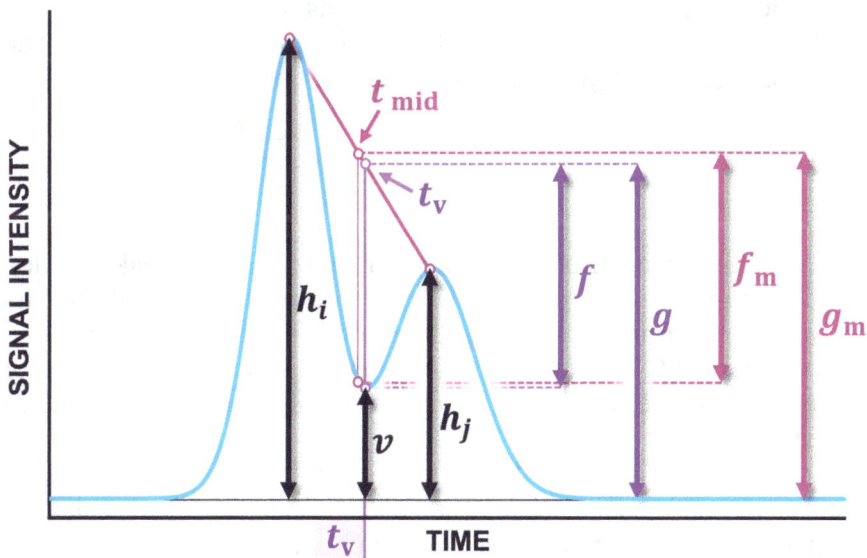

Figure 10.12 Measures used to compute the different peak-to-valley or valley-to-peak ratios.

An important aspect of all peak-to-valley ratios is that they automatically capture any peak asymmetry, and thus – unlike for the resolution criterion – no complicated corrections are needed. A correction for noise may be required, however, as proposed by Wegscheider *et al.*, who modified the peak-to-valley ratio as follows:[10]

$$P = \frac{f}{g + 2N_{pp}} \qquad (10.16)$$

where N_{pp} is the peak-to-peak noise (see Section 9.11.2).

For Gaussian peaks of similar width, an equation can be derived that relates the peak-to-valley ratio to the resolution (see ref. 11, p. 122), *i.e.*

$$P = 1 - 2e^{-2R_{s,i,j}{}^2} \qquad (10.17)$$

Below $R_{s,j,i} \approx 0.59$, this equation is not valid (it yields negative values for P) because there is no valley for such marginally resolved peaks.

10.3.1.3   Fractional Peak Overlap

The fractional peak overlap compares, as its name reveals, the overlap of a peak of interest (A_n) with that of its neighbours. The **fractional peak overlap**, FO, is defined as

$$FO = \frac{A_n - A_{n,n-1} - A_{n,n+1}}{A_n} \qquad (10.18)$$

and is illustrated in Figure 10.13. The fractional peak overlap provides a good indication of the degree of purity of a peak, but its accurate determination strongly hinges on the successful deconvolution of the peaks. This is particularly challenging when the shape of the peaks and the number of co-eluting analytes are unknown (see Section 9.12.3).

Back in 1960, Giddings developed an approximate equation that relates the peak purity (S_{pp}, equal to 1 minus the fractional peak overlap between two peaks) to common parameters,[12] *i.e.*

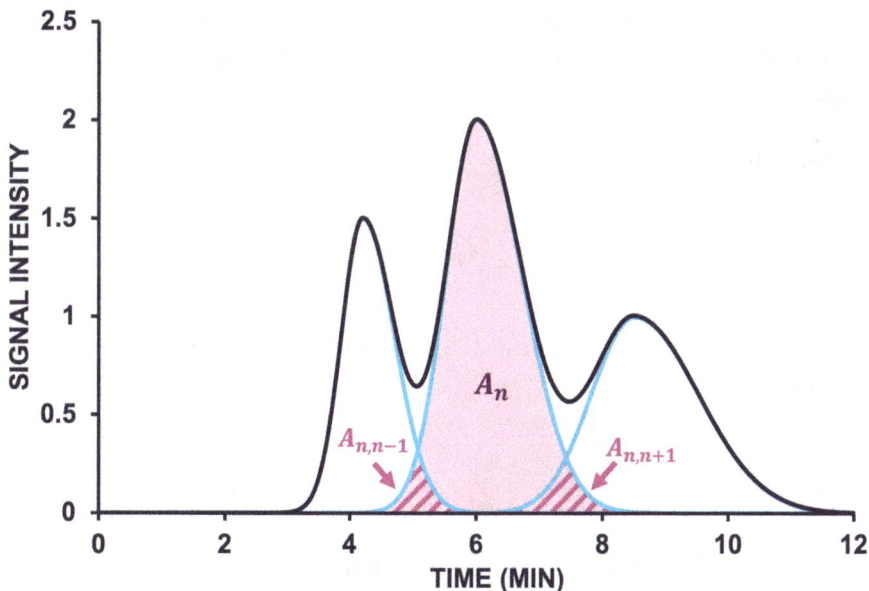

Figure 10.13 Schematic depiction of the fractional peak overlap criterion.

$$S_{pp} \approx 1 - e^{-2F_{j,i}} \qquad (10.19)$$

where

$$F_{j,i} = \frac{\left(t_{R,j} - t_{R,i}\right)^2}{8\left(\sigma_{t,i}^2 + \sigma_{t,j}^2\right)} \qquad (10.20)$$

which is slightly different from $R_{S,j,i}$, which is now a much-more-accepted resolution criterion.

10.3.1.4 Separation Factor

Through eqn (10.3), the resolution ($R_{S,j,i}$) of two Gaussian peaks can be related to fundamental chromatographic parameters, *i.e.* the retention factor (k), the selectivity (α), and the plate number (N). However, it may not always be practical or advisable to optimize all these three parameters simultaneously. For instance, retention and selectivity (k, α) may be optimized by varying the mobile-phase composition (and possibly the temperature) on a given column (*e.g.* from Figure 10.7A to B). After the selectivity-optimization stage, a required number of plates can be established, and the most efficient way to meet this requirement (*e.g.* the column dimensions and flow rate) can be determined in a system-optimization stage (*e.g.* from Figure 10.7B to C).

A practical simplification of the resolution factor is the **separation factor**, which is defined as

$$S_{j,i} = \frac{k_j - k_i}{k_i + k_j + 2} \tag{10.21}$$

and can be further simplified using $k = (t_R - t_0)/t_0$ (eqn (1.13)) to obtain

$$S_{j,i} = \frac{t_{R,j} - t_{R,i}}{t_{R,i} + t_{R,j}} \tag{10.22}$$

Eqn (10.22) allows the separation factor to be directly obtained from the chromatogram, without the need to know t_0 or any assessment of the peak widths. The caveat is the assumption of Gaussian peaks and that the plate count is constant throughout the chromatogram. In addition, $S_{j,i}$ is intended for chromatography under constant (isothermal, isocratic) conditions. Eqn (10.22) can be linked to the number of required plates using eqn (10.2). This yields, if N is constant,

$$R_{S,j,i} = \frac{t_{R,j} - t_{R,i}}{2(\sigma_{t,i} + \sigma_{t,j})} = \frac{t_{R,j} - t_{R,i}}{t_{R,i} + t_{R,j}} \cdot \frac{\sqrt{N}}{2} = S_{j,i} \cdot \frac{\sqrt{N}}{2} \tag{10.23}$$

From this, it follows that to separate a pair of peaks with a required resolution $R_{s,req}$, the required number of plates is

$$N_{req} = \frac{4R_{s,req}^2}{S_{j,i}^2} \tag{10.24}$$

For $R_{s,req} = 1.5$ this yields

$$[N_{req}]_{R_{s,req} = 1.5} = \frac{9}{S_{j,i}^2} \tag{10.25}$$

and for $R_{S,req} = 1$, this yields

$$[N_{req}]_{R_{s,req}=1} = \frac{4}{S_{j,i}^2} \quad (10.26)$$

These equations illustrate that there is a very simple relationship between the separation factor S and the required number of plates under the assumption of Gaussian peaks with equal plate numbers. If a selectivity factor of $S_{j,i} = 0.03$ is achieved, 10 000 plates are needed to realize $R_s = 1.5$. Minimum response functions (see Section 10.3.2.4), such as the minimum resolution ($R_{s,min}$) value or the minimum separation factor (S_{min}), also known as the resolution (or separation factor) of the critical pair ($R_{s,crit}$ or S_{crit}, respectively), can be related directly to system requirements.

10.3.1.5 Comparison

In this section, we attempt to summarize the discussion on elemental criteria by considering the strengths and weaknesses of each approach (see Table 10.3). The resolution (R_a) is clearly defined and well-rooted in chromatographic theory and practice. However, to describe the actual quality of separation, rather complex corrections are needed to reflect the influence of peak properties, such as asymmetry and the ratio of peak heights. We have discussed how resolution can be adapted, but the modified equations cannot always be applied easily and reliably.

The fractional peak overlap (FO) provides an excellent measure of the degree of separation ("peak purity"), but it also raises challenges related to the accurate deconvolution of chromatographic peaks. The deconvolution itself is not highly challenging, but the process hinges on an accurate determination of the number of peaks in a

Table 10.3 Comparison of different elemental criteria.

Property	R_s	P	P_m	P_v	FO	S
Affected by peak asymmetry	Yes	No	No	No	No	Yes
Affected by ratio of peak heights	Yes	Yes	Yes	No	No	Yes
Representative of separation	Varies	Yes[a]	Yes[a]	Yes[a]	Best	Varies
Transferability	Good	Poor	Poor	Poor	Indirect[b]	Good[c]
Robust calculation	Varies	Varies	Varies	Varies	Hard	Easy
Equation	10.2	10.13	10.14	10.15	10.18	10.22

[a]Except for marginal separations (R_s smaller than about 0.5), where no valley can be discerned.
[b]If FO is determined by peak deconvolution and the effects of changes in conditions on the peak-shape parameters are known.
[c]Not intended for use with programmed elution (temperature-programmed GC; gradient-elution LC).

cluster and on the selection of distribution functions that accurately describe the shapes of the peaks.

The separation factor (S) does not require complicated data analysis or corrections, but it is limited to non-programmed (isothermal, isocratic) elution, symmetrical (Gaussian) peaks and a constant plate number. It offers a quick indication of the required plate number, which aids in the transfer of methods to columns of different dimensions.

Peak-to-valley ratios (P, P_m, P_v) inherently take into account the shape of the entire peak (including its width and symmetry), and they suffice to express the extent of separation of the peaks above a certain degree of separation, where a valley becomes visible. Usually, P, P_m, and P_v can be readily determined from a chromatogram. However, it is difficult to predict the effects of changes in operating conditions or column dimensions so that the transferability is quite limited.

Ultimately, Table 10.3 shows that there is no perfect ("one-size-fits-all") elemental criterion and that the selection of a suitable criterion is a case-by-case decision depending on the available information and the optimization process. Fundamental criteria that are rooted in chromatographic theory, such as resolution (R_s) and the separation factor (S), depend on peak properties (*i.e.* peak shape, relative heights or areas, plate numbers), whereas pragmatic criteria, such as P, P_m, and P_v, are more difficult to connect with chromatographic principles and to transfer to other conditions. Therefore, resolution (R_s) and separation factors (S) can be more easily combined with interpretative optimization methods, in which retention factors (and often peak widths and sometimes peak asymmetries) can readily be obtained from the simulated data. Because such optimization procedures are almost invariably computer-aided, using more complex equations is not prohibitive. Resolution (or separation factors) may also guide optimization procedures in regions where severe co-elution occurs (*i.e.* below a resolution of $R_s = 0.6$).

10.3.2   Response Function Operations

The criteria described in Section 10.3.1 only assess the separation of one or two pairs of peaks. Of course, chromatograms may easily contain tens or hundreds of peaks, and comprehensive two-dimensional chromatograms may feature thousands of peaks. To assess the quality of entire chromatograms, the elemental criteria and other aspects are combined in what are commonly known as **chromatographic response functions**. Such a CRF aims to characterize the overall quality of the separation of all peaks throughout the entire chromatogram. Thus, CRFs are of key interest in optimization workflows.

A variety of operations are used to translate the available chromatographic information into a CRF. It is not within the scope of this module to review all possible response-function operations. For this, readers are referred elsewhere.[13,14] Here, we will limit ourselves to some of the most important types of operations.

10.3.2.1 Sum and Product Response Functions

The criteria described in Section 10.3.1 consider the separation of two or, in the case of the fractional peak overlap (FO), perhaps three peaks, and it is difficult to effectively scale such criteria to characterize the separation in an entire chromatogram. Among the simplest operations are summing or multiplying all the assessed criteria across the chromatogram. For the resolution, this results in

$$\text{CRF}_{\text{sr}} = \sum_{i=1}^{n-1} \left(R_{s,i+1,i} \right) \tag{10.27}$$

where n is the number of observed peaks. Here, the chromatographic response function (CRF_{sr}) is the sum (Σ) of all resolution factors for all (*i.e.* $n - 1$) pairs of peaks. Instead of the resolution, we can also take the sum of the values of another elemental criterion, for example,

$$\text{CRF}_{\text{sp}} = \sum_{i=1}^{n-1} \left(P_{i+1,i} \right) \tag{10.28}$$

The sum of fractional overlaps can be approximated using Giddings' equation for peak purity (eqn (1.19)) to obtain

$$\text{CRF}_{\text{sspp}} = \sum_{i=1}^{n-1} \left(1 - e^{-2F_{j,i}} \right) \tag{10.29}$$

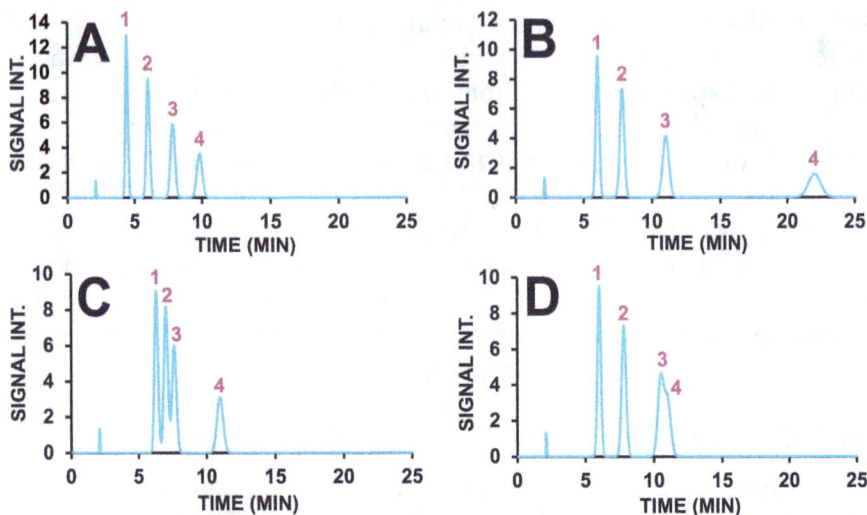

Figure 10.14 Chromatograms with four distinct peaks with different resolutions between them. Summing the resolution yields ΣR_S = 9.75 for the separation in A, and ΣR_S = 15.53 for B. See Table 10.4 for other relevant resolution values.

To enhance legibility, we often write abbreviated versions of the criteria, such as ΣR_S for CRF_{sr}, ΣP for CRF_{sp}, or ΣS_{pp} for CRF_{sspp}.

The main challenge with summing the criteria becomes apparent when we apply eqn (10.27) to the four different separations of four peaks shown in Figure 10.14. The corresponding resulting values are shown in Table 10.4. We observe that ΣR_S = 9.75 for the separation in Figure 10.14A and ΣR_S = 15.53 for Figure 10.14B. The latter is numerically superior, but chromatographers will agree that the separation shown in Figure 10.14A is more time-efficient. Other issues are encountered when taking the product (Π) of the different resolution factors, which is equivalent to the sum of the logarithms and formulated as

$$CRF_{pr} = \prod_{i=1}^{n-1} \left(R_{s,i+1,i} \right) = e^{\sum_{i=1}^{n-1} \ln\left(R_{s,i+1,i} \right)} \tag{10.30}$$

or

$$CRF_{pp} = \prod_{i=1}^{n-1} \left(P_{i+1,i} \right) = e^{\sum_{i=1}^{n-1} \ln\left(P_{i+1,i} \right)} \tag{10.31}$$

Table 10.4 Resolution values for the different peak pairs in the chromatograms shown in Figure 10.14. Various ways to describe the resolution for the full separations are also shown. Truncated implies that R_S values are limited to a maximum of 1.5. See the text for discussion.

Chromatogram	$R_{s,1,2}$	$R_{s,2,3}$	$R_{s,3,4}$	ΣR_s	ΠR_s	$\Sigma R_s/(n-1)$	$\Pi R_s/(n-1)$
A – Normal	3.77	3.19	2.78	9.75	33.51	3.25	11.17
A – Truncated	1.50	1.50	1.50	4.50	3.38	1.50	1.13
B – Normal	3.19	4.17	8.16	15.53	108.7	5.18	36.26
B – Truncated	1.50	1.50	1.50	4.50	3.38	1.50	1.13
C – Normal	1.29	1.01	4.48	6.77	5.81	2.26	1.94
C – Truncated	1.29	1.01	1.50	3.80	1.95	1.27	0.65
D – Normal	3.19	3.61	0.57	7.38	6.58	2.46	2.19
D – Truncated	1.50	1.50	0.57	3.57	1.28	1.19	0.43

The product functions are preferred because they can be used when any one of the R_s or P values equals zero, whereas the sum of logarithms cannot. Multiplication of all resolution values results in $\Pi R_S = 33.51$ for Figure 10.14A and $\Pi R_S = 108.7$ for Figure 10.14B. Truncation to a maximum of $R_{s,max} = 1.5$ makes the numbers more realistic. Both ΣR_S and ΠR_S show that, in terms of resolution, the separation in both chromatograms A and B is optimal. $(\Sigma R_S)/(n - 1)$ and, especially, $(\Pi R_S)/(n - 1)$ provide good indications of inadequate resolution in chromatograms C and D.

Similar results would be obtained if we replace R_s by S to obtain ΣS and ΠS. Neither ΣR_S nor ΣS is very suitable to distinguish between the chromatograms in Figures 10.14A and 10.14B, as their values are largely determined by the largest values of the elemental criteria within the chromatogram. Criteria such as ΣP or ΣFO are more aligned with truncated resolution functions because the underlying elemental criteria are by design constrained to values between 0 and 1. A truncated resolution sum has the advantage that the effects of changes in *NN* (for example, due to changes in flow rate or column dimensions) can easily be predicted. This is clearly not possible for ΣP. Product criteria are largely determined by the lowest values. If a value of 0 is obtained for the separation between any two peaks, the overall product value is also equal to zero. ΠR_S or ΠS equals zero if two (or more) peaks fully co-elute (*i.e.* $t_{R,i} = t_{R,j}$). In the case of ΠP, a result of zero is obtained as long as any two peaks do not show a valley.

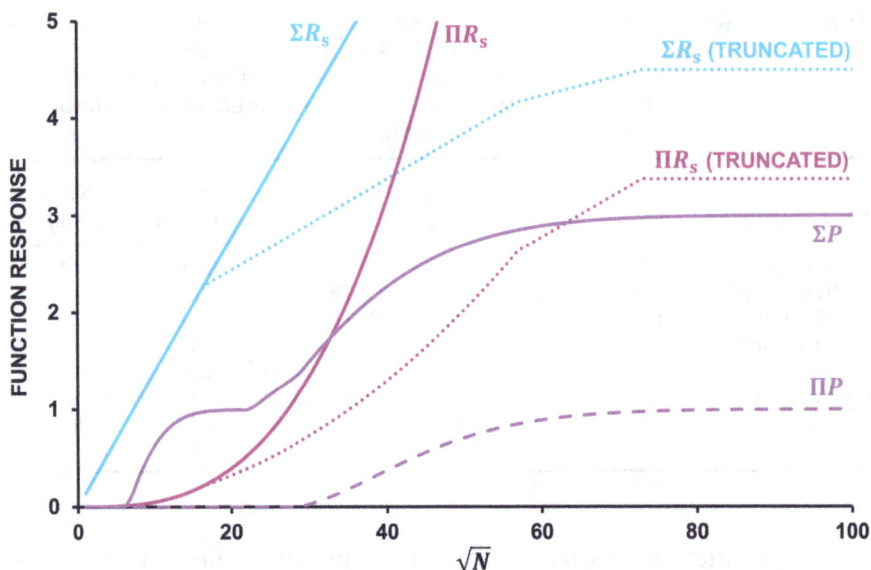

Figure 10.15 Overlay of the response of several chromatographic response functions as a function of $\sqrt{N}$ for the chromatogram shown in Figure 10.14C. Eqn (10.17) was used to estimate P.

It is interesting to study the dependency of CRFs on the column efficiency. Figure 10.15 shows the CRF response for several CRFs plotted against $\sqrt{N}$ for the chromatogram shown in Figure 10.14C. Based on eqn (10.3), we expect ΣR_S to increase linearly with $\sqrt{N}$, which is indeed observed in Figure 10.15. For a chromatogram with four peaks, ΠR_S is expected to increase with $N^{3/2}$. The trend of ΣP is significantly different. As with all other criteria, it starts from the complete overlap of all peaks at $N = 0$. As $\sqrt{N}$ approaches 7, the cluster of the first three peaks starts to become separated from the last peak. The last P value ($P_{4,3}$) starts to increase until ΣP stabilizes at around 1 at $\sqrt{N} \approx 10$. Above $\sqrt{N} = 23$, the first three peaks start to be resolved, and ΣP increases further towards a limiting value of $\Sigma P = 3$ (for three peak pairs) around $\sqrt{N} = 75$. ΠP does not yield any information on the quality of separation until about $\sqrt{N} = 28$ because at least one peak pair is not sufficiently resolved to show a valley ($R_{S,j,i} < 0.59$; see eqn (10.17)). Thereafter, ΠP gradually increases until it reaches a limiting value of 1. In this sense, ΠP may serve as a threshold criterion. Note that plots of ΣS or ΠS would yield a horizontal line in Figure 10.15, as S is independent of N.

In a practical situation, graphs such as Figure 10.15 cannot easily be constructed. If we ignore the possible effects of pressure on retention factors, a limited range of plate heights may be covered by varying the flow rate. Covering a larger range would require studying different columns, which is impractical and futile in the case of transferable criteria, such as R_S.

10.3.2.2 Embedding the Number of Peaks

From the above discussion, we can establish that some of the CRFs treated thus far are strongly affected by the number of peak pairs for which the resolution is computed. For example, if, for chromatogram A, we would only have had the first two peaks, then $\Sigma R_S = 3.77$, whereas for all four peaks, the criterion reads $\Sigma R_S = 9.75$. Three excessively separated peaks with $R_{S,2,1} = 2$ and $R_{S,3,2} = 3$ would yield a higher sum ($\Sigma R_S = 5$) than four equally separated peaks ($R_{S,2,1} = R_{S,3,2} = R_{S,4,3} = 1.5$; $\Sigma R_S = 4.5$). To correct for a possible undesirable effect of the number of observed peaks, n_{obs}, we may use the average resolution, *i.e.*

$$\text{CRF}_{avr} = \overline{R}_s = \frac{\sum_{i=1}^{n_{obs}-1} (R_{S,i+1,i})}{n_{obs}-1} \tag{10.32}$$

or the average P value

$$\text{CRF}_{avp} = \overline{P} = \frac{\sum_{i=1}^{n_{obs}-1} (P_{i+1,i})}{n_{obs}-1} \tag{10.33}$$

At this stage, it is good to note that the number of peaks is itself also a possible CRF

$$\text{CRF}_n = n_{obs} \tag{10.34}$$

Especially when developing a method for separating a very complex sample, the objective of optimization may be to maximize n_{obs}, assuming that no peak splitting effects occur, which would also increase the number of peaks.

Table 10.4 shows the resulting numbers for $\Sigma R_S/(n-1)$ and $\Pi R_S/(n-1)$. We observe a value of $\Sigma R_S/(n-1) = 3.25$ for Figure 10.14A, which is more in line with the appearance of the chromatogram. After correcting for the number of peaks, resolution products

$(\Pi R_S)/(n-1)$ still yield high values, especially for Figure 10.14B, due to the excessive resolution between peaks #3 and #4. As illustrated in Figure 10.15, truncation (*e.g.* $R_{S,max} = 1.5$) may remedy this situation.

10.3.2.3 Normalized Resolution Products

In the previous sections, we have observed that (without truncation) ΣR_S does not reflect the quality of separation very well because it increases with increasing analysis times and because one or more large R_S values may dominate the sum. Using the average resolution (eqn (10.32)) does not mitigate this issue. A chromatogram of four peaks with $R_{S,2,1} = 0.5$, $R_{S,3,2} = 0.5$, and $R_{S,4,3} = 5$, with $\Sigma R_S = 6$ and the average resolution $\overline{R}_S = 2$ is arguably much worse than a chromatogram with $R_{S,2,1} = R_{S,3,2} = R_{S,4,3} = 1.5$, for which $\Sigma R_S = 4.5$ and $\overline{R}_S = 1.5$. It shows ΣR_S (and $\overline{R}_S$) to be more sensitive to changes in the retention factor of the last peak (k_ω) than to changes in the extent of separation. Product criteria are more sensitive to low values of the resolution, as a single occasion of exact co-elution $(R_{S,j,i} = 0)$ renders the entire product zero. For the two above examples, $R_{S,2,1} = R_{S,3,2} = 0.5$, $R_{S,4,3} = 5$, and $R_{S,2,1} = R_{S,3,2} = R_{S,4,3} = 1.5$, the product would amount to 1.25 and 3.38, respectively, indicating the superiority of the second chromatogram. However, if only the last peak is moved in the first chromatogram to a much higher k value, such that $R_{S,4,3} = 14$ or higher, ΠR_S also becomes higher for the first chromatogram than for the second chromatogram.

To correct for this sensitivity to the length of the chromatogram, a normalized resolution product (CRF_{nrp}) can be used, *i.e.*

$$CRF_{nrp} = \prod_{i=1}^{n-1}\left(\frac{R_{S,i,i+1}}{\overline{R}_S}\right) = \left[\prod_{i=1}^{n-1}\left(R_{S,i+1,i}\right)\right]\cdot\left[\frac{\Sigma_{i=1}^{n-1}\left(R_{S,i+1,i}\right)}{n-1}\right]^{(1-n)} \tag{10.35}$$

When we assume Gaussian peaks of equal height and equal plate counts for all peaks, S (eqn (10.21) and (10.22)) is proportional to R_S, so that

$$CRF_{nsp} = \prod_{i=1}^{n-1}\left(\frac{S_{i,i+1}}{\overline{S}}\right) = CRF_{nrp} = \prod_{i=1}^{n-1}\left(\frac{R_{S,i,i+1}}{\overline{R}_S}\right) \tag{10.36}$$

where $\overline{S}$ is the average separation factor.

For our chromatogram with $R_{S,2,1} = R_{S,3,2} = R_{S,4,3} = 1.5$, we find a perfect score of 1 (100%) for CRF_{nrp}, whereas for $R_{S,2,1} = 0.5$, $R_{S,3,2} = 0.5$, and $R_{S,4,3} = 5$, we have $\text{CRF}_{nrp} = 16\%$, and only 3% would be left if we increase $R_{S,4,3}$ to 14. Thus, CRF_{nrp} appears to be a good measure for the distribution of peaks in the chromatogram. It does not differentiate between good and bad average resolutions (nor between short and long retention times). For a chromatogram with $R_{S,2,1} = R_{S,3,2} = R_{S,4,3} = 0.5$, we also find $\text{CRF}_{nrp} = 100\%$.

To ensure that the peaks are not stacking up at higher k values, it has been suggested to include t_0 within the resolution product, as

$$\text{CRF}_{nrp0} = \prod_{i=0}^{n-1}\left(\frac{R_{S,i,i+1}}{\overline{R_S}}\right) = \prod_{i=0}^{n-1}\left(\frac{S_{i,i+1}}{\overline{S}}\right) \tag{10.37}$$

where the average resolution also includes a signal at t_0 ($i = 0$), *i.e.*

$$\overline{R_S} = \frac{\sum_{i=0}^{n-1}\left(R_{S,i+1,i}\right)}{n} \tag{10.38}$$

10.3.2.4 Minimum Response Functions

A completely different operation is to ignore the majority of the separation values and focus exclusively on the smallest value. For instance, in the past, minimum thresholds were set for selectivity screening based on the selectivity α. The CRF then reads

$$\text{CRF}_{sel}: \alpha \geq x \tag{10.39}$$

where x is the minimum threshold. CRF_{sel} yields a binary output: either the criterion is met, or it is not. It could be used to select columns or define regions to be considered for the next stage. It is useful in a sequential optimization workflow, where certain requirements must be fulfilled, after which the next CRF can be employed. Selectivity is not a perfect threshold criterion because a high selectivity at very low k values may not allow separation. A minimum value for the resolution is more sensible. As explained in Section 10.3.1.4, the separation factor S can be directly related to the required number of plates. Setting a requirement of $S_{j,i} \geq 0.03$ allows baseline separation ($R_S = 1.5$) to be obtained with 10 000 theoretical plates. Unfortunately, using $S_{j,i}$ is restricted to non-programmed

Table 10.5 Chromatographic objectives and corresponding response functions. PE denotes programmed elution (temperature programming in GC, gradient elution in LC); NPE denotes non-programmed elution (isothermal, isocratic).

Objective	On a method-development column	On the final column
Shortest analysis time	Minimal final retention factor (k_ω)	NPE: Minimal final retention time ($t_{R,\omega}$) PE: Shortest program duration (t_G)
Minimum required number of plates (N_{req})	Maximal critical resolution: max($[R_{s,j,i}]$min)	N/A
Acceptable separation	Manageable number of plates required ($N_{req} \leq N_{req,max}$)	Adequate resolution ($[R_{s,j,i}]$min $\geq R_{s,req}$) or peak-to-valley ratio ($[P_{j,i}]$min $\geq P_{req}$)
Best distribution of peaks	Normalized resolution product (eqn (10.37))	
Highest sensitivity	NPE: Minimal final retention factor (k_ω) PE: Maximum injectable amount max ($c_{inj}V_{inj}$)	Maximum injectable amount max ($c_{inj}V_{inj}$) Minimal dilution factor (c_{inj}/c_{peak})
Most resolved analytes[a]	Maximal number of detected peaks (n_{obs})	Maximal number of observed peaks (n_{obs})

[a]Especially for non-target analysis.

(isothermal, isocratic) analyses and it assumes Gaussian peaks of similar heights.

10.3.2.5 Response Functions vs. Separation Objectives

Table 10.5 lists a number of sensible goals for chromatographers and their translation into response functions. Some of the functions introduced above appear in this table, while others, such as ΣR_S, do not find a place in the table. The goals of chromatographic method development and optimization are seldom as clear-cut as the objectives listed in the table. This will be discussed in Sections 10.3.3 and 10.3.4.

10.3.2.6 Selectivity Optimization vs. System Optimization

In Table 10.5, a distinction is made between a "method-development" column and a "final" column. It is important to make this distinction because of the many parameters that affect chromatographic separations and the enormous complexity that arises from needlessly considering many parameters at the same time. A great deal of effort is being devoted by chromatographers to creating columns and systems with high separation power. Indeed, the results of those efforts have been discussed at length in this book. However, every sample is different and may require a different solution. If a minimum resolution of 1 is achieved on a method-development column, we may want to double the column length (*e.g.* by using two columns in series) to approach $R_S = 1.5$ because resolution increases with $\sqrt{N}$ or with $\sqrt{L}$. However, this may cause an increase in pressure, which in LC may necessitate lower flow rates or switch to larger particles and even longer columns. All of this can be grasped in kinetic plots (see Module 3.16). In all cases, better results (faster analysis, less dilution) can be obtained if greater selectivity can be achieved, *i.e.* if the required number of plates for the separation (N_{req}) is lower.

The same philosophy applies under programmed conditions (temperature programming, gradient elution). The definition of resolution (eqn (10.2)) remains valid (but eqn (10.3) does not). The resolution obtained on a method-development column can be tuned by using a longer or shorter column. The duration of the elution program (t_G) can be scaled by keeping the ratio t_G/t_0 constant. Especially in gradient-elution LC, care should be taken to correct for the dwell time (see Module 3.4).

In the case of non-target separations, system optimization aims at maximizing the peak capacity of the column (or columns, in the case of multidimensional separations) in combination with the operating conditions (*e.g.* flow rate, gradient parameters).

10.3.3   Composite Response Functions

It is often difficult for chromatographers to formulate the exact goals of a method-development or optimization process. This may be because they do not fully know what they want or, more likely, because they want too much at the same time. The simplest dilemma is the desire for the best possible separation in the short-est possible time. Yet, for computer-aided optimization, a clearly defined CRF is highly desirable. The process of combining different objectives into a single CRF will be discussed in this section.

10.3.3.1   Sum Criteria

One common way to combine two objectives is through a weighted sum of individual criteria. For example, resolution and analysis time can be combined into

$$\text{CRF}_{\text{r\&t}} = w_{\text{r}} \sum_{i=1}^{n-1} \left(R_{\text{s},i+1,i} \right) - w_{\text{t}} t_{\omega} \tag{10.40}$$

where w_{r} and w_{t} are weighting factors for the resolution and time factors, respectively, and t_{ω} is the retention time of the last peak. There are many variations of this approach. For example, peak-to-valley ratios (P) may be used instead of resolution (R_{s}), or the retention factor of the last peak (k_{ω}) may be used instead of t_{ω}. In addition, another factor may be included, such as the number of observed peaks (n_{obs}), with its own weighting factor (w_{p}). The sum of all resolution values can be replaced by a better resolution function (see Section 10.3.2.5). The difficulty with such a chromatographic response function is that it is a typical case of adding apples to oranges. This makes it very difficult to establish sensible weighting factors to avoid an outcome with a very short retention time but too little resolution, or one with a high resolution but a long retention time. It is not really possible to link a sum function as in eqn (10.40) to the objectives listed in Table 10.5.

In a misleading– but not uncommon – version of summation criteria, desired or required resolution values (*e.g.* $R_{s,req}$) and maximal retention times are added (t_{max}). Eqn (10.40) may then be modified to

$$
\begin{aligned}
\mathrm{CRF}'_{r\&t} &= w'_R \cdot \sum_{i=1}^{n-1} \left(\frac{R_{s,i+1,i}}{R_{s,req}} \right) + w'_t (t_{max} - t_\omega) \\
&= \frac{w'_r(n-1)}{R_{s,req}} \sum_{i-1}^{n-1} (R_{s,i+1,i}) - w'_t \cdot t_\omega + w'_t t_{max} \\
&= w''_r \sum_{i=1}^{n-1} (R_{s,i+1,i}) - w'_t \cdot t_\omega + c_0
\end{aligned}
\tag{10.41}
$$

where w''_r is a modified weighting factor and c_0 is a constant, the value of which moves the entire response surface up or down, without affecting the location of the optima. Therefore, the appearance of a required resolution value and a maximum analysis time are illusions in such a criterion.

Product criteria have also been used, such as

$$
\mathrm{CRF}_{r\times t} = \frac{\Sigma_{i=1}^{n-1} (R_{s,i+1,i})}{t_\omega^{w_{te}}}
\tag{10.42}
$$

where the exponent w_{te} takes the place of the weighting factor. However, such $\mathrm{CRF}_{r\times t}$ response factors have the same apples-and-oranges problem as the $\mathrm{CRF}_{r\&t}$-type functions.

10.3.3.2 Product Criteria

Product criteria generally make more sense than sum criteria. Reasonable examples of product criteria are

$$
\mathrm{CRF}_{tcs} = \left(\frac{1}{k_\omega} \right)^{a_s} \prod_{i=0}^{n-1} \left(\frac{S_{i,i+1}}{S} \right)
\tag{10.43}
$$

for optimizing a non-programmed (isothermal, isocratic) separation with (near-)Gaussian peaks on a method-development column,

$$
\mathrm{CRF}_{tcr} = \left(\frac{1}{k_\omega} \right)^{a_r} \prod_{i=0}^{n-1} \left(\frac{R_{s,i,i+1}}{R_s} \right)
\tag{10.44}
$$

for optimizing a programmed separation on a method-development column, or

$$\text{CRF}_{\text{tcp}} = \left(\frac{1}{k_\omega}\right)^{a_p} \prod_{i=0}^{n-1}\left(\frac{P_{i,i+1}}{\overline{P}}\right) \tag{10.45}$$

for optimizing on a final column. However, the balance between the time factor (represented by the retention factor of the last peak, k_ω) and the resolution factor, which is expressed in the exponent a (a_s, a_r, or a_p), is still quite arbitrary.

10.3.3.3 *Threshold Criteria*

A clearer way to connect the objectives of Table 10.5 to response functions is to work with thresholds. This is most sensible in case optimization occurs on the final column. For example, for separation on the final column the response function

$$\text{CRF}_{\text{thr}}: \ \min k_\omega \cap R_{s,\min} \geq x \tag{10.46}$$

can be clearly linked to two objectives in Table 10.5. It aims for the shortest possible analysis time by seeking the lowest value of k_ω, provided that the resolution is adequate (*e.g.* by setting $x = 1.5$).

In the case of gradient elution on the final column

$$\text{CRF}_{\text{thg}}: \ \min t_G \cap R_{s,\min} \geq x \tag{10.47}$$

could be an appropriate threshold criterion, with t_G representing the duration of the gradient.

Generally, much shorter analyses can be achieved by maximizing $R_{s,\min}$ on a method-development column and then optimizing the column dimensions and operating conditions to achieve the required number of plates as quickly as possible.

10.3.3.4 *Hierarchical Response Functions*

It is also possible to combine two (or more) of the objectives from Table 10.5 if the top criterion absolutely dominates the next one – either by itself or after adding weighting factors. For example, a response function based on the normalized resolution product (eqn (10.37))

$$\mathrm{CRF}_{\mathrm{nrs}} = n_{\mathrm{obs}} + \mathrm{CRF}_{\mathrm{nrp0}} = n_{\mathrm{obs}} + \prod_{i=0}^{n-1}\left(\frac{R_{s,i,i+1}}{R_s}\right) \qquad (10.48)$$

or

$$\mathrm{CRF}_{\mathrm{npv}} = n_{\mathrm{obs}} + \mathrm{CRF}_{\mathrm{npp0}} = n_{\mathrm{obs}} + \prod_{i=0}^{n-1}\left(\frac{P_{i,i+1}}{P}\right) \qquad (10.49)$$

is dominated by n_{obs} because the second term has a theoretical maximum of 1. Thus, a chromatogram with one more peak is always better than one with a better spread of peaks. The latter is encouraged by the second term, which becomes the decimal fraction of the response function.

Peichang and Hongxin proposed a response function that fits into this category,[15]

$$\mathrm{CRF}_{\mathrm{lpc}} = 100\,000 n_{\mathrm{obs}} + 10\,000 P_{\mathrm{min}} + (100 - t_\omega) \qquad (10.50)$$

The weighting factors are chosen such that the resolution term (the lowest peak-to-valley ratio[16] observed in the chromatogram) only plays a role in differentiating between chromatograms with equal numbers of peaks, whereas the analysis time (t_ω in minutes) is considered only when both n_{obs} and P_{min} are equal.

Tyteca and Desmet proposed a CRF that is largely a combination of a hierarchical criterion and a combined time-resolution criterion,[13] *viz.*

$$\mathrm{CRF}_{\mathrm{t\&d}} = n_{\mathrm{obs}} + \frac{1}{t_\omega}\left(\frac{R_{s,\mathrm{min}}}{R_{s,\mathrm{req}}}\right)^2 \qquad (10.51)$$

Under most circumstances, the second term is smaller than unity and n_{obs} takes precedence. However, contemporary fast separations (*e.g.* by ultra-high-pressure liquid chromatography) may lead to high values for the second term. There is also an issue with dimensions. These problems can be alleviated by replacing t_ω in eqn (10.51) with t_ω/t_0.

Niezen and Desmet tried to alleviate the disadvantages of sum functions by making the time penalty adapt to the progress of the optimization process.[14] Their $\mathrm{CRF}_{\mathrm{n\&d}}$ behaves as a hierarchical response function. It is based on the ratio between the number of fully resolved peaks (n_{pure}) and the total number of observed peaks (n_{obs}),

$$\text{CRF}_{n\&d} = \frac{n_{pure}}{n_{obs}} + \frac{10^{2\left(n_{pure}-n_{obs}\right)}}{t_\omega} \tag{10.52}$$

As long as not all peaks are resolved, the second term is equal to $1/(100t_\omega)$ or smaller and the first term dominates. In the region where $n_{obs} = n_{pure}$, the criterion becomes $\text{CRF}_{n\&d} = 1 + 1/t_\omega$ and the objective shifts towards minimization of the analysis time. A peak is considered pure when the resolution with its neighbouring peaks exceeds a certain value. This is akin to setting a threshold resolution value (Section 10.3.3.3).

10.3.3.5 Derringer Desirability Functions

The Derringer desirability function was introduced in the field of chromatographic method development by Bourguignon and Massart.[17] It is an elegant way to combine different objectives into a single CRF. The basic principle is that the effect of each variable on the CRF is scaled between 0 (highly undesirable) and 1 (highly desirable). A relevant example is the retention time, which proved difficult to handle in the sum functions of Section 10.3.3.1, where time and resolution were found to be like adding apples and oranges. For a resolution function that scales between 0 and 1, we may use the normalized resolution product, CRF_{nrp0} (eqn (10.37)). To scale the retention time of the last peak, we may use the following set of equations:[17]

$$\text{Der}_{t_\omega} = \begin{cases} 1 & \text{if } t_\omega \leq t_L \\ 1 - \left(\frac{t_\omega - t_L}{t_H - t_L}\right)^{e_t} & \text{if } t_L \leq t_\omega \leq t_H \\ 0 & \text{if } t_\omega \geq t_H \end{cases} \tag{10.53}$$

This function is illustrated in Figure 10.16A for three values of the exponent e_t, i.e. 0.3, 1, and 3, with a lower analysis time (t_L) of 5 min, below which we are fully satisfied, and a higher value (t_H) of 15 min, above which we are dissatisfied. Note that t_L and t_H do not correspond to a minimum and a maximum analysis time. From Figure 10.16A, it appears that an exponent of $e_t > 1$ (e.g. $e_t = 3$) may

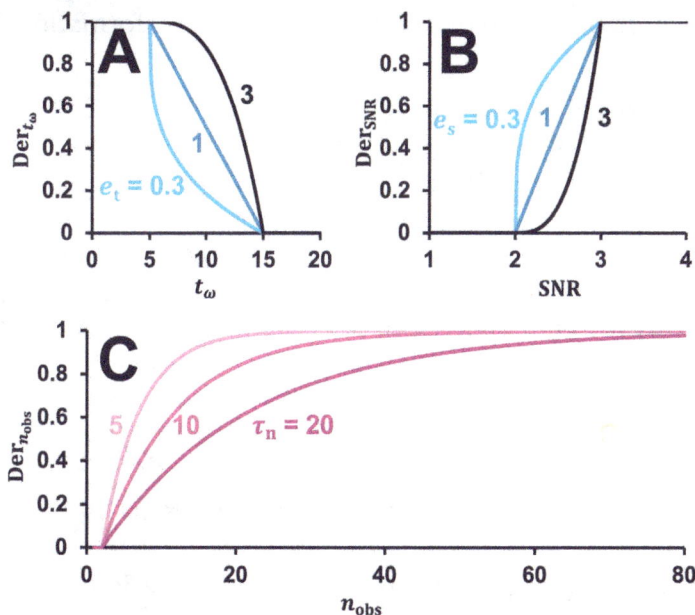

Figure 10.16 Examples of Derringer functions for (A) analysis time (Der$_{t_\omega}$; see eqn (10.53)), (B) the signal-to-noise ratio for the lowest peak (Der$_{SNR}$; see eqn (10.54)), and (C) the number of observed peaks (Der$_n$; see eqn (10.55)).

be appropriate, as we are likely to gradually lose satisfaction once t_ω exceeds t_L.

One property that we may like to see increase is the signal-to-noise ratio for the lowest peak in the chromatogram (SNR$_{min}$). This variable spans a wider range, and therefore, it seems sensible to use its logarithm in a Derringer criterion, *i.e.*

$$\text{Der}_{SNR} = \begin{cases} 0 & \text{if } \ln(\text{SNR}_{min}) \leq \ln(\text{SNR}_L) \\ \left(\dfrac{\ln(\text{SNR}_{min}) - \ln(\text{SNR}_L)}{\ln(\text{SNR}_H) - \ln(\text{SNR}_L)} \right)^{e_s} & \text{if } \ln(\text{SNR}_L) \leq \ln(\text{SNR}_{min}) \leq \ln(\text{SNR}_H) \quad (10.54) \\ 1 & \text{if } \ln(\text{SNR}_{min}) \geq \ln(\text{SNR}_H) \end{cases}$$

This function is illustrated in Figure 10.16B for $\ln(\text{SNR}_L) = 2$ and $\ln(\text{SNR}_H) = 3$ with three different values of the exponent (*i.e.* $e_S = 0.3$, 1, and 3). Considering the logarithmic nature of the variable, $e_S = 1$ may be an appropriate choice.

The number of observed peaks cannot be transformed into a Derringer-type function in the same manner as analysis time or the signal-to-noise ratio because there is no higher value at which the analyst would be fully satisfied. One possible transformation for n_{obs} is as follows:

$$\text{Der}_n = \begin{cases} 0 & \text{if } n_{obs} \leq 2 \\ 1 - e^{\left(\frac{2-n_{obs}}{\tau_n}\right)} & \text{if } n_{obs} \geq 3 \end{cases} \tag{10.55}$$

This function is illustrated in Figure 10.16C for three values of τ_n, *i.e.* 5, 10, and 20. The choice may depend on the complexity of the sample, with a higher value being more appropriate for a more complex sample.

Based on eqn (10.37), (10.53–10.55), the overall Derringer response function may take the form

$$\text{CRF}_{\text{Der}} = \left[\text{CRF}_{\text{nrp0}}\right] + \left[\text{Der}_{t_\omega}\right] + \left[\text{Der}_{\text{SNR}}\right] + \left[\text{Der}_n\right] \tag{10.56}$$

Each of the four terms on the right-hand side is optional. Two or more of the different objectives may be selected, but a greater number of objectives increases the chance that low values for one function may be overshadowed by the other ones. For example, a region where Der_{t_ω}, Der_{SNR}, and Der_n are all (nearly) equal to 1 may yield a high value for CRF_{Der} of approximately 3, even if the resolution is very low ($\text{CRF}_{\text{nrp0}} \approx 0$). Such a risk is lower when only two objectives are combined in a CRF_{Der} function.

10.3.4 Multi-criteria Decision-making

Much of selectivity optimization in chromatography involves multi-criteria decision-making (MCDM), in that several of the objectives identified in Table 10.5 are pursued simultaneously. In Section 10.3.3, we have seen examples of how chromatographers go out of their way to merge these various objectives into chromatographic response functions (CRFs) that ultimately yield a single number to guide the process, with varying degrees of success. The optimization strategies discussed in Module 10.4 often require such a CRF.

Some of the CRFs described earlier, such as hierarchical response functions (Section 10.3.3.4) and Derringer desirability functions (Section 10.3.3.5), are often classified as MCDM, but we have classified them as composite response functions because different objectives are combined in a CRF that yields a single unique num-

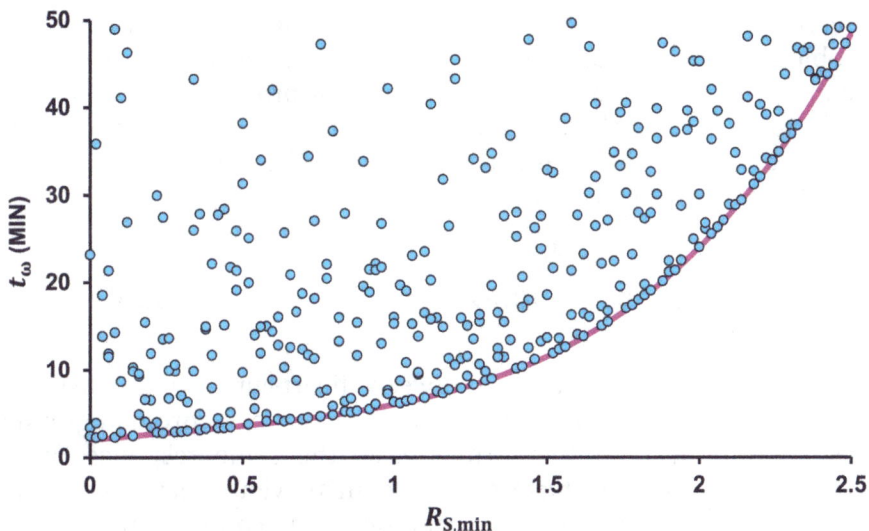

Figure 10.17 Example of a Pareto-optimization plot for minimum resolution and analysis time. The pink line represents the Pareto-optimal (PO) front, and the points falling on this line are PO points, each of which would be a valid outcome of the optimization.

ber. Here, we will discuss Pareto-optimality as the one genuine MCDM method used in chromatographic method development.

10.3.4.1 Pareto Optimality

Pareto optimization is named after Italian scientist Vilfredo Federico Damaso Pareto (1848–1923), and therefore written with a capital P. It is based on inspecting all observations and then making a judicious choice. An example of a Pareto-optimization (PO) plot is shown in Figure 10.17. Each point in the figure represents a set of conditions, for which the elution time of the last peak in the chromatogram (t_ω) and the resolution of the critical pair $(R_{S,min})$ have been computed.

The pink line in Figure 10.17 is the Pareto-optimal (PO) front. Points on this line are considered PO points. For such points, there are no conditions at which a better resolution can be obtained in a shorter time. For all other points, this is the case and these are sub-optimal. The chromatographer is left with a choice between a resolution of 1 in about 6 min, a resolution of 1.5 in 11.6 min, or a resolution of 2 in 24 min – or any other point on the line.

Pareto optimization is a valid choice for many optimization problems, but it requires the analyst to make decisions. It cannot be used in automated method-development processes. It is feasible to decide, for example, to opt for the point where $R_{S,min} = 1.5$ in the shortest possible time, but this would turn the PO approach into a threshold criterion (Section 10.3.3.3).

10.3.5 Response Functions for Two-dimensional Separations

A distinction can be made between heart-cut two-dimensional separations and comprehensive two-dimensional chromatography (see Module 7.1). In the former case, the same response functions can be used as those described in previous sections of this module for one-dimensional separations. In comprehensive two-dimensional separations, some objectives from Table 10.5 remain unaffected, such as minimum analysis time, maximal critical resolution, or highest possible sensitivity, while peak dilution is a specific point of attention, especially in comprehensive two-dimensional liquid chromatography (LC×LC). Distributing the peaks throughout the entire chromatogram deserves special attention in the present section.

10.3.5.1 Peak Capacity

Comprehensive two-dimensional chromatography separations are particularly useful for fingerprinting and non-target analysis of complex samples (*i.e.* samples containing a large number of analytes). To optimize separation systems for use in such cases, peak capacity, as discussed in Sections 7.1.4.1 and 7.1.5.3, is the best response function. A high two-dimensional peak capacity (^{2D}n) increases the probability of separating sample components. The peak capacity of a system is independent of the sample injected. For a specific (type of) sample, the effective peak capacity (*i.e.* the fraction of the available peak capacity that is actually used) depends on the fractional surface coverage ($f_{coverage}$; see Section 7.1.5.1).

10.3.5.2 Surface Coverage

In Figure 10.18, several methods for determining the fractional surface coverage ($f_{coverage}$) are illustrated. The four figures show the apexes of peaks observed in a comprehensive two-dimensional

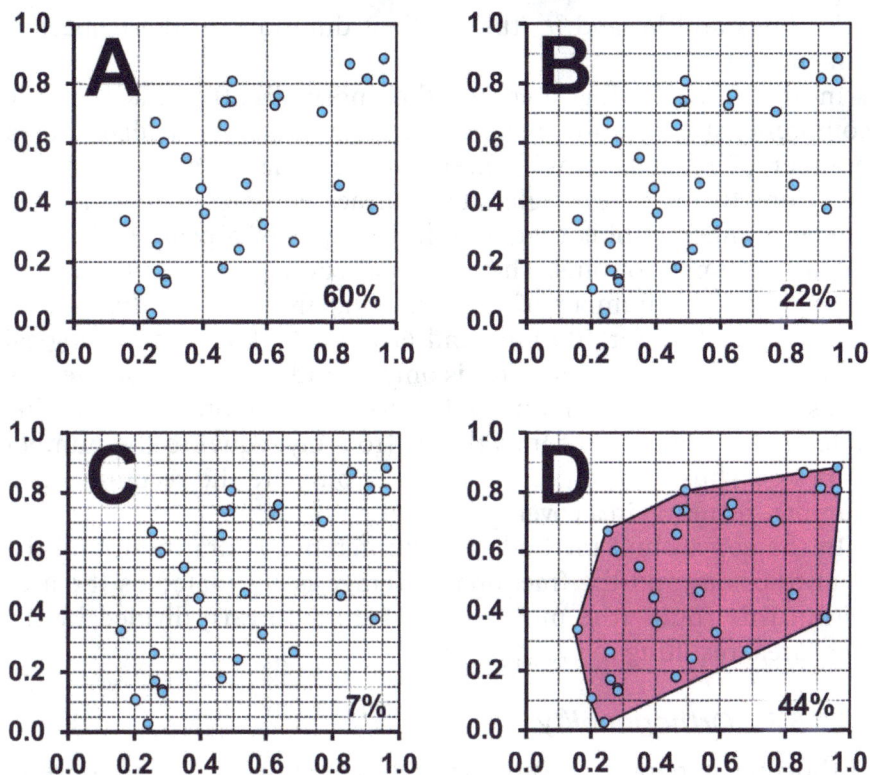

Figure 10.18 Illustration of the bin-counting method (A–C) and the convex-hull method (D) for characterizing surface coverage in comprehensive two-dimensional chromatography; see the text for explanation.

separation as dots. The first- and second-dimension retention times (displayed on the horizontal and vertical axes, respectively) are normalized as follows:

$$\left[{}^{1}t_{R}\right]_{norm} = \frac{{}^{1}t_{R} - {}^{1}t_{0}}{{}^{1}t_{G} - {}^{1}t_{0}} \tag{10.57}$$

$$\left[{}^{2}t_{R}\right]_{norm} = \frac{{}^{2}t_{R} - {}^{2}t_{0}}{{}^{2}t_{G} - {}^{2}t_{0}} \tag{10.58}$$

where ${}^{1}t_{R}$ and ${}^{2}t_{R}$ are the analyte retention times in the first- and second-dimension separation, respectively, and ${}^{1}t_{0}$ and ${}^{2}t_{0}$ are the void times and ^1t_G and ^2t_G the gradient durations in the respective dimensions.

In the bin-counting method, the (normalized) surface of the comprehensive two-dimensional chromatogram is divided into a number of regular "bins". These can be squares, such as shown in Figure 10.18, or rectangles. The surface coverage is now defined as the number of bins that contain the apex of a peak, divided by the total number of bins. The fractional coverage is seen to depend strongly on the number of bins, ranging from 7% in the case of 400 bins to 22% for 100 bins and 60% for 25 bins. This is logical because the number of analytes is only 30 and some of these overlap. Thus, 30 analytes can never fill 400 boxes. Therefore, the guideline is to create a number of bins that is (roughly) equal to the number of peaks. Thus, in the present case, 25 bins are the recommended number, resulting in a coverage of 60%.

In the convex-hull method, a circumference is drawn around the analyte peaks, and the fractional coverage is calculated as the area within this "hull" (the pink area in Figure 10.18D) divided by the total area, resulting in $f_{coverage} = 44\%$.

10.3.5.3   Orthogonality

Orthogonal separations were defined in Section 7.1.5.1 as those in which the retention times in the first and second dimensions were statistically independent. A simple statistical measure would be the correlation coefficient, but this is not very satisfactory, as a low value does not necessarily imply a good spread of peaks across the two-dimensional chromatogram. A better measure can be obtained from the asterisk method devised by Camenzuli et al.[18]. The underlying idea is that if the peaks are well spread, there must be high variances around four different lines in the chromatogram, i.e. a vertical line at $^1t_R = 0.5$, a horizontal line at $^2t_R = 0.5$, and the two diagonal lines $^2t_R = {}^1t_R$ and $^2t_R = 1 - {}^1t_R$. The lines are identified with the Z_1, Z_2, Z_-, and Z_+ parameters, respectively, in Figure 10.19. The standard deviations around the lines can be calculated using normalized retention times as defined in Equations 1.57 and 1.58.

$$\sigma_{Z_1} = \sigma\{[^1t_R]_{norm,i} - 0.5\} \tag{10.59}$$

$$\sigma_{Z_2} = \sigma\{[^2t_R]_{norm,i} - 0.5\} \tag{10.60}$$

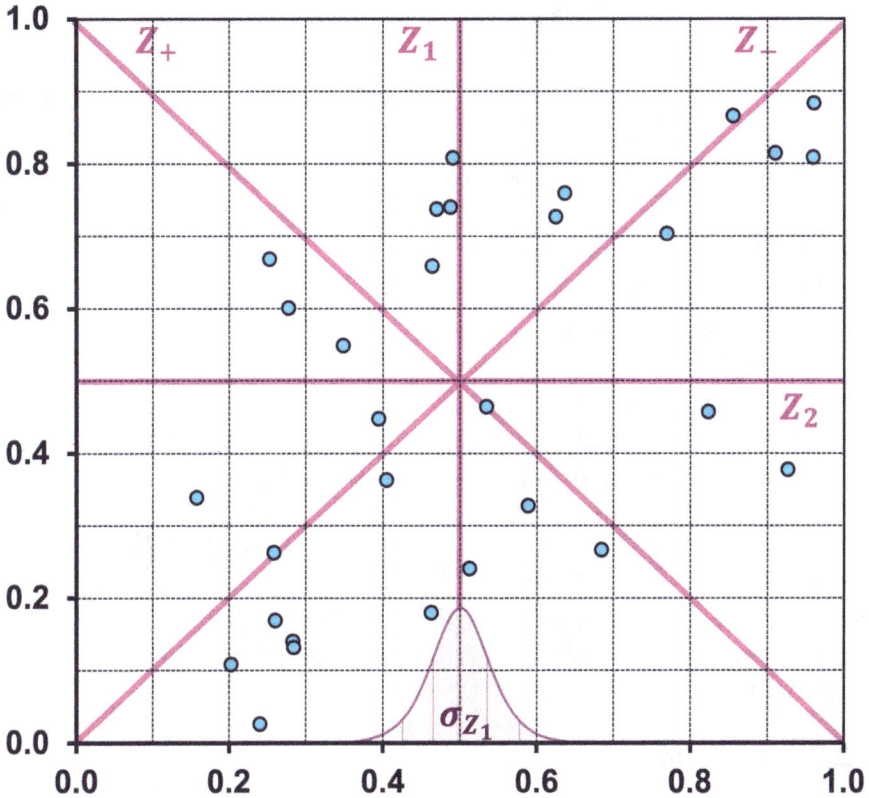

Figure 10.19 Illustration of the asterisk approach to describe orthogonality in comprehensive two-dimensional chromatography. Only the standard deviation of Z_1 is schematically illustrated, but standard deviations are also determined around the other three Z-axes.

$$\sigma_{Z_-} = \sigma\left\{[^1t_R]_{norm,i} - [^2t_R]_{norm,i}\right\} \tag{10.61}$$

$$\sigma_{Z_+} = \sigma\left\{[^1t_R]_{norm,i} + [^2t_R]_{norm,i} - 1\right\} \tag{10.62}$$

These standard deviations are scaled to obtain the Z values that are indicative of how the points are spread in the chromatogram.

$$Z_1 = 1 - 2.5 \cdot \left|2.5\sqrt{2} \cdot \sigma_{Z_1} - 1\right| \tag{10.63}$$

$$Z_2 = 1 - 2.5 \cdot \left|2.5\sqrt{2} \cdot \sigma_{Z_2} - 1\right| \tag{10.64}$$

$$Z_- = 1 - 2.5 \cdot |\sigma_{Z_-} - 0.4| \tag{10.65}$$

$$Z_+ = 1 - 2.5 \cdot |\sigma_{Z_+} - 0.4| \tag{10.66}$$

Finally, the asterisk number (A_0) is calculated as

$$A_0 = \sqrt{Z_1 \cdot Z_2 \cdot Z_- \cdot Z_+} \tag{10.67}$$

A_0 varies between 0 and 1, and it is usually expressed as a percentage orthogonality. For the example shown in Figure 10.19, which contains the same retention data as Figure 10.18, the spread is significant in all directions, viz., $Z_1 = 0.91$, $Z_2 = 0.94$, $Z_- = 0.91$, and $Z_+ = 0.86$, resulting in an orthogonality of $A_0 = 80\%$.

A significant advantage of the asterisk approach is that it can easily be calculated from a table of paired first-dimension and second-dimension retention times. This makes it an easy-to-use criterion, especially in combination with interpretive optimization methods, which rely on computing the retention times of all analytes in both dimensions.

10.3.5.4 Connected Components

Boelrijk et al. adopted a concept from graph theory to establish – and maximize – the number of observed peaks in a comprehensive two-dimensional chromatogram.[19] The concept is illustrated in Figure 10.20. If the resolution between peaks is below a threshold value, they are combined into clusters or **connected components**. An isolated peak also counts as a connected component (with zero connections). The total number of connected components is indicative of the degree of separation achieved. If all analytes yield isolated peaks, the number of connected peaks is maximized. This CRF_{ncc} was used successfully in a Bayesian optimization strategy for comprehensive two-dimensional separations.[19]

10.3.6 Response Functions for Specific Problems

10.3.6.1 Limited Optimization of Target Analytes

There are many cases where a number of target analytes need to be separated in samples in which other irrelevant components are also present. An example is shown in Figure 10.21. Peaks #3 and #6

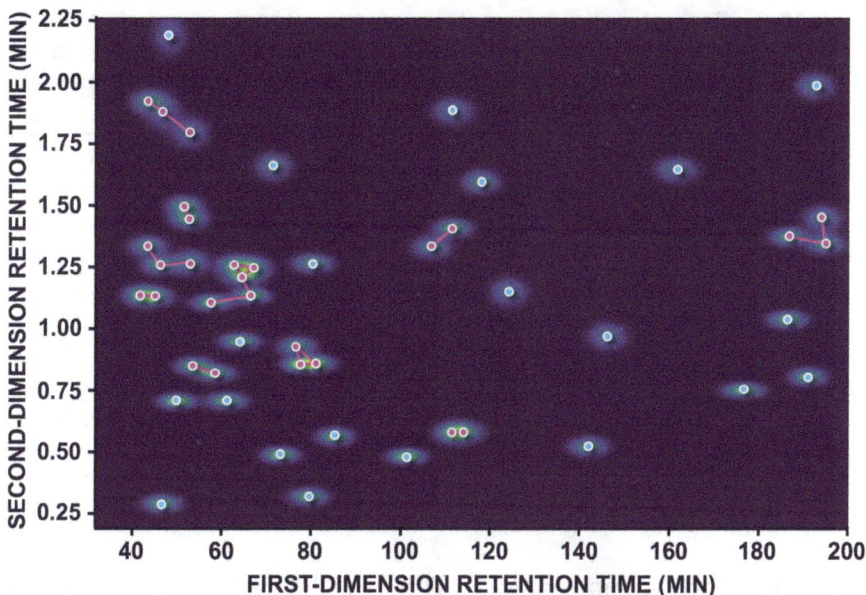

Figure 10.20 Illustration of the concept of connected components. Connected components (clusters) are indicated with pink dots and connection lines. Isolated "connected" components are indicated with blue dots. The total number of connected components (clustered and isolated) can be used as a chromatographic response function (CRF_{ncc}), the maximization of which corresponds to optimizing the separation. Adapted from ref. 19, https://doi.org/10.1016/j.chroma.2021.462628, under the terms of the CC BY 4.0 license, https://creativecommons.org/licenses/by/4.0/.

(blue-shaded numbers) represent the target analytes. Only the four indicated resolution values are relevant because they involve one of the target analytes. $R_{S,3,2}$ and $R_{S,4,3}$ quantify the extent of separation of analyte #3, while $R_{S,6,5}$ and $R_{S,7,6}$ that of analyte #6. Peaks #4 and #5 are poorly separated, but since neither peak is relevant, this does not matter. Peak #1 is totally irrelevant. However, during an optimization process (*e.g.* upon changing the LC mobile phase), the order of elution may change. For example, if the order of peaks #1 and #2 is reversed, $R_{S,3,1}$ becomes relevant instead of $R_{S,3,2}$.

The critical resolution value is the lowest value that involves a relevant peak, which, in this case, is $R_{S,6,5} = 1.82$. When calculating a normalized resolution product, only the relevant resolution factors are used in calculating the product of resolution factors (the numerator), whereas all peaks are used in computing the average

Table 10.6 Characteristic parameters of the chromatogram shown in Figure 10.21. The two σ values represent the leading and tailing sides of the asymmetrical peaks. Relevant retention times and resolution values are printed in bold.

Peak #	t_R	σ		$R_{s,i,i-1}$
		Front	Tail	
0 (t_0)	1.10	0.025	0.050	—
1	3.42	0.038	0.051	13.11
2	4.11	0.049	0.063	3.46
3	**4.98**	0.065	0.080	**3.42**
4	7.49	0.092	0.121	**7.28**
5	7.90	0.108	0.129	0.90
6	**8.82**	0.124	0.152	**1.82**
7	**14.30**	0.192	0.271	**7.95**
—	—	—	$\overline{R}_s$	5.42
—	—	—	CRF_{nrp0}	0.42

resolution (the denominator). This results in lower values for CRF_{nrp0}, except in the imaginary (but ideal) case in which all irrelevant analytes elute at t_0. These calculations are summarized in Table 10.6 for the chromatogram shown in Figure 10.21. Criteria such as

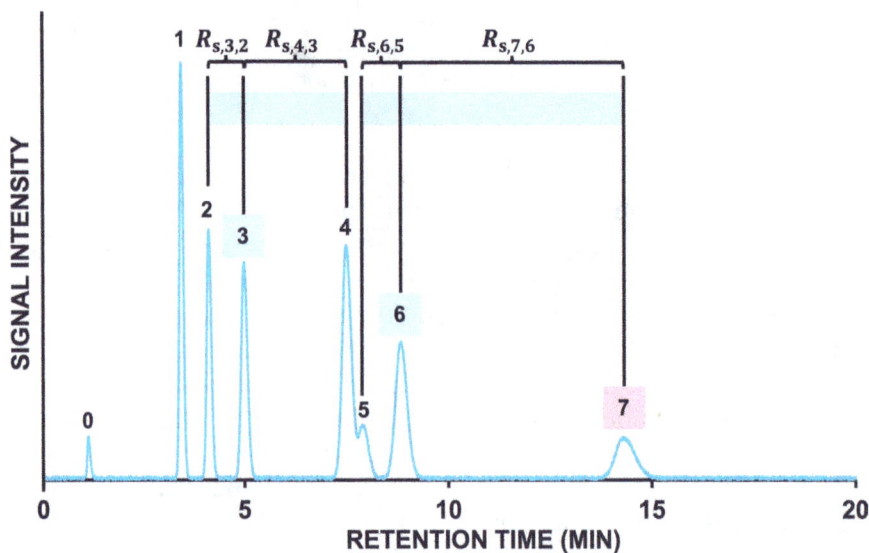

Figure 10.21 Example of a limited optimization of two target analytes (#3 and #6, blue-shaded numbers). Only the four indicated resolution values are relevant. Peak #7 (pink-shaded) is relevant because it determines the analysis time.

the minimum relevant resolution, the distribution of relevant peaks (CRF_{nrp0}), and the retention time of the last peak (t_ω) can be used in threshold criteria (Section 10.3.3.3), hierarchical response functions (Section 10.3.3.4), Derringer desirability functions (Section 10.3.3.5), or in multi-criteria decision-making (Section 10.3.4). Maximizing the observed number of peaks (n_{obs}) is not a meaningful criterion for targeted analysis, as the appearance of more irrelevant peaks is not advantageous.

10.3.6.2 Programmed Elution

Much of the discussion in Module 10.3 is valid for programmed analysis (temperature programming in GC, gradient elution in LC), but care should be taken to use the correct elementary criteria. The definition of resolution (eqn (10.2)) is equally valid in programmed elution as in non-programmed (isothermal, isocratic) elution. However, eqn (10.3) and derived equations are specific to non-programmed elution. The latter is also true for the separation factor S (eqn (10.22)). Peak-to-valley ratios can be used in programmed elution, but their transferability is quite limited. For example, the effect of changes in column dimensions on P values is hard to predict.

In eqn (10.3) and (10.22), it is assumed that peak widths increase with retention time. When using linear temperature programs in GC or linear composition gradients in RPLC, the peak width (of peaks eluting well into the linear gradient and before it ends) is, to a first approximation, independent of the elution time. This implies that the difference in retention times is the key criterion. To achieve sufficient separation, eqn (10.2) yields

$$\sigma_{max} = \frac{[\Delta t_{R,j,i}]_{min}}{4R_{s,req}} \tag{10.68}$$

where σ_{max} is the maximum allowable dispersion to achieve the required resolution ($R_{s,req}$) for the critical (relevant) pair of peaks separated by $[\Delta t_{R,j,i}]_{min}$. The dispersion is determined by the column and flow rate (as in isocratic elution) but is also affected by the slope of the gradient and, especially, the retention factor of the analyte at the moment of elution (k_e). Both σ_{max} and $[\Delta t_{R,j,i}]_{min}$ are affected by the elution program, but the effects are largely predictable (see, for example, Module 3.4).

The distribution of the peaks in the main part of the program, where the peak width is approximately constant, can be expressed in terms of the normalized product

$$CRF_{n\Delta p} = \frac{\Pi_i^{n-1}(t_{R,i+1} - t_{R,i})}{\overline{\Delta t}} \qquad (10.69)$$

where n is the number of peaks and $\overline{\Delta t}$ is the average difference obtained from

$$\overline{\Delta t} = \frac{\Sigma_i^{n-1}(t_{R,i+1} - t_{R,i})}{(n-1)} \qquad (10.70)$$

In the case of a limited number of relevant peaks, only Δt values that involved a relevant peak are included in the product $\Pi\Delta t$, whereas all peaks should be included in $\Sigma\Delta t$ (see Section 10.3.6.1).

10.4   Optimization Strategies (M)

In Section 10.1.3.1, we have explained that retention surfaces are relatively simple, while response surfaces that indicate the quality of separation for the entire chromatogram tend to be highly irregular and complex. If we are to optimize a separation by recording chromatograms, calculating an overall response function, and plotting this against the parameter(s) to be optimized (*i.e.* directly characterizing the pink line in Figure 10.6C), 20 composition points likely will not suffice and 100 points seem a more reasonable number.

In Figure 10.22, we try to indicate what is realistically possible in chromatographic optimization. Chromatographic experiments take time (and consumables), especially if different conditions (and equilibration) are required for each data point. We consider 10 experiments reasonable (indicated by the blue zone at the bottom of the diagram) and 100 experiments quite problematic (entering the light-pink zone). This shows that direct ("non-interpretive") grid-search optimization (without retention models) is already unattractive for a single optimization parameter and unrealistic for the simultaneous optimization of several parameters. If retention surfaces are accurately known, grid search can be performed computationally, and calculating the chromatographic response for a million sets of conditions is not extremely challenging (light-pink

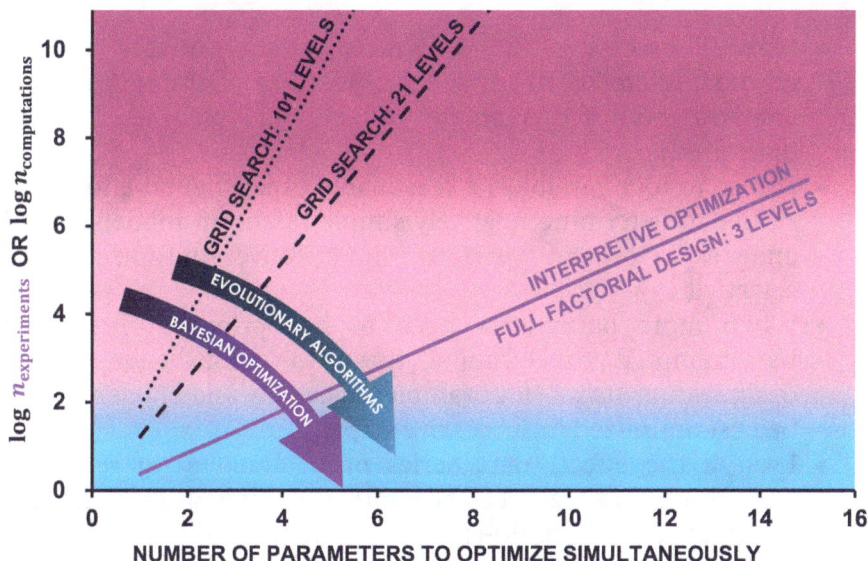

Figure 10.22 Number of experiments or computations needed for chromatographic optimization using grid-search or design-of-experiment (DoE) strategies. Dark-blue lines: grid search of the response surface (no retention model) with 20 (dashed line) or 100 (dotted line) levels per parameter; purple line: interpretive approach, building a retention model based on a full-factorial design with three levels per parameter; blue zone: feasible number of chromatograms; light-pink zone: feasible number of computations; dark-pink zone: unrealistic.

zone), but when we increase the number of parameters above three, we become computationally limited (dark-pink zone). If we are to optimize more parameters simultaneously, search strategies are required that significantly outperform grid-search approaches (*e.g.* evolutionary algorithms, Section 10.4.2.3, or Bayesian optimization, Section 10.4.2.4).

The dark purple line in Figure 10.22 indicates the number of experiments required for an interpretive approach, using a full-factorial design at three levels to determine retention models for each analyte. This approach is reasonable for three parameters (*e.g.* composition, temperature, ionic strength; 27 experiments). In the case of four parameters (81 experiments), the blue colour is already fading. We may reduce the number of experiments by performing a fractional factorial design (see Section 9.8.2). Considering five parameters simultaneously is unrealistic.

We can draw some important conclusions from Figure 10.22.

- Directly assessing the response surface (quality of the chromatogram) is ill-advised for mixtures containing diverse analytes, even for a single variable (*i.e.* parameter to be optimized).
- For up to four variables, a reasonable experimental effort may suffice to determine relatively simple retention models for all analytes in order to perform "interpretive" optimization of selectivity.
- When more parameters need to be optimized, it should be attempted to decouple these to obtain independent optimization stages (for example, selectivity and efficiency may be best optimized in successive steps).
- Even if the effects of a series of parameters on retention and resolution can be accurately predicted (*e.g.* in the optimization of multi-step gradients),[20] the "brute-force" grid-search method soon reaches the limit of computer power.
- There is a need for smarter search algorithms in multi-dimensional space.

10.4.1 The Parameter Space

On the one hand, parameters that have a correlated effect on the outcome (the chromatographic response) should be optimized concurrently (see Section 9.8.1). On the other hand, we have seen above that the number of parameters that can realistically be optimized simultaneously is severely limited. Therefore, it is crucial that the choice of which parameters to optimize is made judiciously. All the relevant information can be found in previous chapters. The resolution was explained in Section 1.5.1, and this led to the all-important eqn (1.39), which described the resolution of two peaks ($R_{S,j,i}$) under constant elution conditions (isocratic, isothermal) as a product of three factors: one for retention ($\bar{k}/(1+\bar{k})$, where $\bar{k}$ is the average retention factor of the two analytes); one for selectivity ($(\alpha_{j,i} - 1)/(\alpha_{j,i} + 1)$, with the selectivity defined as $\alpha_{j,i} = k_j/k_i$), and one for efficiency ($\sqrt{N}/2$, where N is the plate count). Under programmed conditions, the effect of selectivity, in terms of the difference in elution temperatures (in a temperature-programmed GC run) or of elution composition (in a gradient-elution LC run), becomes all important. In terms of elution time, the difference increases with

Table 10.7 Inventory of the main optimization parameters in GC.

Parameter	Effect on selectivity	Effect on peak width and shape	Effect on analysis time
Stationary-phase nature and polarity	Large	Large	Small (if temperature is adapted)
Stationary-phase manufacturer	Small	Small	Small
Stationary-phase firm thickness	None	Large at $\delta_f > 0.3^a$	Significant
Isothermal temperature	Significant	Large	Large
Temperature program	Small	Large	Large
Flow rate	None	Moderate	Minor[b]
Column diameter	None	Large	Minor
Column length	None	Large	Large

[a]Reduced film thickness (see Section 1.7.7)
[b]In programmed analysis.

increasing duration of the gradient, but part of this gain diminishes because the peak width also increases (to a smaller extent).

10.4.1.1 Parameter Space in GC

In Table 10.7, we make an attempt to list the main optimization parameters in gas chromatography (GC). It is instantly clear from this table that the main parameter affecting selectivity is the nature of the stationary phase. The polarity of the stationary phase also strongly affects the width and shape of the observed peaks. This explains the common approach for an unknown sample to try one non-polar column and one polar column and to proceed to the next stage if symmetrical peaks and some selectivity are observed. The starting and final temperatures will then be adapted to the elution range of the sample. Only in special cases, for example, for the separation of enantiomers, a larger number of columns may be tested (see Section 10.2.1). Separations of (pairs of) enantiomers are also among the rare cases in which an isothermal GC separation may be preferred over a temperature-programmed separation. The latter is almost always preferred due to the rapid decrease in chromatographic efficiency with increasing retention factors (see Section 1.7.7).

A larger number of parameters affect the peak width (efficiency) and analysis time, but these effects are quite predictable, and the initial chromatograms will indicate whether, for example, a longer column is required. The flow rate may be used to balance efficiency against analysis time on a given column. Kinetic plots, such as those discussed for LC in Module 3.16, can also be used for establishing the optimal column dimension and flow rate in GC.[21]

10.4.1.2   *Parameter Space in LC*

Establishing the parameter space in LC is much more complicated than in GC, as demonstrated by the size and complexity of Tables 10.8 and 10.9. There are many parameters that affect retention and selectivity, and smart choices have to be made to define a manageable optimization space. For the first phase of method development, *i.e.* selectivity optimization, parameters may be selected that (i) are important for the analytes in the sample (target analytes and potential interferents), (ii) are important for the LC mode selected (based on the information provided in Chapter 3), and (iii) have a large effect on selectivity. It is prudent to limit the selection to a maximum of three parameters (see Figure 10.22 and the related discussion). For some parameters (mobile-phase composition, temperature), it may be argued that two levels allow establishing reasonably accurate retention (and peak-width) models. Other parameters, such as pH, require more levels, and their effects may be more difficult to model. In some cases (notably ion-pair chromatography), there are too many significant parameters to perform a comprehensive optimization. Sensible restrictions should be applied for upper and lower boundaries of the parameter space. For example, for mobile-phase composition in RPLC, almost the entire range may be used. The region between 100% water and (say) 5% of modifier is unattractive because of dewetting phenomena (see Section 3.5.1). The range of temperatures or pH is much more restricted. Temperatures below ambient may not be practical, whereas excessively high temperatures may affect column lifetime. Temperature control in LC is not easy (see Section 3.3.4), and it is ill-advised to try and keep the column temperature constant when small particles and high pressures are encountered (see Section 1.7.6). Both (very) low and (even moderately) high pH values may detrimentally affect LC columns, especially those based on silica (see Section 3.2.2).

The selection of the stationary phase may be aided by the selectivity-screening procedures discussed in Module 10.2. These may be combined with optimizing the mobile-phase composition by running two gradients (with different slopes) on each column. In principle, two gradients may also be run at different temperatures (for a total of four experiments on each column) and with two different modifiers (eight experiments on each column), but the effort multiplies, and the (ideally automatic) optimization of the selectivity on each column becomes rapidly more complicated. Testing several different stationary phases of the same type (*e.g.* ODS-silica columns from different manufacturers) is usually not very productive unless a method needs to be developed in which alternative columns are specified. The surface area and pore-size distribution of the packing material are usually correlated (smaller pores correspond with larger surface areas). Stationary phases that are nominally identical but offer different surface areas, pore-size distributions or particle sizes usually exhibit at least small differences in selectivity.

The mobile-phase composition (in RPLC, NPLC and HILIC) and the ionic strength of the mobile phase (in IEC) tend to be the most important parameters to control retention. These are also the parameters that are typically programmed in the case of a gradient. Retention and selectivity in multi-step gradients can be predicted from retention models,[20] but the robustness of methods relying on very complex gradients may be questioned. If the gradient span (ranging from initial to final conditions) of a linear gradient is chosen broad enough, it is not expected to affect selectivity.

The set of parameters at the bottom of the table (flow rate and length, diameter, and particle size of the column) can be used to optimize the efficiency based on the outcomes of the selectivity-optimization stage. An elegant way to do so is the kinetic plot method (see Module 3.16).

10.4.2 Search Methods

In this section, we discuss computer-aided methods for locating optima on multidimensional response surfaces. As explained in Section 10.1.3.1, chromatographic response surfaces that describe the quality of a chromatogram exhibit (very) many local optima, which is why trial-and-error optimization and direct

Table 10.8 Inventory of the main optimization parameters in LC. RPLC = reversed-phase liquid chromatography; NPLC = normal-phase liquid chromatography; HILIC = hydrophilic-interaction liquid chromatography; IEC = ionexchange chromatography; IPC = ion pair chromatography; SEC = size-exclusion chromatography. ☑ indicates a possible optimization parameter. ☐ indicates a small (possibly undesirable) effect.

Parameter	Neutral analytes	Ionogenic analytes	Ionic analytes	Applies to					
				RPLC	NPLC	HILIC	IEC	IPC	SEC
Stationary-phase nature and polarity	☑	☑	☑	☑	☑	☑	☐	☑	☐
Stationary-phase manufacturer	☑	☑	☑	☑	☑	☑	☐	☑	☐
Ion-exchange capacity	—	☑	☑	☐[a]	—	☐	☑	☐[b]	—
Surface area	☑	—	—	☑	☑	☑	—	☑	—
Pore-size distribution	☐	☐	☐	☐	☐	☐	☐	☐	☑
Nature of the mobile phase (modifier)	☑	☑	☑	☑	☑	☑	☑	☑	☑
Mobile-phase composition[c]	☑	☑	☐	☑	☑	☑	☑	☑	☐
Mobile-phase ionic strength	—	☑	☑	☐	—	☑	☑	☑	☐[d]
Mobile-phase pH	—	☑	—	☑	—	☑	☑	☑	—
Nature and concentration of the ion-pair reagent	—	☑	☑	—	—	—	—	☑	—
Gradient initial hold	☑	☑	☑	☑	☑	☑	☑	☑	—
Gradient span	☑	☑	☑	☑	☑	☑	☑	☑	—
Gradient slope	☑	☑	☑	☑	☑	☑	☑	☑	—

(continued)

Table 10.8 (*continued*)

Parameter	Neutral analytes	Ionogenic analytes	Ionic analytes	Applies to					
				RPLC	NPLC	HILIC	IEC	IPC	SEC
Temperature	☑	☑	☑	☑	☑	☑	☑	☑	☐[e]
Flow rate	☑	☑	☑	☑	☑	☑	☑	☑	☑
Column length	☑	☑	☑	☑	☑	☑	☑	☑	☑
Column diameter	☑	☑	☑	☑	☑	☑	☑	☑	☑
Particle size	☑	☑	☑	☑	☑	☑	☑	☑	☑

[a]Residual silanols may interact with cations or basic analytes.
[b]Ion-exchange capacity is dynamically generated through ion-pair reagent.
[c]Ratio of base solvent and modifier(s).
[d]Large effect for charged polymers.
[e]Temperature affects analyte solubility and may suppress adsorption effects.

Table 10.9 Effects of the main optimization parameters in LC.

Parameter	Effect on selectivity	Effect on peak width and shape	Effect on analysis time	Comments
Stationary-phase nature and polarity	Large	Large	Compensated by mobile-phase composition	Type of stationary phase correlates with selection of LC mode
Stationary-phase manufacturer	Small	Small	Small	Different columns of the same type have smaller effects than changing the mobile-phase composition
Ion-exchange capacity	Small in IEC; significant if secondary mechanism	Significant	Potentially large	
Surface area	None	Small	Compensated by mobile-phase composition	Affects sample capacity
Pore-size distribution	Size-exclusion (or ion exclusion)	Especially for high-MW analytes	None	Pores should be sufficiently large to avoid exclusion (except in SEC, where only total exclusion should be avoided)
Nature of the mobile phase (modifier)	Large (except in IEC and SEC)	Large	Compensated by mobile-phase composition	
Mobile-phase composition[a]	Moderate	Moderate	Very large	Main parameter to control retention
Mobile-phase ionic strength	Mainly in IEC	Moderate	Very large in IEC	Volatile buffers are required for MS (or ELSD) detection

(continued)

Table 10.9 (*continued*)

Parameter	Effect on selectivity	Effect on peak width and shape	Effect on analysis time	Comments
Mobile-phase pH	Large	Large[b]	Compensated by mobile-phase composition	Many (silica-based) columns have limited ranges of pH stability
Nature and concentration of the ion-pair reagent	Large	Significant	Large	Correlated with many other parameters
Gradient initial hold	Small if any	Small	Small	Hold time adds to the analysis time
Gradient span	Moderate[c]	Potentially large	Significant	
Gradient slope	Moderate	Significant	Large	
Temperature	Significant	Significant	Compensated by mobile-phase composition	
Flow rate	Small	Significant	Large	
Column length	None	Large	Large	
Column diameter	None	None	None (as long as the flow rate is adapted)	
Particle size	None (in theory)	Very large	Very large	

[a]Ratio of base solvent and modifier(s).
[b]Especially around the pK_a of the analyte.
[c]Mainly early eluting peaks (affected by initial composition).
[a]See Section 1.7.6.

(non-interpretive) characterization of chromatographic response surfaces are ill-advised. However, once we have obtained retention surfaces (and possibly peak-width and peak-symmetry surfaces) and we can calculate the response using one of the criteria described in Module 10.3, we need search methods to locate the highest or most attractive) of the local optima. This is known as the overall or **global optimum.**

10.4.2.1 Grid Search

Grid search was already mentioned above as the benchmark technique for establishing a response surface. It simply involves imposing a "grid" on the parameter space by dividing the different parameters into a number of (usually equidistant) steps. For example, isocratic composition can range from 100% solvent A to 100% solvent B in steps of 1%, or temperature may range from 30 °C to 60 °C in steps of 1 °C. In a two-parameter space, this yields a grid with 31 × 101 = 3131 different nodes. Again, it is impossible to perform so many experiments, but once retention (and peak-width) surfaces have been established, it is a straightforward task to predict 3131 chromatograms and to compute the corresponding optimization criteria. Grid search is a feasible approach for a one-, two-, or three-dimensional separation space, but more efficient search techniques are required if more optimization parameters are considered simultaneously.[22]

Figure 10.23A illustrates a grid search on a two-dimensional response surface. Although this is obtained for a simple mixture (four analytes), four local optima can be observed. The grid shown is quite coarse (16 levels) for illustration purposes, but a reasonable impression of the response surface can be obtained. A more exact location of the optima and the corresponding chromatograms can be obtained by performing refined searches with a narrower grid in the vicinity of the optimum. Note that a non-interpretive approach with such a coarse grid (256 experiments) would already be highly impractical.

10.4.2.2 Simplex Methods

Simplex methods are aimed to ascend the slope on a response surface to reach the highest point (assuming a response criterion that increases with increasing separation quality). A simplex search starts with a set of experiments (or computations in the case of interpretive optimization) that is one greater than the number of

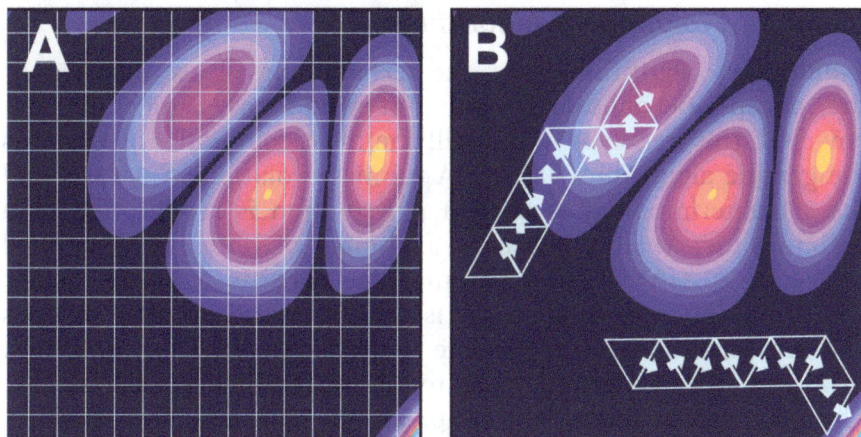

Figure 10.23 Illustration of (A) grid search and (B) simplex optimization.

parameters. In a two-parameter space, such as that shown in Figure 10.23B, this original set forms a triangle (called a simplex). In the simplest simplex algorithm, the point with the lowest response is then discarded and the next experiment or computation is performed at a point obtained by reflecting the triangle away from this lowest point. In this way, the simplex proceeds towards an optimum. A coarse simplex is shown in Figure 10.23B for illustration purposes. In practice, finer designs may be used, or in more-advanced simplex algorithms, the triangle may be contracted or expanded depending on the previously observed values.[23] John Berridge, then at Pfizer (Sandwich, Kent, UK), deserves credit for performing fully automated LC method development using a simplex approach. This was no mean achievement back in 1982.[24] Berridge needed between 15 and 30 experiments to reach an optimum. As illustrated in Figure 10.23B, the simplex algorithm finds an optimum, which is not necessarily the global optimum. The chances of finding the true global optimum are enhanced by restarting the simplex process several times from different locations. For complex mixtures, many more local optima can be expected compared to the four-analyte mixture illustrated in Figure 10.23. The prospect of multiple restarts requiring 15 to 30 experiments makes simplex an unattractive option for a computerized trial-and-error optimization of chromatographic experiments. However, simplex optimization can feasibly be applied to interpretive optimization, locating optima based on retention (and peak-width) surfaces.

10.4.2.3 Evolutionary Algorithms

Evolutionary algorithms – as the term implies – are inspired by the concept of evolution. A number of evolutionary algorithms have emerged, the archetype – and still the most common – of which is the **genetic algorithm** (GA). A GA mimics evolution in nature, and the same terminology is used. Each solution is represented by a chromosome, which consists of a number of genes, as illustrated in Figure 10.24A, where a chromosome consists of twelve genes. For computational purposes, it is easiest to use a string of ones and zeroes, although it is feasible to use letter codes for the genes. In Figure 10.24A, the blue cells represent the binary code for %B. The value can be found by multiplying the gene code by the values listed above (in this case %B = 32 + 8 + 4 + 1 = 45). The last five cells represent the temperature, with a defined offset of 30 °C. The value here is 16 + 4 = 20 °C. The proposed (isocratic) solution is defined by a composition of 45% B and a temperature of 50 °C. Any kind of resolution can be obtained, simply by defining the value of a gene in units of 1% B and 1 °C (as in Figure 10.24) or 0.1% B and 0.1 °C, *etc.* Figure 10.24B shows a population of eight proposed solutions (chromosomes). For each solution, the chromatographic response (or "fitness") is computed from the retention (and peak-width) surfaces. In theory, a chromatogram can be measured corresponding to each solution, but this is highly impractical. One way to enforce the limits of the parameter space is to punish disallowed chromosomes through their response. For example, the fifth proposed solution has 109% B. By assigning a response of zero, it is disqualified.

Based on the response values, we could select the four solutions (chromosomes) with the highest scores to become "parents" of the next generation (rank selection; the four highest ranked solutions are assigned an asterisk in Figure 10.24B). Alternatively, they can each be assigned a probability of becoming a parent that is proportional to their response. For example, the first chromosome gets assigned a probability of 46%, *etc.* In Figure 10.24C, we illustrate the "offspring" from the four parents with asterisks, produced by the crossover at the two indicated points. For example, the first chromosome of the next generation is produced by taking the first five genes of the first parent and the last seven genes of the second parent. The generation may be completed by adding a small number of mutants (not shown), which are produced by making

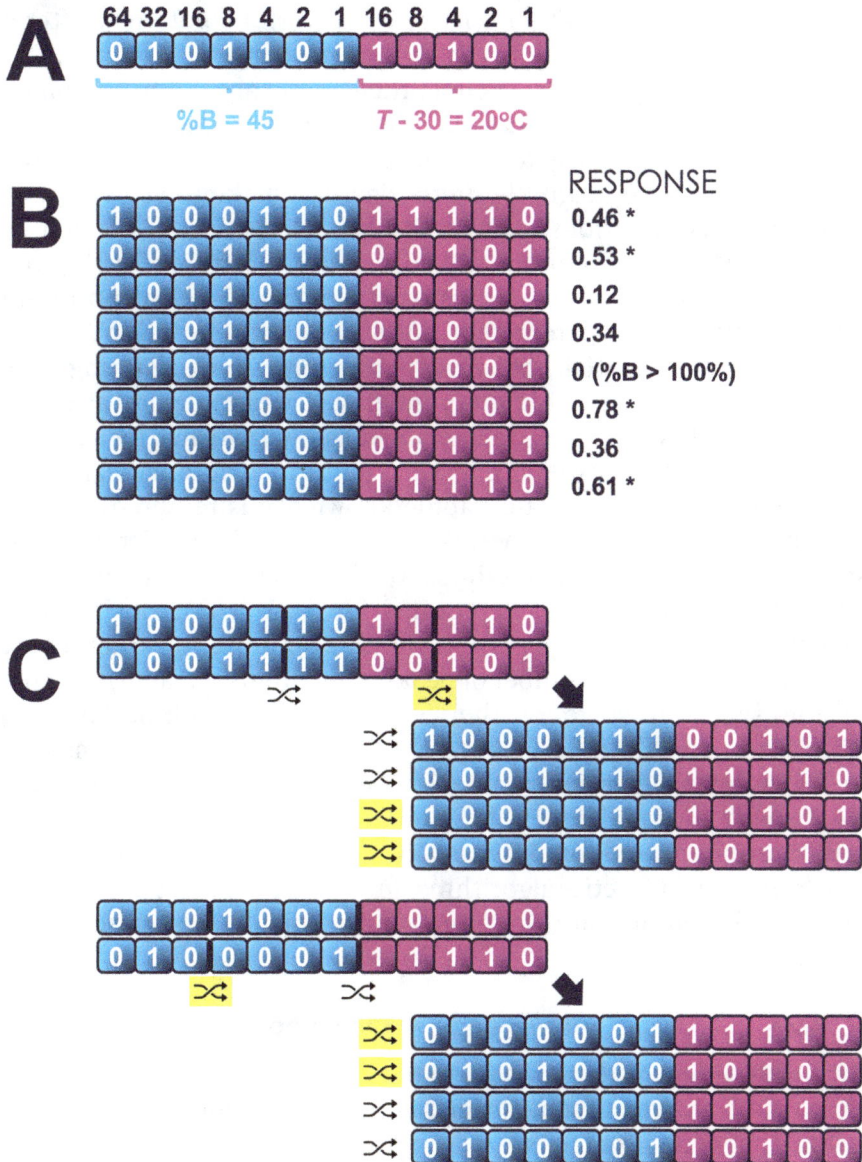

Figure 10.24 Illustration of a genetic algorithm to locate the global optimum in a two-parameter chromatographic search space, spanned by %B (0–100%; blue genes) and temperature (increment above the minimum temperature of 30 °C; pink genes): (A) chromosome (proposed solution) consisting of twelve genes; (B) population consisting of eight chromosomes, with a chromatographic response score assigned to each solution (the highest scoring chromosomes are selected as parents, indicated by an asterisk); (C) reproduction, *i.e.* a new generation produced by crossover of selected parents.

random changes (usually in relatively successful solutions). After a few generations, the GA zooms in on the optimum.

Evolutionary algorithms, such as the genetic algorithm explained above, have proven very successful in locating optima in multidimensional space, are unlikely to get stuck at local optima, and have the potential to deal with multi-criteria decision-making. For example, pareto-optimal solutions (see Section 10.3.4.1) can be selected as parents. O'Hagan *et al.* applied an evolutionary algorithm directly (non-interpretively) on gas chromatograms.[25] They established an optimum for nine optimization parameters (injection volume, injection temperature, split ratio, initial hold time, temperature-programming rate, final temperature, final hold time, and data-acquisition rate) after 120 generations of two chromosomes or chromatograms each, for a total of 240 experiments. All experiments were performed automatically on a single column. No chromatographic knowledge is required, beyond establishing a sensible parameter space. Zelena *et al.*[26] applied a similar strategy to develop LC-MS methods for characterizing human serum, with 130 or 150 LC-MS runs required for establishing positive and negative ESI methods, respectively. Multiple optimization criteria were used simultaneously (number of peaks, resolution, and analysis time) at the selection stage of the evolutionary algorithm. Huygens *et al.*[27] compared the efficiency of several evolutionary algorithms with grid search for the optimization of one-dimensional and comprehensive two-dimensional liquid-chromatographic separations. They used retention models and computed peak widths to perform many different optimizations. Genetic algorithms and more advanced evolution algorithms were found to be much more efficient than grid-search strategies.[27]

10.4.2.4 Artificial Intelligence Approaches

Artificial-intelligence approaches are also finding their way into the domain of analytical separation science. Best known are **artificial neural networks**, which are essentially a modelling approach. A schematic example is shown in Figure 10.25. The neural network connects a number of input nodes (in our case, the chromatographic optimization parameters) with one or more outputs. "Sub-criteria" in Figure 10.25 could be, for example, the number of observed peaks, the minimum resolution (resolution for the critical pair), and the analysis time, whereas a single output relates to a CRF as discussed in Module 10.3. Every line that connects two cells in Figure 10.25 carries a weight. The value of a cell ("neuron") in the hidden layer

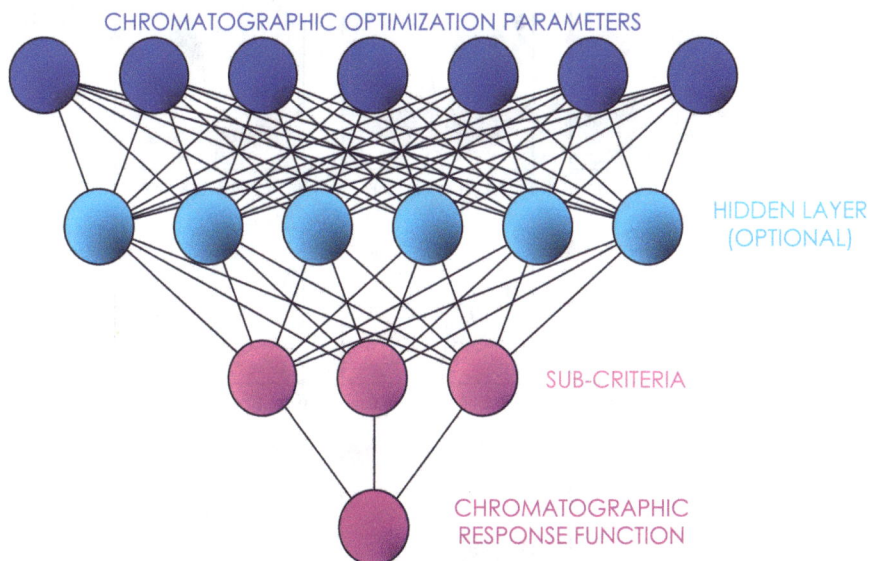

CHROMATOGRAPHIC OPTIMIZATION PARAMETERS

HIDDEN LAYER
(OPTIONAL)

SUB-CRITERIA

CHROMATOGRAPHIC
RESPONSE FUNCTION

Figure 10.25 Schematic artificial neural network for chromatographic method development.

is determined by the values of the operating parameters and the weights. The values of the sub-criteria are determined by those of the hidden-layer cells and their associated weights, and the CRF is obtained from the three sub-criteria. The inclusion of a hidden layer allows generating non-linear models.

Artificial neural networks can connect the output (CRF) to the inputs (chromatographic parameters) without any models, assumptions, or knowledge. The great disadvantage is the large number of experiments needed to train the network. For the network shown in Figure 10.25, $7 \times 6 + 6 \times 3 + 3 = 63$ weights need to be established before any predictions or improvements can be made. Interesting attempts have been made at retention-time prediction using artificial neural networks, but their application for method development has otherwise received little attention thus far.

Bayesian optimization is a much more promising machine-learning method for method development in analytical separation science because it requires a manageable number of experiments. After a parameter space has been defined and a number of initial experiments performed, Bayesian optimization proceeds with fitting a probabilistic function to the chromatographic response function. This implies that at every point in the parameter space, the response is not an exact value but a probability distribution. For example, this

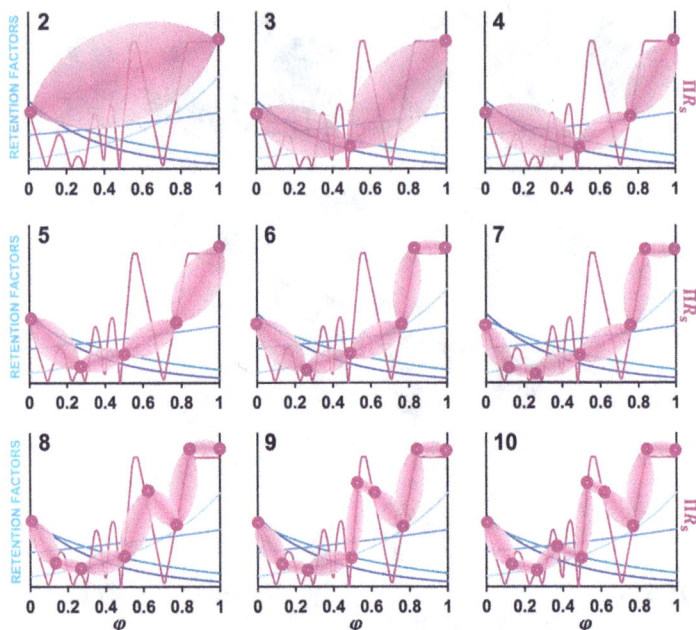

Figure 10.26 Illustration of the application of Bayesian optimization on the response surface of Figure 10.6C. Successive approximations of the response surface are shown from left to right and then from top to bottom, with the number of available data points indicated.

can be a Gaussian function with a mean and a standard deviation. Based on fitting the function to the experimental results, an acquisition function is established to determine where the next experiment will be performed, in a trade-off between **exploitation** (focussing on regions with a high chromatographic response) and **exploration** (focussing on regions where the variance is still high). Figure 10.26 provides a schematic illustration of the application of a Bayesian optimization strategy on the response surface of Figure 10.6C. The process is seen to be akin to a balloon-twisting exercise, with the regions of uncertainty narrowing and the predictions ("balloons") slowly starting to resemble the response surface (drawn pink line) more closely.

Boelrijk *et al.* used Bayesian optimization to establish optimal gradient parameters (initial hold time, gradient duration, and the initial and final compositions) for both the first- and second-dimension gradients in comprehensive two-dimensional LC (LC×LC) separations of four complex mixtures, each containing 50 synthetic dyes.[28] Retention and peak-width models were used in an interpretive optimization. Bayesian optimization was found to greatly outperform grid search (104 *vs.* 11 664 "experiments"). It must be noted that the parameters selected for the optimization do not strongly affect the chromatographic selectivity (see Tables 10.8 and 10.9), which simplifies the response surface. In another study, Boelrijk *et al.* applied Bayesian optimization for the fully automated ("closed-loop") optimization of "multi-linear" LC gradients (*i.e.* gradients consisting of multiple linear segments) for the optimization of dye mixtures. The approach converged on an optimum within 35 experiments.[29] Both single-criteria and multi-criteria optimization were demonstrated.

Reinforcement learning is a branch of machine learning that centres around an "agent", which could represent an analyst or an analyst-by-proxy in a computer program. The analyst takes action in the "environment" (parameter space) and gains information on the "state" (updated response surface). The outstanding attribute of reinforcement learning is that the analyst is rewarded (or penalized) for their actions. The magnitude of the award is determined by the difference between the previous response surface and the updated one. The cycle is repeated until a satisfactory result (the "stop criterion") is reached. The goal of the analyst is to maximize the cumulative reward ("return"). As in Bayesian optimization, a compromise is sought between the exploitation of existing information and the exploration of uncharted territories. The process is illustrated in Figure 10.27. Kensert *et al.* used reinforcement learning algorithms to optimize scouting gradients in LC (see Section 3.17.2) and to derive accurate retention models from these.[30,31]

10.5 Method Validation (M)

The objective of analytical method validation can be formulated as "to demonstrate that the analytical procedure is fit for the intended purpose".[32] The purpose of the method can vary widely. For example, extremely low detection methods may be essential for determining drugs in surface waters, while detection limits are hardly relevant in testing pharmaceutical formulations or seized drugs. Some methods may serve as a "one-off" analysis, such as for solving a particular problem encountered with a product or process. After the problem has been solved (often with the help of analytical measurements), it is advisable to document the methods

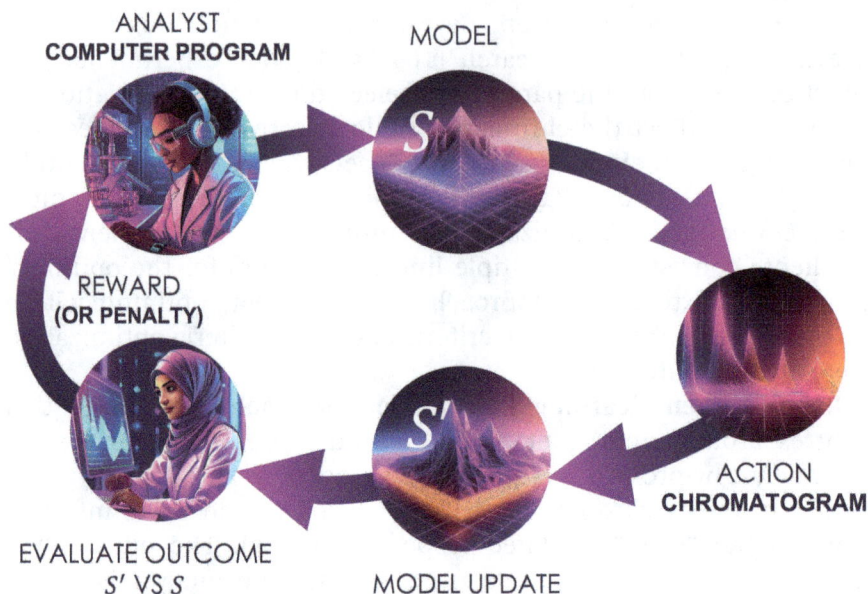

Figure 10.27 Illustration of reinforcement learning.

and the data (chromatograms, spectra) for future reference, but the method may never be needed again. At the other extreme, for submission of a method to a regulatory body, for use as a standardized method, or for approval of, for example, a pharmaceutical product, extensive validation is mandatory. Requirements for method validation are arguably the most strict and best described for human medicines. We refer to the guidelines[32] of the International Council for Harmonisation of Technical Requirements for Pharmaceuticals for Human Use (ICH), as these are intended for global use across all regulatory bodies.

Extensive validation studies follow three stages, *i.e.* formulating a validation protocol, performing the validation tests, and writing a validation report.

The ICH method-validation process is illustrated in Figure 10.28. Table 10.10 lists a number of stages and tests that are part of the method-validation process.

Some **repeatability** testing is always needed in analytical chemistry. If a measurement cannot be repeated at least once, it may be an aberration, and it does not constitute any real information. In the case of a formal evaluation, the number of replicates required is described *a priori* in the validation protocol. A

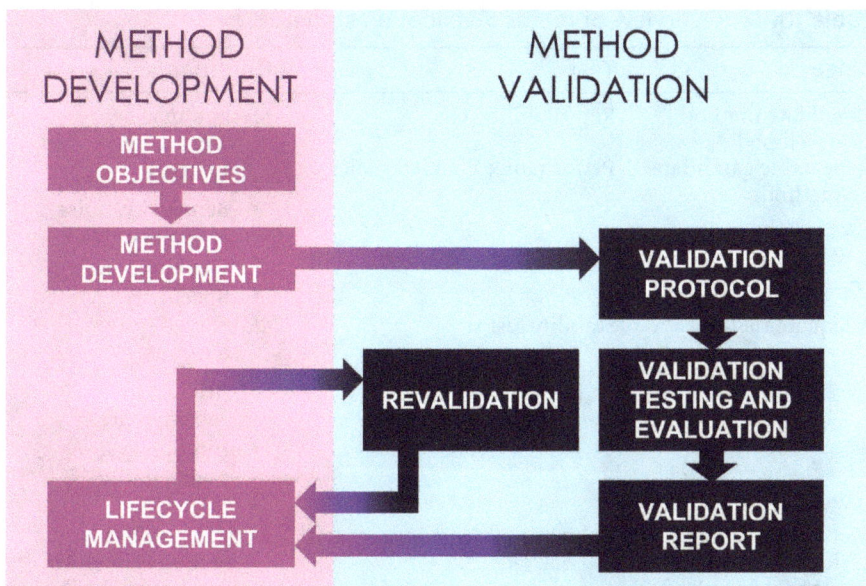

Figure 10.28 Summary of the method-validation process.

typical number of replicates for a formal repeatability test is nine (either nine replicates or three replicates each at three different concentrations). If a method is to be used routinely during a longer period and/or by different analysts, testing the **intermediate precision** becomes appropriate. This includes repeatability tests carried out across multiple days, by multiple analysts, or under varying laboratory conditions (warm/cold, dry/humid, *etc.*). To test these variables simultaneously and efficiently, design of experiments may be used (see Module 9.8). Before including different instruments in such a test, system-suitability tests may be carried out (see below).

Validating the **accuracy** of a method, *i.e.* verifying that the obtained results are correct, should be performed under real method conditions (including sample preparation) and in the presence of a real or realistic matrix. Accuracy testing relies on a comparison of obtained results with expected or, preferably, known values. For this purpose, certified reference materials (CRMs) may be required. For quantifying one or a few components in a relatively simple matrix, pure-component references may be used to create test samples. In other cases, such as determining the aromatics content of a mineral-oil product or the molecular-weight distribution of a synthetic polymer, representative test

Table 10.10   Overview of method-validation stages.

Stage	Test	Comments
Candidate method developed	Repeatability test	Method may be improved if needed
Repeatable candidate method	Performance characteristics	• Method accuracy • Sensitivity, noise, LOD,[a] LOQ[b] • Working range • Specificity
Adequate method	Stability test	Stability of system, solutions
Adequate and stable method	Intermediate-precision tests	• Inter-day repeatability • Inter-analyst repeatability • *Etc.*
Method considered for submission as standard method	Robustness/ruggedness test	Determine whether the method is likely to pass a reproducibility test
Proposed standard or regulatory method	Reproducibility test	Interlaboratory round-robin test
Implementing existing method	System-suitability test	Test whether a specific instrument configuration meets the requirements of the method

[a]Limit of detection.
[b]Limit of quantitation.

samples cannot easily be prepared, and CRMs exist for such applications. A critical requirement for CRMs is their stability. While this may be ensured for oil products or synthetic polymers, the situation is much more complex in complex food, biological, or clinical samples. In such cases, accuracy testing may rely on a standard addition method (see Section 9.6.3.6), applied to the entire analytical method, including sample preparation. Accuracy is often reported as a (percentage) recovery.

The **sensitivity** of a method (*i.e.*, the slope of the calibration curve) or the response (calibration curve), the **limit of detection** (LOD) or detection limit, and the **limit of quantitation** (LOQ) or quantitation limit have been discussed in Section 9.6.5. The working **range** of the method, which is not necessarily a linear range, is typically reported from a lower limit to an upper limit. The range can also be limited by upper and lower values within which the accuracy and precision of the method have been validated. The types of limits that should be reported may depend on the purpose of the analytical method.[32] Ways to fit linear or non-linear models to calibration data have been discussed in Section 9.6.3. In pharmaceutical analysis, at least five different concentration levels across the reported range are recommended to demonstrate the linearity of a calibration curve.[32]

A **stability test**, aimed to validate the stability of the analytical method and required solutions, should not be confused with a stability-indicating method, which is used to test the stability (shelf life) of pharmaceutical products. A stability test may include mobile-phase solutions. Including the stationary phase (column) is more difficult. This latter factor is typically implicitly included in the repeatability and intermediate precision tests.

Testing the **specificity** of the method implies demonstrating that there is no interference of the signals (peaks) of the analytes by any other possible sample constituents. At least all known compounds that may be present, such as known impurities or degradation products, and a range of matrices (if appropriate) should be tested. Impurities may be purposefully generated by stressing (*e.g.* heating, illuminating) a product. Demonstrating that a completely different "orthogonal" method yields identical results may confirm adequate specificity. Sufficient chromatographic (or electrophoretic) resolution (such as $R_s \geq 1.5$) for target analytes is a requirement for demonstrating the specificity of an analytical separation method. If the specificity of the method is not adequate, a second, complementary method may need to be added to meet the requirements for the analysis.

In forensic science, specificity is the all-important criterion. Ideally, it is quantified in terms of **likelihood ratios** using Bayesian statistics. Rather than striving for a given acceptable likelihood, prosecutors may add additional (highly specific) measurements with the idea that independent probabilities can be multiplied, and any alternative explanations can be excluded.

A **robustness test** is not usually a required part of a method-validation protocol for regulation purposes. However, if a method is intended for use at multiple locations or is to be submitted for standardization, it is recommended. It can be seen as a prevention of problems that might have been foreseen. In a robustness test, the effect of small changes in a number of values on the performance of the method is evaluated. Screening designs (Section 9.8.3) may be used to test the main effects of a large number of parameters. The most important parameters may be tested using a factorial design (Section 9.8.2). If design-of experiments (DoE) strategies are used during method development, data on the robustness of the method with respect to the most critical parameters may already be available and can be carried forward when reporting on the robustness of the method. The result of a robustness test provides acceptable ranges for the method parameters and greater confidence that the method will perform at different locations and pass a lengthy and expensive reproducibility test. A robustness test may be combined with a stability test.

A **reproducibility test** constitutes the highest level of method validation. Usually, it is only performed on methods proposed for standardization. It needs the participation of different laboratories, a good deal of administration, and high-level statistics. The term "reproducibility" is often used more lightly in fields other than human medicine, but such (mis)use of the term is discouraged. Although reproducibility refers to the interlaboratory precision of a method, an accuracy component is usually included in a reproducibility test.

A **system-suitability test** (SST) can be seen as part of the initial method-development process, but a formal test may be a part of the method-validation protocol. After a method has been standardized, it used to be essentially frozen, but recently there has been more interest in the lifecycle management of the analytical method.[1] Analytical methods may be used for many years, and changes do happen. For example, new instruments may appear, and these may be applicable following a system suitability test or following instrument performance qualification (PQ) documents produced by the manufacturer. This is more difficult with the emergence of new types of columns, such as core–shell particles (or superficially porous particles, see Section 3.2.2.1) or significantly improved instruments (*e.g.* progressing from high-pressure LC to ultra-high-pressure LC, see Section 3.3.2). If a method is updated along such lines, additional validation is required, but the prospect of a renewed reproducibility test is daunting. Regulatory bodies have been considering a lighter regime for the validation of updated methods, as indicated in Figure 10.28.

10.6 Peak Tracking (A)

10.6.1 Objectives

When comparing two or more chromatograms of identical or similar samples, it is often necessary to locate the different peaks in the different chromatograms. This is needed for

- (quantitative) analysis of a series of chromatograms with small changes in retention times, possibly due to variations in flow rate, temperature, eluent composition (in LC), or dependence of retention on the amount injected;
- comparison of chromatograms recorded on different columns (even if nominally identical), different instruments, or different locations;
- recording of large datasets across a long period of time, for example, for biomarker-discovery studies;
- gradient-scanning experiments in LC (see Section 3.17.2);
- establishing retention models (and, ideally, peak-width and peak-shape models) for the purpose of interpretive optimization.

In the first three of these cases, the expectation is that retention times remain unchanged between chromatograms. Small changes may typically be corrected using the **retention-time-alignment** methods described in Section 9.11.4. Due to the expected stability of the retention times, the assumption is that there will not be much difficulty in tracking the peaks across the chromatograms. Retention time alignment is then also typically conducted for target peaks or, in the case of non-target analysis, based on prior knowledge about the samples.

In the last two cases, the conditions are purposefully changed to alter the selectivity and, possibly, the elution order of the analytes (see Module 10.4). In rare cases, it may be possible to collect the input data for interpretive optimization by injecting all analytes individually, but (all) analytes are often not available in sufficiently pure form. Moreover, the number of experiments (injections) soon becomes prohibitive as the complexity of the sample (number of analytes) increases. Thus, the ability to pinpoint the peaks across the initial chromatograms is crucial, especially in method development (gradient scanning and interpretive optimization). This is referred to as **peak tracking**, the goal of which is to establish the location of the highest possible number of unique peaks, or peak clusters, throughout all chromatograms. It is not necessary to identify any of the analytes, as long as they can be properly tracked. From this information, peak tables can be generated that allow large-scale comparison of retention times as well as retention modelling and interpretive optimization.

The distinction between retention-time alignment and peak tracking is summarized in Table 10.11.

10.6.1.1 Scope of This Module

Peak tracking is facing many of the challenges identified for signal processing in Modules 9.11 and 9.12. Therefore, it is by no means an established process. Some of the peak-tracking methods discussed in the present section are still immature. However, due to their paramount importance for automating method development (see Module 10.4), these methods must be discussed in this textbook. The scope of Module 10.6 is to briefly discuss strategies for peak tracking, their strengths and weaknesses, and their implementation. An overview of published methods is provided in Table 10.12.

10.6.2 Matching Criteria

10.6.2.1 Peak Area and Intensity

The first reported peak tracking algorithm in 1982 employed the peak area as a criterion for matching peaks.[33] The attractiveness and simplicity of the use of peak areas or intensities become apparent from the chromatograms in Figure 10.29. While the two 1D chromatograms on the left show a different distribution of peaks, it is relatively easy to recognize the resemblance of both separations by eye and label (track) peaks 1 through 4 due to the distinct pattern of peak intensities or areas.

However, the use of peak areas or intensities is not very robust, especially not in cases where co-eluting peaks are encountered. We described in Section 9.12.4 that peak detection and integration methods may easily result in a 20–30% error in peak areas for two partially overlapping peaks in a chromatogram. To make things worse, if peak tracking is used for building retention models, the chromatograms are typically obtained under different chromatographic conditions. As a result, the degree of co-elution, and perhaps even the peak order, is likely to change from one chromatogram to another. This variability propagates into the area values that

Table 10.11 Comparison between retention-time alignment and peak tracking.

	Retention-time alignment	Peak tracking
Assumption	The same method has been used and any changes in retention times are accidental.	Different methods have been used and the elution order may have changed.
Aim	Aligning retention times of peaks across chromatograms.	Tracking of peaks across chromatograms.
Conditions	• Conditions kept constant as much as possible • Shifts in retention times are usually small • Elution order does not change	• Different methods can be used for the different chromatograms • Shifts in retention times may be large • Elution order may have changed
Applications	• Correction of small changes in retention times due to minor fluctuations • Comparison of columns • Curation of large-scale long-term data collection	• Establish retention models (*e.g.* based on gradient scanning) • Interpretive optimization
Reference	Section 9.11.4	Module 10.6

are ultimately used for peak tracking. Given that co-eluting peaks with different relative areas may significantly interfere with each other, significant errors may be incurred when using intensities for peak tracking. The background of the signal may complicate peak analysis, and it may thus be prudent to use noise-filtering techniques (Section 9.11.2) and baseline-correction techniques (Section 9.11.3) to mitigate this.

10.6.2.2 Fuzzy Peak Areas

To deal with the variability of the reported peak areas for co-eluting peaks and the effects of varying chromatographic conditions on peak characteristics, Otto *et al.*[37] employed **fuzzy theory**[36] to improve the robustness of peak tracking by considering the uncertainty in the peak areas.

The concept of fuzzy theory is illustrated in Figure 10.30 and will only be briefly explained here. On the *x*-axis, the peak areas are shown for one peak that is suffering from (different degrees of) co-elution

Table 10.12 Overview of different strategies employed for peak tracking. (S) = single-channel data, (M) = multi-channel data (e.g. MS or UV–vis spectra). 1D = one-dimensional chromatography, 2D = comprehensive two-dimensional chromatography.

Peak tracking algorithms

Matching criterion	Strategy	Vulnerable to	Type	Notes
Peak area or signal intensity	Direct comparison;[33,34] signal ratio comparison[35]	Change in response in different mobile phases; shift and/or drift in baseline; co-elution	(S/M); 1D	May benefit from noise filtering and baseline correction
Fuzzy peak areas (based on fuzzy theory)[36]	Comparison of fuzzy peak areas[37]	Any variations in reference chromatogram (including unstable analytes); severe co-elution and/or elution-order shifts	(S/M); 1D	Requires uncertainty in peak areas to be specified; requires assignment of reference chromatogram
UV–vis spectrum similarity	Multivariate comparison;[38] definition of match factors[39]	Effects of chromatographic conditions on spectra; insufficiently distinct spectra; severe co-elution (mixed-component spectra); imperfect peak detection[39]	(M); 1D	Comparison may be improved by Savitzky–Golay noise filtering and baseline correction[39]
Mass-spectrum similarity	Direct comparison;[40] weighed correlation factors;[41] comparison of the 30 most intense peaks[42,43]	Insufficiently intense MS signals; inaccurate peak integration; isobaric analytes; different degrees of	(M); 1D,[40–42] 2D[43]	Ref. 40 requires several datasets from different (orthogonal) methods; ref. 41–43 include candidate-reduction strategy and

(continued)

Table 10.12 (*continued*)

Peak tracking algorithms

Matching criterion	Strategy	Vulnerable to	Type	Notes
Statistical moments	Weighed comparison of peak area, width, asymmetry, and kurtosis[42,43]	co-elution between compared chromatograms; variations in ionization efficiency	(S); 1D,[42] 2D[43]	ref. 42 and 43 adapt to the presence of isobaric analytes and supported by statistical moments
		Poor peak detection; severe co-elution (only for single-channel data); undersampling (only for 2D)		Ref. 42 and 43 supported by mass-spectrum similarity assessment; include a candidate-reduction strategy and adapt to the presence of isobaric analytes
Position of peak relative to its neighbours	Bayesian probabilistic pattern comparison of the surroundings of a peak[44]	Shifts in retention times; dissimilar co-elution and/or retention behaviour between compared chromatograms	(S); 2D	Benefits from baseline correction; a mixture model[45] was used in ref. 44

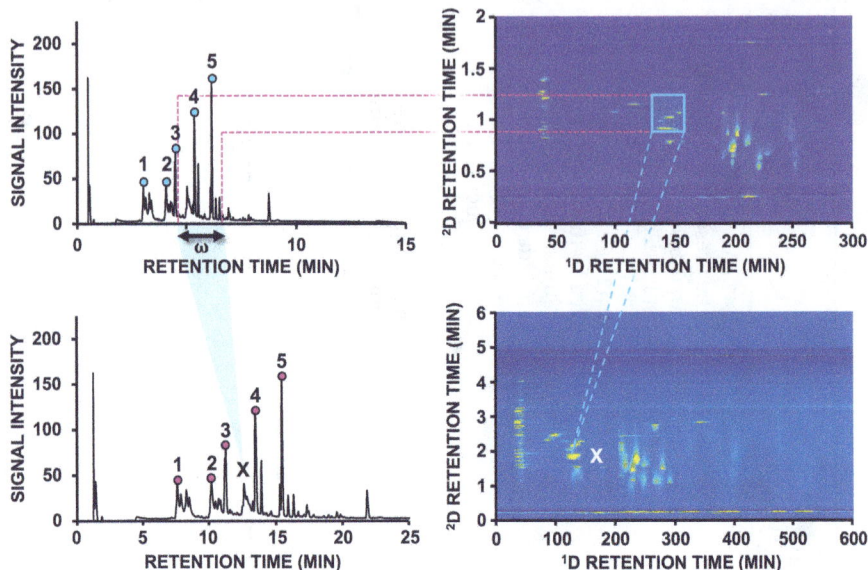

Figure 10.29 Example of two 1D-LC (left) and LC×LC (right) chromatograms on which peak tracking is conducted. The numbered peaks indicate that peak characteristics, such as area and intensity, are seemingly obvious criteria for use during peak tracking. One peak in the bottom chromatograms is marked as X, and the task of peak tracking is to establish its location in the top chromatograms. The number of candidates can be reduced by searching only in limited domains (ω; see Section 10.6.4).

in two different chromatograms (A and B). In Figure 10.30, these are represented as the two blue distributions. Instead of one single value for each peak, the uncertainty is plotted as a distribution across a range of peak area values. In other words, these distributions $a(x)$ and $b(x)$ represent the possible peak areas that can be found for the peak in chromatograms A and B, with the maximum being the most probable value. These distributions are created through the use of so-called membership functions, which are beyond the scope of this book but are explained well in reference[37].

For peak tracking, the essence is that through fuzzy subtraction of $b(x)$ (*i.e.* the uncertainty of the peak area of peak b) from $a(x)$, the total uncertainty in the differences between the two peaks is obtained. This is depicted by the pink dashed line in Figure 10.30. By setting one chromatogram as a reference and comparing the peak area in each chromatogram to that of the reference, the uncertainty profiles of the differences for each pair can be compared. For example, if the uncertainty distributions (*i.e.* the pink line in Figure 10.30) are similar between chromatograms A and B, and between chromatograms A and a third chromatogram C, then peaks *a* and *b* may be tracked through these three chromatograms.

While this approach has been shown to be relatively robust to differences in the degree of co-elution and signal intensity, it remains vulnerable to severe co-elution and hinges on successful peak detection (and preferably deconvolution). The selection of the reference chromatogram is also important If this chromatogram contains peaks pertaining to unstable components, then this may jeopardize the comparison between chromatograms and the peak tracking process.

10.6.2.3 Statistical Moments

Statistical moments are treated in Section 9.12.1. They allow the characterization of a chromatographic peak in terms of area, retention time, width, asymmetry, and kurtosis. In addition to peak area and intensity, the third moment, which characterizes the peak asymmetry, has been shown to be useful for peak tracking.[42] The second moment, which characterizes the peak width, is often not sufficiently discriminative in the case of programmed elution (temperature programming in GC; gradient elution in LC), whereas the fourth moment has limited evidential value. One general remark about statistical moments is that they strongly rely on successful peak detection, with a particular requirement for correctly determining the start and endpoint of the peaks. Peak deconvolution may often be required, as is noise reduction (filtering), because the values obtained for the higher moments, especially, are greatly affected by noise.

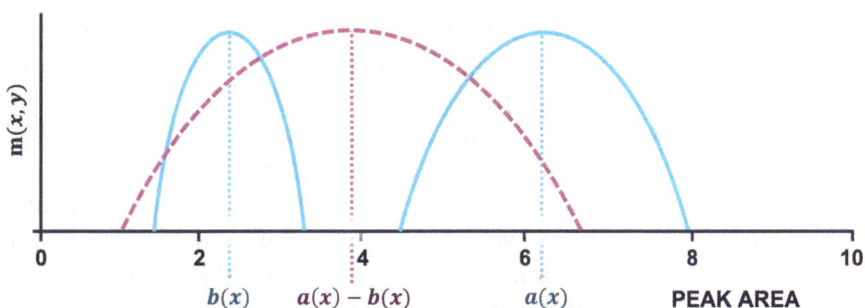

Figure 10.30 Fuzzy subtraction of the area of one peak in one chromatogram ($b(x)$) from that of another chromatogram ($a(x)$).

10.6.2.4 Similarity of Mass and UV–vis Spectra

Given the lack of methods to obtain robust assessments of the peak characteristics, as discussed in the previous sections, most peak-tracking algorithms rely on the use of multi-channel data. This implies that the UV–vis spectrum or mass spectrum is considered a fingerprint of the analyte for peak tracking.

Different approaches have been explored to establish the similarity between the two spectra. A very common method is to use the Pearson correlation coefficient (Section 9.6.2). However, this parameter becomes more difficult to determine reliably when the signal intensity decreases, especially in the case of mass spectrometry. It may then be prudent to weigh the more intense peaks stronger by avoiding the centring of the Pearson correlation coefficient. The comparison of the intensity of the detected peaks can also be included in the similarity assessment.[41] However, this is particularly tricky for MS data when peaks are co-eluting to varying degrees. Another possibility is to only consider the most intense peaks in the mass spectrum. For example, two studies considered the 30 most intense m/z signals in each mass spectrum for similarity assessment. For more than 90% of the peaks, this was found to result in a high evidential value for matching.[42,43] The high information density contained in mass spectra renders them very useful for peak matching, but the presence of isomers may significantly reduce the evidential value of the spectra. The inclusion of relative retention times may result in sufficient evidence to match isomers correctly across chromatograms.[43]

UV–vis spectra contain much less information than mass spectra. However, they have been used successfully in peak-tracking strategies in LC.[38,39] Smoothing of noisy spectra has been found to increase the likelihood of a successful match. Small changes in chromatographic conditions (*e.g.* different concentrations of ACN in water) or the use of different modifiers can cause shifts or changes in the spectrum, and co-elution may contaminate the spectra of nearby peaks. All these factors may put peak tracking at risk.

One interesting study mitigated this problem by exploiting the multivariate character of spectra to perform principal component analysis (PCA; see Section 9.9.1) and factor analysis.[38] Such multivariate approaches are also very common in the -omics and environmental fields for feature analysis of higher-order MS-based datasets. However, such applications are beyond the scope of this book.

10.6.3 Comparison of Multiple Datasets

With the continuous improvement of UV–vis detectors and, especially, high-resolution mass spectrometers, the size of the datasets obtained has increased dramatically. This escalation is even more dramatic for datasets arising from hyphenated systems based on comprehensive 2D chromatography (*e.g.* from GC×GC-MS or LC×LC-MS). While the availability of computational resources has also increased drastically, it is hard to keep up with the increasing complexity of the datasets. Distilling information from the data has required increasingly long computation times.

This is a relevant issue for peak tracking. As the number of chromatograms to be compared increases, tracking all peaks across all gathered chromatograms requires an exponentially increasing number of computations and a concomitant increase in required computational resources. This can be highly challenging for high-resolution separations hyphenated with high-resolution mass spectrometry, but it becomes particularly dramatic for multi-dimensional separation systems with multichannel detection. One example is LC×LC-MS, where one comparison of two chromatograms can easily take several minutes, and a dataset comprising just ten such chromatograms of highly complex samples may take more than an hour to process.

Figure 10.31A illustrates the problem for the comparison of six datasets. Each dataset (chromatogram) is depicted as a numbered circle, and a link between two chromatograms depicts one computational peak-tracking comparison. If all six datasets are immediately available (*e.g.* for establishing retention models or for the retrospective comparison of several datasets), then the number of comparisons C will be given by

$$C = \frac{n_{chrom}(n_{chrom} - 1)}{2} \tag{10.71}$$

where n_{chrom} is the number of chromatograms. The resulting relationship plotted in Figure 10.32 demonstrates the exponential increase in comparisons (and thus computational resources) required as the number of chromatograms increase. For six datasets (as shown in Figure 10.31), the number of comparisons is 15; for twelve chromatograms, it is 132; and for 50 chromatograms, it is 2450. Mitigation of this problem requires the omission of comparisons or links in the graphical depiction of Figure 10.31. Such a linkage problem is a common issue in information sciences, and a linking strategy can be devised to minimize the number of links needed per chromatogram. However, any omission of a link comes with a risk of losing information.

To understand this, the linking strategy can be compared to a game of "*whispers*" or "*telephone*". Suppose that person 1 and person 2 (*i.e.* chromatograms 1 and 2) tell each other the locations of their peaks, and person 2 then shares the location of its peaks and those of person 1 with person 3. Person 2 conveys information on the peaks from person 1 to person 3, without persons 1 and 3 ever communicating with each other. There is a clear danger that person 2 incorrectly transfers the full extent of the information. In this case, incorrect or incomplete information may be due to (i) random variation in the data, (ii) an anomalous signal in the chromatogram, (iii) different degrees of co-elution due to changed method parameters (as would be expected for interpretive optimization), *etc.* Removal of links thus brings risks. The more a chromatogram is compared to other chromatograms, the less likely it becomes that one anomalous chromatogram will jeopardize peak tracking, but also increases the risk of a false positive (Module 9.3).

Figure 10.31B depicts the situation when the number of chromatograms with which each chromatogram is compared, defined as L, changes to 3 (drawn lines) or 2 (pink dashed line), as compared to 5 in Figure 10.31A. The increase in C then becomes linear as

$$C = \frac{L \cdot n_{\text{chrom}}}{2} \tag{10.72}$$

which is also shown by the pink dashed, dotted, and dotted-dashed lines in Figure 10.32 for different values of L. For 2D-LC, an L-value of 3 was found to yield acceptable results.

Of course, different linking strategies are also possible. For example, the chromatograms could be linked in series ($L = 2$). However, such a low L-value requires a strategy for instances when a peak is not found in one of the chromatograms, as this would mean that this same peak is not sought any longer in subsequent chromatograms.

To avoid this, Molenaar *et al.* programmed an algorithm such that each new chromatogram is compared to at least two random prior chromatograms for interpretive optimization, where in every iteration a new chromatogram is recorded based on the peak

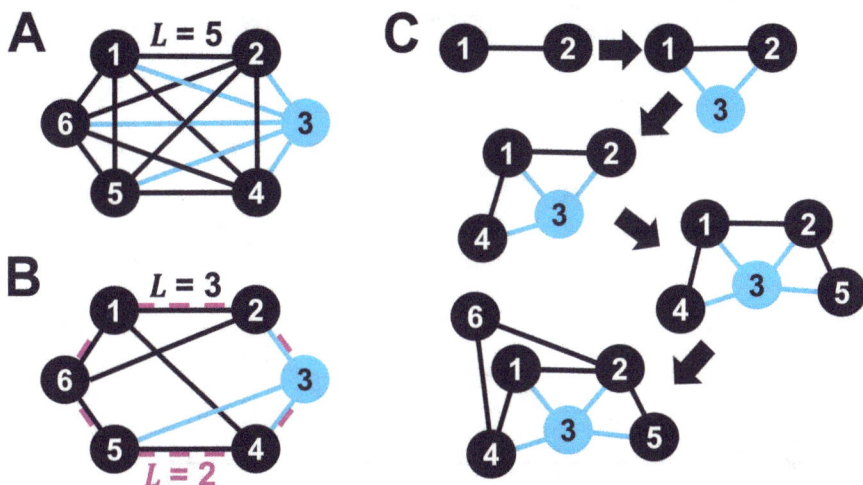

Figure 10.31 Examples of different linking strategies in the comparison of several datasets. Each sphere with a number presents one dataset. Total number of computations (each link) required when comparing six datasets simultaneously when the number of links per dataset is (A) $L = 5$ and (B) $L = 3$. (C) Example of a linking strategy for iterative workflows at the time when a new dataset is added. Adapted from Ref. 46 with permission from Elsevier, Copyright 2023.

tracking data of the previous chromatograms.[46] This strategy is illustrated in Figure 10.31C, and the number of connections needed is plotted in Figure 10.32 (cumulative, light blue drawn line).

10.6.4 Candidate Reduction

A similar problem to the linkage problem described in the previous section also arises when comparing just two chromatograms. Under the assumption that a peak in one chromatogram can exist anywhere in the other chromatogram, a rigorous peak-tracking comparison would require that each candidate peak in one chromatogram be compared with all other peaks in the other chromatogram. This is especially true for selectivity optimization in LC, where the elution order can easily shift.

The context of a peak-tracking operation is thus relevant. If both chromatograms are measured using similar retention mechanisms, then it is unlikely that a peak eluting early in one chromatogram – despite possible changes in elution order – elutes last in the other chromatogram. With this in mind, it can be worthwhile to

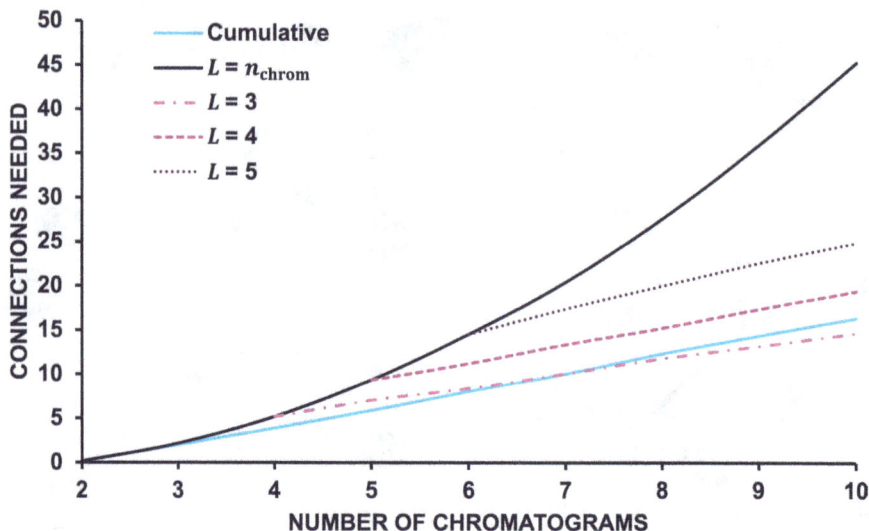

Figure 10.32 The number of required connections plotted as a function of the number of chromatograms for different linking strategies (see the text). The cumulative strategy refers to the approach illustrated in Figure 10.31C, where every new chromatogram is compared to two randomly selected existing chromatograms. Adapted from ref. 46 with permission from Elsevier, Copyright 2023.

define rules for candidate selection. An example is shown for two comparisons of chromatograms measured using RPLC in Figure 10.33. These chromatograms, measured as part of a gradient-scanning strategy (*i.e.* with different gradient slopes; see Section 3.17.2), feature similar separations with possible minor shifts in elution order. The number of candidates can thus be reduced by focussing only on a specific domain (marked as ω) within the chromatogram.

The advantage of candidate selection is illustrated in Figure 10.33A for a hypothetical comparison of two chromatograms. In this figure, the peaks are ranked according to their appearance in the chromatogram, and in either dimension, the number of boxes is equal to the number of peaks. Each box represents a possible combination of a peak detected in the first dimension with one in the second-dimension chromatogram. When the two chromatograms are obtained using similar retention mechanisms, it is reasonable to expect matching peaks to be located in the vicinity of the diagonal. Due to shifts in retention and elution order, this may not be exactly true, and one chromatogram may feature more detected peaks (including background) than the other. In such cases, it may be useful to employ a search window (ω), where all peaks within this window are labelled as "logical" (bright-blue area in Figure 10.33) and the remaining boxes as "illogical" (light-blue area). In such cases, it is important to define the window in units of time to avoid missing peaks in dense clusters. In Figure 10.33B, the retention times of the peaks are plotted on the horizontal axis for chromatogram *i* and on the vertical axis for chromatogram *j*. This strategy is an example of how the number of peak-pair comparisons can be sensibly reduced. If deemed necessary, a second evaluation can follow (see Figure 10.33C and D for Figure 10.33A and B, respectively) to evaluate eventually remaining unpaired peaks.

10.7 Outlook (A)

Compared with 1982, when the first automated "closed-loop" method-development system for LC was published,[24] the need for such systems has increased drastically. This is because the number of highly qualified analytical separation scientists has not kept pace with the immense growth in the number of applications of the various techniques described in this book. At the same time, the complexity of instruments and methods has increased tremendously. There is now an enormous number of options for different elution (and injection) programs, detection and hyphenation, and multi-column or multidimensional separations. Apart from a

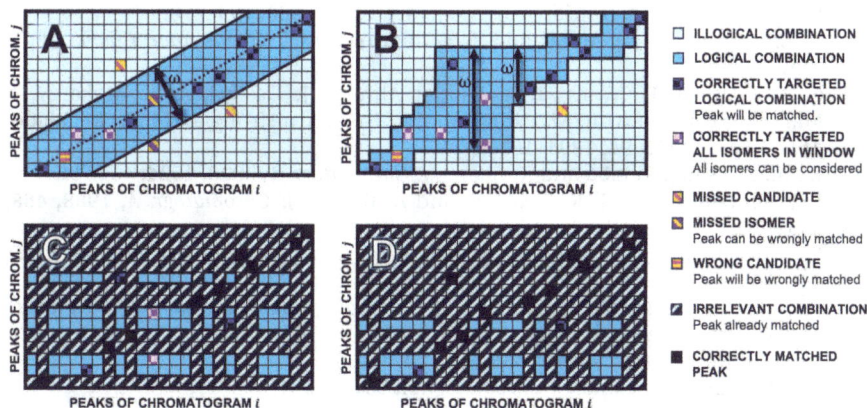

Figure 10.33 Example of a candidate-reduction strategy for comparing the peaks in two chromatograms. Adapted from ref. 42 with permission from American Chemical Society, Copyright 2021.

great need to collect all necessary knowledge in one book, we must also move towards method-development systems that apply all such knowledge correctly and, ideally, automatically.

Recent developments in this direction are exciting. Bos *et al.*[20] developed the AutoLC system that incorporates peak tracking, retention modelling, interpretive-optimization strategies, and unsupervised instrument control in an open-source package. Boelrijk *et al.*[29,47] developed multi-task Bayesian optimization in an attempt to combine the best from interpretive optimization (low number of experiments needed) and Bayesian optimization (an effective way to find optimum settings for many parameters). Developing open-source systems harbours a promise of scientists from different research groups expanding and improving systems with their specialized knowledge, in the same way as communities contribute to the development of open-source programming languages, such as Python or Julia, or to the Wikipedia encyclopaedia.

We will keep you up to date on new developments through the website supporting this book (ass-ets.org).

References

1. European Medicines Agency, in *ICH Q14 Guideline on analytical procedure development*, 2024.
2. T. Hetzel, T. Teutenberg and T. C. Schmidt, *Anal. Bioanal. Chem.*, 2015, **407**, 8475–8485.
3. J. W. Dolan, *LCGC North Am.*, 2018, **36**, 888–890.
4. B. W. J. Pirok, J. Knip, M. R. van Bommel and P. J. Schoenmakers, *J. Chromatogr. A*, 2016, **1436**, 141–146.
5. L. Nováková and M. Douša, *Anal. Chim. Acta*, 2017, **950**, 199–210.
6. G. L. Losacco, J.-L. Veuthey and D. Guillarme, *TrAC, Trends Anal. Chem.*, 2021, **141**, 116304.
7. G. L. Losacco, H. Wang, I. A. Haidar Ahmad, J. Dasilva, A. A. Makarov, I. Mangion, F. Gasparrini, M. Lämmerhofer, D. W. Armstrong and E. L. Regalado, *Anal. Chem.*, 2022, **94**, 1804–1812.
8. J. Lin, C. Tsang, R. Lieu and K. Zhang, *J. Chromatogr. A*, 2020, **1624**, 461244.
9. P. J. Schoenmakers, J. K. Strasters and Á. Bartha, *J. Chromatogr. A*, 1988, **458**, 355–370.
10. W. Wegscheider, E. P. Lankmayr and K. W. Budna, *Chromatographia*, 1982, **15**, 498–504.
11. P. Schoenmakers, in *Optimization of Chromatographic Selectivity - A Guide to Method Development*, Elsevier, 1986, vol. **35**.
12. J. C. Giddings, *Anal. Chem.*, 1960, **32**, 1707–1711.
13. E. Tyteca and G. Desmet, *J. Chromatogr. A*, 2014, **1361**, 178–190.

14. L. E. Niezen and G. Desmet, *J. Chromatogr. A*, 2024, **1727**, 465008.
15. L. Peichang and H. Hongxin, *J. Chromatogr. Sci.*, 1989, **27**, 690–697.
16. L. Peichang, L. Xiuzhen and Z. Yukuei, *J. High Resolut. Chromatogr.*, 1980, **3**, 551–567.
17. B. Bourguignon and D. L. Massart, *J. Chromatogr*, 1991, **586**, 11–20.
18. M. Camenzuli and P. J. Schoenmakers, *Anal. Chim. Acta*, 2014, **838**, 93–101.
19. J. Boelrijk, B. Pirok, B. Ensing and P. Forré, *J. Chromatogr. A*, 2021, **1659**, 462628.
20. T. S. Bos, J. Boelrijk, S. R. A. Molenaar, B. Van 'T Veer, L. E. Niezen, D. Van Herwerden, S. Samanipour, D. R. Stoll, P. Forré, B. Ensing, G. W. Somsen and B. W. J. Pirok, *Anal. Chem.*, 2022, **94**, 16060–16068.
21. S. Jespers, K. Roeleveld, F. Lynen, K. Broeckhoven and G. Desmet, *J. Chromatogr. A*, 2016, **1450**, 94–100.
22. G. B. van Henten, J. Boelrijk, C. Kattenberg, T. S. Bos, B. Ensing, P. Forré and B. W. J. Pirok, *J. Chromatogr. A*, 2025, **1742**, 465626.
23. M. McDaniel, A. D. Shendrlbar, K. D. Relszner, P. W. West, F. D. Pierce, T. C. Lamoreaux, H. R. Brown, R. S. Fraser, M. W. Routh, P. A. Swartz and B. Denton, *Anal. Chem.*, 1977, **49**, 1422–1428.
24. J. C. Berridge, *J. Chromatogr. A*, 1982, **244**, 1–14.
25. S. O'Hagan, W. B. Dunn, M. Brown, J. D. Knowles and D. B. Kell, *Anal. Chem.*, 2005, **77**, 290–303.
26. E. Zelena, W. B. Dunn, D. Broadhurst, S. Francis-McIntyre, K. M. Carroll, P. Begley, S. O'Hagan, J. D. Knowles, A. Halsall, I. D. Wilson and D. B. Kell, *Anal. Chem.*, 2009, **81**, 1357–1364.
27. B. Huygens, K. Efthymiadis, A. Nowé and G. Desmet, *J. Chromatogr. A*, 2020, 461435.
28. J. Boelrijk, B. W. J. Pirok, B. Ensing and P. Forré, *J. Chromatogr. A*, 2021, **1659**, 462628.
29. J. Boelrijk, B. Ensing, P. Forré and B. W. J. Pirok, *Anal. Chim. Acta*, 2023, **1242**, 340789.
30. A. Kensert, G. Collaerts, K. Efthymiadis, G. Desmet and D. Cabooter, *J. Chromatogr. A*, 2021, **1638**, 461900.
31. A. Kensert, G. Desmet and D. Cabooter, *J. Chromatogr. A*, 2024, **1713**, 464570.
32. European Medicines Agency, in *ICH Q2(R2) Guideline on validation of analytical procedures*, 2023.
33. H. J. Issaq and K. L. McNitt, *J. Liq. Chromatogr.*, 1982, **5**, 1771–1785.
34. I. Molnar, R. Boysen and P. Jekow, *J. Chromatogr. A*, 1989, **485**, 569–579.
35. A. C. J. H. Drouen, H. A. H. Billiet and L. De Galan, *Anal. Chem.*, 1985, **57**, 962–968.
36. H. Carter, D. Dubois and H. Prade, *J. Oper. Res. Soc.*, 1982, **33**, 198.
37. M. Otto, W. Wegscheider and E. P. Lankmayr, *Anal. Chem.*, 1988, **60**, 517–521.
38. J. K. Strasters, H. A. H. Billiet, L. de Galan and B. G. M. Vandeginste, *J. Chromatogr. A*, 1990, **499**, 499–522.
39. A. J. Round, M. I. Aguilar and M. T. W. Hearn, *J. Chromatogr. A*, 1994, **661**, 61–75.
40. G. Xue, A. D. Bendick, R. Chen and S. S. Sekulic, *J. Chromatogr. A*, 2004, **1050**, 159–171.
41. M. J. Fredriksson, P. Petersson, B.-O. Axelsson and D. Bylund, *J. Chromatogr. A*, 2010, **1217**, 8195–8204.
42. B.W.J. Pirok, S.R.A. Molenaar, L. Roca and P.J. Schoenmakers, *Anal. Chem.*, 90, 14011–14019.

43. S. R. A. Molenaar, T. A. Dahlseid, G. M. Leme, D. R. Stoll, P. J. Schoenmakers and B. W. J. Pirok, *J. Chromatogr. A*, 2021, **1639**, 461922.
44. A. Barcaru, E. Derks and G. Vivó-Truyols, *Anal. Chim. Acta*, 2016, **940**, 46–55.
45. J. J. de Rooi and P. H. C. Eilers, *Chemom. Intell. Lab. Syst.*, 2012, **117**, 56–60.
46. S. R. A. Molenaar, J. H. M. Mommers, D. R. Stoll, S. Ngxangxa, A. J. de Villiers, P. J. Schoenmakers and B. W. J. Pirok, *J. Chromatogr. A*, 2023, **1705**, 464223.
47. J. Boelrijk, S. R. A. Molenaar, T. S. Bos, T. A. Dahlseid, B. Ensing, D. R. Stoll, P. Forré and B. W. J. Pirok, *J. Chromatogr. A*, 2024, **1726**, 464941.

Appendix

A.1 Kolmogorov–Smirnov Test: Table of Critical Values

Table A.1 Critical values for the Kolmogorov–Smirnov test.

	Kolmogorov–Smirnov test				
n	$\alpha = 0.01$	$\alpha = 0.05$	$\alpha = 0.1$	$\alpha = 0.15$	$\alpha = 0.2$
1	0.995	0.975	0.950	0.925	0.900
2	0.929	0.842	0.776	0.726	0.684
3	0.828	0.708	0.642	0.597	0.565
4	0.733	0.624	0.564	0.525	0.494
5	0.669	0.565	0.510	0.474	0.446
6	0.618	0.521	0.470	0.436	0.410
7	0.577	0.486	0.438	0.405	0.381
8	0.543	0.457	0.411	0.381	0.358
9	0.514	0.432	0.388	0.360	0.339
10	0.490	0.410	0.368	0.342	0.322
11	0.468	0.391	0.352	0.326	0.307
12	0.450	0.375	0.338	0.313	0.295
13	0.433	0.361	0.325	0.302	0.284
14	0.418	0.349	0.314	0.292	0.274

Analytical Separation Science
By Bob W. J. Pirok and Peter J. Schoenmakers
© Bob W. J. Pirok and Peter J. Schoenmakers 2025
Published by the Royal Society of Chemistry, www.rsc.org

		Kolmogorov–Smirnov test			
n	$\alpha = 0.01$	$\alpha = 0.05$	$\alpha = 0.1$	$\alpha = 0.15$	$\alpha = 0.2$
15	0.404	0.338	0.304	0.283	0.266
16	0.392	0.328	0.295	0.274	0.258
17	0.381	0.318	0.286	0.266	0.250
18	0.371	0.309	0.278	0.259	0.244
19	0.363	0.301	0.272	0.252	0.237
20	0.356	0.294	0.264	0.246	0.231
25	0.320	0.270	0.240	0.220	0.210
30	0.290	0.240	0.220	0.200	0.190
35	0.270	0.230	0.210	0.190	0.180
40	0.250	0.210	0.190	0.180	0.170
45	0.240	0.200	0.180	0.170	0.160
50	0.230	0.190	0.170	0.160	0.150
>50	$1.63/\sqrt{n}$	$1.36/\sqrt{n}$	$1.22/\sqrt{n}$	$1.14/\sqrt{n}$	$1.07/\sqrt{n}$

A.2 Lilliefors Test: Table of Critical Values

Table A.2 Critical values for the Lilliefors test.

		Lilliefors test			
n	$\alpha = 0.01$	$\alpha = 0.05$	$\alpha = 0.1$	$\alpha = 0.15$	$\alpha = 0.2$
4	0.417	0.381	0.352	0.319	0.300
5	0.405	0.337	0.315	0.299	0.285
6	0.364	0.319	0.294	0.277	0.265
7	0.348	0.300	0.276	0.258	0.247
8	0.331	0.285	0.261	0.244	0.233
9	0.311	0.271	0.249	0.233	0.223
10	0.294	0.258	0.239	0.224	0.215
11	0.284	0.249	0.230	0.217	0.206
12	0.275	0.242	0.223	0.212	0.199
13	0.268	0.234	0.214	0.202	0.190
14	0.261	0.227	0.207	0.194	0.183
15	0.257	0.220	0.201	0.187	0.177
16	0.250	0.213	0.195	0.182	0.173
17	0.245	0.206	0.189	0.177	0.169
18	0.239	0.200	0.184	0.173	0.166
19	0.235	0.195	0.179	0.169	0.163
20	0.231	0.190	0.174	0.166	0.160
25	0.203	0.180	0.165	0.153	0.149
30	0.187	0.161	0.144	0.136	0.131
>30	$1.031/\sqrt{n}$	$0.886/\sqrt{n}$	$0.805/\sqrt{n}$	$0.768/\sqrt{n}$	$0.736/\sqrt{n}$

A.3 Dixon Q Test: Table of Critical Values

Table A.3 Critical Q_{crit} values for the Dixon Q test.

n	One tail: $\alpha = 0.05$	One tail: $\alpha = 0.01$	Two tail: $\alpha = 0.05$	Two tail: $\alpha = 0.01$
3	0.941	0.988	0.97	0.994
4	0.765	0.889	0.829	0.926
5	0.642	0.78	0.71	0.821
6	0.56	0.698	0.628	0.74
7	0.507	0.637	0.569	0.68
8	0.554	0.683	0.608	0.717
9	0.512	0.635	0.564	0.672
10	0.477	0.597	0.53	0.635
11	0.45	0.566	0.502	0.605
12	0.428	0.541	0.479	0.579
13	0.57	0.67	0.611	0.697
14	0.546	0.641	0.586	0.67
15	0.525	0.616	0.565	0.647
16	0.507	0.595	0.546	0.627
17	0.49	0.577	0.529	0.61
18	0.475	0.561	0.514	0.594
19	0.462	0.547	0.501	0.58
20	0.45	0.535	0.489	0.567
21	0.44	0.524	0.478	0.555
22	0.43	0.514	0.468	0.544
23	0.421	0.505	0.459	0.535
24	0.413	0.497	0.451	0.526
25	0.406	0.489	0.443	0.517
26	0.399	0.486	0.436	0.51
27	0.393	0.475	0.429	0.502
28	0.387	0.469	0.423	0.495
29	0.381	0.463	0.417	0.489
30	0.376	0.457	0.412	0.483

A.4 Grubbs Test: Table of Critical Values

Table A.4 Critical G_{crit} values for the Grubbs test for a single outlier.

	Grubbs test – a single outlier	
n	$\alpha = 0.05$	$\alpha = 0.01$
3	1.155	1.155
4	1.481	1.496
5	1.714	1.764
6	1.887	1.973
7	2.020	2.139
8	2.126	2.274
9	2.215	2.387
10	2.290	2.482
11	2.355	2.564
12	2.412	2.636
13	2.462	2.699
14	2.507	2.755
15	2.549	2.806
20	2.709	3.001
25	2.822	3.135
30	2.908	3.236
35	2.979	3.316
40	3.036	3.381

Table A.5 Critical G_{crit} values for the Grubbs test for outliers of the two largest or two smallest values.

	Grubbs test – two outliers	
n	*α* = 0.05	*α* = 0.01
4	0.0002	0.0000
5	0.0090	0.0018
6	0.0349	0.0116
7	0.0708	0.0308
8	0.1101	0.0563
9	0.1492	0.0851
10	0.1864	0.1150
11	0.2213	0.1448
12	0.2537	0.1738
13	0.2836	0.2016
14	0.3112	0.2280
15	0.3367	0.2530
20	0.4391	0.3585
25	0.5123	0.4376
30	0.5672	0.4985
35	0.6101	0.5469
40	0.6445	0.5862

A.5 Critical Range Method: Table of Critical Values

Table A.6 Critical CR_{crit} values for the critical range method for an outlier.

	Critical range method		
n	$a = 0.1$	$a = 0.05$	$a = 0.025$
2	2.3	2.8	3.2
3	2.9	3.3	3.7
4	3.2	3.6	4.0
5	3.5	3.9	4.2
6	3.7	4.0	4.4
7	3.8	4.2	4.5
8	3.9	4.3	4.6
9	4.0	4.4	4.7
10	4.1	4.5	4.8
11	4.2	4.5	4.9
12	4.3	4.6	4.9
13	4.3	4.7	5.0
14	4.4	4.8	5.1
15	4.5	4.8	5.1

A.6 Sign Test: Table of Critical Values

Table A.7 Critical r_{crit} values for the sign test.

			Sign test	
n	$\alpha = 0.005$ (1T) $\alpha = 0.01$ (2T)	$\alpha = 0.01$ (1T) $\alpha = 0.02$ (2T)	$\alpha = 0.025$ (1T) $\alpha = 0.05$ (2T)	$\alpha = 0.05$ (1T) $\alpha = 0.10$ (2T)
1	—	—	—	—
2	—	—	—	—
3	—	—	—	—
4	—	—	—	—
5	—	—	—	0
6	—	—	0	0
7	—	0	0	0
8	0	0	0	1
9	0	0	1	1
10	0	0	1	1
11	0	1	1	2
12	1	1	2	2
13	1	1	2	3
14	1	2	2	3
15	2	2	3	3
16	2	2	3	4
17	2	3	4	4
18	3	3	4	5
19	3	4	4	5
20	3	4	5	5
21	4	4	5	6
22	4	5	5	6
23	4	5	6	7
24	5	5	6	7
25	5	6	7	7

A.7 Wilcoxon Test: Table of Critical Values

Table A.8 Critical T_{crit} values for the Wilcoxon test.

	Wilcoxon test			
n	$\alpha = 0.005$ (1T) $\alpha = 0.01$ (2T)	$\alpha = 0.01$ (1T) $\alpha = 0.02$ (2T)	$\alpha = 0.025$ (1T) $\alpha = 0.05$ (2T)	$\alpha = 0.05$ (1T) $\alpha = 0.10$ (2T)
5	—	—	—	1
6	—	—	1	2
7	—	0	2	4
8	0	2	4	6
9	2	3	6	8
10	3	5	8	11
11	5	7	11	14
12	7	10	14	17
13	10	13	17	21
14	13	16	21	26
15	16	20	25	30
16	19	24	30	36
17	23	28	35	41
18	28	33	40	47
19	32	38	46	54
20	37	43	52	60
21	43	49	59	68
22	49	56	66	75
23	55	62	73	83
24	61	69	81	92
25	68	77	90	101
26	76	85	98	110
27	84	93	107	120
28	92	102	117	130
29	100	111	127	141
30	109	120	137	152

A.8 Mann–Whitney U Test: Table of Critical Values

Table A.9 Critical U_{crit} values for the Mann–Whitney U test.

Mann–Whitney U test

n_1											n_2								
	2	3	4	5	6	7	8	9	10	11	12	13	14	15	16	17	18	19	20
2	—	—	—	—	—	—	0	0	0	0	1	1	1	1	1	2	2	2	2
3	—	—	—	0	1	1	2	2	3	3	4	4	5	5	6	6	7	7	8
4	—	—	0	1	2	3	4	4	5	6	7	8	9	10	11	11	12	13	14
5	—	0	1	2	3	5	6	7	8	9	11	12	13	14	15	17	18	19	20
6	—	1	2	3	5	6	8	10	11	13	14	16	17	19	21	22	24	25	27
7	—	1	3	5	6	8	10	12	14	16	18	20	22	24	26	28	30	32	34
8	0	2	4	6	7	10	13	15	17	19	22	24	26	29	31	34	36	38	41
9	0	2	4	7	10	12	15	17	20	23	26	28	31	34	37	39	42	45	48
10	0	3	5	8	11	14	17	20	23	26	29	33	36	39	42	45	48	52	55
11	0	3	6	9	13	16	19	23	26	30	33	37	40	44	47	51	55	58	62
12	1	4	7	11	14	18	22	26	29	33	37	41	45	49	53	57	61	65	69
13	1	4	8	12	16	20	24	28	33	37	41	45	50	54	59	63	67	72	76
14	1	5	9	13	17	22	26	31	36	40	45	50	55	59	64	67	74	78	83
15	1	5	10	14	19	24	29	34	39	44	49	54	59	64	70	75	80	85	90
16	1	6	11	15	21	26	31	37	42	47	53	59	64	70	75	81	86	92	98

Mann–Whitney U test

n_1	n_2																		
	2	3	4	5	6	7	8	9	10	11	12	13	14	15	16	17	18	19	20
17	2	6	11	17	22	28	34	39	45	51	57	63	67	75	81	87	93	99	105
18	2	7	12	18	24	30	36	42	48	55	61	67	74	80	86	93	99	106	112
19	2	7	13	19	25	32	38	45	52	58	65	72	78	85	92	99	106	113	119
20	2	8	13	20	27	34	41	48	55	62	69	76	83	90	98	105	112	119	127
21	3	8	15	22	29	36	43	50	58	65	73	80	88	96	103	111	119	126	134
22	3	9	16	23	30	38	45	53	61	69	77	85	93	101	109	117	125	133	141
23	3	9	17	24	32	40	48	56	64	73	81	89	98	106	115	123	132	140	149
24	3	10	17	25	33	42	50	59	67	76	85	94	102	111	120	129	138	147	156
25	3	10	18	27	35	44	53	62	71	80	89	98	107	117	126	135	145	154	163
26	4	11	19	28	37	46	55	64	74	83	93	102	112	122	132	141	151	161	171
27	4	11	20	29	38	48	57	67	77	87	97	107	117	127	137	147	158	168	178
28	4	12	21	30	40	50	60	70	80	90	101	111	122	132	143	154	164	175	186
29	4	13	22	32	42	52	62	73	83	94	105	116	127	138	149	160	171	182	193
30	5	13	23	33	43	54	65	76	87	98	109	120	131	143	154	166	177	189	200

Subject Index